A COURSE IN BIOLOGY

THIRD EDITION

A COURSE
THIRD EDITION

Jeffrey J. W. Baker *Wesleyan University*
Garland E. Allen *Washington University*

IN BIOLOGY

ADDISON-WESLEY PUBLISHING COMPANY Reading, Massachusetts
Menlo Park, California • London • Amsterdam • Don Mills, Ontario • Sydney

This book is in the
Addison-Wesley Series in
the Life Sciences

Cover photo: Redwoods, Sequoia National Park
(Courtesy of David Muench)

Production Editor: Emily Arulpragasam

Designer: Judith Fletcher

Illustrators: Richard Morton
and Oxford Illustrators

Cover Designer: Richard Hannus

Library of Congress Cataloging in Publication Data

Baker, Jeffrey J W
A course in biology.

Abridged ed. of the work published in 1967
under title: The study of biology.
Includes bibliographies.
1. Biology. I. Allen, Garland E., joint
author. II. Title.
QH308.2.B34 1979 574 78-67451
ISBN 0-201-00308-2

0–201–00308–2
ABCDEFGHIJ–HA–79

PREFACE

In preparing the third edition of *A Course in Biology* we have incorporated many suggestions from both instructors and students at a variety of colleges and universities. This has resulted in extensive rewriting, the addition of new material to virtually every chapter, and the reorganization of previously existing material. We have made the third edition of *A Course in Biology* more comprehensive as well as more up-to-date than the previous versions.

New information has been added to the chapters on cellular biology and biochemistry, reflecting recent discoveries and new concepts regarding cell structure and function. Chapters on elementary chemistry and biochemistry have been added to assure that the student has review and reference material easily available. The present edition also contains more discussion of human biology, with entirely new material on human genetics, reproduction, development, and sexuality. The organization of *A Course in Biology* has been changed, with the present 16 chapters grouped into four units: Biology and the Scientific Enterprise, The Structure and Function of Cells, Heredity and Development, and Evolutionary and Population Biology. This unit organization has brought together in a more cohesive way chapters that deal with related topics.

Organizational changes have also been made *within* chapters. First, the main body of the text has been made to read more smoothly by separating it from supplementary material such as analyses of critical experiments, historical vignettes, and discussions of significant biosocial problems. Such supplementary material appears in tinted boxes. This separation of relevant supplementary material will introduce topics of special interest without interrupting the flow of ideas being developed in the body of the text. Second, throughout each chapter we have set off important principles in prominent headlines, a technique which will ensure that the reader focuses clearly on the biological principles being emphasized. Third, each chapter ends with a summary that can be used for study and review.

Important changes in style have also been made. In keeping with the suggestions of many users, we have abandoned (except in Chapter 1) the device of setting off hypothetico-deductive statements by indentation. Although the older format was useful to some. many felt it became distracting and monotonous when carried through the entire text. To aid the student in reading and information retrieval, we have added numerous subheadings to the text.

Although there have been many alterations, deletions, and additions in this revision, the spirit and intent of the previous two editions has been preserved. Chapter 1 explains the important role of scientific methodology in devising and testing hypotheses. Throughout the book, experimental design and detail are included (frequently in supplementary boxes) so that the student can examine some of the evidence on which generalizations and explanations are based. One of the most important opportunities an introductory science course affords the student is the opportunity to learn to think critically. To this end we continue to stress "how we know what we know"—the process and method of scientific investigation. This goal has been maintained even as we have made the book much more comprehensive in its content coverage. To aid the student in acquiring critical attitudes, many of the exercises at the end of the chapters are inquiry-based, challenging the student to interpret scientific data, draw conclusions, and choose between alternative hypotheses.

We have eliminated all vestiges of sexism that inadvertently appeared in previous editions. While such changes are subtle, they reflect evolving social values to which we wish to give full support.

In preparing the present edition we have had the help of a number of individuals. Attila O. Klein of Brandeis University thoroughly reviewed much of the material on chemistry, cell structure and function, and developmental biology. Richard Boohar of the University of Nebraska at Lincoln provided informative evaluation of the material on cell structure and function and on genetics. M. M. Green of the University of California at Davis was most helpful in his critique of the newly added material on human genetics. Alan Covich of Washington University in St. Louis played a major role in the extensive reviewing, revising, and rewriting of the chapter on ecology. Veda Andrus of Wesleyan University made many specific and helpful suggestions on the chapter dealing with human reproduction and development. Carol Lynch, also of Wesleyan University, thoroughly read the chapter on behavior. Without the help of these subject matter experts, the preparation of much of the new material would have been more difficult and less successful. We have not always followed all the advice of these consultants, however, and the authors assume full responsibility for the content and approach selected.

In addition, a number of people have helped in reviewing older chapters for the rewriting process. Among these are Johns W. Hopkins III, David Polcianski, William H. Mason, Barry Mehler, and Marvin Natowicz. Susan Allen undertook the complex and sometimes frustrating job of organizing the permission materials for new photographs. Mike Veith prepared a number of glossy prints, and Josh Tolin helped track down sources of illustrations and assisted in preparation of the Glossary, as did Ruth Balluff of Virginia.

Middletown, Connecticut J.J.W.B.
St. Louis, Missouri G.E.A.
January 1979

CONTENTS

PART THREE
HEREDITY AND DEVELOPMENT

ONE

THE NATURE AND LOGIC OF SCIENCE

1.1 Introduction

How do we know what we know? Almost every field has its accumulated knowledge, the so-called "facts" on which it rests. But how were those facts obtained? And how have they been woven together into beliefs, explanations, or theories?

We feel that one cannot be said to understand a subject without knowing how its ideas are derived and supported. This book is about biology, the study of living things. One way to approach the subject would be to present it as a group of accumulated facts and general principles that biologists today believe to be "true." You would study how evolution works, how living cells get energy, how nerves conduct an impulse, etc. Such a traditional approach to science has a major drawback: it quickly makes what you have learned obsolete. Today's scientific truths are almost as likely as not to be tomorrow's errors.

In science, keeping abreast of changing ideas requires understanding the *process* by which they developed. In short, it means learning how to think. Scientific thinking includes such techniques as framing questions, reasoning logically, and testing ideas against experience. Because this book is about biology, many of these specific techniques—such as designing and analyzing experiments—will be illustrated with biological examples. Yet none of the techniques discussed here is restricted to biology, nor to the natural sciences. The tools of clear thinking are applicable in all aspects of human thought, from nuclear physics and medical research to fixing a car or preparing a dinner. The levels of abstraction may differ, but the methods of approach are very much the same.

Approaching biology as a process rather than a collection of facts requires more participation on the reader's part, but the rewards are far greater. For one thing, people remember what they learn for a longer time, since they understand it more thoroughly. For another, the learning process is more dynamic and thus less tedious and dry.

In an age when human beings are confronted with perplexing social and political problems of profound importance, there is even more reason for learning how to use one's own brain to solve problems. Awareness is dawning that the answers our leaders offer are not always the right ones. Everyone needs to develop confidence in his or her own abilities to find meaningful answers to problems. Mastery of the highly successful methods of problem solving found in the natural sciences can enhance confidence in our own judgment and decisions.

Many factors make the study of biology a particularly appropriate area in which to learn the problem-solving methods. As the study of living things, biology has an immediate interest often lacking in the more abstract sciences of chemistry and physics. We are all living human beings who experience curiosity about ourselves and our biological processes. At the same time, many of today's most immediate problems have an explicitly biological base. Pollution, depletion of our natural resources, drugs, bacterial and chemical warfare, genetic engineering, overpopulation, birth control, abortion, and ideas about the inheritance of intelligence are only a few of many examples. The principles of biology and the methods of deriving those principles have an important bearing on understanding and dealing with these current problems.

1.2 Biology as a Science

First and foremost, biology is a science and the biologist a scientist. Defining science is not easy, for it has meant different things at different places and times in history. We have all been taught that physics, chemistry, and biology are natural sciences, while literature and art are "humanities," and other subjects, including history, sociology, and economics, are called the "social sciences." Are the social sciences really distinct from the natural sciences? If so in what ways? If not, why do we still think of them as different? Further, are both the natural and the social sciences fundamentally the same as, or different from, the humanities? Answers to these questions may suggest more clearly what the term "science" itself means.

To address these questions, it will be well to examine several commonly repeated generalizations about the natural sciences. The natural sciences include all those fields in which study of some aspect of the natural world is carried on. Science is said to be characterized by its use of experimentation and its objectivity; it is said to deal with facts, not opinions. It is regarded as quantitative and exact. Conversely, the social sciences and humanities are characterized as basically observational rather than experimental, subjective rather than objective, qualitative rather than quantitative, and therefore less factual and exact.

None of these distinctions drawn between the natural sciences and the social sciences and humanities are really useful, since they do not help us distinguish science from any other activity. Direct laboratory experimentation is often impossible for astronomers, yet they are recognized as scientists. Psychology, on the other hand, which is not considered a science by some people, often uses quite rigorous experimental procedures, and the social sciences are becoming increasingly quantitative and statistical.

If it is difficult to find methodological features that distinguish the natural sciences from the humanities and social sciences, it is impossible to say that the types of reasoning used in the scientific method are unique to it. In fact, the scientific method is trivially simple, differing not one whit from the thought processes of almost any area of rational thought. To diagnose the ills of an ailing car, a good automobile mechanic uses a variety of scientific means; so does a good cook in preparing a new recipe. A natural scientist does virtually nothing that other people do not do routinely in leading a rational life and trying to solve the myriad of problems confronting them every day.

Even so, the natural sciences do seem different

All human learning involves problem solving. There are fewer differences in problem-solving methods between the "sciences" and the social sciences or humanities than is generally believed.

from these commonplace activities. Part of this is myth—perpetuated by science fiction, advertising, or scientists themselves. The fact that some intelligent people are attracted to science has somehow led to the misconception that *all* scientists are very intelligent people and that science is extremely difficult. This myth places scientists in elevated social positions and imbues scientific activity with an elitism that tends to set it apart in most people's minds from everyday activities.

One of the aims of this book will be to dispel some of the myths about scientists and the scientific method. A major aspect of that method, as we shall see, is problem solving. Natural scientists, like historians or sociologists, are involved in a process of solving problems—of trying to find and communicate answers about the real world. Science, to paraphrase the words of biologist Albert Szent-Gyorgyi, is seeing what everyone else has seen and thinking what no one else has thought. Certainly this epithet would apply to history and sociology as well as to science. The most important difference is that in the natural sciences we can often distinguish between several possible answers to a problem by experimentation. Because of practical and ethical restraints, scientific experimentation is often not possible in the social sciences and humanities.

If it is agreed that the differences traditionally emphasized between the natural sciences on the one hand and the social sciences and humanities on the other are not as great as they are often made out to be, what is the so-called "scientific method" we hear so much about? The scientific method consists simply of logical thinking—a process, as has been pointed out, that can be applied in all realms of thought. What are the elements of scientific observation and logical thinking?

1.3 Observation and Scientific "Facts"

The foundation of all modern science is observation. Our entire view of the natural world depends on the accurate recording of sensory data and the organization of these data into general concepts.

A number of factors influence people's observations. Among these are accuracy of sensory impressions, knowledge of what to look for, the subjective influence of expectation based on previous theoretical constructs, and the operation of such other subjective factors as unconscious bias. These influences complicate the gathering of *reliable* data. In all fields of human endeavor, theoretical constructs depend upon the reliability of the observations from which they are derived. Scientific *reliability,* however, means that *the observation must be repeatable:* different people in different places must be able to observe the same object or phenomenon and record the same data. A good part of biologists' research time is spent determining the reliability of their own and others' observations.

Consider the following example. At the turn of the century, it became an intriguing problem to determine the number of chromosomes present in the body cells of human beings. Chromosomes are rod-shaped bodies found in the nuclei of cells of higher animals and plants. It was known that the number of chromosomes in the cells of any given species was the same: fruit flies have 8, elms 56, chickens 18, sugar cane 80, etc. Since chromosomes were recognized as the bearers of heredity, determining the number present in each species was of considerable interest. Many investigators tried to determine the number of human chromosomes.

To observe chromosomes it is necessary to "fix" and "stain" cell preparations for viewing un-

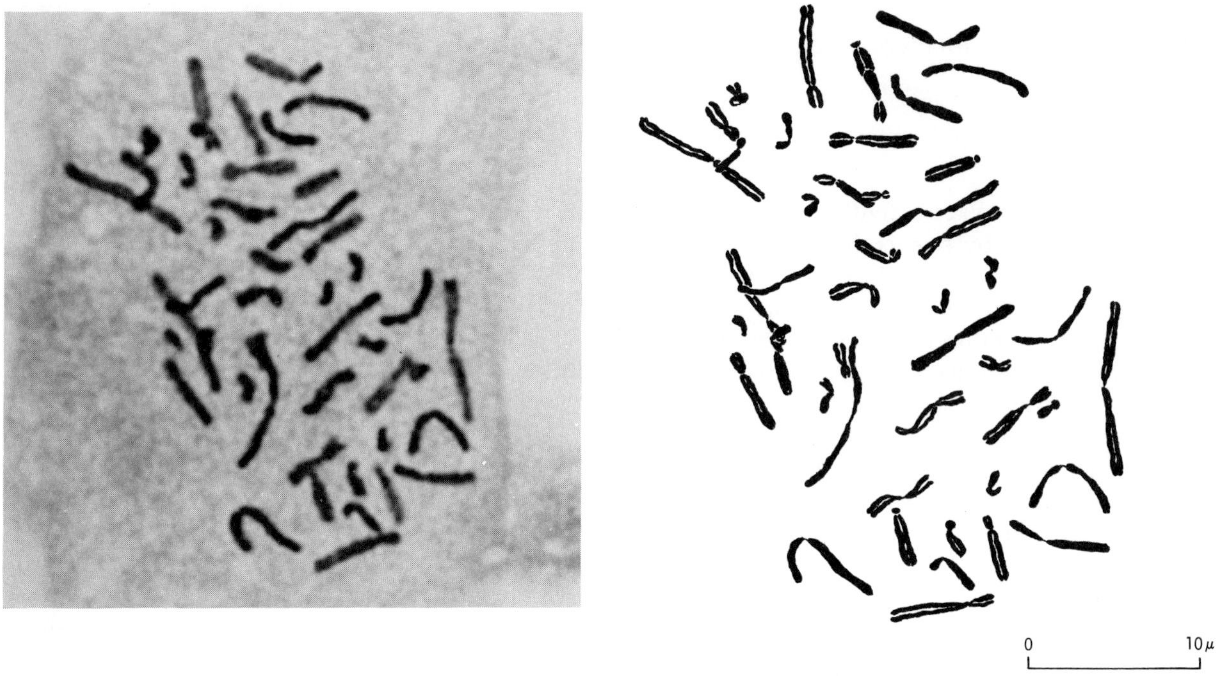

(a) (b)

Fig. 1.1 *(a) Photograph showing chromosomes from a stained spleen cell culture of a four-month-old human fetus. Note how the chromosomes clump, making accurate counts very difficult. (b) Camera lucida drawing of the same group of chromosomes. Such drawings help to elucidate detail more clearly, since they represent a composite of observations at different depths of focus—something no single photograph could provide. (Courtesy T. C. Hsu, from* J. Hered. *43 (1952), p. 168. Reprinted with permission.)*

der the microscope. Chromosomes so treated appear as dark, oblong objects surrounded by other partially stained material in the nucleus. A quick glance at Fig. 1.1(a) will suggest some of the problems observers encountered in making accurate chromosome counts. For one thing, the chromosomes appear clumped together in such a way that it is not always easy to tell where one ends and the next begins. For another, male and female cells seemed to early observers to have a different number of chromosomes, the female having one more than the male. (It is now known that this is due to the sex chromosomes. The female has two *X* chromosomes while males have one *X* and one *Y*, the latter being smaller and far less visible.) Still a third problem is that chromosomes, because they are curled and twisted and lie at all angles to the plane of the microscope stage, appear very different in different preparations. When one part of the chromosome is in focus, another part may not be. Thus it is easy to count as two chromosomes what is really a single chromosome appearing at different focal planes of the microscope.

A camera lucida drawing of the same preparation shown in Fig. 1.1(a) is given in Fig. 1.1(b). It was much easier to count chromosomes in camera lucida drawings than actual chromosome preparations, but the drawings were still not free of ambiguity. A number of subjective decisions often had to be made in tracing the chromosome outlines. For example, different pieces of chromosome come into

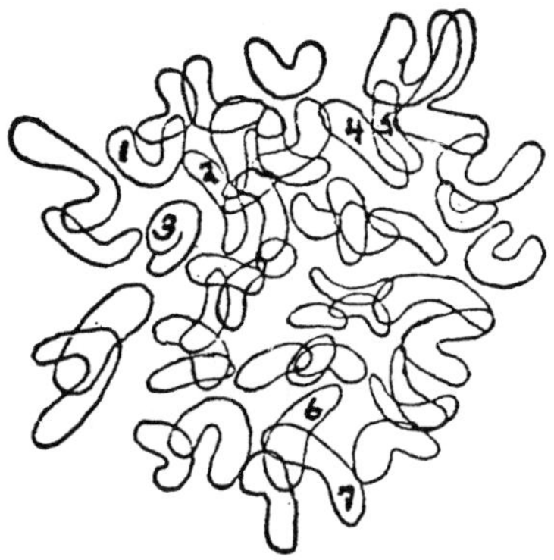

(a)

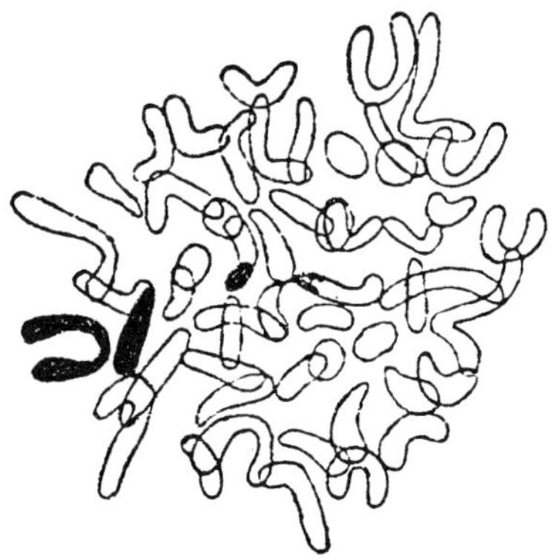

(b)

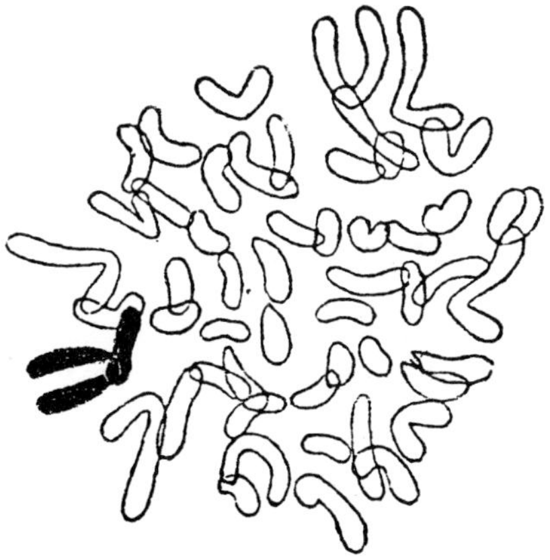

(c)

clear view when the focus of the microscope is changed. It is often difficult to tell whether a piece of chromosome is a separate body or a portion of a chromosome higher or lower in the field of vision. The problem is further compounded by the fact that chromosomes often stick together. The three camera lucida tracings shown in Fig. 1.2 are drawings of the same cell preparation by three different observers. All three were experienced in examining chromosome preparations. Yet they still produced three different representations, which frequently showed up as different chromosome counts. Nevertheless, most biologists eventually agreed that the number of chromosomes in a human cell was 48, and this became the number that appeared in all the textbooks.

Then several new techniques were introduced. One was phase-contrast microscopy (see p. 79), which made it possible to observe *living* cells, in which the chromosomes do not clump together so much. Another involved new methods of fixing and staining cells, allowing the chromosomes to spread out within the nucleus. Still a third innovation was the preparation of **karyotypes.** In this procedure, a chromosome preparation is photographed and the individual chromosomes cut out of the photographic print. The chromosome images are then arranged on a sheet in a systematic fashion, making

Fig. 1.2 *Three different camera lucida tracings of the same chromosome preparation; (a) the preparation drawn by Evans, (b) the same preparation drawn by von Winniwarter, and (c) by Oguma. Note the difficulty in determining whether some parts of chromosomes are attached to, or separate from, other parts. To see how this affects actual counting, observe the chromosome labeled 2 in (a). Evans saw number 2 as a long, crescent-shaped chromosome, while both von Winniwarter and Oguma saw it as two shorter, separate chromosomes. Evans saw chromosome number 3 as a single chromosome, whereas von Winniwarter and Oguma saw it as two separate ones. Such problems greatly confused the issue of human chromosome counts. (From King and Beams,* Anatomical Record *65 (1936), p. 169. Reprinted with permission.)*

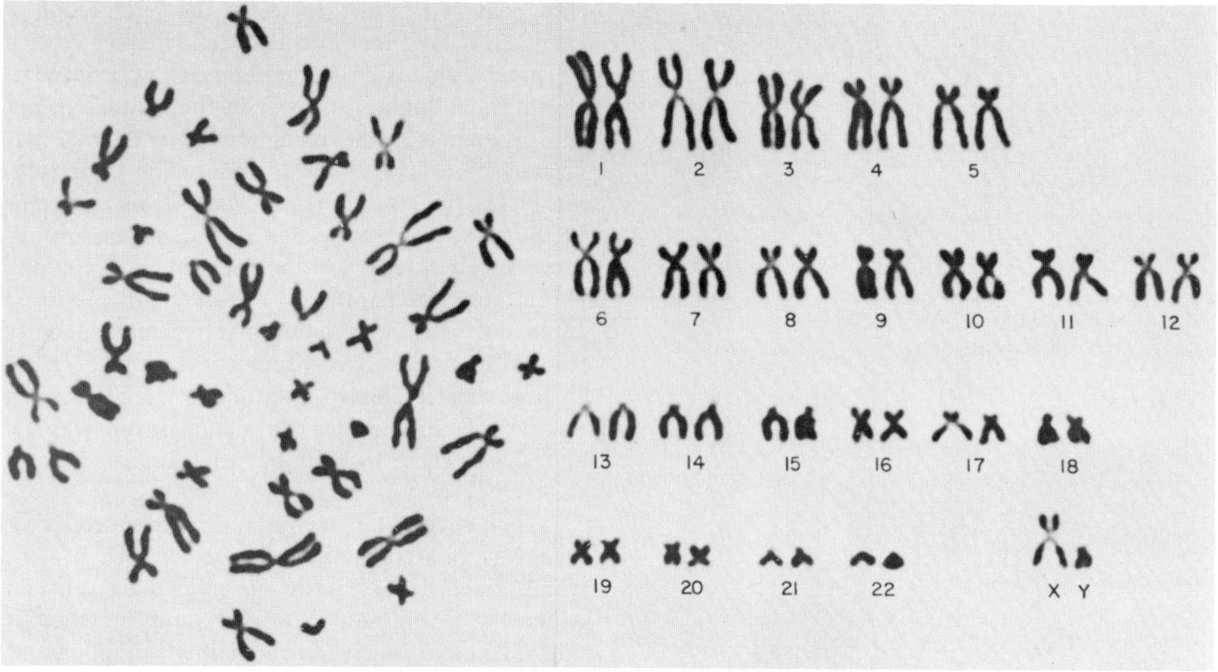

Fig. 1.3 *On the left is a photomicrograph of a full complement of human chromosomes, showing their random positions after fixing and staining. On the right is a karyotype of the same cell, showing the chromosomes lined up in order of decreasing size. Each chromosome pair is given a particular number. The first 22 pairs (44 chromosomes) are the autosomes; the twenty-third pair (the XX or XY pair) represent the sex chromosomes. In the above photograph, the sex chromosomes are X and Y, indicating the cell was from a male. (Photo from A. M. Winchester,* Human Genetics. *Columbus, Ohio: Charles Merrill, 1974.*

it possible to account for each individual chromosome and match it with its partner (see Fig. 1.3).

In 1955, several Swedish researchers were using karyotypes to study the chromosomes of individuals thought to have an inherited disease.* In their chromosome counts, they consistently came up with a total of 46 chromosomes for males and females. They made their counts a number of times to ensure that their observations were not in error. In searching the literature, the Swedish workers found that other medical researchers had also found only 46 chromosomes but had let the matter go. One worker, Dr. Eva Hansen-Melander had been studying human liver tissue the previous spring and had also counted 46 chromosomes. She had discontinued her work, feeling that her "failure" to observe the expected 48 chromosomes reflected poorly on her as an observer.† The Swedish group finally challenged the long-established notion that human

* The karyotype procedure has particular advantages to medicine. Certain conditions such as Down's syndrome (mongolism) are associated with an abnormal number of chromosomes.

† It is tempting to wonder how sociological factors such as age, nationality, sex, etc. influence the development of science. Why did so many workers simply accept the number 48 even when they observed a different number themselves? For example, could Dr. Hansen-Melander's position as a woman in a field long dominated by men have contributed to her lack of confidence in her own observations?

beings have 48 chromosomes. By 1960, after many confirmations of the number 46, the biological community was ready to admit that the older count was wrong.

What can we conclude about the process of observation from this case study?

1. Observations depend upon sensory data. If the material being observed has some ambiguity, such as clumped chromosomes, the observations will reflect that ambiguity in one way or another.

2. Observations contain some subjective input. Determining whether a particular stained mass represents one or two chromosomes often requires the exercise of prejudgment.

3. Even when something different from the expected is observed, the force of accepted dogma may cause investigators to disbelieve their own observations.

4. New techniques often make possible the refinement of observations. New techniques can make observations more accurate and thus add new sensory data for the solution of a problem.

1.4 The "Scientific Method"

The number of scientists in the United States in 1900 was approximately 8000. In 1960 the number was more than 100,000. At least four out of five of all the professional scientists who ever lived are alive today. No one can live in the 1970s and not recognize the tremendous impact this growth in science and the resulting growth in technology have had on the world, both for good and for bad.

The popular, man-on-the-street concept of the scientist and scientific methods is a poor one. The scientist is conceived as having secret means of obtaining knowledge to benefit humankind. It is not widely known that explanations advanced by research scientists may be wrong as often as they are right and that not all of their discoveries benefit us directly (indeed, many seem to be completely useless). This may be due to the fact that wrong guesses are not given much publicity, while right guesses that do not directly benefit us attract less public attention.

Even among scientists, however, there is wide disagreement about what is meant by "the scientific method." Some science textbooks list a series of six or seven steps involved in the scientific method. Such a formal description is quite unrealistic. No research scientist follows any such formalized ritual in performing experiments. Other writers, however, have gone to the opposite extreme in their description of the scientific method. One states that "science is simply doing one's damndest with one's mind with no holds barred." This view certainly conveys that the means used by scientists in solving problems are not necessarily unique to science. As a definition of the scientific method, however, the statement is not very useful. Followed to its logical conclusion, it indicates that artists, philosophers, mechanics, mathematicians, plumbers, and all other persons who work diligently and creatively to solve problems are also scientists. This is not the case.

Induction

Science proceeds by formulating and testing hypotheses. **Hypotheses** are simply tentative explanations put forth to account for observed phenomena. Formulating testable hypotheses draws heavily upon the scientist's creativity and imagination. Attempts to pin down these qualities often seem doomed to failure. All of us know of creative and imaginative people in all professions, yet the attributes that make them so become quite elusive when we attempt to describe them in precise, analytical terms. How does one account for the genius of a scientist like Albert Einstein, whose most brilliant ideas often appeared to be nothing more than an intuitive leap of immense proportions? Perhaps one cannot—or should not—even attempt to do so. Not all scientists are Einsteins, however. For most, the means of hypothesis formulation can be described, at least in general terms. One general pattern of thought often recognized in the creation of hypotheses is known as **induction** or **inductive logic.**

Inductive logic proceeds from the specific to the general; deductive logic proceeds from the general to the specific.

Consider a person who tastes a green apple and finds it sour. Tasting a second, third, and fourth green apple yields the same result. From this experience, the person concludes that all green apples are sour. One physical entity (greenness) has been correlated with another (sourness). The conclusion, "all green apples are sour," is an example of a generalization formed by induction: it is an **inductive generalization.**

Induction is often described as a system of thought proceeding from the specific to the general. Such a description is incomplete since it ignores the fact that some aspects of induction are often highly creative. The description can help us identify induction, however; the tasting of individual apples (the specific) led to the conclusion that all green apples are sour (the general).

The inductive generalization "all green apples are sour" becomes a hypothetical statement when expressed as an "if" statement, the "if" denoting the tentative nature of the hypothesis. But is the hypothesis "true"? We can test it by tasting yet another green apple to see if it is sour; the hypothesis predicts that it will be. This tasting of another apple after forming the hypothesis "all green apples are sour" by induction represents a test of the hypothesis; that is, it is an **experiment.**

Hypotheses that arise from inductive generalizations simply summarize a set of specific data and allow that summary to be tested. For example, observation that the sun rises in the east every day leads inductively to the hypothetical generalization that it always does so. This, in turn, leads to a prediction: the sun will rise in the east tomorrow. Each ensuing day's observation of the sun rising in the east serves as a confirming test of the hypothesis. The second and, to science, far more important type of hypothesis is an **explanatory hypothesis.** Explanatory hypotheses do more than merely generalize from a set of similar observations; they attempt to get at the *causes* for the phenomena in question, the *reasons* behind certain occurrences or processes. For example, the explanatory hypothesis for why the sun rises in the east proposes that it is due to the earth's rotating on its axis from west to east. Thus an explanatory hypothesis attempts to account for one set of observed phenomena in terms of another; in this case it accounts for the sun's rising in the east in terms of the earth's rotation.

Similarly, returning to the green apples example, an explanatory hypothesis goes beyond merely stating that "all green apples are sour" to propose reasons for their sourness. For example, one explanatory hypothesis might propose that the green apples are sour because they contain high concentrations of a particular acid. Such an hypothesis is readily testable by comparing the amount of that acid found in the sour green apples with the amount found in sweet apples. Since it tests the explanatory hypothesis, such a chemical analysis of the apples is, by definition, an experiment. *The primary purpose of scientific experiments is to test hypotheses.* Any hypothesis selected by a scientist

Truth and validity have no necessary connection. Validity is a concept in logic, referring to conclusions derived by deductive syllogisms. Truth is that which agrees with people's experience with the world around them.

must meet a very important requirement: *it must be testable.*

Table 1.1
Comparison of Inductive and Deductive Logic

Inductive	Deductive
Begins with observations; leads to hypothesis	Begins with hypothesis; leads to predictions
Proceeds from specific to general	Proceeds from general to specific
A method of discovery	A method of verification

Deduction

How do experiments test hypotheses? The answer is quite simple. *Experiments test hypotheses by testing the correctness of the predictions that can be derived from them.* This process involves the use of **deduction,** or **deductive logic.**

Deductive logic (often called *if . . . then* reasoning) is the heart and soul of mathematics. It becomes most evident in plane geometry: "*If* two points of a line lie in a plane, *then* the line lies in the same plane." Deduction plays no lesser role in other fields of mathematics: "If $a < b$ and $x \leq y$, then $a + x < b + y$" (the addition law), and "If $x < y$ and $a > 0$ then $ax < ay$" (the multiplication law).

In science (and therefore in biology) deduction is just as vital as it is in mathematics, but there are important differences between the way deduction is used in mathematics and the way it is used in experimental science. Mathematicians generally deal with symbols. They are not concerned with such physical entities as living organisms. Furthermore, the mathematician can manipulate symbols at will. He or she can create situations in proofs which make certain that only one hypothesis is being tested, only one question being asked. Not so the biologist. The organism being studied cannot be so easily manipulated. The biologist can never be absolutely certain that his or her experiment has eliminated all the variables that might influence the results. A major problem in biological research, then, becomes one of experimental design.

The "if" portion of deductive logic represents the hypothesis; the "then" portion, the predictions that *must* follow from acceptance of the hypothesis. The stress placed upon the word "must" emphasizes that once an hypothesis is formulated, the predictions follow automatically.

To illustrate this last point, suppose a student who has been reading an article on automobile safety by consumer advocate Ralph Nader announces to his biology professor, Ms. Smith, that the article has convinced him that anyone who drives a Jonesmobile is an idiot. Ms. Smith replies that she drives a Jonesmobile! The student is trapped; indirectly—and *inescapably*—he has just told his professor that he believes she is an idiot. The situation can be stated hypothetico-deductively as follows:

Hypothesis: *If* all people who drive Jonesmobiles are idiots and *if* Professor Smith drives a Jonesmobile . . .

Prediction: *then* Professor Smith is an idiot.

We have set up the deductive thought pattern into a **deductive syllogism,** The syllogism described here can be visually represented by what in mathematics is called a Venn diagram:

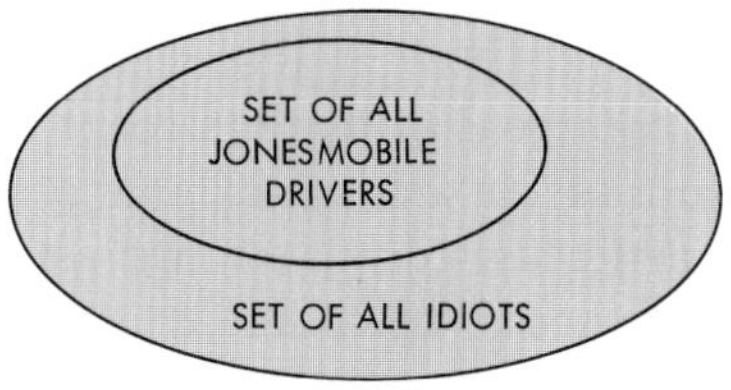

The Venn diagram simply makes it clear that the conclusion that Professor Smith is an idiot is inescapable; we say that it is a valid conclusion. Its "*truth*," on the other hand, is doubtful, considering how unlikely it is that *all* drivers of any one make of automobile, no matter how unsafe, are truly idiots. The example merely illustrates the important point that validity and "truth" have no necessary

A TRUTH TABLE

HYPOTHESIS	CONCLUSION OR PREDICTION
TRUE	TRUE
FALSE	TRUE or FALSE

Fig. 1.4 *The "truth table" shows the relation between an hypothesis and its predictions. Note that a true prediction may be derived from a false hypothesis as well as from a true one. Therefore true predictions do not constitute proof of the truth of an hypothesis.*

connection. Validity is a concept in logic, referring to conclusions derived by deductive syllogisms; truth is that which agrees with people's perception of the world around them.

Testing Hypotheses

Scientific experiments are designed to test hypotheses by determining the correctness of the prediction that follows once an hypothesis is advanced. This implies that there is some sort of consistent relationship between hypotheses and the predictions they automatically generate. Indeed there is. This relationship is shown in the "truth table" (see Fig. 1.4).

An example from scientific research will illustrate how the truth table applies. It has been shown many times that exposure of certain strains of mice to X-ray beams of 600 roentgens or more (a roentgen is a unit of measure of the amount of energy delivered in X-ray beams) causes death within two weeks or less. The death seems to be due to secondary rather than primary effects of the radiation. But the primary cause of death is uncertain at any one time, especially in the period of one to five days after exposure. It was thought that death might possibly be due to bacterial infection resulting from a migration of bacteria through the intestinal epithelium (lining), which histological (tissue) examination showed had been severely damaged by the X-rays. In order to test this hypothesis, antibiotics of various types were administered to the irradiated mice in several different ways to see whether this had any effect on the time of death. No such effect was shown; the mice still died in the same length of time as the control animals, which had been irradiated under the same conditions but given no antibiotics. It was tentatively concluded, therefore, that death in the period tested (from one to five days after exposure) was not due to bacterial infection.

Note the deductive *if . . . then* reasoning here. The experimental logic can be simply stated as follows:

Hypothesis: *If* the deaths of irradiated mice within one to five days after exposure are due to bacterial infection . . .

Prediction: *then* the administration of antibiotics should lower the death rate of mice that receive them.

The experimental results showed the prediction to be false. The mice still died in the same length of time after exposure to the X-rays. Thus we know, barring experimental errors, that the hypothesis explaining the deaths as due to bacterial

A true hypothesis never gives rise to a false conclusion (prediction). However, a false hypothesis can give rise to either a true or a false conclusion (prediction). Hence a false conclusion (prediction) leads to the certain rejection of the hypothesis from which it was derived, but a true conclusion (prediction) does not prove the hypothesis true beyond doubt.

infection is also false and must be either discarded or modified.*

Suppose that the administration of antibiotics *had* caused a lengthening of life. Would this have shown that our hypothesis must be the correct explanation? Absolutely not, although this result would have lent support to the *probability* of its being correct.

Examine again the truth table in Fig. 1.4. Note that the word "conclusion" as used by a mathematician is interchangeable with the word "prediction" as used by a biologist, for predictions that can be made from an hypothesis are simply the conclusions that one must draw from accepting it. In the case of the irradiated mice, it must be predicted that the mice will live longer or recover if the bacterial-infection-as-cause-of-death hypothesis is correct. This did not occur. The mice died as before; the prediction was false.

From the truth table, we see that this automatically means that our hypothesis is false, for *a true hypothesis can never give rise to a false prediction.* In other words, predictions derived from a true hypothesis should never lead to contradictions.

The truth table also shows that we can never *prove* an hypothesis true. While a true hypothesis always gives rise to true predictions, so also may a false hypothesis. The importance of this last fact cannot be overemphasized. It shows that *science can only deal with its "truths" in terms of probabilities and never in terms of certainties.*

In the past, many false hypotheses have been held by scientists and lay people alike, simply because accurate predictions could be made from these hypotheses despite their falsity. Acceptance of the belief that the sun orbits the earth leads one to predict that the sun will rise on one horizon, cross the sky, and set on the other horizon—and so it does.† The fact that this prediction turns out to be correct does not, of course, mean that the sun *does* orbit the earth. In order to demonstrate that this hypothesis is false, other tests must be devised. They would show that it yields false predictions.‡

Although the truth table shows that a true hypothesis never gives rise to a false conclusion (prediction), only in mathematics does obtaining just one false conclusion spell certain death for the hypothesis. Biologists rarely deal with cases in which *every* prediction made by an hypothesis turns out to be correct. The question then becomes one of *how many* or *what proportion* of a given number of predictions must be verified in order to make the hypothesis a useful one. For this reason, experimental data are often subjected to **statistical analyses**, in which mathematics is employed to de-

* Note that it could not be said that death from radiation *in animals* is due to other causes than bacterial infection, for the world "animals" includes many more forms than just mice. Nor could it be stated that death from radiation *in mice* is due to some other cause than bacterial infection, for not all strains of mice were tested. When the research paper is written for publication, the biologist will carefully word the interpretations, limiting them to the precise strains of mice tested and to the time period of death (one to five days after exposure) with which she or he worked.

Despite the care with which experimental results are generally interpreted, biologists often *do* extend their experimental results from one organism to another. Modern drugs, for example, are usually tested first on laboratory animals; if they are successful, their use may be extended to humans. But there is always an element of uncertainty involved. All organisms do not necessarily react the same way to the same drugs.

† It is likely that the observation of this aspect of the sun's behavior contributed to the idea that the sun orbits the earth. This illustrates the fact that hypotheses frequently arise from observations of the very phenomena the hypotheses would predict. It is often difficult to establish which came first.

‡ One such test is to predict the future relative positions of the sun, earth, and other planets, given that the sun does orbit the earth. Such predictions are invariably shown to be false, thus forcing rejection of the hypothesis. On the other hand, accepting the hypothesis that the earth, along with the other planets, orbits the sun leads to very accurate predictions about the relative positions of the sun, earth, and other planets at any particular time.

termine whether deviations from the pattern predicted by the hypothesis are significant.

1.5 Explanations in Science

Like all areas of human problem-solving, the natural sciences seek answers in the form of **explanations.** We talk about explaining things every day. The hometown baseball team loses 10 games in a row; we want an explanation. Inflation continues to rise; various economists try to explain why. The western hemisphere suddenly faces a petroleum shortage; people want an explanation. A friend fails to show up for an appointment; an explanation is in order.

A few years ago biologist Ernst Mayr of Harvard University observed that a warbler living all summer in a tree next to his house in Maine began its southern migration on August 25. This single observation raised a question in his mind. *Why* did the warbler begin its migration on that date?

What does the word "why" mean in this question? Two types of answers can be given to two versions of this "why" question:

1. A teleological answer: *for what purpose* did the warbler begin to migrate?
2. Causal answers: *what caused* the warbler to start migrating?

Scientists try to avoid teleological questions. While it may be appropriate to ask teleological questions about purposeful human behavior, it is futile to ask such questions about natural phenomena. To ask "For what purpose did the warbler begin to migrate?" implies that the warbler made the same kind of conscious, goal-oriented choice on August 25 that a person might make in deciding to go downtown. Teleological explanations imply that the organism knows the goal in advance, such as a comfortable habitat in the south. A teleological hypothesis is untestable.*

In modern science, **causal** answers are sought, since they attempt to explain natural phenomena in terms of the events that may have given rise to them. For example, Dr. Mayr's original question could be rephrased as a second, more precise question: "*What caused* the warbler to begin its migration on August 25?" In most scientific situations, such questions should be posed in another form: "What causes *warblers* to begin their migration *around* August 25?" Seldom in science do questions about individual, particular events find a satisfactory answer. No two situations are absolutely identical. No two warblers are the same. Warbler A might begin its migration on August 25; warbler B might already have departed on August 23; and warbler C might not depart until August 26. In general, scientists are less interested in explaining particular events than in the principles underlying collections of events.

In studying any collection of events in nature, whether dealing with organisms, cells, atoms, molecules, or whatever, it is necessary to deal with **statistically significant** numbers. Statistical significance means that a large enough collection of events or objects is being observed to eliminate the distorting effects of individual differences. If warbler migration were to be studied by examining only two birds, for example, it might be impossible to detect

* Teleological explanations have not been limited to biology. Aristotle's theories of motion, which dominated physics during the classical and medieval periods, were openly teleological. For example, Aristotle explained the falling of stones or the rising of flames and fire by the tendency of things in the universe to "seek" their natural place. A stone's natural place was on the ground with other stones; a flame's natural place was in the heavens with other flames, the sun and stars. The object moved toward its predetermined goal; it knowingly sought its place in the universe.

any common factor influencing both. One bird might begin its migration on August 25, the other on August 26, and there might be a big thunderstorm on August 26. Under such circumstances it could not be determined which bird was more typical or which events (such as the thunderstorm) may have triggered migration. On the other hand, if the sample contained 1,000 birds, and it was noted that about 85% began their migration on August 25, it could be concluded that thunderstorms probably have no causal effect in triggering warbler migration. There are no general formulas for determining what is a statistically significant sample of cases for all questions. Each situation may be different. Observing five birds might be enough to make a generalization about the onset of bird migratory behavior. In general, however, larger samples are necessary, and the greater the number of cases studied, the more likely the hypothetical explanation derived from studying them will be correct.

Causal Explanations

To return now to Dr. Mayr's original question, let us assume that he was indeed interested in what triggered the migratory response of warblers in general, as exemplified by the particular warbler observed in the tree outside his window. In response to the general question of what causes warblers to begin migrating on August 25, Dr. Mayr points out that at least four different kinds of causal explanations are possible.

An ecological explanation. Warblers must move south in response to a decline in food supply. As insect eaters, they would starve to death if they remained in Maine where the insects begin dying out around the end of August. This explanation causally relates warbler migration patterns to general environmental changes, especially those concerned with food supply.

An historical explanation. Warblers begin moving south because through the course of evolution they have acquired an hereditary constitution that programs them to respond to certain environmental changes associated with the end of summer. Other birds, such as screech owls, do not have this same genetic constitution and hence do not respond to end-of-summer environmental stimuli by migrating. This explanation relates warbler migration to genetic causes resulting from evolutionary processes.

An internal explanation. Warblers begin migrating because of their particular physiological state on August 25. The physiological mechanism triggers certain parts of the nervous system, resulting in migratory flight response. This internal physiological explanation is closely associated with the fourth type of explanation.

An external explanation. One or more specific environmental factors activate an intrinsic physiological mechanism in warblers. This explanation puts the emphasis on the external factors that do the triggering, whereas the third explanation deals with the nature of the trigger itself.

These explanations are not mutually exclusive, and the most complete explanation would consist of all of them. They fall into two categories. The last two involve the interaction of environmental factors with the organism's immediate, internal makeup. They can be called the **immediate** or **proximate causes.** The first two explanations, on the other hand, involve interaction of the long-term history of the warblers with their environment, both physical and biological. These can be called **ultimate causes.** They act continuously in the present but have arisen by historical development. The four particular explanations listed above obviously relate only to the particular problem of bird migration.

Facts are very different from hypotheses. Facts are observations that by general agreement have been established as true. Hypotheses are ideas that go beyond the observations and attempt to explain why certain phenomena occur. Two or more people may often develop different hypotheses from the same set of observations.

The existence of two levels of causal explanation, proximate and ultimate, is characteristic of most explanations in the biological sciences. Not every explanation must contain both to be acceptable, however. Sometimes a proximate answer will suffice for a given situation. Suppose the question is "Why is carbon monoxide a poison?" In this case a discussion of the proximate causes (the fact that carbon monoxide molecules interfere with vital respiratory molecules and prevent the delivery of oxygen to body tissues) would probably suffice. It would not be necessary to add that the animal species in question had a genetic constitution that prevented it from synthesizing a respiratory molecule that would not be affected by carbon monoxide.

Cause and Effect

Several other features are worth noting about the general character of explanations as they apply to any field, but especially to biology. For one, explanations tend to be most successful when they relate one kind of particular event to more general processes or events that are already understood. Thus the particular event of warbler migration is "explained" by showing how it relates to ecological, hereditary, physiological, and evolutionary principles. For another, the most successful causal explanations are those that can be tested **empirically:** the particular causal relationship pointed out in the explanation is subject to some kind of experimental test. It is possible, for example, to breed animals and learn something about their genetic constitution; physiological mechanisms can be studied in the laboratory; and determining whether those mechanisms are triggered by one or another external factor is also subject to experimental analysis. This testability criterion for an acceptable scientific explanation is particularly important. As suggested earlier, a causal explanation stating that warblers fly south because of an inherent desire to do so would not be very acceptable, since it could not be tested in any way.

Many cause–effect relationships may be more apparent than real. This usually results from our having no way to sort out the various interactions involved. Suppose, for example, that a cold air mass from Canada arrived in Maine on August 25, just when day length was appropriate to trigger the warbler's migratory response. It might appear that the primary cause for onset of migration was the drop in temperature. In this case, since it is known the actual cause is day length, the drop in temperature would be an example of a **spurious cause.** Such spurious relationships are common, since many events occur simultaneously and thus appear to be related.

In many science textbooks and other sources, the term "theory" is used in place of, or along with, the term "hypothesis." Philosophers of science are not in full agreement on what a theory is, or whether (and how) it differs from or is synonymous with, an explanation. In this book **theory** refers to an hypothesis that has survived being tested a great many times in a number of different ways. Thus, if the hypothesis proposing that day length is the primary cause of bird migration is repeatedly tested and supported, it may eventually gain the status of a theory. This distinction between hypothesis and theory is quite arbitrary and is not necessarily how the terms are distinguished by other authors.

1.6 The Philosophical Side of Biology

For over a century a debate has raged on and off within the biological community between the viewpoints of "vitalism" and "mechanism." This debate has been part of a much larger philosophical dispute that developed during the later nineteenth century, the dispute between philosophical **idealism** and **materialism.** Since certain very different biological assumptions come from each of these philosophical views, it will be important to distinguish between them.

Vitalism

The controversy between vitalists and mechanists has always focused on the question of what causes living organisms to differ from nonliving material. Even the most complex crystal structure is less complex than molecules in living systems. The abilities to move, reproduce, take in and assimilate material from the environment, grow, and have "consciousness" all seem qualitatively different from anything observed in the nonliving world. Vitalists explain this unique feature by postulating, in one form or another, the existence of a "vital force." This force is assumed to be wholly different from other known physical forces and ultimately to be unknowable—not subject to physical and chemical analysis. This vital force departs from a cell or an organism at death. Vitalists do not necessarily deny that chemical analyses of living organisms are valuable. But they feel that "life" involves something more than the principles of physics and chemistry.

Mechanism

Mechanists, on the other hand, maintain that life can be completely understood in physical and chemical terms. They dismiss the idea of a vital force as mysticism and claim that, with proper methods and concepts, it should be possible to explain all the essential processes of living systems. While mechanists admit that modern techniques are imperfect, thus making many questions unanswerable at the moment, they do not believe questions must remain unanswered forever. Essentially, mechanists view organisms as very complicated machines. We can find out how an organism works by taking it apart, studying the components one at a time, and then putting it back together again, so to speak. Basic to the mechanist view is the idea that organisms function according to fundamental laws of physics and chemistry.

The mechanism–vitalism debate has raged periodically throughout the recent history of biology, and the issue has never been entirely settled. Today, some geneticists working primarily on the molecular level claim that the secret of life has been —or soon will be—unraveled. Other biologists, though rejecting a vitalistic viewpoint, feel this is an oversimplification and maintain that reducing an organism to its component molecular parts for study necessarily yields an incomplete picture of the whole.

The conflict between mechanism and vitalism is in many ways an artificial problem for both are limited ways of viewing living systems. Vitalism is limited because it presupposes the existence of a mystical, unknowable force animating all living beings, a force that lies by definition beyond physical and chemical investigation. Since one cannot ask questions about something ultimately unknowable, such a viewpoint limits research. Thus vitalism is an example of **metaphysics**—a mode of thinking that transcends present, material reality to seek nonmaterial causes. Modern biology avoids metaphysical statements as untestable and thus rejects vitalism. Biologists may not know exactly what processes make an organism so unique and different from nonliving matter, but as scientists they assume this can ultimately be determined.

Mechanism has its limitations, too. Biologists, especially, have always sensed that living organisms are so complex that their *organization* is the key to their unique functions. They have attempted to discover that organization by breaking the complex whole down into component parts and studying each part separately. Studies of component parts in isolation do not necessarily give wrong information, but the results are *incomplete*. A limitation of mechanism as a way of viewing living systems is that it often oversimplifies to the extent of error.

The Modern View

Must we choose between two limited perspectives, vitalism or mechanism? Obviously not. Modern biologists accept the basic tenets of mechanism without accepting the analogy of the machine. They view organisms as very different from machines but do not postulate any vital or metaphysical force to account for this difference.

What are the characteristics of this modern view of living systems?

1. All the parts of any whole are interconnected. No part is completely separable from the rest. Thus a change in one part produces various types of changes in all the other parts.

2. Unlike machines, living systems are constantly in a process of change: growth, development, and degradation. This change results from the constant interaction of internal and external forces. For example, a seed sprouts, grows, and matures into a plant, which then flowers, produces fruit, and eventually dies.

3. The internal processes of any living system undergo change as a result of the interaction of opposing forces (see Fig. 1.5). All living organisms, for example, carry out chemical reactions. Thus the growth and development of a seed represents a change in which the overall effect of anabolic (buildup) reactions is greater than that of catabolic (breakdown) reactions. Maturity occurs when the two are balanced, and aging and death result from the dominance of catabolic over anabolic processes. Far from being accidental, this development process is programmed into the genetic and physiological makeup of the organism.

4. The accumulation of many small quantitative changes can eventually lead to a large-scale, qualitative change. For example, the heating of water from 90° to 91°C represents a quantitative, but not a qualitative, change in temperature, since the water has not been fundamentally changed (it is only 1 degree hotter). However, when the temperature goes from 99° to 100°C, the water begins to boil, turning into steam. This represents a qualitative change; water and steam have very different physical properties. The accumulation of many quantitative changes has led to an overall qualitative change.

This principle is as true of the biological as of the physical world. If a nerve attached to a muscle is stimulated with a low-voltage electric shock, the muscle will not respond. If we keep increasing the voltage, we will eventually reach a point called the **threshold**, at which quantitative changes give rise to a qualitative change; *i.e.*, the muscle contracts.

5. In living systems, the whole is greater than the sum of its parts. While this statement may sound metaphysical, it is not. If one of the characteristics of any part is interaction with other parts within the whole, then a description of the part in isolation is not complete. Hence describing each of a collection of parts in isolation will not be equivalent to describing the whole interacting system. Similarly, it is impossible to discuss fully the problem of

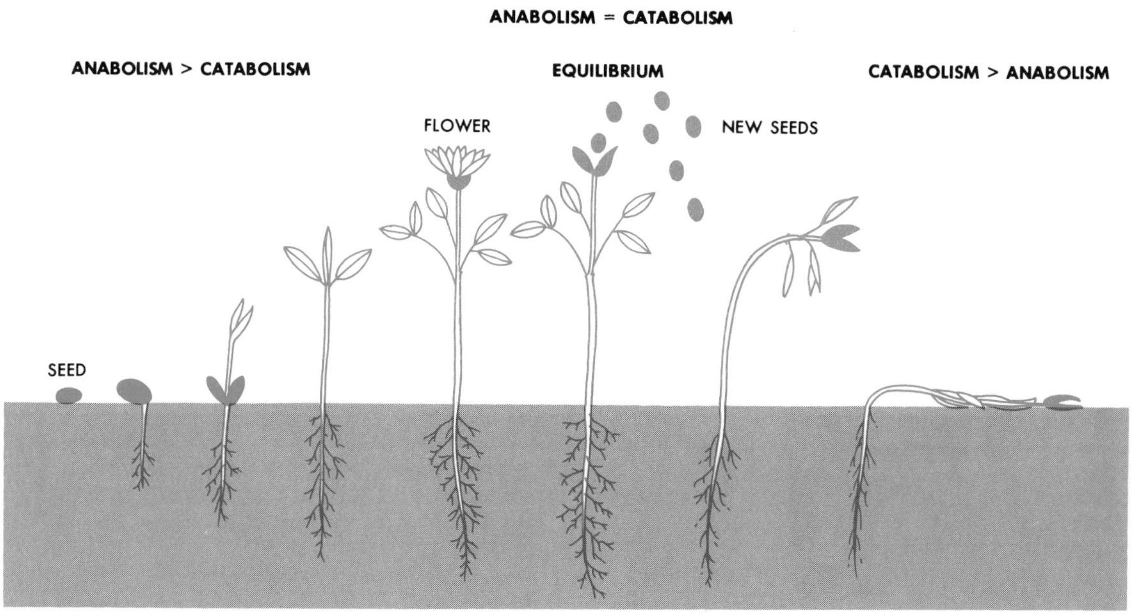

Fig. 1.5 *The developmental processes characteristic of all biological systems are the result of the interaction of many contradictory tendencies inherent in those systems. The life of a plant is shown from seed through death, illustrating the two opposing processes of anabolism (build-up of substances) and catabolism (breakdown of substances). Both go on all the time in all organisms. In the first part of its life, the plant grows and produces seeds because anabolic processes occur in greater number than catabolic processes. In the last part of its life, the plant withers and dies because catabolic processes become the more frequent. This interacting set of processes produces developmental patterns, rather than merely random changes.*

water or air pollution if the industrial needs, labor problems, or general energy needs of the community are ignored. These processes are so intimately connected that the task of cleaning up the environment becomes far more than the sum of solutions to the seemingly isolated problems of water pollution, air pollution, energy needs, unemployment, and the like. A part of the solution of each resides in the solution of all the others.

Realizing that all the parts of a system—living or nonliving—interact and that one cannot accurately consider any one part in isolation, it might appear that studying any complex system is hopeless. How can *all* the parts of a system be studied at once, in all their interactions? There are two answers to this dilemma. First, it *is* still possible to study individual components of a system in isolation if the components are also studied as part of the whole. Second, the computer and the mathe-

---

There are two types of hypotheses in science: generalizing hypotheses (generalizations) and explanatory hypotheses (explanations).

---

matical theory of systems analysis represent important new techniques for studying complex, interacting systems.

These views are characteristic of the way increasing numbers of biologists are coming to see living systems. It is important to be familiar with them, since an investigator's philosophical viewpoint often affects the kinds of questions asked and the kinds of answers considered acceptable.*

1.7 The Limitations of Science

Science is one of our most productive ways of exploring, exploiting, and trying to understand our environment. But it is by no means the *only* way. The historian tries to understand the present (and occasionally to predict the future) by studying the record of the human past. Religion attempts to find certain truths by operating mostly from a platform of faith. Philosophers draw on science, history, religion, and many other fields to consolidate findings in each field and draw meaningful conclusions from them.

Despite its many contributions to human intellectual growth, health, and general welfare, science does have serious limitations. Oddly, one of these stems from one of its greatest strengths. By dealing only with sets of phenomena that can be experienced directly or indirectly through the senses, science is necessarily excluded from other sets of phenomena. Such a qualification rules out any involvement of science in the supernatural. As the biologist George Gaylord Simpson puts it, "This is not to say that science necessarily denies the existence of immaterial or supernatural relationships, but only that, whether or not they exist, they are not the business of science."

The philosopher George Boas has stated that "... what science wants is a rational universe, by which I mean a universe in which the reason has supremacy over both our perceptions and our emotions." Yet what science wants and what science gets are often two different things. For one, nothing in the foregoing statement should be read to imply that scientists as individuals are any more rational, objective, or unemotional than anyone else. Nor is science itself necessarily independent in its thought and action from the society that surrounds and supports it; often the direction in which scientific research swings may be dependent upon the availability of funds allocated by governmental agencies to support that research.

Further, the solutions to purely scientific problems worked out by a scientist may be used by society in a manner that is repugnant to him or her. For example, Yale biologist Dr. Arthur Galston's research on the chemical basis of the shedding of leaves by plants was later used for defoliation of forests in Vietnam. Such occurrences point to a problem in how science is organized and practiced within a country rather than any inherent flaw in the scientific research itself.

Despite what we have outlined above, it is not safe to think that scientists always reason correctly. As a matter of fact, some scientists are notorious for "going off the deep end," particularly when writing in areas outside their own specialty. Furthermore, if scientists can be wrong, so can science.

* The five points outlined above are recognizable as basic tenets of "dialectical materialism." This philosophical school is contrasted with the philosophy of mechanistic materialism, or "mechanism," to which it has been opposed throughout the nineteenth and twentieth centuries.

"In nature we never see anything isolated but everything in connection with something else...."
—Johann Wolfgang von Goethe (1749–1832)

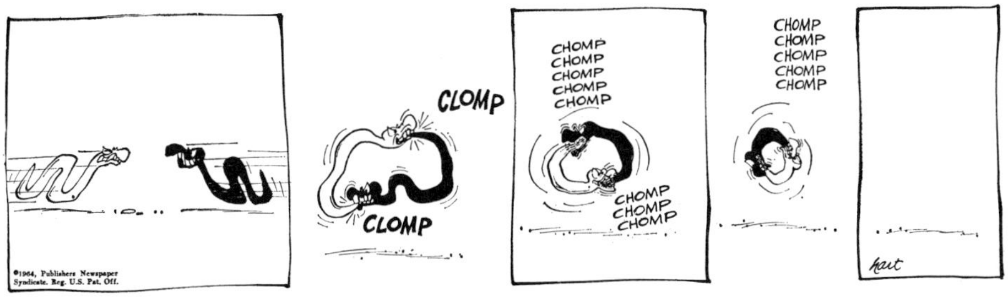

Fig. 1.6 *The consequences of logical thinking do not always lead to "correct" or even logical conclusions. It is a well-known fact of observation that snakes can and sometimes do swallow other snakes, even larger ones than themselves. If it is assumed that the one that is swallowed disappears from sight, the situation depicted is the logical conclusion! (B. C. by permission of Johnny Hart and Publishers Newspaper Syndicate.)*

Ask a physicist about ether, a chemist about phlogiston, or a biologist about Lamarckism. Science has had incorrect theories in the past, it has them now, and it will continue to have them in the future. The strength of science does not lie in any infallibility. Nor does it lie in its logical basis, for the conclusion of a perfectly logical argument can be utter nonsense (see Fig. 1.6). Rather, it lies in the self-criticizing nature of science—the constant search for "truth" by eliminating experimentally established error.

Science is an organized and continually changing body of knowledge based on observation, generalization, and experimentation. It is a disciplined endeavor with a tradition of beliefs that have rational foundations subject to continual review and discussion. Science is separate from the scientists who have contributed to its growth. As an individual, the scientist is only a human being with human emotions and weaknesses.

Despite its limitations, science is a remarkably successful way of accumulating knowledge. In applied research, scientists may use the methods of science for the purpose of developing products to improve human comfort and welfare. In pure or basic research, the scientist searches for knowledge—knowledge for its own sake—regardless of whether or not discoveries will benefit humankind. It is important to note that the results of basic research have contributed as much as or more than those of applied research; indeed, the former often leads to the latter. In science, applications seem to flow naturally from understandings, and the field itself seems to be inherently productive.

SUMMARY

1. Few if any characteristics of the so-called "scientific method" are truly unique to the natural sciences. Like most processes that require imagination, the intellectual creativity of many scientists has defied attempts to break it down into component parts for analysis. Nonetheless, two types of logical thinking, induction and deduction, can be identified in science.
2. Inductive logic is a creative process in which general statements or generalizations are derived from a specific number of discrete observations. A generalization goes beyond a finite number of observations to include all objects or phenomena of that same class. Observing that a few green apples are sour, one might arrive inductively at the generalization "All

green apples are sour." The more specific observations made, the more likely the inductive generalization is correct.

3. Deductive logic, or *if . . . then* reasoning, is the process of inferring from a general hypothetical statement (hypothesis) some specific consequence or prediction that must follow. The "if" represents the hypothesis and denotes its tentative nature as a generalization or explanation. The "then" represents a prediction about what will be true if the hypothesis is true. If all green apples are sour, then the next green apple tasted will be sour. If the prediction is not verified, the hypothesis is false. If the prediction is verified, the hypothesis is supported but not proved.
4. Valid statements are those that follow logically from hypotheses. True statements are those that conform to people's experience. Since a statement may be valid yet still untrue, there is no necessary connection between validity and truth.
5. There are two types of hypotheses. "All green apples are sour" is a generalizing hypothesis. "Green apples are sour because they have a higher concentration of a certain acid than do sweet apples" is an explanatory hypothesis. There is constant interaction between observation and hypothesis in science. Typically, observations lead to the formulation of hypotheses that, in turn, suggest further observations.
6. Mechanism and vitalism are two very different philosophical positions from which to view living organisms. Vitalists claim that some undetectable vital force makes the difference between living and nonliving things. Mechanists claim that nothing unknowable, no special force, resides in living things. The modern view is that though living systems do not contain any special force or vital property, they are not simply organized collections of chemicals. Living organisms are seen as entities containing many separate but interacting parts. This interaction modifies the parts continually, so that all living systems change through time. Living organisms represent a constant interaction of two opposing forces: the constructive process, anabolism, and the destructive process, catabolism.

EXERCISES

1. Distinguish between basic (pure) and applied research. Is it important for both kinds of research to be supported? If so, why?
2. List some of the limitations of science.
3. Explain why the attainment of any absolute truths lies beyond the realm of science.
4. Why must a scientist be careful not to extend experimental conclusions to organisms other than those with which he or she worked?
5. In a recent presidential campaign, some noted scientists and engineers formed a group to support one of the candidates. A spokesperson for this group made the following statement:

   *By the time we were through, any guy in Pittsburgh in a T-shirt with a can of beer in his hand knew that the smartest people in this country considered [our opponent] . . . . to be unfit [for the presidency].**

---

* *Science,* vol. 146, no. 3650, p. 1444.

What are the tacit assumptions made in this statement? Can you see any possible dangers that might arise from acceptance of such reasoning?

6. Explain why the following hypothesis is unacceptable to scientists: Life originated on another planet somewhere else in the universe and came to earth millions of years ago enclosed in a meteorite.
7. Devise an hypothesis to explain each of the following observations. Then outline an experiment to test your hypothesis.
   a) There are more automobile accidents at dusk than at any other time.
   b) When glass tumblers are washed in hot soapsuds and then immediately transferred face downwards onto a cool, flat surface, bubbles at first appear on the outside of the rim, expanding outwards. In a few seconds they reverse, go under the rim, and expand inside the glass tumbler.
   c) It has been noticed that one species of mud-dauber wasp will build its nests from highly radioactive mud, although the radiation received by the developing young may be enough to kill them. Under the same environmental conditions, another species of mud-dauber wasp avoids this mud and selects nonradioactive mud to build its nests.
   d) In mice of strain A, cancer develops in every animal living over 18 months. Mice of strain B do not develop cancer. If the young of each strain are transferred to mothers of the other strain immediately after birth, cancer does not develop in the switched strain-A animals, but it does develop in the switched strain-B animals living over 18 months.

SUGGESTED READINGS

Baker, Jeffrey J. W., and Garland E. Allen, *Hypothesis, Prediction and Implication in Biology*. Reading, Mass.: Addison-Wesley, 1968. Part II contains further analyses of the type applied in this chapter to the work on the precise causes of death in mice exposed to radiation. One of these analyses involves a controversy between researchers. Part III deals with a "Letters-to-the-Editor" debate on the need (or lack of need) for a scientific investigation of racial differences and intelligence.

Bronowski, Jacob, Jr., "The Creative Process." *Scientific American*, September 1958, p. 58. In this article the author tries to show that innovation in science is not different from innovation in the arts or social sciences.

Gardner, M., *Fads and Fallacies in the Name of Science*. New York: Dover Publications, 1957. A well-written, very interesting account of a number of pseudoscientific theories, how they gained adherents, and how most of them can be shown to be invalid. Discusses "Bridey Murphy," "Atlantis," "flying saucers," Lysenkoism, Creationism, and extrasensory perception (ESP).

Hull, David, *Philosophy of Biological Science*. Englewood Cliffs, N.J.: Prentice-Hall, 1974. A concise account of some of the many philosophical problems encountered in theory development and testing in the biological sciences. Not easy reading, this book is a good introduction for the serious student.

Kottler, Malcolm. "From 48 to 46: Cytological Technique, Preconception, and the Counting of Human Chromosomes." *Bull. Hist. Med.* 48 (1974), p. 465. This excellent article gives a fuller account of the case history described in Section 1.3.

Munson, Ronald, *Man and Nature: Philosophical Issues in Biology*. New York: Dell, 1971. This is a collection of essays by many different authors. One of the best is Ernst Mayr's "Cause and effect in biology," which includes our example of the migrating warblers.

Taton, René, *Reason and Chance in Scientific Discovery.* New York: Wiley, 1962. A good account of the role of chance, error, and inspiration in scientific discoveries, as well as an analysis of the relation between science and culture of the times. A very intriguing and thought-provoking work.

Weill, Andrew T., Norman E. Zinberg, and Judith M. Nelson, "Clinical and Psychological Effects of Marihuana in Man." In *The Process of Biology: Primary Sources,* ed. Jeffrey J. W. Baker and Garland E. Allen. Reading, Mass.: Addison-Wesley, 1970. An excellent example of a paper showing the strengths as well as the limitations of scientific investigation.

TWO

MATTER, ENERGY, AND CHEMICAL CHANGE

2.1 Introduction

"Getting and expending energy: that is the basic function of life." This statement captures one of the most important aspects of all living systems. Regardless of what else they do, all organisms must first and foremost provide an energy base for their activities. All organisms must extract energy from raw materials such as sugars, starches, and other foods and convert that energy into useful work. There would be no such thing as life without energy.

Living organisms are constantly involved in **energy flow.** They take in energy from outside themselves and convert it into some useful form. A portion of that useful energy is then released to carry out some life process, after which it leaves the organism once again and is dispersed into the environment. The nature of energy intake is different for different kinds of organisms. Photosynthetic organisms such as green plants and some kinds of bacteria are able to use solar energy directly to produce useful energy. Nonphotosynthetic organisms such as animals, certain bacteria, and plants like yeasts must get their energy in the form of energy-containing substances in the environment (sugars, proteins, etc.). In either case, the organism is part of a flow system in which a certain portion of the energy taken in is extracted and released for the organism's own life processes.

Since all biological phenomena depend upon the energy-getting and energy-using processes in organisms, an understanding of these processes is essential for investigating living systems. It is first necessary to establish some general principles of how chemical reactions occur: the atoms, molecules, or ions involved and the energy exchanges taking place. The present chapter will focus on the interaction of matter and energy.

2.2 Matter and Energy

Matter can be directly defined as anything that has mass and occupies space. The fundamental unit of

All forms of energy are interrelated and interconvertible. The process of interconversion is called the transformation of energy. It always occurs with some loss of usable energy—energy transformations are never 100% efficient.

matter is the **atom,** which in turn consists of particles known as **protons, neutrons,** and **electrons.** **Energy,** on the other hand, must be defined *indirectly* in terms of the movement of matter, i.e., in terms of work.

It is often useful to speak of energy as existing in one of two states, kinetic and potential. **Potential energy** is energy that is stored or inactive. This energy is capable of performing **work** and is frequently referred to as **free energy** (symbolized as G).* A stick of dynamite represents a great deal of potential energy; in quantitative terms it can be said to have a high free-energy value. In a biological context, sugar or fat also represents potential energy, though the free-energy value of either of these substances would be considerably less than that of an equivalent quantity of dynamite.

Kinetic energy is energy in action. It is energy in the process of having an effect on matter, and thus in the process of doing work. A boulder perched on the top of a hill has potential energy. If the boulder is given a slight push and begins rolling down the hill, the potential energy is released as kinetic energy.

There are five forms of energy, each of which exists in either the potential or the kinetic state. These forms are chemical energy, electrical energy, mechanical energy, radiant energy, and atomic energy. The last of these has little direct relationship to the normal functioning of the individual organism, but the others are all directly involved in living systems.

All these forms of energy are interrelated and interconvertible. The conversion of one form of energy to another goes on continually. It is the basis upon which all living organisms maintain themselves. For example, kinetic radiant energy from the sun is converted into potential chemical energy in green plant cells by the food-making process, **photosynthesis.** When an animal eats the plant, it transforms the potential chemical energy of the plant substance into further kinds of chemical energy (by building its own kind of molecule) or into mechanical energy (for movement). Energy transformation is the basis of all life.

If the total amount of radiant energy transmitted to a green plant in a given period of time is measured, it will be found that only a small percentage of the total available energy is captured as potential chemical energy. The same is true of any step in the transformation from one form of energy to another. *The transformation of energy is never 100% efficient.* None of the energy in such transformations is actually lost in the sense of being unaccounted for. Long ago physicists formulated the very important **law of conservation of energy** or, as it came to be called, the **first law of thermodynamics.** This law states that, during ordinary chemical or physical processes, energy is neither created nor destroyed; it is only changed in form.†

* The letters F and G are both used to represent "free energy." F is an older term taken from free energy; G is a more recently used symbol from the name of J. Willard Gibbs (1790–1861), one of the founders of the field of thermodynamics.

† This generalization does *not* apply to atomic fission or fusion reactions, where a very small amount of mass is converted into energy.

2.3 Atomic Structure

Early in this century it was suggested that atoms were composed of a small, dense, central portion—the nucleus—surrounded by various numbers of other particles—the electrons. In 1913, Niels Bohr (1885–1962) suggested that the atom resembled a tiny solar system, with the nucleus representing the sun and the electrons the orbiting planets. Later, the nucleus was shown to contain two types of particles: protons, which are positively charged, and neutrons, which carry no charge. The nucleus was seen to carry a positive charge because of the protons. The electrons circling the nucleus were found to have a negative charge, offsetting the positive charge of the protons. Figure 2.1 shows this general conception of the atom.

Electrons are now conceived of as a negatively charged haze or cloud of particles outside the positively charged nucleus of an atom. Electrons move about the nucleus at varying distances from it, traveling at a high velocity. It is the electrons that are most directly involved in chemical reactions. Electrically neutral atoms have equal numbers of electrons and protons. Under certain conditions, however, an atom can gain or lose electrons. In this way, it acquires a negative or positive charge and becomes an **ion.** When atoms interact with one another by giving up, taking on, or even sharing electrons, a chemical reaction occurs. The exchanges and interactions of electrons among atoms form the basis of chemical reactions and thus of all life processes.

Electrons and Energy

The electron cloud around the nucleus of an atom is composed of electrons in different **energy levels.** Energy levels are roughly analogous to the "planetary orbits" suggested by Niels Bohr. The term "energy level" is an expressive one, since it leads to the idea of electrons as particles possessing certain amounts of potential energy. Electrons can be pictured as moving in specific energy levels outside the atomic nucleus (see Fig. 2.2). The electrons neither absorb nor radiate energy so long as they remain in these energy levels. However, should one or more electrons fall from the energy level they occupy to a lower one, they will *radiate* a precise amount of energy. If energy is *absorbed* by the atom, one or more electrons may jump from a lower energy level to a higher one.

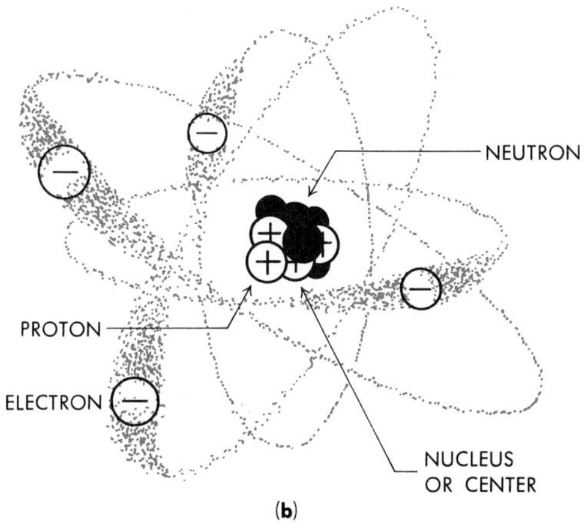

Fig. 2.1 *(a) The modern conception of the atom, showing the dense, centrally located nucleus and the outer haze of electrons. (b) Diagrammatic sketch of the atom showing its parts. Negatively charged electrons circle the nucleus of protons and neutrons. This diagram represents a working model of the atom. It does not, in any real sense, represent a "picture" of the atom.*

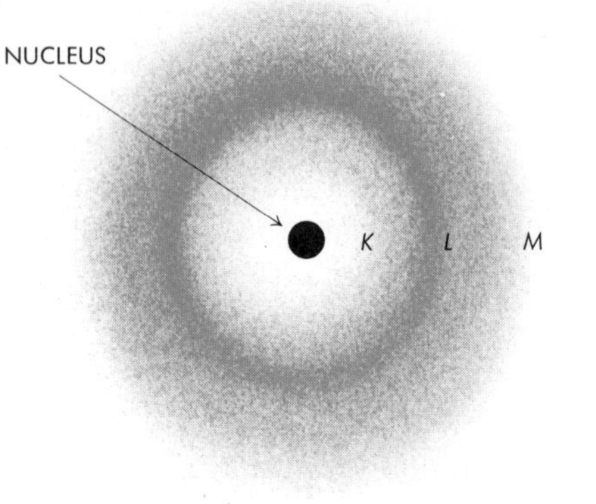

Fig. 2.2 *Diagrammatic representation of an atom, showing three energy levels:* K *represents the lowest energy level,* L *the next highest, and so on. The electrons are shown as a haze because their positions can be defined only in statistical terms.*

An atom can have a large number of energy levels. Indeed, it is possible to recognize *seven* energy levels in which the electrons can be found. These seven energy levels are known as the *K, L, M, N, O, P,* and *Q* levels. Each energy level has a certain maximum number of electrons that it can hold. The seven energy levels are listed in Table 2.1 along with the maximum number of electrons actually found in the isolated, ground-state (i.e., lowest energy state) atoms. More electrons are theoretically possible than are actually found, especially in the *N, O, P,* and *Q* levels; the maximum number of electrons theoretically possible in each energy level is given by the expression $2n^2$, where n is number of the energy level ($K = 1$, $L = 2$, $M = 3$, etc.).

An attractive force exists between the negatively charged electrons and the positively charged nucleus of the atom. This force is greatest at the first energy level and falls off in successively more distant levels. This means that electrons in the outer energy levels are more easily removed from an atom than those close to the nucleus.

Table 2.1

Major energy level	*K*	*L*	*M*	*N*	*O*	*P*	*Q*
Maximum number of electrons	2	8	18	32	32	18	2

Electrons, then, are attracted to the nucleus of the atom. Electrons farther from the nucleus contain more potential energy because energy was required to get them this greater distance from the nucleus. An electron in an outer energy level releases more energy in falling to the lowest inner level than an electron in an intermediate level falling to the same position. Similarly, a satellite 1000 miles from the earth's surface releases more energy in falling to the ground than one 500 miles up.

In an atom electrons can jump to a large number of energy levels, depending on the amount of energy supplied. A small but definite amount of energy will cause an electron to jump only to the next higher level. The right amount of additional energy, however, may cause the electron to jump farther, perhaps so far from the nucleus that it escapes completely from the atom to which it originally belonged. The loss of an electron gives the atom a charge of $+1$, since it now has one more proton than electrons. The electrically charged atom is now an ion. Loss of two electrons would give the atom a charge of $+2$, and so on. Ions can also be formed by the *gain* of electrons. Such ions would be negatively charged. The process of gaining or losing electrons is called **ionization.**

Atoms store potential energy when electrons are raised to higher energy levels; they release energy when electrons fall to lower energy levels.

The first law of thermodynamics (law of conservation of energy) states that in any chemical or physical change, the total amount of energy involved remains the same. Energy is neither created nor lost, but only changed in form.

Raising electrons to a higher energy level produces some gaps below. Some of the lower energy levels are thus incomplete. These are usually filled by other electrons, which fall down from higher energy levels. Just as an electron absorbs energy to jump to a higher energy level, so it releases the same amount of energy in falling back to a lower position.

The movement of electrons to lower energy levels may release energy as X-rays, visible light, or other wavelengths of radiation. The wavelength emitted by an electron in making a downward transition depends on the distance it falls and the type of atom in which the transition occurs (see Fig. 2.3).

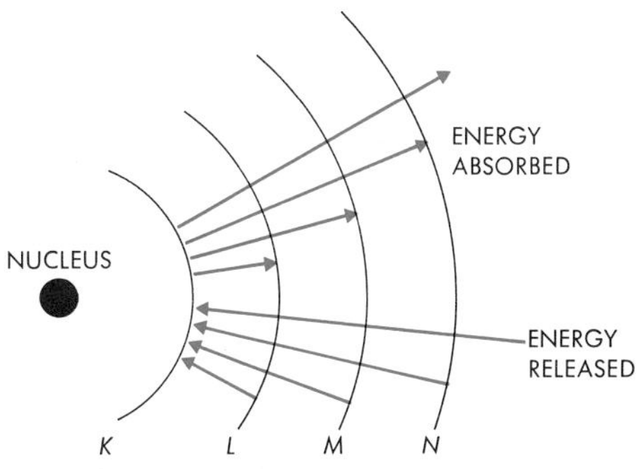

Fig. 2.3 *Electron jumps occur when an atom is supplied with energy. Outward jumps may occur from any energy level to any other, depending on how much energy the electron absorbs. In the diagram, outward jumps are shown only from the K level. When electrons fall inward they give off precise amounts of energy, which appears as X-rays, visible light, and other types of radiant energy.*

When an electron makes a transition, it absorbs or releases distinct amounts of energy. The quantum theory, first proposed by the German physicist Max Planck in 1900, holds that energy can be described as coming in discrete packets, or **quanta.** The light emitted by an incandescent lamp, for example, consists of millions of discrete quanta or **photons** which, because they have some physical properties, can interact with matter. Although the reality of energy "packets" is debated among physicists, the quantum model is useful in understanding how energy can cause or result from electron transitions. If an atom absorbs one or more quanta of radiant energy, electrons at certain energy levels jump to a higher level. The energy is thus temporarily captured by the atom. It may then be released in ways that allow electrons in other atoms to make transitions also. Thus energy can be transferred among groups of atoms by the movement of electrons from one energy level in one atom to a different energy level in another.

Atoms to Molecules

Under ordinary conditions, atoms are often combined with other atoms. Such a combination of two or more atoms is called a **molecule.** When atoms unite to form molecules, the energy levels of the individual atoms interact to form **molecular energy levels.** Electron transitions are possible in the energy levels of molecules, just as they are among energy levels of individual atoms. When a molecule of chlorophyll, for example, absorbs quanta of light energy from the sun, electrons are raised to higher energy levels. The electrons are temporarily lost to the molecule as a whole. But when the electrons make downward transitions, they release the same amount of energy they absorbed. This energy is captured by the plant cell and used to power certain chemical reactions during photosynthesis.

Chemical Bonds

The atoms making up a molecule are held together by **chemical bonds.** A chemical bond is not a physical structure. It is simply an energy relationship between atoms that that holds them together in a molecule.

To understand how chemical bonds are formed, consider the interaction of atoms A and B to form molecule AB. First, A and B must come close enough together for their electron clouds to overlap; rearrangements of electrons between the two atoms is the basis for the formation of chemical bonds. As A and B approach each other, however, they exhibit mutual repulsion as the result of their negatively charged electron clouds (like charges repel, unlike attract). We have to do work (put in energy) to push the atoms still closer together. It is not until the atoms are forced so close together that an electron from one is attracted to the positively charged nucleus of the other that the rearrangements that make the two atoms have a net attraction for each other can take place.

All interactions between atoms to make or break chemical bonds involve the exchange of energy. To break any chemical bond, a certain input of energy (called dissociation energy) is required to overcome the stable state and move the atoms far enough apart so that their mutual repulsion again takes over and sends them on their opposite ways. This energy varies from one type of atom to another. For example, it takes a large amount of energy (104.2 kcal/mole) to break hydrogen-to-hydrogen bonds, but a smaller amount (50.9 kcal/mole) to break sulfur-to-sulfur bonds.*

The above description does not refer to the overall or net energy exchange during chemical reactions, but only to the part leading from the top of the energy hill downward toward stable bond formation.

The overlapping of electron clouds causes rearrangement of electrons in the outermost energy level of each atom. This rearrangement involves one of two possibilities: (1) one atom will *give up* one or more of its electrons to the other or (2) each atom will *share* one or more electrons with the other.

The interaction between outer electrons is the result of a process in which each atom approaches a stable outer electron configuration. This process requires closer examination. It is central to an understanding of current thinking about the formation of a chemical bond.

With the exception of the *K* or innermost energy level, a stable outer electron configuration is achieved with eight electrons. The atoms of any element with eight outer electrons are stable. Neon, argon, krypton, xenon, and radon are all examples of such elements. Such atoms do not generally react with other atoms. Most atoms, however, have fewer than eight electrons in their outer energy level. These atoms tend to reach the stable configuration by giving up, taking on, or sharing electrons. The driving force behind any chemical reac-

* Bond energies are measured usually in terms of kilocalories per mole of reactant. A kilocalorie (kcal) is the amount of heat energy required to raise the temperature of 1.000 g. of water from 15° to 16° C. A mole is a fixed number of molecules, given as Avogado's number, or 6.024×10^{23}.

Chemical bonds represent energy relationships between two or more atoms. These energy relationships involve changed electron orbitals among the atoms involved.

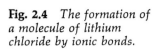

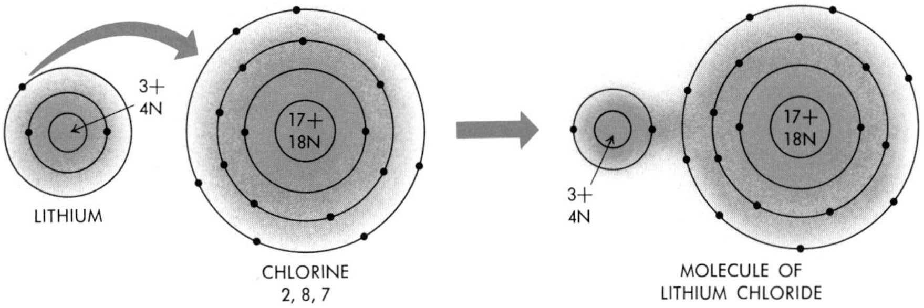

Fig. 2.4 *The formation of a molecule of lithium chloride by ionic bonds.*

tion originates in the tendency of atoms to attain a stable outer energy level.

Ionic Bonds

Two major types of chemical bonds are found in chemical compounds. They are distinguished by the way the stable condition of the outer energy levels is reached.

The first type of chemical bond, found most frequently in inorganic compounds, is the **ionic** or **electrovalent** bond. In the formation of this type of bond, one atom gives up its outermost electrons to one or more other atoms. When this occurs, the outermost energy level of each atom becomes more stable. The formation of lithium chloride from the elements lithium and chlorine is an example of ionic bonding (see Fig. 2.4).

The electron configuration for chlorine, counted from the nucleus outward, is 2, 8, and 7. Atoms of lithium have a single electron in the *L* level. Lithium can reach stability by giving up its one *L* electron to chlorine. This gives the chlorine atom a total of eight electrons in its *M* level—which represents stability for this element. Both atoms now have a stable electron configuration in their outer energy levels.

When lithium gives up its electron, the atom has one less negative charge. Hence it is positively charged (+1). Chlorine, by accepting an electron, now has one more negative charge than positive charges and thus is negatively charged (—1). Since opposite charges attract, the positively charged lithium atom and the negatively charged chlorine atom attract each other. This attraction holds the two atoms together. In this way a molecule of lithium chloride, LiCl, is formed.

There is no 100% ionic bond. Though one atom tends to give its electrons to another, this "handing over" is not complete. The donated electron may still occasionally circle the nucleus of the donor atom.

In the formation of ionic bonds, which atoms will give up electrons and which will receive them? In general, those atoms with fewer than four electrons in the outer energy level tend to lose electrons. Those with more than four tend to gain electrons. Atoms such as sodium, potassium, hydrogen, calcium, and iron possess three or fewer outer electrons. Atoms such as oxygen, chlorine, sulfur, and iodine need one or two electrons to complete their outer energy levels. Thus these atoms tend to take on electrons.

All chemical bonds represent stored chemical energy. Some bonds contain more energy than others. To release this energy bonds must be broken, and this in turn requires the investment of some energy.

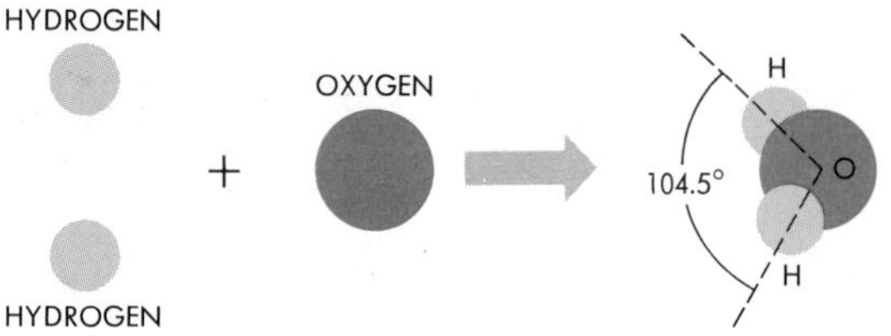

Fig. 2.5 *The formation of water from hydrogen and oxygen. Since the outer electron level of oxygen contains only six electrons and each hydrogen has only one electron to donate, two hydrogen atoms are required to satisfy the stability requirements of the oxygen.*

Covalent Bonds

What about an atom such as carbon, which has four electrons in its outer energy level? Does such an atom tend to give up or take on electrons when combining with other atoms?

Carbon combines with atoms of many other elements by forming **covalent** chemical bonds, a sort of "compromise" between the giving up and the taking on of one or more pairs of electrons between atoms. In such bonding, atoms combine by undergoing a rearrangement of electrons in their outer energy levels. Neither atom loses its electrons completely. Instead the electrons are shared and may orbit the nucleus of any atom in the molecule.

The formation of a molecule of the gas methane illustrates the principle of covalent bonding. Under suitable conditions carbon reacts with hydrogen to form molecules of methane. Four atoms of hydrogen react with each atom of carbon to produce a symmetrical molecule, CH_4:

$$\begin{array}{c} H \\ | \\ H\!-\!\overset{\displaystyle |}{\underset{\displaystyle |}{C}}\!-\!H. \\ H \end{array}$$

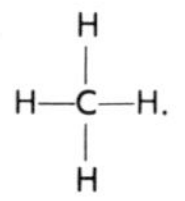

Each line between the carbon atom and a hydrogen atom represents a single pair of shared electrons. The pair consists of one electron from the carbon atom and one from the hydrogen atom. This may be shown more clearly by writing the molecular formula in the following manner:

$$\begin{array}{c} H \\ \overset{ox}{} \\ H \overset{\circ}{\underset{\circ}{\,}} C \overset{\circ}{\underset{\circ}{\,}} H. \\ \overset{ox}{} \\ H \end{array}$$

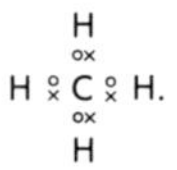

The open dots represent the outer electrons originally in the *L* energy level of carbon. The crosses represent the electrons originally in the *K* level of each hydrogen atom.

Why does this type of bonding take place? The carbon atom has four electrons in its outer energy level. To attain stability, carbon needs eight electrons. Each hydrogen atom has one electron. Hydrogen can reach stability by either losing or gaining one electron. In the formation of the covalent bond between carbon and hydrogen, the electrons in the outer energy levels of each atom orbit the nuclei of both hydrogen and carbon. As a result, each of the four hydrogen atoms has its own electron plus one electron from the carbon to orbit its nucleus. In turn, the carbon atom has not only its own four electrons but also one from each of the hydrogen atoms to orbit its nucleus. This completes the requirements for stability in the outer energy levels of both atoms. The sharing of these outer electrons produces the covalent chemical bond.

When two or more atoms combine to form a molecule, a predictable geometric relationship is established between the atoms involved. The molecule takes on a definite shape. For example, when two atoms of hydrogen combine with one of oxygen to form a molecule of water, the hydrogens always lie at an angle of $104.5°$ apart (see Fig. 2.5). This distance is known as the **bond angle** and is always the same under identical circumstances. Thus molecules of the same kind always have the same geometry.

For the most part, molecules are represented in diagrams as if they were flat, two-dimensional structures. In reality, they are three-dimensional,

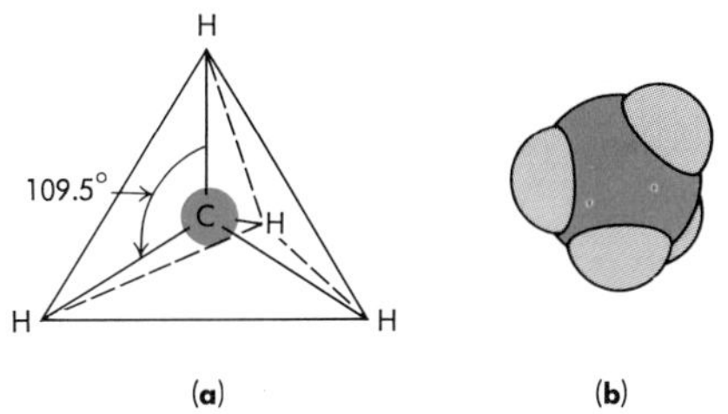

Fig. 2.6 *Two representations of the geometry of a molecule of methans (CH_4). (a) A diagram of the tetrahedral structure, showing the central carbon surrounded by four hydrogens all at equal distances from the carbon and from each other. (b) A space-filling model of methane, showing the actual volumes and geometric relations of the atoms. The atom in the center is carbon. The fourth hydrogen is partly hidden on the other side of the molecule.*

having depth in addition to length and breadth. For example, the organic compound methane, shown in a face on view on page 30, actually forms a solid, four-sided pyramid known as a tetrahedron (Fig. 2.6a). Rather than 90°, the bond angles between the four hydrogen atoms are actually 109.5°. It is therefore desirable to show the three-dimensional structure with a "space-filling" model (Fig. 2.6b). Here the space occupied by the atoms, as well as their orientation within the molecule, is taken into consideration.

The concepts of molecular configuration and bond angles play a role in explaining chemical reactions between molecules. These concepts will be important in later consideration of the larger molecules found in living organisms.

2.4 The Polarity of Molecules

A frequent consequence of the geometric shapes of molecules is a distinct separation of electric charge. This means that one portion of a molecule is positive or negative in relation to another portion of the same molecule. When such an uneven distribution of charge occurs, the molecule is said to exhibit **polarity.** The molecule has a positive and a negative end, separated from each other like the poles of a bar magnet. The charge results because the nuclei of individual atoms in the molecule hold more electrons close to them than do the nuclei of other atoms in the same molecule. Thus the area surrounding these electron-attracting nuclei becomes negatively charged, while the area around the electron-deprived nuclei becomes positively charged.

The water molecule illustrates this point. Although the water molecule as a whole is electrically neutral, it does have a positive and a negative end (Fig 2.7). The geometric configuration of the molecule places both hydrogen atoms at one end. The nucleus of the oxygen atom attracts electrons more than the nuclei of the hydrogen atoms. This results in two positively charged regions on one end of the molecule and a single negatively charged region on the other. The result is a molecule with a positive

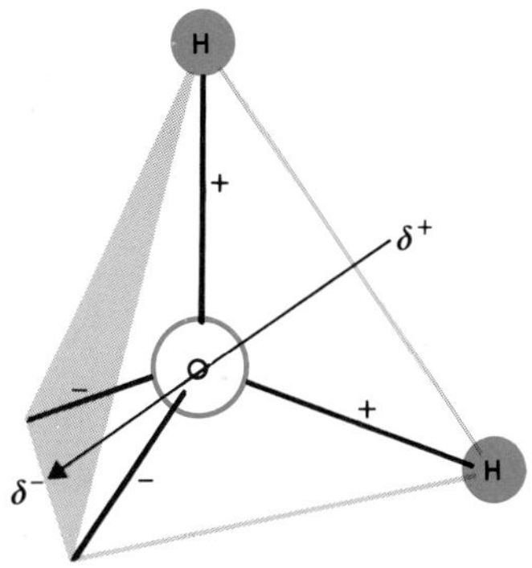

Fig. 2.7 *The geometry of a water molecule contributes to its polarity. Because of the strong force of attraction by the oxygen atom for the four outer electrons (which are unpaired), the centers of negative charge lie on the opposite side of the oxygen from the hydrogen atoms (i.e., to lower left). There is thus a separation of positive and negative charge such that one side of the molecule (the O—H side) is designated δ^+, while the other side is designated δ^-. The arrow pointing from δ^+ to δ^- indicates the direction of the charge distribution, which passes along a gradient from most positive to most negative. The Greek letter δ (delta) indicates "change in." (Modified from Fig. 6.2 of A. G. Loewy and P. Siekevitz,* Cell Structure and Function, *2d ed. New York: Holt, 1969, p. 91.)*

Types of chemical bonds of importance to biological systems include covalent bonds (sharing electrons), ionic bonds (more or less transferring electrons from one atom to another), hydrogen bonds (two negatively charged atoms such as O or N sharing a hydrogen atom), and electrostatic bonds (polar attraction between positively and negatively charged groups).

and a negative end, or two poles; the molecule is therefore polar.

Molecular polarity is significant to the biological sciences in two ways. First, polar molecules tend to become oriented in precise spatial patterns with respect to other molecules (either polar or nonpolar). Because of this, polar molecules are important in certain structural elements of organisms. For example, molecules of fatty acids, found in all living matter, are composed of a nonpolar carbon chain with a polar carbon–oxygen (carboxyl) group (COOH) at one end. When placed in water, the polar ends of the fatty acid molecules are attracted to water molecules, also polar. The nonpolar carbon chains are at the same time repelled by the water. As a result, fatty acid molecules are oriented on the water's surface (Fig. 2.8). Of particular importance to living things is the orientation of **phospholipid** molecules, each of which is a combination of a fat molecule with a phosphate group. Phospholipids are among the most important parts of cell membranes. They tend to become oriented on surface or boundary regions in a manner similar to fatty acids on water. It is partly in this way that cell membranes assume their distinct structure.

Second, polarity is important in understanding both the geometry and the chemical characteristics of large molecules such as proteins and nucleic acids. Proteins are so large they may possess a number of polar groups on one molecule. Polar groups are groups of atoms that as a unit bear a positive or negative charge. Polarity thus tends to

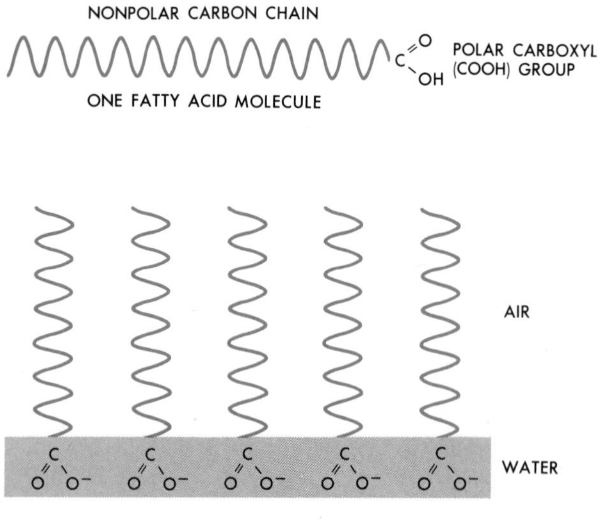

Fig. 2.8 *Polarity of fatty acid molecules determines their orientation at a water–air interface. The carboxyl group is a polar region of the fatty acid molecule. When carboxyl groups touch water, one hydrogen dissociates, leaving the carboxyl group as a whole negatively charged. It thus "dissolves" well in the polar water molecules. The carbon chain is nonpolar and does not dissociate in water. Hence the carbon chains do not dissolve and project out into the air. This is why all fats and oils float.*

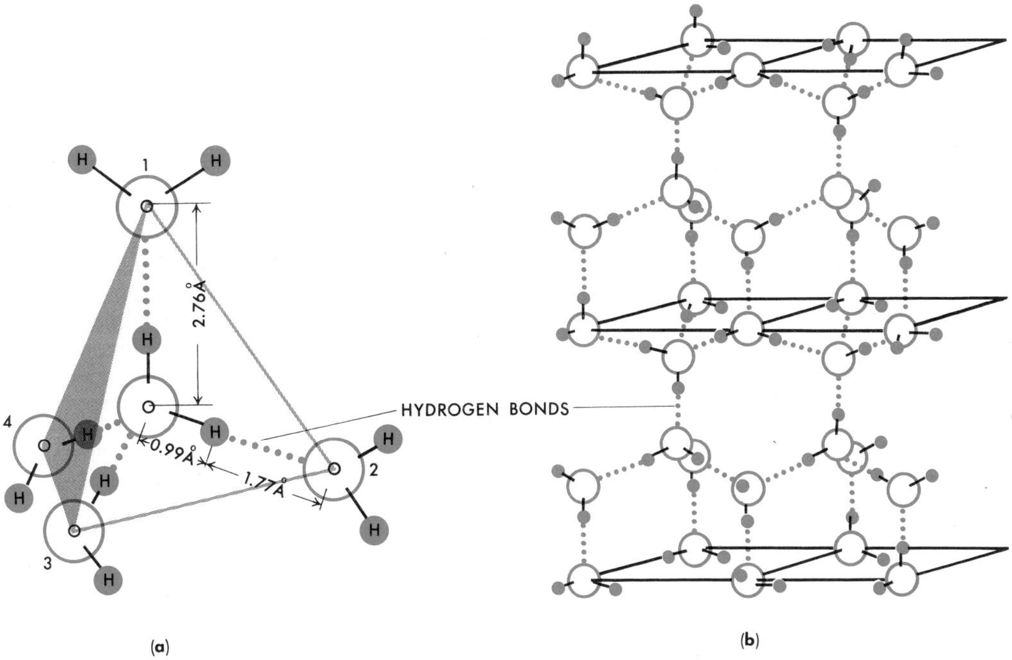

Fig. 2.9 *(a) Hydrogen bonding between five water molecules, showing the precise patterns of bond formation (represented by the dotted line). The five molecules of water form a tetrahedronal lattice. (b) A more complex organization of many water molecules, held in place by hydrogen bonds, forms a crystal of ice. The bonding patterns are identical to the pattern shown in (a) for just five molecules of water. Note the large spaces formed by the latticework. These spaces make the density of ice lower than that of water. (Modified from Fig. 6.3 of A. G. Loewy and P. Siekevitz,* Cell Structure and Function, *2d ed. New York: Holt, 1969, p. 92.)*

bring small molecules, or specific regions of large molecules, into definite geometric relation with each other. In this way, the chemical bonding between individual molecules or between specific regions on the same molecule is brought about more easily.

Hydrogen Bonds

In living organisms, one of the most common types of chemical bond produced by polar attraction is the **hydrogen bond.** Hydrogen bonds are produced by the electrostatic attraction between positively charged hydrogen atoms (protons) on one part of a molecule and negatively charged atoms of oxygen or nitrogen on the same or another molecule. The oxygen and nitrogen atoms tend to be negatively charged because their electrons are held closer to the nuclei and are shared less easily. Because the hydrogen bond occurs between polar regions of a molecule it is, like all polar attractions, relatively weak.

A simple example of hydrogen bonding can be seen between water molecules. The hydrogen atoms of one water molecule form a hydrogen bond with the oxygen atom of the adjacent molecule (Fig. 2.9).

Supplement 2.1

WATER AND LIFE

The living cell is approximately 90% water. Life as we know it occurs in a water medium and could not exist otherwise. The chemical and physical properties of water make it a unique medium for living organisms. In all cases this uniqueness stems from the polar nature of water molecules.

1. *Water is a good solvent.* Water serves as an excellent solvent for a wide range of substances, from uncharged organic compounds to salts such as sodium chloride, which are completely dissociated into ions even in the solid state. Any molecule that contains atoms or groups of atoms that are polar (have unpaired electrons or exposed protons) will attract, and be attracted by, polar water molecules. Hence, water molecules can "snuggle" around or between many other types of atoms and/or molecules. This "snuggling" dissociates like atoms or molecules from one another, just as it can dissociate an atom or group of atoms (as ions) from other groups of atoms. The solvent property of water makes it a universal agent for transporting substances to organisms. Accordingly, water is the liquid medium in which all single-celled organisms and the individual cells of multicelled organisms exist.

2. *Water molecules are highly cohesive.* Because of their polarity, water molecules tend to stick together; they even come to lie in some general patterned relationship to each other, held in position by the hydrogen bonds formed between positive and negative ends of the molecules (see Diagram A). The cohesiveness of water has several effects. One is that water tends to have a high **surface tension.** This is because

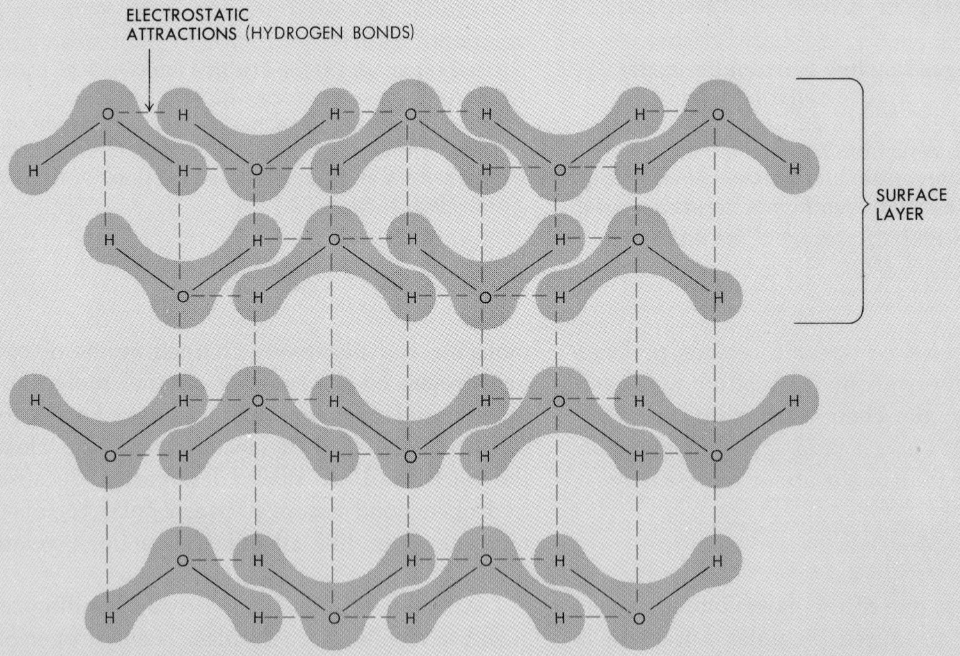

Diagram A

the topmost layer of molecules are pulled from beneath, but not from above. Consequently they pack more densely to the layer beneath them than do other layers of molecules under the surface. This densely packed layer at the top creates the surface tension. Because of this tension, many insects can "walk on water," whereas they cannot walk on other liquids such as alcohol or ammonium. Consequently, the surface layer of bodies of water serves as a habitat for many organisms. Cohesiveness also makes it possible to raise a column of water over greater vertical distances than noncohesive liquids. In a noncohesive liquid, a column (even a thin column) of liquid tends to break apart as it is moved upward. Water columns maintain their integrity as they are moved up the conducting tubes of trees, partly by the force of cohesion.

3. *Water has a high heat capacity and heat of vaporization.* Heat capacity is a measure of the amount of heat required to raise one cubic centimeter (1 cm^3) of a substance one degree Celsius (1°C). The higher the heat capacity of a substance, the more heat it will absorb to show a temperature rise. With its high heat capacity, water can serve to insulate the organism (or cell) from abrupt changes in temperature. Heat of vaporization is defined as the quantity of heat required, per unit mass, for the vaporization of a substance measured at a given temperature (most substances vaporize slowly at any temperature; generally, the higher the temperature, the more rapid the vaporization process). Water requires much more heat to vaporize than most other liquids. This means that water can absorb much more heat, in proportion to the number of molecules passing into the vapor state, than most other liquids. If water had a low heat of vaporization, organisms would lose considerably more water than they do. The high heat capacity and heat of vaporization characteristic of water result from the cohesiveness of the molecules. Because of their electrostatic attraction, it takes more energy to separate the molecules of water than for nonpolar, noncohesive molecules.

4. *Ice is less dense than water.* Because of the particular structure of hydrogen bonds between water molecules, the latticework of the ice crystal has large openings in the center (Fig. 3.10). As these openings are "empty," the ice crystal has less density than the water from which it is formed. This is why ice floats. The advantage of floating ice is considerable to living things. If ice sank, most of the world's rivers, lakes and oceans would be permanently iced in, with only a small surface of melted water in warm weather. With its high heat capacity, the surface water would provide a good insulator, preventing the ice at the bottom of lakes or oceans from ever absorbing enough heat to melt. Such a situation would profoundly alter the pattern of life on earth. Interestingly, other liquids possessing some properties similar to water, such as ammonia, form "ice" that is denser than the liquid. Water is unique in having a solid state less dense than its liquid state.

2.5 Ions and Radicals

At room temperature, the compound hydrogen chloride (HCl) is a gas. If molecules of this gas are dissolved in water, the hydrogen is separated from the chlorine. The separation or ionization occurs in such a way that the hydrogen atom does not take back the electron it loaned to chlorine in forming the bond. Thus the hydrogen atom, now a hydrogen ion or simply one proton, bears a charge of +1. The chlorine retains the extra electron it received from the hydrogen atom. Since it has one more electron than protons, the chlorine bears a negative charge of −1 and is thus a chloride ion. The hydrogen ion is written as H^+, and the chloride ion as Cl^-.

Recall that ions are formed whenever an atom loses or gains electrons. The dissolving of sodium chloride in water results in a separation, or **dissociation,** of the sodium and chloride ions (see Fig. 2.10). Since sodium chloride is an ionic compound, its component atoms are already in the ionic state.

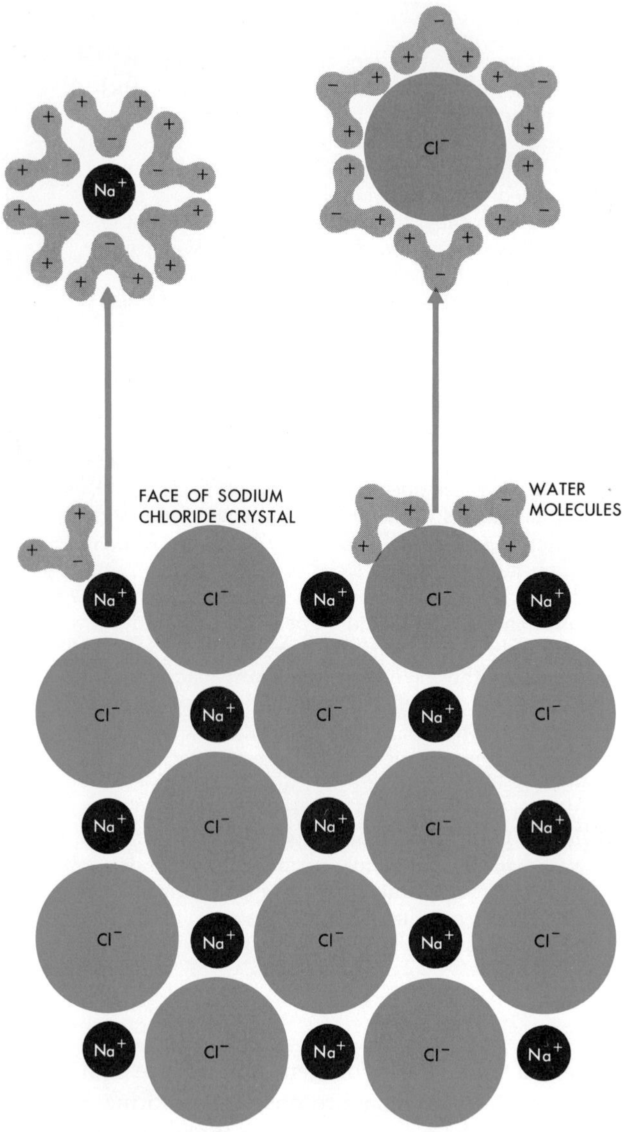

Fig. 2.10 *A representation of sodium chloride dissolving in water. The negative ends of the polar water molecules are attracted to the positive sodium ions and pull them off the crystal lattice. The positive ends of the water molecules are attracted to the negative chloride ions and pull them off. As indicated, each sodium and chloride ion is probably surrounded by at least six water molecules.*

Molecular substances, on the other hand, such as hydrogen chloride, lithium chloride, and acids, undergo ionization when entering into solution. As soon as the water is removed, the oppositely charged particles recombine to form the same molecules or **ion pairs.** The molecule of lithium chloride shown in Fig. 2.4 is an example of an ion pair.

The process of ionization can be represented by an ionization equation. In chemical language, an equation indicates what goes into and what comes out of a certain reaction. For example, the ionization equation for HCl indicates that one molecule of this compound yields, upon ionization, a positively charged hydrogen ion (H^+) and a negatively charged chloride ion (Cl^-):

$$HCl \rightarrow H^+ + Cl^-.\text{*}$$

Similarly, the dissociation of sodium chloride (table salt) gives

$$NaCl \rightarrow Na^+ + Cl^-.$$

When calcium chloride ionizes, it produces two chloride ions for every ion of calcium:

$$CaCl_2 \rightarrow Ca^{++} + 2Cl^-.$$

This equation tells us several things. First, it shows that a molecule of calcium chloride consists of one atom of calcium and two atoms of chlorine. The

---

* Hydrogen ions (H^+) do not exist in solution. As soon as a hydrogen ion (that is, a proton) is removed from a molecule, or dissociates from another ion as in the dissociation of HCl, it is picked up by some other molecule. In dissociation reactions such as that for HCl, the other molecule is inevitably water. This forms a hydronium ion: $H^+ + H_2O \rightarrow H_3O$ (hydronium). The complete reaction for the dissociation of HCl would thus be:

$$HCl + H_2O \rightarrow H_3O + Cl^-.$$

For practical purposes we will usually write hydrogen ions as H^+, but keep in mind that this is convenient notation, not chemical reality.

subscript after the symbol for any atom indicates the number of atoms of that element in the molecule. Second, in writing the ionization equation for this compound, the two atoms of chlorine must be accounted for by showing that there are two chloride ions in the solution. To indicate that these two ions do occur, a 2 is placed in front of the Cl^-. The equation is now balanced. Each of the atoms on the left side of the equation is accounted for on the right.

When some compounds ionize, one of the products is a **complex ion.** Complex ions are associations of two or more atoms which bear an overall positive or negative charge. They are thus collections of two or more atoms that act as a single ion. For example, the ionization of sulfuric acid yields hydrogen ions and a sulfate ion:

$$H_2SO_4 \rightarrow 2H^+ + \underset{\text{(sulfate ion)}}{SO_4^{2-}}.$$

The sulfate ion is composed of one atom of sulfur and four atoms of oxygen. These five atoms have an overall electric charge of −2.

Atoms held together by ionic bonds can separate more easily because one atom has given up electrons while another has accepted them. In this way, the outer energy level of each atom has been satisfied. When such molecules dissociate, no further exchange of electrons is required. Dissociation in this case merely involves overcoming the electrostatic attraction between positive and negative particles. The action of water molecules accomplishes this dissociation (see Fig. 2.10).

In covalent bonds, the outer energy level of each atom is satisfied only as long as the shared electrons revolve about both nuclei. For this to be possible, the atoms must remain close together. It is very difficult to separate one from another if the atoms are forced to assume unstable outer electron configurations. For this reason water molecules generally cannot force covalently bonded atoms apart. Such molecules fail to show ionization in water.

2.6 Oxidation and Reduction: Redox Reactions

Losing electrons during ionization is called **oxidation;** the atom that loses electrons is said to be **oxidized.** Gaining electrons is called **reduction;** the atom that gains electrons is said to be **reduced.**

The process of oxidation does not necessarily involve the element oxygen. The name "oxidation" was originally derived from the class of reactions involving the combination of various elements (mostly metals) with oxygen. Now, however, the term oxidation is used more broadly to refer to any loss of electrons in a chemical reaction, whether or not oxygen is involved.

Oxidation and reduction are useful terms when employed to describe what happens when two atoms, such as sodium and chlorine, combine to form a compound—in this instance, sodium chloride. Sodium atoms undergo a change from a neutral to an electrically charged condition (from 0 to +1) by losing an electron. Chlorine goes from a neutral to a negatively charged condition (0 to −1) by gaining an electron. Thus the sodium is oxidized and the chlorine reduced. The formation of ionic chemical bonds often (though not always) involves an oxidation–reduction reaction.

By contributing electrons that reduce chlorine, sodium acts as a **reducing agent.** By accepting electrons from sodium, chlorine acts as an **oxidizing agent.** An oxidizing agent, then, accepts electrons, while a reducing agent gives up electrons.

When most biological molecules are oxidized, electrons are removed in combination with protons, rather than alone. In other words, a whole

Oxidation is the loss of electrons; reduction is the gain of electrons.

hydrogen atom, in the form of one proton and one electron, is removed during biological oxidation. Thus biological oxidation is frequently associated with hydrogen removal, or "hydrogen transfer," as it is sometimes called. This fact should not obscure the important point that the process is still one of oxidation—the loss of electrons.

2.7 The Collision Theory and Activation Energy

The collision theory is derived from the idea that all atoms, molecules, and ions in any system are in constant motion. Particles that are to interact chemically must first come into contact so that electron exchanges or rearrangements are possible. The collision of any two particles is considered to be a completely random event. If two negatively charged particles approach each other, each will mutually repel the other and direct collision is not likely. The same will be true of two positive particles. If a positive and a negative particle approach each other, however, a collision is more likely. Furthermore, this collision may be "successful" in the sense that it produces an interaction and hence a chemical change.

Not every collision between oppositely charged particles will produce a chemical interaction. Several other factors are involved. First, the average velocity of the particles determines what percentage of collisions will be successful for any given kind of reactants. The more rapidly the particles travel, the more likely that they will yield successful collisions.

Second, particles of each element or compound have their own minimum energy requirements for successful interaction. Imagine a system in which molecules of A and B interact to produce C and D. For any collision between A and B to produce a reaction, each molecule must have a certain minimum kinetic energy. This energy is usually referred to in terms of particle velocity. Greater kinetic energy of a particle means greater velocity. Greater velocity means greater probability that a collision will be successful. If the average kinetic energy of a system is increased, the number of successful collisions will also generally be increased.

Third, molecular geometry plays a role in determining whether or not a collision is successful. If a molecule collides with an atom or another molecule in such a way that the reactive portion of the molecule is not exposed to the other particle, no reaction will occur. This is true in spite of the fact that the particles may have possessed the proper amount of kinetic energy. For this reason, molecular geometry, though a factor in any chemical reaction, is particularly important in reactions between very large molecules. Here the relative position of two colliding molecules is crucial to successful interaction. Living systems have developed means of holding large molecules in specific positions which aid in exposing the reactive portion of the molecule. This is one of the main functions of organic catalysts, or **enzymes** (Section 3.3).

The minimum kinetic energy required by any system of particles for successful chemical reaction is known as the **activation energy.** Activation energy is a characteristic of any reacting chemical system. If the average energy of the particles is below this minimum, the reaction will proceed slowly or not at all. If the average is above the minimum, the reaction will proceed more rapidly.

As an example, the velocities of the molecules in a gas will vary considerably with some showing relatively low velocity, some very high, while others are somewhere in between. Depending upon the level of activation energy of a particular reaction, perhaps only those molecules possessing the highest velocity will be able to participate. In Fig. 2.11, we can see that a hypothetical reaction requiring an activation energy greater than that impacting a molecular velocity of 2000 meters per second could not occur at a temperature of 270° Kelvin, but could for some of the molecules at the higher temperatures of 1273°K and 2273°K, since some (though considerably less than half) of the molecules possess the required velocity. The graph in Fig. 2.11 emphasizes the fact that *chemical interac-*

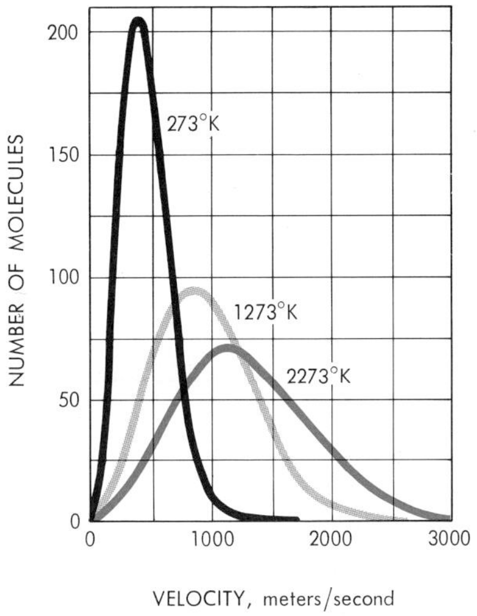

Fig. 2.11 *The distribution of molecular velocities (Maxwell-Boltzmann distribution) in nitrogen at given temperatures. Only those molecules possessing sufficient velocity have enough activation energy to enter into chemical combination. The peak of each curve represents the most probable velocity of the molecules.*

tion between atoms or molecules can be discussed only in terms of probability. The rate of a chemical reaction is influenced by factors that increase or decrease the probability that collisions between particles will be successful.

2.8 Free-Energy Exchange and Chemical Reactions

All chemical reactions involve an exchange of free energy. On the basis of these exchanges, chemical reactions can be divided into two classes. Reactions that absorb more energy than they release are called **endergonic** reactions. Those that release more free energy than they absorb are called **exergonic** reactions.

Endergonic and exergonic reactions can be compared in terms of the energy hill analogy. Endergonic reactions occur in an uphill direction. Exergonic reactions occur in a downhill direction. This means that, like rolling a stone uphill, endergonic reactions require an input of free energy. And, like a stone rolling downhill, exergonic reactions release free energy (see Fig. 2.12).

Change in Free Energy as ΔG

As pointed out earlier, the energy available in any particular chemical system for doing useful work is known as free energy (G). If a net change in free energy occurs during a chemical reaction, the system has either more or less free energy after the reaction than before. An exergonic reaction always involves the loss of free energy. We can say that such a system shows negative free-energy change, $-\Delta G$, where Δ means "change." An endergonic reaction takes in free energy. Thus it shows an increase in free energy, $+\Delta G$. It is possible, therefore, to show whether a given reaction involves an overall increase or decrease in free energy simply by putting the symbol — or + after the equation.

Like the energy in chemical bonds, free energy exchange in reactions is measured in kilocalories per mole of reactant. For example, in the reaction between hydrogen and oxygen to produce water, we find

$$H_2 + \tfrac{1}{2}O_2 \longrightarrow H_2O,^* \qquad \Delta G = -56.56 \text{ kcal/mole.} \tag{2–1}$$

The reaction has a negative ΔG and hence gives off energy. The more negative the value for ΔG, the more energy the reaction releases. Consider the re-

* In equations where energy equivalents are given, the numbers before each molecule refer to numbers of moles. The one-half O_2 thus means one-half mole of oxygen.

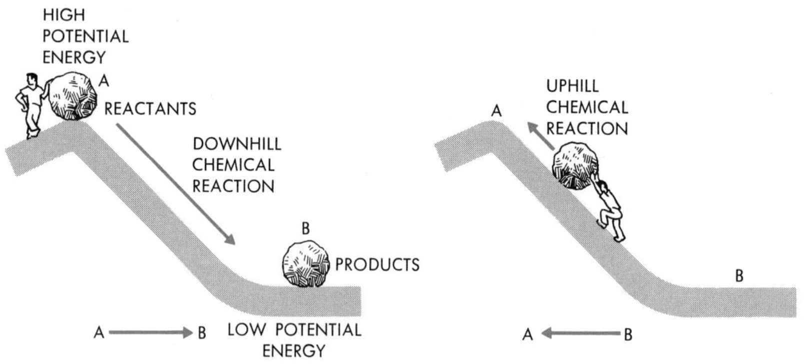

Fig. 2.12 *In this analogy, a chemical reaction that releases free energy is compared to a stone rolling down a hill. When the stone has reached the bottom, it has less potential energy than when at the top. To go in the reverse direction back up the hill, the stone will have to absorb the same amount of energy that it released while rolling down. Under natural conditions, absorbing this amount of energy is quite unlikely. Only very rarely, if at all, would a stone ever get back to the top unless it was pushed. The same is true of chemical reactions that release large amounts of energy. They are considered to be irreversible. Unless energy is supplied from the outside, the reverse reaction will not occur.*

action of the sugar glucose and oxygen, which releases in several steps the energy for many life processes. These steps can be summarized in the following equation:

$$C_6H_{12}O_6 + 6O_2 + 6H_2O \rightarrow 6CO_2\uparrow + 12H_2O, \qquad \Delta G = -690 \text{ kcal/mole.} \tag{2–2}$$

This overall reaction releases a great deal more energy than the reaction shown in Eq. 2–1).

In a similar manner, the numerical value for reactions with a positive ΔG indicates how much energy the reaction requires. In Eq. (2–3), iodine reacts with hydrogen to form the compound hydrogen iodide:

$$\tfrac{1}{2}I_2 + \tfrac{1}{2}H_2 \rightarrow HI, \qquad \Delta G = +0.315 \text{ kcal/mole.} \tag{2–3}$$

This reaction requires a small amount of energy, as shown by the low positive value for ΔG. On the other hand, the process of photosynthesis, in which green plants produce carbohydrates from carbon dioxide and water, requires a large intake of energy. This energy is supplied by light. The overall process can be written as

$$6CO_2 + 12H_2O \rightarrow C_6H_{12}O_6 + 6H_2O + 6O_2\uparrow, \qquad \Delta G = +690 \text{ kcal/mole.} \tag{2–4}$$

Knowing the value of ΔG makes it possible to compare the amounts of energy that various reactions absorb or release.

All exergonic reactions show an overall loss of free energy. Many of these same reactions, however, require an energy input to get them started. If left to themselves, many reactants will never show any measurable chemical activity. However, if the right amount of energy is supplied, the reaction begins. It then goes to completion without the addition of more energy from the outside.

How can this be explained? In such chemical systems, the reactants have relatively high activation energies. The addition of energy gets a larger percentage of particles in the system up to the required kinetic energy. In absorbing this energy, the particle becomes activated. When particles are in an activated state, a successful reaction is much more probable.

Chemical reactions can occur only when two or more interacting atoms collide with enough "energy of activation." This energy is necessary to overcome the natural tendency of the negatively charged electron clouds of the atoms to repel one another.

A specific example will clarify this point. Formic acid, HCOOH, is the pain-causing substance in wasp and bee stings. Under certain conditions, formic acid decomposes into carbon monoxide (CO) and water, with a slightly positive ΔG. For this reaction to occur, however, a formic acid molecule must first absorb enough energy to become activated. In being activated, the molecule undergoes a rearrangement of one hydrogen atom. The molecular structure is changed, and along with it the stability of the whole molecule. It splits into two parts, carbon monoxide and water:

$$HCOOH \rightarrow CO + H_2O. \qquad (2\text{–}5)$$

Graphing Energy Exchanges

Energy exchanges in chemical systems are often given on a graph, which shows the changes in potential energy during the course of reaction. These changes are then compared with the time it takes the reaction to go to completion. Such a graph for the formic acid reaction is shown in Fig. 2.13.

Analysis of this graph reveals some important things about this chemical reaction. The graph describes the changes in energy for one molecule as that molecule undergoes the decomposition reaction shown in Eq. (2–5). Before reacting, an individual molecule is in a relatively low energy state. By absorbing energy, this molecule passes to a higher potential-energy level. It is now in an activated state. The appropriate rearrangement occurs, and the product molecules are formed. Note that the product molecules are at a slightly higher energy state than the original molecule of formic acid. This indicates that the overall reaction absorbed a small amount of energy. The reaction is endergonic.

The distance *h* on the graph indicates the energy of activation for this chemical system. The height of the graph line can thus be considered an **energy barrier:** a "hill" over which the molecule has to climb before it can roll down the other side

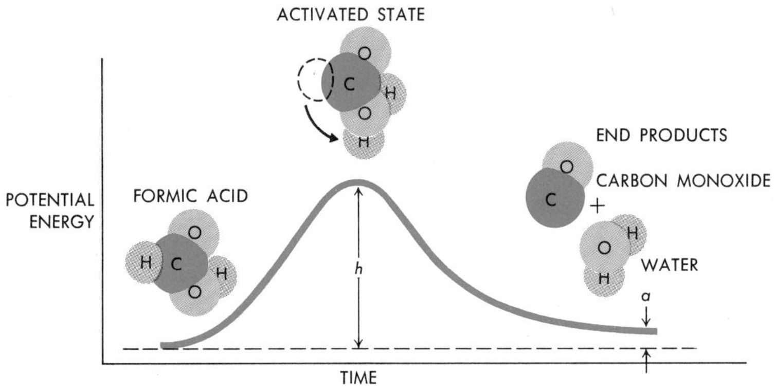

Fig. 2.13 *Changes in potential energy during the decomposition of formic acid. The original molecule, to the left, is in a relatively low energy state. By collision with another molecule of high kinetic energy, this molecule becomes activated. The potential energy of such an activated molecule is greater. During activation, a molecular arrangement occurs and the molecule splits. The end products, carbon monoxide and water, are at a higher energy state (*a *on the graph) than the original molecule. Distance* h *represents the height of the energy barrier.*

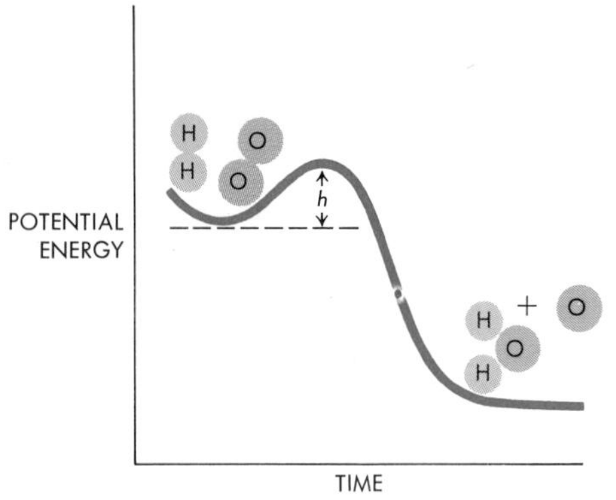

Fig. 2.14 *A spontaneous exergonic reaction, the formation of water from H_2 and O_2. The end products are at a lower energy state than the two starting reactants, indicating that the reaction releases energy. This reaction has a small energy barrier, as indicated by the size of* h.

to completion. After absorbing the required activation energy the reaction proceeds spontaneously, just as a stone rolls down a hill once it is pushed over a rise at the top. Figure 2.14 shows an energy diagram for a spontaneous exergonic reaction. The products of exergonic reactions are always at lower potential-energy states than the reactants.

Hydrogen and oxygen normally exist together in a single container without the least indication of reacting to produce water. The molecules simply do not have the necessary energy of activation. However, if a small electric spark is introduced into the chamber, an explosive reaction takes place. Thus hydrogen and oxygen react quickly to form water if given the necessary push to get them started. The fact that a spark is all that is needed to provide this push shows that the energy barrier for this reaction is not very high. Passing the spark through hydrogen and oxygen provides enough energy to put some molecules of each in the activated state. The activated molecules spontaneously react to form molecules of water. From Eq. (2–1) we see that this reaction releases energy. The energy released from one reaction is enough to get several other molecules of each element over the energy barrier. The spark provides an initial push. The rest of the reaction occurs by a chain-reaction effect, just as a house of cards collapses when one card is disturbed.

Energy Exchange and Chemical Bonds

How does the energy involved in chemical reactions relate to the formation and breaking of chemical bonds? In exergonic reactions, the end products are at a lower energy state than the reactants; in endergonic reactions, the end products are at a higher energy state. Some exergonic reactions result in the formation of chemical bonds, while others result in the breaking of bonds. There is no necessary correspondence between exergonic reactions and the breaking of bonds or between endergonic reactions and the building of bonds. However, quite frequently those reactions that result in the building or synthesis of a large molecule from smaller components are, indeed, endergonic, while those that result in the breaking down of larger molecules into smaller parts are exergonic.

2.9 Rates of Reaction

The rate of any chemical reaction is defined as *the amount of reaction in a given period of time.* The amount of reaction is generally measured in terms of the change in concentration of reactants or products. The basic relationship between amount of reaction and time can be expressed as a word equation:

$$\text{RATE OF REACTION} = \frac{\text{CHANGE IN CONCENTRATION}}{\text{CHANGE IN TIME}}. \quad (2\text{–}6)$$

The concept of rate in chemical reactions is vital to an understanding of chemical equilibrium. In addition, knowledge about reaction rates allows a clearer understanding of the mechanisms by which a particular chemical process occurs. Under given conditions, the rate of a reaction is a predictable characteristic of chemical systems. Such systems can thus be described in terms of reaction rates as well as direction or energy exchange.

A specific example will help in understanding the concept of rate as it applies to chemical reactions. Consider a reaction in which molecules of A and B combine to yield the product C and D:

$$A + B \rightarrow C + D. \quad (2\text{–}7)$$

If we begin with molecules of A and B only, the rate of reaction at the outset is very high. Rate in this case can be measured in terms of how rapidly the reactant molecules A and B disappear in specific units of time. If we were to stop the reaction every thirty seconds and determine the amount of A and B present, these data plotted on a graph would give a line like that shown in Fig. 2.15(a).

We can see that the rate at which A and B disappear from solution changes with time. For instance, the concentration of A and B decreases most rapidly in the first two minutes and begins to level off about the sixth minute. By the ninth minute nearly all of A and B have been used up in the reaction. At this point, the rate at which A and B combine to yield C and D is almost zero.

If we also plot the rate of appearance of products C and D in this reaction, the curve we obtain is just the opposite of that for the disappearance of A and B, as shown in Fig. 2.15(b). This is not surprising, since the rate at which C and D appear depends directly on the rate at which A and B interact.

The molecular explanation for the change in rate of chemical reactions goes back to the collision

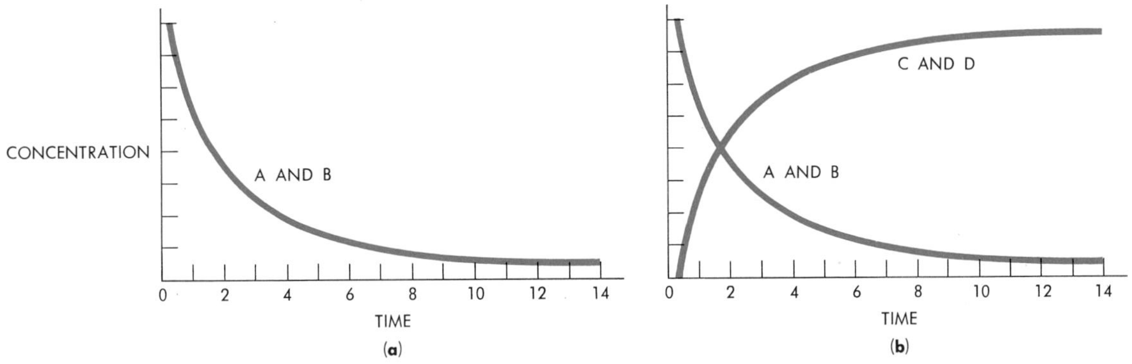

Fig. 2.15 *Changes in concentration of reactants A and B and products C and D during the course of a chemical reaction. This relationship between change of concentration and time indicates the rate of reaction. Graph (a) shows only change in concentration of reactants. Graph (b) also shows the change in concentration of products. Change in concentration of either reactants or products is most rapid during the first few minutes of reaction.*

theory. The rate at which a chemical reaction progresses toward completion depends on the number of effective collisions between reacting molecules or atoms. This number is determined for any given reaction by the concentration of reactants. The more molecules or atoms of reactants, the greater the number of effective collisions.

During the course of any chemical reaction in a closed system, the concentration of reactants decreases as the chance of collision between a molecule of A and a molecule of B decreases. At the same time, the concentration of product molecules C and D is increasing. This means that collisions between C and D molecules will become more frequent. The effects of this increase in the frequency of collisions on the course of chemical reactions will be discussed in Sections 2.10 and 2.11.

Many factors influence the rate of chemical reactions: temperature, concentration of reactants, pH (hydrogen ion concentration), and the presence or absence of catalysts. Each of these factors influences the rate of chemical reactions by increasing or decreasing the number of effective collisions. *Temperature* affects reaction rates by changing the average velocity of particles in a chemical system. It also affects reaction rates by changing the fraction of molecules possessing the minimum energy for reaction. Higher temperatures mean faster velocities and a greater fraction of molecules possessing the minimum energy; lower temperatures mean slower velocities and a smaller fraction of molecules possessing minimum energy. This works as follows: change in velocity means a change in the possibility of collision; the faster the velocities of the reacting particles, the greater the frequency of collision. In general, for every rise of 10°C, the rate of a chemical reaction is doubled; conversely, for a fall of 10°C, the rate is cut in half.

The concentration of reactants has a similar effect. The greater the concentration of one or both reactants, the greater the chance that a collision will occur. The smaller the concentration of reactants, the less chance of collision, hence the slower the rate of reaction. **Catalysts** affect the rate of reaction by increasing the effectiveness of any given collision once it has occurred or by increasing the chances of the reaction occurring by providing a surface on which the reactants can meet. Catalysts are molecules (or sometimes atoms) that facilitate a particular reaction without themselves being permanently changed in the reaction. Catalysts do not cause a reaction to occur that would not occur on its own. They do, however, speed up the rate of reactions that would occur anyway.

2.10 Reversible and Irreversible Reactions

It follows from the collision theory that chemical reactions are **reversible:** the reaction can go in either direction. In any chemical system, then, two reactions are usually taking place:

$$A + B \rightarrow AB \qquad (2\text{–}8)$$

and the reverse:

$$AB \rightarrow A + B. \qquad (2\text{–}9)$$

The forward and reverse equations can be combined into one, with the reversibility indicated by double arrows:

$$A + B \rightleftharpoons AB. \qquad (2\text{–}10)$$

Equation (2–10) indicates that at the same time reactants A and B on the left are combining to form product AB on the right, product AB is decomposing to yield the two reactants again. The forward and the reverse reactions are occurring simultaneously within one test tube.

In Eq. (2–10) the two arrows are of equal length. This shows that the forward reaction occurs just as readily as the reverse. In some reactions, however, this is not the case. Equation (2–11) shows a longer arrow to the right than to the left:

$$A + B \xrightleftharpoons[]{\quad} AB. \qquad (2\text{–}11)$$

This indicates that the reaction occurs more readily in the forward direction than in the reverse direction. Since the forward reaction is favored in this case, the reaction is shifted to the right. The relative size of the two arrows indicates the general direction of reversible reactions.

In principle, all chemical reactions are reversible. There is no reaction known which, under suitable conditions, cannot proceed (however slowly) in the reverse direction. Yet under ordinary conditions some reactions are far less reversible than others. In these cases, the reaction from right to left occurs so slowly that its rate is barely detectable. Such reactions are said to be **irreversible.**

2.11 Chemical Equilibrium

Within a certain period of time, reactions reach a state of equilibrium. When this condition is reached, the *proportion* of reactants in relation to products remains the same. Notice that this does *not* mean that the *amounts* of reactant and of product are necessarily equal.

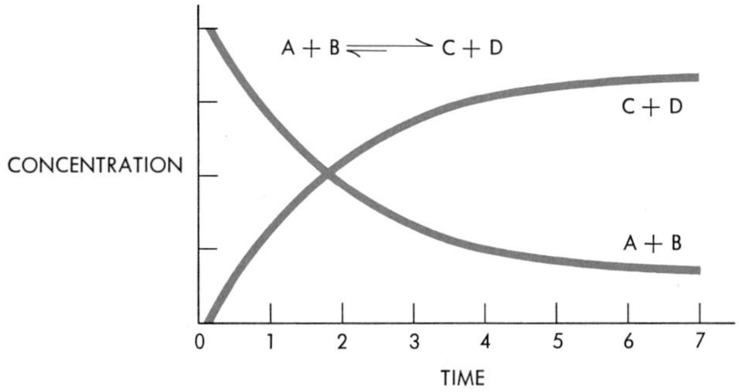

Fig. 2.16 *Graph showing the change in concentration of reactants and products in a reversible reaction. The point of chemical equilibrium is reached where the two curves level off (at about the fifth minute). At this point, the rate of the forward reaction is equal to the rate of the reverse reaction, the defining criterion of chemical equilibrium.*

An example will illustrate this point. Consider the reaction in which molecules A and B yield products C and D:

$$A + B \rightleftharpoons C + D. \qquad (2\text{–}12)$$

The different lengths of the arrows indicate that the conversion of A and B to C and D occurs more readily than the conversion of C and D to A and B. In other words, the energy barrier is lower for a successful interaction of A and B than for C and D. This means that, in a given period of time and at the same concentrations, more A and B will react with each other than will C and D. Under the same conditions, the initial rates of the two reactions are different.

If the reaction begins with only molecules of A and B present, it will occur at first only to the right. The graph in Fig. 2.16 plots change in concentration of reactants and products. It is apparent that the change in concentration ceases after about the fifth minute. Beyond this point, there is no net change. When these curves level off, the concentration of C and D is greater than the concentration of A and B. The reaction is thus directed to the right, as the relative lengths of the arrows indicate. The energy of activation for the forward reaction is less than that for the reverse reaction.

As long as molecules of reactant and product are still present in a system, chemical reactions never cease. In the above reaction, C and D accumulate because of the relatively high activation energy required to convert them back into A and B. Eventually the number of collisions between molecules of C and D will be higher than the number of collisions between molecules of A and B. As a result, the rate of the reverse reaction will increase, in spite of its higher energy of activation. At the same time, the rate of the forward reaction will decrease because of the decreasing concentrations of A and B.

Eventually a point will be reached at which the forward rate equals the reverse rate. When this condition is reached, we say that a state of **chem-**

Chemical equilibrium is established when the forward rate of any chemical reaction is equal to the reverse rate. The equilibrium point of any reaction is measured as the ratio of product to reactant after equilibrium is established.

ical equilibrium exists. The important feature of equilibrium is that *the rates of forward and reverse reactions are the same.* This means that as many molecules of A and B are being converted into C and D as molecules of C and D are being converted into A and B.

Those reactions that are completely reversible can be symbolized by arrows of equal lengths, pointing in both directions, as in the equation

$$A + B \rightleftharpoons C + D. \tag{2–13}$$

A graph showing rates of reaction for completely reversible systems can be seen in Fig. 2.17. The converging of the two lines indicates that the concentration of reactants and products is equal after the reaction reaches equilibrium.

When a point of equilibrium is reached in a chemical reaction, *the rate of reaction in one direction is equal to the rate of reaction in the other.* The reaction is still occurring both to the right and to the left. Since the rates of these reactions are equal, however, the concentration of reactants and products remains the same.

2.12 Acids, Bases, and the pH Scale

The concept of acids and bases is important in a discussion of chemical reactions in living systems, so important that a special scale has been devised to conveniently indicate the acidic or basic character of solutions.

Acids and Bases

An **acid** is defined as any substance that can donate a proton. This property is represented by the generalized equation:

$$HA \rightarrow H^+ + A^- \tag{2–14}$$

where HA stands for any acid, H^+ the proton that the acid can release, and A^- the negative ion to

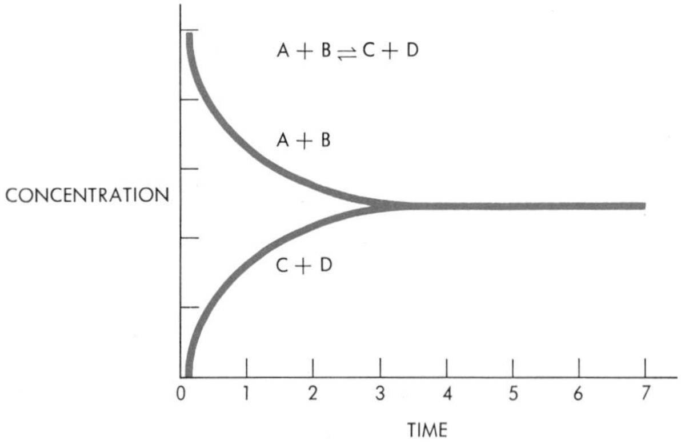

Fig. 2.17 *A graph showing changes in concentration over a period of time for a completely reversible reaction. Note that an equilibrium is established about 3½ min after the reaction has begun. At this point, the concentration of reactants equals the concentration of products.*

which the proton is bound in the acid molecule. Especially common as proton donors in living systems are carboxyl groups, written —COOH. Found in many organic molecules, —COOH groups dissociate to yield a proton and a negatively charged COO⁻ group:

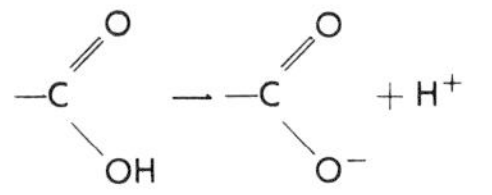

$$-C\overset{O}{\underset{OH}{\diagup}} \longrightarrow -C\overset{O}{\underset{O^-}{\diagup}} + H^+ \qquad (2\text{--}15)$$

A **base** is any substance that accepts protons. Bases are thus the opposite of acids in their chemical properties. Their dissociation reactions can be written in general as:

$$B^- + H^+ \longrightarrow BH. \qquad (2\text{--}16)$$

It can be surmised from this equation that whatever is designated as B^- in reality is identical to what we called A^- in Eq. (2–14), since the product of combining a hydrogen ion with B produces a molecule that can release the hydrogen ion again. Many substances known as hydroxides (because they have a —OH group attached to them) are bases because they dissociate in water into hydroxyl ions, which are very powerful hydrogen ion acceptors:

$$BOH \longrightarrow B^+ + OH^- \qquad (2\text{--}17)$$

$$OH^- + H^+ \longrightarrow HOH \text{ (that is, } H_2O) \qquad (2\text{--}18)$$

Water acts as both an acid and a base and consequently mediates most acid–base reactions. In fact, when an acid gives up a hydrogen ion, it always gives it up to water first. When an acid is added to water, the acid dissociates and gives its hydrogen ions to a water molecule, creating the hydronium ion (H_3O^+). If there is a stronger base around than water (one that has a stronger affinity for hydrogen ions than water does), the hydronium ion will pass the extra hydrogen to the stronger base. If no stronger base is around, the number of hydronium ions increases in the solution.

Because acids and bases are defined as opposites, it is apparent that when put together in the same solution they would tend to counteract each other's effects. Chemists say that acids neutralize bases, and vice versa. One of the neutralization products is a salt, and the other is water. Consider the neutralization of hydrochloric acid (HCl) by sodium hydroxide (NaOH):

$$HCl + NaOH \longrightarrow NaCl + H_2O. \qquad (2\text{--}19)$$

The products in this case are a salt, sodium chloride (table salt), and water.

The pH Scale

So far we have discussed acids and bases in general, qualitative terms. However, scientists have devised a scale to measure how acidic or basic a given solution is. Called the **pH scale**, it is based upon the concentration of hydrogen ions* in a liter of solution.

As shown in Fig. 2.18, the pH scale runs from 0 to 14. The lower numbers refer to acid solutions. The higher numbers refer to basic solutions. The midpoint in the scale is 7, the pH of water. At this point the concentration of hydrogen ions equals the concentration of hydroxide ions. Any solution with a pH of less than 7 has more hydrogen than hydroxide ions in solution. Conversely, any solution with a pH of more than 7 has fewer hydrogen than hydroxide ions in solution.

The pH scale is based on actual calculations of the number of hydrogen ions in a solution. The

* Traditionally, pH has been defined in terms of the "concentration of hydrogen ions" per liter of solution. Since hydrogen ions are protons, pH could just as well be discussed in terms of "concentration of protons" per liter of solution. To be in line with other descriptions of pH found in the literature, the term "hydrogen ion" will be used throughout this section. The pH scale, of course, actually measures the concentration of hydronium ions.

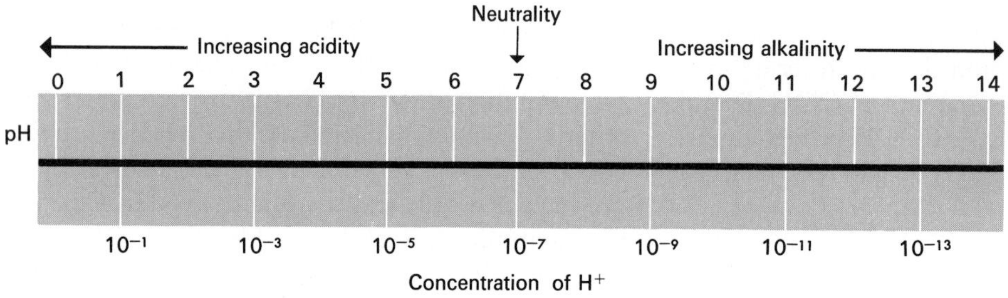

Fig. 2.18 *The pH scale. The pH value is at the top. At the bottom is the actual concentration of hydrogen ions expressed in moles per liter.*

concentration of hydrogen ions in solution is expressed as moles of hydrogen ion per liter.

The following statements summarize how pH scale values relate to the actual strength or weakness of acidic and basic solutions:

1. The pH scale ranges from 0 to 14. Solutions having a pH of less than 7 are acidic; those with a pH greater than 7 are basic.
2. The midpoint of the scale, pH 7, represents neutrality. Here the number of hydrogen ions equals the number of hydroxide ions. The pH of water is 7.
3. The pH scale is a logarithmic one. A change of one unit in pH corresponds to a tenfold change in hydrogen ion concentration.
4. The pH scale is a standardized means of expressing the acidity or alkalinity of any solution. It provides a frame of reference by which the concentration of hydrogen ions in various solutions can be judged.

It is quite possible to have a pH outside of the normally encountered range of 1 to 14. These extremes rarely occur in biochemical work, however. They are certainly not found in living organisms. As a matter of fact, the large majority of pH measurements within living organisms lie well between 6 and 8. It is true that the stomach, with large quantities of hydrochloric acid, has a pH of 1 or 2, and that certain bacteria require a very acid medium in which to live. But these are exceptions to the rule. The majority of plants and animals are restricted to an internal environment that varies only slightly to one side or the other of neutrality.

SUMMARY

1. Living organisms are highly organized entities, and as such they represent thermodynamically improbable states. They are able to maintain their improbable condition only by enormous and continual expenditure of energy. The study of biochemistry is largely an investigation of the principles by which organisms get and expend energy.
2. Matter has weight and occupies space. Matter is composed of atoms, one type of atom for each of the known elements. Energy is measured as the movement of matter, or work.
3. Atoms are composed of electrons, protons, and neutrons; the protons and neutrons, located in the nucleus, determine the element's atomic number (protons) and weight (protons plus

neutrons) and thus its basic physical properties (density). Electrons, located in a cloud outside the nucleus, determine the atom's chemical properties, the ways it interacts with other atoms. Since the number of electrons equals the number of protons for any atom, the atom is electrically neutral. Loss or gain of an electron creates a positive or negative (respectively) charge for the atom as a whole; such charged particles are called ions.

4. Atoms can store energy by upward transitions of electrons; they release energy by downward transitions. Electron transitions occur in discrete steps.

5. Chemical bonds represent electron redistribution among the atoms involved. Bonds contain energy. Two or more atoms joined together by such bonds are called molecules.

6. The various kinds of chemical bonds of importance to living systems include covalent bonds (electrons shared), ionic bonds (electrons essentially given by one atom to another), hydrogen bonds (between hydrogen and atoms such as oxygen or nitrogen), and electrostatic bonds (attraction between negatively and positively charged atoms and molecules). The latter two types of bonds are called weak interactions since they can be readily broken.

7. Electronic structure of atoms determines molecular geometry, and molecular geometry determines chemical reactivity and specificity.

8. Polar molecules are those that have differential charge distribution so that some regions are more positively and others more negatively charged. Several important properties of water are due to molecular polarity.

9. Gain of electrons by an atom or molecule is called reduction; loss of electrons is called oxidation. The loss or gain of one or more electrons produces an overall net charge (+ or −, respectively) on the atom or molecule. Such charged atoms or molecules are called ions.

10. Endergonic chemical reactions are those that absorb more energy than they liberate; exergonic reactions are those that give off more energy. Endergonic and exergonic refer to the net overall energy balance in a chemical reaction.

11. All chemical reactions are reversible. Reversibility means that the reaction proceeds in two directions: from reactants to products, and back from products to reactants. For practical purposes, some reactions may be considered irreversible, since their reverse rates are infinitesimally small compared to their forward rates.

12. Chemical reactions reach equilibrium when the rate of forward reaction is equal to the rate of reverse reaction. Equilibrium point is a numerical value expressing the ratio of product to reactant at equilibrium.

EXERCISES

1. Describe the difference between an *endergonic* and an *exergonic* reaction. How does this feature relate to the equilibrium of the reaction (is there any relation between whether the equilibrium is shifted to the right or left, and whether the reaction is endergonic or exergonic)? How does it relate to how spontaneous the reaction is (spontaneous reactions require very little to get them started)? How does this relate to whether the reaction involves synthesis (the building of chemical bonds) or breakdown (the breaking of chemical bonds)?

2. What is a chemical bond? Why is the term "bond" misleading?
3. Discuss the three graphs in Fig. 2.19 below in terms of the chemical events they represent. What does the change in concentration of reactants and products indicate? Relate these graphs to the concept of chemical equilibrium.

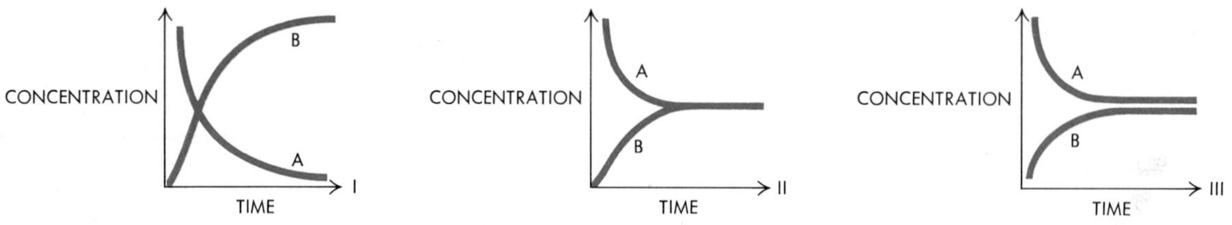

Figure 2.19

4. What is the difference between an acid and a base? What is important about the concept of acid and base in relation to living systems?

SUGGESTED READINGS

Suggestions for further reading about topics discussed in this chapter appear at the end of Chapter 3.

THREE

CHEMISTRY AND LIFE

3.1 Introduction

The study of chemical reactions involving the element carbon is known as **organic chemistry.** Although it might seem that, because it deals only with reactions containing carbon, organic chemistry would be a far more restricted field than inorganic chemistry, such is not the case. Carbon is capable of forming more varieties of compounds with more different elements than almost any other atom in the periodic table.

That branch of organic chemistry specifically concerned with the chemical reactions within living organisms is called **biochemistry.** Both biochemistry and organic chemistry as a whole involve the principles of chemical reactions, bond formation, and energy exchange described in Chapter 2. There is nothing fundamentally different in principle about the chemical reactions within living organisms and those in nonliving systems. In general, the chemical reactions within living organisms are more complex and involve many more individual steps and specialized energy sources than those that occur in the inorganic or the nonliving organic world. This greater complexity is important, for it imparts to living systems their unique abilities to organize and regulate chemical reactions in a way that produces the characteristics of life.

3.2 Classes of Molecules in Living Systems

In all animal life, from bacteria to humans, there are three major types of compounds involved in metabolic reactions. These are **carbohydrates, lipids (fats),*** and **proteins.** All three types of molecules

* The terms "lipid" and "fat" are often used interchangeably. They are not identical, however. Fats are a specifically defined class of molecules, composed mostly of fatty acids combined with glycerol. Lipid is a more general category of substances having no structural features in common but made up of molecules soluble in nonpolar solvents. Thus all fats are lipids, but not all lipids are fats.

Carbohydrates are a major quick-energy source for living systems. The most immediately available form is sugars; many carbohydrates are stored as polysaccharides such as starches. Some polysaccharides also serve as structural elements, as in plant cell walls.

contain the elements carbon, hydrogen, and oxygen, with proteins containing other elements, such as nitrogen and sometimes sulfur, as well. The ratios of these elements differ between one type of molecule and the next, as do the three-dimensional arrangements of the atoms that form the molecules.

Carbohydrates

Carbohydrate is a general category containing two major types of substance, sugars and starches. Sugars serve as the primary fuel molecules for all living cells. Starches, composed of many simple sugar units linked together, serve as reserve fuels. Carbohydrates generally contain hydrogen and oxygen in the ratio of 2:1. A generalized carbohydrate formula can thus be written as $(CH_2O)_n$, where the subscript n means that molecules of carbohydrates are multiples of this basic unit. The formula for glucose, one of the most common carbohydrates in living systems, is $C_6H_{12}O_6$ ($n = 6$). Another common carbohydrate is ribose, whose formula is $C_5H_{10}O_5$ ($n = 5$). A structural formula, showing the basic arrangement of atoms in a molecule of glucose, a simple sugar, is as follows:

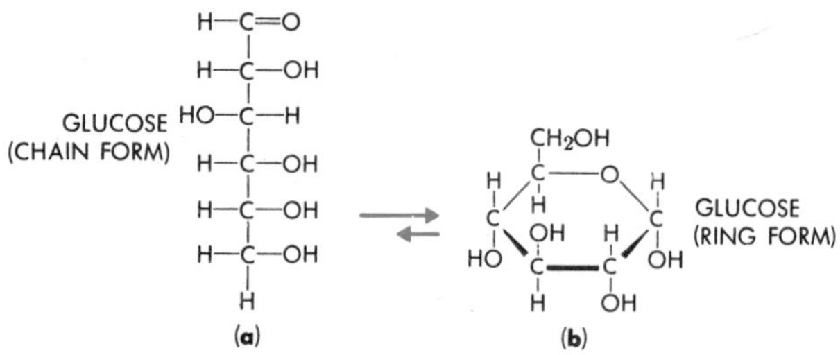

Glucose, ribose, and several other five- or six-carbon sugars are the basic building blocks of all types of larger carbohydrate molecules. Glucose is accordingly called a simple sugar, or **monosaccharide.** Two simple sugars joined together (such as glucose and glucose, or glucose and fructose) are called double sugars, or **disaccharides.** Large carbohydrate molecules, such as starch or cellulose, are composed of the simple glucose units joined end-to-end, forming very large, complex carbohydrates called **polysaccharides.** Large starch molecules can have molecular weights up to 500,000 or more. The term **macromolecule** is sometimes used to refer to any type of large molecule.

The following example illustrates how glucose units can be joined together to produce a larger molecule, in this case, the disaccharide maltose. The equation shows that during this process one molecule of water is eliminated:

$$C_6H_{12}O_6 + C_6H_{12}O_6 \rightarrow \underset{\text{(maltose)}}{C_{12}H_{22}O_{11}} + H_2O.$$

This process of joining, with the elimination of a water molecule, is called **dehydration synthesis** and is shown with the following molecular formulas:

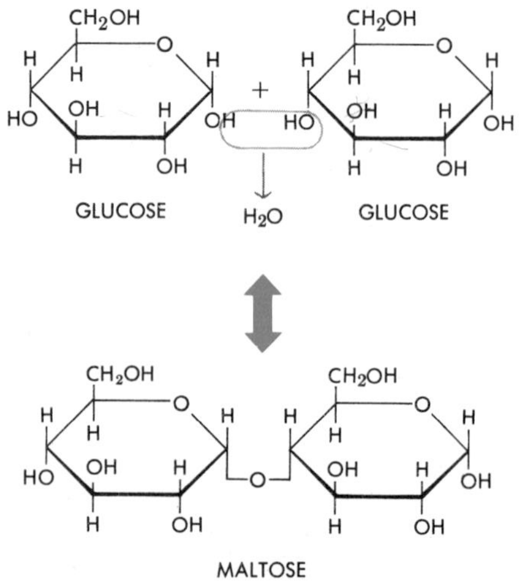

The reverse process, whereby maltose is broken down by the chemical *addition* of water between two units, is called **hydrolysis.** Both dehydration synthesis and hydrolysis occur in the formation and degradation of all major macromolecules in the living cell. The chemical bond that joins two glucose units (or glucose and a maltose) is known as an α-**glycosidic** bond. Some disaccharides (such as lactose, to the right) have a β-**glycosidic** bond. This bond is formed in the same way as the α-bond, but it has a different orientation in space. Whereas the α-glycosidic bond projects below the plane of the two rings, the β-bond projects above it:

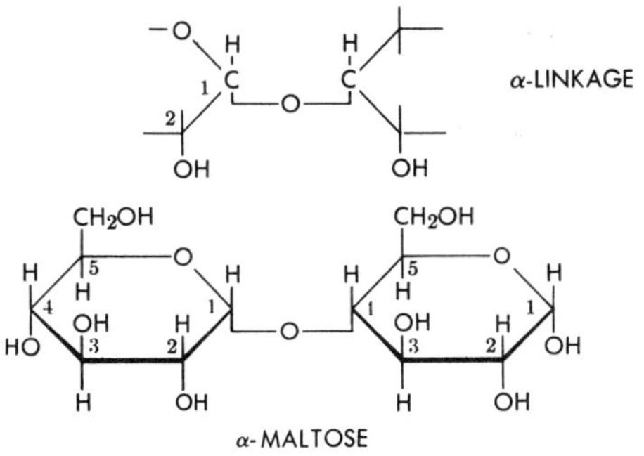

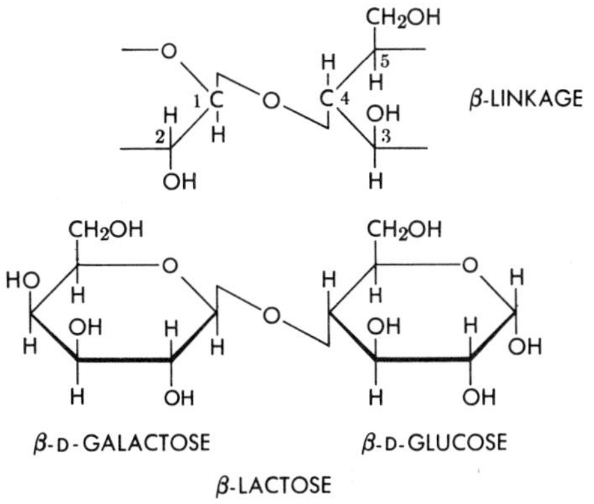

Disaccharides bound together by α-glycosidic bonds thus have a different overall three-dimensional shape from those bound together by β-bonds; accordingly, the two types of molecules behave differently in cellular metabolism.

Lipids

The lipids are a group of organic chemical compounds including the fats, oils, and sterols. They are an important tissue component and a major foodstuff. A lipid molecule, like that of a carbohydrate, contains the elements carbon, hydrogen, and oxygen. However, unlike the carbohydrates, the ratio of hydrogen to oxygen is far greater than 2 : 1.

The true fat molecule has two parts. These are (1) *alcohol* (usually glycerol) and (2) a group of compounds known as *fatty acids.* Fats are broken down into these two parts during digestion. The glycerol portion of the molecule has the following structural formula:

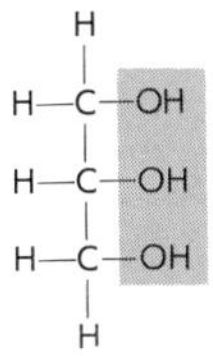

The shaded portion, the alcohol groups, indicates the region where fatty acids can be attached.

There are many kinds of fatty acids, which differ mostly in molecular size and in the degree to which the carbon atoms are completely bonded to or **saturated** with hydrogen atoms. The following is the formula for a saturated fatty acid, stearic, $C_{17}H_{35}COOH$:

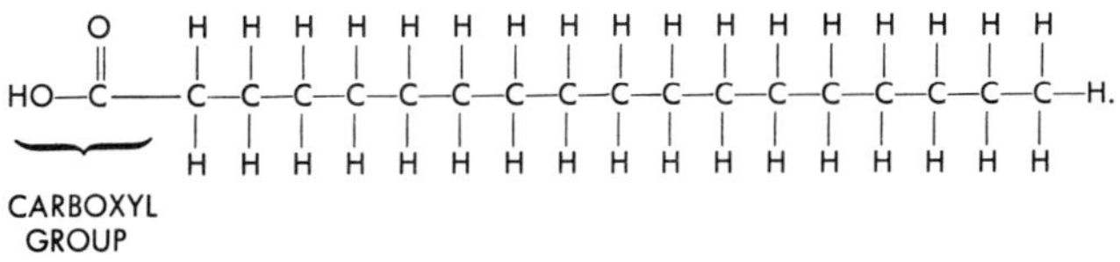

The following is the formula for a singly unsaturated fatty acid (one in which two hydrogens are replaced by a double bond between two adjacent carbons), oleic, $C_{17}H_{33}COOH$:

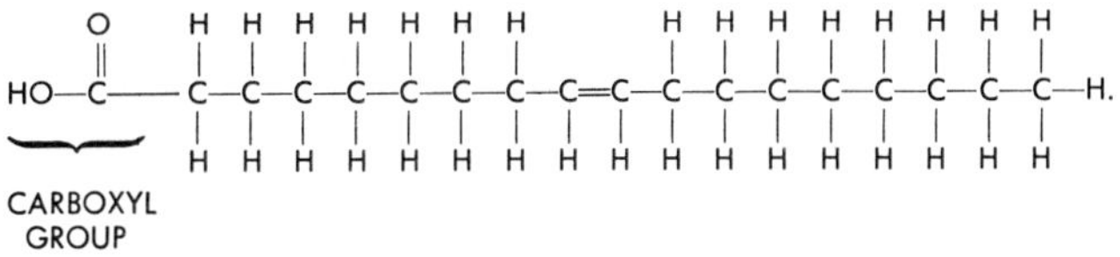

All fatty acids consist of a **carboxyl** group (—COOH) attached to varying numbers of carbon and hydrogen atoms. The carboxyl group is found in all organic acids. The joining of a fatty acid to the glycerol molecule is accomplished by the removal of an H^+ from the glycerol molecule and an OH^- group from the fatty acid. The H^+ and OH^- unite to form water. Since glycerol has three OH^- groups available, three fatty acid molecules can be attached, as follows (the R represents the hydrocarbon region of the fatty acid):

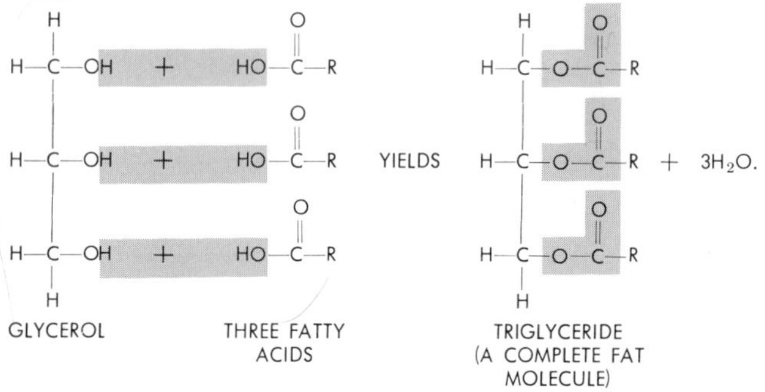

The glycerol and fatty acid are joined by dehydration synthesis, yielding in this case an **ester bond.** The splitting of fats during human digestion involves breaking the ester bond, and the reaction is catalyzed by an enzyme (pancreatic lipase). The important lipids found in biological membranes are composed of two fatty acids joined to glycerol phosphate, producing a **phospholipid.**

Fats are an organism's most concentrated source of biologically usable energy. Most of them provide up to twice as many calories per gram as do carbohydrates. The chemical reasons for this are evident from a comparison of molecular formulas. Fats contain considerably more hydrogen per molecule than carbohydrates. These hydrogen atoms supply electrons for the energy-releasing chemical processes in cells (removal of electrons is called oxidation; oxidation is the basic process in energy release). The oxidation of fats can create considerably more total energy than oxidation of even large carbohydrates. Fats have hydrogens where carbohydrates have some oxygens. Relative to fats, therefore, carbohydrates are partially oxidized and thus yield less energy upon completion of the oxidation process.

Proteins

Proteins play the most varied roles of any molecules in the living organism. As enzymes, proteins serve to keep all the various chemical reactions within a cell operating smoothly and continuously. As structural elements, proteins serve in such places as the contractile fibers of muscle, the spongy supporting tissue between bones, and in hair, nails, and skin.

The fundamental building block of protein is the **amino acid.** Amino acids are nitrogen-containing compounds with an amino group (NH_2) and a carboxyl group (COOH). Amino groups give basic properties to amino acids, while carboxyl groups give acidic properties.

A diagram of a generalized amino acid is shown below. Attached to a carbon atom in every amino acid is a characteristic group of atoms, symbolized as R.

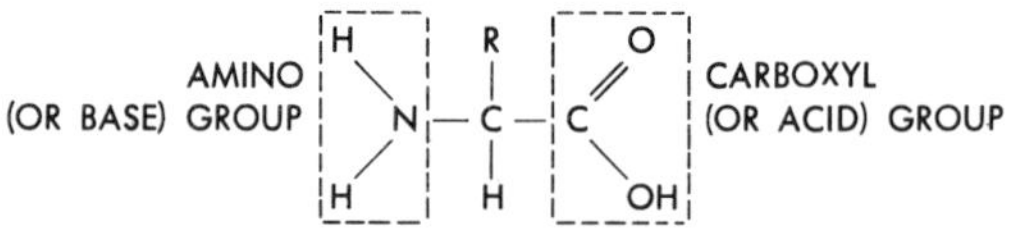

It is in the number and arrangement of atoms comprising the R group that one amino acid differs from another.

From about twenty different amino acids, all the proteins known to exist in plants and animals are constructed. Amino acids are to proteins as letters of the alphabet are to words. A group of amino acids can be joined together in a specific order to produce a given protein, just as a group of letters can be arranged to form a specific word. For this reason, amino acids are often referred to as the "alphabet" of proteins.

Amino acids are linked in end-to-end fashion to form long chains. *The variety found among proteins is the result of the types of amino acids composing each and the order or sequence in which these types are arranged.*

The comparison of amino acids to letters of the alphabet is helpful in representing how a small change in a protein molecule can completely change its chemical properties. Changing one letter in a word may cause that word to become meaningless. For example, the word "skunk" conveys one idea, while the word "skank" means nothing at all. Likewise, a change or substitution of one amino acid for another may make an entire protein molecule "meaningless" to the cell. The molecule is no longer able to carry out its function.

In one way, however, the comparison of proteins to words falls short. This is in the matter of length. Words are generally composed of relatively few letters. Even the name of the New Zealand village Taumatawhakatangihangakoauauotamateapokaiwhenuakitanatahu only approaches the complexity and length of a small to average protein. Proteins are macromolecules, often consisting of several hundred to over a thousand amino acids. Their molecular weights range from 6000 to 2,800,000.

The great size of proteins gives them added versatility in cell chemistry. They can take on a variety of shapes and sizes, each of which may serve very specialized functions. For this reason, biochemists speak of **chemical specificity** as being characteristic of many proteins. It is through the use of such chemically specific proteins as enzymes that living organisms are able to carry out their many different reactions so efficiently.

When amino acids unite to form proteins, the amino end of one amino acid molecule forms a chemical bond with the carboxyl end of the other, with the removal of one molecule of water:

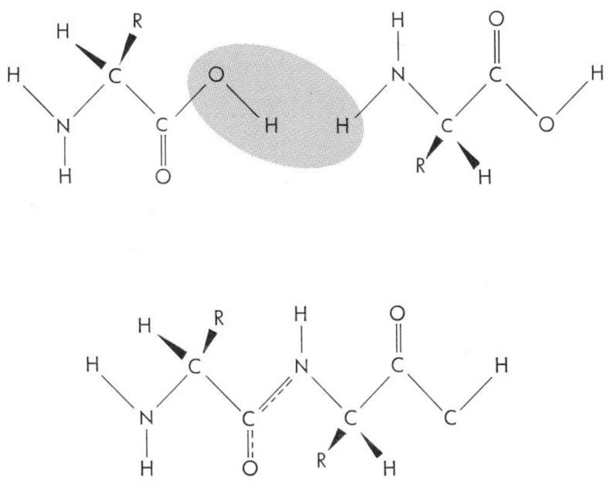

The result is the formation of a connecting link between the two, much like the connection of railroad cars. This linkage process continues until all the amino acids necessary to form the protein are joined together in the characteristic order for that particular molecule. Note in the diagram above that this process involves the loss of one water molecule between each two amino acids and is thus a dehydration synthesis. The resulting linkage is called a **peptide bond.** The peptide bond consists of the C—N linkage, thus:

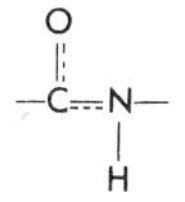

and involves a partial double bond between C and N and C and O (indicated by a solid and dotted bond line). This so constrains the atoms in its vicinity that the whole region of the linkage forms a flat plane (shown in color in the diagram presented below). The flatness of this plane is the same regardless of what two amino acids are joined together in the peptide bond. As we shall see later, this has considerable importance for our understanding of protein synthesis and breakdown.

The joining of amino acids in this manner forms a larger unit called a **peptide.** Peptides may contain from two to thirty or more amino acids. The molecular structure of a peptide consisting of four amino acids is shown below:

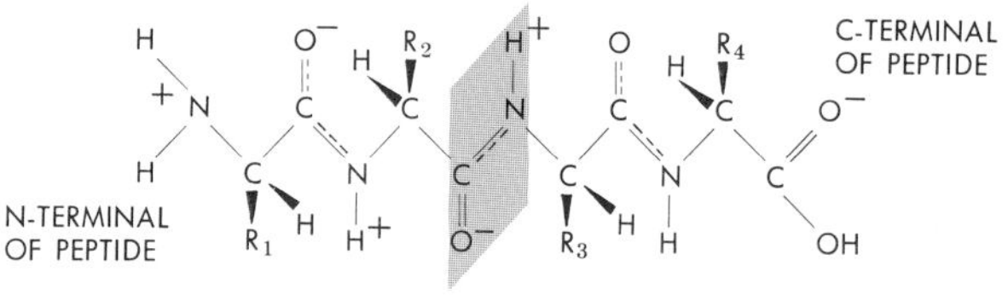

Note that the polypeptide bears numerous positive and negative charges along its backbone. At the left and right ends of the chain (called the N-terminal and C-terminal, respectively) the charges are a result of the interaction of those groups (the amino and carboxyl, respectively) with water. The NH_2 terminal tends to pick up a hydrogen ion from hydronium (H_3O^+) to produce an overall net positive charge on the amino group. The carboxyl terminal tends to lose a hydrogen ion to water, creating H_3O^+ and an overall net negative charge at the C-terminal. The peptide bond region also shows charged groups. Because of the partial nature of the double bond between C and O and C and N, the oxygen gets more of the shared electron pairs, while the hydrogen gets less. Consequently the C=O group is negative, and the C=N=H group is positive.

The sequence of amino acids formed from a series of covalent peptide bonds may be considered the most fundamental level of organization of proteins, their "backbone."* This sequence of amino acids is responsible, in part, for the uniqueness of each type of protein.

Few proteins exist as extended polypeptide chains stretched out like a rope. Most polypeptide chains are coiled or twisted in a variety of ways by the formation of hydrogen bonds between adjacent portions of the molecule. The most prominent form produced by hydrogen bonding is the **alpha helix.**

An alpha helix is produced by spiral twisting of the amino acid chain. To visualize the geometry of a helix, think of a ribbon as a straight-chain polypeptide. A helical structure can be formed by twisting the ribbon several times around a pencil. The spiral that remains after the pencil is removed has the general shape of an alpha helix. The alpha helix is held in position by the formation of hydrogen bonds between amino acids on one part of the chain with those on another part of the same chain.

The hydrogen bonds are formed between the C=O group of one amino acid and an N—H group nearby. Hydrogen bonding is the result of an electrostatic attraction between an unshared electron pair of one atom and the positively charged hydrogen end of a polar molecule. Only a few atoms will form hydrogen bonds, and three of these—oxygen, sulfur, and nitrogen—are found in proteins. Thus, for example, a hydrogen bond can be formed when the two groups shown below approach each other (the dashed line represents the hydrogen bond):

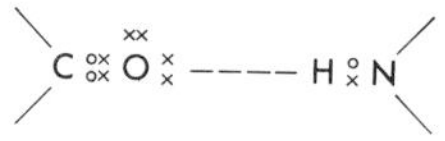

* Currently, the terms *primary, secondary,* and *tertiary* structure of proteins are often used to refer, respectively, to: (1) the amino acid sequence; (2) the coiling of the polypeptide chain into the *alpha helix,* or the interaction of two polypeptides to produce the *beta configuration;* and (3) the folding of the alpha helix into various shapes to produce a more or less globular protein molecule (see following sections). However, the terms primary, secondary, and tertiary structure actually refer to types of forces stabilizing a protein molecule, and not to any actual geometric shape. Thus "primary structure" refers to the covalent bonding of peptides within a protein. "Secondary structure" refers to hydrogen bonding, as well as to various ionic and so-called salt bonds. "Tertiary structure" refers to ionic bonds, interactions between atoms placed extremely close to each other, and the like, such as disulfide bonds. In order to avoid discussing the numerous types of bonds that contribute to the overall stability of a protein molecule, we have chosen to discuss only the major geometric patterns proteins may take.

Fig. 3.1 *A diagram of a peptide backbone wound into an alpha-helix configuration. The heavy dashed curve traces the helical structure. The broken lines going from the C—O to the N—H groups represent the hydrogen bonds that hold the chain in the alpha-helix form. (After R. B. Corey and Linus Pauling,* Proc. Intern. Wood Textile Research Conf. **B. 249,** *1955).*

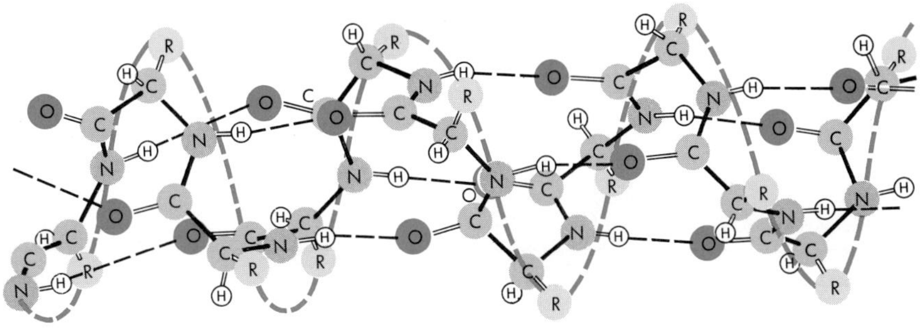

It is difficult to overestimate the importance of hydrogen bonding to our present concept of protein structure. Although hydrogen bonds are quite weak individually, many hydrogen bonds reinforce each other to produce a relatively stable structure (see Fig. 3.1).

Hydrogen bonds can be broken by many physical and chemical means. For example, a change in pH or temperature is very effective in weakening hydrogen bonds. When the pH or temperature of a protein solution is raised, a sudden change occurs in the number of protein molecules existing in helical form. At certain critical pH or temperature values, the weak forces such as hydrogen bonds break, and the molecule unrolls or unwinds; it is then said to be **denatured.** Up to a limit, such a process is reversible. If the pH or temperature is returned to its original value the hydrogen bonds re-form, and the helical structure spontaneously re-forms itself. However, at certain extremes of pH (low or high), or at certain critically high temperatures, denaturation becomes an effectively irreversible process, because other forces than hydrogen bonds have been destroyed. A hard-boiled egg is an example of irreversibly denatured protein (the egg white, or albumin). The white does not become liquid again upon cooling. The effects of denaturation are represented in Fig. 3.2.

While the alpha helix is frequently found in proteins, virtually no protein has its full length of polypeptide chain coiled into the alpha helix configuration. Protein chains fold back on themselves, producing very elaborate configurations. The folding is specific and precise because the position of amino acids in a polypeptide is always precise. Because the amino acid sequence is different for each type of protein, the folding patterns are always unique for that protein type. The sizes and shapes of three proteins are represented in Fig. 3.3.

Changes in the folding of an alpha helix can

Proteins play the most varied roles of all macromolecules in living systems. Normally they serve as enzymes for all biochemical reactions; they are important structural components in cell membranes, skeletal structures, hair, nails, muscle, and connective tissue such as tendons and ligaments; under extreme conditions they can be oxidized to yield energy.

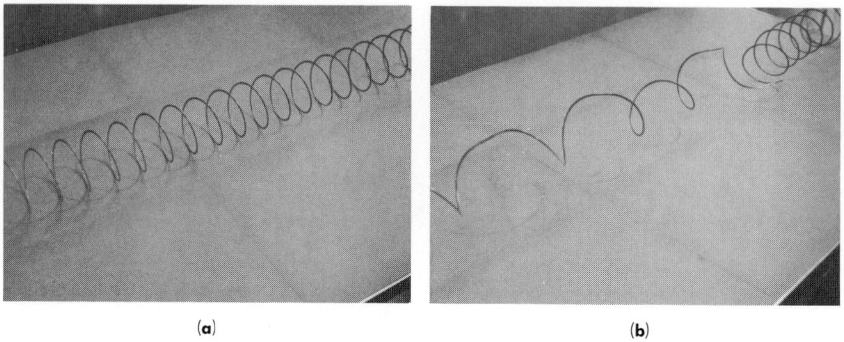

Fig. 3.2 *(a) A coiled spring toy known as a "slinky." This represents the alpha-helix structure of certain proteins. When the slinky is pulled out of shape, as shown in photograph (b), it cannot recoil into the original helical structure. It has been irreversibly altered. This is analogous to denaturing a protein by such physical means as heating or by such chemical means as placing the protein in a concentrated solution of urea.*

alter the **biological activity** of a protein. When we speak of biological activity of a protein we mean that each protein will catalyze the reaction involving only one type of molecule or a small group of other types of molecules. Apparently molecules of various substances are able to fit onto the surface of the protein molecules with which they normally react. The fit must be a good one so that appropriate atoms of the reacting substance and the protein are brought close enough together for a reaction to occur. This depends on the surface configuration of the protein. Even slight changes in any one of the three levels of organization in a protein could render that protein nonfunctional.

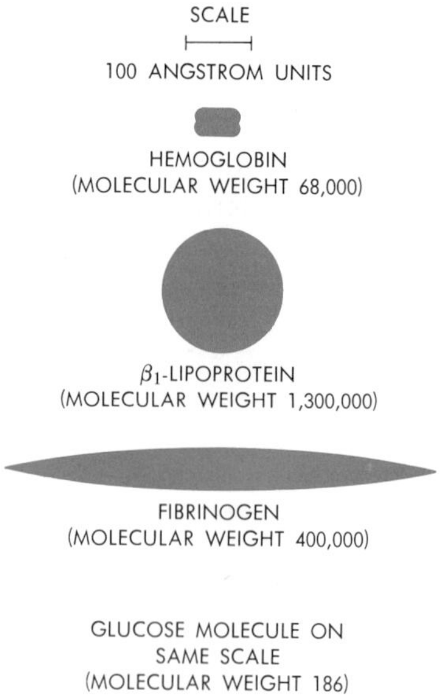

Fig. 3.3 *The various sizes and molecular configurations for several types of protein found in living systems. Hemoglobin is the oxygen-carrying protein of the blood. It consists of four polypeptide chains. β_1-lipoprotein is found in the liquid portion of blood and in the tissue fluids of higher organisms, such as mammals. It is composed of lipid elements bound to protein and serves in maintaining the stability of certain enzymes. Fibrinogen is also found in the liquid portion of blood and is involved in forming clots. The glucose molecule is included for size comparison.*

3.3 Enzymes

There are many varieties of proteins in the bodies of living plants and animals. One group—the enzymes—deserves special attention. Enzymes are involved in virtually all the chemical reactions within living organisms. Without these specialized proteins, life as we know it could not exist.

Enzymes are organic catalysts. Catalysts, you will recall, are molecules that speed up a chemical reaction without themselves being chemically changed in the process. After participation in one reaction, a catalyst can go on to participate in a second, a third, and so on. As catalysts, enzymes participate in almost all the chemical reactions that keep organisms alive. They enable a human being to support vigorous chemical activity at only 98.6°F. They enable Antarctic fish to remain alive and active at near 0°C. Furthermore, since they come out of chemical reactions unchanged, a few enzyme molecules go a long way. This means that the organism does not have to expend a great deal of energy in order to constantly resynthesize enzymes at a rate proportional to the rate of the reactions they catalyze.

Enzyme Kinetics

Much can be learned about the nature of enzyme-catalyzed reactions by studying the kinetics of reactions: the various changes in rate during the course of a reaction. For example, it is possible to measure the amount of product formed in an enzyme-catalyzed reaction from the moment the reactants (called the **"substrate"**) and enzyme are brought together until the reaction has stopped (reached an end-point, or equilibrium).

If the amount of product formed is measured at 1-min intervals and this quantity is plotted against time on a graph, curves like those shown in Fig. 3.4 are obtained. This graph shows data for two different reactions, designated here as A and

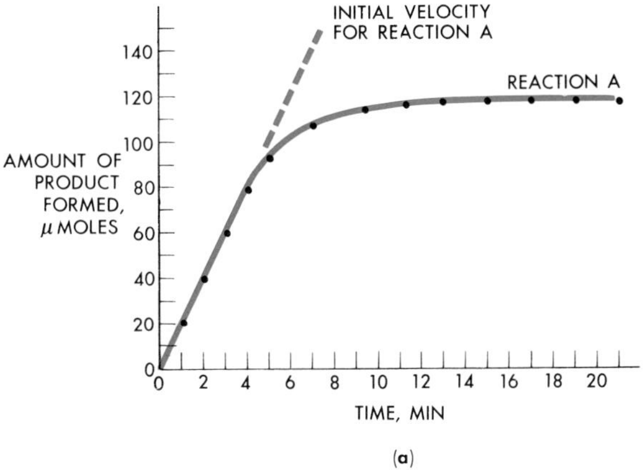

Fig. 3.4 *The relationship between total amount of product formed and the time period for two different enzyme-catalyzed reactions. Graph (a) represents reaction A, catalyzed by enzyme α; graph (b) represents reaction B, catalyzed by enzyme β. The initial rates of reactions A and B are different, as indicated by the different slopes of the part of each curve extended by the dotted line. As the graphs indicate, enzyme α acts more effectively on its substrate, affecting more molecules per unit time, than enzyme β. In time, both reactions will reach an equilibrium point and the curve will level off. For comparative purposes, it is necessary that both reactions begin with equivalent excesses (in molar quantities) of substrate.*

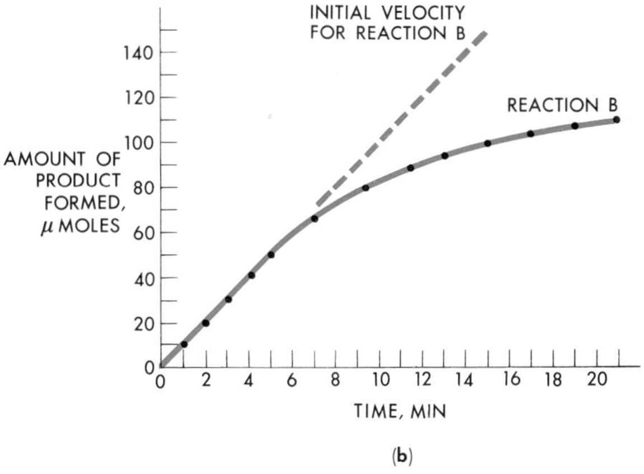

B. As an example, observe the solid line for reaction A. This reaction can be written symbolically: Reactant $\overset{\text{Enzyme}}{\rightleftharpoons}$ Product. At time 0 there is no product detectable in the reaction system; this represents the beginning of the reaction. After 1 min, 20 μmoles have been formed; after 2 min, 40; after 3 min, 60; after 4 min, 80. The rate of the reaction could be given as 20 μmoles of product formed per minute for this initial period. Note, however, that by the fifth minute this rate has begun to slow down. Instead of a total of 100 μmoles of product formed by the 5-min mark, only about 92 have been formed. In the interval between the fourth and the fifth minute only about 12 μmoles, rather than 20, have been formed. During the first 4 min, the rate has been constant and the change in rate has been zero. But from the fourth minute on through about the twelfth minute, the rate is changing; i.e., it is slowing down. For each successive minute, the amount of product formed in that interval is less than in the preceding minute. From the twelfth minute onward, the reaction rate again becomes constant; no more product is being formed, and the change in rate is again zero.

Using the principles of chemical kinetics discussed earlier, how can we explain these changes? During the first 4 min of reaction A, the number of substrate molecules greatly exceeds the number of enzyme molecules (this is usually true for all catalyzed reactions). This means that every enzyme molecule is working at its maximum capacity. There are so many substrate molecules around that each time an enzyme finishes with one substrate molecule, it is confronted with others. After a period of time, however (in the present case, after 4 min), the number of substrate molecules begins to dwindle and the concentration is lowered. As the concentration of reactants (substrate) is lowered, the chance of successful collisions is reduced (see Section 2.9). In the present case we are concerned not only with successful collisions between the reactants, but also between reactants (substrate) and the enzyme. As time goes on, the substrate concentration becomes less and less since more substrate is being converted into product. The leveling off of the curve after 12 min indicates that the total amount of product is remaining constant. The reaction system has reached equilibrium.

A similar curve can be seen for reaction B. The difference in the shape of this curve represents the difference in effectiveness of enzyme α and enzyme β in acting on their respective substrates. Given the same initial concentrations of both enzyme and substrate, it is apparent that enzyme β acts less rapidly on its substrate than enzyme α. Since the initial substrate concentrations in the two reaction systems are the same, the total amount of product formed in B, given enough time, will eventually be equivalent to that formed in A, as long as the two reactions start with the same concentration of reactants. In this example both reactions reach equilibrium with the ratio of reactants and products the same.

In comparing the kinetics of reactions A and B we need a common reference point. For example, do we want to compare the two reactions at the 2-min mark, the 12-min mark, or later? It is obvious that comparisons made at different times will give different values for the total product formed in reactions A and B. The reason for this is that we are faced with two variables (given equal concentrations of enzyme molecules): (1) the rate at which the two different enzymes can act on their substrates (which is a characteristic of the enzyme molecule), and (2) the constantly changing sub-

---

Enzymes can catalyze the reaction for a large number of substrate molecules in a very short period of time—in some cases up to 10,000,000 molecules of substrate per enzyme molecule per second!

---

strate concentrations during the course of reaction. There is, however, a means of eliminating one of these variables. In the first few minutes of the reaction, the number of substrate molecules is so large compared to the number of enzyme molecules that changing the concentration does not, for a period at least, affect the number of successful collisions. Note that during this early period the rate of change is constant; i.e., the enzyme is acting on substrate molecules at a constant rate. The slope of the graph line during these early minutes defines what biochemists call the **initial velocity** of the reaction. The initial velocity of any enzyme-catalyzed system is determined by the characteristics of the enzyme molecule and is always the same for the enzyme and its substrate as long as temperature and pH are constant and substrate is present in excess. As shown in Fig. 3.4, the initial velocities of reactions A and B are different and represent a difference in effectiveness between the two enzymes. This difference is a measure of the maximum rate at which an enzyme can act on substrate molecules in a given period of time. It is important to understand why substrate should be present in excess to measure the true initial velocity of an enzyme-substrate reaction.

Suppose that we run a series of reactions with enzyme system A, in which the concentration of enzyme is held constant but the starting concentration of substrate is varied. If we measure the initial velocity (say, for the first 2 min) of each reaction and plot this value on a graph against substrate concentration, we get a curve like that shown in Fig. 3.5. Note that at low substrate concentrations the initial velocity of the reaction is low. This means that when there are few substrate molecules in solution to begin with, the enzyme can never reach its maximum rate of conversion (since the frequency of successful collisions becomes the limiting factor). In other reactions, where the starting concentration of substrate is greater, the initial velocity is greater. As the graph line shows, the increase in initial velocity with increasing substrate concentration is linear, but only up to a point. By the time the starting concentration of substrate has reached 0.1 mole per liter, the curve has begun to level off. This suggests that the enzyme is approaching its maximum initial velocity. In other words, beyond about 0.5 moles per liter, an increase in substrate concentration does not affect the initial velocity of the reaction. Enzyme molecules are working as fast as they can; the presence of more and more substrate at the start of the reaction will not affect the initial velocity. Thus, at this point we can measure true initial (maximum) velocity because one variable—substrate concentration—has been eliminated. Rate of conversion now depends solely on characteristics of the enzyme molecule.

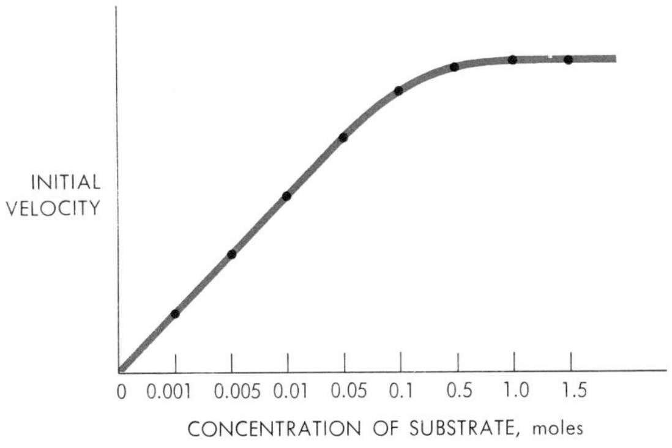

Fig. 3.5 *Graph showing the relationship between substrate concentration and initial velocity of an enzyme-catalyzed reaction. The reaction volume and enzyme concentration are held constant. Each point on the graph represents the measured initial velocity of a specific reaction where substrate concentration is the only variable.*

The above phenomena make sense in terms of our general knowledge of the kinetics of any chemical reactions, whether enzyme-catalyzed or not. However, enzyme-catalyzed reactions have some characteristics that are distinct from non-enzyme-catalyzed reactions. Understanding the nature of some of these characteristics will help us answer an important question: By what mechanism do enzymes increase the rate of biochemical reactions?

Effect of Temperature

The rate of enzyme-catalyzed reactions is greatly affected by temperature. If the initial velocity of a specific enzyme-catalyzed reaction is measured at a number of different temperatures (with enzyme and substrate concentrations held constant), a curve like that shown in Fig. 3.6 is obtained. Note that at low temperatures the rate of reaction is quite slow and that the rate increases with increasing temperature, up to about 36°C. Beyond 36°C, however, the rate begins to slow down again even though the temperature is raised. From our knowledge of all chemical reactions, we can explain the first half of this curve: the higher the temperature, the more rapidly the reacting molecules (in this case, substrate and enzyme molecules) move about, and the greater the fraction of molecules that possess minimum energy of activation.

Should not this principle apply to temperatures above 36°C as well? In non-enzyme-catalyzed reactions, initial velocity does indeed increase as the temperature is raised beyond 36°C. Why, then, are enzyme-catalyzed reactions so sensitive? To answer this question we must develop a model to explain how enzymes function, a topic that will be taken up in the next section.

The initial velocity/temperature curve shown in Fig. 3.6 varies from one type of enzyme to another. Every specific enzyme has its so-called **optimum temperature:** that temperature at which the enzyme achieves its maximum rate (initial velocity). For the particular reaction shown in Fig. 3.6, 38°C is the optimum temperature; for another enzyme-catalyzed reaction, 25° or 40° might be the optimum.

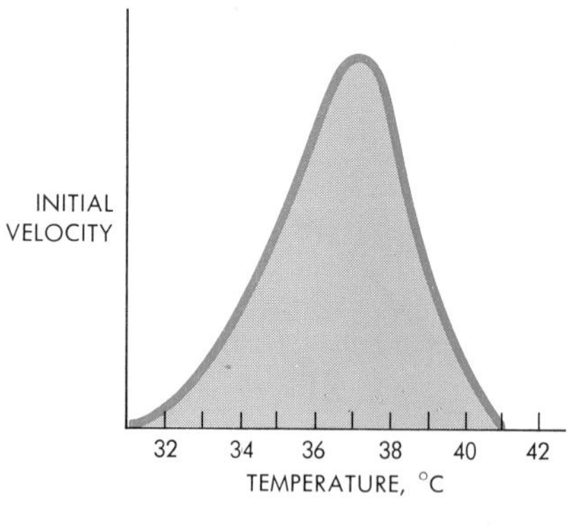

Fig. 3.6 *The effect of temperature on the activity of one enzyme.*

"Poisons"

Enzymes can be "poisoned." Certain compounds, such as bichloride of mercury or hydrogen cyanide, are deadly poisons to all living organisms. They exert their effect by inactivating one or many enzymes. Hydrogen cyanide blocks one of the enzymes involved in the chemistry of respiration. The way in which this is believed to occur will be considered shortly.

Specificity

Enzymes are specific in their action. This is one of the most distinctive characteristics of enzymes. They will often catalyze only one particular reaction. For example, the enzyme sucrase will catalyze only the breakdown of sucrose to glucose and fructose. It will not split lactose or maltose. Lactase and maltase, respectively, must be used as the catalytic agent for these two sugars.* With other enzymes, specificity of action is not quite so obvious. Trypsin, for example, is a **proteolytic** (protein-splitting) enzyme that acts upon many different proteins.

Why should some enzymes work on only one compound while others will work on several? Enzymes act upon a specific chemical linkage group. In the case of trypsin, only those peptide linkages of protein molecules formed with the carboxyl group of the amino acids lysine or arginine are acted on by trypsin:

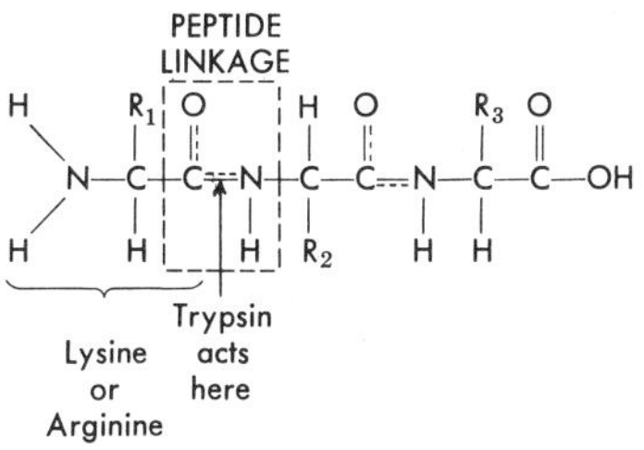

Since peptide linkages involving the carboxyl groups of lysine and arginine are found in many proteins, it is not surprising that trypsin will act on a broad range of protein molecules.

3.4 Theories of Enzyme Activity

We now come to a crucial question in the study of enzymes: By what mechanism do enzymes operate?

The most important hypothetical model proposes that enzymes have certain surface configurations produced by the three-dimensional folding of their polypeptide chains. On this surface, there is an area to which the substrate molecule is fitted. This area is called the **active site** of the enzyme. It is thought that when the substrate molecule becomes attached to the enzyme at this site, the internal energy state of the substrate molecule is changed, bringing about the reaction (Fig. 3.7).

A helpful analogy for visualizing how enzymes work is to picture the substrate as a padlock and the enzyme as the key that unlocks it (the "lock and key" model). The notched portion of the key thus becomes the active site, since it is here that the "reaction," or the unlocking of the padlock, takes place. The padlock comes completely apart, just as a molecule is broken apart by enzyme action. The key serves equally well, however, to run the reaction in the reverse direction, i.e., to lock the padlock again. The key comes out unchanged and ready to work again on another padlock of the same type; similarly, the enzyme is ready to catalyze another reaction of the same type. In light of this analogy, trypsin becomes a sort of "skeleton key" enzyme. It can open several types of padlocks (proteins) as long as they have similar engineering design (certain peptide linkages).

Inhibition of Enzyme Activity

Especially interesting in light of the lock-and-key analogy is the effect of inhibitors, or "poisons," on enzyme activity. Consider the following example. One step in the breakdown of sugars involves the conversion of a four-carbon molecule, succinic acid, to another four-carbon molecule, fumaric acid, with the removal of two electrons (and two protons) (see Fig. 3.8a). This reaction is catalyzed by the enzyme succinic dehydrogenase, which is highly specific for its substrate, succinic acid. If the reaction is carried out in a test tube with just enzyme

---

* The *-ase* ending indicates that the compound is an enzyme. Other enzymes, such as trypsin, end with *-in*. This signifies that they, like all enzymes, are proteins. Enzymes ending in *-in* were discovered and named before an international ruling was made in favor of the *-ase* ending. A few changes have been made. For example, the mouth enzyme ptyalin is now called salivary amylase. Enzymes are also named after the compounds they attack. Thus peptides are attacked by peptidases; peroxides by peroxidases; lipids by lipases; ester linkages by esterases; hydrogen atoms are removed by dehydrogenases; and so on.

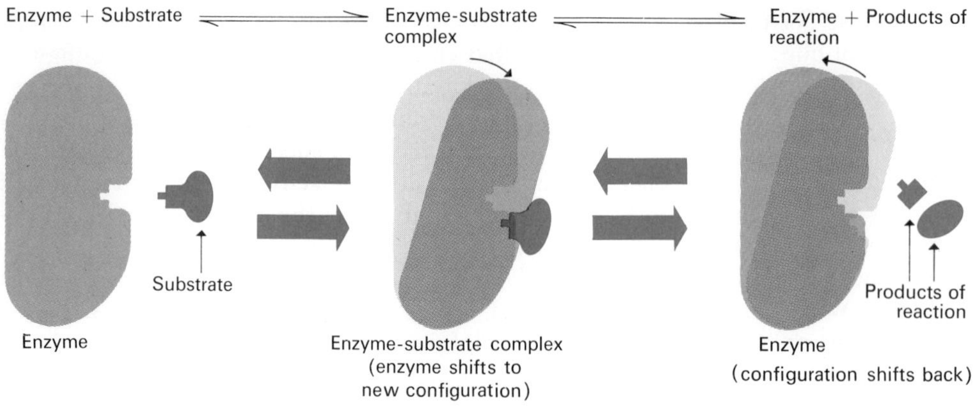

Fig. 3.7 *Schematic version of the interaction of enzyme and substrate. The specificity of enzymes for certain substrates is thought to be due to the surface geometry of the enzyme and substrate molecules, which allows them to fit together in a precise manner. Once the substrate molecule is situated in the appropriate surface region of the enzyme, certain groups of atoms in each molecule interact in such a way that the substrate is permanently changed. Enzymes can speed up the rate of breakdown (or synthesis) of substrate molecules. The enzyme molecule is only temporarily altered by interaction with the substrate, so it can catalyze other substrate molecules as soon as it finishes with one.*

and substrate, a particular initial velocity can be observed. If along with succinic acid malonic acid is added to the test tube, the rate of formation of fumarate is greatly reduced. Malonic acid is like a key that fits into a lock but is just different enough from the proper key not to be able to turn. For a moment it "jams" the lock. Malonic acid molecules can fall out of the active site, however. If many succinic acid molecules are around, it is possible for them to enter the site once it is free. Malonic acid is a **competitive inhibitor:** it competes with the normal substrate for the enzyme's active site. But it does not permanently deactivate the enzyme.

Certain molecules can also affect enzyme molecules more or less permanently. These are called **noncompetitive inhibitors,** two examples of which are carbon monoxide and cyanide. Carbon monoxide molecules attach to the active site of certain oxygen carriers (such as hemoglobin) and certain respiratory enzymes (such as the cytochromes). When either of these inhibitors is attached to a respiratory protein, normal function is impossible (the cytochrome or hemoglobin cannot interact with oxygen). Unlike malonic acid, however, neither carbon monoxide nor cyanide becomes detached from the active site of the protein. As a result, the effect of noncompetitive inhibitors is permanent; this is why both carbon monoxide and cyanide are such deadly poisons. The distinction between competitive and noncompetitive inhibition is diagrammed in Fig. 3.8.

Small Molecule Requirements for Enzyme Function

Besides requiring environmental conditions such as proper temperature, pH, etc., many enzymes also need the presence of certain other substances before they will work. For example, salivary amylase

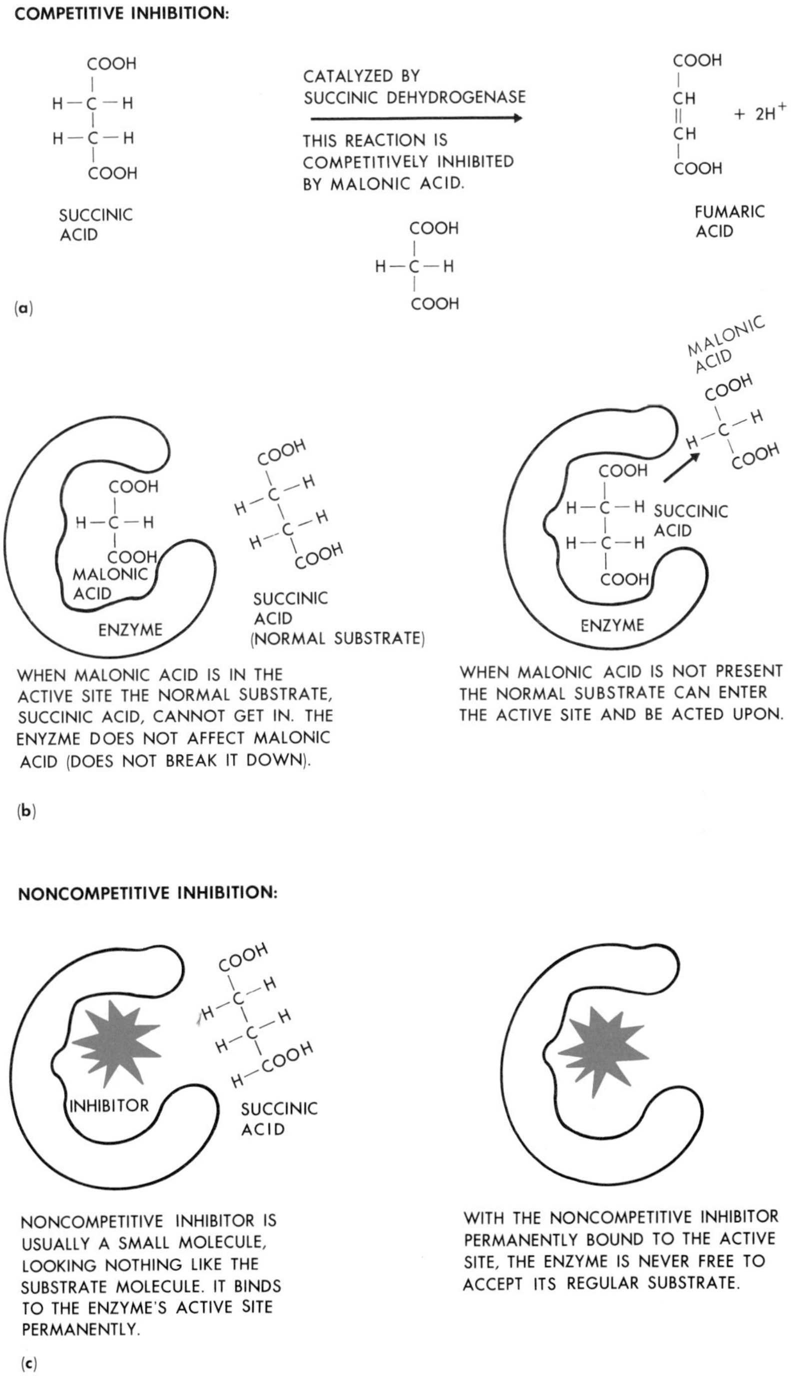
COMPETITIVE INHIBITION:
COOH
H—C—H
H—C—H
COOH
SUCCINIC ACID
CATALYZED BY SUCCINIC DEHYDROGENASE
THIS REACTION IS COMPETITIVELY INHIBITED BY MALONIC ACID.
COOH
CH
CH
COOH
+ 2H+
FUMARIC ACID
COOH
H—C—H
COOH
(a)
COOH
H—C—H
COOH
MALONIC ACID
ENZYME
SUCCINIC ACID (NORMAL SUBSTRATE)
WHEN MALONIC ACID IS IN THE ACTIVE SITE THE NORMAL SUBSTRATE, SUCCINIC ACID, CANNOT GET IN. THE ENYZME DOES NOT AFFECT MALONIC ACID (DOES NOT BREAK IT DOWN).
MALONIC ACID
SUCCINIC ACID
ENZYME
WHEN MALONIC ACID IS NOT PRESENT THE NORMAL SUBSTRATE CAN ENTER THE ACTIVE SITE AND BE ACTED UPON.
(b)
NONCOMPETITIVE INHIBITION:
INHIBITOR
SUCCINIC ACID
NONCOMPETITIVE INHIBITOR IS USUALLY A SMALL MOLECULE, LOOKING NOTHING LIKE THE SUBSTRATE MOLECULE. IT BINDS TO THE ENZYME'S ACTIVE SITE PERMANENTLY.
WITH THE NONCOMPETITIVE INHIBITOR PERMANENTLY BOUND TO THE ACTIVE SITE, THE ENZYME IS NEVER FREE TO ACCEPT ITS REGULAR SUBSTRATE.
(c)

Fig. 3.8 *Comparison of competitive and noncompetitive inhibition of enzymes. (a) Chemical reaction involving conversion of succinic acid to fumaric acid, as catalyzed by the enzyme succinic dehydrogenase. This reaction can be competitively inhibited by malonic acid, whose molecular structure is similar to succinic. (b) Diagrammatic representation of competitive inhibition by malonic acid. Because of its similar shape, malonic acid can fit into the enzyme's active site in place of succinic. It is not acted upon, however, and can fall out of the active site. Competitive inhibitors do not permanently block the site. (c) Noncompetitive inhibition occurs when some inhibitor, usually a small molecule, fits into the active site and permanently binds to it. This keeps the normal substrate out and inactivates the enzyme. The noncompetitive inhibitor does not necessarily resemble the normal substrate in molecular shape.*

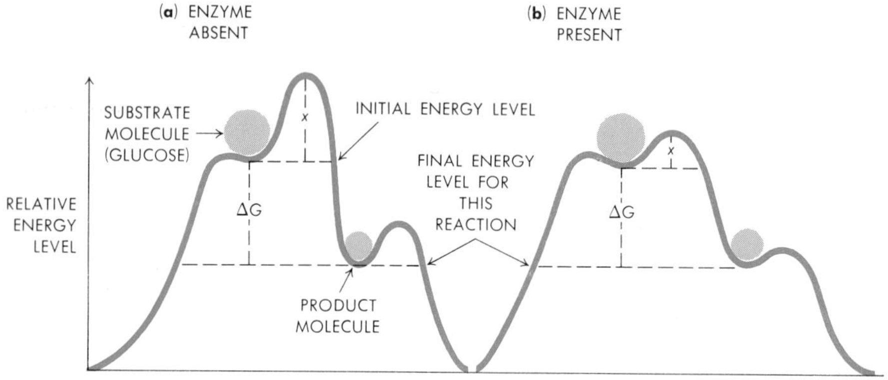

Fig. 3.9 *The effect of an enzyme on the activation energy requirements of a molecule undergoing chemical degradation. The amount of activation energy required is represented by the distance* x. *The net gain in free energy is symbolized by* ΔG *(it is assumed that the molecule falls to the lowest energy level). Note that the uncatalyzed reaction (a) has a far higher activation energy barrier than the catalyzed reaction (b) on the right. The new molecule must overcome another energy barrier, requiring more activation energy, before it can be broken down and release more energy. In a living organism another enzyme, specific for this second reaction, would be needed.*

will work on amylose (a sugar) only if chloride ions (Cl^-) are present. Magnesium ions (Mg^{++}) are needed for many of the enzymes involved in the breakdown of glucose.

Some enzymes require another organic substance in the medium in order to function properly. In a few cases, enzymes actually consist of two molecular parts. One of these is a protein, called an **apoenzyme.** The other molecular part is a smaller, nonprotein molecule called a **coenzyme.** Its name signifies that it works with the main apoenzyme molecule as a coworker in bringing about a reaction.

In an apoenzyme–coenzyme case, the two molecular parts are chemically bonded to each other. In other cases, the coenzyme is combined only briefly with the enzyme. In either case, the coenzyme must be present for any catalytic activity to take place.

Chemical analysis of the smaller coenzymes has shown that they often contain a vitamin as part of the molecule. This finding has led to the idea that vitamins serve as coenzymes. This would explain why the absence of certain vitamins causes such remarkable physical effects on the organism. When the coenzyme is missing, an entire series of important physiological reactions may be blocked.

The Energetics of Enzyme Catalysis

Since most biochemical reactions are to one degree or another reversible, it is not surprising that enzymes can catalyze reactions in either direction. In a completely reversible reaction, the enzyme can catalyze the reverse reaction as readily as the forward. This observation means that enzymes do not make reactions occur that would not occur on their own, but only increase the rate at which the reactions take place. Enzymes speed up reactions by changing the energy requirements for getting the reaction started. They do not affect the net energy changes (the ΔG) of any reaction.

How do enzymes change these energy requirements? The energy hill analogy, discussed earlier in the book, will help to elucidate this point. The glucose molecule at the top of the energy hill shown in Fig. 3.9 represents a certain amount of potential chemical energy. However, glucose may be stored indefinitely without ever releasing its chemical energy. In terms of the energy hill, there is nothing to lift it over the hump at the top and start it rolling down.

This is where enzymes fit into the picture. Their presence lowers the amount of activation energy (x) needed to start the reaction. Enzymes accomplish this by putting the substrate molecule into a constrained position where the *probability* that certain electron exchanges will take place is greatly increased. Enzymes bring particular chemical bonds in a substrate molecule into close proximity with specific reactive groups of the enzyme. This process converts a relatively unlikely event into a likely one. By expending only a small amount of energy, a living organism can now release the chemical energy available in the glucose molecule.

3.5 Enzymes and the Control of Biochemical Reactions

Regulatory Enzymes and Feedback Pathways

If a given reaction occurred all the time at a high rate, living cells would waste a considerable amount of energy. If the rate could never be speeded up, the organism would be at a disadvantage when increased activity was required. The ability to speed up and slow down rates of individual chemical reactions spells greater efficiency of operation for the organism as a whole. Let us consider how enzymes are crucial to controlling rates of reactions in living systems.

A classic example of an enzymatic control system is illustrated by the biochemical reactions involved in the synthesis of cytosine triphosphate (CTP), an essential building-block of nucleic acids, from carbamyl phosphate and aspartic acid. This reaction is diagrammed in Fig. 3.10. The initial step in the reaction involves the interaction of carbamyl aspartate. This reaction is catalyzed by the enzyme aspartate transcarbamylase (ATCase for short). Carbamyl aspartate is further transformed through additional reactions into cytosine triphosphate. Such a reaction series is called a **biochemical pathway** because it consists of a number of sequentially interrelated chemical reactions. How are such pathways regulated? Apparently the concentration of the cellular end product has a profound effect on the rate at which new CTP molecules are formed. CTP acts as some sort of inhibitor to ATCase, the enzyme catalyzing the initial step in the reaction series. An enzyme, such as ATCase, capable of being regulated by its end product is

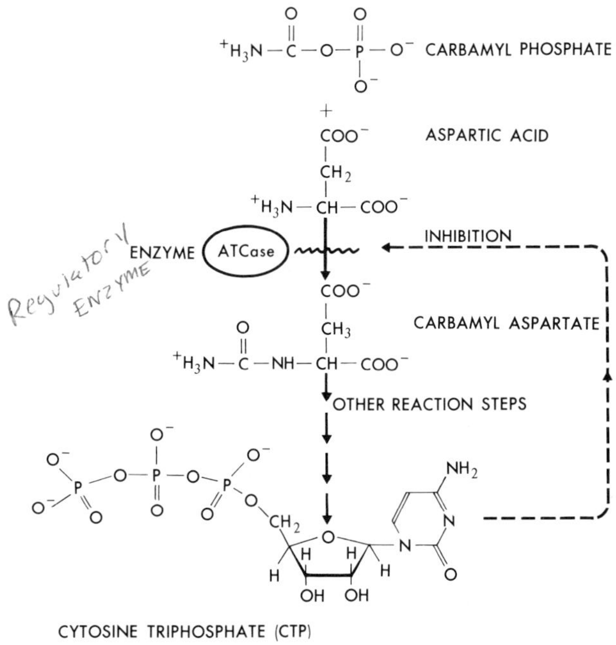

Fig. 3.10 *The biochemical pathway for cytosine triphosphate synthesis from carbamyl phosphate and aspartate transcarbamylase. The end product, CTP, acts as an inhibitor for the enzyme catalyzing the first step in the pathway, aspartate transcarbamylase. Note that the molecular geometries of end product and original substrates are quite different, implying that CTP does not inhibit ATCase competitively.*

called a **regulatory enzyme.** Where the regulatory enzyme is inhibited by one of its own end products, the process is known as **negative feedback inhibition** (the effect of the end product is a negative one on the enzyme).

In the biochemical pathway in our example, as the amount of cytosine triphosphate builds up, increasing numbers of ATCase molecules are inhibited, so that the pathway as a whole is slowed down. Conversely, as the amount of aspartic acid and carbamyl phosphate build up, ATCase begins to catalyze more reaction. The rate at which the enzyme acts is thus greatly affected by the amount of either substrate or end product available. The control process is far more finely tuned than could be accounted for by the equilibrium effects of increasing concentration of end product or reactants. Something else must be involved.

In the middle and late 1950s the following information was available regarding regulatory enzymes and feedback pathways:

1. Inhibition by the end product is reversible; as more substrate is made available, the negative effects of the end product can be overcome.
2. No structural relationship could be found between substrate and end-product molecules. They often appeared to be very different in size and molecular structure.

The first observation suggests that the end-product molecule might be a competitive inhibitor for the enzyme. If that were true, however, from what we know about competitive inhibition the end-product

Supplement 3.1

CHARACTERISTICS OF REGULATORY ENZYMES

1. The enzyme molecule is composed of several subunits—usually two or four, depending upon the particular enzyme. The subunits are usually of two different types, represented by two different polypeptide chains.
2. They contain two different active sites, usually on the separate types of subunits. One active site, the catalytic site, is specific for the substrate; the other, the regulatory site, is specific for the regulatory molecule.
3. End-product molecules act to inhibit the enzyme's activity by combining with the regulatory site.
4. Each subunit, and thus the enzyme molecule as a whole, can shift between two configurational states. In one state the catalytic site is opened up while the regulatory site is closed; in the other state the opposite is true. Binding of either substrate or end product in their respective active sites stabilizes the subunit (and whole molecule) in the appropriate configurational state for a moment of time.
5. Regulatory enzymes are usually found after branch points in a sequential biochemical pathway.

Organisms gain economy and flexibility by constructing large proteins out of several identical subunits. Less genetic information is needed to code for the amino acid sequences, and a protein composed of such subunits is more flexible in the geometric configurations it can assume than one consisting of a single long chain.

molecule would be expected to be very similar to the original substrate. The second observation contradicts that prediction, so it is clear that simple competitive inhibition is not involved.

Allostery

In the early 1960s, the French investigators François Jacob, Pierre Changeaux, and Jacques Monod proposed a model to explain how regulatory enzymes function. They developed the concept of **allostery** (or allosteric proteins—i.e., enzymes), in which an enzyme has two active sites per molecule. One active site, called the "catalytic site," is specific to the substrate; the other, called the "regulatory site," is specific to the end product. Later studies on protein structure indicated that this idea was correct; moreover, it appears that for most enzymes consisting of subunits, the catalytic site is contained on one subunit, while the regulatory site is contained on another.

Based on our knowledge of protein structure, the allosteric model can be stated as follows: at any moment each subunit of the enzyme can exist in either of two conformational states (geometric configurations). Both positions are thermodynamically stable and the subunit can shift back and forth between them. For convenience, we will call these two states the R state and the T state. Since all the subunits of an enzyme molecule are linked together, a shift in the conformational state of one tends to cause a shift among the others in that same direction. Thus, the molecule as a whole will always tend to be in either the R or T state. Within any one molecule, no mixture of states among the subunits (some in R, others in T) will be stable. When a substrate binds to the catalytic site of a subunit, that subunit shifts into the R state. This shift causes the other subunit to shift into the R configuration also; as a result, the regulatory site on the second subunit will be in a distorted state. That is, it will be physically altered so that it will not readily bind to an end-product molecule.

After substrate catalysis takes place, the product(s) fall away from the enzyme. This leaves the substrate active site free; the subunits can then shift to the T state, a process which occurs spontaneously and more or less randomly. In the T state, the active site of the second subunit is not distorted and can receive the end-product molecule. This momentarily stabilizes the whole molecule in the T state. Under these conditions, the catalytic site of one subunit is distorted. It cannot accept substrate molecules and hence cannot act as a catalyst. The important point to keep in mind is that with neither substrate nor end product present, the subunits (and thus the whole enzyme molecule) will constantly shift back and forth between the R and T states. With plenty of substrate and very little end product, the enzyme will tend to be more frequently in the R state and thus catalyze substrate reactions. As catalysis proceeds, end-product molecules will build up and attach with increasing frequency to the regulatory site on the second subunit. As a result, in the T state the enzyme will be unable to bind to substrate, reducing the overall rate of reaction. Regulation is thus brought about by the relative abundance of substrate and end product acting on allosteric enzymes.

SUMMARY

1. The three major classes of large molecules (macromolecules) of great importance to living systems include carbohydrates (sugars and starches), lipids, and proteins. Carbohydrates serve primarily as fuel molecules (energy sources) but also as structural elements; lipids serve as reserve fuel and also take part in membrane structure; proteins serve as struc-

tural components (hair, nails, skeletal systems, muscle, and connective tissue) and as enzymes.

2. Carbohydrates, lipids, and proteins are all constructed by piecing together various special subunits or component parts. A great variety of each type of molecule is achieved by the variable arrangement of essentially a few basic elements.

   a) The basic structural unit of carbohydrates is the sugar molecule (three-carbon, four-carbon, five-carbon, or six-carbon sugars), joined together by glycosidic bonds.
   b) The basic structural units of lipids are fatty acids and glycerol, joined together by ester bonds.
   c) The basic structural units of proteins are the amino acids (of which there are approximately 20 types), joined together by peptide bonds.
   d) The glycosidic, ester, and peptide bonds are formed by dehydration synthesis (removal of a molecule of water between two units); the bonds are broken by hydrolysis (addition of water between two units).

3. Enzymes serve as organic catalysts. They are highly specific for their reactants, or substrates. The enzyme molecule consists of one or more polypeptide chains folded into more or less globular shapes. The substrate fits into a particularly structured region on one part of the enzyme molecule, known as the active site. Fitting together of enzyme and substrate causes a shift in three-dimensional structure of the protein (known as a conformational or configurational shift), which brings certain active groups of the enzyme into contact with specific regions of the substrate. This weakens certain chemical bonds and promotes a reaction.

4. The rate at which enzymes catalyze reactions among their substrates is often subject to allosteric control. Allosteric enzymes have two types of active sites, the catalytic site for substrate, the regulatory site for end product and other regulatory molecules. When the end-product molecule fits into its active site, a conformational shift occurs that causes distortion of the substrate active site. Allosteric enzymes usually consist of two or four subunits (of two different types), with each type of active site on a different type of subunit. Conformational shifts by one subunit are transferred to other subunits as well, since all subunits are tightly bound by electrostatic forces.

EXERCISES

1. The reaction $A + B \rightarrow AB$ takes place slowly at 20°C unless either compound x or y is present. Compound x is a metallic catalyst and y is an enzyme; both compounds catalyze the reaction. Ten milliliters of a solution of A and B are placed in each of four test tubes, to which varying amounts of x or y are added, as shown below:

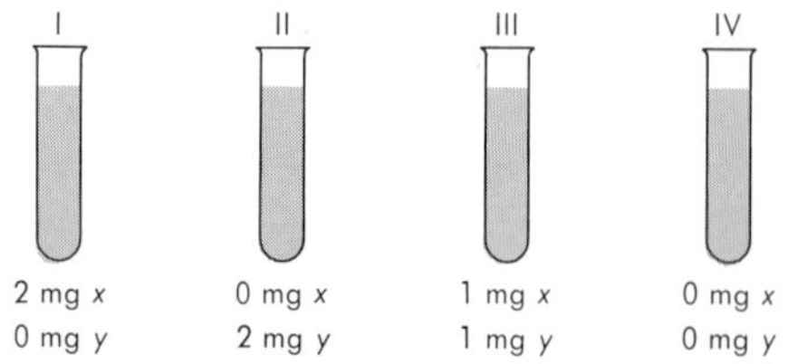

a) Predict which tubes would show the greatest and which the least rate of reaction at 20°C. Explain your reasons.

b) If A and B are heat-stable at 100°C, in which tube(s) would the reaction rate be greatest at this temperature? Least? Why?

c) Increasing the temperature from 20° to 30° will probably double the reaction rate in which tube(s)? How do you know?

d) If the reaction is allowed to reach equilibrium, in which tube(s) will the amount of AB be greatest? Least? What are your reasons?

e) The contents of tubes I, II, III, and IV are poured into separate dialyzing sacks made from cellophane. The sacks are placed in separate containers of distilled water. The reaction slows down in sacks I and III but continues at the previous rate in sacks II and IV. If A and B are added to the distilled water from outside of sacks II and IV (each tried separately), the reaction proceeds at a very slow rate. When A and B are added to distilled water from outside sacks I and III, the reaction speeds up. Explain the results (assume that A, B, and AB are all large molecules).

2. What are the major differences between carbohydrates, fats, and proteins in terms of (a) molecular weight, (b) molecular geometry (size and shape), and (c) general biological uses (in cell metabolism)?

3. Explain the lock-and-key model of enzyme function. How is this model useful? In what ways could it be misleading?

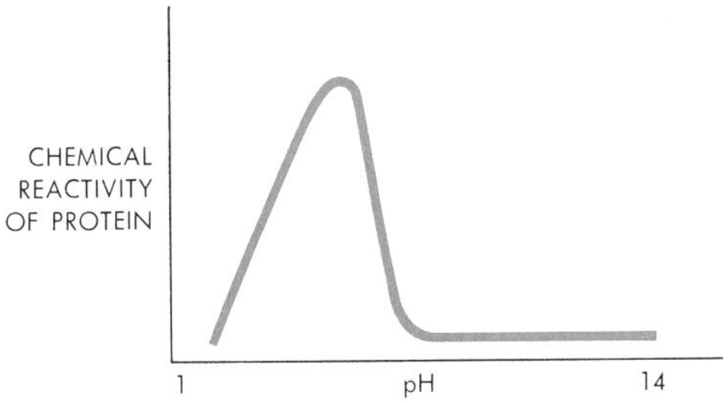

Figure 3.11

4. Proteins are said to display optimal ranges of activity within any environment, for example, within a range of pH values, temperatures, etc. In terms of the graph in Fig. 3.11, explain how changes in pH could affect the chemical reactivity of protein molecules. Include in your answer a discussion of the following points:

a) Where on a protein molecule (and at what level of organization of the molecule —primary, secondary, or tertiary structure) changes in hydrogen ion (H^+) concentration in the medium have their effect.

b) How changes in the protein brought about by change in H^+ concentration could affect chemical reactivity of the protein.

c) Why is the curve relatively steep on either side of the optimal peak?

5. The following data represent optical density readings for three enzyme-catalyzed reactions.

Table I

Enzyme + Substrate

Time, min	Optical density
0	.030
15	.064
30	.095
45	.125
60	.150

Table II

Enzyme + Substrate + Inhibitor

Time	Optical density
0	.011
15	.030
30	.049
45	.068
60	.088

Table III

Enzyme + Inhibitor + Overabundance of Substrate

Time	Optical density
0	.033
15	.055
30	.085
45	.110
60	.135

Assuming that the more reaction that occurs, the greater the optical density of the solution, interpret the data in light of current knowledge of enzyme structure and function. Include in your answer the following:

a) Why does the curve have the shape it does for Table I? Why does the change in optical density occur as it does? Explain this in terms of interaction between enzyme and substrate molecules.

b) What is happening in Tables II and III? What kind of inhibition is shown here? On what evidence is your answer based?

c) How can these graphs be explained in terms of the concept of active site or the lock-and-key analogy of enzyme function?

It was further found that when the enzyme and inhibitor were first heated together in solution, the inhibitor subsequently removed, the protein solution slowly cooled, and a normal amount of substrate added, the reaction occurred in much the same way as indicated in Table I. Explain this phenomenon in terms of your knowledge of the nature of the forces involved in determining protein structure.

SUGGESTED READINGS

The following books contain more detailed treatment of atomic structure, chemical bonding, and formation of molecules:

Baker, J. J. W., and G. E. Allen, *Matter, Energy, and Life,* 3d ed. Reading, Mass.: Addison-Wesley, 1974.

Cheldelin, Veron H., and R. W. Newburgh, *The Chemistry of Some Life Processes.* New York: Reinhold, 1964. An introduction to organic chemistry with special reference to biochemical processes. Assumes more background in chemistry and atomic structure (bonding, etc.) than *Matter, Energy, and Life,* but covers some biochemical processes in great detail.

Hendrickson, James B., *The Molecules of Nature.* Menlo Park, Calif.: W. A. Benjamin, 1965.

Hoffman, Katherine B., *Chemistry of Life.* Washington, D.C.: National Science Teachers' Association, 1964. A very readable introduction to some basic biochemical processes. Amply illustrated.

Ryschkewitsch, George E., *Chemical Bonding and the Geometry of Molecules.* New York: Reinhold, 1963. A simple introduction to types and properties of chemical bonding in terms of orbital theory.

Sisler, Harry H., *Electronic Structure, Properties, and the Periodic Law.* New York: Reinhold, 1963. An up-to-date treatment of orbital theory and its relation to bond formation. Requires some simple mathematical background.

White, Emil, *Chemical Background for the Biological Sciences.* Englewood Cliffs, N. J.: Prentice-Hall, 1964. This paperback serves well as the text for a one-semester course in organic chemistry. It is primarily concerned with a variety of organic rather than biochemical reactions.

The following books contain some valuable information on enzymes and the principles of biochemistry as applied to living systems:

Avers, Charlotte J., *Basic Cell Biology.* New York: Van Nostrand, 1978. Chapter 3 of this text contains a useful summary statement on the modern theories of enzyme structure and function. A good, modern reference in relatively simplified form.

Bernhard, Sidney, *The Structure and Function of Enzymes.* Menlo Park, Calif : W. A. Benjamin, 1968.

Koshland, D. E., "Protein shape and biological control." *Scientific American* 229 (no. 4, October 1973), p. 52. A good discussion of the way enzymatic control is obtained through shifts in conformational structure of the protein.

Lehninger, A. L., *Biochemistry.* New York: Worth, 2d ed., 1975. This is probably the single best biochemistry textbook available at the present time. Very thorough and detailed; excellent for reference.

Loewy, Ariel, and Philip Siekovitz, *Cell Structure and Function.* New York: Holt, 2d ed., 1969. Not too detailed, a good general textbook for review.

Moss, D. W., *Enzymes.* London: Oliver and Boyd, 1968. This small paperback is an excellent simple introduction to enzymes, their variety, structure, and function.

FOUR

CELL STRUCTURE AND FUNCTION

4.1 Introduction

Within the last 20 years, biologists have increasingly been turning their attention to problems at the cellular level. In present-day terms, the problems of heredity, embryonic development, immunity, and cancer have all been studied on the level of the cell and its components. What are cells? What different kinds of cells are found throughout the animal and plant kingdoms? What are the structural components of cells? How do cells function in chemical and molecular terms? And finally, how do cells interrelate to form higher levels of organization in multicelled animals and plants?

4.2 The Cell Concept

The "cell concept" as viewed today can be summarized in the following four propositions:

1. *Cells are the structural units of virtually all organisms.* From simple one-celled organisms like an *Amoeba* or bacterium to multicelled organisms like a human being or tree, all organisms are composed of basic structural units called **cells.** Cells are the building blocks in the architecture of life.

2. *Cells are the functional (and dysfunctional) units of virtually all organisms.* All the chemical reactions necessary to the maintenance and reproduction of living systems take place within cells. The chemical processes (metabolism) that provide the energy for contraction of a muscle cell, for example, take place within the muscle cell itself. The same is true for the processes of cell reproduction: they all occur within cells. At the same time, the failure of functions, or sicknesses, are also cellular in nature. Cancer, for example, is the result of a failure of certain cells to regulate the reproductive process.

3. *Cells arise only from preexisting cells.* Cells do not spontaneously generate. A multicellular or-

ganism grows by duplication of its individual cells. By special cell divisions, some organisms produce **gametes,** or specialized sex cells such as eggs or sperm, capable of generating a whole new organism. The idea of **biogenesis**—that all living cells originate only from preexisting living cells, or that life originates from life—is fundamental to the modern cell concept.

4. *Cells contain hereditary material* (**nucleic acid**) through which specific characteristics are passed on from parent cell to daughter cell. The hereditary material contains a "code" that ensures continuity of species from one cell generation to the next.

The modern cell concept (sometimes called "the cell theory") is not the work of any one person. Cells were first observed and described by some of the early seventeenth-century microscopists. Robert Hooke's *Micrographia* (1665) contains some of the first clear drawings of plant cells, made from observations of thin sections of cork.* Hooke coined the term "cell" to refer to the boxlike structures he saw. The fact that all living organisms are composed of cells was not recognized, however, until the nineteenth century. The most important generalized statement (our first proposition) about the cellular nature of living organisms was made by two German biologists, Matthias Schleiden and Theodor Schwann, in 1838 and 1839. Schleiden, a botanist, and Schwann, a zoologist, studied many types of tissue in their respective fields. Both came to the conclusion that the cell was the basic structural unit of all living things. The second and third propositions were added by the German pathologist and statesman Rudolf Virchow (1821–1902). In his work *Cellular Pathology* (1858), Virchow spoke of the cell as the basic metabolic as well as structural unit. In this same work he emphasized the continuity in living organisms with the statement: "*omnis cellula e cellula*"—"all cells come from [preexisting] cells." The last proposition in the modern cell concept is a more recent addition.

All cells, whether living individuals as in a one-celled plant or animal, or collectively in a multicelled organism, face certain common problems. As a result, they have numerous structural and functional features in common. Although we will discuss many of the differences between cell types throughout this book, in the next three chapters we emphasize the numerous features they have in common. All cells have a surface covering called a **cell membrane** regulating what passes in and out of the cell. All cells contain some form of hereditary information, coded in molecules of **nucleic acid.** In some types of cells nucleic acid is contained within a special, membrane-surrounded region called the **nucleus;** in others, it is more loosely distributed throughout the cell. In addition, cells contain various kinds of smaller structures called **organelles.** Organelles are "little organs" representing the cell's machinery. Some organelles are responsible for energy release, others for protein synthesis, packaging of storage products, and transportation of substances throughout the cell. Not all types of cells have all the various organelles. Some of the most fundamental taxonomic categories among living organisms are based on presence or absence of certain cell organelles.

Presence or absence of a well-defined nucleus is the basis for one of the most fundamental classi-

* The term "cork" refers to a portion of the bark, or outer covering layer, of any woody plant. The cork used for thermos bottles is part of the bark of the cork oak tree, found mostly in Spain.

Cells are the structural, functional, and reproductive units of virtually all organisms. They are the seat of disease as well as health.

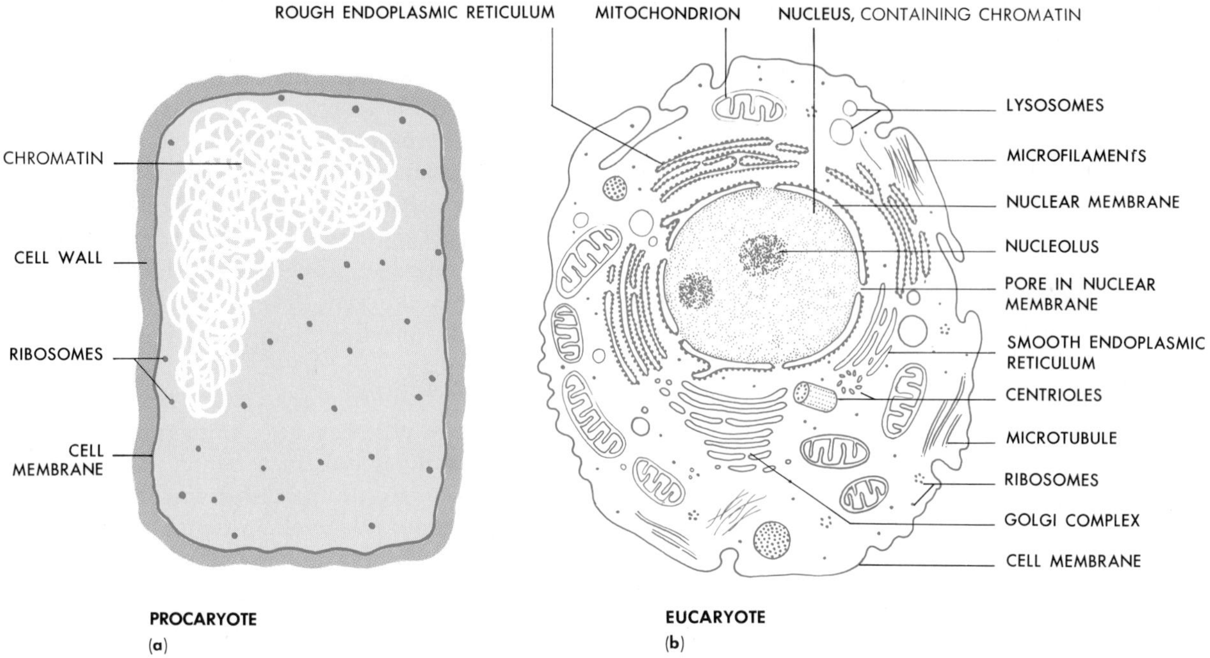

Fig. 4.1 *Comparison between a procaryotic and eucaryotic cell. Although the procaryotic cell possesses a high degree of organization, it is not as complex and does not show as high a degree of specialization as the eucaryotic cell. (After E. O. Wilson* et al., Life on Earth. *Sunderland, Mass.: Sinauer, 1973.)*

fications in the animal and plant worlds: eucaryotic and procaryotic cells. **Eucaryotes** (Greek, *eu* = true, *cary* = nucleus) have a definite nucleus. Eucaryotic cells are usually larger and more complex than procaryotic cells; they contain many more types of organelles than procaryotes. The cells of all higher plants and animals are eucaryotic. The cells of bacteria and blue-green algae are **procaryotes.** Because they lack both a well-defined nucleus and most of the cell organelles found in eucaryotes, procaryotes are unable to form the complex interactions that make possible the development of multicellular organisms. Figure 4.1 compares a procaryote and a eucaryote.

4.3 Tools and Techniques for Studying Cells

Advances in our knowledge of cellular anatomy are tied directly to improvements in the techniques and instruments used to examine cells. The nineteenth and early twentieth centuries saw the introduction and refinement of certain techniques, such as staining and microtoming. In **staining,** parts of cells and tissues are colored with dyes, causing them to stand out in sharp contrast to other structures left unstained or stained different colors. In **microtoming,** cells and tissues are cut into thin sections so that they may be examined under a microscope (see Fig. 4.2).

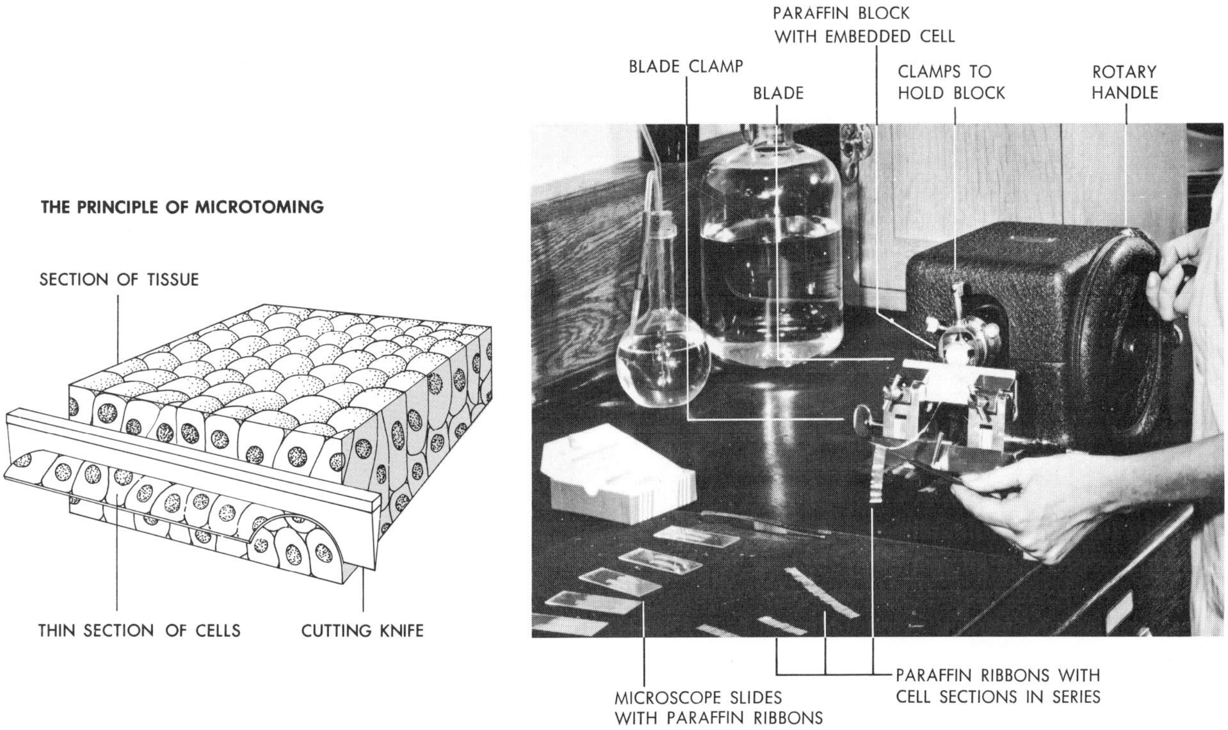

Fig. 4.2 *A rotary microtome in use. The cells are permeated with paraffin to hold them rigid when cut by the blade. When the wheel is turned, the cells are pushed against the blade and a section is cut. The next turn of the wheel moves the cells forward a predetermined number of microns, and another section is cut. The entire section series is then stained and studied under the microscope. From such studies, it is possible to recreate the cell in three-dimensional perspective. (Photo courtesy Douglas C. Anderson, Department of Fisheries and Forestry, Ontario, Canada.)*

The Light Microscope

The primary instrument for the study of cell structure is the light microscope (Fig. 4.3, a and c). From the seventeenth century to the 1920s, detailed knowledge of cellular anatomy increased as the resolving power of the light microscope was increased. **Resolving power** is a microscope's capacity for separating to the eye two points that are very close together.* Resolving power is inversely proportional to the wavelength of radiant energy used, (as well as such factors as the opening size of the microscope lens). This means that as wavelength *decreases* (using light more toward the violet end of the spectrum), resolving power *increases*. The

* The naked human eye has a resolving power of approximately 0.1 mm. Thus two lines placed closer together than 0.1 mm will appear as a single line to the naked eye, no matter how close to them the observer may get. Since most cells are smaller than 0.1 mm across, the need for a microscope is obvious.

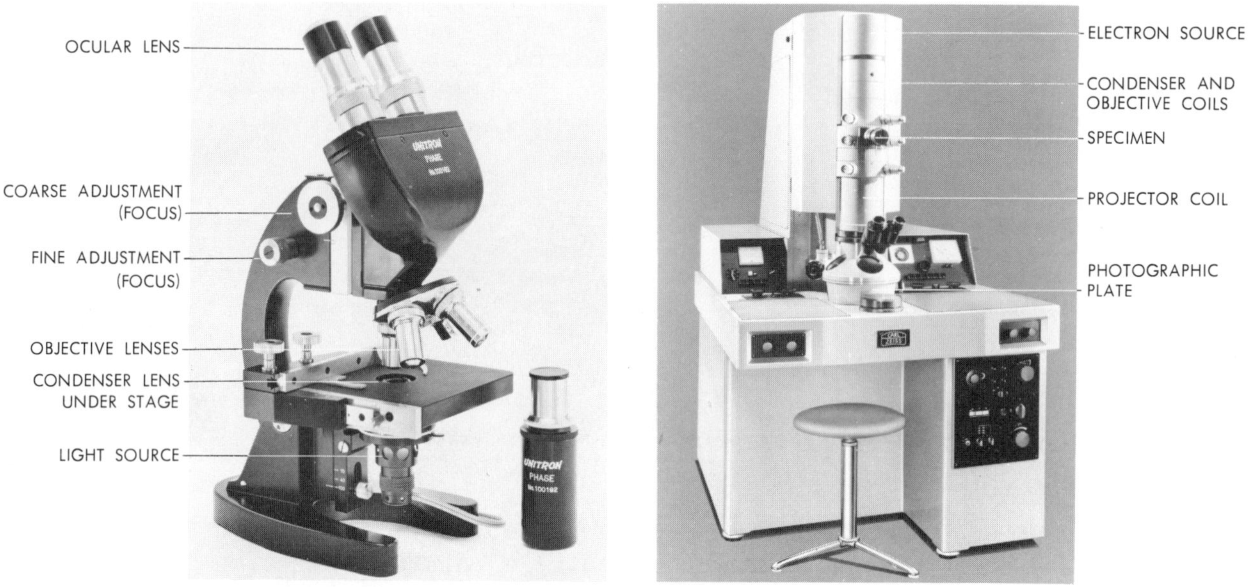

(a) **LIGHT MICROSCOPE**

(b) **ELECTRON MICROSCOPE**

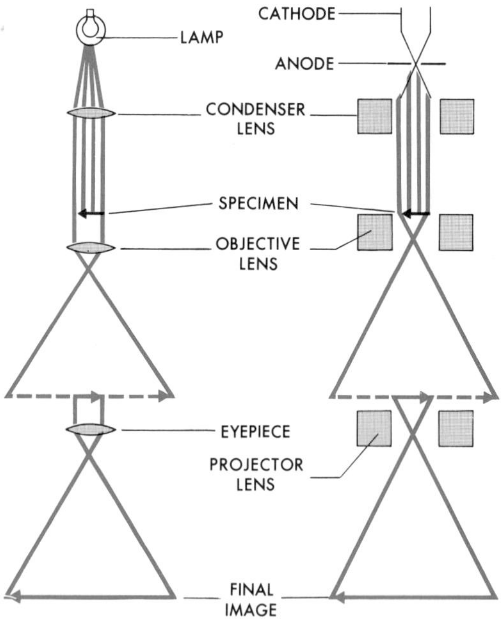

(c)

(d)

Fig. 4.3 *Comparison of light (a and c) with electron (b and d) microscopes. The principles on which they work are very similar. The light microscope uses radiation within the visible spectrum, whereas the electron microscope uses beams of electrons at much shorter wavelengths. For purposes of comparing the optics of these microscopes, the light microscope is inverted from its normal position in (c). (Photo of light microscope courtesy Unitron Instrument Company, Newton Highlands, Massachusetts. Photo of electron microscope courtesy Carl Zeiss, Inc.)*

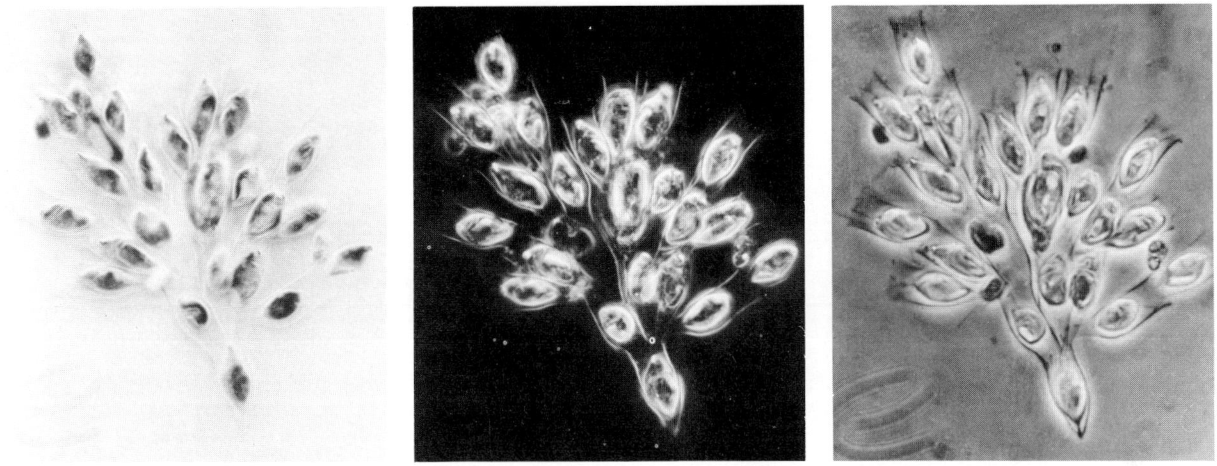

(a) BRIGHT-FIELD (b) DARK-FIELD (c) PHASE-CONTRAST

Fig. 4.4 *Three microscope techniques are shown here in viewing the same organism,* Dynobryon sertularia. *Bright-field microscopy directs light up through the specimen toward the viewer. Dark-field microscopy sends light in at an angle from the side. Thus only light bent upward by the specimen is seen in the field. (Photos courtesy Dr. Earl Hanson, Wesleyan University.)*

best modern light microscopes have resolving powers of 0.17 μ, or 1700 Å, using violet light; two dots less than 0.17 μ apart will appear as a single dot when viewed under violet light. If light of longer wavelengths is used, for example white light (a mixture of many wavelengths, most of which are longer than violet), the resolving power is limited to 0.25 μ. By using very short wavelengths of radiation (such as ultraviolet "light"—invisible to the human eye but visible to photographic plates) and quartz lenses, light microscopes can obtain a resolving power of 0.1 μ, but no more.

Other techniques have been employed to increase the amount of detail a light microscope can disclose other than simply size of object. One method is known as **dark-field microscopy.** In this process, light rays are bent by the lenses in such a way that only the light passing through the specimen passes upward to the viewer's eye; the background remains dark. This contrast makes the object being viewed, and especially its boundaries, more distinct, as shown in Fig. 4.4(b).

Another technique is known as **phase-contrast microscopy** (Fig. 4.4c). The human eye responds best to wavelengths of light that pass easily through water; much of the cell's content is composed of water. Hence, the majority of cell components appear transparent to the human eye. This difficulty can be overcome by taking advantage of the fact that in *all* light mircoscopes, light waves undergo slight changes in **phase:** a part of the beam of light passing through the cell (or any object being studied) is momentarily slowed down and thus rendered out of phase with the rest of the beam. In the normal light microscope this retardation is so slight that it cannot be detected by the human eye. In the phase-contrast microscope, the retardation is amplified by a set of mirrors so that it becomes visible. Hence, light passing through the

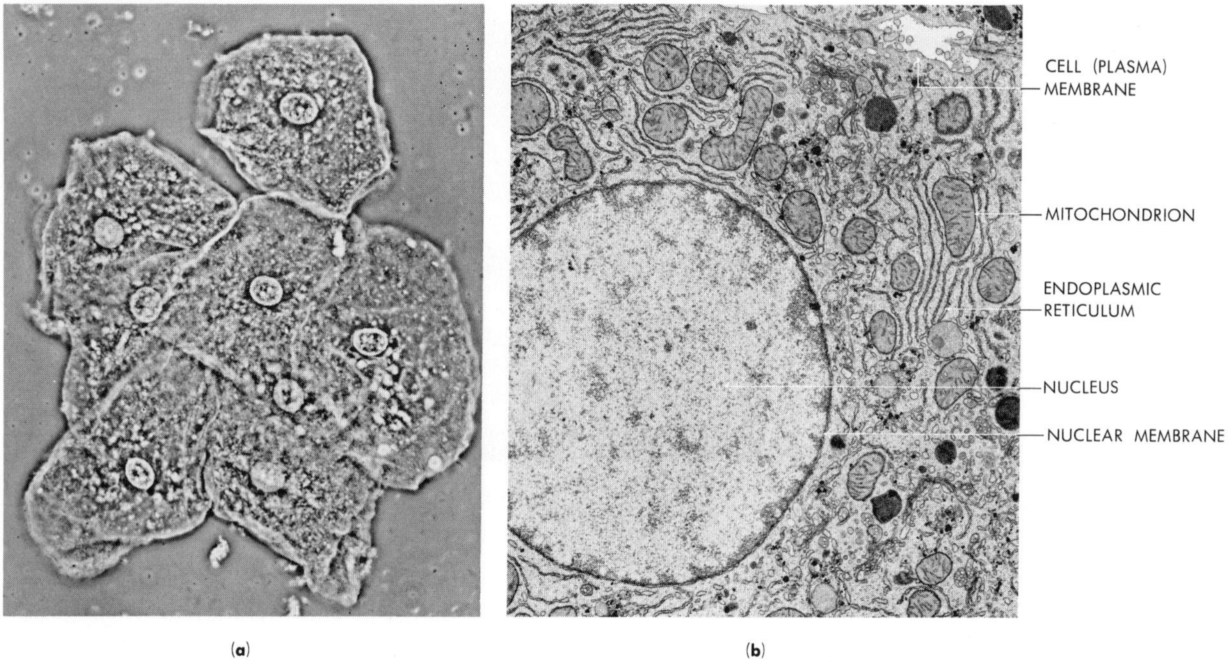

Fig. 4.5 *View of two mammalian cells under light (a) and electron (b) microscopy. (a) A human cheek cell viewed under phase-contrast light microscopy (×400). Note that the only part clearly visible inside the cell is the nucleus. The cytoplasm appears granular and can be seen to contain numerous particles, but few details are visible. (b) Electron micrograph (×17,000) of a rat liver cell. Note the large, dominant nucleus, the many mitochondria, the cell and nuclear membranes, and the endoplasmic reticulum. (Electron micrograph courtesy Albert L. Jones, M.D.)*

cell and portions of the cell reaches the eye at different times. The thicker the portion of cell the light passes through, the more those waves are slowed down. The result is that details become visible because each structure slows the light down (makes it out of phase with other parts of the light beam) in a very specific way.

The Electron Microscope

In 1924, the physicist Louis de Broglie theorized that electrons had a wave nature. The length of that wave was found to be only 0.05 Å (one angstrom unit equals 10^{-8} cm or 1/100,000,000 cm). Therefore, it was reasoned, if a beam of electrons were substituted for a beam of light, the shorter wavelength of the electrons should cause a corresponding increase in resolving power. This prediction proved to be correct. In 1934, an electron microscope was used to obtain pictures with greater resolution than those that could be obtained through the light microscope. The electron microscope uses a beam of electrons rather than a beam of light. Magnetic coils serve to focus the beam, much as the objective lenses focus light beams in the light microscope (see Fig. 4.3, b and d). The electron microscope has revealed more new knowledge of cell structure in 30 years than was discovered with the light microscope in the previous 300.

As mentioned above, the limit of resolving power of the light microscope is about 0.17 to 0.25 μ (depending on wavelength used); that for the electron microscope is about 0.0005 μ (5 Å). Furthermore, while the light microscope can magnify an object only at a maximum of about 2000 times, an electron microscope can magnify up to 1,000,000 times! It is obvious why the electron microscope gives an enormous advantage to modern biologists.

Figure 4.5 shows a comparison between a light microscope picture and an electron microscope picture, called a **photomicrograph,** of an animal cell. The difference in the detail between the two is striking. The greater detail revealed by the electron microscope is a result of both its greater magnifying power and its greater resolving power.

From a distance, a forest-covered mountain appears a solid green or bluish color. Close to the mountain, however, one can see that this apparently solid green blanket is highly diverse, composed of individual trees differing widely in size and kind. The development of the electron microscope enabled the biologist to view the cell "up close." Like the blanket of trees, the cytoplasm showed itself not to be a uniform substance, but a highly structured region, with intricate design and detail.

The Scanning Electron Microscope

Recently a new instrument called the **scanning electron microscope** has been used to study the fine structure of cells and organisms. (The type of

Fig. 4.6 *The scanning electron microscope allows high-magnification three-dimensional viewing of objects such as cells or, in this case, pollen grains. Shown below are pollen grains of two different species of plants: the small herb* Cosmos bipinnatus *(left),* ×*6500; and the lily* Lilium longiflorum *(right),* ×*2000. (Photos courtesy J. Heslop-Harrison, University of Wisconsin.)*

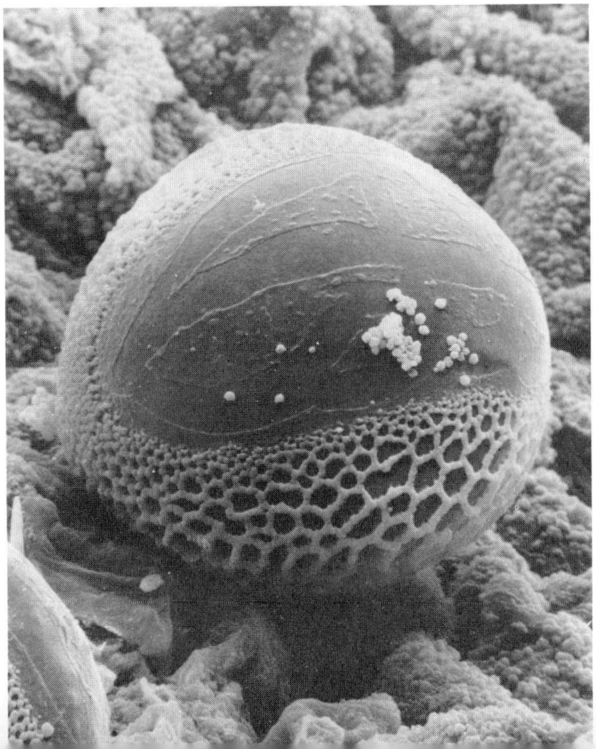

electron microscope already discussed is technically called a transmission electron microscope.) In the scanning electron microscope, a beam of electrons scans and bounces off the surface of an opaque specimen, rather than passing through an extremely thin section). This new instrument greatly facilitates the study of whole specimens or surface structures, such as those found in pollen grains, red blood cells, kidney tubule cells, bone and teeth surfaces, and cells in culture (grown outside the organism in a special medium in the laboratory). Scanning electron micrographs appear in Fig. 4.6.

4.4 The Eucaryotic Cell

The "typical" cell pictured so often in biology texts does not exist. To get some idea of the internal organization of cells, it will be necessary to consider both plant and animal cells. There are many remarkable similarities between animal and plant cells. Throughout this section we will compare and contrast the two, discussing the organelles they share in common as well as those they do not.

Representative animal and plant cells are shown in Figs. 4.7 and 4.8. Note that for both types of

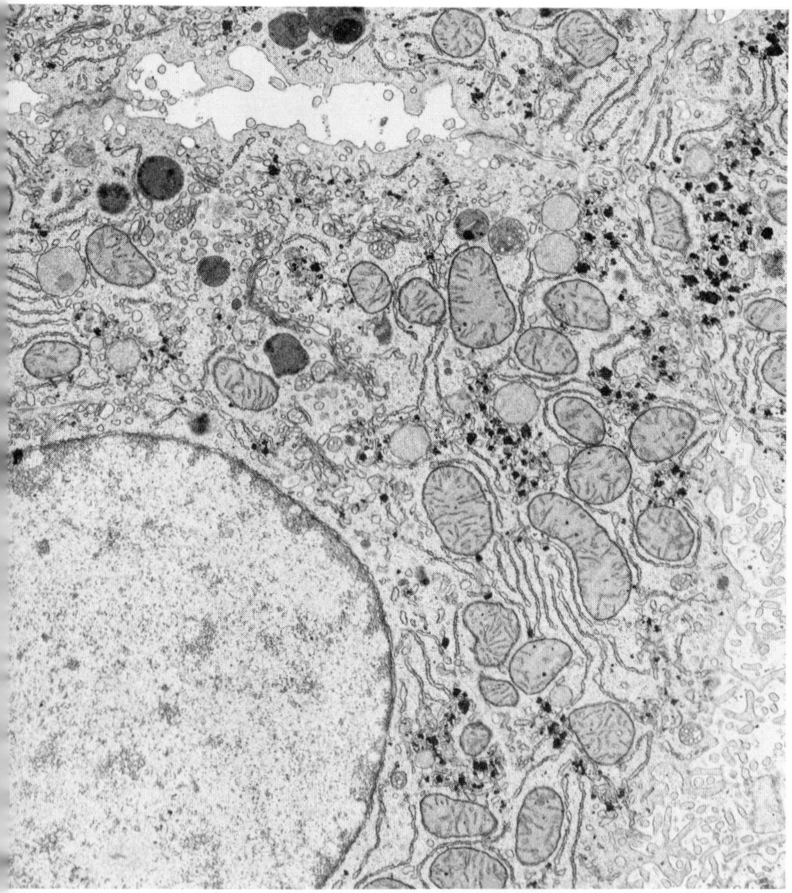

(a)

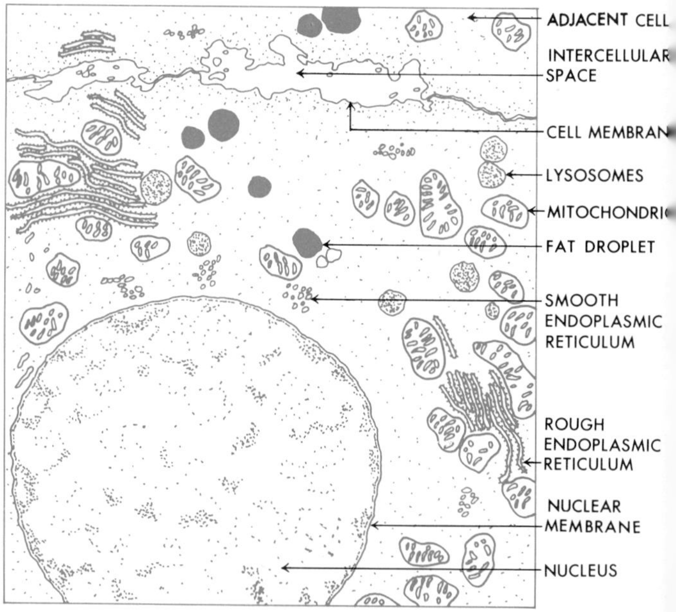

(b)

Fig. 4.7 *Electron micrograph (a) and diagram (b) of a representative animal cell. Like Fig. 4.5 (b), this cell is from a rat liver. The dark globules are fat droplets; the smaller dark granules are glycogen, a polysaccharide known as "animal starch," which is synthesized and stored in liver cells. (Electron micrograph courtesy Albert L. Jones, M.D.)*

Supplement 4.1

THE UPPER LIMIT ON CELL SIZE: A MATTER OF GEOMETRY

Cells vary widely in size and shape. The difference in size between the smallest and largest is comparable to the difference in size between a mouse and a whale. The smallest cells, known as pleuropneumonialike organisms (PPLO), or mycoplasmas, range in size from 0.1 to 0.25 μ. They are responsible for a highly infectious pneumonia in cattle and also for some forms of human pneumonia. The largest free-living cells in terms of volume include protozoans such as the giant amoeba *Chaos chaos,* which reaches a length of 5 cm. In terms of length, in multicelled organisms some nerve cells run the length of an entire arm or leg: for an elephant such a cell may be more than 8 ft long. Diagram A offers a range of cell sizes drawn to scale. In life, the large amoeba is easily visible to the naked eye; the human cheek cell is barely visible; the others are invisible without a microscope.

How large can cells get? The answer to this question comes partly from solid geometry. All cells depend for their continued existence on the passage of materials (such as nutrients and wastes) in and out through their boundaries, the cell membrane. From solid geometry we know that as a sphere becomes greater in size, its volume increases as the *cube* of the radius. The surface area, on the other hand, increases only

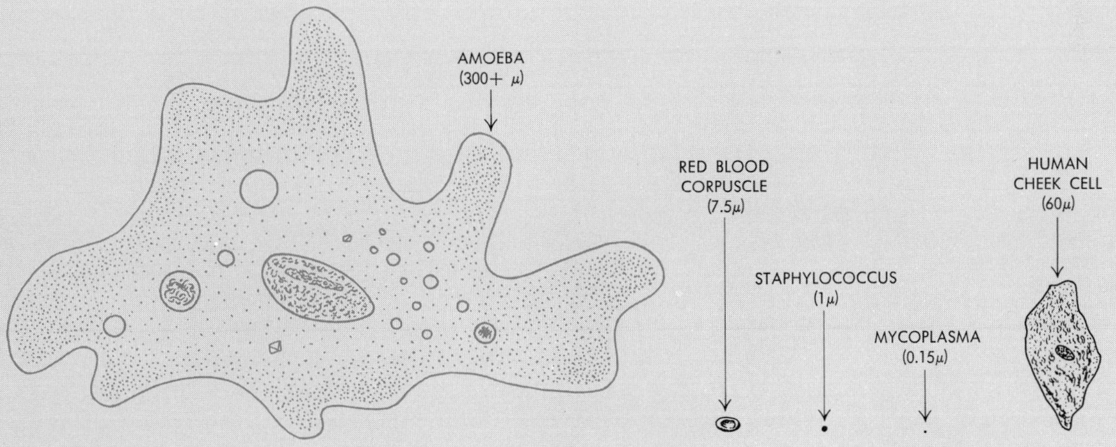

Diagram A

as the *square* of the radius. This simple principle
applies to the other three-dimensional shapes
a cell may assume. The surface–volume ratio
of solid bodies is shown in Diagram B.

 A cell may become no larger than the maximum size at which it can successfully carry out
its life processes. That is, its volume cannot ex-
ceed the capacity of the surface area to exchange
materials between inside and outside. If its size
exceeds this limit, the cell must either divide
(thereby restoring a surface-area-to-volume ratio
compatible with life) or die.

$r = 1$ mm

SURFACE AREA $= 4\pi$
VOLUME $= \frac{4}{3}\pi$

$r = 2$ mm

SURFACE AREA $= 16\pi$
VOLUME $= \frac{32}{3}\pi$

$r = 3$ mm

SURFACE AREA $= 36\pi$
VOLUME $= \frac{108}{3}\pi$

1 mm

SURFACE AREA $= 6$ mm^2
VOLUME $= 1$ mm^3

2 mm

SURFACE AREA $= 24.$mm$^2 =$ FOURFOLD INCREASE
VOLUME $= 8$ mm$^3 =$ EIGHTFOLD INCREASE

3 mm

SURFACE AREA $= 54$ mm$^2 =$ NINEFOLD INCREASE
VOLUME $= 27$ mm$^3 =$ TWENTY-SEVENFOLD INCREASE

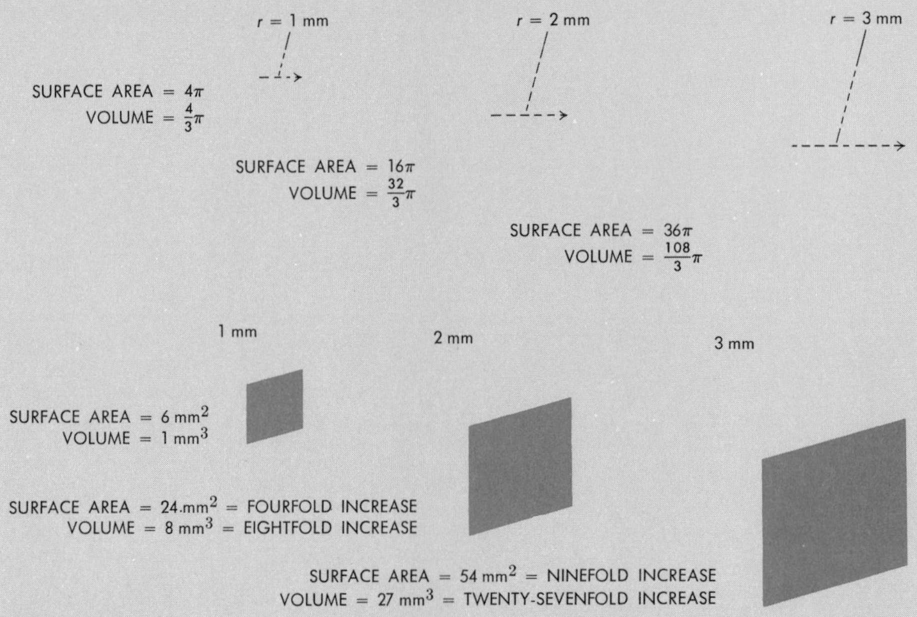

Diagram B

Two great developments have occurred in the evolution of cells: the procaryotes, without a distinct nucleus, and the eucaryotes, with a membrane-bounded nucleus. Cells of most plants and animals are eucaryotic; cells of bacteria, mycoplasmas, and blue-green algae are procaryotic.

cells, there is a general division between cytoplasm and nucleus. Most of the cell's organelles are located in the cytoplasm, which we may think of as the cell's "living machinery." The nucleus is the center for hereditary information; it programs and guides the overall activities of the living machinery. We will consider first the outer boundary of the cytoplasm, the cell membrane or plasma membrane. Then we will turn to the cytoplasm and its various component organelles. Finally, we will consider the organization of the cell nucleus. As you read, refer to the whole-cell electron micrographs (Figs. 4.7 and 4.8) as well as the more detailed electron micrographs of individual cell structures.

The organelles discussed in this section are not the only ones found in animal and plant cells. We may later have occasion to refer to a variety of

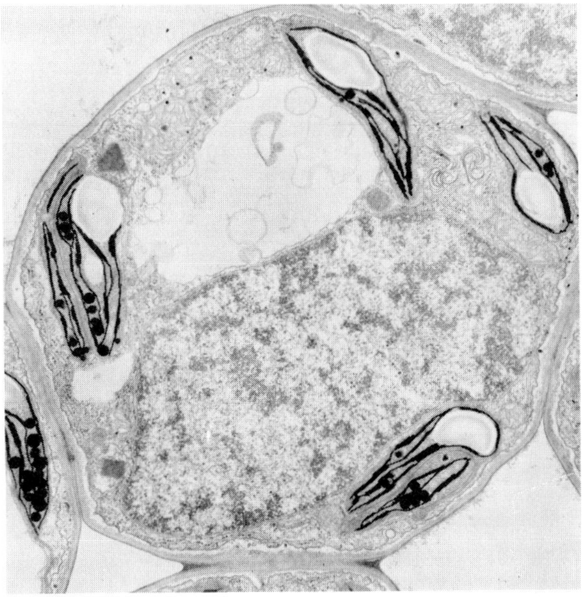

(a)

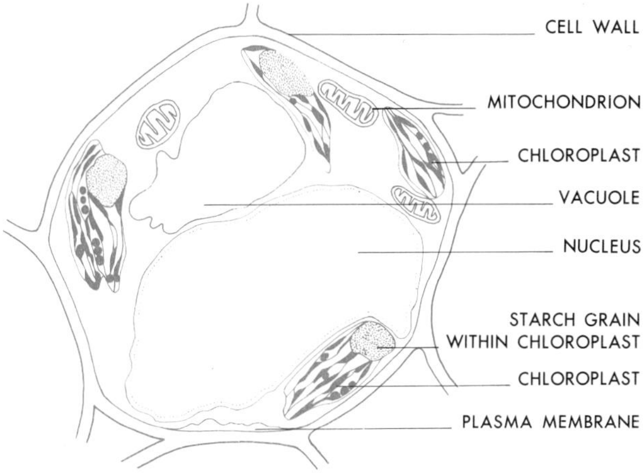

(b)

Fig. 4.8 *Electron micrograph (a) and drawing (b) of a representative plant cell. Note the presence of a thick cell wall, in addition to the plasma membrane, at the boundary of the cell. Note also the large vacuole, typical of most kinds of plant cells. (Electron micrograph courtesy H. J. Arnott, University of Texas at Arlington.)*

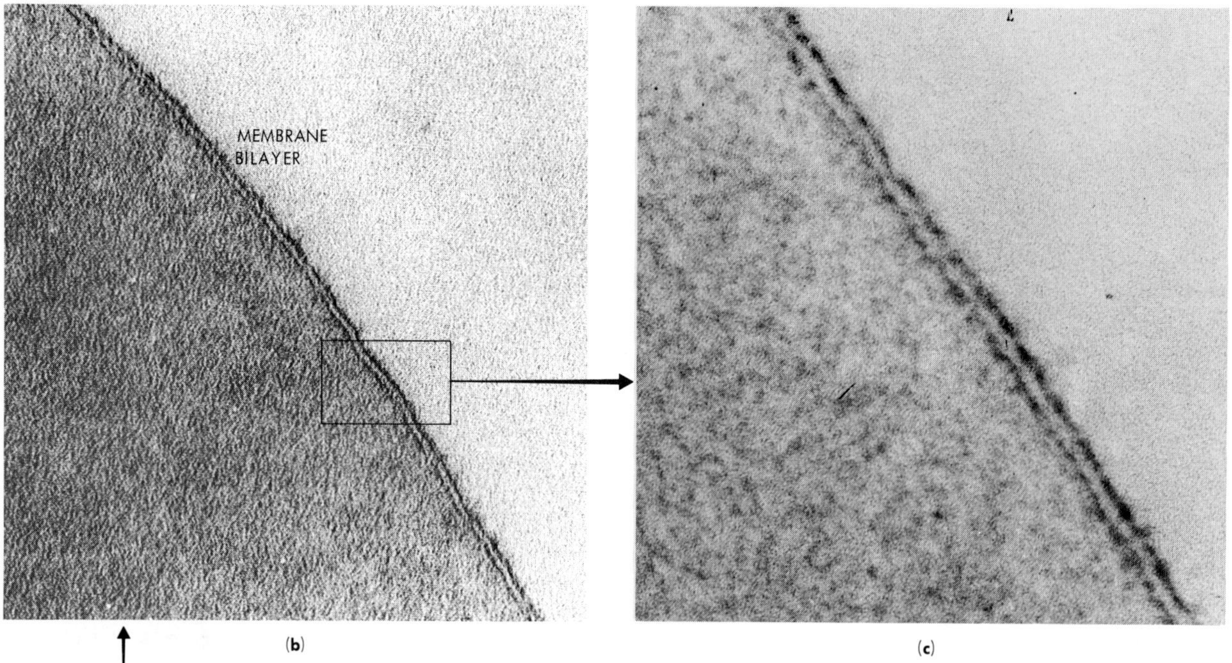

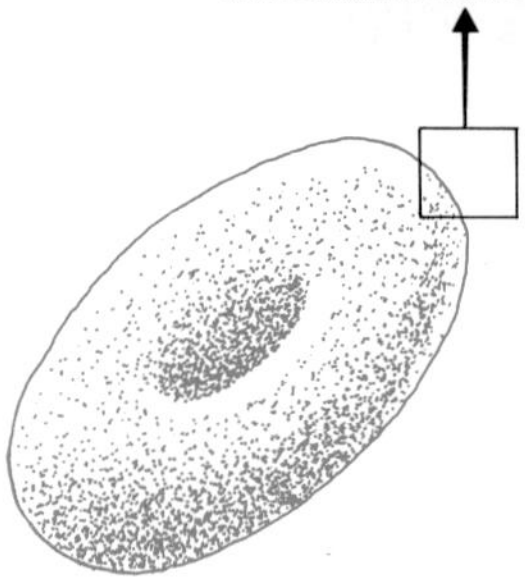

Fig. 4.9 *The cell membrane from a single red blood cell. (a) Drawing of a whole red blood cell. (b) Electron micrograph of a part of the cell membrane showing the unit membrane bilayer. Denser layers (dark) are composed of glycerol and protein portions; less dense areas (light) in between are largely fatty acid chains (×360,000). (c) Higher magnification of membrane bilayer (×758,000). (Photos from* Biochemical Society Symposium *16 (1959), courtesy of J. David Robertson and Academic Press.)*

others, found in some but not all cell types. The organelles we will discuss here constitute the main ones found in nearly all eucaryotic cells.

The Cell Membrane: An Overview

Structure. Materials pass in and out of the cell through the cell membrane or plasma membrane (Fig. 4.9). Chemical studies of cells show their isolated plasma membranes to be composed mainly of phospholipids and proteins. Phospholipids consist, at one end, of a long chain of atoms with nonpolar properties, and at the other end of a shorter group with polar properties, as shown in Fig. 4.10. Molecules with polar properties tend to dissolve in water. Thus phospholipid molecules in

---

The cell membrane is a dynamic association of lipid (phospholipid and glycolipid) and protein. It is the site of many biochemical reactions in the cell.

---

contact with water tend to line up with their polar ends in the water and their lipid ends away from it.

The basic structure generated by the polarity of phospholipids is thought to account for the general appearance of membranes. Note in the enlargement shown in Fig. 4.11 that each membrane looks like a railroad track, consisting of two parallel lines bordering a clear space. The darker lines shown in the electron microscope photomicrographs

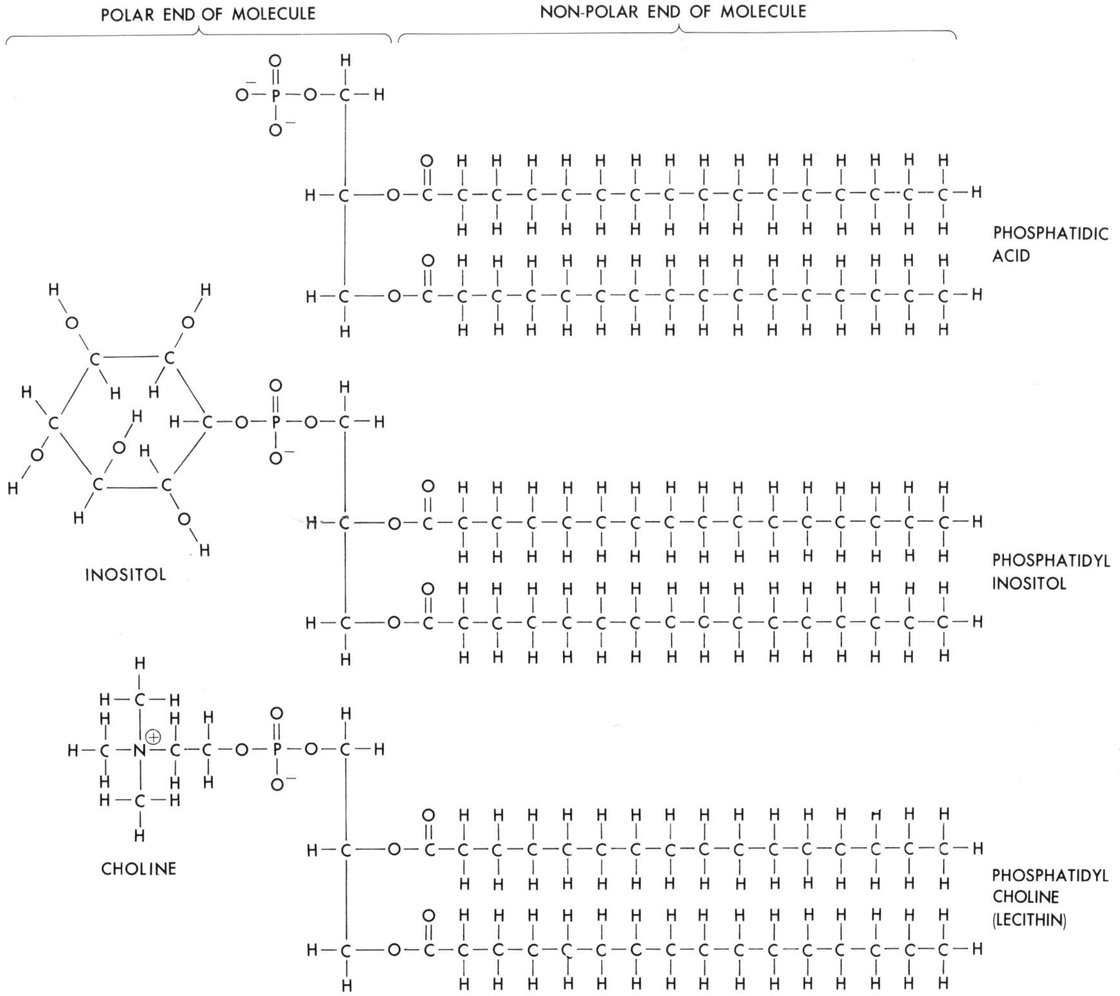

Fig. 4.10 *Three prominent phospholipids found in many types of animal and plant membranes. All three molecules consist of two hydrocarbon chains (fatty acids) bound to a glycerol. In place of a third fatty acid chain, phospholipids have a phosphate compound bound to carbon 1 of the glycerol.*

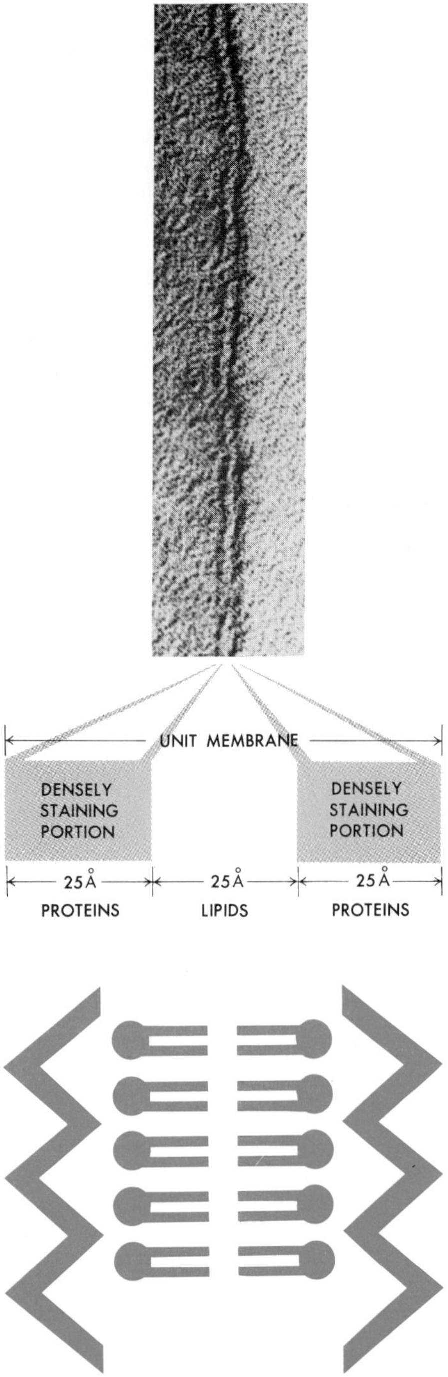

are areas of greater density, while the light space between them is an area of lesser density. In 1937, H. Davson and J. F. Danielli suggested that the polar, gylcerol-phosphate end of the lipid molecules make up the dense region, while the nonpolar fatty acid chains make up the less dense region (Fig. 4.11). According to the Davson-Danielli model, the membrane appears to be a kind of bilayer (two-layer) whose basic structure derives in large part from the chemical nature of its phospholipid components. The three regions making up a bilayer are known as a **unit membrane.**

However, other molecular components of the membrane are also involved in the unit membrane, and a simple bilayer model for its structure is too simple. More details of the molecular basis for membrane structure will be discussed later in this chapter. In particular, there are some problems with the Davson-Danielli model, which will be discussed further in Section 4.7. It can be stated, however, that the general bilayer nature of the membrane appears to be characteristic of *all* membranes, including those surrounding intracellular components as well as the plasma membrane.

Fig. 4.11 *The basic structure of cell membranes as proposed by the Davson–Danielli model. The upper illustration shows a high-power electron micrograph of a membrane (belonging to a cell in the intestine), showing the basic unit-membrane structure. The middle and bottom illustrations show schematically the relationship between the membrane structure as it appears in the electron microscope and its molecular composition. The densely staining region of each unit membrane consists of the polar head of phospholipids and surrounding protein. The less densely staining region in the middle consists of the hydrocarbon chains of the fatty acids. The Davson–Danielli model has been refined and expanded in recent years, though much of its basic idea remains intact. [Electron micrograph (×24,000) courtesy Professor E. De Robertis; reproduced from E. De Robertis,* Cell Biology, *6th ed. Philadephia: W. B. Saunders, 1975.]*

Function. The plasma membrane has several very important functions. Structurally, it provides a boundary that keeps cell components together. Without the membrane, nothing would confine the various organelles in a single space with close proximity to each other. Functionally, the membrane serves as a doorway through which substances enter and leave the cell. More importantly, it serves as a *selective* doorway: it allows some substances to pass through while restricting the passage of others. With astounding precision, cell membranes can distinguish between units as small as ions. The membrane can also allow certain types of very large molecules to pass through and prevent others from passing through.

Transport. Passage of materials through membranes can occur in one of two general ways. The first, known as **passive transport,** is based on the simple principles of diffusion. Diffusion always occurs from an area of greater concentration (of the substance doing the diffusing) to an area of lesser concentration (of that substance). Thus, passive transport cannot be used to create a greater concentration of something on one side of a membrane than exists on the other side. But many cells contain far higher concentrations of certain substances on the inside than exist outside, and they accomplish this by moving the materials into the cell. This process is called **active transport,** since it involves moving materials from an area of lower concentration to one of higher concentration. Active transport requires the expenditure of energy on the cell's part; the membrane is intimately involved in this process.

Many cells in both single-celled and multicelled organisms (principally animals) are capable of a process called *bulk transport,* in which substances do not pass through the plasma membrane by passive or active transport. **Bulk transport** moves large masses, such as droplets of water, grains of food, or globules of protein, into or out of the cell by first surrounding them in a kind of membrane-limiting vesicle. An example of this process is the classic behavior of the protozoan *Amoeba* in its capture and ingestion of prey (see Fig. 4.12). The organism forms a pseudopod (literally, "false foot") that extends around the prey and completely encloses it with membrane. Once trapped in the vesicle, the prey is literally inside the *Amoeba,* where it can be digested. White blood cells in the human body perform a similar function when they engulf

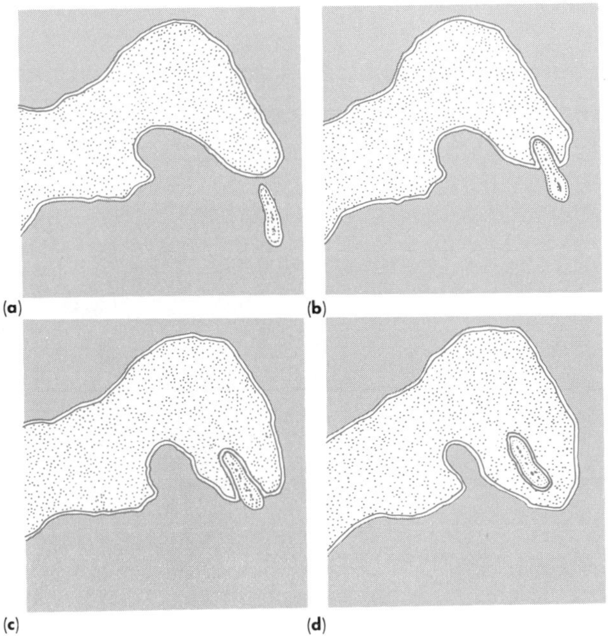

Fig. 4.12 *Phagocytosis in the giant amoeba,* Chaos chaos. *In more common terms, the amoeba is said to "engulf" the prey, a smaller protozoan called* Paramecium. *The amoeba extends parts of its cell body (a pseudopod) around the prey, enclosing it. The cell membrane then fuses to form a space, the vacuole, with the* Paramecium *inside. Digestion takes place by the passage of enzyme molecules from the prey's cytoplasm into the vacuole. At bottom right, the* Paramecium *is actually inside the vacuole, although the perspective makes it appear to be in the cytoplasm.*

Fig. 4.13 *Electron micrograph of pinocytosis, showing infolding of the membrane. The cell shown here is very thin and narrow, and it is located in the wall of the tiniest blood vessels in the animal body, the capillaries. (Photo courtesy Dr. George E. Palade, ×315,000.)*

▼

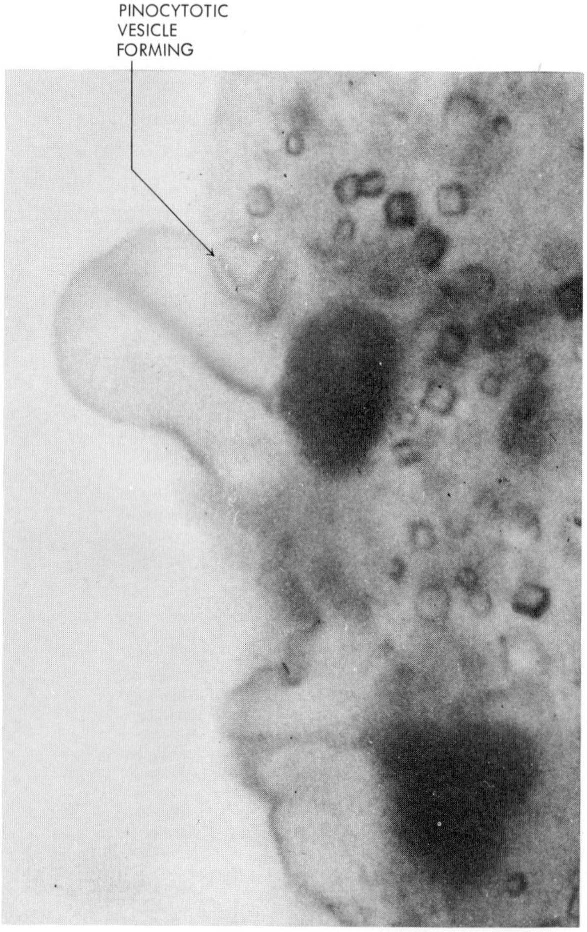

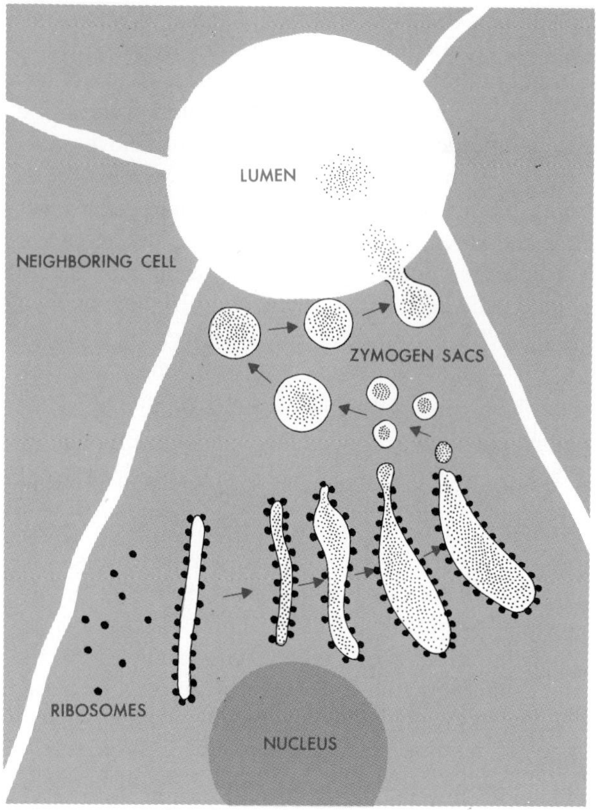

▲

Fig. 4.14 *Reverse phagocytosis is depicted here as this pancreatic cell secretes protein. Protein is synthesized in the cell cytoplasm (around membranes containing the small organelles called ribosomes). The protein (zymogen, a precursor of a digestive enzyme) is concentrated within the sacs in the cytoplasm near the bottom portion of the cell. Small sacs are pinched off each large sac; these small sacs fuse together forming zymogen-containing sacs that are eventually extruded into the duct at the upper end of the cell. (After Hokin, Lowell E. and Mabel R. Hokin, "The chemistry of cell membranes,"* Sci. Am., *October 1965, 78–86.)*

bacteria and other infectious agents they encounter in the body tissues. The process of bulk transport of solid materials is called *phagocytosis* ("cell eating"). The process of bulk transport of liquids is called *pinocytosis* ("cell drinking") and is shown in Fig. 4.13. A good example of bulk transport out of the cell by reverse phagocytosis (which might be termed "cell regurgitation") appears in Fig. 4.14.

Vacuoles

Many kinds of nonliving inclusions are found within the cytoplasm. Plant cells, for example, contain bubblelike structures called **vacuoles** (see Fig. 4.15). These vacuoles often serve as reservoirs, holding sap or waste products. A **vacuolar membrane** separates the contents of the vacuoles from the surrounding cytoplasm. The vacuolar membrane has a structure similar to the outer plasma membrane of the cell, and it regulates the passage of materials into and out of the vacuole and the cytoplasm.

Fig. 4.15 *(a) A region of the cytoplasm of a rat hepatocyte (an epithelial cell of the liver) showing several membrane-surrounded storage regions for newly synthesized substances, waste materials, or newly absorbed materials. [Electron micrograph (a) courtesy Dr. E. Sue Lumb, Department of Biology, Vassar College (×55,600); (b) courtesy Dr. William A. Jensen (×6000).]*

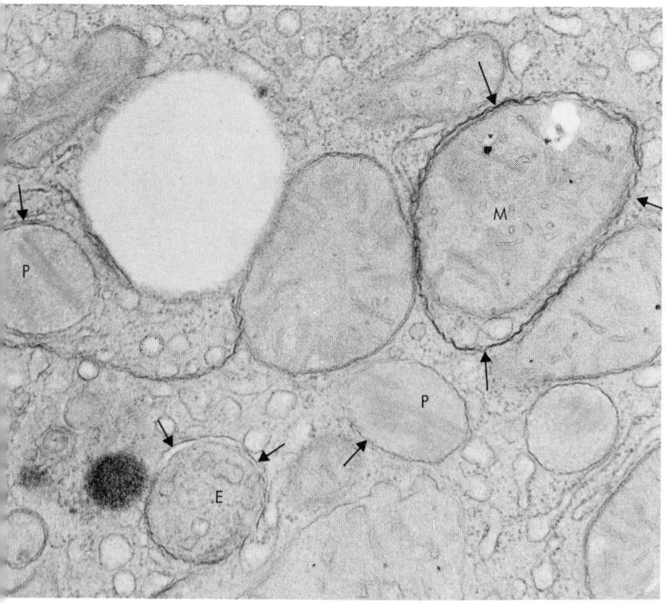

(a)

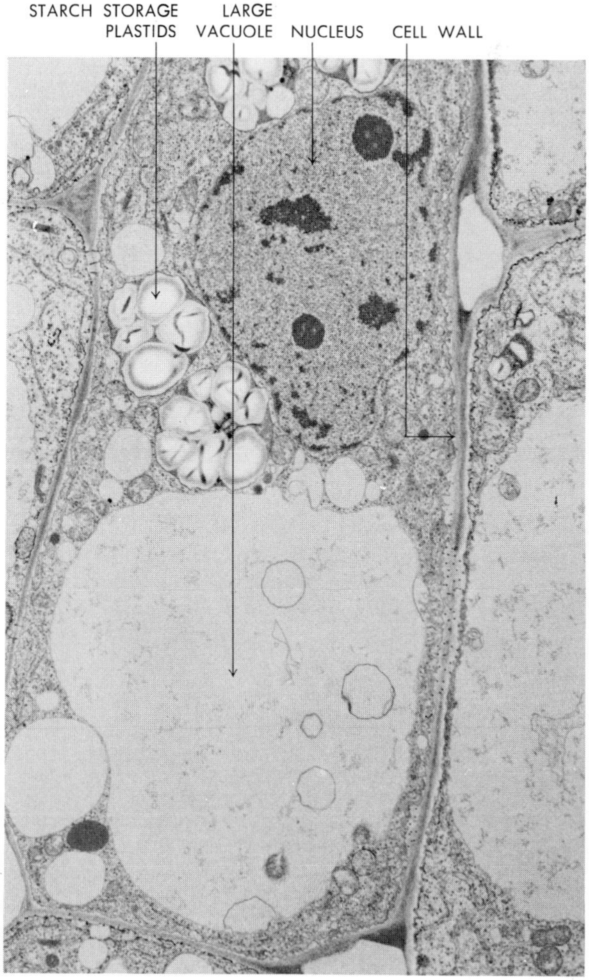

(b)

Fig. 4.16 *Electron micrograph of a cell from the root tip of a plant, showing a number of cell organelles, including a prominent plastid* (P) *(×28,000). (Photo courtesy Dr. M. C. Ledbetter, reproduced from T. P. O'Brien and M. McCully,* Plant Structure and Development, *New York: MacMillan, 1969, p. 7.)*

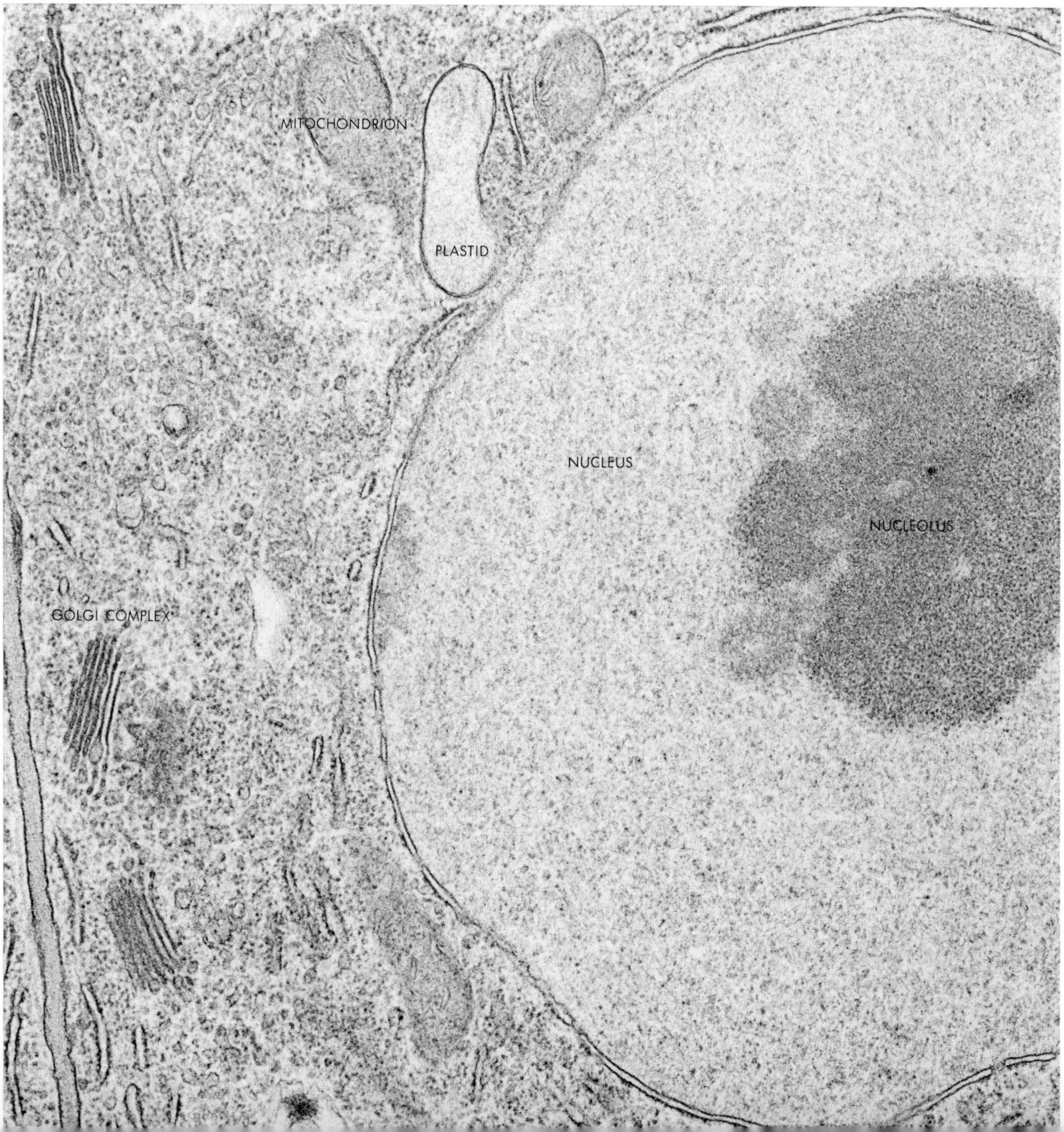

Plastids

Many plant cells have small bodies called **plastids** (Fig. 4.16). **Chloroplasts** are perhaps the most important kind of plastid and are discussed in detail below (see Fig. 4.23). Chloroplasts contain the pigment **chlorophyll,** which gives the plant its green color and is essential to its manufacture of food. **Leukoplasts** are believed to serve as centers for storage of starch (Fig. 4.17) in the cytoplasm.

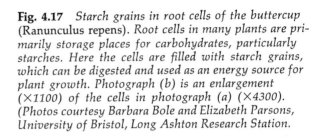

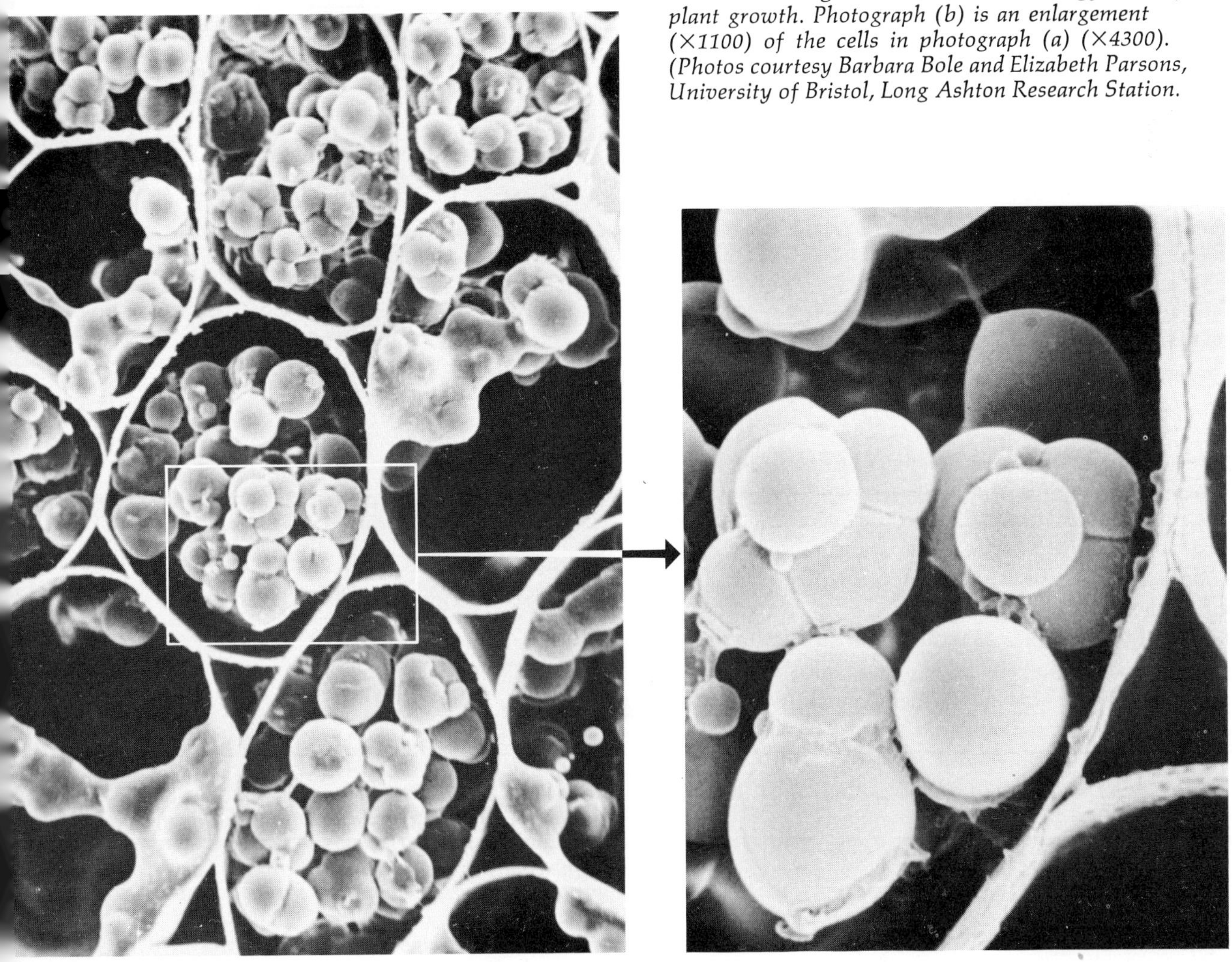

(a) (b)

Fig. 4.17 *Starch grains in root cells of the buttercup* (Ranunculus repens). *Root cells in many plants are primarily storage places for carbohydrates, particularly starches. Here the cells are filled with starch grains, which can be digested and used as an energy source for plant growth. Photograph (b) is an enlargement (×1100) of the cells in photograph (a) (×4300). (Photos courtesy Barbara Bole and Elizabeth Parsons, University of Bristol, Long Ashton Research Station.*

Chromoplasts, because of the pigments they contain, are the plastids that give color to the flowers and fruits of many plants. Plastids do not occur in animal cells.*

Inclusions

Many cells, especially plants, also contain a number of small structures called **inclusions.** Inclusions are not organelles but consist of various kinds of organic or even inorganic structures such as starch granules (Fig. 4.17), crystals, globules of lipid, and the like. These inclusions often stain differentially so that they can be detected by simple microscopic observation.

Inclusions represent storage forms of various substances, usually of such large size or insolubility that movement in and out of the cell or across plastid or vacuolar membranes would be impossible. For starch to be removed, for example, it must first be broken down into its soluble components, simple sugars, or monosaccharides.

Centrioles

Centrioles (Fig. 4.18) are small, deeply staining particles seen lying near the nuclear membrane in cells that are not dividing. Under the electron microscope the centriole appears as a small hollow cylinder, the walls of which are formed by nine sets of triplet microtubules. In cells that have two centrioles, the centrioles separate from each other and move toward opposite sides of the cell shortly before the cell begins to divide. In cells with only one centriole, a new centriole appears to grow from the side of the other, and the two resulting centrioles separate. It has been hypothesized that the

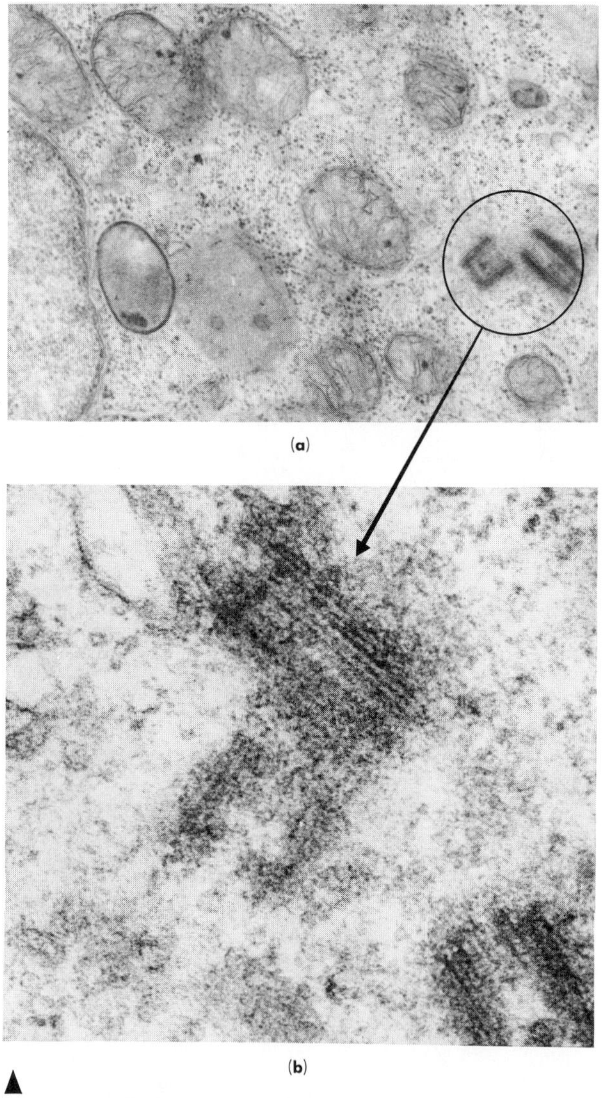

▲

Fig. 4.18 *Electron micrographs of centrioles. (a) Longitudinal section of a pair of centrioles from an epithelial cell of chicken duodenum. (Courtesy S. P. Sorokin.) (b) Higher magnification of centriole in longitudinal section (×120,000). (Courtesy J. André.) (c, to the right) Cross section of one of the barrel-shaped components, showing the filamentous tubules that run the length of the centriole. Tubules are arranged as nine groups of three units each (×180,000). (Courtesy J. André.)*

* Certain marine animals, called slugs, are known to contain functional chloroplasts within the cells of their body tissues. However, the chloroplasts are obtained by feeding on algae; not all the chloroplasts are digested and some end up inside the slug's body cells.

centriole may play a role in the formation or assembly of fiberlike structures called **spindle fibers,** which are involved in the movement of chromosomes during cellular division. The centriole does not seem to be essential for cellular division, however, since many higher plant cells manage to divide without them.

Microtubules

During the 1960s, detailed electron microscope studies of cells verified the existence of **microtubules** in the cytoplasm. Microtubules are long, straight cylinders about 230 to 270 Å in diameter (Fig. 4.19). In cross section, microtubules appear to be composed of 13 globular protein subunits arranged around a 45-Å central core. In longitudinal section the subunits appear to be lined up in 13 longitudinal rows. Microtubules are very abundant immediately adjacent to the plasma membrane and

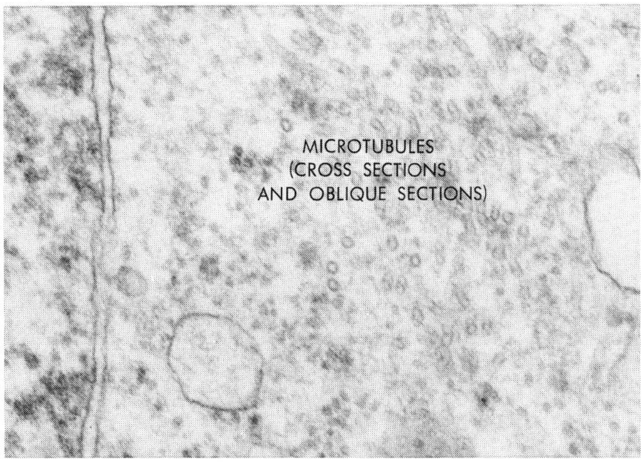

(a)

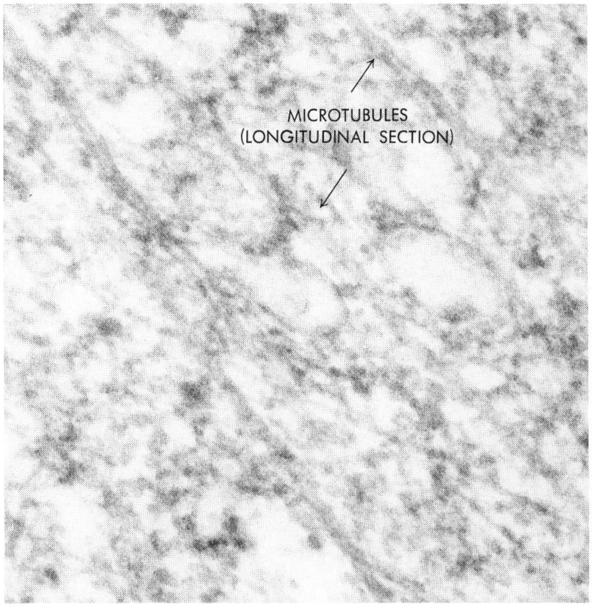

(b)

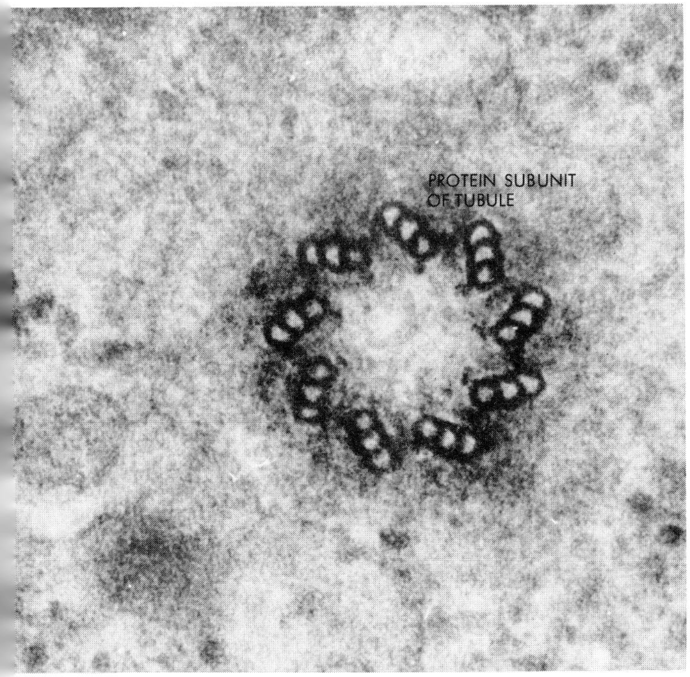

(c)

▲

Fig. 4.19 *Microtubules shown in two electron micrographs. (a) Shown in cross section from sperm-producing cells of the scorpion (×39,000). (b) Shown in longitudinal section from human kidney fibroblast cell (fibroblasts are cells that produce collagen and other structural proteins, especially for supportive and connective tissue) grown in culture (×47,000). (Photo courtesy Dr. David M. Phillips.)*

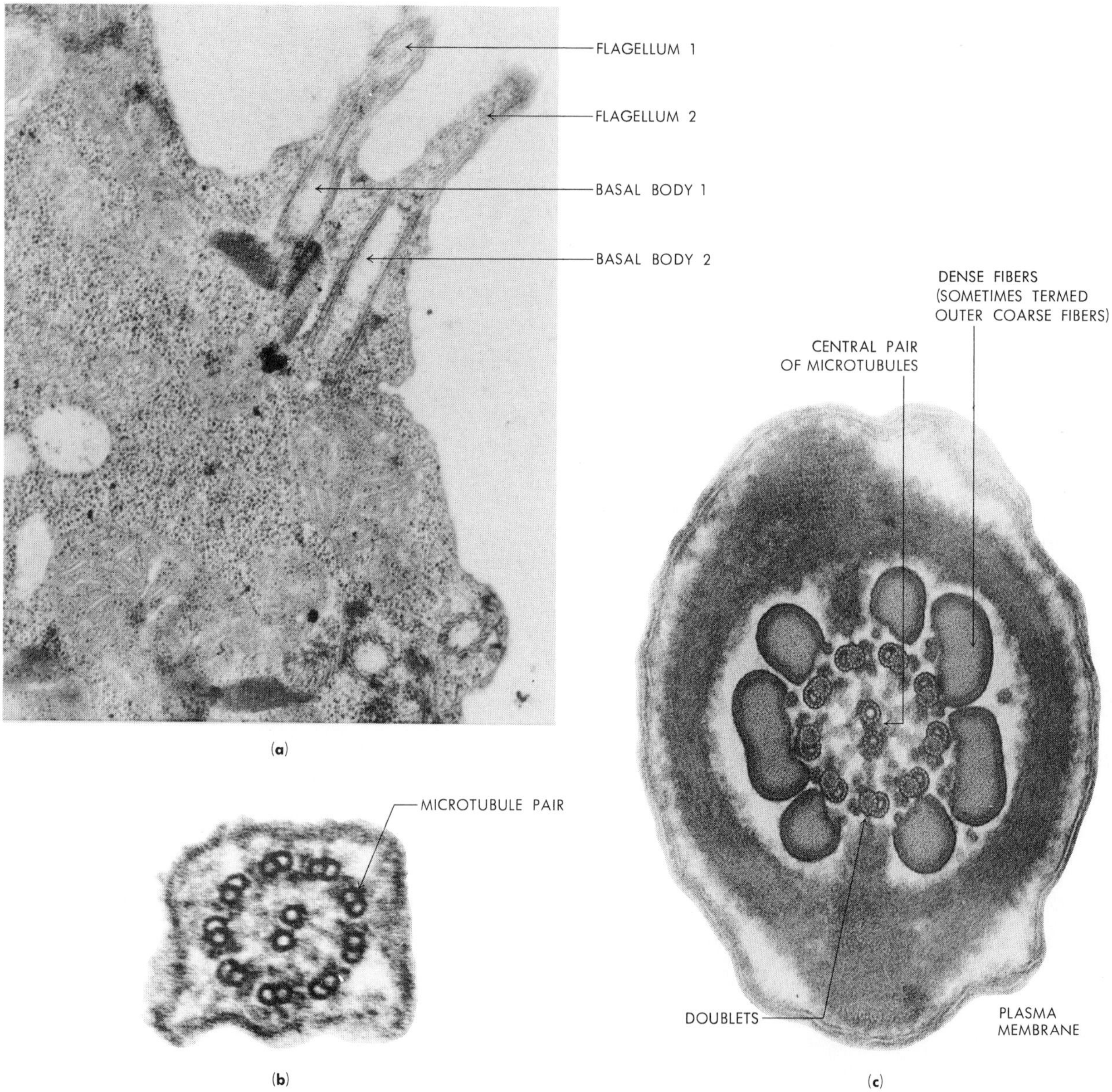

Fig. 4.20 *Flagella, showing the microtubular composition. (a) The basal bodies (two centriole-like structures) that go to form a flagellum in the one-celled protozoan* Naegleria. *This species has a double flagellum, as shown here. Note the similarity to centriole structure (×28,000). (b) Cross section of* Naegleria *flagellum, showing characteristic 9 + 2 arrangement of microtubules (×120,000). (c) Cross section of sperm flagellum of the rat (×255,000), showing the 9 + 2 structure. [(a) and (b) courtesy A. D. Dingle, "Control of flagellum number in* Naegleria," J. Cell Sci. *7 (1970); (c) courtesy Dr. David M. Phillips.]*

The basic function of many organelles is to enclose in a limited space a set of chemical reactions that are interdependent.

Fig. 4.21 *Electron micrograph of microfilaments from mouse fibroblast (cells that form collagen and other structural proteins) grown in culture (×23,000). The microfilaments are shown in longitudinal section (lengthwise) and appear as long strands, almost hair-like. They seem to function in certain kinds of cell movements, such as the contraction of heart muscle cells. (Photo courtesy Dr. David M. Phillips.)*

▼

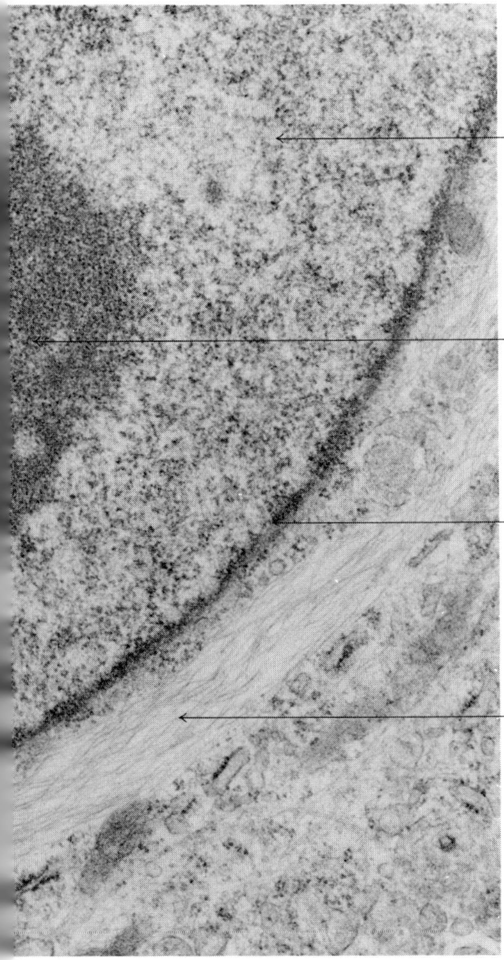

appear to be involved in the determination of cell shape. There is some evidence to support the hypothesis that these structures may also be involved in the flowing of cytoplasm within the cell (cytoplasmic streaming). In wheat, a band of microtubules encircles the nucleus just before the cell begins to undergo division. During cell division, some of the microtubules become attached to each of the chromosomes and apparently help to move the chromosomes during the division process. Microtubules are also found in *cilia* and *flagella*, the hair-like cytoplasmic extensions of many microorganisms. Flagella are generally longer than cilia, but the structure and function of both are very similar. Capable of various kinds of whiplike motion, they serve to move the one-celled organism around. As Fig. 4.20 shows, cilia and flagella are characterized by the so-called "9 + 2" arrangement of microtubules. This means that there are nine groups of two tubules arranged in a circle, with one group of two tubules in the center. The universality of this arrangement, from the flagella of sperm to the cilia of cells lining the nasal passages, implies some ancient paths of common descent among widely diverse organisms.

Cells also contain structures called **microfilaments** (Fig. 4.21). Microfilaments are bundles of protein threads, and make up the contractile elements. They are found in many kinds of cells and appear to aid in a variety of types of movement. Like microtubules, microfilaments consist of subunits of protein. Unlike microtubules, microfilaments do not consist of subunits arranged in a "tubule" form, but appear as fine, threadlike structures.

Mitochondria

Mitochondria usually appear as spherical or sausage-shaped structures scattered throughout the cytoplasm (see Fig. 4.7a for an electron micrograph and Fig. 4.22 for a schematic drawing). They are usually found in the region of the cell that shows the greatest metabolic activity, and they appear in their greatest numbers in the most metabolically active cells, such as liver or muscle cells. If the mitochondria are separated from the rest of the cellular parts by centrifugation, their metabolic characteristics can be studied. Such isolated mitochondria can break down certain organic compounds to carbon dioxide and water. In so doing, they release energy in a form that can be used by organisms. Thus the mitochondria seem to be the cellular centers of respiratory activity, and in them originate the carbon dioxide and water exhaled in breathing.

The electron microscope reveals the structure of the mitochondrion in considerable detail. Seen in section, the innermost of the mitochondrion's two membranes shows infoldings. These infoldings, called **cristae,** form partitions extending into the cavity of the mitochondrion. The presence of cristae increases surface area within a mitochondrion. It is on cristae that collections of enzymes and other types of molecules, collectively called **respiratory assemblies,** are located. Within each respiratory assembly, molecules resulting from the breakdown of sugar and other carbon compounds react during respiration. As might be predicted, there is a direct relationship between the number of cristae and the activity of the cells from which the mitochondrion comes. Mitochondria from active cells show a large number of cristae. This greatly increases the surface area on which the energy-releasing reactions can take place.

Mitochondria contain some of the genetic material deoxyribonucleic acid, or DNA, and their own ribosomes. The DNA codes largely for the mitochondria's ribosomal ribonucleic acid, or RNA. Mitochondrial ribosomes are involved in the synthesis of some proteins, though it is not known which ones.

Fig. 4.22 *Mitochondrion from a pancreas cell (a), shown in longitudinal section (×20,000). Mitochondria are covered by an outer membrane and lined by an inner membrane that is infolded into numerous partitions, called cristae. Schematic drawing (b) illustrates the overall structure of the mitochondrion in three-dimensional perspective. (Photo courtesy Dr. George E. Palade, Rockefeller University.)*

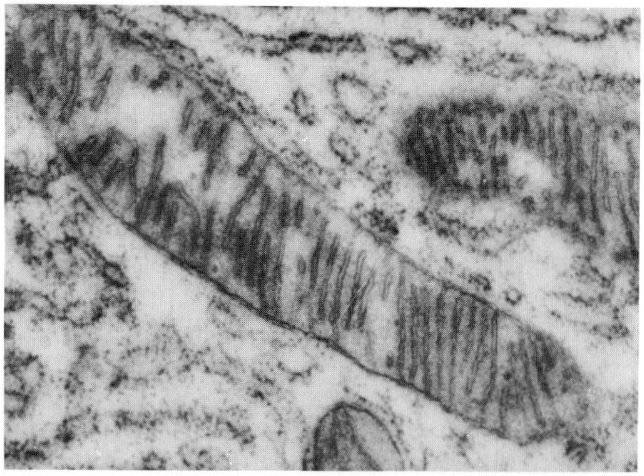

(a)

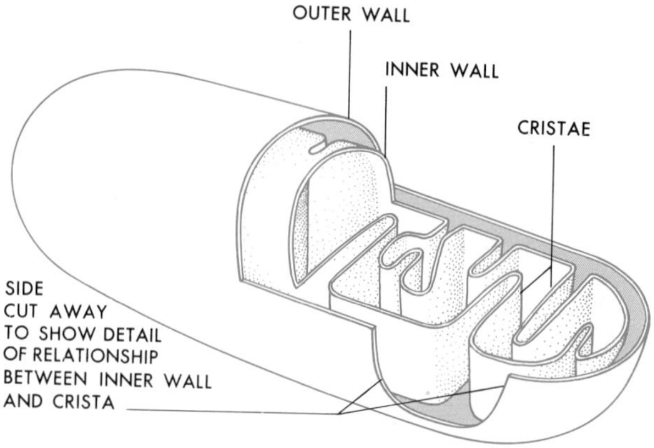

(b)

Supplement 4.2

BIOLOGICAL SELF-ASSEMBLY

Microtubules and microfilaments can be taken apart and rebuilt by simple chemical treatments. For example, microtubules can be broken down either by physical disruption of the cell or by addition of high concentrations of calcium ions. Under these conditions the microtubule breaks down into subunits of a protein called, appropriately, tubulin. These subunits are all the same size and molecular weight for any given type of organism. If isolated tubulin subunits are placed in a calcium-free medium and supplied with a phosphate compound such as adenosine triphosphate (ATP) or guanosine triphoshate (GTP), the subunits reaggregate and form a section of intact microtubule. The structure can assemble itself if provided with the appropriate chemical agents.

Microfilaments can be disaggregated by a substance known as cytochalasin B, which does not appear to alter any other cell components. If cytochalasin B is subsequently removed, the microfilaments reform. Aggregation and disaggregation are reversible. More important, the fact that reaggregation occurs spontaneously indicates that the subunits carry within them the "instructions" for their own assembly. It is enough to produce the subunits with a specific amino acid sequence, and the larger structure (microtubule) will form spontaneously.

Knowledge of disaggregation and reaggregation of tubulin subunits makes it possible to learn something of the role microtubule structures may play in cell life. When cytochalasin B is applied to dividing cells, the furrow that normally develops to cut the two parts of the cell in half never appears. We conclude that the furrow is formed by the action of microfilaments. Application of chemicals known specifically to disrupt the structure of microfilaments makes it possible to pinpoint the areas of cell activity in which these structures are involved.

Similar disruption experiments have been carried out with microtubules. Low temperature or the chemical colchicine disrupts microtubules in much the same manner in which cytochalasin B disrupts (disaggregates) microfilaments. When colchicine is applied to dividing cells, the duplicated chromosomes do not separate from each other. Formation of pseudopods in organisms such as *Amoeba* and the familiar streaming of cytoplasm within plant cells are also interrupted by colchicine. Removal of colchicine usually restores these functions. It is thus possible to conclude that there may be some distinct relationship between microtubular function and these various motions observed within cells. It is important to keep in mind that such evidence does not suggest an unequivocal cause–effect relationship. It might be that colchicine is disrupting a process that affects both microtubule structure and various cell processes. More and different kinds of evidence would be necessary before we could conclude that microtubules cause chromosome movements, pseudopod formation, or cytoplasmic streaming.

Chloroplasts

One of the most important types of plastid is the chloroplast, found in all green plant cells. Chloroplasts resemble mitochondria in many ways, and it is useful to compare the two organelles. A photograph and diagram of a chloroplast are shown in Fig. 4.23.

As Fig. 4.23 shows, chloroplasts are membrane-bound organelles containing stacks of internal membranes, called **lamellae,** running along the major axis of the plastid. Periodically these membranes form closed, disc-shaped structures called **thylakoid discs.** The closely packed regions of the discs are called **grana** (Fig. 4.23, a and b). Attached to the lamellae are the light-absorbing molecules of **chlorophyll.** In the process of photosynthesis, chloroplasts absorb light energy, con-

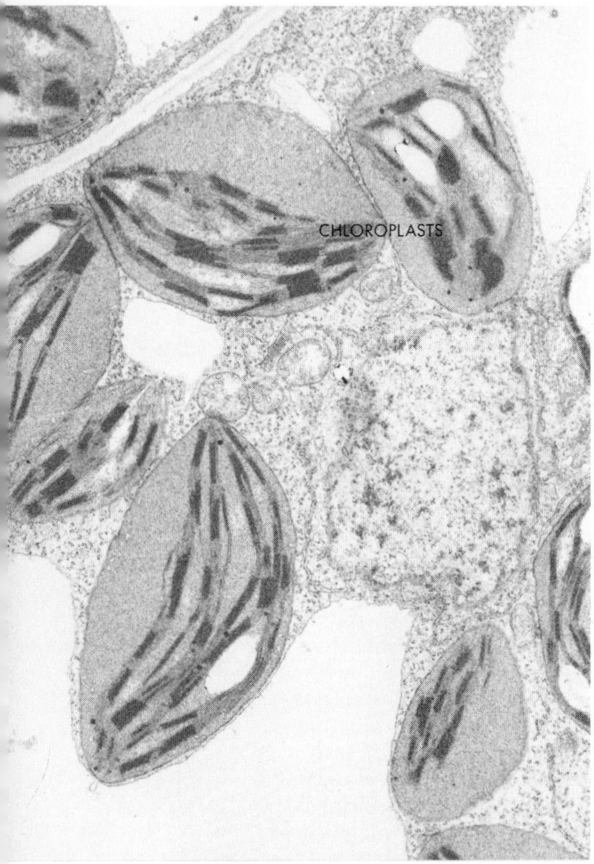

(**a**)

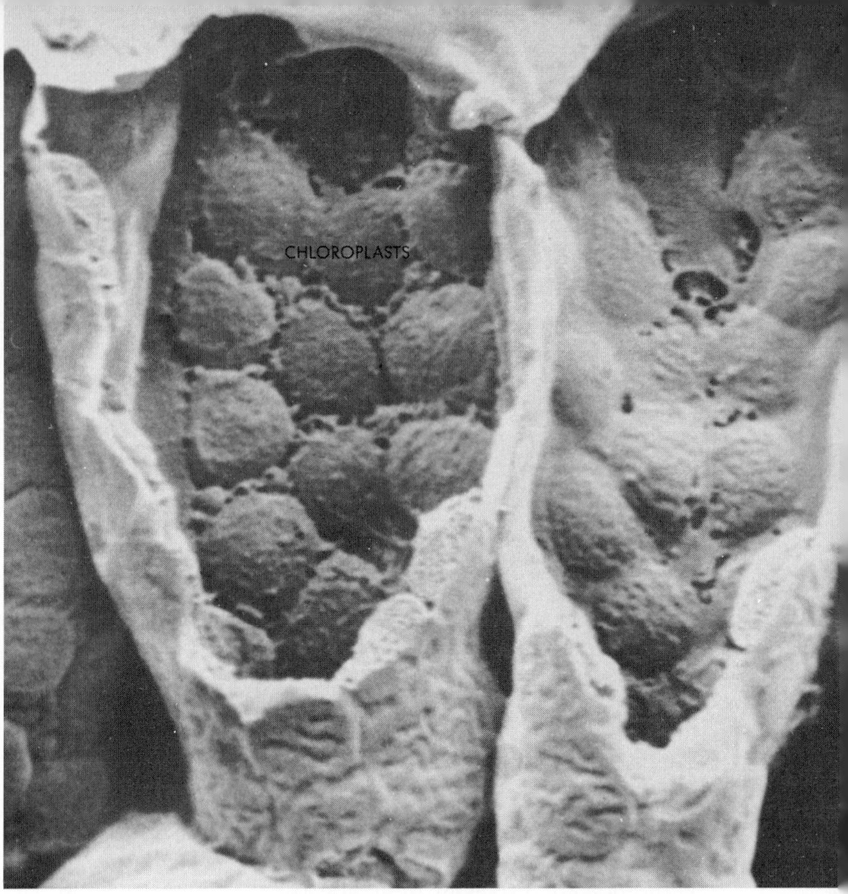

(**b**)

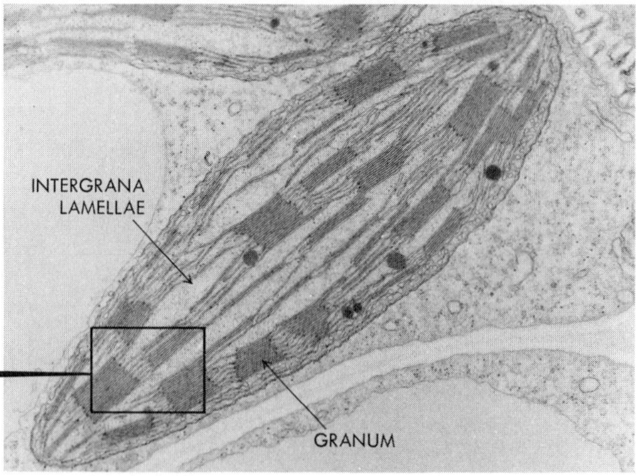

(**c**)

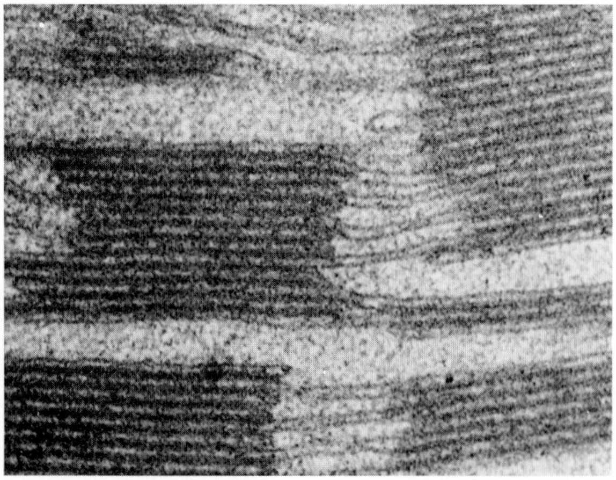

(**d**)

verting it into a form useful to the cell by a process known as **photophosphorylation.** In photophosphorylation, light energy is used to convert the low-energy compound adenosine diphosphate (ADP) into a high-energy compound, adenosine triphosphate (ATP), which is used for cell work. This process will be discussed in greater detail in Chapter 6. It is important to emphasize here the fact that photophosphorylation is very similar to the energy-releasing chemical processes that occur in mitochondria. The main difference between the two is the source of energy: light in one and chemical oxidation in the other. Furthermore, like mitochondria, chloroplasts contain their own DNA and ribosomes. Chloroplast DNA appears to code for certain of the enzymes involved in the chemistry of photosynthesis, and for chloroplast ribosomal RNA. Exact determination of the genetic information coded in the chloroplast DNA has yet to be made.

Fig. 4.23 *Chloroplasts. (a) Electron microscope view of chloroplasts in cells of the bean plant,* Phaesolus *(×1700). The chloroplasts are the football-shaped bodies within the cell boundaries. (Photo courtesy W. M. Laetsch). (b) Scanning electron microscope view of chloroplasts from the leaves of the pea plant (×2000). Note how tightly packed the chloroplasts are in these cells, compared to those shown in (a). (Photo courtesy Barbara Bole and Elizabeth Parsons, University of Bristol, Long Ashton Research Station). (c) Electron micrograph of a thin section of a single chloroplast from a leaf of* Froelichia *(×45,550). Note how the internal membranes, the lamellae, run the length of the chloroplast. Periodically the lamellae are stacked together, producing a denser array of membranes known as grana. The space between lamellae is fluid-filled and is called the stroma. (Photo courtesy W. M. Laetsch). (d) Enlarged electron micrograph of a number of grana, showing the stacked lamellae from a single chloroplast of* Phyllorpadin *(×61,750). (Photo courtesy W. M. Laetsch).*

The structural similarity between chloroplasts and mitchondria is also striking. Both consist of internal membranes on the surfaces of which chemical reactions occur. Functionally, both organelles are involved in the production of high-energy compounds such as ATP, though they use different energy sources to accomplish this task. The structural and functional similarity of chloroplasts and mitochondria has led many biologists to speculate about the possible evolutionary relationship between these organelles (see Supplement 13.2).

Like animal cells, plant cells have mitochondria in addition to chloroplasts. The function of mitochondria is the same in both plants and animals. The function of choroplasts is unique to plants. It allows the plant to convert the energy contained in light (as photons) into usable chemical form. Animals are unable to do this and must depend upon chemical energy, in the form of food, for all their energy needs.

Endoplasmic Reticulum and Ribosomes

The electron microscope also reveals a complex network of channels extending throughout the cytoplasm. These channels are infoldings, often continuous with the plasma membrane. They also appear to connect with the nuclear membrane, but agreement on this point is not universal. The entire system of channels is called the **endoplasmic reticulum** (Fig. 4.24). It is thought that the endoplasmic reticulum may speed the transport of materials from the extracellular environment into the cytoplasm, as well as from the nucleus to the cytoplasm. Support for this hypothesis comes from the observation that tiny bodies called **ribosomes** (Fig. 4.25) are often found along the outer surface of a channel.

Ribosomes are so named because they contain high concentrations of RNA. By "tagging" with radioactive atoms the smaller molecules (amino acids) of which proteins are composed, it can be shown that the amino acids go first to the ribo-

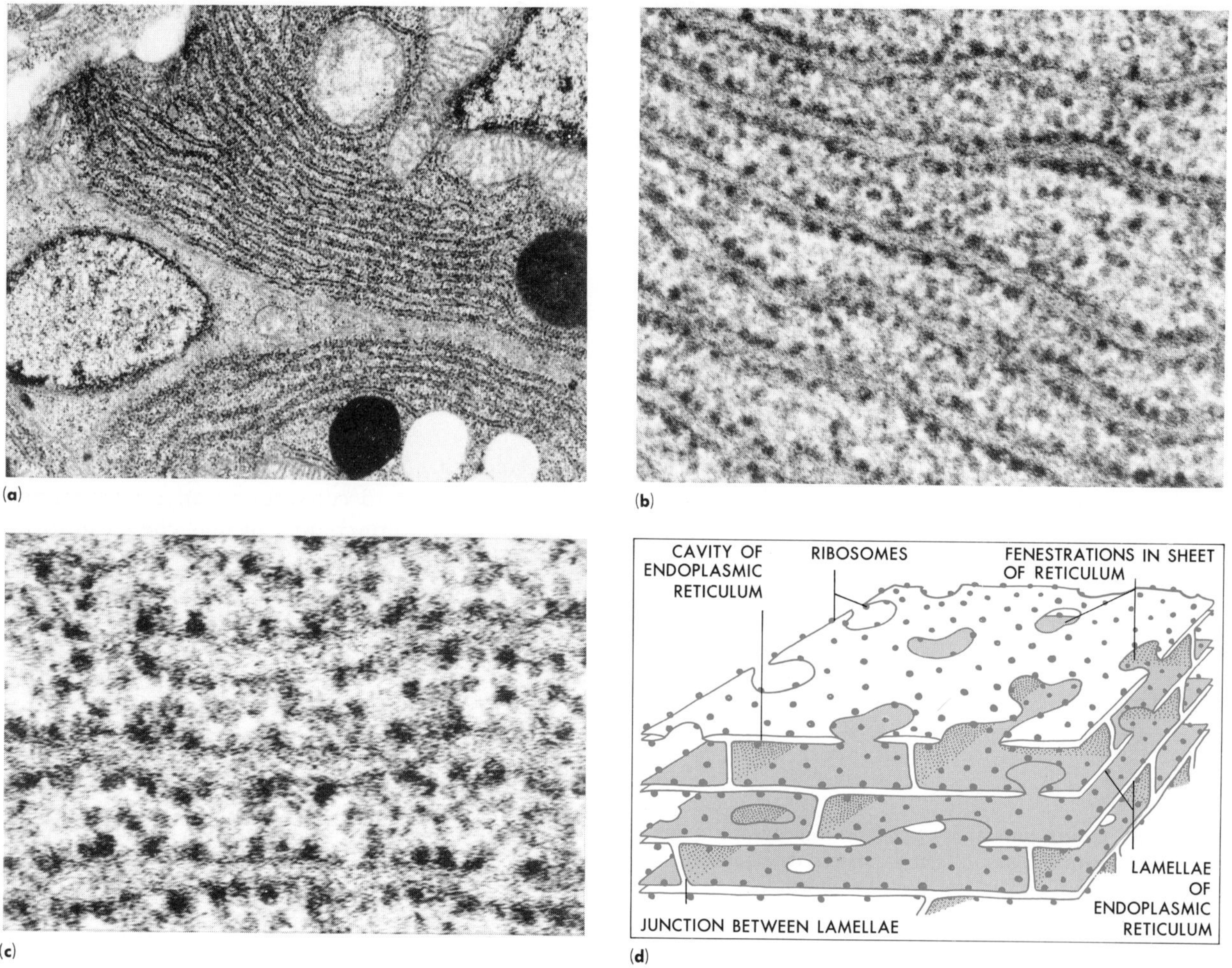

Fig. 4.24 *Electron micrographs of sections through chick embryo pancreas, showing endoplasmic reticulum at three different magnifications—(a) ×15,400, (b) ×75,600, (c) ×115,500. At the higher powers, the double nature of the lamellae and their associated ribosomes is clearly visible. Compare with the diagram (d). (Photos courtesy Elizabeth Johnson, Wesleyan University. Diagram after S. Hurry,* The Microstructure of Cells. *Boston: Houghton Mifflin, 1964, p. 10.)*

somes and later show up in protein molecules. Thus it is reasonable to deduce that proteins are synthesized either in or on the ribosomes. Since virtually all living cells synthesize proteins, ribosomes are one of the few subcellular particles that must be present in all living things. The hypothesis that ribosomes are the site of protein synthesis would predict their presence in larger numbers in cells that are very active in protein synthesis than

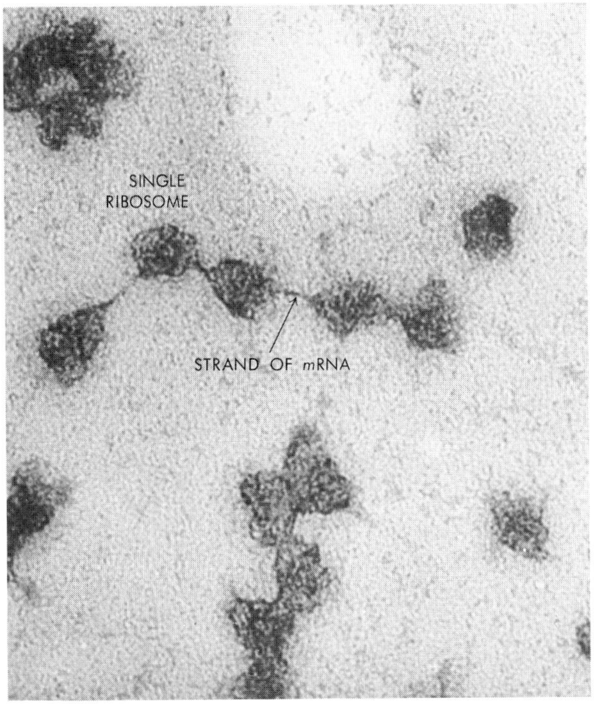

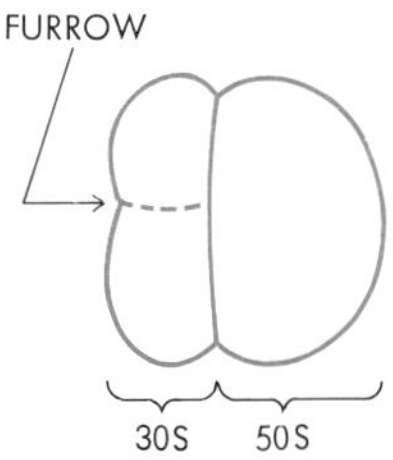

Fig. 4.25 *Ribosomes are cell organelles responsible for protein synthesis. (a) Several ribosomes with what is thought to be a molecule of* mRNA *strung between them. Messenger RNA contains genetic information (from DNA) specifying amino acid sequence. Ribosomes are assembly units that "read" the genetic information. In the ribosomes, specific amino acids are joined into polypeptides. Groups of ribosomes joined together by a strand of* mRNA *are called polysomes. (b) Diagram of a ribosome showing the two-subunit structure (30S and 50S). [Electron micrograph courtesy Alexander Rich, J. R. Warner, and C. E. Hall, from A. B. Novikoff and E. Holtzman,* Cells and Organelles. *New York: Holt, Rinehart and Winston, 1970, p. 69.]*

in cells that are less active. This proves to be the case; a typical bacterium, growing rapidly (and thus synthesizing large quantities of protein), may contain some 15,000 ribosomes.

Electron microscope examination of bacterial ribosomes shows that they have a diameter of about 200 A, whereas those from mammalian cells are a few angstroms larger. Unfortunately, the electron microscope does not provide the detail necessary to study the *precise* structure of ribosomes. What is now known about ribosome structure has been established by indirect experimental evidence. It has been shown that the bacterial ribosome consists of two subunits. These subunits are designated 50S and 30S respectively, because of their sedimentation (settling out) characteristics in a centrifuge.* Neither subunit can synthesize protein by itself, but it remains a mystery why this two-subunit structure is essential to ribosomal functioning. The subunits themselves differ in structure, with the 50S bacterial ribosome subunit consisting of 23S ribosomal RNA and about 35 different proteins, and the furrowed 30S subunit consisting of 16S ribosomal RNA and about 20 different proteins. The proteins appear to be present as globular spheres with a diameter of 32 to 40 A. Ribosomes are often found associated in groups known as **polyribosomes** or **polysomes.**

Golgi Complex

The Golgi complex (Fig. 4.26) is a cluster of flattened, parallel, smooth-surfaced sacs found within the cytoplasm. The sacs, often referred to collectively as the **dictyosome,** contain numerous smaller infoldings or pockets. At times the Golgi complex resembles the endoplasmic reticulum without its

* The S stands for Svedberg units. A Svedberg unit is a sedimentation constant—a measure of the rate of sedimentation of, for example, a ribosomal subunit in a centrifuge.

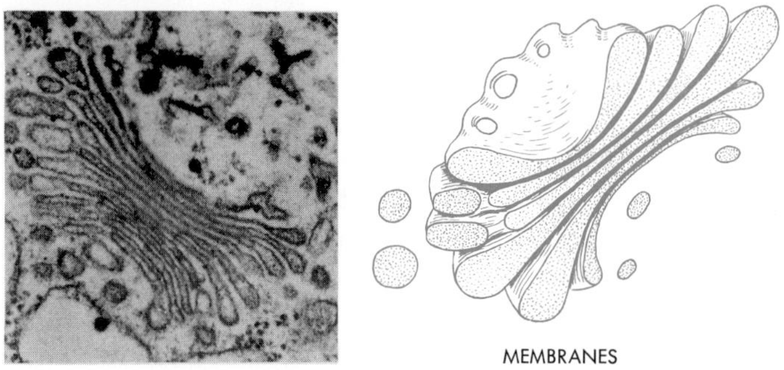

Fig. 4.26 *Golgi complex, showing the stack of folded membranes called collectively the dictyosome (×43,000). (Photo courtesy Dr. George E. Palade, Rockefeller University.)*

ribosomes. There is evidence to indicate that the Golgi complex may be concerned with excretion and the transport of particles into and out of the cell, particularly lipid materials. Golgi complexes are especially prominent in glandular cells involved in secretions of protein enzymes and other substances. Golgi material also functions in the synthesis of cell wall and cell plate material. More information has recently been learned about the Golgi because new methods have been developed for isolating Golgi dictyosomes in a purified form.

Lysosomes

The electron microscope reveals another structure in the cytoplasm that rather resembles a mitochondrion. This is the **lysosome** (Fig. 4.27). The

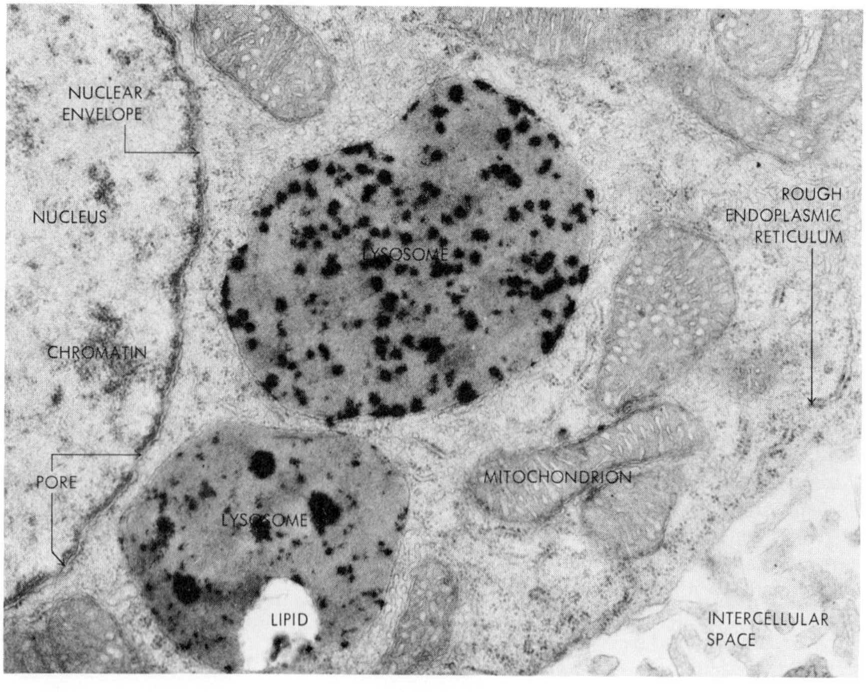

Fig. 4.27 *Electron micrograph (×38,000) of a portion of a functional kidney tubule cell of a 13-day-old chick embryo. The dark spots in the lysosome are the sites of lead deposits that resulted from chemical tests for the activity of one of the enzymes present in the lysosome. Also labeled are the nucleus, nuclear envelope and pore, endoplasmic reticulum, and intercellular space into which cell processes extend. (Courtesy Dr. E. Sue Lumb, Department of Biology, Vassar College.)*

lysosome is a saclike structure containing enzymes that catalyze the breakdown of molecules such as fats, proteins, and nucleic acids into smaller molecules. These smaller molecules can then be used as energy sources. It may be that the lysosomes serve to isolate these digestive enzymes from the cell cytoplasm, thereby keeping the cell from digesting itself. This hypothesis is supported by the fact that when the lysosome membrane is ruptured, the cell undergoes chemical breakdown, or **lysis.** There is experimental evidence linking the lysosomes to the muscle atrophy occurring after surgical denervation or disease-caused nerve paralysis. It has further been noticed that at death the mitchondria and lysosomes are broken down. It may be that the breakdown of the lysosomes contributes to the early, irreversible changes that occur after death.

It is significant to note that lysosomal enzymes are among the most stable components of living matter. Some lysosomal enzymes can remain stable at temperatures that would denature almost any other large protein. We can hypothesize why this might be an important characteristic of lysosomal enzymes. Since these enzymes live in an environment of highly concentrated, compartmentalized digestive enzymes, they must have various structural means (perhaps many covalent bonds holding the polypeptide chain of the enzyme in its tertiary form) to keep from being digested by each other. Lysosomal enzymes are therefore adapted to exist in "hostile" environments.

Nucleus

The nucleus (Fig. 4.28) is usually the most obvious anatomical feature of a cell. Most cell nuclei can easily be seen under the light microscope, often standing out as rounded bodies slightly darker than the surrounding cytoplasm. A nucleus has a limiting membrane, the **nuclear membrane** or **envelope,** which separates the nuclear contents, the **nucleoplasm,** from the surrounding cytoplasm. This membrane bears the same relationship to the

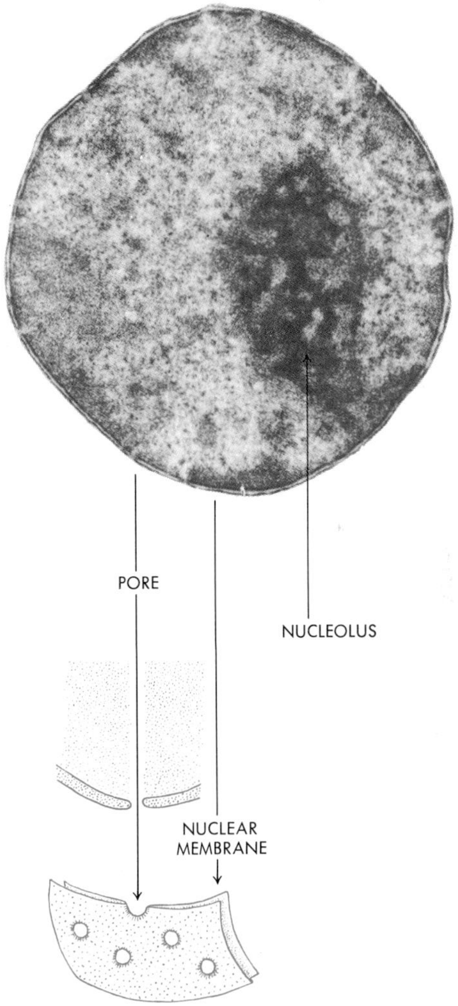

Fig. 4.28 *The cell nucleus, showing the prominent nucleolus and the cell membrane containing pores (×12,000). (Courtesy Dr. George E. Palade.)*

nucleus as the plasma membrane bears to the cytoplasm: it serves as a regulatory device through which materials pass in and out of the nucleus. The electron microscope reveals that the nuclear membrane has distinct pores through which substances may pass going from the nucleus into the cytoplasm and vice versa. The membrane is a double one, consisting of two unit membranes.

When we stain a living onion cell with iodine, the nucleus takes up the reddish-brown stain far more readily than does the cytoplasm. Thus the nucleus stands out well against the cytoplasm. Closer examination reveals that only certain portions of the nucleoplasm have taken up the stain. Particularly visible are one or more **nucleoli,** which are small, spherical bodies. The nucleoli are extremely variable structures, often changing in size and shape. When a cell is dividing, the nucleoli disappear, then reappear shortly after division is completed. Their precise function is not entirely understood. But we know that nucleoli contain the genetic information, coded in molecules of DNA, for the synthesis of ribosomal RNA. The synthesis occurs in the nucleolus itself.

Darkly staining regions within the nucleus indicate the presence of "chromatin" (from the Greek word for "color"). Chromatin has a selective affinity for certain stains and thus is particularly "color-prone." Chemical analysis reveals chromatin to be composed of nucleic acid and proteins. In chromatin this nucleic acid is deoxyribonucleic acid, or DNA. When cells begin the reproductive process, the existence of the chromatin in the form of rod-shaped bodies called **chromosomes** becomes evident. One of the so-called giant chromosomes from salivary gland cells of the fruit fly is shown in Fig. 4.29.

The nucleus functions as the main information and control center of cellular activity. The nucleus appears to be responsible for the continuing life of

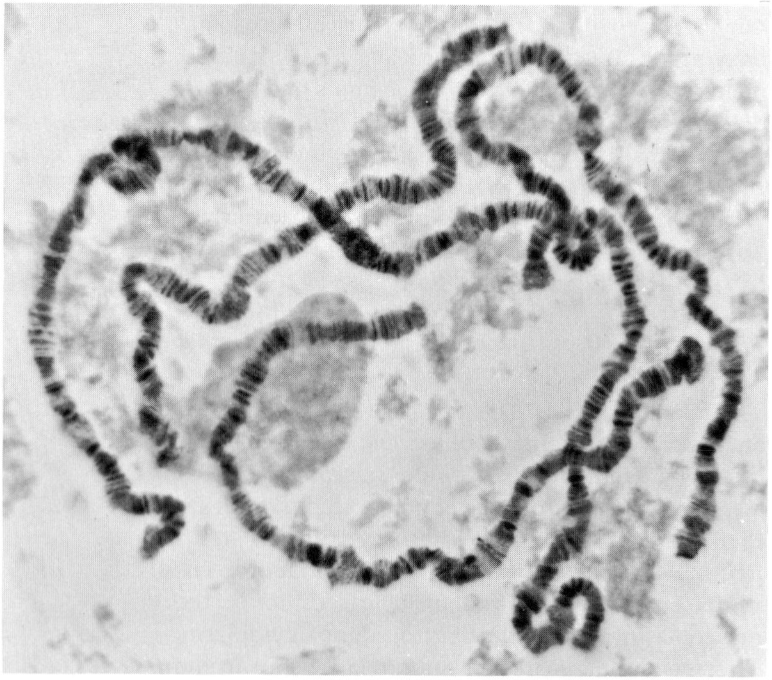

Fig. 4.29 *Photo of chromosomes from the salivary gland cells of the fruit fly. The dark bands represent areas that preferentially take up stain. Chromosomes are composed mostly of protein and deoxyribonucleic acid (DNA). (Courtesy CCM: General Biological, Inc., Chicago.)*

the cell and is necessary for its reproduction. Cells that have no nucleus when fully developed (such as most mammalian red blood cells) are unable to reproduce. Nor do such cells live for very long: They may live up to 3 or 4 months, but seldom any longer. They must be produced by other groups of cells that are found to retain their nuclei when mature. In the case of red blood cells, they are produced by cells found in the bone marrow. The nucleus appears to carry out its functions by interaction with the cytoplasm, principally through the agency of various forms of ribonucleic acid. Certain types of RNA synthesized in the nucleus pass out into the cytoplasm and direct chemical events there. The nucleus is also responsible for the transmission of hereditary information, as DNA, from one cell generation to the next. DNA can be found in the chromatin. This material must duplicate and be equally divided between the two products (daughter cells) of cell division.

Eucaryotic Cellular Reproduction

Eucaryotic cells generally reproduce by a process known as **mitosis** (Fig. 4.30a). In mitosis the chromosome number is doubled, and then the cell divides into two parts by pinching in at the center. One complete set of chromosomes goes to each daughter cell, so that each is in most ways identical to the parent cell and to its "sibling." The complex chromosome movements involved in mitosis will be discussed in more detail in Chapter 7. Mitosis is a process of **asexual** reproduction since it involves only one parental cell (or organism).

Most eucaryotic cells also engage in another process called **meiosis** (sometimes called "reduction division"; see Fig. 4.30b). In meiosis, the chromosome number is reduced by one-half of what it was at the start of meiosis. In most meiotic divisions the chromosomes in a cell are duplicated, and then the cell undergoes two successive mitoses. The net result of meiosis, wherever it is found, is that the daughter cells have half as many chromosomes as the starting parental cell. The division process is not random, however. It occurs in such a way that each daughter cell has one complete copy of all genetic information. Meiosis occurs most generally in the animal and plant kingdom in the formation

Fig. 4.30 *Schematic diagram illustrating the basic difference between sexual and asexual reproduction. The symbols ♂ and ♀ stand for male and female, respectively.*

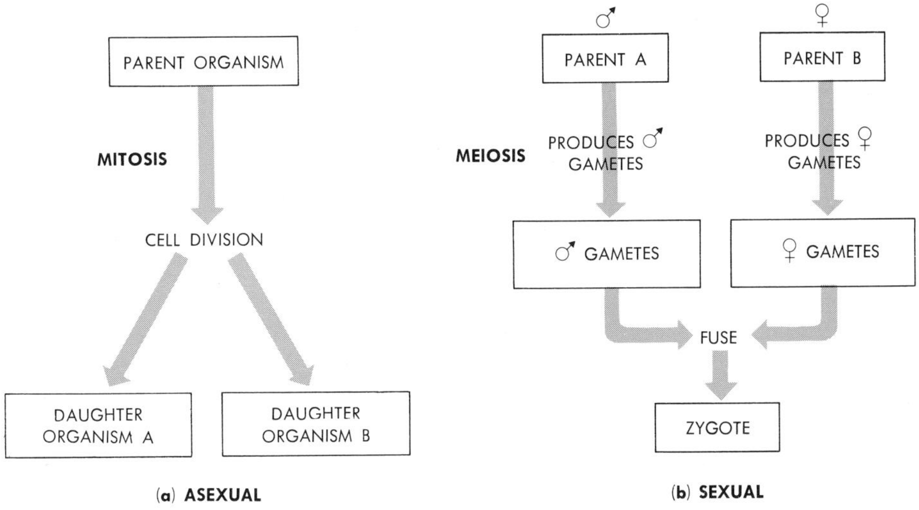

Supplement 4.3

THE FUNCTION OF CELL PARTS: HOW DO WE KNOW?

The electron microscope is extremely useful in studying cell structures. But how do we know what functions the various components—membranes, organelles, and the like—serve? Here the electron microscope is less useful. Cell biologists use the centrifuge as one tool for studying the function of cell components. A batch of cells of a particular type (kidney cells, liver cells, etc.) are put into a blender so that the cell membranes are broken down. The cell contents are thus spilled out and can be suspended in liquid. This suspension is poured into a tube and spun in a high-speed ultra-centrifuge. A centrifuge uses the power of centrifugal force to force particles in suspension to settle out, or sediment, at the bottom of the centrifuge tube. A centrifuge can effectively increase by many thousands the force of gravity that causes particles to settle out of a suspension if allowed to sit undisturbed for a period of time (see Diagram A). Most particles the size of cell organelles would never settle out by simple gravitational force.

Increasing the speed of centrifuge rotation increases the centrifugal force. The denser the particle, the less centrifugal force is necessary to cause it to settle out of suspension. Thus a suspension of whole cell parts might first be spun at low speeds, forcing to the bottom of the tube only the densest particles. The fluid remaining at the top, the **supernatant,** is poured off, and the precipitate is resuspended. The resuspended precipitate can either be centrifuged again, to refine the separatory process (there will be mixed in with the dense particles some less dense particles that started out at the bottom of the tube), or studied biochemically. The first supernatant is then centrifuged at a higher speed, and the precipitate obtained from this procedure is resus-

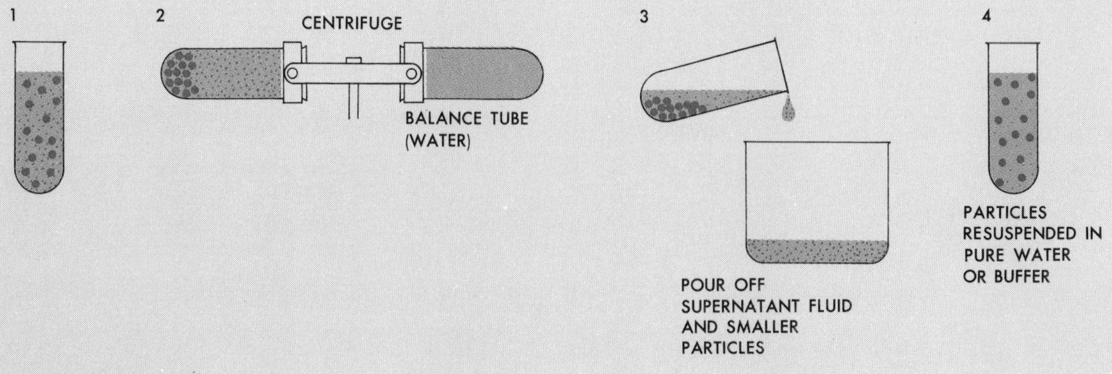

Diagram A

pended and subjected to biochemical tests. This process can be repeated a number of times until the final precipitates (containing the least dense particles) are obtained. In this way the cell can be taken apart and each fraction separated from other fractions and studied separately.

For further refinement of this technique, the cell components can be suspended in a specially prepared liquid called a **sucrose density gradient,** simply a sucrose solution that shows increasing density from the top of the tube to the bottom. The cell components are then added to the top of this tube as shown in Diagram B. In a liquid of variable density, particles will tend to settle out at various densities of the medium that more closely resemble their own density. (Sucrose is commonly used to form the density gradient because it is readily available and is not usually metabolized directly by most cell components.) After cell components are spun in a sucrose density gradient, particles of different density become separated into relatively homogeneous layers. The density gradient thus allows the precise separation of numerous components in a single spin, as shown in Diagram C. The layers can then be drawn off one at a time by a special separatory process.

Once a relatively pure fraction is obtained, it can be subjected to biochemical tests to determine its function. The mitochondrial fraction, for example, has been found to metabolize pyruvate or acetate to produce carbon dioxide. This suggests that the function of mitochondria is to carry out cell respiration. The microsomal fraction (which consists of ribosomes attached to membrane fragments of the endoplasmic reticulum) is found to incorporate radioactively labeled amino acids into polypeptide chains. This suggests that the function of ribosomes is protein synthesis. Using these and other techniques, we can determine the role of each cell component.

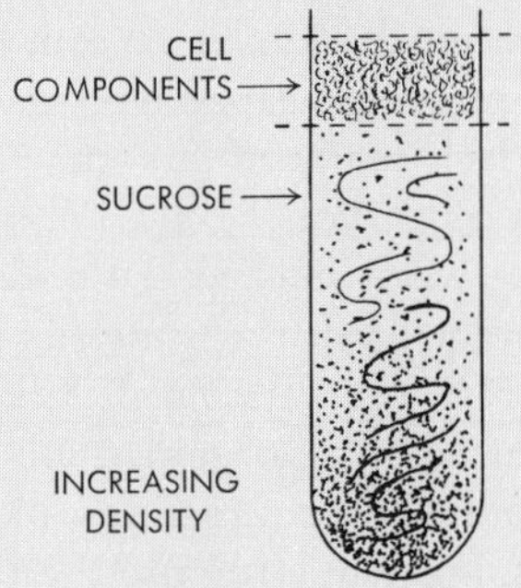

Diagram B

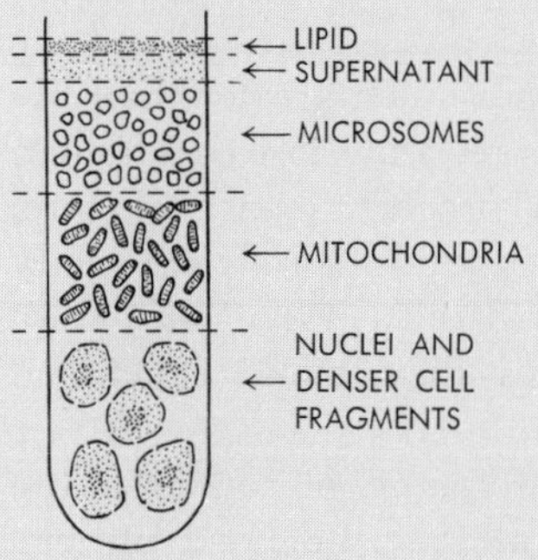

Diagram C

of sex cells, or **gametes,** such as egg and sperm. It is thus an important part of **sexual** reproduction, since it involves the union of gametes produced by two different parents (see Fig. 4.30). Meiosis is a biological necessity: without it, the chromosome number in the species would double with every generation of sexual reproduction.

4.5 The Procaryotic Cell

Procaryotic cells seem to be much simpler in comparison to eucaryotic cells. Although capable of independent existence, they lack many of the characteristics associated with eucaryotic cells. The most obvious difference between eucaryotes and procaryotes is that the latter lack a distinct nucleus. All procaryotes contain DNA, are capable of the synthesis and breakdown of complex organic molecules, divide, and carry out all the functions of life. While they are much simpler in structure than the eucaryotes, procaryotes are very much living organisms. Their simplicity limits their collaborative behavior, however, for few procaryotic cells are capable of forming multicellular associations in which cell specialization is observed. Although procaryotes often associate in certain very simple ways, each cell functions almost totally on its own at all times.

It has been said that the structural differences between the most complex procaryotic cell and the simplest eucaryotic cell are much greater than between the cells of an oak tree and those of a human being. In cellular detail, organizational structure, and functional capacity, procaryotic cells are vastly different from eucaryotic cells. What are these differences, and how do they affect functional capacity?

One of the main differences, of course, is the absence in procaryotes of a well-defined nucleus surrounded by a membrane. The electron micrograph of a typical bacterium in Fig. 4.31 shows this clearly. The lightly stained region is nucleic acid, the dark stained region the cell cytoplasm. While there is no definite nucleus, it is clear that the nuclear material is not necessarily scattered randomly throughout the cell. In most bacterial and other procaryotic cells, nuclear material tends to be localized in the cell.

Fig. 4.31 *Electron micrograph of a dividing bacterium,* Bacillus subtilis, *showing localization of DNA and absence of membrane and a distinct nuclear membrane. Plasma cell wall are also visible (×120,000).*

A second main difference between procaryotic and eucaryotic cells is their respective sizes. On the average, eucaryotic cells are much larger than nearly all the procaryote types. Animal cells range in size (on the average) between 10 and 60 μ; eucaryotic plant cells range from 50 to 100 μ; bacteria tend to average about 2 to 5 μ, while mycoplasmas range from 0.1 to 0.25 μ. This difference in size is related (though not causally) to a third important difference: complexity. Complexity, however, is a two-way street. It makes possible, *but also requires,* greater size. Therefore, most eucaryotic cells are larger in size than procaryotic. Going back to the geometric relationship between the volume of a sphere and its surface area, recall that while surface area increases by the square of the radius, volume increases by the cube. Consequently, larger cells encounter greater problems in exchanging materials between the inside and outside; the larger they get the less surface area they have per unit volume inside. Beyond a certain volume, larger cells must have special organelles (endoplasmic reticulum, vacuoles) and processes (bulk transport, cytoplasmic streaming) for taking in and releasing substances. All these processes require energy, so that special energy-generating organelles (mitochondria and chloroplasts) are also required.

From these considerations still another major difference between eucaryotic and procaryotic cells

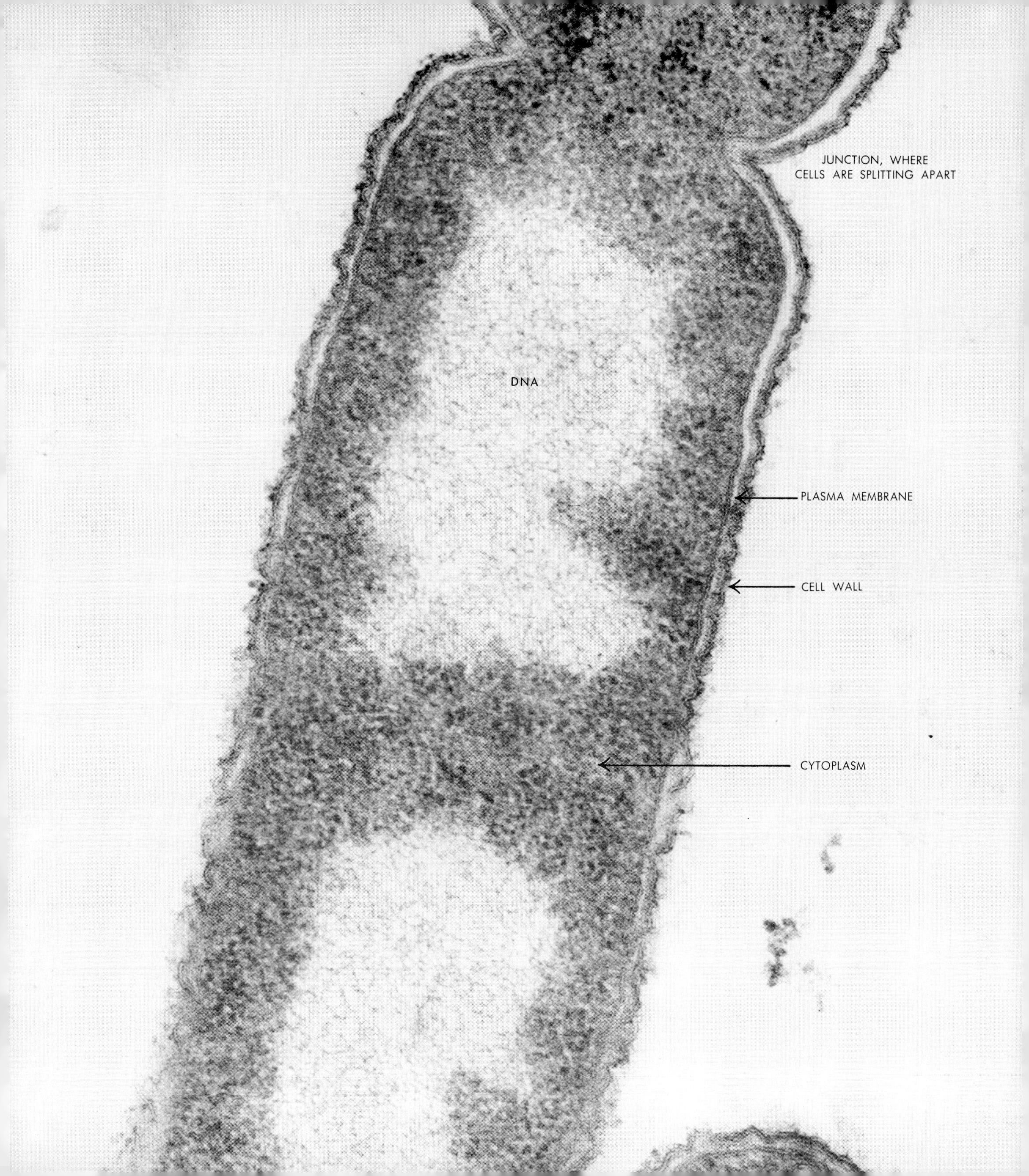
JUNCTION, WHERE
CELLS ARE SPLITTING APART
DNA
PLASMA MEMBRANE
CELL WALL
CYTOPLASM

emerges: the presence or absence of organelles. By and large, eucaryotic cells (whether of plant or animal) have many more types of organelles than procaryotic cells. Most procaryotes lack (in addition to a well-defined nucleus) mitochondria, endoplasmic reticulum, chloroplasts, and Golgi apparatus. Many procaryotes (especially among the bacteria) contain flagella. All procaryotes contain ribosomes and a plasma membrane (usually of the bilayer, unit-membrane structure).

What are the varieties and types of procaryotes? The most widespread groups are the bacteria and the blue-green algae. A less diverse group includes the mycoplasmas, the smallest known living cells.

While the structure and function of these cell types will be discussed in greater detail in a later chapter, it is helpful to summarize some characteristics of these organisms now to obtain a general insight into the nature of all procaryotic cells.

Bacteria

While bacteria can be seen by the light microscope, most of their detail has been revealed only by the electron microscope. Bacteria appear in overall shape as rods, spheres, or spiral structures. They have a very heavy cell wall covering their plasma membrane (see Fig. 4.31). The cell wall is made of polysaccharide and is often highly resistant to acids and other agents. The cell wall serves as a protection for the delicate bacterial cell within. Most kinds of bacteria are like animals—they cannot synthesize their own food photosynthetically but must obtain it from organic matter in the environment. Some bacteria, however, are capable of photosynthesis, although they carry out the process in slightly different ways from higher plants. Because of the presence of a cell wall, and the fact that some species are photosynthetic, the bacteria have been traditionally classified as plants. However, modern taxonomists increasingly prefer to classify bacteria, along with a variety of other one-celled forms, as Monera.

Bacteria generally reproduce asexually by cell division, or **fission.** Many bacteria have all their genetic information in one long single-stranded DNA molecule, the bacterial "chromosome." This molecule duplicates, and the cell splits in half. Periodically, however, most strains of bacteria undergo conjugation—a form of sexual reproduction. Two cells come to lie side-by-side, their cell walls break down at one adjacent point, and the DNA of one cell passes into the other cell. The recipient cell is thus temporarily in possession of two complete "chromosomes." The donor cell simply deteriorates. The recipient cell soon undergoes fission, in which one chromosome goes to each daughter cell. Conjugation allows the exchange of genetic information. The two chromosomes can break and exchange parts, producing a new chromosome with a combination of hereditary information that it did not possess before. The conjugation process takes several hours to complete. Fission requires less than half an hour for many species.

Some species of bacteria produce spores from time to time. Sporulation (spore formation) is not a reproductive process, but one that helps to sustain the cell during unfavorable conditions. The electron micrograph and drawing in Fig. 4.32 show the process of sporulation. When conditions become unfavorable, the bacterial cell manufactures a very strong, virtually impermeable membrane—the spore coat—from materials in its cytoplasm. This coat surrounds the nucleic acids and a small part of the cytoplasm. The remaining parts of the bacterial cell disintegrate, leaving the well-protected spore containing the organism's DNA. Spores can withstand temperatures up to 100°C (boiling) for several hours; nonsporulated bacterial cells of the same species are killed instantaneously by such temperatures.

Despite the differences between bacterial and eucaryotic cells, there are numerous similarities. All bacteria are able to carry out some of the same stages in oxidation of carbohydrates (cell respiration), and some bacteria can carry out all the steps. Bacteria synthesize protein with ribosomes in a

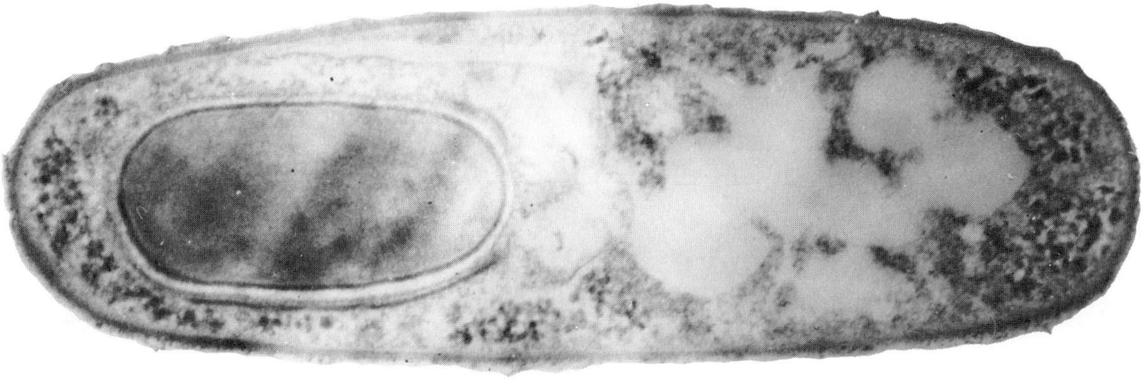

(a)

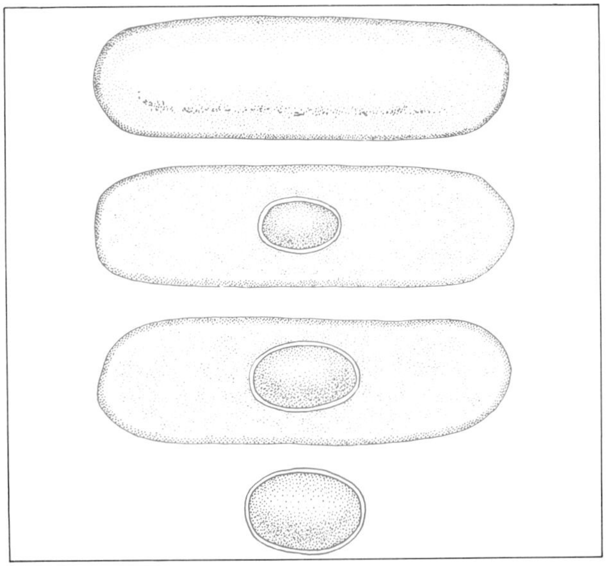

(b)

Fig. 4.32 *The process of sporulation in bacteria. The electron micrograph in (a) shows the large spore being formed toward one end of the cytoplasm of the bacterial cell. The spore has a very tough protective coat enabling the cell to survive through unfavorable conditions (×46,000). (b) Drawing of sporulation process. The spore is formed from elements in the cytoplasm and encased in its coat while still in the cytoplasm of the original cell. (Electron micrograph courtesy Dr. George B. Chapman and the American Society for Microbiology.)*

process very similar to that in eucaryotes. Furthermore, bacteria actively transport molecules across their cell walls and plasma membranes in a process similar to that which operates in cells of higher organisms.

Blue-Green Algae

Blue-green algae share many characteristics with bacteria. As in the bacterial cell, the DNA of blue-green algae cells is not contained within a true nucleus, but is localized in certain regions (see Fig. 4.33a). Like bacteria, blue-green algae have ribosomes and lack most of the other organelles found in encaryotic cells. And, like bacteria, blue-green algae cells are surrounded by a protective carbohydrate cell wall. Unlike most bacteria, however, all the 1500 or so species of blue-green algae are able to carry on photosynthesis. This process is accomplished without the chloroplasts typically found in cells of higher plants. Blue-green algae have chlorophyll *a* (and none of the other forms of chlorophyll found typically in higher plant cells) and several other pigments (such as phycocyanin, a bluish pigment) involved in the photosynthetic process. Lacking chloroplasts, blue-green algae cells have chlorophyll bound to cytoplasmic membranes (Fig. 4.33b). These folded membranes give the internal structure of the blue-green algae the appearance of considerably greater complexity and orga-

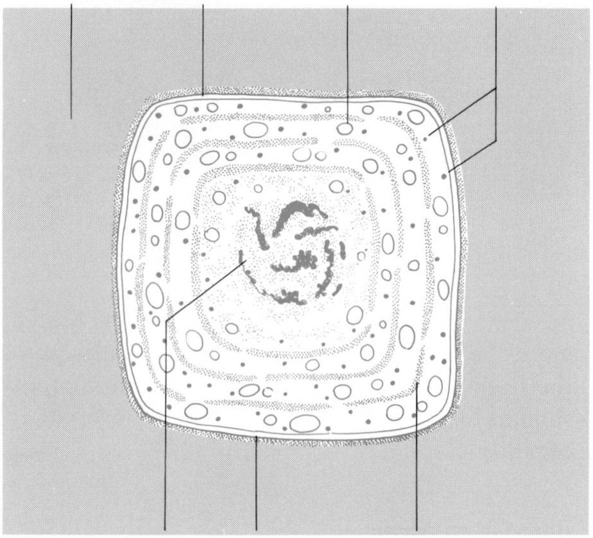

(a)

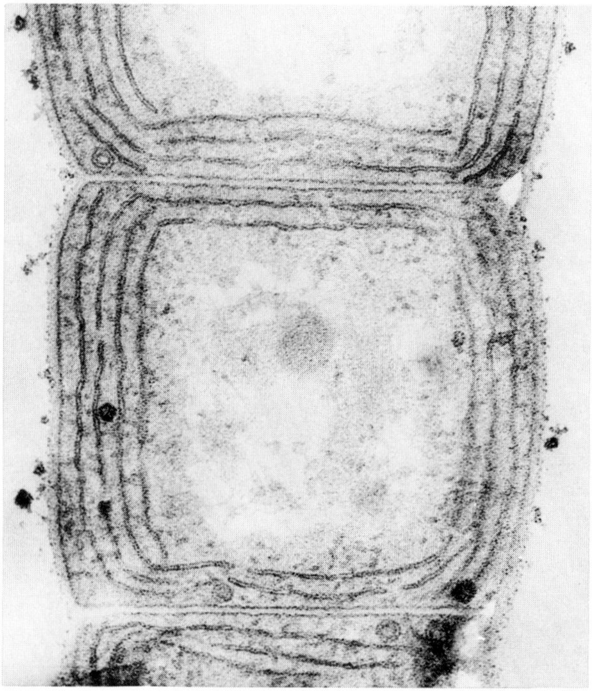

(b)

Fig. 4.33 *Blue-green algae are similar in many organizational features to bacteria. Nuclear material is not localized in an enclosed nucleus. There are few organelles, with the exception of ribosomes and a cytoplasmic membrane system for photosynthesis. The cell is surrounded by a thick cell wall of carbohydrate; outside this there is sometimes another, more amorphous layer called a "sheath." The sheath is usually a semi-fluidlike material, similar to thick jelly. It prevents water loss from the algal cell, as well as affording increased protection. The electron micrograph shown in (b) gives a good indication of the cytoplasmic membrane system in one alga,* Anabaena *(×6800). (Photo courtesy William T. Hall, Electro-Nucleonics Laboratories, Inc.)*

nization than the bacterial cell. In reality, the difference is not great. With the exception of the cytoplasmic membrane system and the presence of photosynthetic pigments, the blue-green algae are organized in a manner very much like bacteria. Many biologists think the two groups share a close evolutionary relationship.

Many, though not all, species of blue-green algae exist as groups of cells arranged in long filaments (see Fig. 4.34). Some of these live in the moist earth, others in water. A few species can endure very high temperatures—as high as 70°C—such as those found in hot springs. The variety of colors observed in the hot springs of Yellowstone National Park is due to the accessory pigments of many species of blue-green algae that live in the springs. Like the bacteria, the blue-green algae are very widely distributed throughout the world.

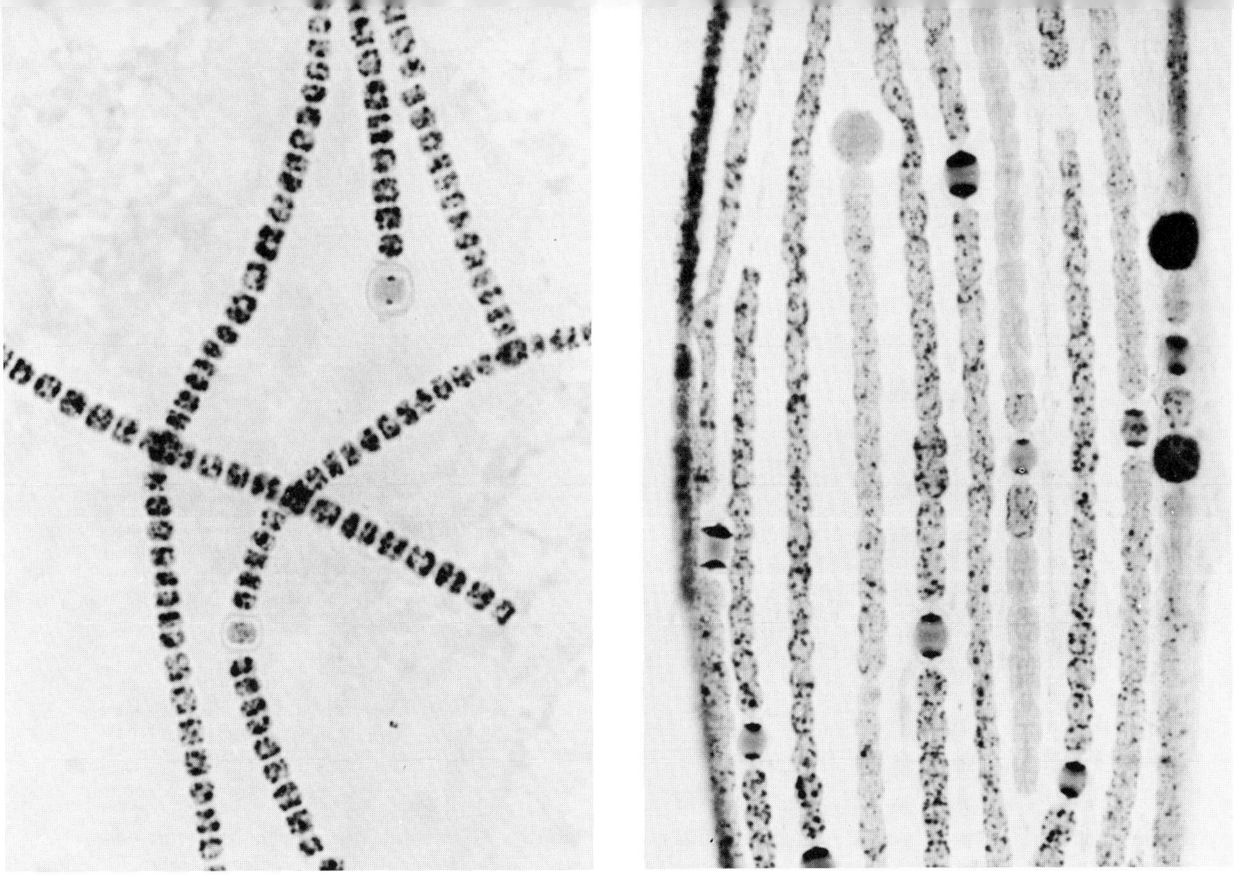

(a) (b)

Fig. 4.34 *Two filamentous species of blue-green algae. (a)* Nostoc, *showing the beadlike cells arranged in short filaments with a larger reproductive cell, the heterocyst. Heterocysts develop from ordinary vegetative cells, and are spore-like in nature. They appear to function in reproduction, as they can germinate under special conditions. (b)* Anabaena, *showing special reproductive cells interspersed at intervals along the filament. Each cell in such filamentous forms is independent of the others and can live by itself if separated from the filament. (Courtesy Carolina Biological Supply Company.)*

Mycoplasmas

Mycoplasmas are the smallest known cells. About thirty different species have been observed, ranging in diameter from 0.1 to 0.25 μ (about one-tenth the size of average bacteria). They are frequently referred to as PPLO, or "pleuropneumonia-like organisms." Their existence was first suspected in the nineteenth century in connection with a highly contagious disease of cattle, pleuropneumonia. But they were not identified at the time because they were so small that they could not even be caught in the porcelain filters used to remove bacteria from blood. In 1898 it was shown that PPLO could grow on a culture medium and did not require living host cells. Finally, in 1931, a special filter capable of trapping these tiny organisms was developed. It was only then that they could be observed, their

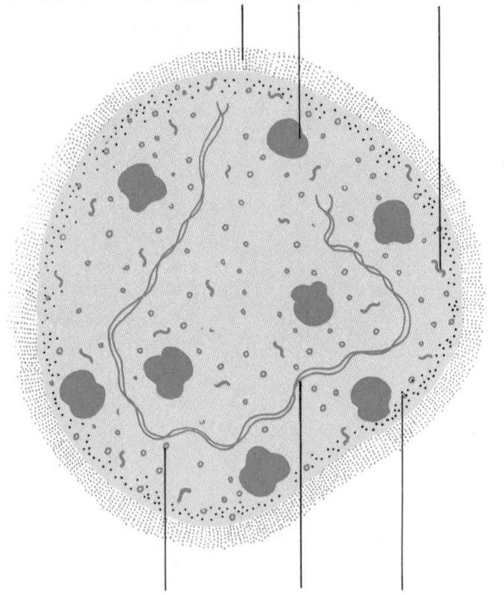

Fig. 4.35 *Diagram of a typical mycoplasma, the PPLO (pleuropneumonia-like organism). These extremely small cells contain all the types of molecules found in most living things (carbohydrate, lipids, protein, and nucleic acid) as well as ribosomes. Note the much larger relative size of ribosomes shown here; this indicates something of the tiny size of PPLO. Some mycoplasmas like PPLO are highly pathogenic in humans and other mammals. (After* Biology Today. *New York: CRM/Random House, 1972, p. 157.)*

actual size adequately estimated, and something of their structure determined (see Fig. 4.35).

PPLO cells lack most of the organelles that characterize eucaryotic cells, but they contain most components found in procaryotic cells. They have what appears to be a typical lipid–protein membrane surrounding the cell. They have DNA loosely organized as the hereditary material within the cytoplasm. They have ribosomes for protein synthesis. Their internal organization appears even less highly structured than that of bacteria. PPLO, for example, do not appear to have their nuclear material localized in the way that is common for most bacteria.

PPLO reproduction involves a complex life cycle that takes two days to complete (see Fig. 4.36). The species of PPLO that has been studied most intensively forms small "elementary bodies" that grow in size for several days until they are the size of a bacterium (about 1 μ in diameter). The large cell may divide into two daughter cells, or it may form elementary bodies within itself that are eventually released. PPLO may not repre-

Fig. 4.36 *Generalized life cycle of a mycoplasma. Elementary bodies grow into intermediate cells and eventually large cells. These large cells either reproduce without forming elementary bodies or form elementary bodies within themselves to be released and start the cycle again. The process requires about two days. (After* Biology Today. *New York: CRM/Random House, 1972, p. 156.)*

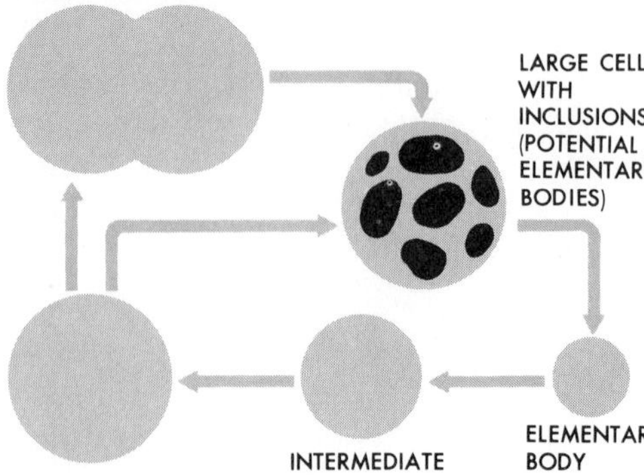

Supplement 4.4

THE LOWER LIMIT ON CELL SIZE: A MATTER OF SPACE

Earlier in this chapter we discussed the question of how large cells could be in terms of the geometry of volume versus surface area for solid figures. Just as the upward limit on cell size is determined by ratio of surface area to volume, the lower limit is determined by the sizes of cell components and macromolecules. The plasma membrane of cells is anywhere from 70 to 100 Å thick; with such a membrane on all sides, the external measurements of a cell cannot be less than about 300 to 400 Å (0.03 μ) at an absolute minimum. Even that would severely limit the available space inside the cell. Although we do not often think of units as small as molecules taking up space, every such molecule down to the smallest ion has dimensions. Only so many molecules can fit into a given volume of space. The most basic biochemical reactions characteristic of all living cells (replication of DNA, synthesis of proteins, breakdown of sugars to produce usable energy) require a large number of different kinds of molecules. Thus, according to recent calculations, a cell containing the minimum number of ribosomes, enzyme molecules, and DNA could hardly be smaller than 0.4 μ, or about one-tenth the size of mycoplasmas. Biologists doubt that any cells have reached this minimum possible size.

Summary of the Characteristics of Animal, Plant, and Bacterial Cells

Eucaryotic		**Procaryotic**
Animal	**Plant**	**Bacteria**
Size: 10 to 60 μ*	Size: 50 to 100 μ	Size: 2 to 5 μ (some up to 100 μ)
No cell wall	Cell wall	Cell wall
Centriole present	Centriole in lower plants only	Centriole absent
Small vacuoles	Large vacuoles	No vacuoles
Organelles:	*Organelles:*	*Organelles:*
Mitochondria	Mitochondria	Ribosomes
No chloroplast	Chloroplast	
Endoplasmic reticulum	Endoplasmic reticulum	
Ribosomes	Ribosomes	
Nucleus (+ membrane)	Nucleus (+ membrane)	
Golgi apparatus	Golgi apparatus	

* 1 μ = 1/1,000,000th of a meter, or 10^{-6} meters

sent the very smallest existing cells—there may still be smaller ones yet undetected. However, it seems unlikely that there will be discovered structures qualifying as true cells that are very much smaller than PPLO (See Supplement 4.4).

4.6 Viruses

In addition to procaryotes and eucaryotes, there is another group of "organisms" that are not cellular in nature. **Viruses** are small particles composed of two types of substances: protein and nucleic acid. The nucleic acid carries a program of information for producing more viruses; the protein forms a coat around the nucleic acid "core." Figure 4.37 shows one type of virus, the so-called bacteriophage, which attacks bacterial cells. Viruses function by invading living cells (from bacteria to those of higher animals and plants). Inside the host cell, the virus reproduces itself by using the cell's chemical machinery. Often (but not always and depending upon the type of virus and pattern of infection) this destroys the host cell.

Viruses cannot reproduce outside of living cells, no matter what raw materials are available. Furthermore, viruses can be crystallized, placed in a bottle, and stored for many years. When the crystals are resuspended in solution, they are capable

Fig. 4.37 *Electron micrograph of a group of T2 bacteriophages. Each phage consists of a head, neck, and tail region. The head outer coat and the neck and tail consist of protein. Inside the head is DNA carrying the virus's genetic information. Phages without the thickened neck-piece are "triggered," meaning they have released their DNA. DNA is shot out through the neck and tail into a bacterium to which the phage has attached itself. [Photo courtesy Dr. Sidney Brenner and Dr. R. W. Horne,* Journal of Molecular Biology. *Copyright © 1959 by Academic Press (London) Ltd.]*

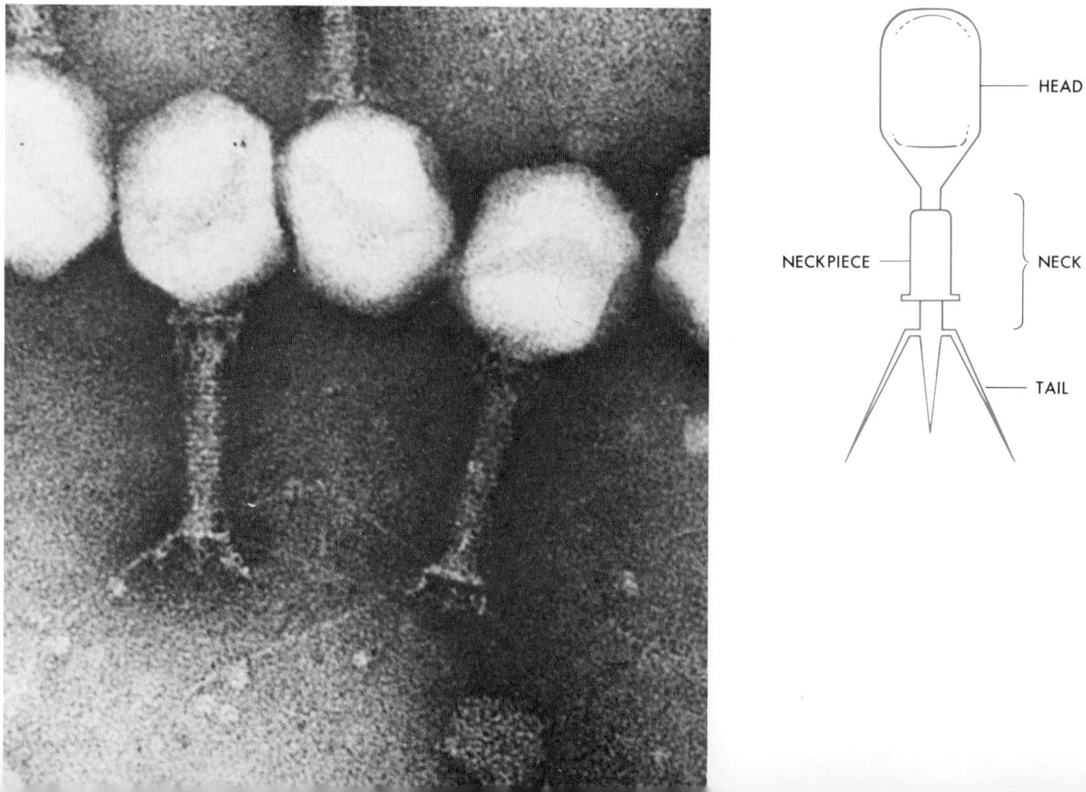

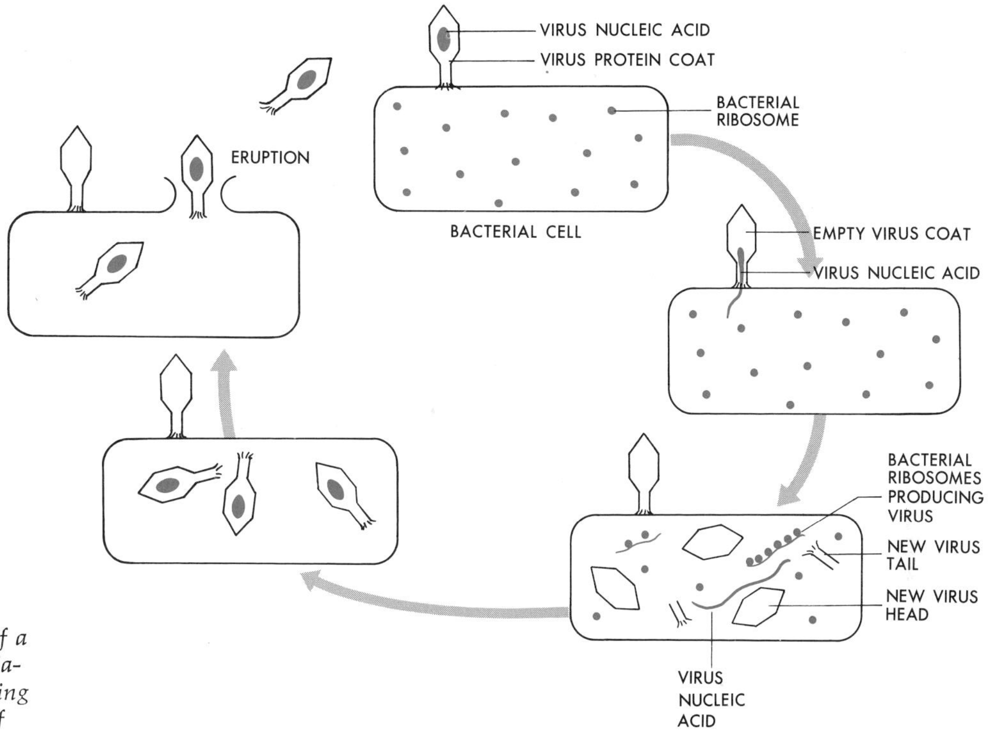

Fig. 4.38 *Life cycle of a bacteriophage (bacteria-infecting virus) showing the typical features of viral life cycles.*

of infecting cells once more, but while they are crystals, they do not carry out any of the chemical reactions associated with living cells. In view of these characteristics, many biologists have found it impossible to classify viruses as either living or nonliving. The terms simply have no meaning when applied to them. If the ability to reproduce constitutes "living," then viruses are alive. On the other hand, if the ability to carry out respiration and synthesize one's own proteins is considered "living," then viruses are not alive. Functioning only with the aid of intact living cells, they are like a set of blueprints with no factory of their own where their plans can be put into action. Yet, once inside a living cell, viruses can direct the synthesis of proteins, reproduce virus parts, and assemble those parts into new, intact viruses.

Since all viruses live within already-existing cells, they are for the most part harmful to the host cell. Viruses are major causes of disease in both animals and plants. A mottled, deteriorating condition affecting tobacco plants is caused by the tobacco mosaic virus, TMV. Smallpox and polio are caused by viruses, as are many varieties of the common "cold." Viruses are particularly difficult to combat because common antibiotics do not usually affect them. The most effective medical approach to viral diseases is prevention of the disease through immunization (e.g., vaccination). Exposure of the human (or animal) body to a small dosage of virus allows the body's defenses against that specific virus to build up without the organism's succumbing to the disease. Then when the virus is encountered in the environment, it can be eliminated from the body early because the body "recognizes" it as foreign. Once the virus has invaded the body's cells in an infection, it is much more difficult to eliminate.

The general reproductive cycle for most viruses is diagrammed in Fig. 4.38. The diagram shows a bacteriophage invading a rod-shaped bacterial cell. This entire cycle can be completed in less than half an hour under favorable conditions. The virus nucleic acid is injected into the host-cell-to-be, leaving the protein coat attached to the outside of the cell wall. Inside the host cell, virus DNA commandeers the bacterial ribosomes, amino acid, and carbohydrate pools and begins to direct the production of virus protein. Viral nucleic acid is also replicated prodigiously. When components have been completed, they self-assemble in the host cell's cytoplasm. The cell is ruptured and the new virus particles erupt to infect other cells nearby.

Some viruses do not destroy the host cell immediately after invading it. Certain viruses enter a host cell and simply deposit their nucleic acid in the cell but do not replicate new viruses. The viral nucleic acid either exists independently of the host cell's nucleic acid or can actually be incorporated into it. In either case, viral nucleic acid replicates at exactly the same rate as the host cell's. At some indefinite time in the future (for viruses in human cells this may be from a few months to 20 or more years), the viral nucleic acid does begin to take over the cell and produce new viruses. Thus there can be a long delay between infection and manifestation of viral disease. It is thought, for example, that some forms of cancer are caused by viral nucleic acid that gets into a person's cells early in life but is not activated until middle age or later.

4.7 A Case Study in Structure and Function: The Cell Membrane

Using radioactive tracers and very delicate microchemical techniques, biologists have devoted considerable attention to the problem of how substances pass through membranes in living cells. The experiment diagrammed in Fig. 4.39 shows an artificial membrane (nonliving, but functionally similar in many ways to living membranes) inside of which has been placed a solution of blood plasma (proteins), a monosaccharide sugar (glucose), and an inorganic salt (sodium chloride). The membrane, with its enclosed substances, is then lowered into distilled water.

After two hours have passed, the distilled water outside the membrane is chemically analyzed for the presence of plasma, glucose, and salt. The test for plasma is negative; none has passed through the membrane to the distilled water. The tests for glucose and salt, however, are positive. Evidently these compounds have passed through the membrane into the water. These results show this membrane to be **permeable** to glucose and sodium chloride, but **impermeable** to protein. A membrane that is partially permeable to solvent but not permeable to solutes is said to be **semipermeable.** A membrane allowing one substance (solute or solvent) to pass through more easily than another substance (solute or solvent) is called **differentially permeable.**

Note, however, that something else happened in the experiment. The membrane bag became swollen, indicating that something had moved in from the outside. Since nothing but water is present in the beaker, we can only conclude that as molecules of glucose and ions of sodium chloride moved out, molecules of water moved in. This membrane, then, is permeable to water molecules.

If the amounts of salt and glucose present in the distilled water outside the membrane are quantitatively compared with the amounts of salt and glucose on the inside, they will be found to be equal. Molecules and ions in the solution are in a constant state of random motion. For example, if the salt copper sulfate is dissolved in distilled water, the copper and sulfate ions will spread out among the water molecules. In time, every area of the container will contain the same amount of copper and sulfate ions. This is because *all molecules, atoms, or ions in a solution tend to move from an area of greater concentration to one of lesser concentration.* We call such movement **diffusion.** When all the copper and sulfate ions are equally dispersed throughout the water, an *equilibrium* has been reached. This does *not* mean that the ions have

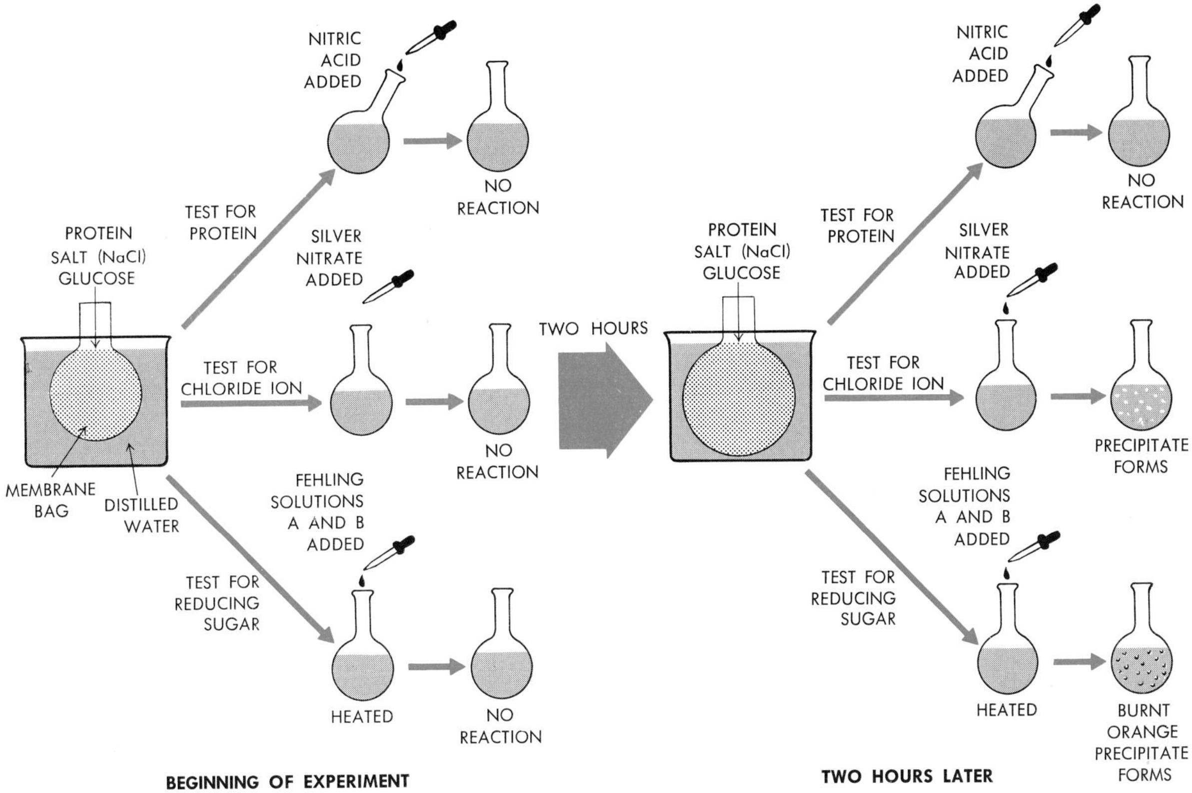

Fig. 4.39 *A series of experiments performed to investigate the nature of membranes in relation to the solutions within them. Note the change in size of the membrane bag after two hours.*

stopped moving. It means that for every ion that moves from the left side of the container to the right, another ion moves from the right to the left. The net effect is as if the ions had not moved at all. The equilibrium in such a situation is thus a **dynamic equilibrium.**

It is easy to hypothesize an explanation for the results of the experiment involving the artificial membrane filled with blood plasma, glucose, and sodium chloride. Molecules of glucose and ions of sodium and chlorine must have passed through the membrane into the distilled water by diffusion. At first, more molecules and ions passed out through the membrane into the distilled water than passed in. This is because the *concentration* of these molecules and ions was greater inside the membrane than outside. For every molecule and ion that passed in through the membrane, perhaps 1000 molecules and ions passed out into the distilled water. As diffusion continued, however, the concentration of molecules and ions inside the membrane became closer to the concentration outside, and the rate of diffusion slowed down. Eventually the number of molecules passing out through the membrane became equal to the number passing in. At this point,

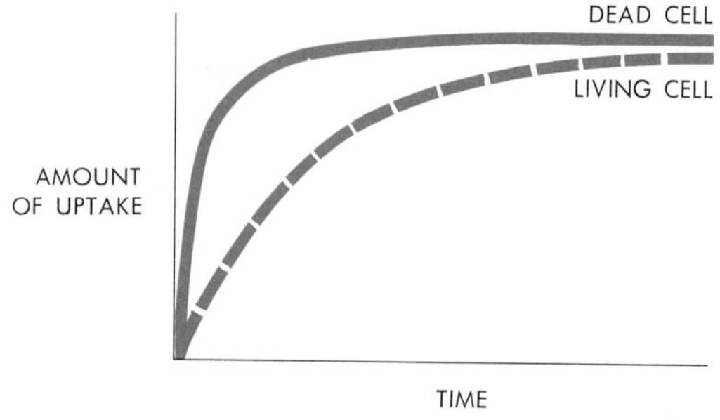

Fig. 4.40 *This graph shows the uptake of ions introduced into a liquid medium surrounding dead and living cells. Note that the living cells are able to offer some resistance to diffusion of ions into the cell. A graph of the respiratory rates of such cells shows that they must expend energy to accomplish this.*

a dynamic equilibrium was attained. No such equilibrium was attained with the protein molecules in the blood plasma, since they were unable to pass through the membrane. The swelling of the membrane can be accounted for by assuming that the water molecules entered more rapidly than glucose or the sodium and chloride ions escaped.

The results obtained in this experiment with the membrane are duplicated in experiments using individual cells. In general, proteins are not passed through the plasma membrane of such cells. It is logical to wonder why this should be so. One hypothesis is that the plasma membrane is perforated by pores so small that they must be measured in angstrom units. We might hypothesize, then, that only those molecules small enough to pass through the pores in the plasma membrane will pass into the cell. This hypothesis is supported by the fact that small molecules, such as those of water, generally pass freely in and out of cells, while larger molecules, such as proteins, generally do not.

Other observations, however, contradict this hypothesis. Ions of sodium are kept out of resting nerve cells, yet such ions are far smaller than the smallest pore in the plasma membrane. Conversely, under certain conditions large molecules of nucleic acid will pass into some cells. Thus the pore hypothesis may account for the movements of some substances in and out of cells, but it cannot account for others. The cells of many different organisms can concentrate molecules or ions to a far higher degree than the surrounding medium. Many of these molecules and ions are of such small dimensions that they could easily pass out of the cell, and since diffusion occurs from a region of greater to a region of lesser concentration, it seems they would do so. But they do not. We might ask, then, why the experiment with the membrane resulted in equal concentrations of glucose molecules and sodium and chlorine ions on both sides of the membrane when such an equilibrium is not necessarily attained with other types of cells.

The answer is suggested by additional experiments. It has been shown that the respiratory rate of cells that are concentrating ions or molecules against the normal concentration gradient is far higher than the respiratory rate of similar cells that are not concentrating the same molecules or ions. It seems evident, then, that these cells maintain a high concentration of dissolved substances within themselves by expending energy. A similar group of cells, intact but nonliving, is unable to maintain a high concentration of salts. Not being alive, these cells are unable to expend the energy necessary to hold the ions and molecules within their membranes. The laws of diffusion have the upper hand, and the concentration of substances inside and outside these nonliving cells soon equalizes (Fig. 4.40).

As reported by one investigator, the pores in the plasma membranes of red blood cells have a diameter of 7 Å. Yet glucose molecules 8 Å across are able to pass into the cells. Here again, experiments show that energy is being expended by the red blood cells; they must perform work in order to get the glucose inside. The process is spoken of as one of **active transport,** as opposed to the passive transport of substances into and out of the cell by diffusion. In the latter process, the cell need expend no energy.

As yet, no single hypothesis has explained all the types of movements of materials known to occur across the plasma membrane. It has been suggested that the polarity of the phospholipid molecules composing part of the plasma membrane, as well as the polarity of the molecules passing through it, may determine whether the substance gets through. In the light of our present knowledge, it seems best to view the plasma membrane as an active boundary layer rather than as a rigid, inflexible structure. It might be pictured as similar to a line of men standing a few feet apart. Their duty is to guard a certain area, making certain that only properly identified persons or vehicles get through. If a large vehicle must pass through the line, the men can move aside to make room. If a small person without the proper credentials approaches the line, the men can move together to prevent his passage. We cannot use this example in the case of nonliving cells. Instead, we can picture the plasma membrane of the nonliving cell as similar to a line of posts stuck in the ground around a guarded area. In this situation, the openings between the posts cannot be adjusted, and dynamic equilibrium is attained between the ions or molecules of only those substances able to pass between the posts.

The hypothesis pictured by this analogy views the plasma membrane as more than a passive, inactive structure. Such an hypothesis helps to explain why a balanced equilibrium is not always attained between similar substances inside and outside the cell. The hypothesis also pictures the cell as playing an active role in determining which substances enter it by diffusion and which do not, as well as how much of a substance may do so. Occasionally, therefore, cells must expend energy in obtaining or holding needed ions or molecules.

Revision of the Davson–Danielli Model

The Davson–Danielli unit membrane concept, discussed in Section 4.4, has been remarkably successful in accounting for the electrical and passive-permeability characteristics of plasma membranes. Further, artificial membranes made on the basis of the model closely resemble real plasma membranes in both structure and function. Yet it is difficult to see how the protein components of the membrane can perform all the functions they must surely perform during active transport if, as the Davson–Danielli model indicates, they are merely spread out over the lipid surface.

Evidence that the Davson–Danielli unit membrane model needs considerable revision comes from several observations:

1. Electron microscope examinations of membranes after treatment with a lipid-digesting enzyme, phospholipase, were carried out by A. Otholenghi and M. Bowman at Ohio State University. These examinations show that the membrane is not digested all over, but only in patches. Presumably these patches are areas where exposed phospholipid molecules are concentrated.

2. M. Glazer and his colleagues working at the University of California at La Jolla have shown that when red blood cell membranes are heated, the areas assumed to be those attacked by phospholipase are left intact, but areas sensitive to temperatures of the range that affects proteins are changed.

These observations have necessitated a revision of the Davson–Danielli model. The revision is based on data from electron microscope studies, combined with biochemical analysis of membrane properties. Such studies have focused particularly on the membrane of red blood cell ghosts (cells that have been treated in such a way as to lose their

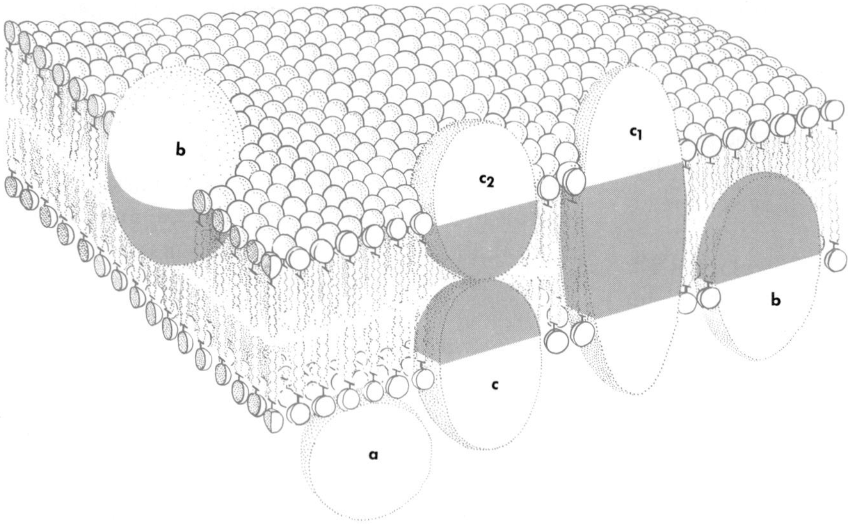

Fig. 4.41 *Newer view of membrane structure showing protein molecules "floating" in the liquid lipid bilayer. Three different types of protein, classified according to their relation to the lipid bilayer, are shown here. (a) Extrinsic or surface proteins, which do not penetrate the bilayer. (b) and (c) Intrinsic or immersed proteins, which penetrate the bilayer to some degree. Type (b) proteins penetrate the bilayer partially. Type (c) proteins either penetrate the bilayer as a single molecule (c_1) or as two subunits (c_2). Shaded sections of each intrinsic protein represent the region of the molecule in which nonpolar groups predominate. (From Roderick Capaldi, "A dynamic model of cell membranes,"* Sci. Am., *March 1974. Copyright © 1974 by Scientific American, Inc. All rights reserved.)*

hemoglobin contents but retain an intact plasma membrane), and the inner membranes of mitochondria. The following observations have emerged from these studies:

1. Plasma membranes as well as membranes internal to the cell (such as endoplasmic reticulum and mitochondrial membranes) all have the same unit membrane structure. In addition, their chemical compositions, while showing many small differences, are also very much alike.

2. All membranes appear to consist of three kinds of molecules: phospholipids, glycolipids (lipids bound to a carbohydrate unit instead of phosphorus), and protein. Lipids account for approximately one-half of the mass of most membranes, proteins the other half.

3. The arrangement of the lipid and protein components follows a modified version of the Davson–Danielli bilayer model. The lipid bilayer is a fluid structure at body temperature (because the fatty acids are unsaturated and have a lower melting point than saturated fatty acids) in which the protein molecules "float" (see Fig. 4.41). The nature of flotation depends upon the protein type (mo-

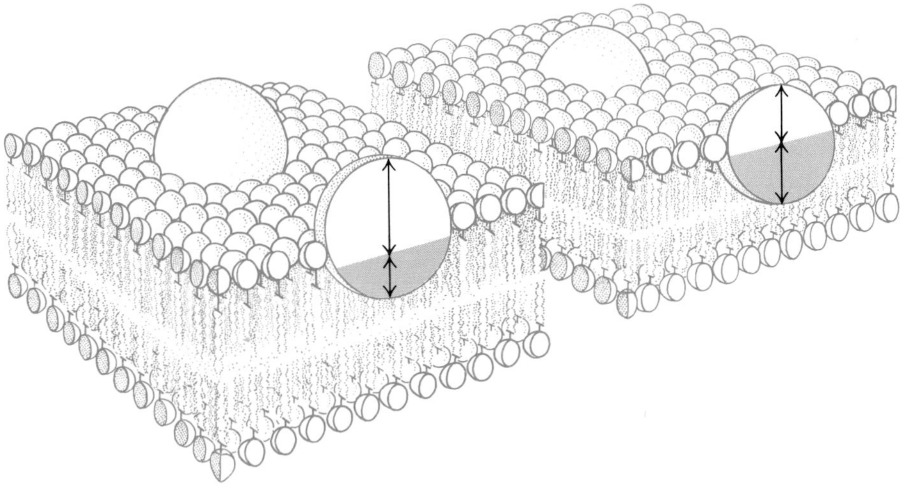

Fig. 4.42 *Changes in degree of penetration of intrinsic proteins in the lipid bilayer of membranes are illustrated with the molecule rhodopsin (spherical unit). Rhodopsin is a visual pigment, responsible for receiving photons of light in animal retinas and converting it to electrical energy within retinal cells (called rods). In the dark, rhodopsin penetrates the cytoplasmic membrane (to which it is attached) about one-third of the diameter of the molecule. When exposed to light (right), the molecule sinks deeper into the lipid bilayer. The chemical function of the molecule is intimately related to its structural position in the lipid bilayer of the membrane. (From Roderick Capaldi, "A dynamic model of cell membranes,"* Sci. Am., *March 1974. Copyright © 1974 by Scientific American, Inc. All rights reserved.*

lecular weight, charge distribution), but it generally occurs in two patterns:

a) Proteins lying at the external surface of the membrane that do not penetrate the lipid bilayer to any degree. These are called extrinsic or external proteins.

b) Proteins that penetrate the lipid bilayer to some degree (called intrinsic or internal proteins). Some intrinsic proteins penetrate the lipid layer only partially, others may extend completely through it. It has also been found that the degree of penetration of the lipid bilayer by any individual intrinsic protein varies with chemical changes at the membrane level. For example, the light-sensitive pigment rhodopsin is an intrinsic protein attached to cytoplasmic membranes in visual cells of the eye (in the retina, at the back of the eye). When retinal cells (called "rods") are in the dark, rhodopsin is only one-third immersed in the lipid bilayer. When the cell is illuminated by a photon of light, the rhodopsin molecules become about one-half submerged (see Fig. 4.42). The membrane is thus intimately involved, structurally and functionally, with many biochemical reactions in the living cell.

Supplement 4.5

MODELS IN SCIENCE

Models are often used in science as an aid to explanations. They differ somewhat from hypotheses and are not necessarily a part of every scientific explanation. Yet they have become such an integral part of many explanations, especially in contemporary biology, that the term has become commonplace. What do we mean by the term "model"? The Davson–Danielli model for membrane structure provides an example.

Models are like analogies in that they attempt to represent the unfamiliar in terms of the familiar. Models usually involve some sort of physical representation of a complex process or structure whose details cannot be directly observed. An electron micrograph of a cell is not a model—it is a direct visual representation of the cell structure. The Davson–Danielli concept of membrane structure *is* a model because it attempts to represent the physical structure of something we cannot observe directly. Similarly, the Bohr concept of the atom as a miniature solar system is a model.

Most models attempt to relate their physical components to known functions. The Davson–Danielli membrane model postulates a physical structure that was in agreement with known functional properties of membranes at the time it was published (1937). Partly because they attempt to represent complex phenomena and structures with physical components, most models are admitted oversimplifications. By simplifying, and by finding physical components to represent what would otherwise be abstract notions, models can bring considerable coherence and easy comprehension into scientific theories.

Models differ from theories in that they are generally more concrete and less all-encompassing. The Davson–Danielli model of membrane structure is a part of what we can call the overall theory of membrane structure and function. The theory would encompass many more elements than the Davson–Danielli model. For instance, it would include concepts of ion diffusion (physical chemistry) and electrical properties of cells. Any broad theory is likely to encompass at least several models within its overall framework.

One of the chief functions of models is to aid in making predictions. The Davson–Danielli model predicts that dissolving away the lipid portion of a membrane should totally destroy the membrane's structure and function. This prediction can be (and was) put to the test. Because the results were negative, the model could be rejected, at least as it stood. The value of models is that they can be tested.

Like theories and hypotheses, models are not permanent. They are constantly being tested, modified, and sometimes rejected. Again, the Davson–Danielli model is a good example. Because it failed to predict certain phenomena accurately, the model had to be changed. This change did not reject all elements of the original model, but it did significantly alter the model. The new model (it does not really have a single name) accounts far more precisely than the old model for observational and biochemical data about membrane structure and function.

A persistent danger plagues the use of models in science. Sometimes workers or students tend to take the models too literally. The Bohr model of the atom is a good case in point. Planets in the solar system travel in specific orbits. It is possible to predict where a planet will be at almost any given moment for a long time in the future. Physicists have found that such predictions are not possible with electrons in atoms. The idea of electrons traveling in fixed orbits like planets became a hindrance to understanding how atoms are really put together. The quantum theory, which replaced Bohr's planetary model, is conspicuous in the absence of a mechanical model for atomic structure. Physicists now claim it is impossible to develop any model of atomic structure in physical terms. That is, they hold that no model (as we have been using the term) is possible. Most physicists now view atomic structure in mathematical and probabilistic terms.

4. Membranes are not passive "walls" acting only as containers or filters. They are directly involved in many chemical reactions occurring at the cell surface (both the inner and the outer side of the plasma membrane) as well as within cells. The proteins that make up a large part of all membranes are not merely structural units. They also have a distinct function, as illustrated with rhodopsin. As the chief photochemical pigment in most vertebrate (and many invertebrate) eyes, rhodopsin's biochemical role is to convert the energy of photons of light into chemical energy to serve as the nerve cell impulse. This chemical process is closely related to the change in position of the rhodopsin with respect to the lipid bilayer of the membrane in which it "floats." Structural changes in the membrane proteins mirror functional changes in the cell's biochemistry.

By virtue of the new view of membranes that has emerged in the last few years, biologists are beginning to understand a large number of biological phenomena in terms of their relation to membranes. Processes such as the immune response, cancer, photosensitivity, active transport, and embryonic development, as well as many specific biochemical pathways, are all coming to be understood as membrane or cell surface phenomena. The field of membrane research promises to gain increasing importance in the future.

With the newer model of membranes now in mind, let us return to the problem of how materials enter and leave cells. At the beginning of this section we discussed the movement of molecules and ions in terms of the general principles of diffusion. At that time we defined diffusion as the tendency of substances to move from an area of greater to one of lesser concentration. Of course diffusion can occur without passage across membranes, but here we are concerned with cells, where membranes are involved. Diffusion of dissolved molecules and ions across a membrane is known as **dialysis;** diffusion of water across a membrane is called **osmosis.** The presence of the membrane affects the process of diffusion in both these cases.

Osmosis

Osmosis is the passage of a solvent through a semipermeable membrane from a region of greater to a region of lesser concentration. In a living organism, the solvent is usually water and the semipermeable membrane is usually that of the cell. Since water molecules are small, the cell is somewhat limited in the amount of direct control it can impose on their passage across the membrane.

Figure 4.43(a) illustrates the diffusion of water and sugar molecules (sucrose) across a membrane permeable to both molecules. At the start, water is present on both sides of the membrane, but sucrose is present only on the right. Since there are only water molecules on the left side of the membrane, while the right side has both water and sucrose molecules, the greater concentration of water is on the left side of the membrane. Thus water molecules will tend to move through the membrane from the left side to the right by simple diffusion, and sucrose molecules will move from the right to the left. Each compound is passing from its area of greater to its area of lesser concentration. As time passes, the concentration of sucrose molecules on the left will increase until it balances the concentration on the right. At the same time, the concentration of water molecules on the right will increase until it balances the concentration on the left. Eventually, the ratio of sugar molecules to water molecules will be the same for both sides of the membrane. After this point has been reached, both sucrose molecules and water molecules will continue to pass through the membrane, but the concentration on both sides will remain constant. The system will have attained dynamic equilibrium for both kinds of molecules.

Now consider the situation diagrammed in Fig. 4.43(b). In this case, the membrane is semipermeable. In other words, this membrane allows free passage of the small water molecules but restricts

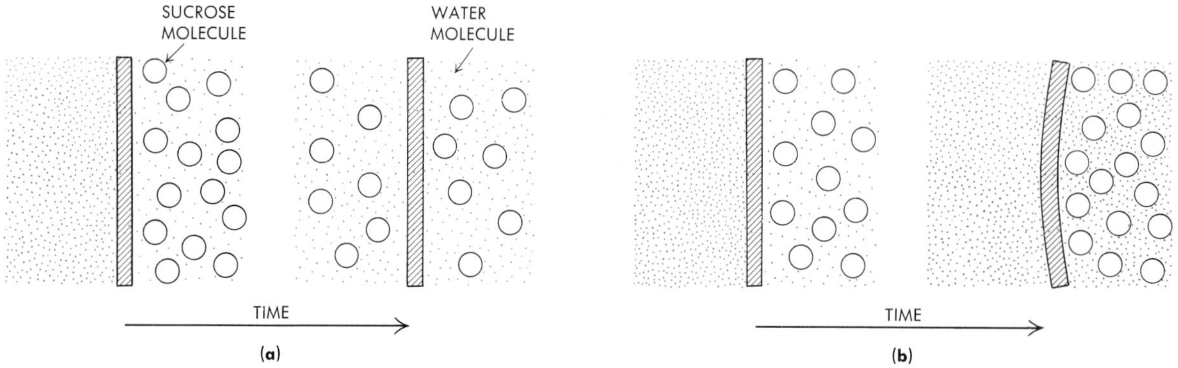

Fig. 4.43 *(a) Diffusion of sucrose and water molecules through a membrane permeable to both. Eventually, the rates of each kind of molecule passing through the membrane from either side become equal. (b) A membrane impermeable to sucrose but permeable to water. The water molecules pass from left to right, since their concentration is greater at left. The resulting osmotic pressure exerts a force on the membrane.*

the passage of the larger sucrose molecules. At the start, sucrose is in greater concentration on the right and water is in greater concentration on the left. In this system, the sucrose molecules cannot pass from an area of greater to lesser concentration; the membrane is impermeable to them. The water molecules, on the other hand, can pass freely through the membrane. Thus they pass from an area of greater to an area of lesser concentration, i.e., from left to right. Of course, as a result of random motion, water molecules are also passing from right to left. However, since the concentration of water molecules is greater on the left side of the membrane than on the right, far more molecules pass in from left to right. The passage of these water molecules is an example of osmosis.

Figure 4.43(a) showed a system that attained dynamic equilibrium. The system shown in Fig. 4.43(b) could never attain that condition, and here is why: The concentration of water on the left side is 100%. Now, the sugar molecules must stay on the right side, since they cannot pass through the membrane; therefore, no matter how many water molecules move from the area of greater concentration (the left) to the area of lesser concentration (the right), the concentration of water on the right can never equal that on the left (that is, 100%).

In the diffusion of water across the membrane in Fig. 4.43(b), do the molecules pass from left to right at the same rate from the beginning of the experiment to its end? No. The rate of osmosis *decreases* with time; the longer the process goes on, the slower the rate at which it occurs. Thus, if 25 units of water molecules pass through the cell membrane the first minute, 15 may pass through the second minute, 8 the third minute, and so on. The passage of water through the membrane constantly changes the relative concentrations of both water and sugar molecules on the right side. Since osmotic rate is directly proportional to concentration differences between the substances involved, the osmotic rate is constantly changing. At the beginning of the experiment represented in Fig. 4.43(b), there are many water molecules passing from left to right, and we can say that the system has a high osmotic rate.

On first thought, one might well be tempted to view osmosis as simply a special case of diffusion by which water (or a solvent) passes through submicroscopic pores in semipermeable membranes. Certainly it is true that the direction and equilibrium of osmotic processes can be accurately predicted by an hypothesis that interprets osmosis in this manner. Yet careful measurements, using isotopically labeled H_2O^{18}, uncover a contradiction to this hypothesis: water traverses a porous semipermeable membrane faster than would be predicted, even if the theoretical maximum rate of diffusion is assumed to prevail. Current hypotheses concerning osmotic flow are based on concepts of bulk flow resulting from hydrostatic pressure differences working in conjunction with diffusion. Thus even the seemingly "simple" processes occurring in living matter often turn out to be far from simple when subjected to more thorough analysis.

Dialysis

Dialysis is the diffusion of a dissolved substance through a differentially (semi-) permeable membrane. Dialysis takes place as long as the membrane is permeable to the molecule or ion in question. If the membrane is impermeable to a molecule, the molecule cannot diffuse across it. Obviously, then, the properties of the membrane are crucial in determining what dissolved substances pass through (all biological membranes are permeable to water).

Membranes are not passive agents, however, like a wall with doors or other openings. Study of living cells has revealed that membranes play a crucial role not only in selecting what molecules will be allowed to pass, but in actually *aiding* the physical movement of certain substances into or out of the cell by the processes of **facilitated transport** and **active transport,** which are associated with living cell membranes. These forms of transport provide a good example of interrelationship between structure and function in biological membranes.

Facilitated and active transport have many features in common. They also have some important differences. Their common features include: (1)

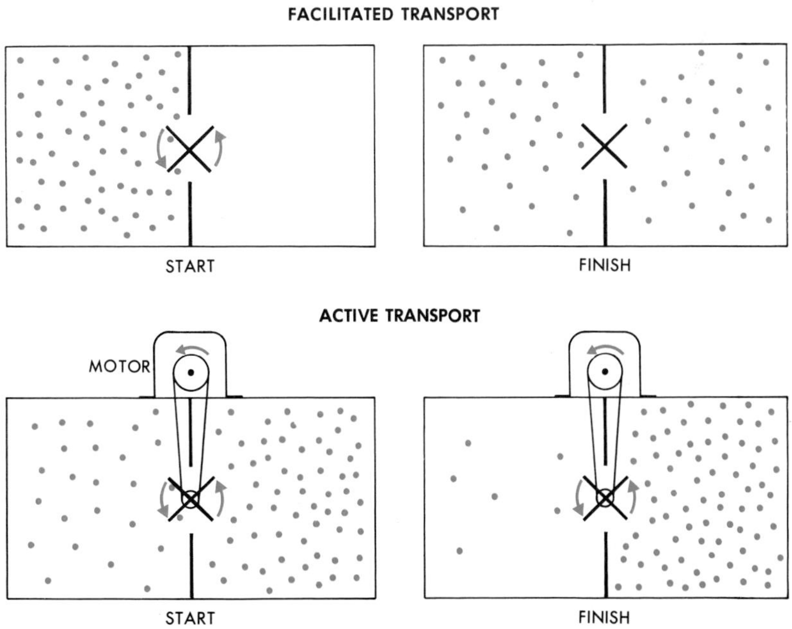

Fig. 4.44 *General model showing the chief differences between facilitated and active transport. Active transport involves the expenditure of energy (as in running the motorized paddlewheel in the diagram) to move substances against a concentration gradient. Facilitated transport does not involve expenditure of energy on the cell's part.*

movement of molecules and ions across membranes at rates faster than would occur by simple dialysis alone, and (2) a high degree of specificity; both are selective in terms of which molecules they transport. Their difference lies mainly in the fact that active transport requires the expenditure of energy by the cell, whereas facilitated transport does not. Active transport can actually move substances across a membrane *against* a concentration gradient (that is, from an area of lower to one of higher concentration), whereas facilitated transport cannot. Two models showing this distinction between facilitated and active transport appear in Fig. 4.44.

Facilitated transport. A number of membrane-bound transport enzymes have been identified that aid in moving specific molecules across cell membranes, particularly plasma membrane. For example, glucose enters some animal cells by the aid of specific enzymes at the cell surface. No energy is expended, so the process is one of facilitated transport. Yet sugars similar in size, structure, and properties to glucose are not transported in the same way. This indicates the highly specific nature of facilitated transport. The specificity is found to lie in the transport enzymes. Active transport also involves highly specific membrane proteins. In both cases the membrane proteins may act as carriers, facilitating the movement of molecules from one side of the membrane to another.

A hypothetical model has been developed to explain both facilitated and active transport in terms of membrane-bound carrier proteins. Figure 4.45 shows two generalized schemes for facilitated and active transport. Consider facilitated transport first. A molecule to be transported (S) must first come into contact with the membrane—specifically with a carrier protein component of the membrane. Carrier proteins are of the "intrinsic" type; that is, they are partially immersed in the lipid layer. When the specific transportable S molecule contacts a carrier protein, the latter binds to it much like any enzyme binds to its substrate. The binding process causes a conformational shift in the protein, such that the protein may sink deeper into the lipid bilayer, or perhaps even flip completely over, undergoing a 180° change in position. The result is that the S molecule being transported is now on the inside of the membrane rather than the outside. The carrier protein simply ushered the S molecule through the lipid region of the molecule. Once on the other side of the membrane, the S molecule is released in a manner possibly similar to the release of a substrate from its enzyme after a chemical reaction.

Active transport. Active transport (Fig. 4.45b) may follow the same general pattern. In fact, it may even be that the same carrier proteins are involved in both facilitated and active transport. It is known, for example, that such sugars as glucose and galactose can be moved by both facilitated and active transport. Hence, it is likely that their transport may involve the same carrier molecule. However, since active transport can move molecules across membranes against a concentration gradient, the process has several differences from facilitated transport, although the binding and translocation steps may be very similar. If the concentration of S is greater inside the cell than outside, the release step is different. With a high concentration of S inside, the new S molecule would have a tendency to stay bound to the carrier. Energy is used in the form of the fuel molecule adenosine triphosphate (ATP) to remove S from the carrier protein. This frees the carrier, which can then pick up another S molecule at the outer surface of the membrane. In this way, the cell does work to increase the concentration of substances inside it to a degree considerably greater than the outside concentration.

Table 4.1 shows the enormous differences in concentration that can be achieved through active transport. The table illustrates several points. Active transport is selective—potassium (K^+) and calcium (Ca^{+2}) are selectively transported into the cell

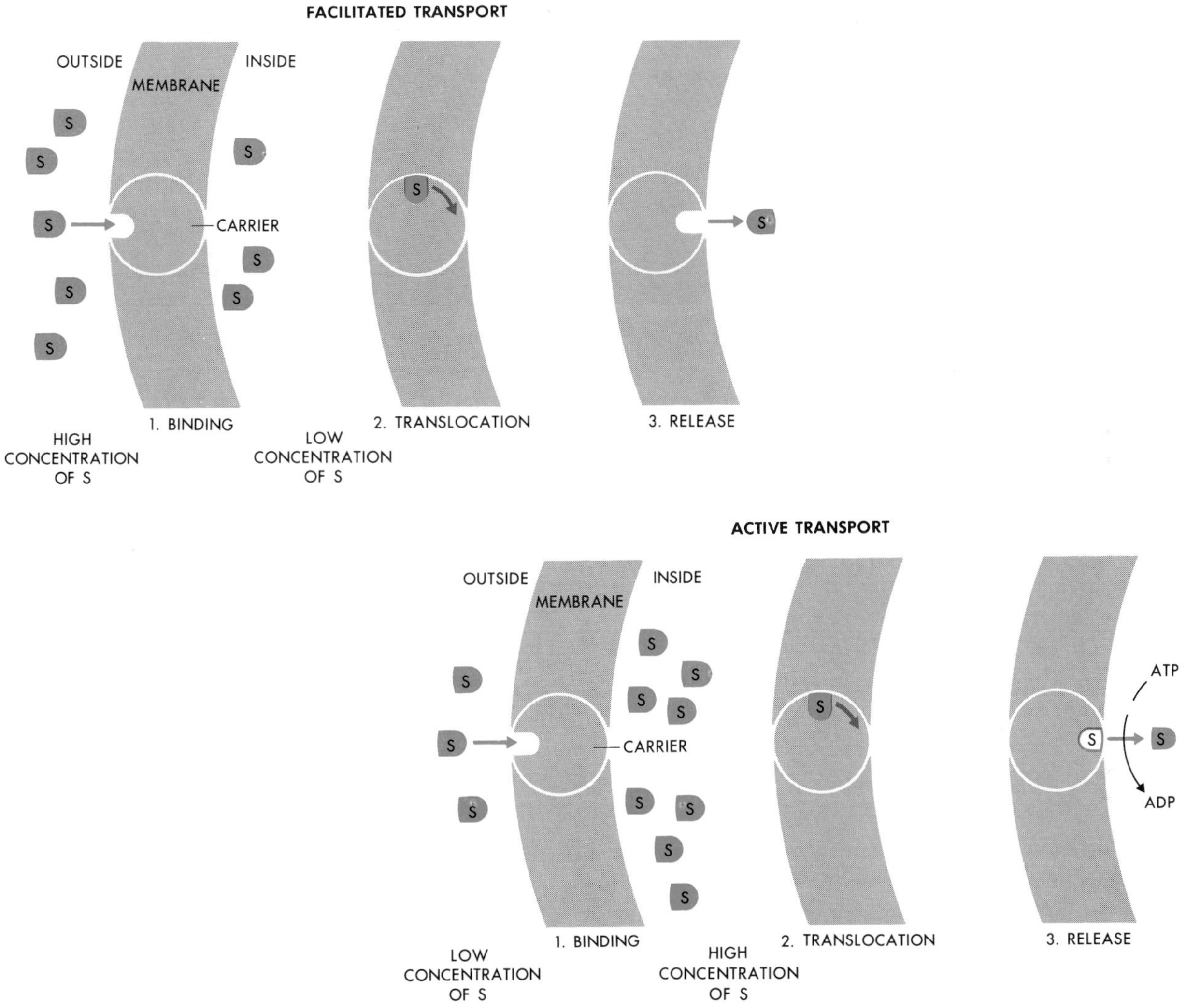

Fig. 4.45 *Molecular model for facilitated and active transport. In active transport, energy must be expended in the form of molecules of ATP (which are converted to the less energy-rich compound ADP) to split S from the carrier. This is because from the beginning there is a higher concentration of S inside the cell than outside. (After A. G. Loewy and P. Siekevitz,* Cell Structure and Function, *2d ed. New York: Holt, Rinehart and Winston, 1969, p. 443.)*

Table 4.1
Ionic Concentrations in the Red Blood Cell and the Surrounding Blood Plasma

	Ions, concentration in milliequivalents per liter	
	Red blood cell	Blood plasma
K^+	150	5
Na^+	26	144
Cl^-	74	111
Ca^{+2}	70.1	3.2

against very high concentration gradients, whereas sodium and chloride (Cl^-) are not. But, equally important, note that sodium and chloride are actively transported *out* of the cell. Active transport works in both directions, bringing substances into the cell and moving substances out of the cell—in both cases against a concentration gradient.

Active transport is sometimes referred to as a cell pumping system (for example, the "galactose pump" or the "sodium pump"). The term "pump" should not be taken too literally. It does signify, however, the fact that work is being done and that something is being moved in an "uphill" direction.

The molecular details of both facilitated and active transport are very poorly understood. That carrier proteins in the membrane exist is clear. Exactly how they work involves a great deal of speculation and model-building. Most of the ideas expressed in the models in Fig. 4.45, for example, have relatively little direct evidence behind them. Using the model it is possible to make certain predictions: for example, that applying cyanide to cells (which stops all energy-generating reactions in cells, and hence interferes with production of ATP) would soon bring a stop to all active transport but would not affect facilitated transport. The predictions are borne out. Soon after application of cyanide to red blood cells, the ionic concentrations inside and outside come to equilibrium. Yet much research still needs to be done in attempting to isolate and characterize specific carrier proteins and to understand the actual process by which they translocate molecules across the membrane. The field of membrane research is indeed an open and exciting area for future investigation.

SUMMARY

1. The cell "theory" maintains that (a) cells are the structural units of virtually all organisms; (b) cells are also the functional units of all organisms and thus the seat of both health and disease; (c) cells arise only from preexisting cells and are not spontaneously generated from inorganic material; (d) cells contain hereditary material (nucleic acid) through which specific characteristics are passed on to the next generation of cells.
2. All cells, single-celled organisms or cells in multicelled organisms, face certain common problems: they must carry out metabolism (gain and use energy), reproduce, transport substances in and out of the cell, quite frequently show some form of motion (intracellular as well as overall cell motion), exhibit growth and development, and maintain themselves in the face of constantly changing external conditions. It is not surprising to find that most cells have many of their parts, such as molecules and organelles, in common.
3. Biologists generally recognize two major types of cells: eucaryotic (possessing a true nucleus surrounded by a nuclear membrane) and procaryotic (lacking a true nucleus, with no nuclear membrane). The cells of all protozoa and most types of algae and the higher plants and

animals are eucaryotic. The cells of bacteria, blue-green algae, and mycoplasmas are procaryotic. Viruses are not cellular in nature, being composed only of nucleic acid surrounded by a protective coat of protein. Viruses are able to reproduce themselves inside living cells. Outside living cells, they behave very much like the crystals of any large molecule. Hence they cannot be grouped with either eucaryotes or procaryotes.

4. Tools for studying cell structure include the microtome (for preparing thin tissue slices), the light microscope, and the electron microscope. Light microscopes can only magnify to a maximum of about 2000 times; electron microscopes can magnify up to 1,000,000 times. Using the electron microscope, cell biologists have been able to gain an accurate picture of the structure of many very small cell components.

5. Techniques for studying cell function include: (a) correlating structure with function; for example, by noting that ribosomes appear to be heavily concentrated in cells that secrete proteins one could conclude that ribosomes are involved in protein synthesis; (b) breaking cells apart, separating out the various components, and performing biochemical tests on each fraction; this process involves use of the centrifuge and the ultracentrifuge to separate one fraction from another.

6. Eucaryotic cells contain a number of structures, or organelles, each of which has its own particular function(s):

   a) Cell membrane. This is a lipid bilayer in which a variety of proteins float. The cell membrane functions partly to determine what substances go in and out of the cell, and partly as a surface structure on which many reactions are localized. Membranes in living cells are physiologically very active, not the passive "barrier" structures that have been described in the past. Because of the lipid bilayer structure, membranes have a distinct polarity: polar regions forming each surface of the membrane (glycerol and phosphate regions of the lipid molecule), with a nonpolar region in between (fatty acid chain region of the lipid molecules). This polarity determines the arrangement of proteins in the membrane as well as its structural and chemical properties.

   b) Microtubules. These are tubules composed of protein subunits, running throughout the cytoplasm in many cell types. Microtubules appear to function in relation to various cell movements: pinching in of a cell, cytoplasmic streaming in certain types of plant cells, and especially the motion of cilia and flagella.

   c) Centriole. The centriole is a structure consisting of two sets of microtubular units resembling barrels, placed at right angles to each other. Centrioles are found in most cells except those of many kinds of higher plants. They appear to have a role in the process of cell division.

   d) Microfilaments. These are bundles of protein threads found in certain cells. They appear to be related in some way to certain types of cell movements, such as the contraction of heart muscle.

   e) Mitochondria. Mitochondria are sausage-shaped structures located throughout the cytoplasm of all eucaryotic (but not procaryotic) cells. They are usually found in areas of the cell showing the greatest metabolic activity. Mitochondria have an outer membrane surrounding an inner membrane that is folded inward to form partitions called cristae that increase the inner surface area. Enzymes involved in the respiratory process are bound to the inner membrane. Mitochondria are the centers of capture of usable energy from oxidation of foodstuffs. Within the mitochondria, energy-rich molecules of adenosine triphosphate (ATP) are generated from energy released when glucose and other substances are oxidized.

f) Chloroplasts. These are a kind of plastid found in the cells of all green plants from algae to oak tree. Chloroplasts are bounded by an external membrane and contain numerous internal membranes called lamellae. Lamellae are stacked at certain intervals into compact units called grana. Chlorophyll molecules are bound to lamellae. Chlorophyll absorbs light and converts it to a useful form of energy (ATP) for synthesizing carbohydrates. The process of energy capture and carbohydrate synthesis is known as photosynthesis; chloroplasts are the site of photosynthesis in all green plants.

g) Endoplasmic reticulum. This consists of a series of membrane-lined channels running through the cytoplasm of most cells. The endoplasmic reticulum facilitates transport within the cell; its membranes also serve as sites for many chemical reactions. There are two types of endoplasmic reticulum: rough endoplasmic reticulum, in which ribosomes are bound to the membranes; and smooth endoplasmic reticulum, in which there are no ribosomes attached to the membranes.

h) Ribosomes. These small structures are the site of protein synthesis in all cell types, procaryotic as well as eucaryotic. In eucaryotic cells ribosomes are usually attached to endoplasmic reticulum; in procaryotic cells they are freely dispersed in the cell cytoplasm. Ribosomes consist of two subunits of different densities, which can be separated by chemical means (such as changing the ionic concentration of the medium).

i) Golgi complex. The Golgi complex is a cluster of flattened, parallel sacs found within the cytoplasm of many cell types. It appears to be involved in the transport, packaging, and excretion of materials, particularly lipids.

j) Lysosomes. These are spherical or rod-shaped organelles surrounded by a unit membrane boundary. Lysosomes contain a variety of enzymes, mostly those that break down ("digest") macromolecules such as proteins, carbohydrates, and lipids. Damage to lysosomes can lead to release of their powerful enzymes into the cell cytoplasm and hence to generalized degradation.

k) Nucleus. The nucleus is the center of hereditary processes in the cell. The nucleus contains nucleic acid as deoxyribonucleic acid (DNA) and ribonucleic acid (RNA). DNA is passed on from generation to generation and transmits the cell's hereditary information. RNA is involved in various aspects of translating the information contained in DNA into protein structure. In the nucleus of eucaryotic cells, DNA is organized into a network of filaments called chromatin. Just before cell division, the chromatin strands condense into visible, darkly staining rods called chromosomes. Procaryotes have chromatin and even visible chromosomes, often localized in a central region of the cell. They have no distinct nucleus.

7. All eucaryotic cells undergo the processes of mitosis and meiosis. Mitosis involves replica-

tion of the existing set of chromosomes and subsequent division of these chromosomes equally between the two daughter cells. Mitosis preserves the chromosome number from one cell generation to the next. Meiosis is a process in which the chromosome number is reduced by half. The process occurs in such a way that each daughter cell gets one complete set of chromosomes. Meiosis is sometimes called the reduction division. Meiosis is particularly involved in the sexual phase of reproductive cycles in one-celled and many-celled organisms. Meiosis prevents the repeated doubling of the chromosome number that would otherwise result from sexual reproduction.

8. Procaryotes reproduce by a process known as fission, which is asexual. Fission is analogous to mitosis, since it involves only a duplication of the nucleic acid and division of the two copies between daughter cells. The process is less complex than mitosis. Procaryotes also engage in sexual reproduction, during which they exchange genetic material and subsequently undergo a division similar to meiosis.

9. Cells transport materials (water, ions, and small and large molecules) in and out by a variety of different processes. These include:

   a) Diffusion. A physical process in which materials tend to move from an area of greater to one of lesser concentration. This can occur passively across cell membranes.

   b) Osmosis. A physical process in which water diffuses across a semipermeable membrane. The cell membrane appears to be relatively passive in osmotic flow.

   c) Bulk transport. The process by which large, macroscopic portions of liquids or solids are moved into or out of cells. The membrane is intimately involved in bulk transport. Bulk transport involves at least two distinct kinds of processes: (i) Phagocytosis: the taking in (engulfing) of solid material by formation of a membrane enclosure around it, which becomes incorporated into the cytoplasm as a vacuole. (ii) Pinocytosis: the taking in of bulk quantities of water (or liquid). Both phagocytosis and pinocytosis appear to occur in the elimination of materials from the cell as well.

   d) Facilitated transport. The movement of materials across the cell membrane from areas of higher to areas of lower concentration, but at a rate faster than could occur by diffusion alone. Facilitated transport does not require expenditure of cellular energy. It is particularly involved in the transport of sugars into the cell.

   e) Active transport. The movement of materials across the cell membrane from areas of lower to areas of higher concentration, against a concentration gradient. This process involves the expenditure of energy by the cell. Both facilitated and active transport involve membrane-bound carrier proteins as agents for translocating molecules across the membrane.

10. The structure of cells serves to facilitate their biochemical functions. Most biochemical reactions in cells are localized: either on membrane surfaces, or in areas (such as lysosomes or vacuoles) surrounded by membranes. Cells are not simply bags of molecules floating loosely about; they are highly structured entities in which most types of molecules have distinct places. It is partly for this reason that cells can carry out so many different chemical reactions with such efficiency.

EXERCISES

1. Discuss briefly the functions of the following parts of a cell: (a) plasma membrane, (b) cell wall, (c) ribosomes, (d) endoplasmic reticulum, (e) mitochondria, (f) centriole, (g) plastids, (h) lysosomes.
2. What is significant about the fact that most cells contain almost all the same structures?
3. Describe the plasma membrane of a living cell. In what way is the structure of the membrane closely correlated with its function?
4. If electron micrographs show that mitochondria are grouped around a particular structure or region of the cell, what might you infer is happening in this region?
5. Table 4.2 shows the results of a series of differential centrifugations of a cell homogenate. At each step the pellet (or, in the case of the last step, the final supernatant) is resuspended and studied for its biochemical and structural properties (see last column on the right). Study this table and answer the questions posed below. Answers should be as brief as possible to get the point across; one- or two-word answers will suffice in many cases. Please number each part carefully according to the numbering system in the table.

   For the pellet fraction whose properties are described in the last column of Table 4.2, list which organelle or organelles (or parts of organelles) might be ex-

Sample	Centrifugation speed (× force of gravity)	Time centrifuged	Physiological and/or chemical characteristics of pellet fraction resulting from centrifugation
1. Cell homogenate	900 g	10 min	Contains much DNA and RNA, and *large* lipid-containing fragments.
2. Supernatant from No. 1	15,000 g	20 min	Is able to oxidize glucose and other carbohydrate intermediates. Consumes oxygen and produces CO_2. Contains particles that, if ruptured, are capable of digesting away most other cell components.
3. Supernatant from No. 2	25,000	30 min to 1 hr	Lipid and narrow tubular-shaped membranes; sometimes has ability to produce full-scale proteins from amino acids.

Sample	Centrifugation speed (× force of gravity)	Time centrifuged	Physiological and/or chemical characteristics of pellet fraction resulting from centrifugation
4. Supernatant from No. 3	50,000	30 min to 1 hr	Some lipid and tubular shaped membranes; seldom shows ability to produce proteins from amino acids.
5. Supernatant from No. 4	100,000	2 hr	Small lipid fragments
6. Supernatant from No. 5	[No centrifugation. Supernatant is studied directly.]		Contains particles with RNA and protein. Able to produce proteins from amino acids under proper conditions.

pected to appear in that fraction. Explain briefly what evidence leads you to identify particular organelles with particular fractions. In general, you should make use of two kinds of data supplied in the table: (1) the relative densities of the organelles as indicated by the centrifugation speeds at which they settle out, and (2) the biochemical, physiological, and structural information given in the last column on the right. It is possible for an organelle, or fragments of large organelles, to appear in more than one fraction. Taking Fraction 1 as an example, your answer might run like this:

1. *Probably nuclei (containing nucleoli), and large fragments of plasma or nuclear membranes. Nuclei are very large and are among the densest bodies in the cell; they contain DNA (in chromosomes) and RNA in nucleoli. Cell membranes contain much lipid, and large fragments have a relatively high density.*

6. Distinguish between osmosis and diffusion.
7. Explain what is meant by calling a membrane permeable, semipermeable, or impermeable.
8. A particular membrane is said to be permeable to water and sodium chloride, yet impermeable to glucose molecules. Explain how this might be possible.

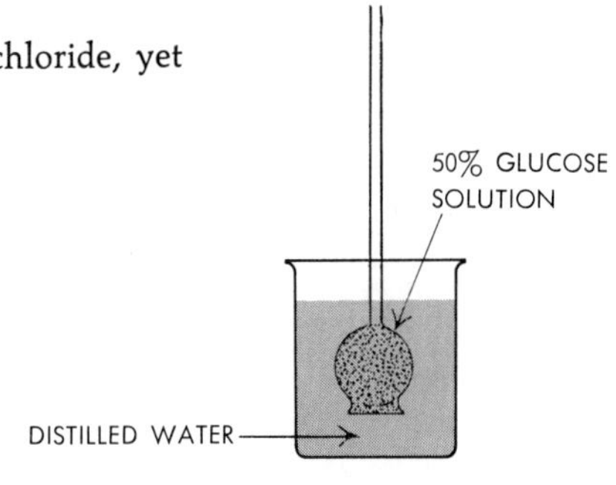

Figure 4.46

9. Figure 4.46 is a diagram of an osmometer similar to those used to demonstrate osmosis. At the beginning of an experiment there is a 50% glucose solution within

the membrane, which is impermeable to glucose. Draw a graph representing the rate of osmosis in this system, indicated by the rate at which the water climbs up the tube.

10. Which graph in Fig. 4.47 would most accurately represent change in osmotic rate over time in a semipermeable, nonliving, enclosed membrane system? Explain your choice.

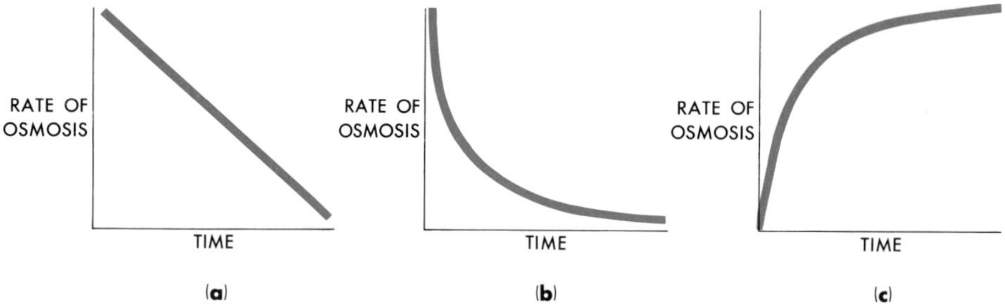

Figure 4.47

11. Indicate on the graph you drew for Exercise 9 what the line would look like if the glucose concentration within the membrane were only 10% at the beginning of the experiment.
12. A snail can be killed by throwing salt on it. You get very thirsty (indicating that your body cells are low on water) after eating a salty meal. In growing living cells in test tubes, physiologists are very careful to use a medium known as Ringer's solution, in which the salt concentration is exactly that found in body and cellular fluids. Explain all three of these facts in terms of osmosis and dialysis.
13. Plasmolysis is a term used to refer to the collapse of a cell from excessive loss of water. Describe the conditions that probably exist if a cell undergoes plasmolysis when it is placed in a solution.

For Exercises 14–18, read the following directions carefully, and refer to Fig. 4.48.

Necessary data: At the beginning of the experiment the solutions in the two arms of the tube are as pictured. They are separated at the bottom of the tube by a differentially permeable membrane. The volumes on either side of the tube are the same, and thus the level of the liquid in both arms is also the same. The apparatus is allowed to stand for several days.

14. During the experiments what will happen to the water level?
    a) It will rise in side A, since water (not the substances in it) will tend to pass from its area of greater concentration to its area of lesser concentration.
    b) It will rise in side A, since the water (not the substances in it) will tend to pass from its area of lesser concentration to its area of greater concentration.
    c) It will remain the same, because atmospheric pressure is equal on both sides of the system.
    d) None of the above.

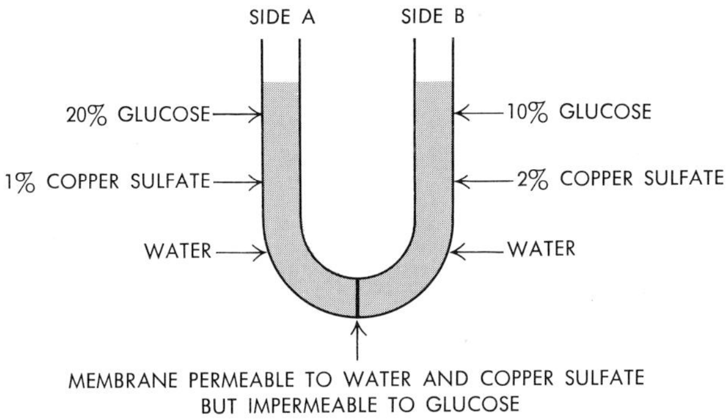

Figure 4.48

15. What will happen to the glucose solution of side A?
    a) It will become more concentrated, and that on side B will become less concentrated, since water moves from A to B.
    b) It will become more concentrated, and that on side B will become less concentrated, since water passes from B to A.
    c) It will become less concentrated, and that on side B will become more concentrated, since water passes from A to B.
    d) It will become less concentrated, and that on side B will become more concentrated, since water passes from B into A.
16. Which of the following best describes what will happen to the copper sulphate as the experiment proceeds?
    a) There will be no passage of copper sulphate because solutes do not go through semipermeable membranes.
    b) There will be a slight passage of this substance, but the passage will be restricted by the size of the pores in the membrane.
    c) There will be a slow movement, since the concentrations are nearly equal.
    d) There will be no passage, since copper sulphate is insoluble.
17. Osmotic pressure will be greatest on:
    a) Side A at the beginning of the experiment.
    b) Side A at the end of the experiment.
    c) Side B at the beginning of the experiment.
    d) Side B at the end of the experiment.
18. Which of the following substances show(s) the phenomenon of dialysis in this experiment? (a) water (b) glucose (c) copper sulphate.

The experiment shown in Fig. 4.49 was performed on the multinucleated protozoan *Pelomyxa*. For the purpose of this experiment, the effect of many nuclei is no different from that of a single nucleus. The amount of radiation in all cases was equal to a lethal dose.

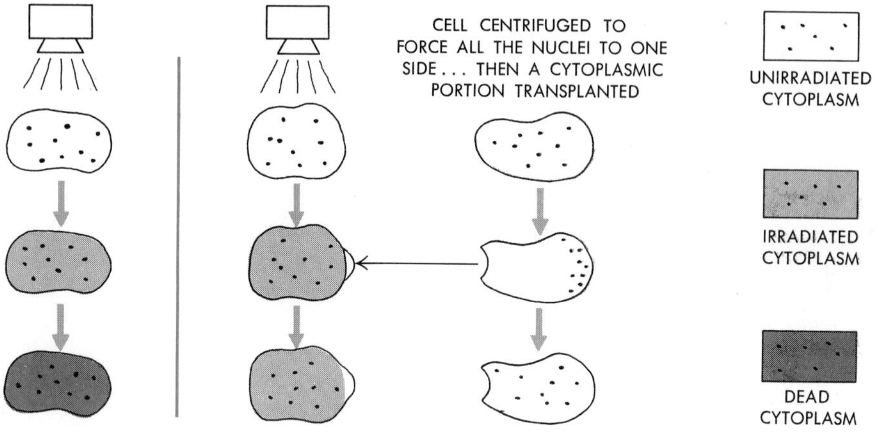

Figure 4.49

Examine separately each statement in Exercises 19–28. If the statement is possible on the basis of the experimental results given, write A opposite that number on your paper. If the statement cannot be evaluated on the basis of the experimental results, write B. If the statement is contradicted by the evidence given, write C.

19. Under normal conditions, radiation (a dosage equal to the lethal dose) kills a cell.
20. Radiation undoubtedly affects the nucleus of a cell more than the cytoplasm.
21. The cells on the left-hand side of Fig. 4.49 died because the nucleus was unable to govern protein synthesis.
22. The mechanical effect of transplanting a portion of cytoplasm could counteract the effects produced by radiation.
23. Unirradiated cytoplasm will enable the cell to survive in the presence of radiated cytoplasm.
24. The transplanted, unirradiated cytoplasm may produce a repair substance that enables the nucleus to recover from the radiation.
25. On the right-hand side of Fig. 4.49, the critical damage from the radiation may have been restricted to the cytoplasm.
26. The transplanted portion of cytoplasm may have contained all the necessary components for the continuing existence of the cell.
27. Centrifuging a cell causes its death.
28. Radiation does not kill cells.
29. Offer an hypothesis to account for the results shown in Fig. 4.50. Consider closely the comparative effects of the radiation (which was of equal dosage in all cases) on the nucleus and the cytoplasm. Does the radiation affect the cytoplasm and the nucleus equally? Explain your hypothesis completely.

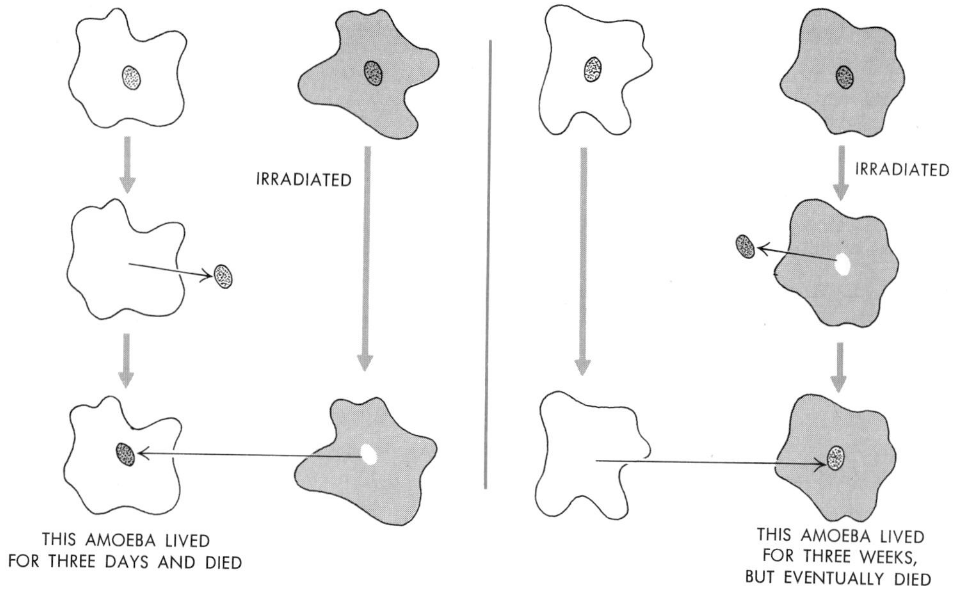
IRRADIATED
IRRADIATED
THIS AMOEBA LIVED
FOR THREE DAYS AND DIED
THIS AMOEBA LIVED
FOR THREE WEEKS,
BUT EVENTUALLY DIED

Figure 4.50

SUGGESTED READINGS

A number of useful paperback and hardback books now exist summarizing recent findings in cell biology. Only a few are listed here as guides for more detailed information on many aspects of cell structure and function.

Avers, Charlotte J., *Basic Cell Biology*. New York: Van Nostrand, 1978. A complete, up-to-date, and thorough introduction to cell biology and biochemistry. Useful as a simplified reference source.

Branton, D., and D. W. Deamer, *Membrane Structure*. New York: Springer-Verlag, 1972. An advanced monograph on membrane structure including experimental evidence for many conclusions.

De Robertis, E. D. P., *Cell Biology*. Philadelphia: W. B. Saunders, 5th ed., 1970. A comprehensive textbook of cell structure and function.

Dyson, Robert D. *Cell Biology: A Molecular Approach*. Boston: Allyn and Bacon, 1974. A well-illustrated, authoritative account of many aspects of cell structure and function. It is quite complete and assumes some background on the part of the reader.

Fawcett, D. W., *The Cell, Its Organelles and Inclusions: An Atlas of Fine Structure*. Philadelphia: W. B. Saunders, 1966. An outstanding collection of electron micrographs, arranged by organelle type, with thorough descriptions of each organelle's functions.

Kimball, John W., *Cell Biology*. Reading, Mass.: Addison-Wesley, 2d ed., 1978. This paperback introduces the general principles of cell and molecular biology.

Loewy, Ariel, and Philip Siekevitz, *Cell Structure and Function*. New York: Holt, Rinehart and Winston, 2d ed., 1969. Contains good discussions of selected topics in cell structure and function; presupposes relatively little background.

Novikoff, A. B., and E. Holtzman, *Cells and Organelles*. New York: Holt, Rinehart and Winston, 3rd ed.,

1977. Simple and well-presented descriptions of cell structures and functions. Contains a large section on nerve cells as membrane models.

Porter, K. R., and M. Bonneville, *An Introduction to the Fine Structure of Cells and Tissues.* Philadelphia: Lea & Febiger, 4th ed., 1973. An excellent collection of electron micrographs of animal cells, accompanied by detailed commentaries.

Thomas, L., *The Lives of a Cell: Notes of a Biology Watcher,* New York: Viking, 1974. Thoughtful and stimulating short essays that broaden the reader's appreciation of all living things.

Wolfe, Stephen, *Modern Cell Biology.* Belmont, California: Wadsworth Publishing, 1972. Not as detailed or authoritative as the De Robertis text but handsomely illustrated.

Articles on various cell components are listed below for further reference:

Bretscher, Mark, "Membrane structure: some general principles." *Science* 181 (August 1973), p. 622. A good review article, somewhat detailed and technical.

Capaldi, Roderick A., "A dynamic model of cell membranes." *Scientific American,* March 1974, p. 26. An interesting and up-to-date treatment of membrane structure.

De Duve, Christian, "The lysosome." *Scientific American,* May 1963. An older but well-written article by a pioneer in the study of lysosomal structure and function.

Loewenstein, Werner R., "Intercellular communication." *Scientific American,* May 1970, p. 30. A discussion of the mechanism by which specific signal molecules travel between adjacent cells of a tissue to enable all the cells of that tissue to act in concert.

Mahler, Henry S., "Mitochondria: molecular biology, genetics, and development," No. 1 in the Addison-Wesley Module in Biology (Reading, Mass.: Addison-Wesley, 1973). A 50-page booklet summarizing much of the current information available about mitochondria, including structure and biochemical function.

Morowitz, Harold J., and Mark E. Tourtellotte, "The smallest living cells." *Scientific American,* March 1962, p. 117. A discussion of PPLO, a mycoplasma.

Satir, B., "The final steps in secretion," *Scientific American,* October 1975, p. 28. A discussion of the interaction of cell organelles in the synthesis and secretion of various substances.

Satir, Peter, "How cilia move." *Scientific American,* October 1974, p. 44. An interesting article on the role of microtubules in ciliary movement. An essay in the interrelationship between structure and function.

Stein, Wilfred and William Lieb, "How molecules pass through membranes." *New Scientist,* January 10, 1974, p. 77. A simplified review article dealing largely with facilitated and active transport.

Swanson, Carl P., and Peter L. Webster, *The Cell.* Englewood Cliffs, N.J.: Prentice-Hall, 4th ed., 1977. A comprehensive, yet simplified introduction to cell biology. Written for beginning biology students at the college level.

FIVE

CELL METABOLISM I: ENERGY RELEASE AND METABOLIC PATHWAYS

5.1 Introduction

Cells are chemical machines. They consume fuels (e.g., sugars) and release waste products (e.g., carbon dioxide and water). The energy they extract from their fuel is used to perform **work,** defined as the force required to move matter through a distance. Work is involved whether the matter being moved is as small as a proton or as large as a rock or an automobile. All cells perform general types of work, such as building up or synthesizing molecules, or actively transporting various substances across cell membranes. In addition, some cells also do specialized work, such as physical movement (e.g., the movement of an amoeba or the contraction of a muscle cell) or electrochemical work (e.g., the conduction of an impulse by a nerve cell).

Like machines, cells release energy from fuels in an orderly way to accomplish specific tasks. In both cases, the release of energy is accomplished by breaking down fuels in a process known as "oxidation." But here the comparison between the cell and the machine begins to fail. When looked at more closely, the details of energy extraction and use show the cell to be a far more complex and delicate unit than any existing machine. In the long run, the differences between cell and machine are far more important than their similarities in adding to our understanding of the nature of life.

To understand these differences, consider a cell and an internal combustion engine, such as an automobile engine. The comparison will be limited here to just two points: (1) the *mechanism* and (2) the *rate* of energy release. In the automobile engine, energy is released from the fuel (gasoline) by a series of rapid, single-step oxidations, each of which is a single explosion in a cylinder. All the available energy from a given quantity of fuel is released in one step at each ignition. The power from these small explosions pushes the pistons up and down, generating forward motion. Work is

Metabolism is the sum total of biochemical reactions by which cells obtain and use energy to do work and maintain themselves. The particular sequence of biochemical reactions leading from one set of reactants through a series of intermediates to one or more final products is called a metabolic pathway.

performed, since a mass (the car) has been moved a distance. Much of the energy, however, is lost as heat and thus wasted. Living cells, on the other hand, release energy from their fuel substances (sugars, fats) in a number of successive but interconnected oxidation reactions, each of which releases a small amount of energy. In place of the one-step, rapid burning process that occurs in the automobile engine, the cell breaks down its fuels by a multistep, "slow burning" process.

The rate at which an internal combustion engine operates is usually controlled by an outside agent (the driver) through a mechanical device called the accelerator. But some engines also have an internal, self-controlling mechanism called a "governor," a mechanical device designed to retard the motor's rate of activity. The faster the motor goes, the more the governor is brought into action to slow the motor down. A governor is an example of an internal, self-regulating device, in contrast to the driver, who is an *external*, autonomous regulator. Cells have no counterpart to the "driver" of an automobile. Yet, to respond to changing external conditions and use their energy most efficiently, cells must be able to control the rate at which their internal processes occur. To accomplish this, cells have a variety of internal chemical mechanisms. Some of these mechanisms regulate what substances come into or go out of the cell; some regulate the rate of one particular step in a long series of biochemical reactions. All the controls are internal—analogous to the governor in an automobile engine. The chief difference between control in cells and control in machines lies in the ability of the former to exercise high levels of precision not through the regulation of gross mechanical processes, but by regulating small-scale chemical reactions.

The totality of the processes by which cells obtain and use energy to do work and maintain themselves is known as **metabolism.** All metabolic activity is ultimately chemical in nature; i.e., it involves the interaction of atoms and molecules. The metabolism of all cells is carried out by many series of interconnected chemical reactions known as **metabolic pathways.** Any particular metabolic pathway usually consists of many individual reactions whose total effect is the synthesis or degradation of molecules and the use or release of energy.

Here are a number of generalizations that apply to all biochemical pathways:

1. All pathways consist of a number of discrete steps, each of which produces a particular change in the reactants or intermediates involved.

2. Each step is highly specific in that it involves only a single kind of change. As a result, the end product of the pathway is also highly specific.

3. Particular biochemical pathways are localized within certain regions of a cell: some occur within organelles like the mitochondria or chloroplasts, while others occur at the internal surface of the cell membrane. Some pathways are not so strictly localized, but occur throughout the cytoplasm.

4. Every pathway as a whole and every discrete step of a pathway involves energy exchange. The overall pathway may be exergonic or endergonic

and individual steps may be one or the other. The net energy requirement for the pathway as a whole is a result of the sum total of exergonic and endergonic steps.

5. Every discrete step in a metabolic pathway is catalyzed by a specific enzyme. The ability of the pathway as a whole to function depends upon the catalytic function of each of the individual enzymes.

6. Biochemical pathways are controlled by various input and feedback mechanisms that constantly adjust the reaction rates according to the cell's state.

7. All metabolic pathways within a cell are interrelated to varying degrees. The pathways for carbohydrate breakdown are connected to the pathways for fat synthesis and breakdown or amino acid synthesis and breakdown. The rate at which one pathway occurs can often affect many other pathways within the cell.

To illustrate the major characteristics of metabolic pathways, it is helpful to look in some detail at the series of reactions involved in the breakdown of sugar and the consequent release of energy in cells. Sugars are fuels and provide the basic energy source of all living processes. Furthermore, the reactions involved in sugar breakdown are quite similar in all cells, from bacteria to humans. Since the getting and expending of energy is one of the most important and crucial life processes, understanding the specific reactions involved in the release of energy not only furnishes some information about how cells themselves function, but also provides a means of showing some general characteristics of all metabolic pathways.

5.2 ATP: The Energy Currency in Cells

Living cells do not use directly the energy released from the breakdown of a fuel molecule, but rather convert that energy into a molecular "currency" that can be "spent" at a later time for any energy-requiring process. Within each cell, energy released from the breakdown of fuel molecules is captured in the form of **high-energy phosphate bonds.** These high-energy bonds serve as the major energy source in all living things. High-energy phosphate bonds are found primarily in a molecule called **adenosine triphosphate,** or **ATP.** This consists of a molecule known as adenine, a five-carbon sugar (ribose), and three phosphate groups. The ATP molecule in diagrammatic form is:

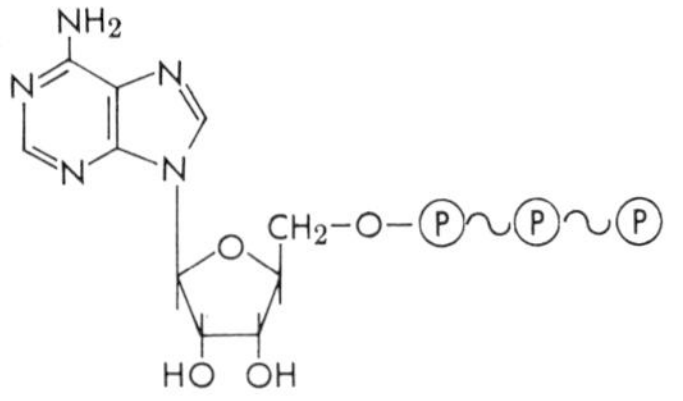

The two wavy lines between the end phosphate groups of the ATP molecule indicate high-energy bonds. The bond of the first phosphate linked to the ribose is a low-energy bond. The difference between high- and low-energy phosphate bonds lies in the overall amount of energy that each makes available to biological systems upon hydrolysis. High-energy phosphate bonds release about 7 kilocalories (kcal) per mole. Low-energy phosphate bonds release approximately 3 to 4 kcal per mole.

ATP is the energy currency of the cell. It is the molecule that directly energizes virtually all biochemical reactions in the cell.

High-energy phosphate bonds result from internal rearrangements of electrons between phosphorus and oxygen atoms. The same numbers of atoms and electrons are involved in both low-energy and high-energy phosphate bonds. The difference in the amount of energy released when each is broken is the result of a difference between the electron configurations in the respective bonds.

Molecules of ATP, which contain two high-energy phosphate bonds, serve as the major packets of energy in living cells. When energy is required for any cellular activity (e.g., the synthesis of a protein), one or two of the three phosphate groups from an ATP molecule are removed. This removal process leaves a compound less rich in energy: either **adenosine diphosphate (ADP)** or **adenosine monophosphate (AMP),** according to the particular reaction. Molecules of ADP and AMP are rebuilt into ATP by the energy released from the breakdown of glucose molecules. Thus a constant cycle of breakdown and reformation of ATP occurs in living systems (Fig. 5.1).

The amount of energy released by breaking one high-energy phosphate bond in ATP is slightly more than is necessary to drive most endergonic reactions. About 5.5 kcal are required to build a glycosidic bond, about 5 kcal to form an ester bond, and about 3.5 kcal to form a peptide bond. Thus ATP can provide enough energy for all the major synthetic reactions within a cell.

Energy can be transferred from the processes involved in the breakdown of one type of molecule to the processes involved in the synthesis of another by means of **coupled reactions.** For example, the formation of a glycosidic bond between two simple sugars is coupled to the reaction in which ATP breaks down to form ADP. This reaction provides a good example of the principle on which coupled reactions take place.

Consider the overall reaction in which glucose and fructose join together to form the double sugar sucrose:

$$\Delta G = +5.5 \text{ kcal/mole}$$

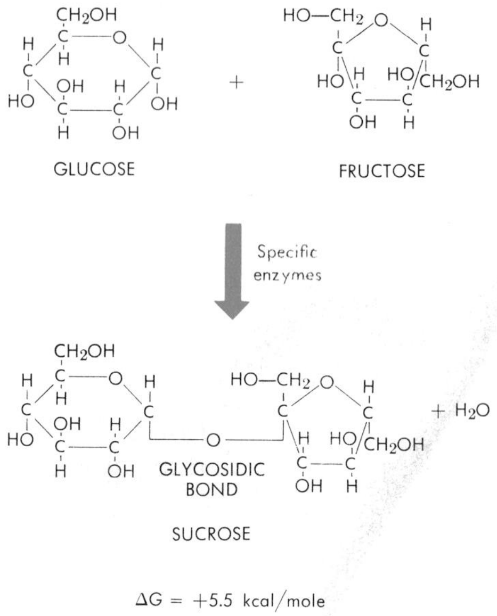

Two important details are missing from this overall representation of the reaction, however: (1) ATP is involved, and (2) the reaction takes place in two separate steps. The two steps and the exact role of ADP are shown in Fig. 5.2. In the first reaction a molecule of ATP interacts with a molecule of glucose; during this reaction one phosphate group is transferred from the ATP to form glucose-1-phosphate, leaving ADP. Glucose-1-phosphate is now an energy-rich intermediate, containing some of the energy originally held in the high-energy phosphate bond of ATP. In the second reaction, hydrolysis of the phosphate bonds of glucose-1-phosphate yields about the same amount of energy as required to form the glycosidic bond. Thus the second reaction proceeds spontaneously as long as a supply of glucose-1-phosphate and fructose is present. In the transfer of the terminal phosphate of ATP to glucose, electrons are rearranged around carbon 1 of the glucose; the result is the formation of a relatively high-energy phosphate bond. The rearrangement of electrons forming this new chemical bond represents potential chemical energy.

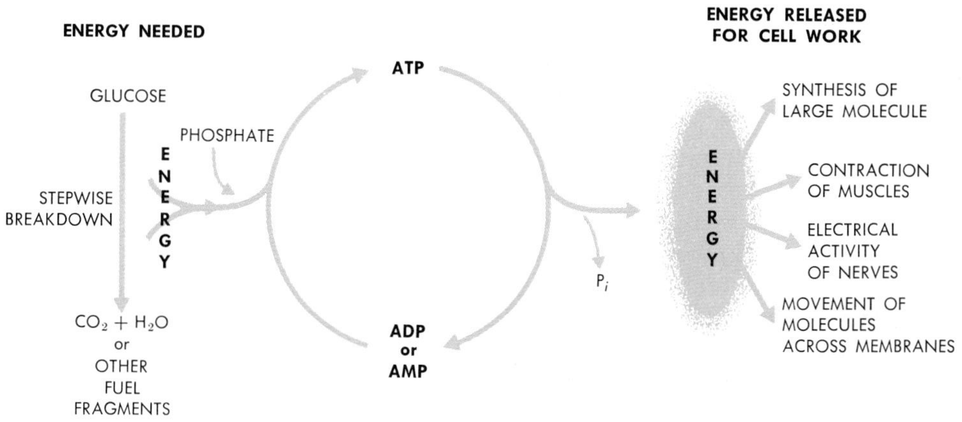

Fig. 5.1 *The ATP cycle. Molecules of ATP, the "small-change currency" of the cell, give up one or two phosphate groups, releasing the energy locked in the high-energy bonds to perform cell work. Some of the types of work for which energy is needed are shown to the right. The "energy-poor" compounds ADP and AMP are built back up into ATP by addition of one or two phosphate groups. The energy for this build-up is derived from the breakdown of glucose*

Fig. 5.2 *The two-step process by which glucose is activated by picking up a phosphate group from ATP and subsequently reacts with fructose to form the double suger, or disaccharide, sucrose. The reaction shows an overall increase in free energy obtained from the breakdown of ATP to ADP. The exergonic breakdown of ATP is coupled to the endergonic formation of the glycosidic bond between glucose and fructose.*

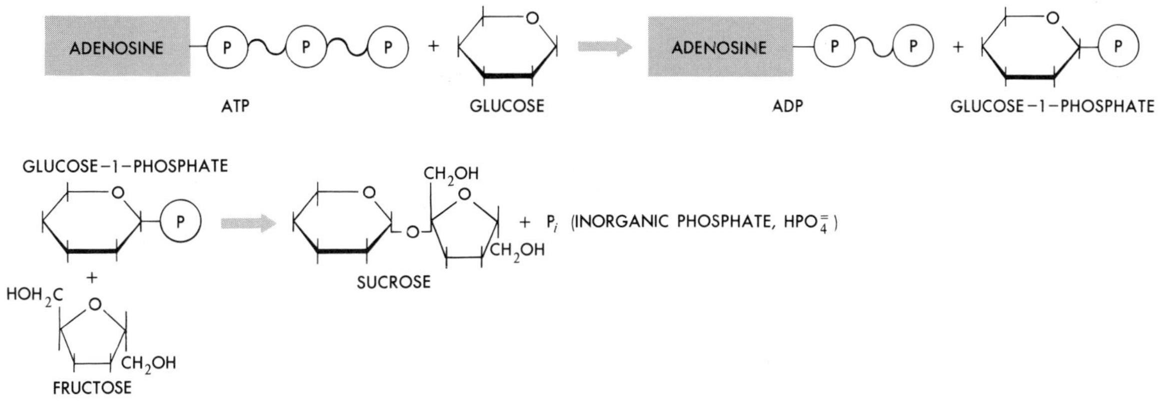

This potential energy is realized as kinetic chemical energy when the glucose-1-phosphate interacts with fructose to form a new chemical arrangement (the glycosidic bond). Coupled reactions, then, involve the creation of a high-energy intermediate by rearrangement of atoms within at least one reactant. This first step in the process is exergonic. The "activated," high-energy intermediate now contains the energy that directly drives the second, or endergonic, step. When exergonic and endergonic reactions are coupled in this way, relatively little of the total potential energy is lost.

ATP is not the only molecule that can provide energy for biochemical reactions. Other molecules, similar in structure to ATP, may also have high-energy phosphate bonds, e.g., uridine triphosphate (UTP), cytidine triphosphate (CTP), and guanosine triphosphate (GTP). In addition, high-energy sulfur bonds are involved in a few biochemical reactions. Nevertheless, ATP is rightly considered the main energy source of the living cell. It serves in far more biochemical reactions than any other single molecule.

5.3 Electron Transport and Energy Release: The Generation of ATP

Cells constantly fight an uphill battle to remain alive. Most of the chemical reactions necessary for the maintenance of life are endergonic. ATP is constantly being used to drive these uphill reactions. Where, then, does the ATP come from? How does the energy-poor compound, ADP, go back to form the energy-rich compound, ATP? In other words, what exergonic reactions are necessary to drive the uphill processes of ATP formation?

ATP is generated from AMP or ADP by a general process known as **phosphorylation.** Phosphorylation involves the addition of a phosphate group onto a molecule such as ADP, creating a high-energy phosphate bond in the process. There are two major pathways by which phosphorylation of ADP occurs: **substrate phosphorylation** and **oxidative phosphorylation,** the latter involving electron transport. Substrate phosphorylation will be discussed later in this chapter. At present, we will focus on oxidative phosphorylation from electron transport, the process by which eucaryotic cells obtain most of their ATP.

Oxidative Phosphorylation

In Chapter 2 we saw that energy is captured in the formation of chemical bonds by the transition of electrons from lower to higher energy levels in atoms and molecules. Similarly, energy is released in the breaking of chemical bonds by the downward transition of electrons. The energy to move electrons upward in the formation of ATP comes from the release of energy in the downward transition of electrons from other molecules. These other molecules are the fuels (primarily carbohydrates and lipids) that come to the cell as its food. A main function of fuel molecules in living systems, therefore, is to provide a source of electrons that, through downward transition, can be coupled to the formation of particular chemical bonds in ATP. In living cells, electrons are removed from fuel molecules and passed through a series of "acceptor" molecules, coming to occupy successively lower energy levels as the process continues. The relation between electron transport and the degradation of fuel molecules is diagrammed in Fig. 5.3.

As electrons are passed from one acceptor to another, their potential energy levels are lowered and some of the released energy is bound into high-energy phosphate bonds of ATP. At least five different acceptor molecules are involved in the electron tranport system, the ETS (Fig. 5.3). At three points in the process the drop in energy level during electron transfer is sufficient to generate a high-energy phosphate bond. Since electrons are usually transported in pairs, the complete passage of a pair of electrons along the electron transport system results in the generation of three ATP

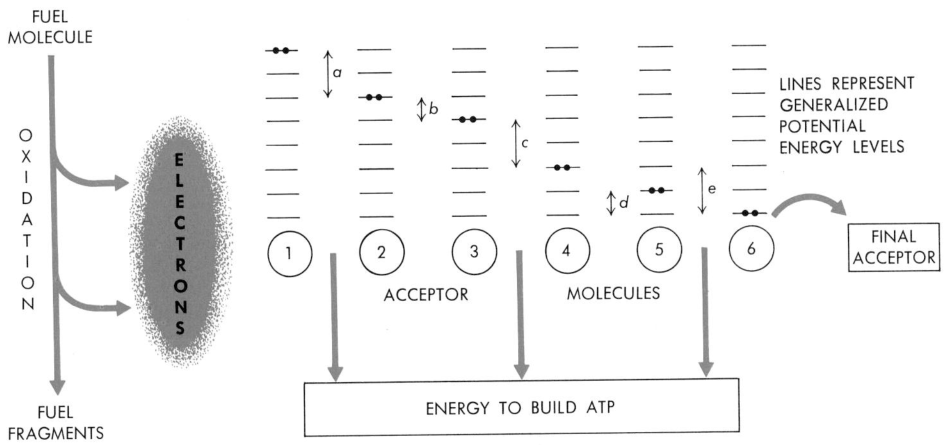

Fig. 5.3 *Generalized diagram showing the principle involved in biological oxidation and the capturing of energy as ATP. The circular structures labeled 1 through 6 represent specific electron transport molecules. The electron drops in potential-energy level as it passes from one transport molecule to another. These drops occur within the electron cloud of each transport molecule and are indicated by the distances* a, b, c, d, *and* e. *The drop in energy level with each transfer results in the release of small amounts of energy. This energy is captured in the formation of ATP.*

molecules from three molecules of the ADP and inorganic phosphate (P_i). Thus the endergonic reactions associated with ATP formation are coupled to the exergonic reactions in which electrons are removed from fuel molecules.

The Role of the Cytochromes

What molecules act as acceptors in the electron transport system and what properties allow them to serve this special purpose? Of particular importance in the electron transport system are the **cytochromes.** In Fig. 5.3 the cytochromes are represented by the last four molecules in the electron transport system (numbers 3, 4, 5, and 6). The first two acceptors in the transport system are coenzymes known respectively as nicotinamide adenine dinucleotide (NAD) and flavo protein (FP). The cytochromes, sometimes referred to as "respiratory enzymes," are iron-containing molecules with certain molecular features in common. Each is composed of a basic ring structure (called a porphyrin ring) in the center of which is an iron atom bonded to four nitrogens. The porphyrin ring structure is also found in the hemoglobin and myoglobin molecules. In cytochromes, the porphyrin ring is bonded at several points to a polypeptide chain. The protein segment of the molecule stabilizes the active site in the region of the iron atom.

Reversible changes involved in cytochrome activity result from successive oxidations and reductions. When any tissue is active and using the energy from the breakdown of sugar to produce ATP, electrons are being transported at a rapid rate along the electron assembly. Each transport molecule (e.g., one of the cytochromes) picks up an electron and thus becomes reduced. It passes the electron on to the next cytochrome molecule in the transport system and thus becomes oxidized. The ability of cytochromes to successively undergo oxidation and reduction in a reversible manner is a function of the electron configuration about the central iron atom. Iron is one of several types of atoms that can exist in one of several valence states: it can

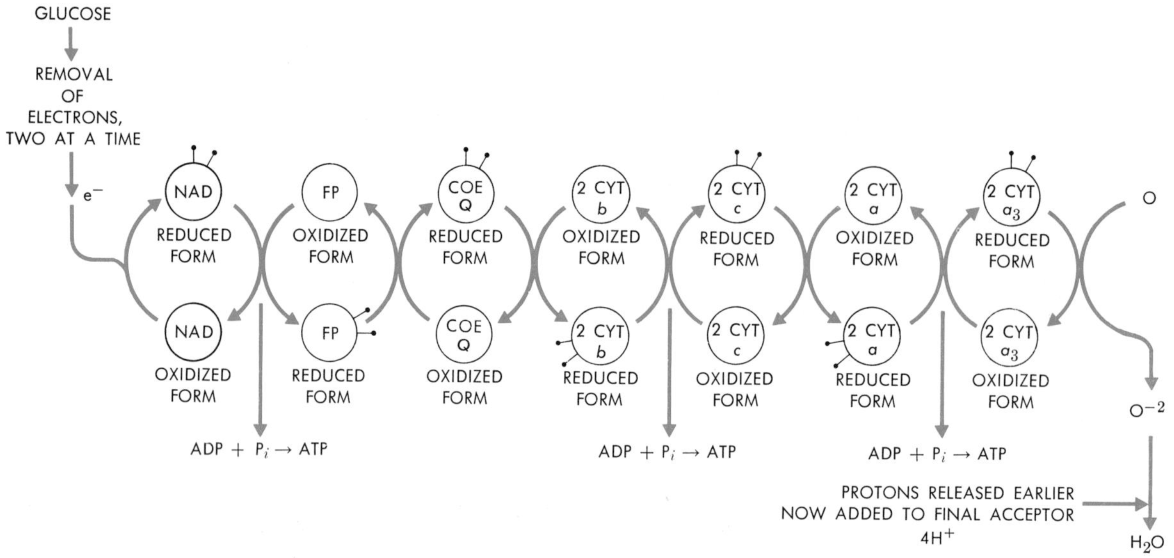

Fig. 5.4 *Diagram of the various electron transport molecules involved in the passage of electrons from a fuel molecule to a final acceptor. During the course of this process, the energy from the electrons is used to add inorganic phosphate (PO_4) onto ADP molecules, forming ATP. The sites of ATP production are shown by the arrows leading downward from three transfer points. P_i stands for inorganic phosphate.*

give up either two or three electrons and exist as either the ferrous ion (Fe^{++}) or the ferric ion (Fe^{+++}). The ferric ion can pick up an electron and thus become reduced to the ferrous ion. The ferrous ion, in turn, can give up the electron and thus be oxidized back to the ferric ion. The ability of cytochrome molecules to pick up and release electrons easily allows them to function efficiently in the electron transport process.

The complete electron transport assembly is shown diagrammatically in Fig. 5.4. Electrons are removed two at a time from glucose and picked up first by oxidized NAD or NAD^+ (which then becomes reduced NAD, symbolized $NADH + H^+$).*

* This way of symbolizing reduced NAD emphasizes that what each NAD molecule picks up as a carrier is not two complete hydrogen atoms, but rather two electrons and one proton; the extra proton (the H^+) is given off to the medium where it combines with a water molecule to form the hydronium ion (H_3O^+). H_3O^+ can give up its extra proton to oxygen at the end of the electron transport system to yield water, as shown in Fig. 6.4. The equation would be: $2H_3O^+ + O^{-2} \rightarrow 3H_2O$.

Reduced NAD passes its electrons to oxidized FP, which in turn becomes reduced FP (abbreviated FPH_2). FPH_2 then passes electrons on to the cytochromes, first to *b*, which passes them on to *c*, *a*, and a_3.

A "Final Acceptor"

What is the ultimate fate of the electrons on cytochrome a_3? Obviously, if the electrons cannot be passed somewhere, the electron transport system will soon grind to a halt, since all the transport molecules will be in the reduced form, or "filled up." There must, therefore, be a "final acceptor." This is the function of oxygen in cellular respiration. As shown in Fig. 5.4, oxygen picks up electrons from reduced cytochrome a_3 and thereby

Supplement 5.1

CYTOCHROMES: HOW DO WE KNOW ABOUT THEM?

The existence of the cytochromes and the nature of their cellular function were first noted in 1925 by the British biochemist David Keilin. Keilin started out to study the chemical composition of different kinds of living tissues by a process known as microspectroscopy. In this process, a piece of living tissue is placed on a microscope slide and light is passed through it. The ocular lens of the microscope is replaced by a prism, however, so that light emerging from the tube is diffracted and a spectrum produced. Such spectra show a broad, continuous background ranging in color from red on one end to violet on the other. This presents the spectrum of light emitted by the light source. Onto this continuous background are superimposed a number of dark absorption lines, representing various types of molecules in the tissue being observed. Each type of molecule absorbs light at different specific wavelengths. Thus by examining the absorption spectrum it is possible to identify some of the major components of living tissue.

Keilin was interested in studying a particular group of molecules (known at the time as "myohaematin") found in all types of living cells. In particular, Keilin was studying the characteristics of these molecules in the thoracic muscles (which control wing movements) of the bee. He placed the entire organism on a microscope slide and allowed light to pass through the wing muscle at the joint between the wing and thorax (see Diagram A).

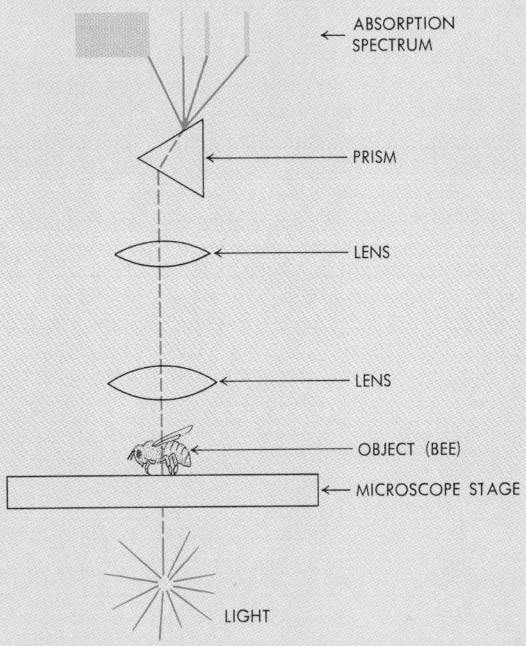

Diagram A

He noted that when the insect was quiet no particular absorption bands could be observed. When the insect moved its wings, however, several distinct absorption bands appeared on the spectrum, as shown in Diagram B.

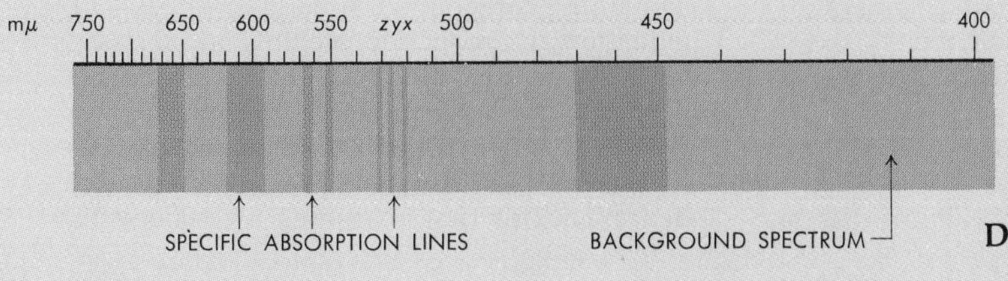

Diagram B

The continuous spectrum forming the background is produced by the light source and ranges from red on the left to violet on the right. The dark lines represent specific wavelengths of light that molecules in the tissue have absorbed. Each type of molecule absorbs at specific wavelengths. If the molecule is absent from the tissue, its corresponding lines on the spectrum are also absent.

Keilin's studies on myohaematin in yeast cells showed yet another interesting result. When examined with the microspectrograph, yeast growing normally in a suspension showed the major absorption bands. When oxygen was bubbled through the yeast culture, the absorption bands disappeared. If nitrogen was bubbled through the culture, the absorption bands became intense. These observations suggested that myohaematin underwent reversible oxidation and reduction and that the absorption bands were characteristic of the reduced state and their absence characteristic of the oxidized state.

One more interesting piece of information was added to Keilin's observations. He found that the appearance and disappearance of absorption bands was markedly affected by the presence of such agents as carbon monoxide and cyanide. As was pointed out in Chapter 3, these same agents strongly affect the respiratory pigment hemoglobin. Keilin knew that hemoglobin was an iron-containing compound that could both pick up and discharge oxygen molecules, thereby serving as the oxygen-transporting molecule of the blood. When carbon monoxide or cyanide is added to a hemoglobin solution, however, the hemoglobin molecules are permanently reduced and thus rendered unable to pick up oxygen. Similarly, when carbon monoxide or cyanide is added to a solution of myohaematin, the dark absorption bands occur and persist, even if oxygen is added later. Oxygen is unable to reverse the effects of cyanide or carbon monoxide on myohaematin, just as it fails to reverse the effect of these molecules on hemoglobin.

Keilin drew the following specific conclusions from his observations:

1. The oxidation–reduction reactions of myohaematin are normally reversible: reduced myohaematin appears when oxygen is absent; it disappears when oxygen is present.

2. With specific poisons such as cyanide or carbon monoxide, molecules of myohaematin behave in some ways quite similar to those of hemoglobin. Thus myohaematin could be thought of as an iron-containing molecule with many of the structural properties of hemoglobin.

3. The existence of a number of different dark bands on the absorption spectrum of reduced myohaematin suggested that several different kinds of this molecule were present in the cell.

Fifteen years later, Keilin and one of his coworkers, Edward F. Hartree, showed that the dark lines of the absorption spectrum were produced by several different molecules rather than the single substance myohaematin. They renamed those molecules **cytochromes.** Each set of bands turned out to represent one of several types of cytochromes, called, in order of their sequence in the electron transport process, cytochrome *b*, cytochrome *c*, cytochrome *a*, and cytochrome a_3 (or cytochrome oxidase).

Note the kind of reasoning involved in Keilin's and Hartree's work. Since it was difficult to isolate pure cytochrome, and since no techniques were then available for determining its exact molecular structure anyway, they were forced to reason by analogy. *If* hemoglobin shows an absorption spectrum characteristic for reversible oxidation and reduction of iron, and *if* myohaematin shows similar absorption bands, *then* it is possible that myohaematin is an iron-containing compound that undergoes reversible oxidation and reduction. Keilin's initial studies did not provide direct evidence for this chemical property of myohaematin. Only more detailed studies on purified samples of myohaematin (cytochrome) itself would provide *direct* evidence. Lacking that, reasoning by analogy provided a start.

The electrons that pass through the electron transport system come from removal of electrons from glucose (in other words, oxidation of glucose). During their subsequent transport through the cytochrome system, electrons pass successively from higher to lower energy levels.

becomes negatively charged (O^{-2}). This negatively charged oxygen ion then picks up protons from the medium to produce water. Here is one of the chief differences between the use of gasoline as a fuel in an engine and the use of sugars as a fuel in a living organism. Both processes generally consume oxygen and release carbon dioxide. When gasoline burns, however, oxygen reacts directly with the molecules of the fuel. In cellular oxidation, oxygen does *not* unite directly with molecules of sugar. Rather, electrons are removed in successive stages from the fuel molecules and ultimately transferred to oxygen.

Although both cellular oxidation and the oxidation of fuel in an engine may produce the same end products (CO_2 and H_2O), the chemical mechanisms by which these processes occur are very different. This difference is vitally important in understanding the energy-harnessing process involved in the rebuilding of ATP. Release of energy in small steps, such as occurs in the cell, allows more energy to be captured in a usable form. The one-step oxidation of fuel through burning produces a great deal of heat. The many-step oxidation of sugars in the cell produces a minimum of heat. For organisms, as for the machine, heat represents wasted energy. Thus the stepwise breakdown that occurs in cells, although more intricate, is ultimately more efficient.

The electron transport system has a structural as well as a functional basis in cells. Found in virtually all types of cells, it is usually associated with membranes—in bacteria as parts of the cell membrane, in cells of higher organisms as part of the mitochondrial and (as we shall see later) chloroplast membranes as well. Recent evidence has shown that the inside walls or cristae of mitochondria contain many small globular units believed to represent groups of respiratory enzymes called **respiratory assemblies.** Each is thought to contain one set of cytochrome molecules, arranged in a specific order on a membrane. It is also probable (though not definitely established) that at least FP is also bound to the membrane with the cytochromes. The specific structural arrangement of the cytochrome molecules makes possible the effective passing of electrons from one acceptor to the other. Like workers lined up on an assembly line, the electron transport molecules are arranged in a physical order corresponding to their function in transporting electrons. The generation of ATP by electron transport, then, occurs on the inner membraneous walls of the mitochondria.

Why is ATP generated at only three points during electron transport? As we have seen, energy is released as electrons pass from one carrier to the next, as a result of the stepwise, downward transition of electrons from higher to lower energy levels. If the amount of energy given off at each step of the transfer process is measured, data such as those presented in Supplement 5.2 on electron flow are obtained. To drive the uphill reaction involved in ATP formation, 7 kcal per mole are required. According to Table 5.1 only three specific points in the electron transport process generate more than 7 kcal per mole of electrons transported. These points occur in the transfer from NAD to a flavin containing protein, from cytochrome *b* to cytochrome *c*, and from cytochrome *a* to a_3 to molecular oxygen.

Supplement 5.2

ELECTRON FLOW: A ONE-WAY STREET

How are electrons passed in a specific order along the electron transport system? Why, for example, does cytochrome *c* pass electrons only to cytochrome *a*, and not back to cytochrome *b*? Furthermore, why does cytochrome a_3 pass electrons to oxygen rather than back to cytochrome *a*? The answer to these questions is obtained by measuring the so-called oxidation–reduction potentials in each molecule of the respiratory assembly. Essentially, oxidation–reduction potentials measure the affinity of a particular atom or molecule for electrons. The more positive the attraction for electrons in a given atom or molecule, the greater potential that atom or molecule possesses. When oxidation–reduction potentials for NAD, FAD, and the cytochromes are measured, data similar to those shown in the table are obtained. These data indicate that each

Oxidation–Reduction Potential for Various Electron Acceptors

Electron acceptor	Oxidation–reduction potential
NAD	−0.3
FAD	−0.1
Cyt *b*	+0.1
Cyt *c*	+0.28
Cyt *a*	+0.29
Cyt a_3	Data not available
Molecular oxygen	~+0.80

successive molecule in the electron transport system has a greater affinity for electrons than the molecules preceding it (i.e., has a stronger oxidation–reduction potential). Thus, as soon as any one electron transport molecule picks up electrons, they will be pulled away from it by the adjacent molecule, which possesses a more powerful oxidation–reduction potential. In this way the flow of electrons through the respiratory assembly is always maintained in one direction.

Table 5.1
Potential Energy Released as Electrons Are Passed through Each Step of the Electron Transport System
(FAD = flavin adenine dinucleotide)

Electron passage	ΔG, kcal/mole
NAD → FAD	−12.4
FAD → Cyt *b*	− 4.1
Cyt *b* → Cyt *c*	−10.1
Cyt *c* → Cyt *a*	− 1.3
Cyt *a* → Cyt a_3 → O_2	−24.4

The stepwise release of potential energy during electron transport allows the capture of a significant percentage of the total potential energy as high-energy phosphate bonds.

5.4
The Release of Electrons: Pathways for the Breakdown of Sugars

Having seen the complex forms of energy release and use in cell processes, we can now look at the ultimate source of energy for the cell: the sugars, starches, and fats. How are sugars oxidized, step by step, to release usable energy? What changes must the sugar molecule undergo during this process so that only two electrons are removed at a time, and how are these changes specifically controlled by enzymes? How efficient is the cell in extracting energy from glucose? Where does the process of sugar breakdown occur in cells, and how is it regulated? How is the metabolism of sugars related to the metabolism of other molecules in the cell, such as proteins and the nucleic acids? These and other questions will form the basis for a closer look at the metabolic pathways involved in biological oxidation.

In most cells the breakdown of sugars occurs in two main stages:

1. **Glycolysis or fermentation** is the partial breakdown of sugars into simpler components without

the presence of molecular oxygen. Some electrons are removed from the sugars during glycolysis, though only a small amount of the total potential energy of the sugar is extracted. In certain specific cells the end product of glycolysis, pyruvic acid, is converted into alcohol in a process called alcoholic fermentation (for example, in yeasts), or lactic acid fermentation as in some bacteria or muscle cells.

2. **Aerobic respiration** is the series of changes by which the products of glycolysis undergo further oxidation and ultimate degradation to CO_2 and H_2O. During aerobic respiration, many pairs of electrons are removed from the sugar fragments, with oxygen acting as the final electron acceptor.

The general reaction for glycolysis can be written as

$$\text{FUEL MOLECULE} \rightarrow \text{FUEL FRAGMENTS} + \text{ENERGY.} \qquad (5\text{–}1)$$

In particular, the equation for glycolysis can be written as

$$\underset{\text{(glucose)}}{C_6H_{12}O_6} \rightarrow \underset{\text{(pyruvic acid)}}{C_3H_4O_3} \begin{cases} \nearrow 2C_2H_5OH \text{ (alcohol)} + 2CO_2 + \text{ENERGY (yeast)} \\ \searrow 2C_3H_6O_3 \text{ (lactic acid)} + \text{ENERGY (bacteria, (muscle cells)} \end{cases} \qquad (5\text{–}2)$$

A generalized reaction for aerobic respiration can be written as follows:

$$\text{FUEL MOLECULE} + O_2 \rightarrow \text{FUEL FRAGMENTS} + H_2O + \text{ENERGY.} \qquad (5\text{–}3)$$

In more specific terms, the equation for respiration can be given as

$$C_6H_{12}O_6 + 6O_2 + 6H_2O \rightarrow 6CO_2 + 12H_2O. \qquad (5\text{–}4)$$

The differences between the outcomes of glycolysis and respiration reflect differences in individual reactions, in the enzymes involved, and in the total energy extracted. Most cells of higher organisms are able to carry out both glycolysis and respiration. The cells of some lower organisms, however, are able to carry out only glycolysis. We will discuss their metabolic pathways in a later section. For the moment, let us consider the complete pathway of carbohydrate metabolism, including both glycolysis and respiration, as it occurs in the cells of higher plants and animals. The process of sugar breakdown involves three phases: (1) mobilization of reserve carboyhdrates, (2) glycolysis, and (3) aerobic respiration.

Mobilization

Mobilization entails preparing various carbohydrate stores in the cells for entry into the glycolytic

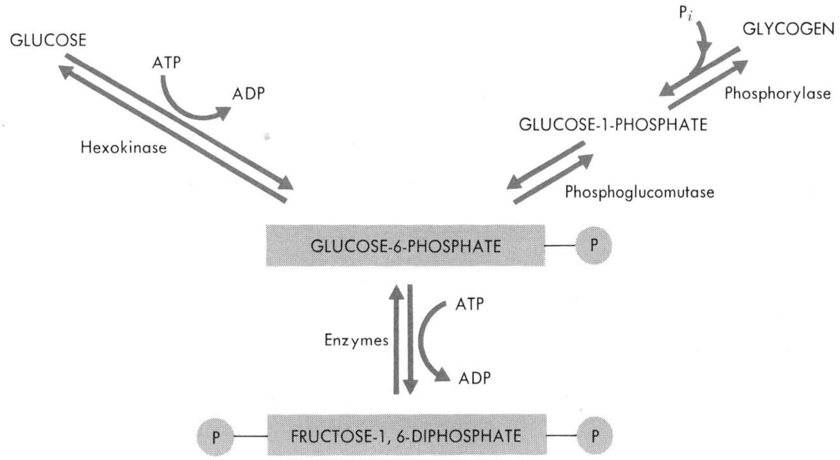

Fig. 5.5 *Carbohydrate breakdown, phase I: mobilization. In this phase an activated six-carbon unit, fructose-1,6-diphosphate, is produced, starting with either free glucose or glucose bound up as glycogen. Two ATP molecules are used to convert two glucoses into fructose-1,6-diphosphate. The cell thus spends energy to mobilize its carbohydrates for further oxidation. This can be considered an initial "investment."*

pathway. The starting point of mobilization can be either single, free glucose units or glucose bound together as animal starch or **glycogen** (see Fig. 5.5). Both glucose and glycogen are converted to glucose-6-phosphate and ultimately to fructose-1,6-diphosphate. The mobilization of glucose requires the expenditure of ATP, while that of glycogen does not. The expenditure of ATP is necessary to form glucose units into glycogen in the first place. In the long run, whether the starting point is free glucose or glycogen, the creation of glucose-6-phosphate requires the expenditure of one ATP molecule at some point along the way.

As Fig. 5.5 indicates, a second ATP molecule is added to the glucose-6-phosphate, converting it into the compound fructose-1, 6-diphosphate. This step marks the end of the mobilization phase. Fructose-1, 6-diphosphate is an "energized" molecule with more potential energy than the original glucose. Since fructose-1, 6-diphosphate represents a shove *up* the energy hill, mobilization is an endergonic reaction. Only a small amount of energy has been invested by the cell up to this point. As the energized molecule now enters the second stage of respiration, glycolysis, the original investment is paid back—along with several dividends: extra ATP's that give the cell more energy than that with which it started.

Glycolysis

The details of the next two phases of respiration—glycolysis and aerobic respiration—are summarized in Figs. 5.6 and 5.7, while their interrelationships with the generation of ATP in the cell appear in Fig. 5.8. Full discussion of the many intermediate chemical steps are presented in a number of textbooks, some of which are listed at the end of this chapter. Students with special interest in the biochemistry of respiration should consult these sources. For our purposes only a general summary will be necessary, since the main points about cellular respiration can be made without entering into excessive biochemical detail.

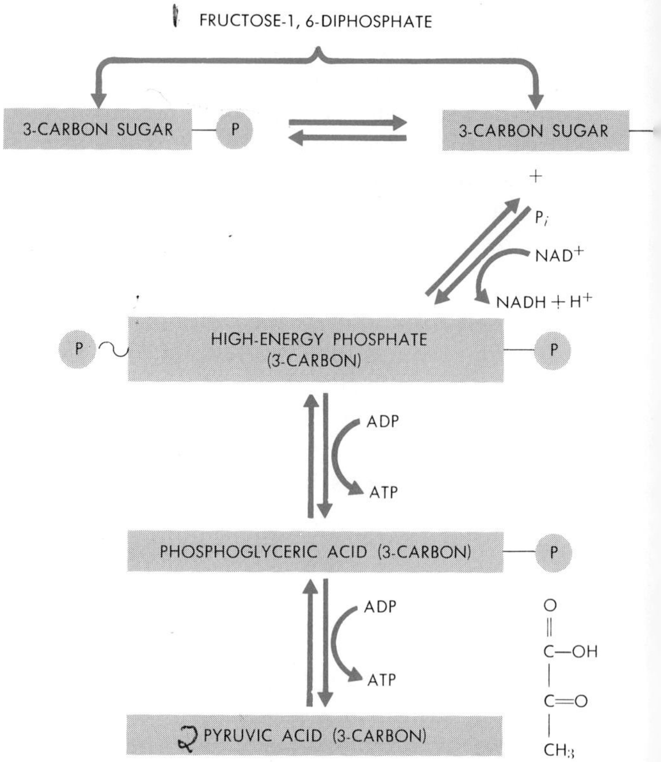

Fig. 5.6 *Carbohydrate breakdown, phase II: glycolysis. Once glucose has been mobilized as fructose-1, 6-diphosphate, the six-carbon sugar is broken down into two three-carbon sugars. These three-carbon sugars are interconvertible between themselves and exist in equilibrium. Thus both pass through the glycolytic pathway shown on the right. The ultimate products of glycolysis are pyruvic acid, ATP, and reduced NAD (written as $NADH + H^+$).*

Fructose-1, 6-diphosphate is eventually converted into two three-carbon molecules of pyruvic acid during glycolysis. In completing this series of reactions, more energy is extracted from the molecule than went into producing it from the original glucose. The additional energy comes, of course, from the potential energy stored in the chemical

Supplement 5.3

THE ECONOMIC VALUE OF FERMENTATION

Over the centuries fermentation has become a valuable process in the preparation and preservation of foods, the manufacture of such alcoholic beverages as wine, beer, and whiskey, and the manufacture of certain chemicals. The different end products of various fermentation processes depend on the type of substrate (sugar, starch, etc.) and the type of organism used to carry out the fermentation.

The most familiar fermentation process is that employed in making alcoholic beverages. For making wine, the substrate is fruit juice. The most common fruit is grapes, though any fruit juice with high sugar content will work. The fermenting organism is yeast. The fruit juice is first placed in a very large vat or crock and seeded with a desirable strain of yeast. Fermentation is allowed to proceed for a few days, during which the vat bubbles with carbon dioxide. During this period the sugar is consumed and ethyl alcohol is produced. When the alcohol content reaches about 14%, the yeast cells can no longer survive, and the process eventually comes to a halt. The contents of the vat are then poured into storage containers that allow limited access to air. The raw wine sits in these containers for a month or more, during which time many impurities (including tannins and the dead yeast cells) settle out. After several transfers the wine is bottled.

The sweetness of a wine is largely determined by the amount of sugar added at the beginning—or after the initial fermentation has been completed. The color of the wine depends on whether red or white grapes are used. The flavors of various wines, on the other hand, are a result of many factors: the strain of yeast used, the type of grape, and environmental factors such as the soil in which the grapes were grown and the climate that particular year. This accounts for the interest shown by wine connoisseurs in the national and/or regional origin of the wine as well as the year it was produced. In order to produce wine of consistent taste and quality, winemakers must learn to control these variables as accurately as possible.

One of the most serious problems winemakers face is contamination of the fermentation vats with bacteria. Since bacteria ferment sugars to lactic acid rather than alcohol, their presence gives the wine a very sour taste. When contamination of wine by bacteria precipitated the great wine calamity in France during the mid-nineteenth century, the wine industry called in a young organic chemist, Louis Pasteur. Pasteur demonstrated that the contaminating substance was lactic acid. He went on to show that living organisms, bacteria, produced the lactic acid. His solution involved the introduction of sterile

bonds of the glucose itself. Thus, mobilization primes the glucose molecule so that more of its own potential energy can be released in a systematic way.

On its way to the production of pyruvic acid, fructose-1, 6-diphosphate is first broken down into two three-carbon sugars, each containing a phosphate group (Fig. 5.6). These two sugars are interconvertible: they exist in chemical equilibrium with one another in solution. Consequently both will eventually pass through the entire glycolytic pathway to pyruvic acid.

During glycolysis two very important types of chemical events take place, each relating to the extraction of energy from the three-carbon phosphate sugar:

1. Electrons are removed from the substrate by an enzyme in conjunction with the electron ac-

procedures for handling the grape juice and inoculating it with yeast. These procedures, later applied to milk as the well-known "pasteurization" process, made it possible to avoid bacterial contamination while still allowing yeast to ferment the grape juice. Pasteur's discoveries about the role of bacteria in infections grew out of his early work with the French wine industry.

Beer is produced much like wine, except that the starting substrate is malt rather than fruit juice. Malt is derived from the seeds of barley plants, which are softened by soaking in water to stimulate germination. The malt is fermented by a special strain of yeast known as "brewer's yeast." The flavor of beer is produced by adding hops (the tops of certain flowers of the mulberry family, which produce a bitter substance) to the fermentation vats. After a few days the fermented liquor is drawn off and, like wine, stored for a period of time in closed containers. It is important in making both wine and beer to be sure that too much air does not become available during the fermentation process. This would stimulate the yeast cells to carry out aerobic respiration, consuming the sugar and producing only carbon dioxide and water rather than alcohol.

Hard liquor, such as bourbon, scotch, or gin, is produced by a more complex process. In each case, a different substrate is used for the initial fermentation: for bourbon, corn mash, and for gin, potatoes. Fermentation proceeds in large open vats for a few days in a manner similar to that for wine and beer. The liquor is then poured off and allowed to settle for a period of time to remove the least tasty impurities (mostly tannic acid). The liquid is then poured into distillation vats, where it is heated until it evaporates. Since alcohol vaporizes at a much lower temperature than water, it is driven off first and collected in condensing coils (distillation). By distillation the alcoholic content of the final product can be made much greater than is possible for either wine or beer. Alcoholic content of hard liquors is reckoned as "proof," 100-proof being 50% alcohol. Further distillation of the fermented liquid drives other substances into the condensing apparatus as well, adding to the pure alcohol flavors derived from the original fermented liquid.

After distillation the liquor is stored in wooden vats or barrels. The kinds of wood used and the way it has been processed determine the final flavor of the liquor. Bourbon, for example, is usually stored in maple barrels whose insides have been charred. The liquor leaches out various substances from the charred wood, giving bourbon a characteristic taste. The materials leached from the wood also give bourbon and scotch their color. Since gin is not stored in wooden vats, it remains clear.

The preparation of commercial alcohol involves complete distillation of a fermented mixture, giving relatively pure ethyl alcohol. Commercial alcohol can be 200-proof, or 100% alcohol. "Denatured" alcohol is merely ethyl alcohol with an additive to make it undrinkable.

Lactic acid fermentation, so injurious to the wine and beer industries, has been used beneficially in the preparation and preservation of certain foods. For instance, the bacterium *Lactobacillus casei* is employed in the production of sour cream from milk. Other kinds of bacteria are used to produce cottage cheese, yogurt, and cheeses. *Lactobacillus plantarum* is used as a preservative in the preparation of pickles and sauerkraut.

Controlled fermentations are also used to produce chemicals on an industrial scale. The yeast *Citromyces* ferments sugars to citric acid, and *Aspergillus niger* ferments sugars to oxalic acid. Citric acid is used commercially in producing carbonated soft drinks and in blueprinting. Oxalic acid is used in the leather and textile industries to condition raw materials, as well as in the manufacture of dyes.

ceptor molecule, NAD. This process occurs simultaneously with the addition of another phosphate group to the three-carbon phosphate sugar, creating a three-carbon diphosphate sugar. This removal of electrons is an actual oxidation process similar to those associated with cellular respiration. A consequence of oxidation is that one of the phosphate groups on the three-carbon sugar is converted from a low-energy to a high-energy phosphate bond.

2. Catalyzed by specific enzymes, ADP molecules from the medium now interact with the high-energy phosphate sugar produced in the first step to generate ATP, the high-energy phosphate group being transferred from the sugar to the ADP. The process of generating ATP by direct interaction of ADP with a high-energy phosphate sugar is an example of the **substrate phosphorylation** mentioned earlier in Section 6.3. Substrate phosphorylation is the only method of ATP generation that occurs in glycolysis.

From one starting glucose molecule, the end products of glycolysis are two molecules of pyruvic acid, two molecules of reduced NAD (NADH + H^+), and four molecules of ATP. Since a total of two ATP's were used to start the process, the net energy gain for the cell is two ATP's. In other words, the cell has spent two ATP's and gotten back four.

We should emphasize that the net gain of two ATP's does not mean that cells defy the laws of thermodynamics. Through the process of respiration cells do not "create" energy from nothing. The energy represented in the two additional ATP's comes solely from the potential energy of the glucose molecule itself. Through the enzyme-catalyzed, highly specific reactions of glycolysis, the cell is able to rearrange electrons within the glucose molecule so that electrons already at higher potential energy levels in the molecule are lowered, just as, in a tower of blocks, the input of only a small amount of additional energy (such as removing one block at a key point) can release much of the potential energy in the tower. Glycolysis does this in such a way that the glucose "tower" only collapses a very little—less than 20% of its total capacity. Thus pyruvic acid, the end point of glycolysis, does *not* represent the lowest potential energy stage to which glucose can be brought.

The Citric Acid Cycle

The *complete* oxidation of pyruvic acid, with the consequent extraction of considerably more energy, occurs during the final stage of respiration, the **citric acid cycle.** This series of reactions is also known as the **Krebs cycle** (for the English biochemist Hans Krebs, who first worked out the chemical details of the pathway) and the **tricarboxylic acid cycle** (since a number of three-carbon sugars are involved).

A stepwise pathway. Some of the major stages in the citric acid cycle are shown in Fig. 5.7. Specific enzymes are involved in the conversion of the three-carbon pyruvic acid into a two-carbon acetic acid molecule. During this conversion, one carbon dioxide molecule is removed and two electrons are picked up by NAD, forming a complex that combines immediately with coenzyme A, which is always present in the cell. This reaction forms acetyl coenzyme A (called acetyl CoA), which is able to enter directly the pathways involved in the citric acid cycle. The two-carbon acetyl CoA combines

---

The oxidation of glucose occurs in a stepwise fashion. At each specific step, NAD^+ removes two electrons at a time. These electrons furnish the energy for ATP generation.

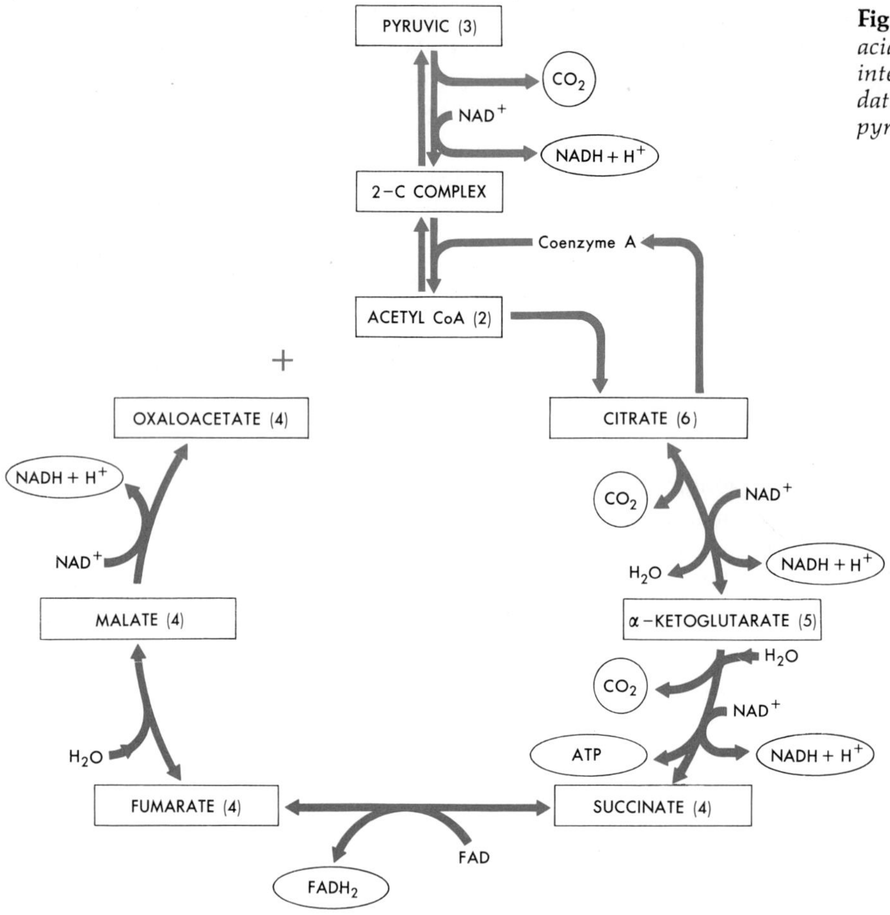

Fig. 5.7 *Diagram of the citric acid cycle, showing several of the intermediate stages in the oxidation and decarboxylation of pyruvic acid.*

with the four-carbon oxaloacetic acid, also abundant to the cells, to yield the six-carbon citric acid. In this reaction the two-carbon acetate molecule is separated from coenzyme A, which is then returned to the medium and can be used for combination with another acetate. As a result of this process, two $NADH + H^+$ molecules are generated, one from each pyruvic acid.

The six-carbon citric acid molecule undergoes a series of oxidations and decarboxylations (removal of carbon dioxide molecules) as shown in Fig. 5.7. After the removal of each carbon dioxide or each pair of electrons (equivalent to hydrogen atoms), internal rearrangements of the molecule are necessary to preserve its stability. Note that during the progress of the citric acid cycle, the six-carbon molecule is converted successively into a five-carbon molecule (α-ketoglutarate) and a series of four-carbon molecules (succinate, fumarate, malate, and oxaloacetate). Thus each turn of the cycle begins with a four-carbon compound combining with a two-carbon compound to yield a six-carbon intermediate which is immediately broken back down to the original four-carbon compound by a stepwise series of reactions. Because one molecule of oxaloacetic acid is regenerated from each molecule of oxaloacetate that originally combined with acetyl CoA, this series of reactions is truly a cycle.

The gain in ATP. Two sources of potential chemical energy are created during the citric acid cycle. One is the direct production of ATP by substrate phosphorylation, such as that found in glycolysis. A second is the production of reduced coenzymes, including NADH and $FADH_2$. Through the process of electron transport, these molecules can generate a considerable amount of ATP. Each $NADH + H^+$ molecule yields three ATP's (except the two produced during glycolysis, each of which generates only two)* by passing its two electrons along the complete electron transport chain (as shown in Fig. 5.4). In one case where electrons are picked up from succinate by FAD, one step (from NAD to FAP) in the electron transport system is bypassed. Thus each pair of electrons passing from $FADH_2$ down the chain generates two ATP's instead of the customary three.

Each glucose molecule that enters the breakdown process generates two pyruvic acid molecules. Thus the complete oxidation of a single glucose results in two turns of the citric acid cycle. To obtain an accurate tally of the input and output of glucose oxidation, it is necessary to remember that every reaction after the breakdown of fructose-1, 6-diphosphate occurs twice for each original starting glucose. The relationship among glycolysis, the citric acid cycle, and the generation of ATP by electron transport is shown diagrammatically in Fig. 5.8.

ATP is thus generated in the breakdown of glucose both by substrate phosphorylation and by electron transport. Through both of these processes, but particularly through electron transport, an enormous amount of potential energy of glucose is harnessed as the potential energy of high-energy phosphate bonds. Table 5.2 shows the precise quantity of ATP generated by each process, and the particular steps where energy is harnessed. As the data show, for each starting glucose molecule, 6 ATP's are generated by substrate phosphorylation and 32 ATP's are generated by electron transport. Two ATP's were required to get the process going, so that from the complete oxidation of a single glucose molecule, a net total of 36 ATP's is produced. Or, put another way, from a mole of glucose molecules, 36 moles of ATP become available to the cell for work. Considering that only 2 ATP's were spent to get the process started, this represents a substantial return in potential energy.

The number of ATP's generated by complete oxidation of a single glucose molecule has been a matter of dispute for some years in textbooks and among biochemists. The total number produced has been claimed by some to be 38 and by others to be 40. The reason for such discrepancy lies in the existence of a so-called "shuttle" system by which electrons removed from glucose in the cyto-

* Cytoplasmically generated $NADH_2$ from glycolysis cannot pass across the mitochondrial membrane and hence does not enter the electron transport system directly. Cytoplasmically generated $NADH_2$ reduces a three-carbon sugar (dihydroxyacetone phosphate) into another compound (glycerol phosphate) that can penetrate the mitochondria. Glycerol phosphate can be oxidized by one of the electron acceptor molecules (usually cytochrome Q), but at a point after the first ATP-generating site. Thus for each molecule of $NADH_2$ generated in the cytoplasm during glycolysis, only two ATP's are produced.

ATP is generated during glycolysis and the citric acid cycle by two means: (1) substrate phosphorylation and (2) oxidative phosphorylation during electron transport.

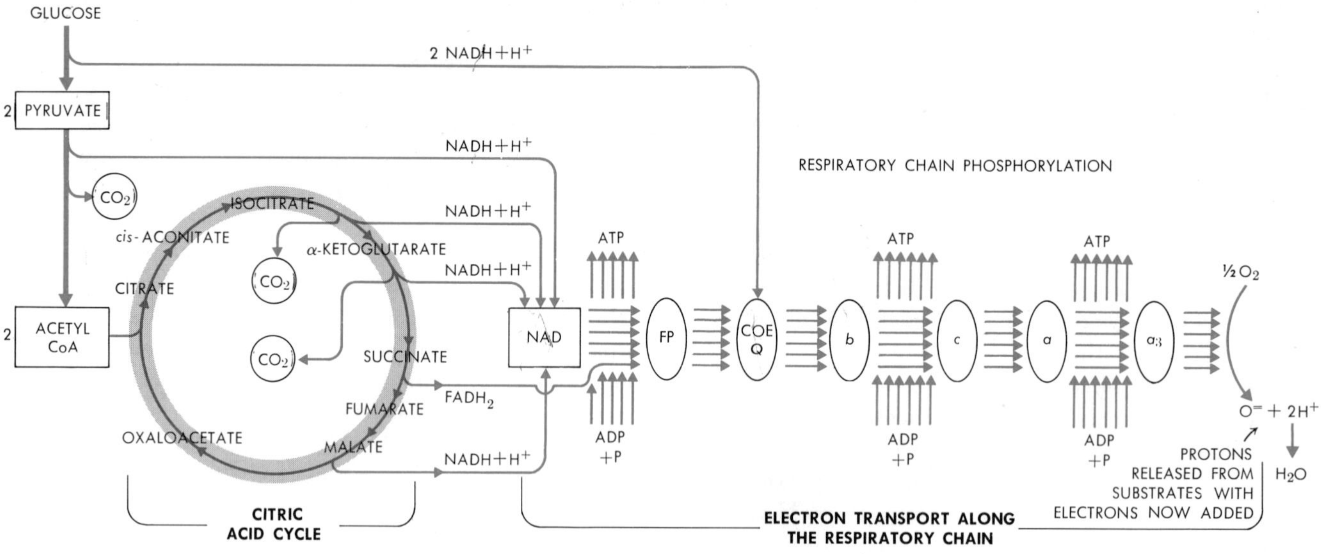

Table 5.2
Energy Harnessed in ATP's during Complete Glucose Oxidation

By substrate phosphorylation		
Glycolysis:		
High-energy phosphate (hexose) sugar → 2 Phosphoglyceric acid		2 ATP
2 Phosphoglyceric acid → 2 Pyruvic acid		2 ATP
Citric acid cycle:		
2 α-Ketoglutaric acid → Succinic acid		2 ATP
Total from substrate phosphorylation		6 ATP
By electron transport		
Glycolysis:		
2 Three-carbon sugar phosphate → High-energy sugar diphosphate	2 NADH →	4 ATP
Citric acid cycle:		
2 Pyruvic acid → Acetic acid	2 NADH	6 ATP
2 Citric acid → α-Ketoglutaric acid	2 NADH	6 ATP
2 α-Ketoglutaric acid → Succinic acid	2 NADH	6 ATP
2 Succinic acid → Fumaric acid	2 FADH	4 ATP
2 Malic acid → 2 Oxaloacetic acid	2 NADH	6 ATP
Total from electron transport		32 ATP
Total from substrate phosphorylation		6 ATP
Total		38 ATP

◀ **Fig. 5.8** *The relation between glycolysis, citric acid cycle, and generation of ATP by electron transport. Every NADH + H^+ molecule passes its electrons along the complete electron transport assembly, generating three ATP's. In the one site in the cycle where FAD is the direct electron acceptor, each $FADH_2$ passes its electrons directly to FP, bypassing NAD, and thus generating only two ATP's. For each glucose molecule mobilized at the beginning, two of each of the reactions shown here take place. This is a result of the breakdown of the original six-carbon sugar into three-carbon fragments early in the pathway.*

plasm during glycolysis are brought into the mitochondria and thus through the electron transport system. As mentioned in the footnote on page 161, the mitochondrial membrane is impermeable to $NADH_2$ produced in the cytoplasm during glycolysis. Electrons from cytoplasmically generated $NADH_2$ are thus "shuttled" across the membrane by passing to a three-carbon sugar which in its reduced form can pass into the mitochondrion. The reduced three-carbon sugar (known as glycerol phosphate) passes electrons to the electron transport system. Electrons shuttled into the electron transport system via glycerol phosphate enter at cytochrome Q; thus each pair of electrons shuttled in generates only two, rather than three, ATP's. Prior to knowledge of this shuttle system, it was assumed that cytoplasmically generated $NADH_2$ produced three ATP's per molecule; hence the disparity in ATP count.

The metabolic context. Note that in many of the reactions diagrammed above, various intermediates and other molecules are constantly exchanged between the pathway and the surrounding medium. It is very important to realize that cells contain pools of molecules of various sorts that are constantly available for interactions with one or another intermediate during a particular metabolic reaction. For example, coenzyme A is plentiful in the cell and is available to react with acetic acid to produce acetyl CoA. Coenzyme A is also used in other metabolic reactions, but it is always returned to the medium. As another example, water is constantly being used in various reactions of the citric acid cycle. The water comes, of course, from the liquid medium that comprises a good part of the cell mass, and it is "repaid" to the medium by the formation of water at the end of the electron transport process.

Like whole organisms, cells are "open systems." This means that they must continually exchange materials with the external environment. Cells cannot produce ATP without oxidizing some fuel molecule, and that fuel must come from somewhere. For animals, it comes from the food they eat: plants and other animals. Like animals, plants also oxidize carbohydrates via glycolysis and the citric acid cycle. Unlike animals, however, plants can produce their own carbohydrates internally by photosynthesis. The energy to drive that uphill reaction is obtained externally from sunlight. Though more self-sufficient than animals, plants are still open systems in the sense that they require an external energy source. The ultimate source of all energy in living systems is external, whether in the form of specific fuel molecules that can be oxidized to regenerate ATP, or in the form of sunlight for the synthesis of carbohydrates. Without this continual input of energy from the environment, life would quickly come to a halt.

If the glucose molecule were completely oxidized all at once (as in burning), its free energy would be released in an uncontrolled way and hence could not be captured by the cell for the generation of ATP.

5.5 Life without Air: Anaerobic Oxidation

While human liver cells can exist for several hours without an oxygen supply, brain cells will cease to function properly if oxygen is cut off for only five or six minutes. In fact, most cells (especially in multicellular organisms) live on such a high-energy budget that constant electron transport is necessary to generate a sufficient supply of ATP to keep the cells alive. These cells are thus dependent on oxygen to keep the electron transport system in operation. Such cells are known as **aerobes,** which term indicates that they require a supply of oxygen-containing air for continued existence. Some types of cells, however, have a different type of electron transport system and do not require oxygen to generate ATP. These cells, called **anaerobes,** produce ATP solely by substrate phosphorylation.

There are two types of anaerobic cells. **Obligate anaerobes** completely lack an electron transport system and cannot use molecular oxygen whether it is present or not. Since obligate anaerobes never use oxygen, they exist on a very low ATP budget. **Facultative anaerobes** can exist anaerobically for relatively long periods of time, even indefinitely, if oxygen is not present. But given a supply of oxygen, facultative anaerobes can carry out electron transport (because they have the electron transport molecules) and thus generate larger quantities of ATP. Certain kinds of bacteria are obligate anaerobes; not only does their growth not require oxygen, but in some instances it is actually *hindered* by oxygen. Yeasts and the muscle cells of higher organisms are examples of facultative anaerobes. If oxygen is absent, yeast cells carry out the process of fermentation, producing ATP only by substrate phosphorylation. Similarly, when muscle cells are exercising very rapidly, the oxygen supply brought by the blood to the tissue is not sufficient to completely oxidize glucose to ATP. Muscle cells exist anaerobically during these pe-

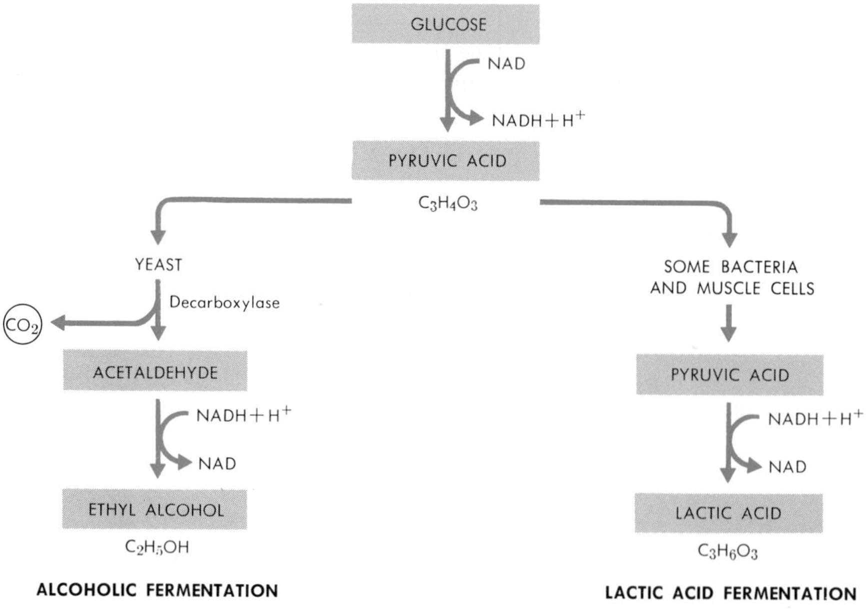

Fig. 5.9 *Diagrammatic representation of two possible pathways of anaerobic oxidation. Since the electron transport system is not used in either pathway, some means must be available for disposing of the NADH + H$^+$ generated during the breakdown of glucose to pyruvate. This is accomplished by passing the electrons back to pyruvate to generate lactic acid, as in bacteria and muscle cells, or by passing the electrons to an intermediate produced from decarboxylating pyruvate, to generate ethyl alcohol, as in yeast.*

riods of vigorous contraction, producing ATP by substrate phosphorylation. When a sufficient oxygen supply becomes available to either type of cell, much more ATP is generated by electron transport.

The end products of anaerobic breakdown are slightly different in different types of cells. Two common pathways of anaerobic respiration, one for yeasts and one for bacteria and muscle cells, are shown diagrammatically in Fig. 5.9. In its initial stages, anaerobic oxidation follows the pathway of glycolysis, during which glucose is converted to pyruvic acid. In this process, two NADH + H^+ molecules and two pyruvic acid molecules are produced for each starting glucose. Anaerobes share the glycolytic pathway with aerobic organisms, even to the extent of having most of the same enzymes involved in mobilization and initial oxidation of the glucose molecule. It is in the fate of the pyruvic acid molecule that anaerobes differ markedly from aerobes.

When no oxygen is available, bacterial and muscle cells dispose of the electrons picked up by NAD^+ by passing them (and their accompanying protons) back to the pyruvic acid, producing lactic acid:

$$\underset{(C_3H_4O_3)}{\text{PYRUVIC ACID}} + \text{NADH} + H^+ \longrightarrow \underset{(C_3H_6O_3)}{\text{LACTIC ACID}} + \text{NAD}^+ \qquad (5\text{–}5)$$

Lactic acid is the end product of this form of anaerobic oxidation as far as the facultative anaerobe is able to carry the oxidative process without molecular oxygen, or as far as the obligate anaerobe can carry the process under any circumstances. The accumulation of lactic acid by anaerobic bacteria also produces the characteristic taste of sour milk. Because lactic acid is the end product, this pathway is sometimes called **lactic acid fermentation.**

In yeast cells, the pyruvic acid molecule is decarboxylated to form an intermediate compound known as acetaldehyde. The two-carbon acetaldehyde picks up electrons (and accompanying protons) from reduced NAD to produce ethyl alcohol, as shown in the following equation:

$$\underset{(C_3H_4O_3)}{\text{PYRUVIC ACID}} \xrightarrow{\text{decarboxylase}} \underset{(C_2H_4O)}{\text{ACETALDEHYDE}} + CO_2 \xrightarrow[\text{NADH} + H^+ \;\curvearrowright\; \text{NAD}^+]{\text{alcohol dehydrogenase}} \underset{(C_2H_5OH)}{\text{ETHYL ALCOHOL}} \qquad (5\text{–}6)$$

Because ethyl alcohol is the end product of this anaerobic pathway, the process is sometimes referred to as **alcoholic fermentation.**

Anaerobic organisms get their ATP by substrate phosphorylation during the conversion of glucose to pyruvic acid. The reactions from pyruvic acid onward are simply means of disposing of the end products of the earlier reactions (i.e., pyruvic acid and NADH + H^+). People have taken advantage of the anaerobic process of alcoholic fermentation to produce a variety of alcoholic beverages; the wine, beer, and distilling industries all depend on this fermentation process (see Supplement 5.3).

In facultative anaerobic cells, such as human muscle tissue, lactic fermentation represents only a temporary pathway that is brought into play when the oxygen supply is low. As lactic acid accumulates, muscles become less and less able to carry out normal contraction. This occurs primarily because the ATP supply is gradually diminishing. During a period of rest, muscle cells can reoxidize the lactic acid through the electron transport system and generate more ATP for future use.

From an energy point of view, it is clear that anaerobic cells gain far less ATP from the oxidation of glucose than do aerobic cells. Fig. 5.10 summarizes the total free-energy change (ΔG) in oxidation of glucose and compares anaerobic and aerobic processes. The total ΔG for the oxidation of a mole of glucose molecules to a mole of carbon dioxide and water is 686 kcal. The ΔG for oxidation of a mole of glucose to a mole of pyruvic acid is 56 kcal. Since anaerobic cells must reduce pyruvic acid to lactic acid or ethyl alcohol, the net ΔG for glycolysis is less than 50 kcal per mole. Thus anaerobic cells must use a great deal more sugar

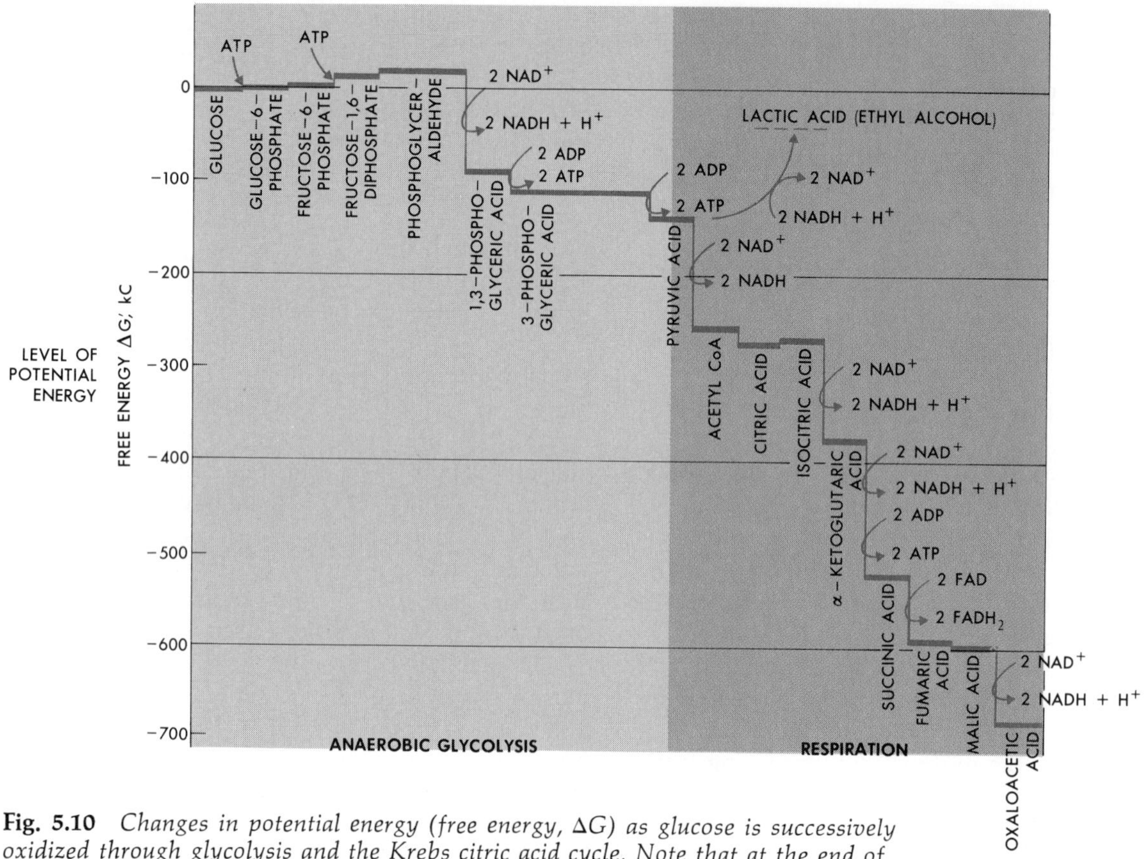

Fig. 5.10 *Changes in potential energy (free energy, ΔG) as glucose is successively oxidized through glycolysis and the Krebs citric acid cycle. Note that at the end of glycolysis only a small amount of the total energy of a glucose molecule has been released. The release of energy in successive steps occurs through electron rearrangements, a few at a time; at each step a few electrons are lowered from a higher to a lower energy level within the molecular orbitals composing the glucose and intermediate substances. The electrons picked up by acceptors such as* NAD^+ *and FAD are eventually passed to lower energy levels through the electron transport system associated with the cytochromes. (After E. O. Wilson* et al., Life on Earth. *Sunderland, Mass.: Sinauer Associates, 1973, p. 161.)*

Aerobic organisms gain far more of the free energy available in glucose oxidation than do anaerobic organisms. Aerobic organisms capture about 41% of the total free energy available in complete oxidation of glucose, while anaerobic organisms capture only about 2% of the total available in the glucose molecule.

to get the same amount of energy as aerobic cells: about 12 to 15 times as much! Because of their relative inefficiency, anaerobic cells face a constant "energy crisis" that aerobic cells do not face.

5.6 Characteristics of Metabolic Pathways as Illustrated by the Oxidation of Sugars

Near the beginning of this chapter a series of questions was asked about the basic characteristics of metabolic pathways. Each of these questions can now be considered in relation to the specific examples of aerobic and anaerobic respiration.

The stepwise nature of pathways. Both glycolysis and the citric acid cycle involve a number of individual steps by which the original glucose molecule is broken down into smaller fragments and the potential energy extracted. As we have seen, the stepwise nature of the pathway ensures that maximum energy can be extracted in a usable form from the fuel molecule. For example, only a few electrons can be removed at a time from the molecule without creating a condition of instability. In order to extract energy in the most controlled (and therefore efficient) manner, the molecule is dismembered a few atoms at a time. Since life itself depends on the efficient use both of the substances present in a cell and of the available energy, the stepwise breakdown or buildup of molecules is essential.

The specificity of individual steps in metabolic pathways. Each step in the metabolic pathways of glycolysis and the citric acid cycle is catalyzed by specific enzymes that ensure that only certain chemical changes occur. For example, NAD can pick up electrons only from certain intermediate products in the citric acid cycle: it can interact with malate to form oxaloacetate and NADH + H^+, but it cannot interact with fumarate. The conversion of malate to oxaloacetate (with the production of NADH + H^+) is catalyzed by a specific enzyme (NAD-malic dehydrogenase). If this enzyme is not present, the reaction will not occur and the entire cycle can be stopped.

Similarly, malate must be formed from fumarate in order for the subsequent reactions of the citric acid cycle to proceed. This is accomplished through the action of an enzyme (fumarase) that catalyzes only this particular reaction. Every step throughout the citric acid cycle and glycolysis depends on such specific reactions. It is obvious that if fumarate, for example, could be converted into a very large number of intermediates other than malate, the specific production of ATP would be greatly retarded. All metabolic pathways depend on a similar high level of specificity of each component step to yield specific products.

Another aspect of specificity is apparent in glucose oxidation. Given the same starting reactant, the differences between two metabolic pathways are the result of the differences in enzymes present. For example, aerobic organisms can oxidize pyruvic acid to CO_2 and H_2O, with the production of a considerable amount of ATP, because they possess enzymes for the citric acid cycle and electron transport molecules, the cytochromes. Obligate anaerobes lack all these enzymes and consequently cannot carry out the citric acid pathways. Similarly, yeast and bacteria differ in the pathways by which they act upon pyruvic acid. Yeast has an enzyme (decarboxylase) that converts pyruvate to acetaldehyde. Acetaldehyde is then converted by another specific enzyme into ethyl alcohol. Some bacteria, on the other hand, lack decarboxylase and convert pyruvic acid directly into lactic acid. Again, the specific fate of any molecule in a metabolic pathway depends on the specific enzymes present.

Localization of pathways. In the cells of most organisms, pathways are localized to some extent in various parts of the cell. The example of complete glucose oxidation illustrates this principle clearly. The glycolytic pathway occurs in the cytoplasm of cells and is catalyzed by soluble enzymes not bound to any particular cell organelles. The enzymes for

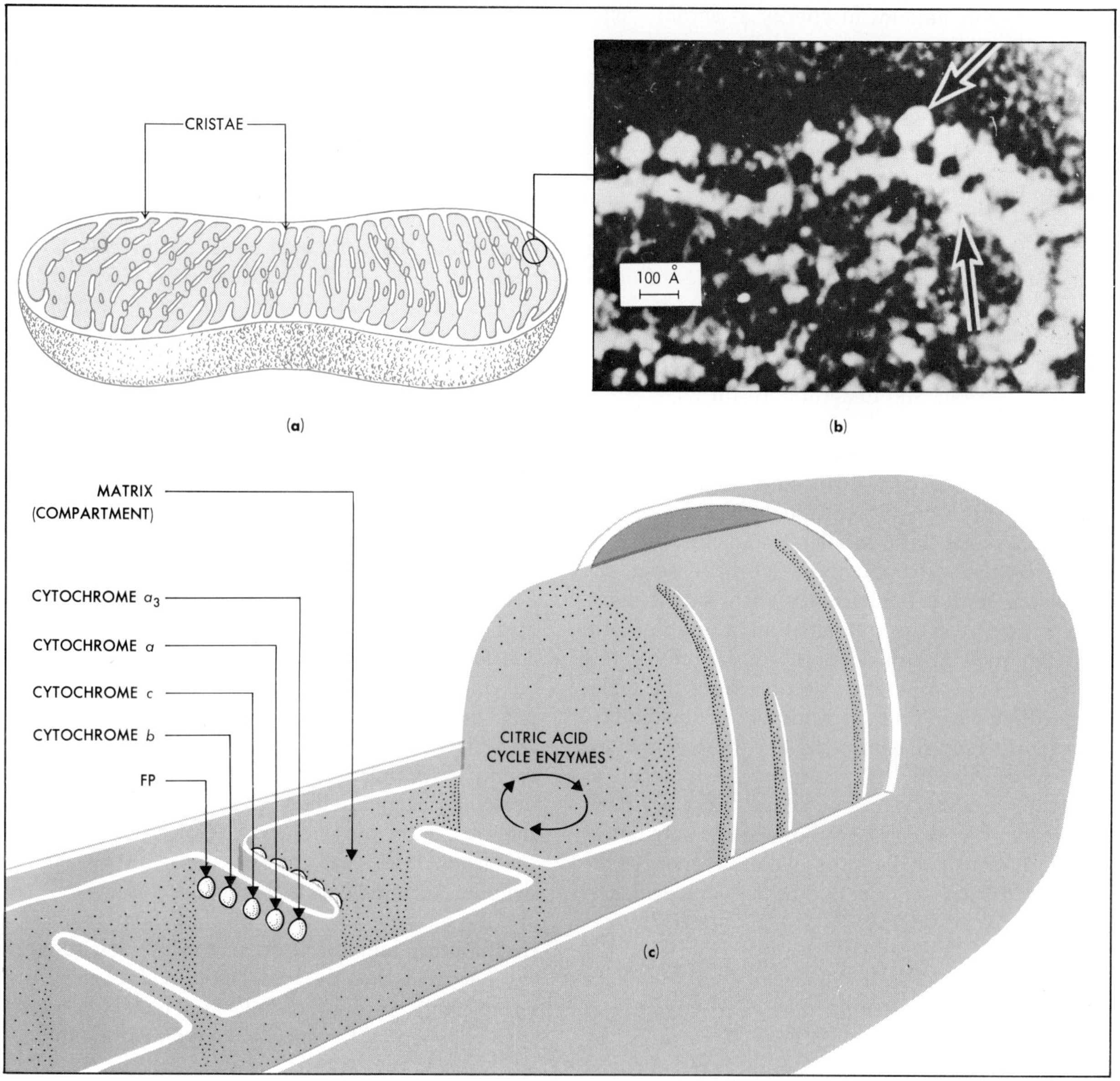

Fig. 5.11 *Successively enlarged views of the internal structure of mitochondria, showing the localization of pathways. Note the intimate relationship between the mitochondrial membrane and certain stepwise enzymatic reactions, oxidative phosphorylation and electron transport in particular [Photo (b) courtesy Institute for Enzyme Research, University of Wisconsin; diagram (c) after* Biology Today. *New York: CRM/Random House, 1972, Interleaf 6.1.]*

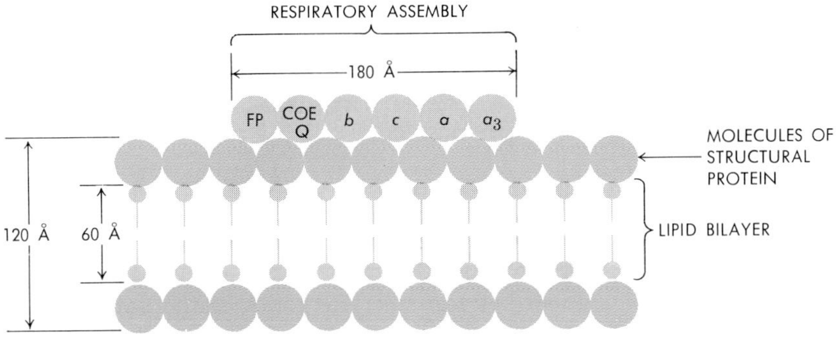

Fig. 5.12 *Cross section of a mitochondrial wall, showing ordered molecules of a single respiratory assembly attached to the wall. Note that the electron transport molecules of the assembly are arranged in the precise order in which they receive and pass on electrons. (Adapted from A. L. Lehninger,* Bioenergetics. *Menlo Park, Calif.: W. A. Benjamin, 1965, p. 113.)*

Fig. 5.13 *Schematic summary of localization of various pathways in the complete oxidation of glucose in aerobic cells. The reactions of glycolysis occur in the cell cytoplasm by enzymes located there. The pathway of the citric acid cycle occurs through soluble enzymes located in the liquid matrix inside the mitochondrion. Electron transport, during which ATP is generated from ADP, occurs in the respiratory assembly, a group of molecules attached to the inside wall of the mitochondrion.*

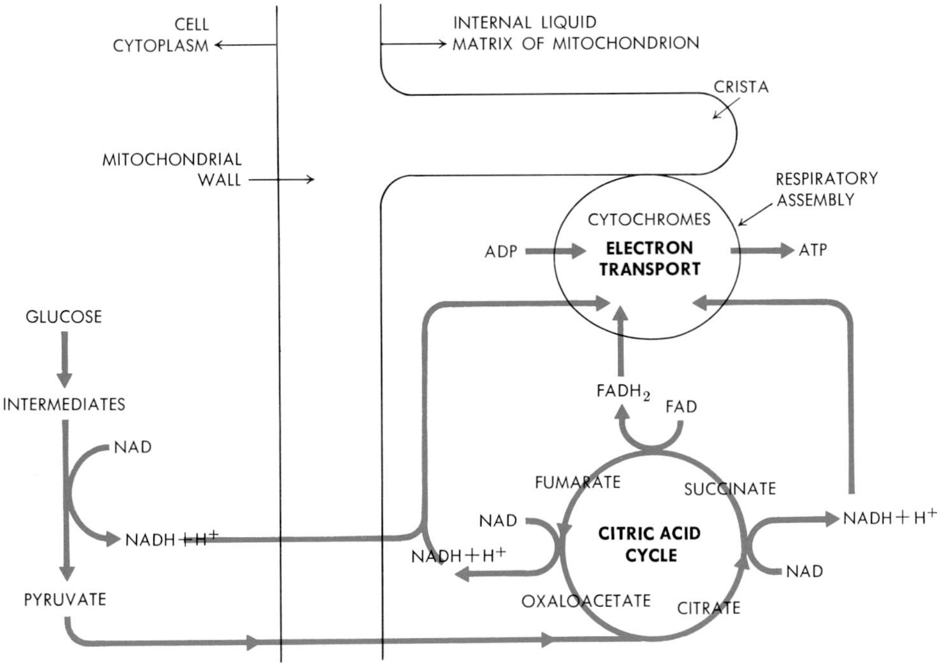

the citric acid cycle, however, are not free in the cytoplasm but dissolved in the liquid matrix inside mitochondria. These enzymes are apparently not bound to the mitochondrial membranes, but they are too large to pass through the walls of the mitochondria into the cytoplasm. The electron transport molecules (the cytochromes, FP to some extent, and possibly NAD) are generally bound to the cristae (Figs. 5.11, 5.12 and 5.13). Pyruvic acid, generated in the cytoplasm by glycolysis, can pass through the mitochondrial wall and come into contact with the enzymes of the citric acid cycle. As soon as electrons are removed from intermediates in the citric acid cycle, they are carried by NAD (or FAD) to the respiratory assembly where electron transport can take place. The by-products of the citric acid cycle (CO_2, water, and ATP) can pass out of the mitochondria into the cytoplasm for elimination or involvement in other metabolic pathways.

We see here, then, three levels of localization. Glycolysis is the least localized of the pathways, occurring throughout the cytoplasm. The citric acid cycle is much more localized, being confined to the liquid medium within the rather small mitochondria. Electron transport is the most highly confined of all the pathways, taking place within a respiratory assembly less than 50 Å in diameter. The advantages of this localization are obvious. When all the reactants and intermediates as well as the enzymes of a particular pathway are confined in a small space, the rate of reaction is markedly increased and thus efficiency improved. The electron transport system shows one of the most highly developed localizations in all of biochemistry. Not only are the molecules tightly bound together in close physical proximity, but they even seem to be arranged in an order appropriate to their functioning in the electron transport process. A diagram of one respiratory assembly, attached to the mitochondrial wall, is shown in Fig. 5.12. With this highly ordered spatial arrangement, electrons can be transmitted rapidly from one carrier to the next in line. In the reactions for glycolysis, the citric acid cycle, and electron transport, the relation between structure and function on a subcellular level is clearly evident. A summary of the localizations involved in glucose breakdown is given in Fig. 5.13.

Energetics of metabolic pathways. In looking at any metabolic pathway, it is important to distinguish between the energetics of the pathway as a whole and the energetics of the individual steps in that pathway. For example, some of the steps in the oxidation of glucose are endergonic, as is the conversion of glucose to glucose-6-phosphate or the conversion of glucose-6-phosphate to fructose-1, 6-diphosphate, both of which require the input of ATP. Yet the overall oxidation of glucose is strongly exergonic. In looking at the energetics of glucose oxidation, we are concerned with (1) the total free-energy change and (2) the efficiency of the entire process—that is, the amount of energy recaptured in a usable form compared to the total free-energy change. In the case of glucose oxidation, the amount of work accomplished is equivalent to the number of high-energy phosphate bonds built for a given amount of glucose oxidized. As seen earlier, the total free-energy change in the complete oxidation of a mole of glucose is 686 kcal. This figure represents the total potential energy released during the process. Information given in Table 5.2 reveals that a total of 38 moles of ATP are generated per mole of glucose oxidized. At a value of 7 kcal per mole of ATP, roughly 266 out of 686 kcal are regained by the cell. Recall, however, that two moles of ATP are required to mobilize a mole of glucose for the oxidative pathways. The cell, therefore, shows a net gain of 36 moles of ATP, or 252 kcal per mole of glucose oxidized. Since 252 is approximately 37% of 686, glycolysis and the citric acid cycle combined extract more than one-third of the energy released during glucose oxidation. This is a remarkable degree of efficiency when compared to the best internal combustion engines, which usually achieve less than 25% efficiency.

Supplement 5.4

LOCALIZATION WITHIN MITOCHONDRIA: THE BASIS FOR OUR KNOWLEDGE

Figures 5.11 and 5.12 illustrate the rather precise knowledge obtained about localization of respiratory pathways within mitochondria. Three separate pathways, for oxidizing glucose in the citric acid cycle, electron transport, and oxidative phosphorylation, have been localized in three separate parts of the mitochondrion.

How is it known that such precise positioning in the geometric space of the mitochondrion actually exists? As in all such biochemical questions, direct observation is not possible. True, the electron microscope has revealed some information about the fine structure of the inner regions of the mitochondrion. But the electron microscope cannot "see" molecules, nor can it ever reveal direct information about function. It is necessary to correlate structural observations with functional observations through experimental methods.

One such experimental approach is the disruption and reconstitution procedure illustrated in Diagram A.

If intact mitochondria, separated from other cell components, are subjected to osmotic treatment so that they take in water and swell, the inner membranes can be seen to contain numerous tiny, sphere-shaped particles, each on a short stalk. The osmotic treatment appears to cause the particles (called F_1 particles) to stick up out of the inner membrane surface where they are normally imbedded. Further treatment of the mitochondria to sonic disruption (very high-frequency sound) breaks the inner membranes down into small vesicles. These vesicles are membrane bilayers from which the F_1 particles protrude.

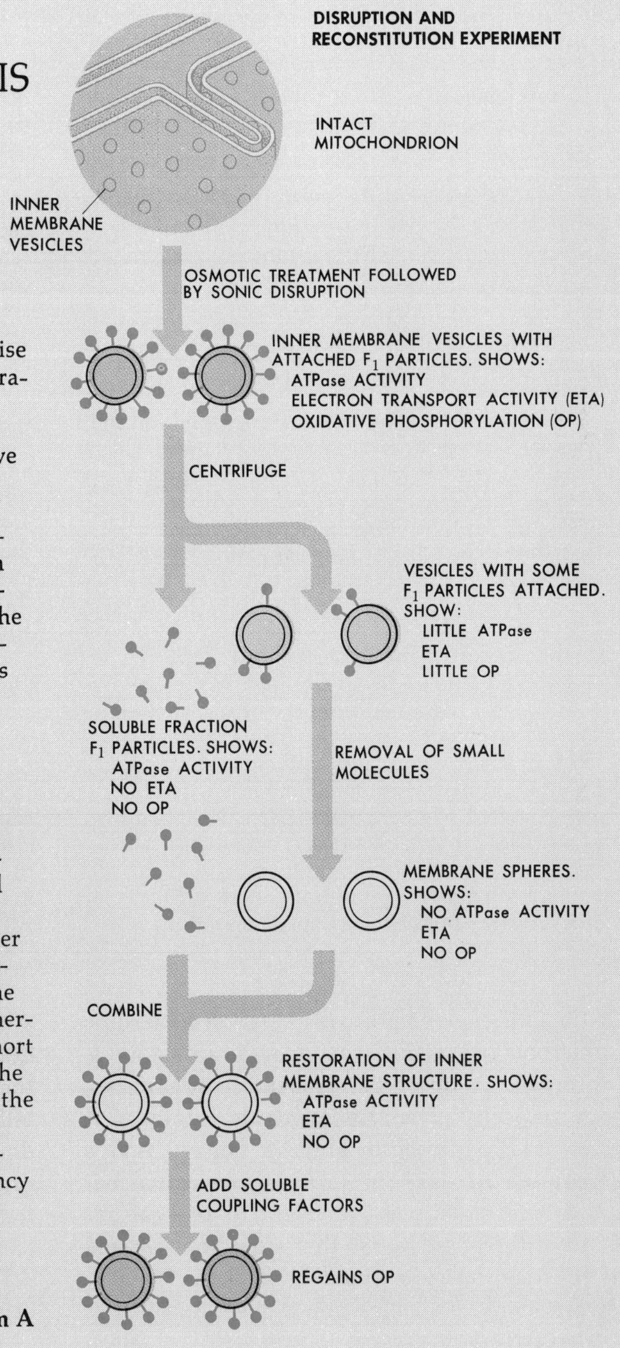

Diagram A

Biochemical tests of the vesicles and attached F_1 particles indicate that the intact units have the following properties:

1. They show electron transport activity (that is, they have the electron transport system, the ETS, composed of the cytochromes and NAD^+).

2. They are capable of carrying out oxidative phosphorylation (they can effect the formation of ATP from ADP and inorganic phosphate), designated as OP (oxidative phosphorylation).

3. They show ATPase activity (they contain the enzyme ATPase, adenosine triphosphatase, which couples electron transport to oxidative phosphorylation). In other words, ATPase activity is necessary for the energy derived from electron transport to be joined to the formation of ATP from ADP.

If the vesicles and attached F_1 particles are further sonicated (subjected to high-frequency sound) and centrifuged, the F_1 particles can be removed and separated from the membranes. The separation of F_1 particles from the membranes by sonication takes time, so that after only a short exposure, not all the F_1 particles are separated. Membrane vesicles with only a few F_1 particles are capable of carrying on electron transport (have ETS), can perform a small amount of oxidative phosphorylation (OP), and show the presence of a small amount of ATPase. Continued sonication removes all the particles. Membrane vesicles without F_1 particles have the ETS, but they show no ATPase activity nor any OP. The F_1 particles show a great deal of ATPase present, but no ETS and no OP. These data suggest that the membranes (which represent the inner membranes of the mitochondrion) contain the electron transport system, while ATPase (the so-called "coupling factor") and other enzymes for oxidative phosphorylation are located on the F_1 particles.

The above hypothesis can be confirmed by reconstituting the intact vesicle-F_1 units (putting them back together). If the membrane vesicles and free F_1 particles are put together in a test tube and the ionic concentration adjusted in a very precise way, the F_1 particles recombine with the membrane to produce the intact respiratory unit. This unit now has full restoration of its three functions: it has ATPase, contains the ETS, and is capable of full OP activity.

Such reconstitution experiments are important steps in confirming the results of disruption experiments, for while it is important evidence to show that removal of a part *eliminates* the function of that part, it is still more convincing to show that replacing the part *restores* the lost function. Such experiments have shown that different parts of the mitochondrion are the sites for different parts of the biochemical pathways involved in glucose oxidation and generation of ATP.

If the efficiency of glucose oxidation in aerobic and anaerobic organisms is compared, an astounding difference is observed. Anaerobic organisms, carrying out only glycolysis, generate a total of 4 moles of ATP per mole of glucose. Since 2 ATP's were required to start the process, the actual net gain is 2 ATP's. This is equivalent to 14 kcal per mole out of the possible 686 that would be released if the glucose molecule could be broken down to its lowest energy states, carbon dioxide and water. Anaerobic organisms are thus able to extract less than 2% of the energy theoretically available to them in the glucose molecule. This figure does not tell the whole story, however. Anaerobic organisms do not break glucose down nearly as far as aerobic organisms do. The total energy released in the partial breakdown of glucose to alcohol or lactic acid is 56 kcal per mole. At a net gain of 2 moles of ATP per mole of glucose oxidized, anaerobic cells can recapture 25% of the energy they release.

Table 5.3
Efficiency of Anaerobic versus Aerobic Respiration

Process	**Anaerobic respiration (glycolysis)**	**Aerobic respiration (glycolysis plus citric acid cycle)**
Total free-energy change during reaction	56 kcal	686 kcal
ATP synthesized (net gain)	2	36
Total free energy stored as high-energy phosphate bonds	14 kcal	252 kcal
Efficiency of recapturing usable energy from total energy released	25%	37%
Fraction of total available free energy recaptured as ATP	2%	37%

Thus, for the reactions they *do* carry out, anaerobic organisms are only slightly less efficient than aerobic. These calculations are recorded in Table 5.3.

Every metabolic pathway has a relative degree of efficiency which can be calculated in a similar fashion. It is particularly important to understand that the relatively high efficiency of cellular metabolic pathways is brought about by (a) their stepwise nature, (b) the fact that they are frequently localized in specific regions of the cell, and (c) the role of enzymes in reducing the amount of activation energy necessary to get the reactions going.

Control of metabolic pathways. Controls affect both the overall direction and the rate at which metabolic reactions take place. Such controls are usually exerted on specific steps of the pathway, and they can occur in a number of ways. One is by the **law of mass action** or **Le Châtelier's principle,** which states that the rate and direction of a reaction (or series of reactions) are affected by the relative concentrations of reactants and products. In a reversible reaction, the greater the concentration of reactant and the less the concentration of products, the more the reaction is shifted to the right. Conversely, the greater the concentration of products and the less the concentration of reactants, the more the reaction is shifted to the left. Concentration of either reactants or products is determined by several factors. One is the availability of reactants, in turn determined by what can get into and leave the cell. A second is the fate of the end products—whether they are removed from the site of the reaction, packaged up and stored in the cell, broken down, or simply transported out of the cell. In aerobic respiration, for example, CO_2 diffuses out of the cell, thus "pulling" the reaction toward more sugar breakdown. An alternative fate is that the end products build up and accumulate in the cell. All these factors affect both the direction and the rate at which a pathway operates by exerting "pushes" and "pulls" on both reactants and products.

A second means by which metabolic pathways are controlled is through the energy requirement—the degree of reversibility—of the individual steps of the reaction series. For example, most of the individual reactions in the citric acid cycle are reversible, having relatively low free-energy changes in either direction. However, from Fig. 5.7 it can be seen that at least two steps—conversion of oxaloacetate and acetyl CoA to citrate, and the conversion of α-ketoglutarate to succinate—are shifted heavily to the right (i.e., in the forward direction). This means that the energy requirement for the forward reaction is considerably less than for the reverse. Thus, although it is theoretically possible for succinate to go back and reform α-ketoglutarate or for citrate to go back and reform acetyl CoA and oxaloacetate, neither of these reactions is very likely to occur. From the standpoint of equilibrium and energy requirements, both reactions are virtually nonreversible. Thus the citric acid cycle is kept moving in a single direction, even though most of the steps are individually reversible. A few essen-

The rate at which the respiratory pathways function is controlled largely by feedback signals to allosteric enzymes located at key points along the pathway.

tially irreversible steps in a metabolic pathway act in a manner analogous to a series of one-way turnstiles in a subway station: they prevent the whole reaction series (the passage of people) from backing up by stopping the reverse flow at specific points. The existence of irreversible steps does not change the *rate* of the overall reaction, which is controlled by other factors, including the concentration of reactants and products. But it does keep the reaction series moving in one direction.

Another example of how directionality is controlled occurs in the electron transport system. Each successive transport molecule has a greater affinity for electrons than the molecule preceding it in the series. There is thus little chance for the electrons to move in a reverse direction and slow down or to avoid being passed on in the sequence necessary to generate ATP.

The rate at which metabolic reactions take place can be controlled by a third and highly subtle device: controlling the activity of enzymes operating on specific steps in the pathway. This control can be exerted in two ways: (1) by controlling the rate at which existing molecules are synthesized, and (2) by controlling the rate at which existing molecules catalyze specific reactions with their substrate. The first method is essentially a genetic mechanism (i.e., controlling the rate at which enzymes are synthesized by the genes) and will be dealt with in more detail in a later chapter. The second process operates by allosteric enzyme controls as described in Chapter 3. This interaction can serve either to increase (activate) or decrease (inhibit) the rate at which the enzyme acts on the substrate.

A single example will illustrate both activation and inhibition of enzymes by products of the pathway. The conversion of citric acid to α-ketoglutaric acid involves several steps, one of which is catalyzed by the enzyme isocitrate dehydrogenase, as follows:

$$\text{CITRATE} \rightleftharpoons \rightleftharpoons \xrightleftharpoons[]{\text{isocitrate dehydrogenase}} \alpha\text{-KETOGLUTARATE} + \text{NADH} + \text{H}^+ + \text{CO}_2.$$

As part of the citric acid cycle, this reaction leads to the production of usable energy (NADH + H^+ or, ultimately, ATP). Chemical investigation has shown that both these molecules are capable of inhibiting the action of the enzyme isocitrate dehydrogenase. In other words, in the presence of these two molecules the enzyme greatly slows down the conversion of citrate to α-ketoglutarate. Thus, as the cell builds up a greater concentration of energy-rich compounds, the pathway for energy production is signaled to slow down. Conversely, it has been found that molecules of ADP or AMP stimulate isocitrate dehydrogenase to convert citrate into α-ketoglutarate. Thus, as the energy supply of the cell begins to run low, enzymes in the energy-producing pathway are signaled to speed up their action.

Other control points can be found in the pathways of glucose oxidation (Fig. 5.14). Note that control occurs both by inhibition and activation of enzymes. With this number of sensitive "control spots," subject to either activation or inhibition by different reaction products, it is apparent that an extremely sensitive mechanism exists by which cells can constantly monitor and regulate their metabolic pathways. Such subtle control is highly advantageous to cells, which must always work to conserve as much energy as possible. A control system on metabolic processes allows the cell (1)

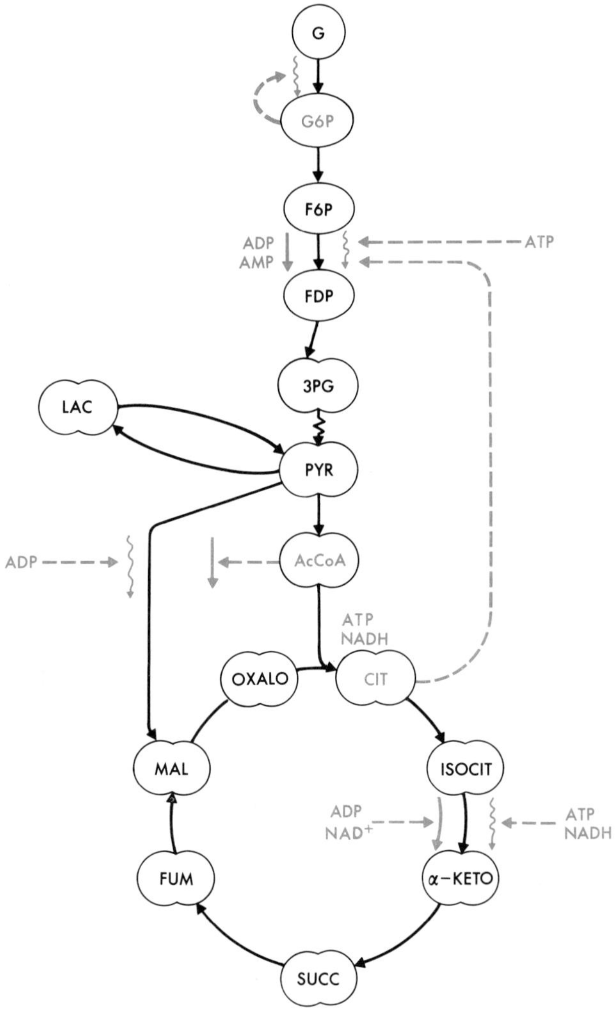

Fig. 5.14 *The many control points in glucose oxidation. Each control point indicated here functions through enzyme inhibition or activation. Allosteric effector molecules are shown in color; heavy colored arrows indicate enzyme activation at that particular point in the pathway; wavy arrows indicate enzyme inhibition by the effector molecule. Dashed arrows point out the site of a control mechanism.*

to speed up necessary and vital reactions and thus not get "caught short" on crucial substances such as ATP, and (2) to slow down the production of unnecessary substances and thus not waste energy and valuable raw materials. Since the external conditions to which cells are exposed are constantly changing (especially for one-celled organisms), survival depends on being able to adjust internal biochemical reactions to meet new needs. The capacity to vary enzyme reaction rates brings about the most effective controls within cells.

The interrelationship between metabolic pathways. Thus far, glycolysis and the citric acid cycle have been considered primarily as pathways for the breakdown of glucose and the production of energy. However, nearly all the intermediates in these two pathways are starting reactants, intermediates, or end products in other metabolic pathways within the cell. For example, citric acid cycle intermediates form connecting points with pathways for the synthesis (and breakdown) of all the major types of molecules found in the cell: fatty acids, amino acids, steroids, carbohydrates, and nucleotides (Fig. 5.15a). This relationship is shown still more specifically in Fig. 5.15(b). Consider, for example, the very important intermediate acetyl coenzyme A, the specific form in which the oxidation products of glucose enter the citric acid cycle. Acetyl CoA represents an intermediate in the breakdown of carbohydrates. In addition, the condensation of several acetyl CoA molecules (each containing two carbons) can produce the hydrocarbon chains, which are major structural components of fatty acid molecules. Thus, when fatty acids are broken down to be used as fuel, they can enter the citric acid cycle as acetyl CoA. Acetyl CoA can also be converted, by a series of reactions, into one of several amino acids. Furthermore, it is the starting point for the synthesis of steroids, molecules that form the basis of several important hormones.

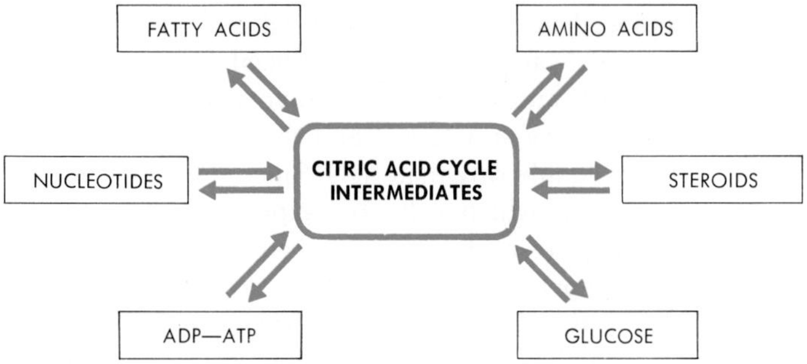

(a)

Fig. 5.15 *The metabolic pathways of glucose oxidation seen as a "metabolic hub." (a) The many other substances that are synthesized from citric acid cycle intermediates. (b) A number of specific branch-points from glucose oxidation, where pathways for synthesis and degradation of other substances enter or depart.*

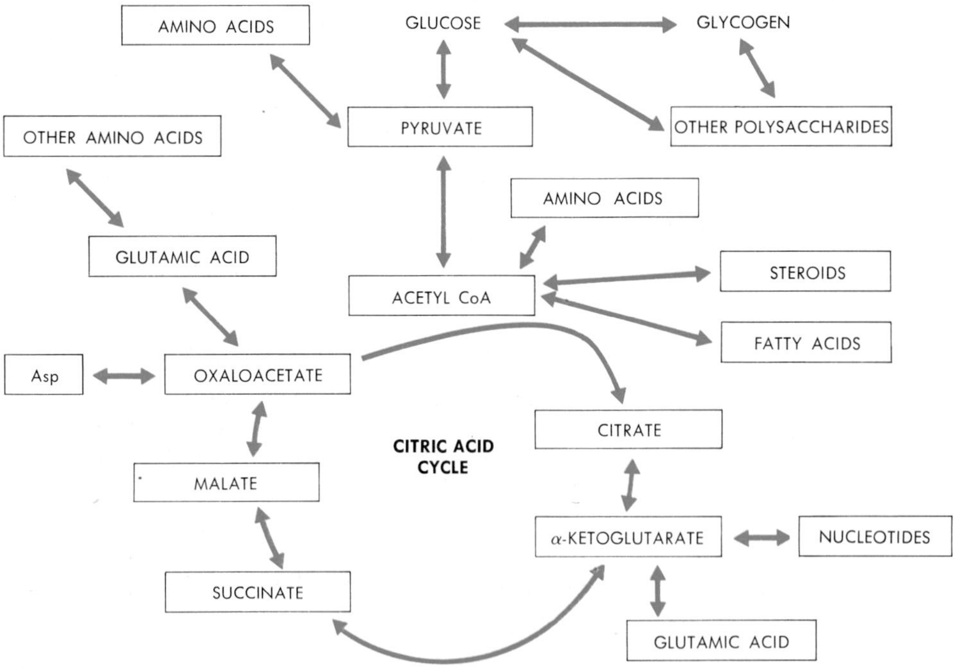

(b)

Supplement 5.5

HOW THE STEPS IN COMPLEX BIOCHEMICAL PATHWAYS ARE INVESTIGATED

The series of reactions involved in glycolysis and the citric acid cycle are numerous and intricate. How do we know the intermediate steps involved and the sequence in which they occur? Several examples will illustrate the way in which biochemists have learned about the respiratory reactions.

In 1935 the Hungarian biologist Albert Szent-Gyorgyi studied the respiration of isolated pigeon breast muscle. He found that the rate of oxygen uptake, quite rapid at first, fell off slowly with time. Pigeon breast muscle respires very rapidly, producing little lactic acid. Its by-products are CO_2 and H_2O. Szent-Gyorgyi noted something else of importance: the fall in rate of respiration paralleled the rate at which succinic acid disappeared from the muscle. He found, however, that the respiratory rate could be restored to its normal level by the addition of small amounts of succinic or fumaric acid. Szent-Gyorgyi concluded that these substances must somehow be involved in the oxidation of carbohydrate. Later studies showed that if succinic, fumaric, or oxaloacetic acid were added to muscle preparation, a large amount of citric acid would be produced. This gave such workers as Hans Krebs the idea that something like the following must be taking place:

$$\left.\begin{array}{l}\text{SUCCINIC ACID}\\ \text{FUMARIC ACID}\\ \text{OXALOACETIC ACID}\end{array}\right\} \rightarrow \text{CITRIC ACID}$$

However, the exact sequence of reactions was not at all clear.

The task of determining the exact reaction series was complicated by the fact that several of these substances are chemically interconvertible. For example, succinic acid can be converted into fumaric acid, and vice versa. This **reversible reaction** is catalyzed by the enzyme succinic dehydrogenase. When succinic acid is added to isolated muscle, fumaric acid and citric acid accumulate. When fumaric acid is added, succinic acid and citric acid accumulate. There is, therefore, no conclusive way to distinguish between the two alternative hypotheses for the sequence of reactions:

Hypothesis I: Succinic acid → fumaric acid → citric acid.

Hypothesis II: Fumaric acid → succinic acid → citric acid.

One experimental technique does make it possible to decide between these alternatives, however. The enzyme succinic dehydrogenase acts only in the chemical reactions by which succinic acid is converted to fumaric acid, and vice versa. We know from an earlier chapter that a substance called malonic acid selectively inhibits succinic dehydrogenase. This means that when malonic acid is present in the system, only molecules of succinic dehydrogenase will cease to function. On the basis of the two hypotheses given above, it is possible to make two predictions. **Hypothesis I:** If the correct sequence of reactions is succinic acid → fumaric acid → citric acid, then addition of malonic acid along with succinic acid should *not* produce an increase in the amount of citric acid. **Hypothesis II:** If the correct sequence of reactions is fumaric acid → succinic acid → citric acid, then addition of malonic acid along with succinic acid *should* produce an increase in the amount of citric acid. Performing the experiment shows that citric acid does not accumulate in the muscle tissue. These results thus support the predictions of hypothesis I and contradict those of hypothesis II.

The foregoing examples illustrate two major types of techniques involved in working out biochemical reaction series. The first technique involves the addition of large amounts of a suspected intermediate to a biochemical system, measurement of its disappearance rate, and identification of the new substance that will then begin to build up. This was the technique applied by Szent-Gyorgyi. The second technique involves the addition of some substance known to inhibit a certain enzyme. The alteration of the rate at which certain products accumulate will indicate at which point in the reaction series the particular enzyme acts. For example, consider this series:

BLOCKED WITH METABOLIC POISON (at enzyme 3)

$$A \xrightarrow{\text{enzyme 1}} B \xrightarrow{\text{enzyme 2}} C \xrightarrow{\text{enzyme 3}} D \xrightarrow{\text{enzyme 4}} E \text{ (END PRODUCT).}$$

If enzyme 3 is put out of commission by a known inhibitor, the intermediate products B and C will accumulate while D disappears (it is converted to E). This indicates that C precedes D in the reaction series. It also indicates that enzyme 3 catalyzes the conversion of C to D. Several substances, such as cyanide and carbon monoxide, are known to block the action of certain cytochromes. By employing these poisons, it has been possible to work out the sequence of reactions in electron transport.

There are, of course, many other techniques frequently used in biochemical analysis. One such technique involves radioactively labeled molecules. These are fed to a biochemical system and the reaction series is stopped at various places. The distribution of radioactive atoms in various intermediate products gives an idea of the pathway taken by the series of reactions. An example of a labeling experiment will be discussed later with regard to the pathways of photosynthesis.

Note also from Fig. 5.15(b) that the conversion of acetyl CoA to each of these other types of molecules occurs by reversible pathways, a very important factor in the regulation of cell metabolism. If a greater amount of carbohydrates comes into the cell than is needed to build ATP, much of the carbohydrate can be shunted off through acetyl CoA to form fatty acids. Fats are an important storage form of fuel molecules and can be held in reserve until needed. The existence of this pathway is responsible for the buildup of fat in human beings on a carbohydrate-rich diet.

As Fig. 5.15(b) shows, many intermediates in the citric acid cycle, e.g., α-ketoglutarate and oxaloacetate, can be converted into other types of molecules, principally amino acids. Pyruvate can enter a pathway for the formation of several types of amino acids, and glucose can be converted into a variety of polysaccharides including, most prominently, glycogen. The existence of these many interconvertible steps between the citric acid cycle and other metabolic pathways allows for maximum efficiency in the use of molecules in the cells. If every pathway started with its own reactants, proceeded through its own particular set of intermediates, and produced its own individual set of products, far more types of molecules would be needed in a cell than are actually found there. That the end product of one pathway can serve as an intermediate or a reactant for another pathway conserves both the number and the different types of molecules necessary and reduces the amount of work the cell must do. Cells are small, and space and energy are usually at a premium. Thus the interconvertibility of pathways not only allows the exercise of more effective control but also increases the efficiency of metabolic activity.

Cells are amazing biochemical systems. With their high degree of structural organization and the close interrelationships between internal cellular structure and biochemical function, cells operate at a level of activity and efficiency that the most elaborate machine human beings have devised cannot even approach. This does *not* mean that cells exhibit mystical or nonphysical properties defying explanation. It *does* mean that cells are far more intricate and subtle units of structure and function than people have sometimes thought.

In this chapter we have discussed the interrelationships between structure and function in the processes by which all types of cells, from simplest to most complex, gain usable energy by running the downhill or exergonic reactions of glucose oxidation. In the next chapter we will explore the other side of that energy hill: how cells of certain types of organisms, green plants, use the energy of sunlight to synthesize carbohydrates in an endergonic process, photosynthesis.

SUMMARY

1. Cells resemble internal combustion engines in that they oxidize fuels to produce energy for work. They are unlike machines in that their internal organization is far more complex, with more subtle interaction of parts, than is found in any machine.
2. Metabolism is the sum total of chemical reactions involved in the degredation and synthesis of substances in the cell. Metabolism involves the reactions in which glucose is oxidized to produce energy, as well as the reactions by which carbohydrates, fats, proteins, nucleic acids, and other important molecules are synthesized or degraded.
3. Adenosine triphosphate (ATP) is the energy currency of the cell. By hydrolysis of a phosphate group (yielding ADP + P_i), 7 kcal of energy are released. This is slightly more energy than is needed to synthesize a peptide, glycosidic, or ester bond. The reaction by which ATP is hydrolyzed to ADP (or in some cases to AMP by loss of the last two phosphate groups) is coupled to the reaction in which the energy from ATP is used. Coupling is enzyme-catalyzed and usually involves an intermediate substrate to which the terminal phosphate group from ATP is attached. In this way, ATP provides virtually all the energy for cell reactions.
4. In animal cells, ATP can be generated in two main ways: (1) through electron transport, in which three ATP's are normally generated for every pair of electrons passing down the acceptor chain, a process known as oxidative phosphorylation, and (2) through substrate phosphorylation, the process by which a phosphate group is added to an intermediate in a reaction series and then converted into a high-energy phosphate bond by electron rearrangements within the substrate. This high-energy phosphate group is then transferred to ADP, producing ATP. In cells of higher plants, a third method of ATP production is available: the use of light energy to join P_i onto ADP to produce ATP. This process is called photophosphorylation (see Chapter 6). The cells of

some lower organisms, particularly among the bacteria, may lack the electron transport system and hence produce ATP only through substrate phosphorylation.

5. The electron transport system (ETS) consists of two acceptors, nicotinamide adenine dinucleotide (NAD) and flavine adenine dinucleotide (FAD), and the cytochromes (cytochromes *b*, *c*, *a*, and a_3). The sequence of electron flow through this system is: NAD → FAD → Cyt *b* → Cyt Q → Cyt *c* → Cyt *a* → Cyt a_3. The final electron acceptor is oxygen, producing water. The cytochromes have a central iron atom that undergoes successive reduction and oxidation as electrons are taken up and passed on, respectively. The flow of electrons goes in one direction, because each acceptor along the chain has a higher affinity for electrons (a higher electronegative potential) than the molecule preceding it. The electron transport system is bound to the inner membranes of mitochondria and is thus a highly localized system.

6. The ETS is coupled, through the enzyme adenosine triphosphatase (ATPase), to the generation of ATP from ADP and P_i. Energy is released during electron transport because the electrons pass, successively, from higher to lower energy levels in moving from one acceptor molecule to another. These downward transitions in energy levels release small quantities of energy, some of which are just large enough to be harnessed to produce the high-energy phosphate bonds of ATP.

7. There are three phases in the oxidation of glucose to yield energy for ATP production: (1) mobilization, during which glucose (or its storage form glycogen and other starches) is "primed" by addition of two phosphate groups; (2) glycolysis, during which the primed sugar (called fructose-1, 6-diphosphate) is broken down into two three-carbon sugars, which through oxidation (i.e., the removal of a pair of electrons) and internal rearrangement are converted into pyruvic acid. By substrate phosphorylation, four ATP's are generated, two from each three-carbon sugar; (3) citric acid cycle, during which pyruvic acid is further oxidized through a series of intermediate products that include citric acid; during the citric acid cycle, 28 ATP's are produced by electron transport, and 2 by substrate phosphorylation. The total ATP produced by all processes of electron transport and substrate phosphorylation is 38. Since the cell spends 2 ATP's to "mobilize" the glucose, the net gain is 36 ATP's per glucose molecule oxidized.

8. Anaerobic respiration is a process of glucose breakdown involving only mobilization and glycolysis. Anaerobic respiration occurs without the process of electron transport; hence it does not require oxygen. There are two kinds of anaerobic organisms: obligate and facultative. Obligate anaerobes lack the cytochrome and other molecules of the electron transport system; they must gain all their energy from substrate phosphorylation. The electrons removed from the substrate during glycolysis are passed back to pyruvic acid to produce a waste product such as butyric acid or lactic acid. Facultative anaerobes are those that can switch from aerobic to anaerobic respiration when the oxygen supply runs low. Facultative anaerobes possess the electron transport system and can carry out the citric acid cycle. One of the most common facultative anaerobes is yeast; when respiring anaerobically, yeasts transfer the electrons picked up during glycolysis to pyruvic acid, eventually producing ethyl alcohol.

9. Aerobic respiration refers to the complete oxidation of glucose, beginning with mobilization and ending with the citric acid cycle. The final products of aerobic respiration are carbon

dioxide and water, along with the ATP generated by substrate and oxidative phosphorylation.

10. Because aerobic respiration oxidizes glucose to its lowest free-energy level, this process produces far more ATP than anaerobic respiration. The aerobic process releases 686 kcal of energy per mole of glucose oxidized. Anaerobic respiration releases 56 kilocalories, or about 8% of the total free energy of glucose. The efficiency of each process is measured as the total energy recaptured by the cell (as ATP) from the energy released. Aerobic organisms recapture 252 kcal per mole of glucose (a net gain of 6 ATP × 7 kcal), and about 37% efficiency (252/686 = 37). Anaerobic organisms recapture 14 kcal per mole of glucose oxidized to pyruvic acid (a net gain of 2 ATP's × 7 kcal), for an efficiency of about 25% (14/56 = .25).

11. The complex pathways involved in glucose oxidation are localized within specific parts of the cell. Mobilization and glycolysis occur in the cytoplasm and are mediated by enzymes not attached to membranes. The citric acid cycle occurs within the mitochondria, in the liquid matrix between the internal partitions, or cristae. These enzymes, too, appear to be unattached to membranes. Electron transport is carried out by the electron transport system, whose molecular components (NAD, FAD, and the cytochromes) are bound to the inner walls of the mitochondrion. In addition, the enzyme adenosine triphosphatase (ATPase), which couples electron transport to oxidative phosphorylation, is bound with the cytochromes to the cristae.

12. The pathways for glucose oxidation are interconnected to the pathways for synthesis and breakdown of most of the other molecular components of the cell. Thus fats (including steroids) are synthesized from acetyl CoA and are broken back down to acetyl CoA; certain amino acids and nucleotides are synthesized from and broken down to α-ketoglutarate.

13. Control of the rate of metabolic pathways, such as glucose oxidation, comes about through three processes: (1) the law of mass action, or the concentration of reactants and products, (2) the degree of reversibility of individual steps; highly reversible reactions are especially sensitive to mass action effects, while those less reversible are sensitive only in the direction toward which their equilibrium point is shifted, and (3) the feedback control exerted by allosteric enzymes. The latter is by far the most subtle and important of the control processes. Details of the mechanism of allosteric control are found in Chapter 3.

EXERCISES

1. Explain why the equation

$$C_6H_{12}O_6 + 6O_2 + 6H_2O \rightarrow 6CO_2 + 12H_2O$$

is a more accurate representation of the process of respiration than the equation

$$C_6H_{12}O_6 + 6O_2 \rightarrow 6CO_2 + 6H_2O.$$

What is the role of water in the process of aerobic respiration?

2. Give several reasons why energy-releasing reactions in living cells occur in a number of steps.

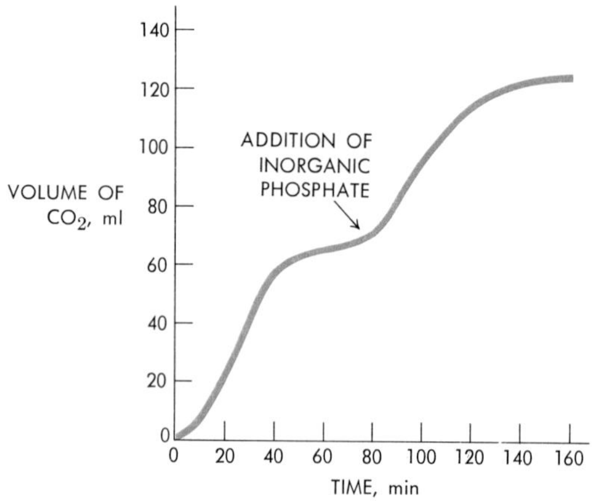

Figure 5.16

3. Figure 5.16 relates the amount of carbon dioxide produced by yeast fermentation to time. Answer the following questions on the basis of the information contained in this graph.
   a) Why does the addition of inorganic phosphate at about the 75-min mark increase the amount of CO_2 produced? What does this show about the role of phosphate in the process of respiration?
   b) Compare the *rates* of CO_2 production for the first 25 min with those for the second 25 min [rate of respiration is determined by dividing the total amount of reaction (in this case measured as the number of ml of CO_2 produced) by duration of the reaction (25 min for each period)]. Are the *rates* the same or different? Is this expected in terms of the role we think phosphate plays in the process of respiration? Why or why not?

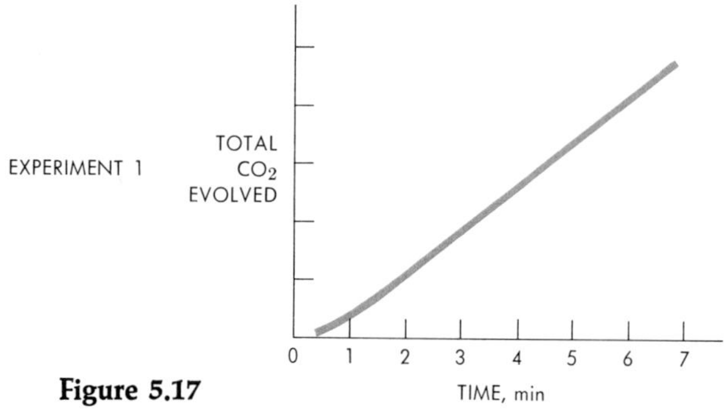

Figure 5.17

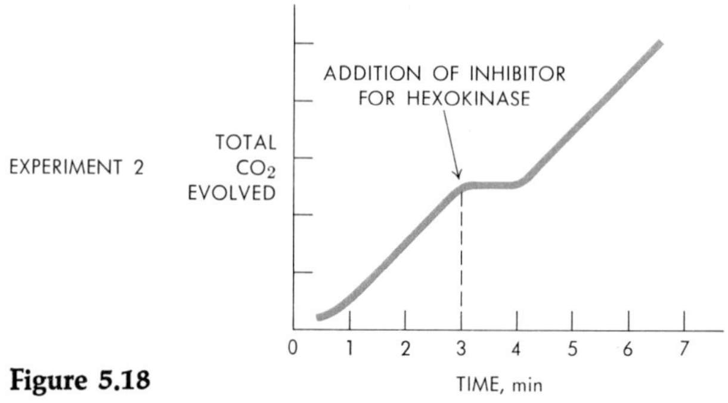

Figure 5.18

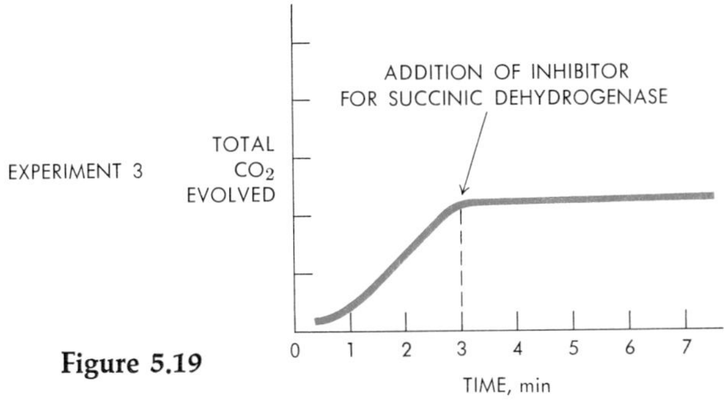

Figure 5.19

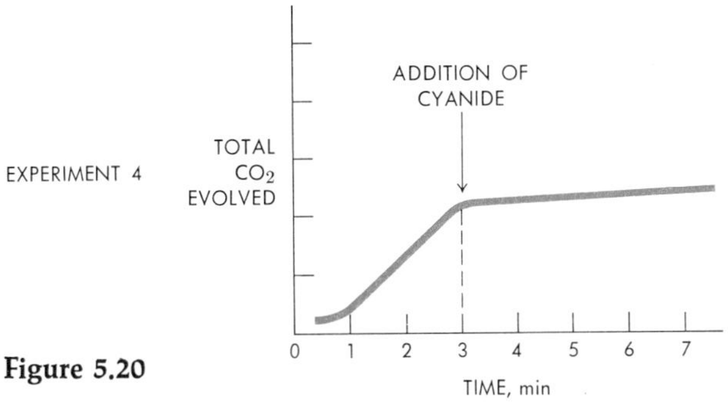

Figure 5.20

4. A slice of liver tissue (an aerobic type of cell) was prepared with cells intact and allowed to carry on respiration. The culture medium in which the cells were incubated contained all the necessary precursors for respiration: glucose, oxygen, ADP, and inorganic phosphate (P_i). Respiration rate was measured by the total amount of carbon dioxide evolved in given periods of time. As long as the culture was carefully maintained, respiration occurred normally and the graph for CO_2 evolution was as shown in Fig. 5.17.

   The enzyme hexokinase catalyzes the conversion of glucose into glucose-6-phosphate. If a competitive inhibitor for hexokinase is added to the incubating liver cells, the curve for CO_2 evolution is as shown in Fig. 5.18.

   The enzymatic succinic dehydrogenase catalyzes the conversion of succinic acid to fumaric acid. When an inhibitor for this enzyme is added, the curve for CO_2 evolution is much the same as shown in Experiment 2 (see Fig. 5.19).

   If cyanide is added to the incubating cells (no other inhibitors added), the curve for CO_2 evolution appears as in Fig. 5.20.

Assume that oxygen is readily available to all the cells and that its delivery does not depend on the presence of hemoglobin or myoglobin. Answer the following questions based on these data:

a) Explain the difference between the graph lines for Experiments 2, 3, and 4 and that for Experiment 1. Discuss each experiment separately, showing the metabolic basis for the observed effects. Relate your answers to occurrences on the molecular level in the various pathways involved. Discuss the molecules affected by the various inhibitors and how these molecules affect equilibrium in the pathway.

b) Dinitrophenol (DNP) is a drug that uncouples ATP production from the electron transport system, thereby interfering with ATP production but allowing electron transport to occur. Suppose that DNP is added to each of the cultures in Experiments 2, 3, and 4 at time period 3 (the same time point at which each of the inhibitors is added). Predict what the resulting line for CO_2 evolution will be like. Draw the graph for each experiment. Explain your reasons for each graph.

c) Predict the changes, if any, in the graph lines if pyruvate were added to the culture system for Experiments 2, 3, and 4. Give your reasons.

5. How is it that the anaerobic respiration of glucose yields different end products in yeast and animal tissues? What must be the principal differences in the two types of cells?

6. Tables 5.4 and 5.5 show the amounts of carbon dioxide produced and oxygen consumed during the respiration of germinating seedlings. To compare the respiratory processes of different organisms, these data can be treated in two ways. They may

Table 5.4
Data on Germinating Wheat Seedlings

CO_2 produced, ml	O_2 consumed, ml
9.2	9.6
14.8	17.1
15.6	14.9
16.6	13.8
17.0	17.4
20.1	19.9
21.6	21.2

Table 5.5
Data on Germinating Castor Bean Seedlings

CO_2 produced, ml	O_2 consumed, ml
5.1	7.4
13.1	17.4
9.1	13.2
2.5	4.4
5.0	6.6
11.3	15.1
18.8	26.7

be plotted on a graph (ml of O_2 used on the horizontal axis and CO_2 produced on the vertical) and the slope of the lines compared, or the **respiratory quotient** for each type of organism may be determined. Respiratory quotient is determined by dividing the amount of oxygen consumed by the amount of CO_2 produced. This can be done by averaging the amounts of each gas, then using these mean values for calculation. On the basis of this information, answer the following questions:

a) What are the respiratory quotients for the wheat seed and the castor bean seed?

b) Plot the data for each table on a graph. How do the slopes for the two lines compare to each other? How do the slopes relate to respiratory quotient?

c) A difference in respiratory quotient indicates that one organism uses more O_2 during its respiration than the other. Knowing that the oxygen used in this process unites with hydrogens from a foodstuff, can you account for the *differences* between the respiratory quotients of wheat and castor bean?

d) What additional information would you need to verify your answer for (c)?

7. Observe the experiments described by Fig. 5.21.

a) Which of the graphs *best* illustrates the temperature–time curves of the four flasks?

b) Explain why the line for each flask (on the graph you chose) should have the shape that it does.

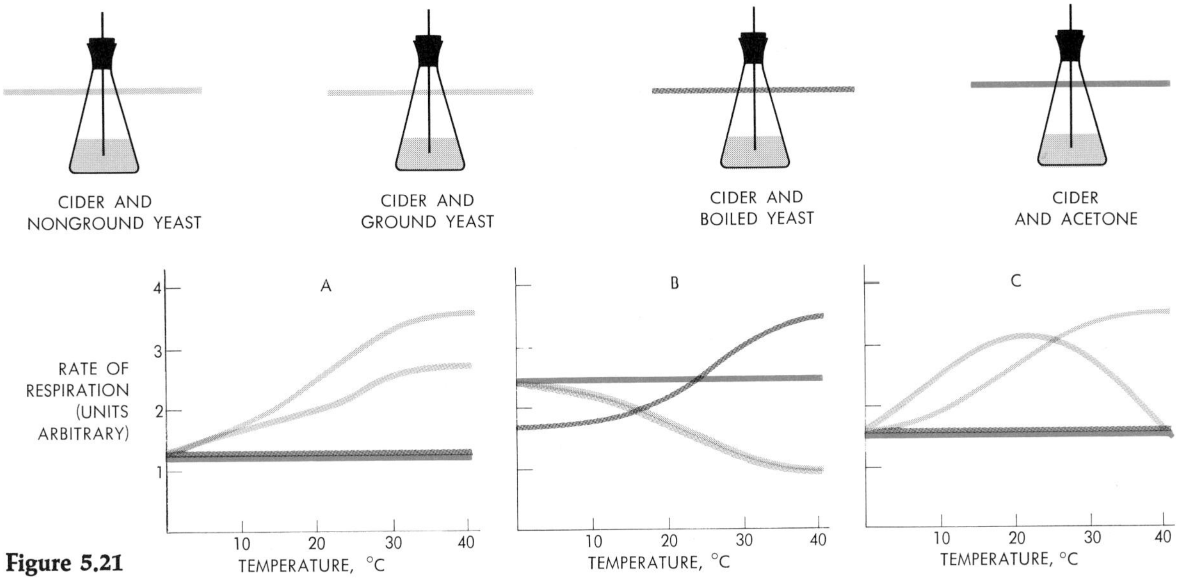

Figure 5.21

c) On the molecular level, why should rate of respiration in yeast cells vary with temperature?

8. Why are carbohydrates said to be fattening?
9. In what sense is the citric acid cycle a central metabolic hub in the biochemistry of living cells?
10. Inorganic phosphate (P_i) is a competitive inhibitor for the enzyme alkaline phosphatase (APase). APase catalyzes the removal of a terminal phosphate group from several kinds of compounds by a reversible reaction of the type shown below:

$$\text{GLUCOSE-6-PHOSPHATE} \xrightleftharpoons{\text{APase}} \text{GLUCOSE} + P_i.$$

If this reaction is allowed to occur in a system with excess substrate, and a concentrated solution of P_i is added as an inhibitor at some point after the reaction has started, the graph in Fig. 5.22 is obtained. If a concentrated solution of sodium arsonate ($NaHAsO_4$), also a competitive inhibitor for APase, is added to a second reaction in place of the P_i, the graph of Fig. 5.23 is obtained.

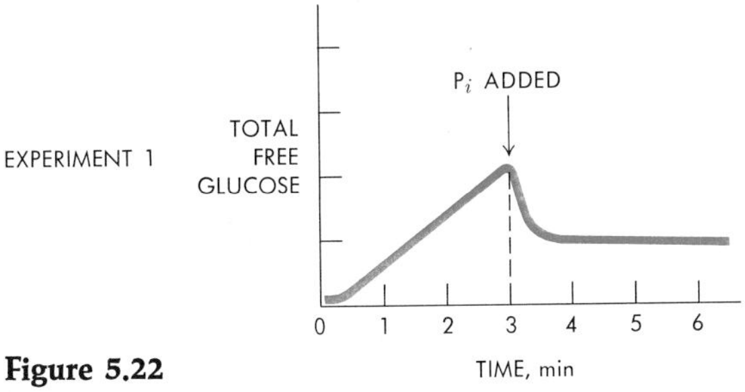

Figure 5.22

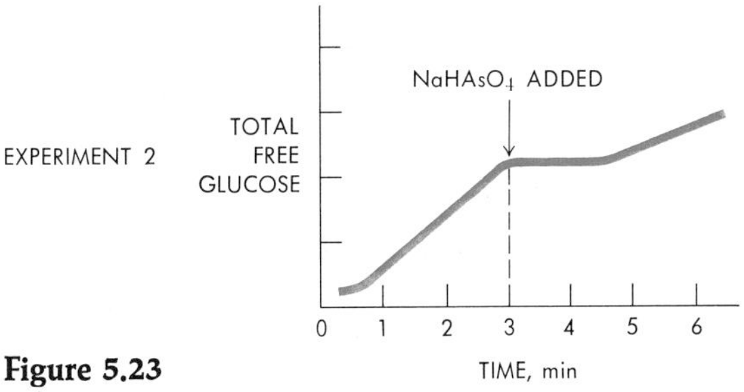

Figure 5.23

Explain the differences between these two graphs in terms of molecular interactions between substrate, enzyme, and inhibitor. Discuss the following points:

a) In both experiments, why does the graph line stop climbing after inhibitor is added?

b) In Experiment 1, why does the graph line fall to a lower level after the addition of inhibitor, whereas the line remains horizontal for Experiment 2?

c) The effect of each inhibitor on the equilibrium of the reaction.

11. A scientist was trying to study the events involved in *intracellular* packaging and transport of proteins that eventually culminated in the secretion of these proteins outside the cell. To do this she studied the secretory process in pancreatic cells, inasmuch as their primary function is the secretion of enzymes and hormones. The secretion of these materials is caused by the fusion of secretory vesicles with the plasma membrane and the release of their contents into the extracellular space. In pancreatic cells these vesicles are also called zymogen granules. So as to perform these studies she made use of electron microscopic and autoradiographic procedures. Her technique was as follows:

1. She gave the pancreatic cells radioactive precursors of proteins (i.e., radioactive amino acids) for 3 min. Thus, *any proteins made in these 3 min would be radioactive.*

2. After 3 min she "chased" away all the radioactive precursors by supplying the cells with nonradioactive precursors. Hence all proteins synthesized after the initial 3-min radioactive pulse feeding would be nonradioactive.

3. She then cut thin sections of the pancreatic tissue at various time intervals, including 7 min after the initial radioactive pulse feeding, 17 min after, and so on.

4. By placing photographic film over the electron microscopic sections, she was able not only to see the pancreatic cells in great detail, but could also see where the radioactivity was localized within the cell. This is because the radioactive protein molecules within the cells caused marks called grains on the photographic film.

The scientist made two statistical treatments of her data. In interpreting these data it will be useful to know that the forming face of the Golgi apparatus are those cisternae of the Golgi that are closest to the endoplasmic reticulum and the nucleus, and that condensing vacuoles are the parts of the Golgi apparatus closest to the plasma membrane. Also, zymogen granule is the name of the final secretory vesicle. Finally, one may interpret the Golgi apparatus as part of the smooth endoplasmic reticulum fraction of the cell.

Based on the data in Fig. 5.24 and Table 5.6, interpret the intracellular pathway of proteins involved in secretion in pancreatic cells. That is to say, what is the flow (from where to where) of the proteins involved in pancreatic secretion? A short qualitative answer is sufficient.

12. The graph in Fig. 5.25 shows the effect of carbon monoxide on the respiration of yeast cells.

a) Offer an hypothesis to explain the difference in respiratory rates of yeast cells treated with carbon monoxide (CO).

b) What seems to be the effect of light on the inhibitory effect of CO?

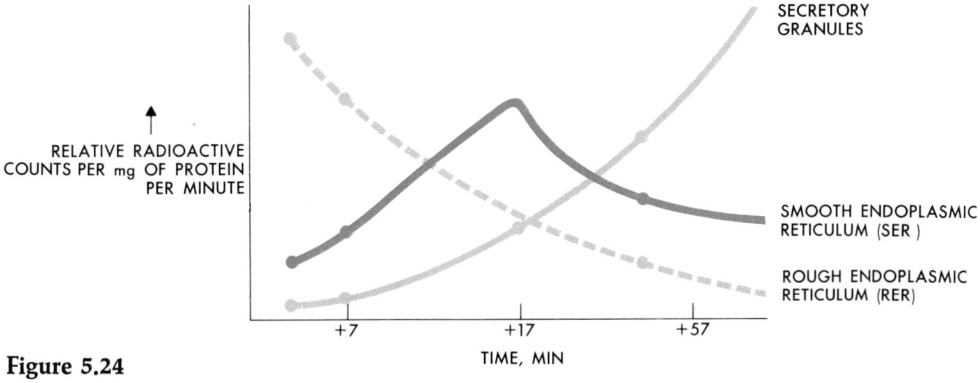

Figure 5.24

Table 5.6
Percent Grains Observed after Pulse Feeding of Cells

Cell Part	3 Min	7 Min	37 Min	117 Min
Rough endoplasmic reticulum	86	40	25	20
Golgi apparatus				
Forming face	3	43	15	4
Condensing vacuoles	1	4	49	8
Zymogen granules	3	5	11	54

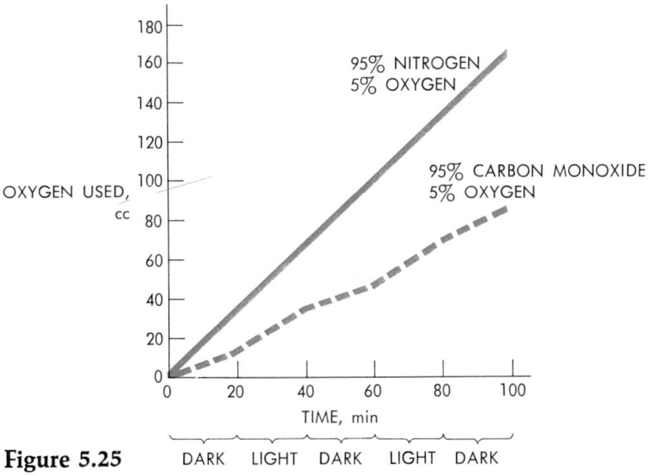

Figure 5.25

13. Which substance would you take for quick energy, glucose or sucrose? Why? Why is chocolate candy said to be a source of quick energy?

SUGGESTED READINGS

A great deal of material has been published recently concerning cellular metabolism, ATP, electron transport, and the like. Below are listed some of those articles and books that might be most useful in providing additional background and recent information in the field.

Baker, J. J. W., and G. E. Allen, *Matter, Energy, and Life,* 3d ed. Reading, Mass.: Addison-Wesley, 1974. A more detailed account of the background material discussed in this chapter. Various chapters of the book deal with such topics as the structure of matter, formation of molecules, the course and mechanism of chemical reactions, acids and bases, the chemical composition of living matter, carbohydrates and lipids, proteins, enzymes, and nucleic acids. The book assumes no formal knowledge of chemistry or physics.

Hinkle, Peter C., and Richard E. McCarty, "How cells make ATP." *Scientific American,* March 1978, p. 104. A review of recent studies on generation of ATP by factors as different as light and oxidation.

Lehninger, A. L., *Bioenergetics,* 2d ed. Menlo Park, Calif.: W. A. Benjamin, 1973. An excellent book, though difficult reading. Presupposes some college-level background in chemistry and biology, though it is written clearly and for the most part is comprehensible to the diligent beginning student.

Loewy, Ariel, and Philip Siekovitz, *Cell Structure and Function,* 2d ed. New York: Holt, Rinehart and Winston, 1969. In Part Four, Chapters 11, 12, and 13 contain a valuable summary of metabolism and energetics. The diagrams are very helpful.

Rosenberg, Eugene, *Cell and Molecular Biology.* New York: Holt, Rinehart and Winston, 1971. Chapter 3 deals with metabolic pathways. This book is well written and aimed at the beginning student, though it is not "watered down." Provides easier access to additional detail on metabolism than Lehninger.

SIX

CELL METABOLISM II: THE UPHILL ENERGY PATHWAY—PHOTOSYNTHESIS

6.1 Introduction

Photosynthesis, the manufacture of energy-rich compounds in plants, is one of the most important chemical processes on earth. Experimental investigations of this process extend back over several hundred years, indicating the long-standing interest human beings have had in a subject on which their very existence depends.

In this chapter we will take an approach different from that followed up until now. Instead of presenting the most up-to-date view of photosynthesis first, we will begin with a brief history of how our knowledge of photosynthesis came about over the past several hundred years. In each example of work carried out, consider what kinds of questions the investigators were asking, what theoretical framework they were using, and what methods they employed to answer their questions.

Approaching the development of scientific ideas historically is not a new idea. It represents one kind of problem solving, allowing the reader to confront a problem, and the nature of the evidence available, gradually. If the strong and weak points of each step are understood along the way, the entire process becomes easier to grasp at the end.

The historical method also allows us to glimpse the logic involved in the work of investigators at earlier periods of history than our own. Often their approach, which at first glance might seem antiquated by modern standards, is quite sophisticated when understood in its own historical context. There is sometimes the tendency to think that all, or at least most, past scientific work is inferior to what goes on in today's laboratories. An historical approach can help dispel such myths.

The topic of photosynthesis is not the only one that could be approached in this manner. Many other subjects—respiration, genetics, evolution, in fact almost any area of biology—could be treated similarly. For reasons of space, we are not employing the historical method extensively with any other topic. Photosynthesis lends itself particularly well to such a treatment. Most of the historical material is found in Section 6.3. Details of the modern view are presented in Sections 6.4 through 6.6.

6.2 Reading Case Histories

When reading about the ideas held by the early natural scientists, one is often tempted to scoff at their work. One may wonder how they could ever have held certain beliefs. It is all too easy, however, to look back from the twentieth century to an earlier period and point out scientific mistakes or false hypotheses; hindsight is much more accurate than foresight. We have at our disposal a great deal of factual knowledge entirely unknown to earlier scientists.

In the early seventeenth century, when the experimental study of plant physiology began, there was little evidence to show how chemical reactions occurred. Investigators noted that some substances went into a chemical reaction and other substances came out. From such observations, the **transmutation theory** evolved. Transmutation was believed to be a process by which one substance could change into another. For example, early investigators noted that when water was boiled away in a flask, a small residue of crystals remained. This they interpreted to mean that some of the water had been transmuted into "earthy material." Many of these early experimenters were called alchemists. Although alchemists are frequently referred to as those who sought to change base metals into gold, only a small number of them actually put much serious effort into the attempt. The early development of chemistry owes a great deal to alchemists. By their many investigations with a variety of inorganic and organic substances, alchemists collected a great many empirical data. Some of these data formed a foundation for the development of the first chemical theories, of which the theory of transmutation is a prime example.

In the seventeenth century, little or nothing was known about the nature of gases or, as they were then called, "aeroform substances." Many believed them to be spirits released when certain solids were heated, while others considered them to be actually alive. Furthermore, if there was little understanding of the nature of gases, there was certainly no more comprehension of their significance to living organisms. As we shall see, this lack of knowledge played an important part in the early investigations of photosynthesis.

The "phlogiston theory" was as important to the seventeeth-century scientist as the atomic theory is to chemistry today. The phlogiston theory was advanced to explain many phenomena, especially burning. It was thought that the flames leaping upward and away from a burning object represented something escaping from the object. This unknown something was called "phlogiston." Today it is known that far from giving anything up, a burning substance unites with oxygen. If all the products of the burning process are measured (including the gaseous compounds released), the weight is greater after burning than before. The additional weight can be accounted for by the amount of oxygen used in the burning process. The seventeenth-century scientist was also aware that a burning substance gained rather than lost weight, and one might think that this fact would raise serious questions about the phlogiston theory. To those early scientists, however, this posed no problem. Phlogiston was simply assigned a negative weight! In other words, due to the loss of the negatively weighted phlogiston, a substance would weigh more after burning than before.

The phlogiston theory actually accounted for a considerable number of observable phenomena. It is a prime example of a false hypothesis that gives rise to many true conclusions. For example, the early scientists noted that a candle burned under a sealed bell jar eventually goes out. This, of course, is due to the fact that there is no more oxygen available in the air within the jar. To adherents of the phlogiston theory, however, the air was simply referred to as being "phlogisticated." Such phlogisticated air was said to be "fixed" and no longer capable of supporting burning. The air present under the bell jar before the candle burned was referred to as being "dephlogisticated air." The phlogiston theory thus adequately explained the phenomenon of burning without running into any contradictions that might disprove it. For this particular event, the burning of a candle, the phlogiston theory is just as good as the modern theory involving oxygen.

A belief in transmutation, a lack of any concept of the nature and significance of gases, and an acceptance of the phlogiston theory mark the mental perspective from which early plant physiologists approached their investigations of photosynthesis.

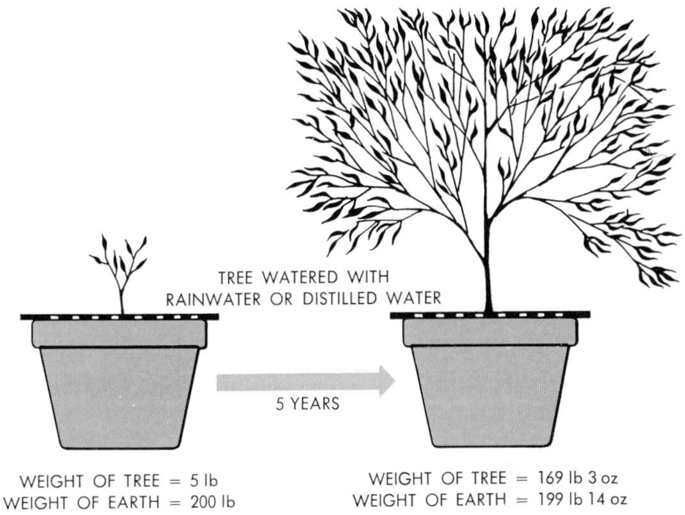

Fig. 6.1 *Van Helmont's experiment. He concluded that the gain in weight shown by the plant was due entirely to the water he had given it over the 5-year period.*

6.3 Some Early Hypotheses and Experiments

The problem of plant nutrition may seem a simple one. How does a plant grow? Where does it get the materials to build more plant matter? It seems easy to understand how animals may grow. They are seen devouring food which, it is assumed, they use to build more animal matter. With rare exceptions, however, plants do not feed in this manner.

The Experiments of van Helmont

One of the first men to study this problem was Jan Baptista van Helmont (1577–1644). While most of his studies were of a chemical nature, van Helmont performed a significant experiment with a willow tree. In this experiment he intended to discover the source of plant nutrients.

After drying and preweighing soil, van Helmont planted in the soil a willow tree stem which he watered and watched grow. Five years later, the tree had gained approximately 164 pounds, and the soil, which he again dried and weighed, had lost only two ounces (Fig. 6.1). It seemed to him that the plant matter accounting for the gain in weight had come from the water alone, since the earth had not lost any appreciable weight. He assumed that water had been transmuted into wood, which he regarded as an "earthy" material.

To van Helmont, this conclusion seemed both logical and inescapable. It agreed well with the predictions of the transmutation hypothesis, widely accepted in his day. Thus, in the light of the time at which this experiment was performed, van Helmont's conclusions are both understandable and reasonable. The experiment itself was a rather good one. (At that time, the use of a control was not considered essential). What of the gases that surround the plant in the atmosphere? Van Helmont did not consider them capable of adding to the plant material. Although van Helmont here overlooked an extremely important variable, he must still be given credit for beginning an investigation that others were stimulated to carry much farther.

The Work of John Woodward

Does the water itself actually provide all the matter used by the plant in making its stems, leaves, and roots? This conclusion of van Helmont's was questioned during the 1690s by John Woodward, a physician and professor at Cambridge University.

From his own observation, Woodward knew that even the clearest water is very far from being pure, that it contains an abundance of "terrestrial particles." To determine whether it was the water or the "earthy matter" to which plants owed their growth, Woodward placed measured amounts of water solutions into glass containers which were closed to prevent evaporation. After carefully weighing each plant, he inserted it through a small hole in the covering of a container so that the roots were immersed in the water solution, and he added measured amounts of water, as necessary, during the experiment. All plants received equal exposure to air and sunlight. Woodward noted that the gain in weight of the plants over a period of time was infinitesimal compared to the amount of water used. He concluded that most of the water is simply passed through the plant, escaping from the leaves. Since the containers used as controls (those containing only water solution) seemed to have a larger quantity of "terrestrial matter" than those with plants, Woodward concluded that it was the "terrestrial matter" of the earth, not the water, that constitutes plants.

Nehemiah Grew and Stephen Hales: Interaction of Plants and Air

In the late seventeenth century, development of the microscope and its use in natural science led to the discovery that the leaves of plants have many openings (stomata) on their surfaces. One early microscopist, Nehemia Grew, reported that the skins of plants, like those of animals, contained orifices whose function might be either to aid in the evaporation of superfluous sap or to admit air. Still other microscopists identified the leaves of plants as digestive organs and the functioning pores as outlets for digestive waste products. But Grew had hit on a very striking possible function for these openings. Could it be that they allowed exchange of substances between the plant and the atmosphere?

The English clergyman Stephen Hales (1677–1761) became interested in the whole problem of the flow of materials through plants. He seems to have been one of the first to note (although indirectly) that water and earth are not the only substances that might be involved in plant nutrition.

Hales set up two chambers as follows: He inverted two glasses in a vessel containing soil and water. In one chamber he rooted a peppermint plant, in the other, nothing. In spite of barometric pressure changes which caused fluctuations in the water levels of both chambers, after two months the level of the water in the chamber containing the plant was higher than that in the chamber without a plant (Fig. 6.2). Some of the air originally present must have been taken in by the plant—this was the only variable that could account for the reduced volume of air in that chamber.

Hales's experiment is significant for several reasons, one of which is that he very skillfully employed the use of a control, the empty container. Had he used only the system in which the mint was growing, it would have been very difficult for him to tell which of the fluctuations of water level were due to increases in barometric pressure and which were due to changes brought about by the plant. In Hales's experiment, however, any changes in *both* systems would obviously not be the result of the plant, but of some outside force (in this case, atmospheric pressure). Thus the use of a control helped Hales to define clearly what effects his plant had upon the atmosphere in its container.

Hales noted another important factor. Placing a fresh mint plant into the jar from which the original plant had been removed, Hales noticed that the

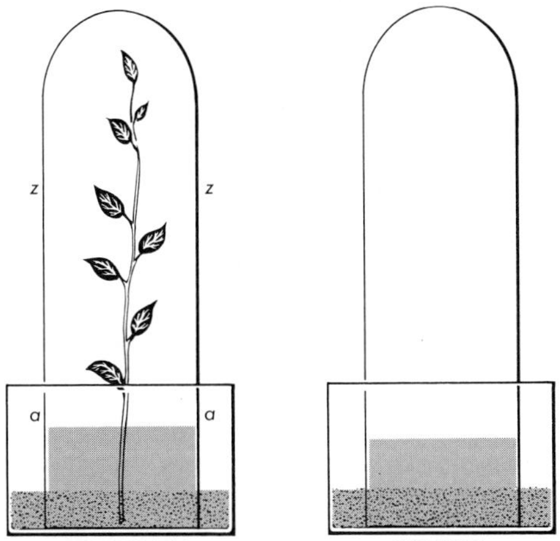

Fig. 6.2 *Stephen Hales's experiment. He concluded that plants remove something from the air. He also showed that the plant changed the composition of the air, although he was not sure what this change was.*

new plant "faded in four or five days." This allowed him to make a very significant conclusion. He could see that the first plant had caused some *change* in the air. From this experiment, then, Hales was able to draw some very important conclusions:

1. Plants interact with the atmosphere. As a result, there was a smaller volume of gas in Hales's glass after the reaction than there was before it.

2. Plants in some way affect the condition of the atmosphere with which they come in contact. The atmosphere in the glass was slightly changed from its original composition.

Hales did not find it easy to interpret the results of his experiments. He saw two possible reasons why the water level in the glasses might have risen. First, it might have been because the leaves imbibed gases. Second, it might have been because the plant exhaled a substance which, on combining with particles already in the air, caused a shrinkage. (For example, when equal volumes of alcohol and water are added together, the total volume is less than the sum of the two.) Hales leaned towards this last explanation, though he did not understand how it could occur. Through Hales's work, the interaction of plants and the atmosphere became established (something that van Helmont had completely overlooked), but the exact nature of this interaction remained obscure.

Joseph Priestley: The Effects of Animals and Plants on the Atmosphere

The British chemist Joseph Priestley (1733–1804) also became interested in investigating gases involved with plant life. By experimentation, Priestley had discovered that a candle will burn in an enclosed space (such as an inverted bell jar) for only a certain period of time. He also noticed that mice soon suffocate when placed in a similar situation. He recognized, therefore, that animals and burning candles somehow "damaged" the air, rendering it no longer capable of supporting life.

Priestley maintained that ". . . there must be some provision in nature" for the purpose of restoring air "rendered noxious" by animals and burning flames. He was aware of the need for a balance in nature as far as the atmosphere was concerned. It seemed evident to him that life could not long exist if there were not some such restorative function, for he could see how even a small candle flame contributed heavily to the "damaging" of its surrounding atmosphere. Priestley was, indeed, coming very close to the heart of the matter. At first, however, he was unaware of just how the balance he envisioned as a necessity was accomplished.

One day, quite by accident, Priestley discovered something: Vegetation is nature's way of restoring air which has been "injured" by burning candles. Earlier, he had hypothesized that both plants and animals affect the atmosphere in the same manner. To test this hypothesis, he did an ex-

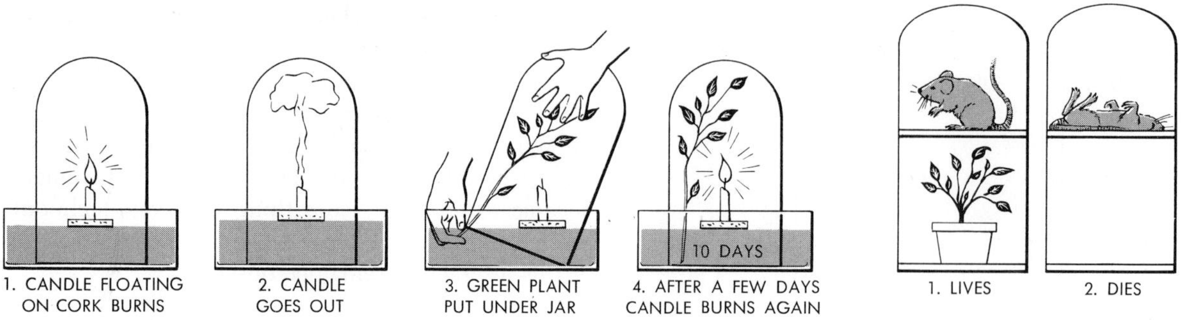

Fig. 6.3 *Priestley's experiments. From his results, he concluded that plants "reverse the effects of breathing."*

periment similar to that performed by Hales; he put a sprig of mint into a glass jar which stood inverted in a vessel of water. He was surprised to see it continue to grow there for several months, for this result contradicted his hypothesis. He also found that the air surrounding this plant allowed a candle to burn and a mouse to live.

These discoveries led Priestley to perform another experiment (Fig. 6.3). He put a sprig of mint into air in which a wax candle burned out, and twenty days later found that another wax candle burned perfectly well in it. Next he decided to divide into two parts a quantity of air in which a candle had burned out. In one part he placed a plant, in the other he did not. After five or six days, a candle always burned in the chamber with the plant, but never in the other chamber.

Up to this point, Priestley's work with plants involved air that had been "vitiated," or made unfit for a living animal, by a burning candle. He next turned to air that had been exposed over long periods of time to decaying plant and animal materials in a water-sealed vessel. Such materials also vitiated the air. He found that almost exactly the same thing happened—the introduction of living plants purified the air.

Priestley thus came to the conclusion that plants "reverse the effects of breathing." He concluded that animals affect the atmosphere by adding something to it, while plants subtract something from it when they purify the atmosphere. His findings were not accepted by many other workers. Attempts to repeat his experiments with the mint often failed, and Priestley was unable to explain why. However, his early experiments had laid a strong foundation on which others would build. Though he himself later questioned them, his experiments gave clear evidence to other investigators that plants *do* have a "purifying" effect on the air.

Ingenhousz Sees the Light

Priestley's works were read carefully by a Dutch physician, Jan Ingenhousz (1730–1799). Ingenhousz carried on investigations of his own at a remarkable pace. By performing variations of the experiments done by Priestley, Ingenhousz made several important discoveries. He found that plants absorbed air from the atmosphere, confirming the discovery of Priestley that plants somehow restore the air. However, he also saw plants as constantly absorbing gases from the atmosphere and changing them into dephlogisticated air. (Recall that a burning candle or respiring animal was thought to release phlogiston into the air and thus render it unfit for breathing. In light of the phlogiston theory, then, plants removed phlogiston from the air, thereby making it fit to support the burning candle or a living animal again.) Ingenhousez concluded that plants give off purified air that diffuses through the atmosphere. Most important, he observed that the process of purification takes place only during daylight, while the sun is above the horizon. At sunset, the process ceases. Here, indeed, was a major observation. For the first time, light was seen as necessary for the photosynthetic process.

But Ingenhousz went on to make yet another contribution, noting "that this office is not performed by the whole plant, but only by the leaves and the green stalks that support them." He remarked that all plants contaminate the air at night and in shaded places, and that flowers, fruits, and roots removed from the ground do so by both day and night. Thus Ingenhousz discovered a variable overlooked by Priestley. He saw that not *all* of the plant contributes to restoring the atmosphere. Some portions act just like burning candles or animals: they give off carbon dioxide and use oxygen. It is now known, of course, that *all* living organisms carry on respiration, and that plants are no exception. Plants, however, contribute more oxygen to the air by their photosynthetic activity than they remove by their respiratory processes.

Ingenhousz did not understand *why* light was important. Nor did he recognize the significance of the green coloring. However, he did see two possibilities that would account for the production of pure air by plants. First, the pure or dephlogisticated air arose from the plant, which had drawn in phlogiston from the atmosphere. Second, Ingenhousz suggested that plants completely immersed in water produced dephlogisticated air because of a vital force that, under the influence of light, transmuted water or other substances into plant materials. This explanation sounds quite similar to van Helmont's hypothesis of transmutation of water into plant material. Nonetheless, it was Ingenhousz's final working hypothesis. Its value was far less than that of the observations he made and the experiments he performed. It remained for later workers to properly interpret the results of those experiments.

Berthollet and Senebier: The Source of Oxygen from Photosynthesis

By the late eighteenth century, quite independently of work on plant physiology, the phlogiston theory was slowly abandoned. The element oxygen was isolated and many of its physical and chemical properties were described. In the light of these advances, the steps of the photosynthetic process were suspected to be something like the following:

SOMETHING FROM AIR + WATER —(LIGHT + GREEN COLORING OF PLANTS)→ PLANT MATTER + "RESTORED AIR"

Note that no distinction was made between the actual final product of photosynthesis (carbohydrate) and other plant growth materials, such as protein. Today, we write the process in a simplified form:

CARBON DIOXIDE + WATER → GLUCOSE + OXYGEN

$$CO_2 + H_2O \rightarrow C_6H_{12}O_6 + O_2 \quad \text{(unbalanced)}$$

Since only oxygen is released as a waste product of this reaction, the problem boils down to discovering what parts of the water and carbon dioxide molecules go into making new plant materials. There are three possibilities:

1. The carbon of the carbon dioxide (CO_2) unites with the water (H_2O) to form glucose ($C_6H_{12}O_6$), the oxygen of the CO_2 being released into the atmosphere.

2. The hydrogen of the H_2O unites with the CO_2, the oxygen of the H_2O being released into the atmosphere.

3. Both of these reactions occur, the oxygen being released from both the CO_2 and the H_2O.

These three alternatives actually serve to point up the main question: Where does the oxygen released by the plant come from, the water, the carbon dioxide, or both? The answer to this question indirectly tells us where the "plant matter" comes from, for whatever is left over after the oxygen is removed must go into the glucose.

Conflicting hypotheses were put forth and tested by two men. The French scientist M. Berthollet (1748–1822) decided that the released oxygen came from the water molecules. Berthollet's reasoning can be expressed in the following deductive way: if the oxygen that plants release into the air comes from the water molecules, then plants grown in a medium free of all dissolved hydrogen should still contain hydrogen in their tissues.

Berthollet performed an experiment to test his hypothesis, growing plants in a medium that did not contain dissolved hydrogen. He then chemically analyzed the plant material for the presence of this element. Hydrogen *was* found in the tissues. Berthollet saw that the hydrogen could have come only from the water, since carbon dioxide contains no hydrogen. By taking the hydrogen, Berthollet reasoned, the plant releases the oxygen in the water molecules into the air. Therefore, he said, *the oxygen released by the plant must come from the water molecule.*

Another Frenchman, Jean Senebier (1742–1809), took issue with Berthollet's conclusion. Senebier reasoned as follows: if the oxygen comes from the water molecule (as Berthollet believed), then the leaves should give off oxygen when immersed in only water (no CO_2 present). However, Senebier's data contradict this prediction; leaves immersed in water give off oxygen only when carbon dioxide is present (Fig. 6.4). This indicated to Senebier that *the oxygen released must come from the carbon dioxide molecule.*

Senebier eliminated a possible variable by a second experiment (Fig. 6.5). After the leaves used in the first experiment had stopped releasing oxygen, he replaced them with fresh leaves. The fresh leaves, however, released no oxygen. This showed that it was a lack of carbon dioxide rather than a "tiring" of the leaves that caused oxygen release to stop. When fresh water containing carbon dioxide was added, oxygen release began again. Thus, the results of this second experiment appeared to support Senebier's hypothesis that carbon dioxide molecules were the source of the released oxygen.

Senebier's experimental data also seemed to show that the amount of oxygen produced by submerged leaves is directly related to the amount of fixed air (carbon-dioxide-laden air) available to those leaves. In other words, the more fixed air supplied, the more oxygen released.*

Thus, by the early nineteenth century the following was known about the photosynthetic process:

1. Plants use both water from the soil and carbon dioxide from the air to manufacture a product, carbohydrate.

2. A by-product of photosynthesis is oxygen, released by plants into the atmosphere.

3. Only the green parts of plants are capable of photosynthesis. Nongreen parts of plants, like animals, can only use oxygen and produce carbon dioxide in the respiratory process.

4. Sunlight is necessary for photosynthesis to occur.

5. Rate of photosynthesis is dependent upon the amount of both light and carbon dioxide available.

Still, there were unsolved problems about photosynthesis. These were:

1. Nothing was really known about the chemical events taking place within plants by which carbon dioxide and water were converted into carbohydrates. In other words, it was known what went

* While performing these experiments, Senebier made an interesting discovery. He noted that shredded bits of leaves, when immersed in water and irradiated with light, released oxygen just as well as whole leaves. These results indicated that the photosynthetic process is not performed by the leaf as an organ. Today it is known that photosynthesis is carried on within the chloroplasts found in living green plant cells.

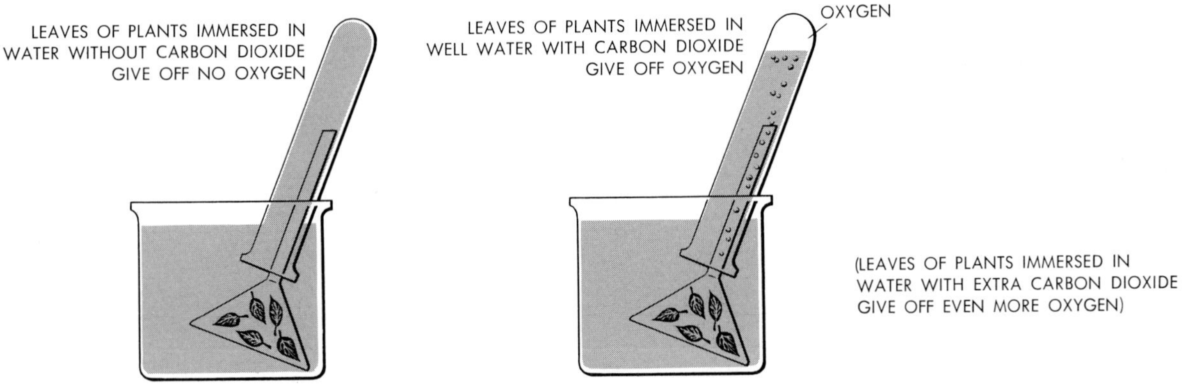
LEAVES OF PLANTS IMMERSED IN
WATER WITHOUT CARBON DIOXIDE
GIVE OFF NO OXYGEN
LEAVES OF PLANTS IMMERSED IN
WELL WATER WITH CARBON DIOXIDE
GIVE OFF OXYGEN
OXYGEN
(LEAVES OF PLANTS IMMERSED IN
WATER WITH EXTRA CARBON DIOXIDE
GIVE OFF EVEN MORE OXYGEN)

Fig. 6.4 *Senebier's first experiment.*

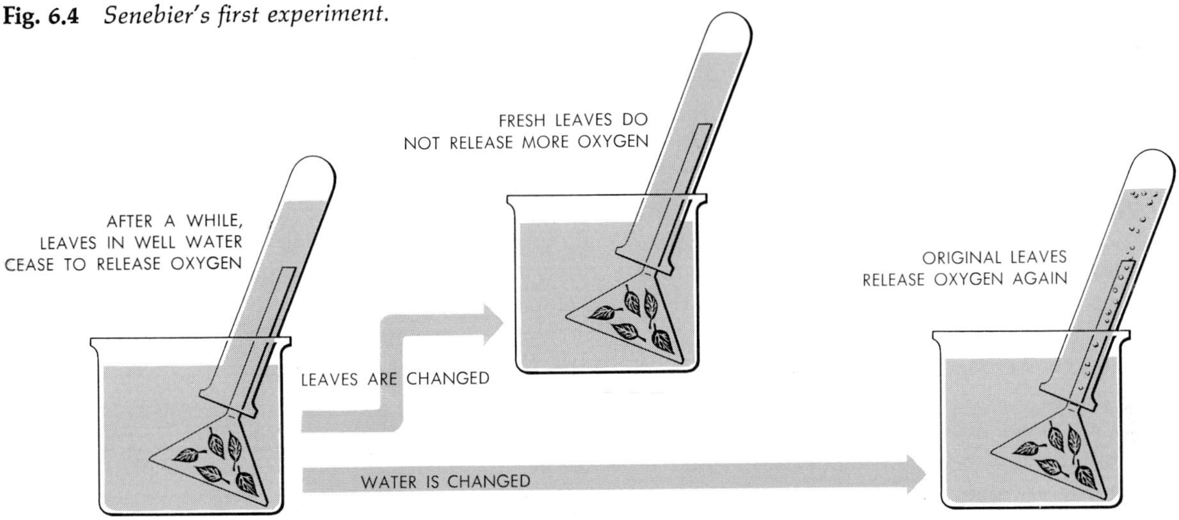
AFTER A WHILE,
LEAVES IN WELL WATER
CEASE TO RELEASE OXYGEN
FRESH LEAVES DO
NOT RELEASE MORE OXYGEN
LEAVES ARE CHANGED
ORIGINAL LEAVES
RELEASE OXYGEN AGAIN
WATER IS CHANGED

Fig. 6.5 *Senebier's second experiment.*

into and came out of the photosynthetic process, but *how* the process actually occurred was completely unknown.

2. The source of the oxygen generated during photosynthesis was a matter of debate. As we have seen, two schools of thought developed, one claiming that oxygen originated from the water, the other that it originated from the carbon dioxide.

One of the most important steps taken in the modern study of photosynthesis involved the attempt to solve this latter problem. A look at how this question was answered will bring us up to the present-day understanding of photosynthesis as an aspect of cellular metabolism.

The Oxygen Problem Is Resolved

Clues leading to the solution of a problem in one area of biological research are often provided by investigation in a slightly different area. So it was with plant photosynthesis. In the 1930s the Dutch microbiologist C. B. Van Niel had studied photosynthesis in purple sulfur bacteria. Like the chlorophyll-containing cells of green plants, these bacteria use light energy to synthesize carbohydrate materials. However, purple sulfur bacteria use hydrogen sulfide (H_2S) instead of water. This fact suggests a deduction that might identify the origin of the oxygen given off by green plants during photosynthesis: if the oxygen released by plants during photosynthesis comes from the carbon dioxide molecule, then the purple sulfur bacteria will release oxygen as a result of their photosynthetic activity. On the other hand, if the oxygen released by plants during photosynthesis comes from a water molecule, then purple sulfur bacteria should release sulfur as a result of their photosynthetic activity.

From his observations, Van Niel already knew the answer. Photosynthesizing purple sulfur bacteria release sulfur, not oxygen, as a waste product. The process can be summarized as follows:

$$CO_2 + 2H_2S \text{ — } (CH_2O)_n + H_2O + 2S.$$

Van Niel reasoned that light decomposes the hydrogen sulfide into hydrogen and sulfur. The hydrogen atoms are then used to reduce the carbon dioxide to carbohydrate. It was an easy step to suggest that the same process occurs in green plants, except that water, rather than hydrogen sulfide, is decomposed by light. The hydrogen atoms so released can then be used to reduce carbon dioxide, while the oxygen is given off. Van Niel hypothesized, therefore, that the oxygen given off by green plants during photosynthesis comes from the water molecules, and not those of carbon dioxide.

There is one basic premise here, of course. It is assumed that, other than the raw materials involved, there is no difference between the photosynthetic process carried out by the purple sulfur bacteria and that performed by green plants. Purple sulfur bacteria are quite primitive organisms; most green plants are not. Therefore, in terms of evolutionary status, it might seem unlikely that the food-making process in both forms is similar.

On the other hand, the conversion of light energy to chemical energy is truly a remarkable feat for a living system to perform, and so it is even less likely (though by no means impossible) that more than one way to accomplish such a conversion would have evolved. This consideration lends strength to Van Niel's extrapolation from the photosynthetic processes of the purple sulfur bacteria to the same processes in green plants. The basic premise is still present, however, with all its accompanying doubts, and prevents Van Niel's observations of the purple sulfur bacteria from being conclusive.

It often happens that the crucial experiment that would resolve the issue between two conflicting hypotheses cannot be performed until the proper experimental instrument or technique is developed. Such was the case here. The introduction of tracer experiments by George Hevesy (1885–1966) in 1923 opened many new experimental pathways in biology. Hevesy used radioactive isotopes of lead to trace the pathways through which materials moved from place to place in plants.

In 1941, a team of scientists including Samuel Ruben and Martin Kamen at the University of California performed the crucial experiment that determined the source of oxygen released in photosynthesis. They exposed the green alga *Chlorella* to water which had been labeled with O^{18}. This isotope can be detected by a technique known as mass spectrometry. The experiment showed that the O^{18} turned up in the gas released during photosynthesis, while there was no O^{18} in the carbohydrate produced. A second experiment seemed obvious. The carbon dioxide was labelled (CO_2^{18}) and the water was not. In this case, Van Niel would predict that the O^{18} would appear in the carbohydrate rather than in the oxygen released during photosynthesis. The experimental data verified this prediction also. It was now clear that the oxygen released during photosynthesis came from the water, not the carbon dioxide.

Van Niel's conclusion was deduced initially from his observation that purple sulfur bacteria do not produce oxygen. Yet this seemed curious to Van Niel, since it was an exception to what seemed like a generally accepted idea. Photosynthetic organisms release oxygen as a by-product of photosynthesis. Observing that purple sulfur bacteria do not release oxygen, Van Niel had to decide whether or not his observation invalidated the general concept. He decided that the sulfur bacteria were simply a special case within the limits of the general theory of photosynthesis. This was admittedly a guess. However, it is sometimes easier to accommodate unexpected observations into a theoretical framework than to reject the entire theory. A third alternative, of course, would have been to reject the observations as somehow faulty, which he was confident was not the case. Van Niel's choice to try to accommodate these new observations into the general theory led him to recognize the parallels between purple sulfur bacteria and green plant photosynthesis.

If the oxygen released in green plant photosynthesis comes from the water, then why should such previous workers as Senebier have found that plants do not give off oxygen when placed in water that has no dissolved CO_2? The answer to that question lies in the concept of limiting factors in metabolic processes. For a plant to synthesize carbohydrate from CO_2 to H_2O (in the presence of light), both reactants must be present in quantities proportional to their rate of consumption in the reaction. Thus, in boiled water with all the dissolved CO_2 driven off, plants could not carry on photosynthesis. Because one ingredient was lacking, the entire process could not proceed. The CO_2 would be called a **limiting factor** in this case. Light also proves to be a limiting factor in the rate of photosynthesis. In the dark there is no photosynthesis. In very bright light, all other factors being equal, photosynthesis proceeds at a maximum rate. To follow Senebier's reasoning, the observation that oxygen release stops in the dark would lead to the conclusion that the oxygen came from the light!

Another relationship can be observed here. As the quantity of any necessary factor in a biochemical reaction is increased, given that the quantities of all the other factors are sufficient, the rate of reaction increases proportionately. There is a limit, however, to which increase in the quantity of any one factor can continue to increase the rate of the reaction as a whole. There are certain upward limits to the amount of CO_2 concentration or the brightness of light that will continually accelerate photosynthesis. Beyond these limits, increase in quantity of either CO_2 or light will not increase the rate of photosynthesis. In fact, too much CO_2, or too bright a light (primarily because it produces too much heat), causes the rate of photosynthesis to *decrease*. The rate of a biochemical reaction is proportional to the quantity of any reactant or factor *only within certain limits*.

A modern follow-up of the story indicates the difficulty of getting conclusive evidence in any field of science. The isotopic tracer studies by Ruben, Kamen, and others in 1941 contained one source of error: if cells are supplied with O^{18} as either carbon dioxide (CO_2^{18}) or water (H_2O^{18}), it is possible for there to be an exchange of oxygen atoms between the two molecules. That is, some of the O^{18} supplied in carbon dioxide can be exchanged with the normal oxygen (O^{16}) of the water, and vice versa. Thus, even though isotopic oxygen is administered to plant cells as either CO_2^{18} or H_2O^{18} (though never as both simultaneously), there is no way to guarantee that some exchange does not occur. As a result it is impossible to ensure that isotopic oxygen is being fed to cells from only one source.

Only in 1975 did a way around the above difficulty emerge. Biochemists Alan Stemler in Illinois and Richard Radmer in Maryland knew that the bicarbonate ion (HCO_3^-) plays a critical role in the initial stages of photosynthesis. So critical is this role, in fact, that when the ion is removed more than 90% of the oxygen evolution by the cells is stopped. Bicarbonate is a source of CO_2, since the ion interacts with water to produce CO_2. Stemler and Radmer decided to use isolated, broken chloroplasts rather than whole cells for their studies for two reasons: (1) Breaking up the chloroplasts allowed them to remove the enzyme carbonic anhydrase, which catalyzes the exchange of oxygen between CO_2 and H_2O. (2) It also allowed them to remove all bicarbonate ions already present in the chloroplasts. They then supplied the chloroplast fragments with isotopically labelled bicarbonate (HCO_3^{18-}). Oxygen evolution began almost immediately—but the oxygen was not labeled! Even though bicarbonate was the only source of CO_2, it was not the source of the evolved oxygen. The conclusion from this experiment seems inescapable: the source of oxygen evolved during photosynthesis must be water. Van Niel's hypothesis was thus verified in a conclusive way. The oxygen released in photosynthesis by green plants comes from the splitting of water molecules. The hydrogens from the water are picked up by NADP (yielding NADPH + H^+) and used to reduce carbon dioxide to carbohydrate.

6.4 Photosynthesis: The Modern View

Photosynthesis consists of two separate but interdependent processes. These are known as the light reaction and the dark reaction. The end product of photosynthesis is carbohydrate in the form of glucose. Glucose contains a considerable amount of potential chemical energy that, as we saw in Chapter 5, can be released by a series of oxidative steps. The functions of the two phases of photosynthesis are geared toward the production of energy-rich glucose. The light reaction serves to convert the radiant energy of light into a usable chemical form that has a high reducing capacity. The dark reaction involves the reduction of carbon dioxide to glucose. Reduction, the addition of electrons to the carbon dioxide molecule, requires energy and reducing agents. Both of these are generated during the light reaction. In studying the overall process of photosynthesis, it is important to keep in mind that the light reaction yields energy and reducing agents which then drive the reduction of carbon dioxide to carbohydrate.

Phase I: The Light Reaction

Light is a form of radiant energy. Physicists have proposed two models to explain the behavior of light in certain situations. The first model is the **wave theory** of light. This model pictures light as traveling in waves, similar to the waves on the surface of a body of water. As with water, the **wavelength** of light may vary from one type of wave to another. Figure 6.6 shows how wavelengths are measured. *The longer the wavelength, the less energy conveyed. The shorter the wavelength, the more energy conveyed.*

Various wavelengths of light are detected by the human eye as different colors. At one end of the visible spectrum is red light, which has a long wavelength. At the other end is violet light, which has a short wavelength. Violet light, of course, conveys more energy than red light.

There are other wavelengths of light to which the human eye does not respond, since they lie beyond the range of the visible spectrum. For example, ultraviolet light has a wavelength shorter than that of violet, and it is not detected by the human eye. Infrared light waves, which are longer than those of red light (and so convey less energy) are also invisible to humans. Thus the human eye is selective in that it reacts to certain wavelengths of light while other wavelengths pass by it undetected.

Chlorophyll. Absorption of light energy in the plant cell is accomplished by photosynthetic pig-

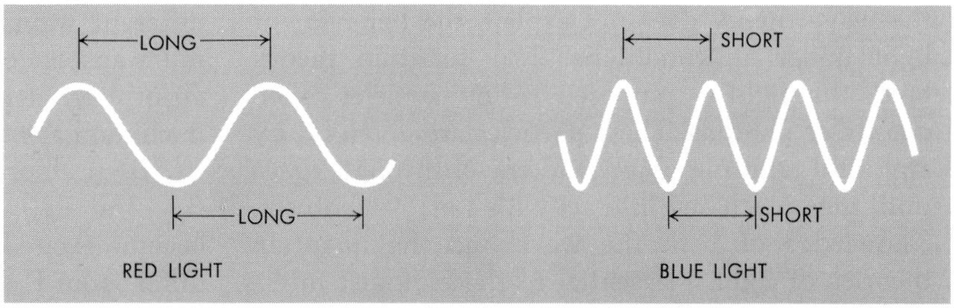

Fig. 6.6 *Light energy travels in waves, the longer wavelengths conveying less energy than the shorter ones.*

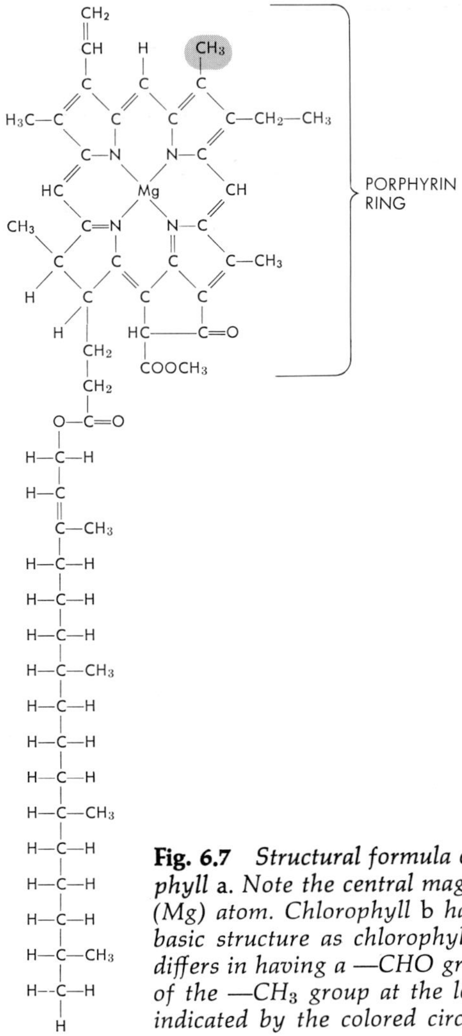

Fig. 6.7 *Structural formula of chlorophyll* a. *Note the central magnesium (Mg) atom. Chlorophyll* b *has the same basic structure as chlorophyll* a, *but differs in having a —CHO group instead of the —CH_3 group at the location indicated by the colored circle.*

ments. In higher plants the most common of these are the **chlorophylls,** which confer on these plants their green color. In addition, higher plant cells have another set of pigments, the carotenoids, which in a solution by themselves appear orange-yellow. Lower plants, such as the algae, contain a series of photosynthetic pigments such as phycocyanin, phycoerythrin, and fucoxanthin, which are chemically different from chlorophyll. The green algae contain chlorophylls like the higher plants. To understand the general function of these pigments in the photosynthetic process, consider green plant chlorophyll.

Chlorophyll bears a distinct chemical resemblance to the cytochromes and to hemoglobin in that it is based on the porphyrin ring structure (see Fig. 6.7). Instead of an atom of iron in the center of the porphyrin ring, chlorophyll has an atom of magnesium. There are actually several kinds of chlorophyll, known collectively as chlorophyll *a* and chlorophyll *b*. As we shall see shortly, both have very similar functions. Chlorophyll molecules are bound to membranes within the chloroplasts and serve to trap the energy of light and begin its transduction, or transformation, into usable chemical energy. Chlorophyll is absolutely essential in this initial step in the uphill reaction of photosynthesis. How, exactly, does chlorophyll act to absorb light and to aid in its transformation? To answer this question we must return, for a moment, to the physics of light.

Although it has been useful and predictive,

the wave theory does not explain the behavior of light under all conditions. The **quantum theory** holds that light is composed of tiny particles called quanta or photons. These particles are given off by any light-emitting object and travel through space until they reach and interact with a material object. Knowledge of both the wave and the quantum theories of light is essential to the biologist interested in photosynthesis, for while light absorption by chlorophyll can be explained in terms of the wave theory, the quantum theory must be employed to explain what happens after light is absorbed. In other words, light seems to travel as a wave, but it interacts with matter as a particle.

Photosystems I and II. Figure 6.8 shows schematically the major steps in the light reaction. The reaction begins with the two pigment systems, known as photosystems I and II. Both pigment systems consist of chlorophyll *a* and one or more other pigment molecules. The pigment molecules are arranged in complexes, or groups, which stay together when isolated chloroplasts are broken down and their contents centrifuged. The complexes are associated into larger unit structures called quantosomes attached to the membrane system within the chloroplast (for more structural details see Section 6.5, and Fig. 6.12). A complex of photosystem I consists of one molecule of a form of chlorophyll *a* known as P_{700}, approximately 200 molecules of regular chlorophyll *a*, and about 50 molecules of carotenoid. A complex of photosystem II consists of one molecule of another form of chlorophyll *a* known as P_{684}, about 200 molecules of regular chlorophyll *a*, and about 200 molecules chlorophyll *b*. Although the special molecules P_{700} and P_{684} comprise less than 1% of the chlorophyll molecules in either photosystem complex, they play a crucial role in the absorption of light energy. It is important to emphasize that both P_{700} and P_{684} are considered molecules of chlorophyll *a*. Their differences are characterized by the different wavelengths at which each shows maximum absorption. P_{700} has a maximum light energy absorption at a wavelength of 700 mμ. P_{684} has a maximum absorption at wavelength 684 mμ.

Both chlorophyll *a* and its two variant forms, P_{700} and P_{684}, can absorb light quanta and thus become excited. However, the regular chlorophylls differ from P_{700} and P_{684} in terms of how that excitation leads to the capture of energy in a usable form. Molecules of regular chlorophyll (both *a* and *b*) act like antennae, absorbing light quanta by exciting electrons that are then raised to higher energy levels. This "excitation energy" is then funneled to molecules of P_{700} or P_{684}, depending on the photosystem involved. The molecules of P_{700} or P_{684} become excited in such a way as to actually lose electrons and thus initiate the major energy-capturing steps of the light reaction.

To understand both the uniqueness and efficiency of the light reaction phase, it will be helpful to consider the functioning and interaction of the two photosystems in a little more detail. The following discussion will refer to Fig. 7.8.

Both photosystems absorb light energy simultaneously. Let us start with photosystem I. Electrons from photosystem I are picked up by an acceptor labeled *Z* (or in some schemes labeled *X* or *FX*). They are then passed by *Z* to another acceptor, the iron-containing ferredoxin. The electrons can now follow two courses. They can be picked up by an electron acceptor known as NADP (nicotinamide adenine dinucleotide phosphate), a molecule very similar to NAD, one of the electron acceptor molecules in respiration. By this process, NADP becomes reduced to NADPH + H^+. Reduced NADP is ultimately used in the dark reactions of photosynthesis to reduce carbon dioxide and to produce carbohydrate.

On the other hand, the electrons may pass from ferredoxin through an alternative pathway by way of an acceptor known as cytochrome b_6. In this case they pass on to another series of acceptors, including cytochrome *f* and plastocyanin, eventu-

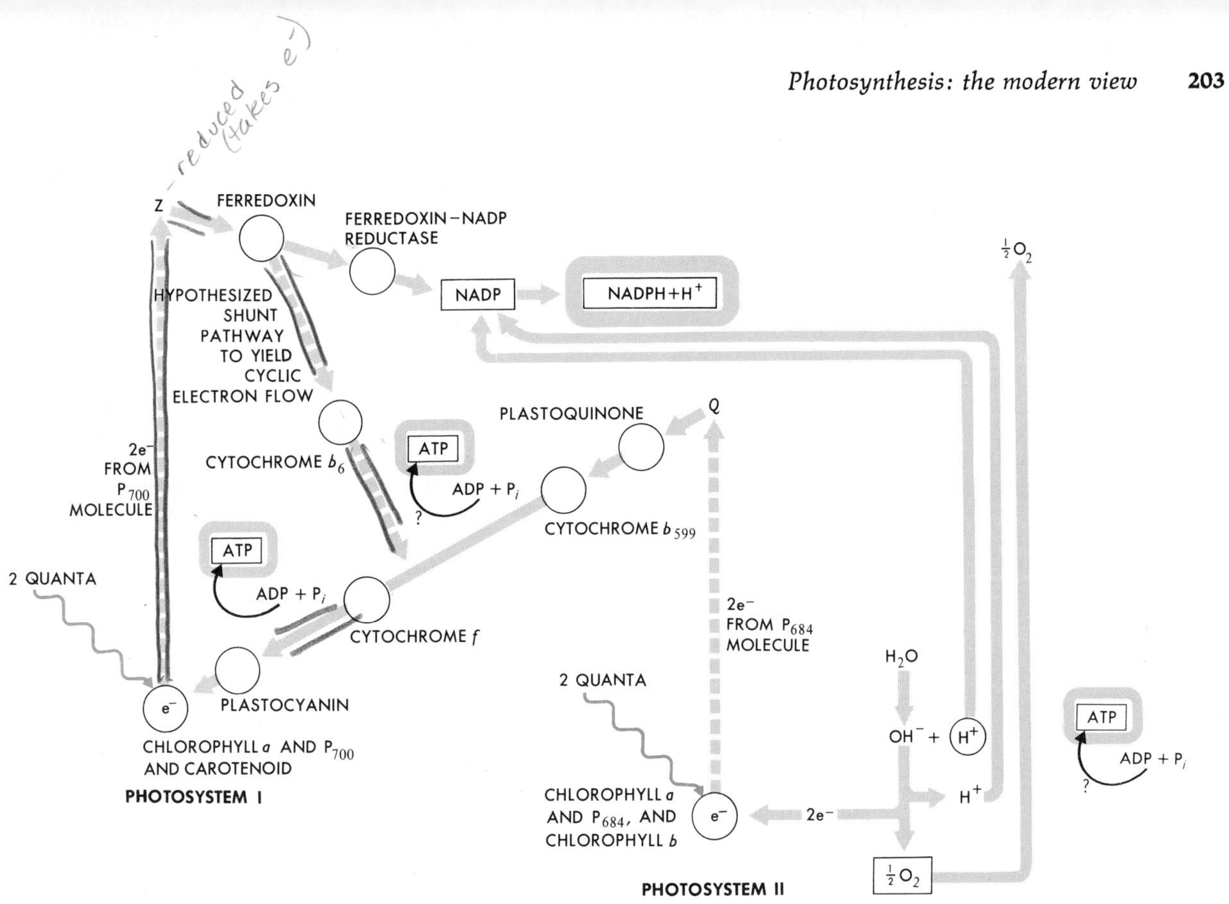

Fig. 6.8 *Postulated pathways of electron flow during the light reaction of photosynthesis. Photosystem I can carry out either noncyclic electron flow (leading to production of NADPH + H^+) or cyclic electron flow (leading directly to ATP generation). The sites of ATP production are currently in dispute. Most researchers now agree that one ATP is generated in the passage of an electron pair from cytochrome* f *to plastocynanin. The other ATP is generated either during the passage of an electron pair from cytochrome* b_{599} *to cytochrome* f *or somewhere in the process of splitting water. Cyclic flow returns the electrons back to chlorophyll* a *and* P_{700}*, the two pigments of photosystem I. Photosystem II is capable of only one-way electron flow from chlorophyll* b *to chlorophyll* a*. The end product of the light reaction is ATP and an energy-rich reducing compound such as NADPH +* H^+*. Molecular oxygen,* O_2*, is also given off. (After A. L. Lehninger,* Bioenergetics, *2d ed. Menlo Park, Calif.: W. A. Benjamin, 1973, p. 112.)*

ally being returned to P_{700}. In passing through this pathway, the electron flow is also coupled to the phosphorylation of ADP, producing some ATP. Photosystem I has built into it a cyclic pathway by which electrons are returned, through an electron transport system, back to pigment P_{700}.

Chloroplasts are biological energy transducers. They take the energy from one transmission system, light, and transmit it to another transmission system, involving electron transport and photophosphorylation.

Supplement 6.1

CHLOROPHYLL AS THE ENERGY ABSORBER IN PHOTOSYNTHESIS

What evidence justifies the claim that chlorophyll is the energy absorber powering the photosynthetic process? How do we know that the selective absorption of certain wavelengths of light by chlorophyll has any direct connection to photosynthesis? A simple experiment performed as long ago as 1881 by the German plant physiologist T. Engelmann (1843–1909) provides an answer.

Engelmann had been interested in the ability of certain aerobic bacteria to move toward areas of high oxygen concentration within a liquid medium. He had also carried out some studies on photosynthesis in algae. From this work, Engelmann knew that bacteria will concentrate themselves in a region of the medium where oxygen is plentiful—in fact, in direct proportion to the amount of oxygen. The more oxygen present, the more bacteria will assemble there. He thus hit upon an ingenious way to measure the amount of photosynthesis green plant cells can carry on in different wavelengths of light.

Engelmann took a filament of the green alga *Cladophora* and placed it on a microscope slide. He then passed the light source of the microscope through a small prism, so that the white light was spread out into a spectrum. He focused the prism so that it fell along the length of the *Cladophora* filament. Thus, different portions of the filament were selectively exposed to different wavelengths of light. Knowing that photosynthesis releases oxygen as a by-product, Engelmann reasoned that if different wavelengths of light were absorbed differently by the photosynthetic apparatus within the cells, then different parts of the filament should produce different amounts of oxygen. Oxygen production should be in direct proportion to the amount of the absorbed light of a given wavelength. By introducing bacteria onto the slide, Engelmann provided himself with an observable measure of oxygen production.

After exposing the *Cladophora* filament to the spectrum for some period of time in the presence of bacteria, Engelmann obtained the results shown in part (a) of Diagram A.

Comparing these results with the absorption spectrum of the pigments found in the *Cladophora* filament, Engelmann noted their exact correspondence (Diagram A, part b). In other words, where the bacteria accumulated in the greatest concentration were found the regions of the spectrum where the pigments were known to absorb most. Conversely, where the bacteria were least concentrated, were found regions of the spectrum where the pigments absorbed least.

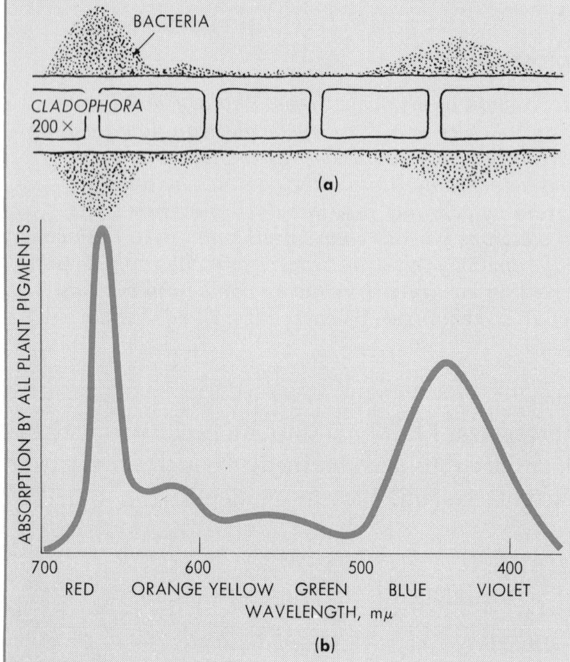

Diagram A

This exact correspondence provided striking evidence to suggest that the plant pigments, chiefly the chlorophylls, are directly involved in absorbing the energy to power photosynthesis.

More recent research has confirmed and extended Engelmann's work. Diagram B shows the spectra for three different photosynthetic pigments, the chlorophylls (*a* and *b*), the carotenoids, and phycocyanin. The carotenoids and chlorophylls are present in the cells of all higher plants and the green algae. Phycocyanin is present in some groups of the algae, especially the blue-greens. Note that for the higher plants, the presence of both chlorophyll and carotenoids means that the cells can absorb light energy over a considerable range of the spectrum. If just one of the pigments were involved, the wavelengths that could be effectively absorbed would be greatly restricted. Note that what with the three pigments shown in Diagram B, almost all of the visible spectrum is covered by an absorbing pigment. It is possible that such a distribution of pigments is especially suited to photosynthesis in the oceans. Marine algae account for the largest part by far of the earth's photosynthesis. In the oceans, there is a mixture of many kinds of algae: green, blue-green, red, brown, etc. With their specialized distribution of pigments, these algae can collectively make maximum use of the solar energy striking the ocean surface.

Diagram B *Absorption spectra for several important plant pigments found in the green alga* Chroococcus. *Note that the three major groups of pigments have absorption peaks at three different areas of the visible spectrum. Together, these pigments can absorb light energy from a considerable range of wavelengths. These pigments are found, in varying proportions, in most of the green plants.* (*After* Biology Today, *New York: CRM/Random House, 1972, p. 96.*)

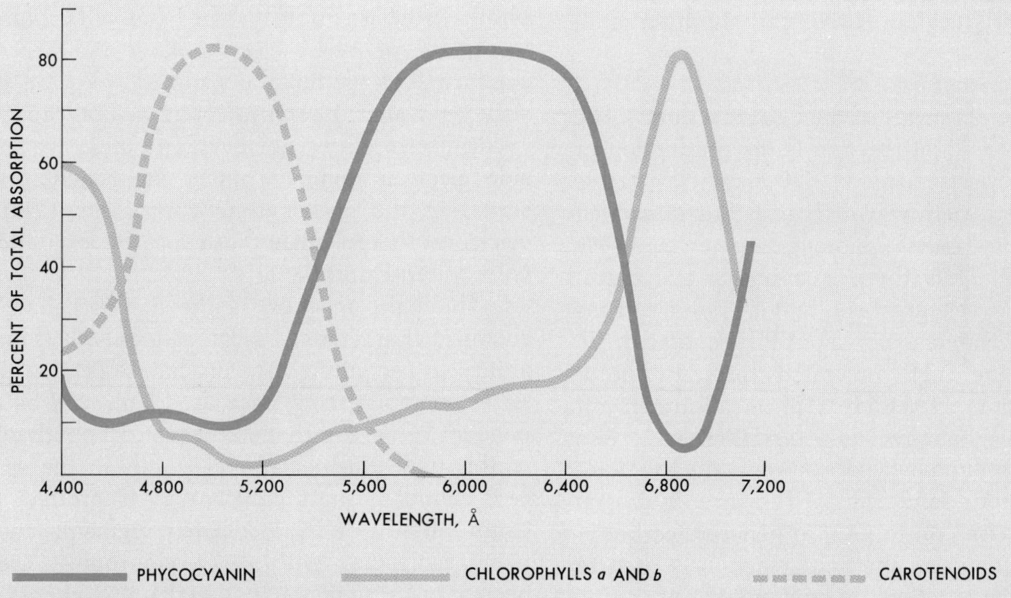

Photosystem II does not have a cyclic flow of electrons open to it. Rather, electrons lost from P_{684} are picked up by the acceptor Q, from which they pass through a series of acceptors among which are plastoquinone, cytochrome b_{599} (different from cytochrome b_6), cytochrome *f*, and plastocyanin. Electrons removed from photosystem II are thus used to replace those lost from photosystem I. But how are electrons replaced for P_{684} molecules of photosystem II? The electrons come from water. Ionized molecules of water, which occur in approximately one out of every million, can donate electrons to chlorophyll *b*, as shown in the lower right-hand portion of Fig. 7.8. This process produces hydrogen ions and free oxygen—the oxygen released into the atmosphere as a by-product of photosynthesis. It is partly to get electrons back into the pigment systems that water is split (ionized) at greater-than-normal rates during the light reaction of photosynthesis. The splitting of water also supplies electrons to reduce NADP, which, on picking up two hydrogen ions, becomes $NADPH + H^+$. These hydrogens are later added to CO_2 to produce the basic carbohydrate unit, $C(H_2O)_n$.

The unique achievement of the light reaction. The net result of the light reaction is the production of energy in a form (reduced NADP and ATP) usable for biosynthetic pathways. For each pair of excited electrons from either photosystem, two ATP's are produced. Thus there is a one-to-one correspondence between quanta of light energy, electrons excited, and the amount of ATP generated. In addition to the ATP generated through photophosphorylation, $NADPH + H^+$ is an important product of the light reaction. $NADPH + H^+$ is a reducing compound whose electrons (in the company of protons, and thus in essence hydrogen atoms) are passed on to CO_2 to generate carbohydrate. The ATP from the light reaction is used to drive this latter reaction, an endergonic one, up its energy hill. We can write the equations for the light reaction in terms of the input and output of photosystems I and II:

Photosystem I

$$\text{2 QUANTA} + \text{NADP} + \text{Pigment System I} \rightarrow \text{NADPH} + \text{H}^+ + \text{Pigment System I}^{++} \qquad (6\text{–}1)$$

OR:

$$\text{2 QUANTA} + \text{2ADP} + 2\text{P}_i + \text{Pigment System II} \rightarrow \text{2ATP} + \text{Pigment System II}^{++} \qquad (6\text{–}2)$$

Summarizing, we get:

Photosystem II

$$\text{4 QUANTA} + \text{NADP} + \text{2ADP} + 2\text{P}_i + \text{Pigment System I and II} \rightarrow \text{NADPH} + \text{H}^+ + \text{2ATP} + \text{Pigment System I and II} \qquad (6\text{–}3)$$

As these equations show, photosystem II is responsible for the process of oxygen evolution in photosynthesis. Photosystem I (i.e., chlorophyll *a*) is capable of cyclic electron flow and hence can produce ATP without any outside electron donor such as water. Photosystem II is not capable of cyclic electron flow, however, and must use an outside electron donor. This is the function water serves in the photosynthetic process: it supplies electrons to replenish those lost by excitation of chlorophyll *b* and P_{684}.

The light reaction is the distinguishing biochemical characteristic of green plants. It is in their ability to use sunlight to produce ATP that plants differ so much from animals. Plants can be independent of an external source of carbohydrate. No matter how long an animal stands in the sun, its cells cannot harness light energy to produce ATP. Chlorophyll and its associated pigments and enzyme molecules within the chloroplast are the necessary ingredients that make this energy conversion possible.

Supplement 6.2

THE LIGHT REACTION: HOW DO WE KNOW?

Several lines of evidence have led over the years to our current understanding of the events in the light reaction.

One of the earliest experiments was that of Robin Hill in England in the mid-1930s. Hill found that isolated chloroplasts extracted from spinach leaves could, when illuminated, cause certain dyes like methylene blue to change color. Methylene blue is an oxidation–reduction indicator; it is blue when oxidized and green when reduced. Moreover, the chloroplasts evolved oxygen while reducing the dye. If the dye was not present in the medium, no oxygen was evolved (see Diagram A). From Hill's experiment came several important conclusions:

1. Chloroplasts contain all the necessary elements for splitting water and producing oxygen. The rest of the plant cell is not necessary for this phase of photosynthesis.

2. When illuminated, chloroplasts have the ability to chemically reduce available electron acceptors. Without these acceptors, no oxygen is evolved.

3. There must be naturally occurring reducing agents in the chloroplast that act analogously to methylene blue.

In the 1950s a team of researchers headed by Daniel Arnon, at the University of California at Berkeley, began more detailed studies of Hill's isolated chloroplast system. Arnon found that isolated chloroplasts could carry out complete photosynthesis (the evolution of oxygen and the reduction of CO_2 to carbohydrate) if supplied with the following substances: ADP, inorganic phosphate, (P_i), NADP, CO_2, and light. This demonstrated for the first time that chloroplasts are the site of the complete photosynthetic process within cells.

Using this knowledge, Arnon then tried to dissect the various reactions involved by omitting one component at a time and observing the effects. For example, in his first experiment, Arnon supplied the chloroplasts all components except CO_2 (see Diagram B). Without CO_2, chloroplasts could carry out all aspects of photosynthesis except the production of carbohydrate.

HILL'S EXPERIMENT (1937)

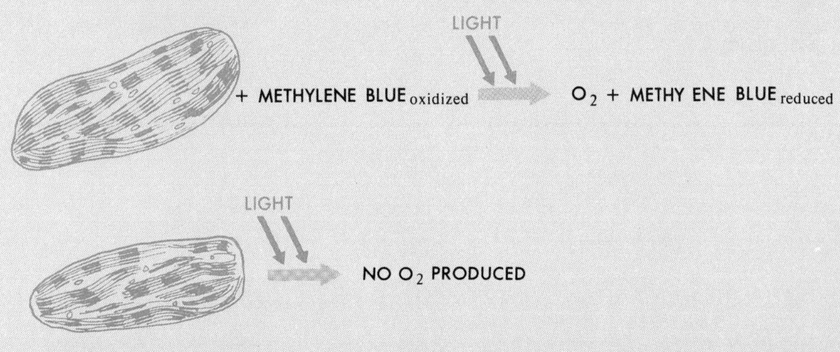

Diagram A

Diagram B

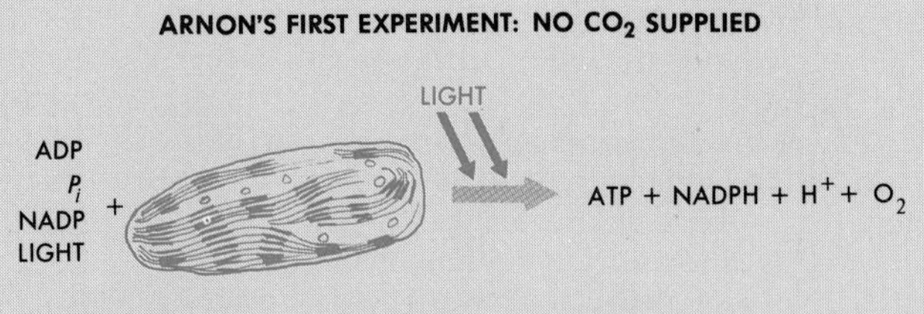

On the basis of these results, Arnon and others proposed that the photosynthetic process can be divided into two main sets of reactions: the light reaction, involving the splitting of water and the generation of ATP and reduced NADP (that is, NADPH + H^+), and the dark reaction, reduction of CO_2 by products of the light reaction.

Arnon then went on to reason that if photosynthesis was divided into a light and a dark phase, then some additional predictions could be made. For example, if both CO_2 and NADP were withheld, then the chloroplasts ought to be able to generate only ATP. The results are shown in Diagram C. The prediction was verified. Chloroplasts supplied with only ADP, P_i and light could generate ATP but could produce neither oxygen nor reduced NADP. This experiment suggested to Arnon that the interaction of light with chloroplasts alone was enough to generate ATP. In terms of electrons, this meant

Diagram C

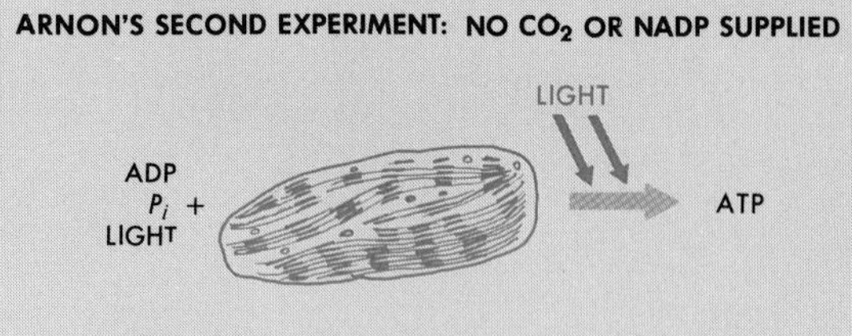

Light energy is absorbed into the biological system when electrons on chlorophyll are raised to higher energy levels and subsequently passed through an acceptor system. Downward transitions during electron transport release energy that is captured as high-energy phosphate bonds of ATP.

there might be some cyclic flow, since no terminal acceptor such as NADP was available. The generation of ATP by illuminated chloroplasts was thus seen to be independent of the generation of NADPH + H^+.

Now Arnon made another prediction—one with far-reaching implications. If the function of the light reaction was to produce ATP and NADH + H^+, then chloroplasts kept in the dark ought to be able to produce carbohydrates if supplied with CO_2, ATP, and NADH + H^2. He tried the experiment illustrated in Diagram D. The prediction was again borne out: chloroplasts kept from the light, but supplied with products of the light reaction plus the raw material CO_2, were still able to produce carbohydrates. This experiment seemed to confirm that the light reaction generates ATP and reduced NADP and that these products are used to reduce CO_2 to carbohydrate during the dark reaction.

Diagram D

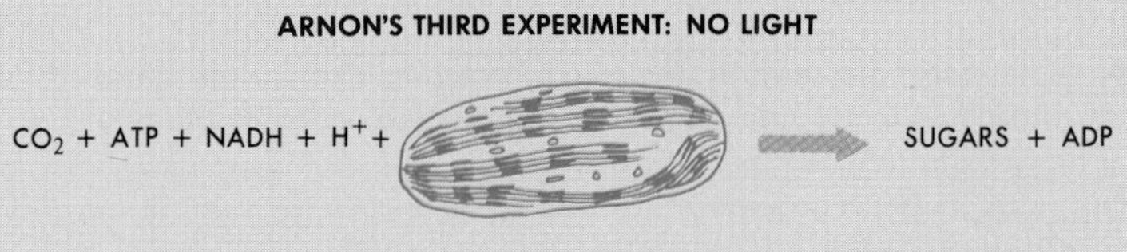

A further implication of Arnon's work pointed to the possible biochemical unity between plant and animal cells. Arnon knew that animal cells are capable of synthesizing certain kinds of carbohydrates from simple precursors by running the pathways of glycolysis and the citric acid cycle in reverse. Liver cells, for example, specialize in producing carbohydrates for storage (glycogen) by such a pathway. Arnon thus supplied liver cells with CO_2, ATP, and NADPH + H^+ (recall that animal cells have only NAD as an electron carrier). He found that the live cells *could* synthesize simple sugars from these basic raw materials: animal cells were capable of carrying out the dark reaction of photosynthesis.

It is thus only in the first phase of photosynthesis—the light reaction—that plants differ so markedly from animals. Otherwise, the cells of each have enough of the same enzymes to carry out similar reaction pathways. All animal cells need is a supply of reduced NADP and they, too, can produce their own carbohydrates!

Phase II: The Dark Reaction

What happens to the energy generated during the light reactions? How is it used to synthesize glucose and other carbohydrates from CO_2? The pathways in which carbohydrate synthesis takes place are known as the **dark reaction.**

The dark reaction of photosynthesis is a multistep biochemical pathway. Like the pathways involved in glucose oxidation, part of the dark re-

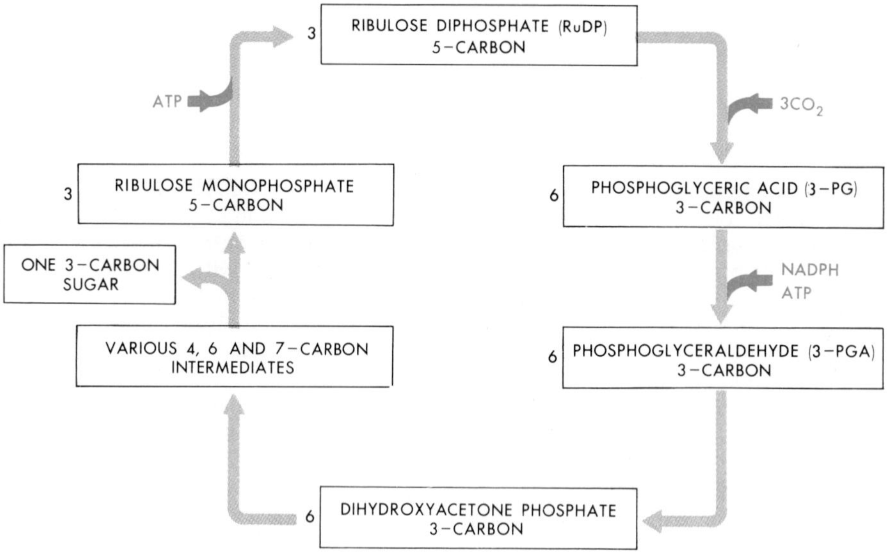

Fig. 6.9 *Schematic drawing of the major steps in the Calvin–Benson cycle (the C-3 cycle) of photosysthesis. The names of the various intermediates are given in the boxes. The designation underneath the name, such as "5-carbon" or "3-carbon," indicates the number of carbon atoms in one molecule of the substance. The numbers outside the boxes, to the left, indicate the number of molecules of that intermediate involved in a complete "balancing" of the cycle's equation. Thus to produce one molecule of a three-carbon sugar, which is the net output of the dark reaction, three molecules of RuDP must combine with three molecules of CO_2 to produce six molecules of PGA, etc. The net photosynthetic product, one three-carbon sugar, can combine with another molecule like itself to produce the six-carbon glucose molecule. The cycle "begins" with RuDP combining with CO_2. RuDP is returned again at the end of the cycle.*

action is cyclical. The main reactions in the dark phase of photosynthesis in most plants consist of what is commonly called the **Calvin–Benson cycle,** named for the two investigators who worked out its details. Prominent in the Calvin–Benson cycle are a series of three-carbon intermediates; hence the cycle is referred to as a C-3 cycle. In the last few years, an alternative pathway has been studied in certain plants, particularly those, like cactus, found in hot and dry habitats. This pathway involves a number of prominent four-carbon intermediates and hence is referred to as a C-4 cycle.

The Calvin–Benson cycle. As we have seen, the pathway of CO_2 reduction is powered by NADPH, which supplies both energy and hydrogen, and ATP, which supplies only energy. But how are the hydrogens from NADPH joined onto CO_2? The events of the Calvin–Benson cycle, in which the

The dark reactions of photosynthesis (also called the Calvin-Benson, or the C-3, cycle) constitute an endergonic pathway in which energy from ATP and NADPH drives the reduction of CO_2 to carbohydrate.

reduction of CO_2 takes place, are outlined in Fig. 6.9. A molecule of CO_2, taken in from the atmosphere, is joined onto an already-existing, five-carbon compound in the cell, ribulose diphosphate (RuDP). This reaction is catalyzed by the enzyme RuDP carbozylase. The product is two three-carbon sugars, **phosphoglyceric acid** (3-PG), and 3-PG is then reduced to **phosphoglyceraldehyde** (3-PGA) by NADPH, an endergonic reaction driven by ATP. Thus phosphoglyceraldehyde, the simplest carbohydrate, is a more energy-rich compound than phosphoglyceric acid. Some of the phosphoglyceraldehyde is converted into another three-carbon sugar (dihydroxyacetone phosphate), which combines with some of the remaining phosphoglyceraldehyde to make the six-carbon, fructose-1, 6-diphosphate. Approximately one-third of the fructose-1, 6-diphosphate is converted into glucose —the end product of the whole synthetic process. The remainder of the fructose-1, 6-diphosphate and the 3-phosphoglyceraldehyde are involved in a complex set of cyclic reactions that eventually generate more ribulose diphosphate, which can then combine with another molecule of CO_2. The pathway is thus a true cycle, because it regenerates itself even while producing a net product that is subsequently funneled off.

The Calvin–Benson cycle has many similarities to glycolysis. In fact, a close comparison indicates that many of the same intermediates and enzymes are involved in both pathways. The difference lies in the fact that the pathways run in opposite directions. Whereas glycolysis releases energy, the Calvin–Benson cycle consumes it. For instance, an important energy-yielding phase of glycolysis is the oxidation of 3-phosphoglyceraldehyde to 3-phosphoglyceric acid. Similarly, the production of 3-phosphoglyceraldehyde from fructose-1, 6-diphosphate in glycolysis is the reverse of what happens in photosynthesis.

This duplication of reactions can be understood in the context of the total metabolism of green plant cells. All green plant cells have mitochondria and carry out glycolysis and the citric acid cycle *in addition to* carrying out photosynthesis. The two processes are separate only to the extent that they are localized within different cell organelles, and to the extent that glycolysis involves NAD while the Calvin–Benson cycle involves NADP. Since both processes evolved together in green plant cells, it is not unusual to find that they have employed much cell machinery in common. (In fact it is highly efficient to have done so. Any chemical reaction will go in reverse if supplied with enough energy. The cell economizes on building materials and energy if it simply runs one set of pathways in reverse, rather than building enzymes for two wholly different pathways.)

An alternative pathway. As mentioned earlier, a second pathway of CO_2 fixation has recently been discovered in desert plants, although it appears to occur in some other types of plants as well. This pathway, referred to as the C-4 pathway, utilizes a different set of intermediates, prominent among which is phosphoenol-pyruvate, or PEP. Consequently the C-4 cycle is also called the PEP cycle, or the PEP system. The PEP cycle is particularly

Supplement 6.3

THE DARK REACTION: HOW DO WE KNOW?

The dark reaction of photosynthesis has been described as a cycle involving a number of intermediates that pass through a specific and predictable series of transformations. How has the sequence of events making up the Calvin–Benson cycle been determined? Much of the work was pioneered and carried out by Melvin Calvin and his colleagues at the University of California at Berkeley.

Calvin carried out his studies on the one-celled alga *Chlorella,* which is easy to culture and photosynthesizes at a great rate. *Chlorella* are grown in a large tank, attached to a flow-coil system as shown in Diagram A. CO_2 is bubbled into the tank, which is constantly exposed to light. Liquid from the tank, containing algal cells, is made to flow through the coils at a fixed rate. At a certain point along the coils, radioactive CO_2, containing the C^{14} isotope of carbon, is injected into the flowing liquid. The cells absorb this CO_2 during the remainder of the flow through the coils. At the bottom, the cells are plunged into a beaker of boiling methanol, which kills them instantaneously. The cells can

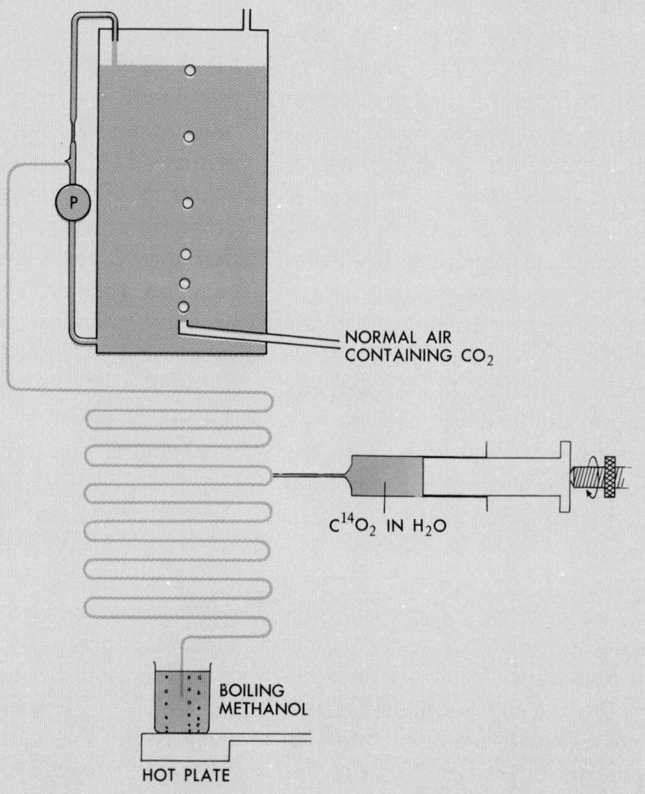

Diagram A

then be homogenized and their contents separated by paper chromatography. The compounds containing C^{14} carbon dioxide can then be identified by placing the dried chromatographic paper in the dark next to an X-ray film. The film is then developed, giving a typical "autoradiograph" such as that shown in Diagram B. The identity of each spot is determined by comparison with a standard or known chromatogram for each particular type of compound involved. The autoradiograph below, which involved an exposure of *Chlorella* cells to radioactive CO_2 for about 1 min, identifies the major intermediates

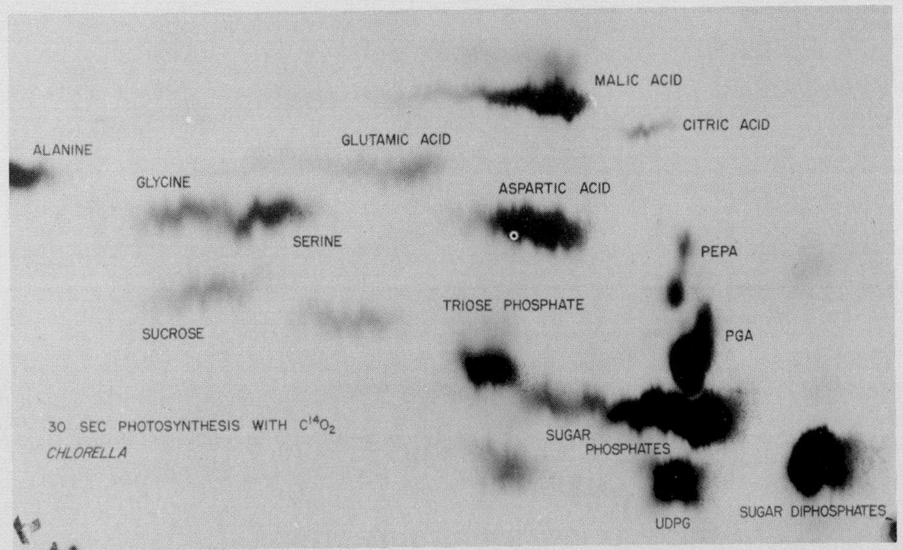

Diagram B *Paper "autoradiograph" of intermediate products of the Calvin–Benson cycle (dark reaction) of photosynthesis. (Courtesy Dr. James A. Bassham.)*

involved in the dark reaction. It does not, however, reveal anything about the *sequence* in which these intermediates were formed.

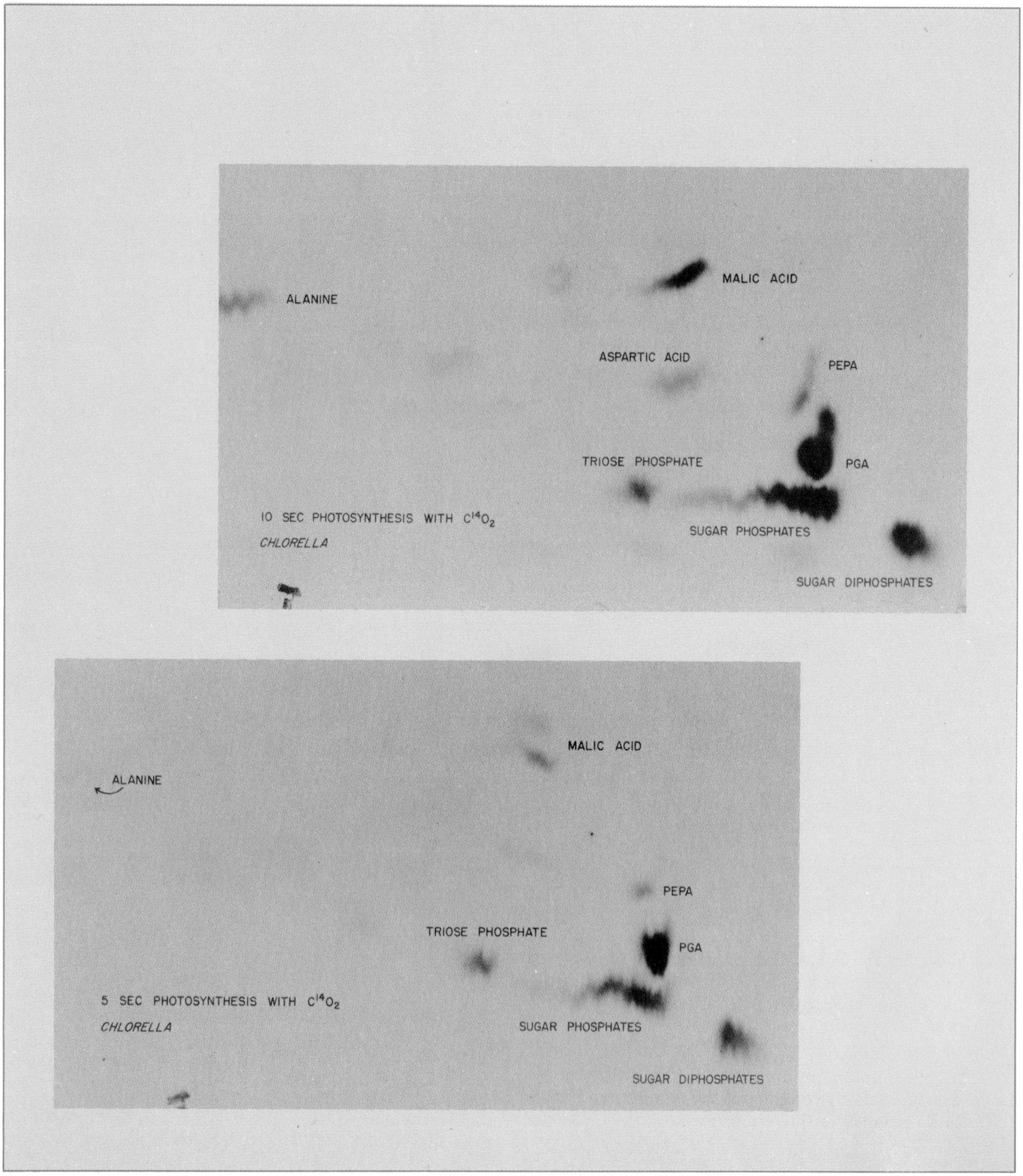
MALIC ACID
ALANINE
ASPARTIC ACID
PEPA
TRIOSE PHOSPHATE
PGA
10 SEC PHOTOSYNTHESIS WITH $C^{14}O_2$
CHLORELLA
SUGAR PHOSPHATES
SUGAR DIPHOSPHATES
MALIC ACID
ALANINE
PEPA
TRIOSE PHOSPHATE
PGA
5 SEC PHOTOSYNTHESIS WITH $C^{14}O_2$
CHLORELLA
SUGAR PHOSPHATES
SUGAR DIPHOSPHATES

Diagram C *Autoradiographs of intermediates in the Calvin–Benson cycle at 5-sec (top) and 10-sec (bottom) exposure of algal cells to radioactive* $C^{14}O_2$. *(Courtesy Dr. James A. Bassham.)*

Calvin and his associates realized that if the rate of flow of the *Chlorella*-containing medium through the coils was varied, they could increase or decrease the amount of time that cells were exposed to radioactive CO_2. When the rate of flow was increased so that the cells were exposed to CO_2 for only 30 sec, for example, the number of intermediates appearing on the autoradiographs decreased. When exposure was lowered to only 5 sec, still fewer spots appeared (see Diagram C). These spots were identified as the three-carbon sugars, RuDP in greatest quantity, and phosphoglyceric acid in somewhat less quantity. By varying the flow rate, and by using metabolic inhibitors that block certain steps in the process, Calvin and others were able to deduce the stages of incorporation of CO_2. It was obvious, for example, that one of the first compounds into which radioactive CO_2 was incorporated was RuDP. Using such techniques over the course of many years, a number of investigators have been able to piece together the complex series of reactions we designate as the Calvin–Benson cycle.

Note from examining the autoradiographs in Diagrams B and C that many of the substances into which radioactive C^{14} finds its way are familiar to us in other contexts than photosynthesis. For example, there are a number of amino acids (aspartic or glutamic) and a number of intermediates in the oxidative pathways of the citric acid cycle (malic, citric, and pyruvic acids). Since molecules entering a cell's biochemical pathways during photosynthesis soon turn up in the building blocks of proteins, carbohydrates, or lipids, it seems clear that nearly all biochemical pathways are interrelated.

adapted to the functioning of a desert plant and works in conjunction with the Calvin–Benson cycle. Desert plants, such as members of the Goosefoot family (*Atriplex*) of the southwestern United States, face two difficult and contradictory needs during the daylight hours. On the one hand, desert plants need to restrict water loss by keeping their stomata closed, since most water loss occurs by evaporation through these stomatal openings. When the stomata are closed, however, very little CO_2 can enter the leaf from the outside; hence, the rate of photosynthesis is greatly retarded. Under such conditions most plants, if they could survive at all, would have so low a concentration of CO_2 in their leaf cells that photosynthesis would proceed at a snail's pace.

Yet *Atriplex*, with its stomates closed, photosynthesizes at a maximum rate in the hottest part of the day. This is possible because of the remarkable characteristics of the PEP cycle.

What CO_2 is available is initially picked up not by RuDP (as in C-3 plants), but by phosphoenol-pyruvate, as shown in Fig. 6.10. This reaction is catalyzed by the enzyme PEP carboxylase. Here is the first place we find an interesting feature of the C-4 system. PEP is far more reactive with CO_2 than is RuDP. In addition, RuDP carboxylase is somewhat inhibited by high oxygen concentration, while PEP carboxylase is not. This means that the PEP system is more effective at picking up CO_2 than the RuDP system when the proportion of carbon dioxide is low and that of oxygen is high. Together, these two features of the C-4 system mean that in low CO_2 concentrations, such as would prevail when the stomata are closed, PEP would take up CO_2 whereas RuDP would not.

However, none of the products of the C-4 cycle are useful in photosynthesis. In fact, careful examination of Fig. 6.10 shows that if allowed to make a full turn, the cycle releases the CO_2. Thus, the C-4 cycle itself does not lead to photosynthetic fixation of CO_2. The C-4 cycle *does* facilitate the incorporation of CO_2 into the Calvin–Benson cycle. Fig. 6.11 (a and c) shows the general structure of the leaves of two types of plants: (a) *Atriplex*

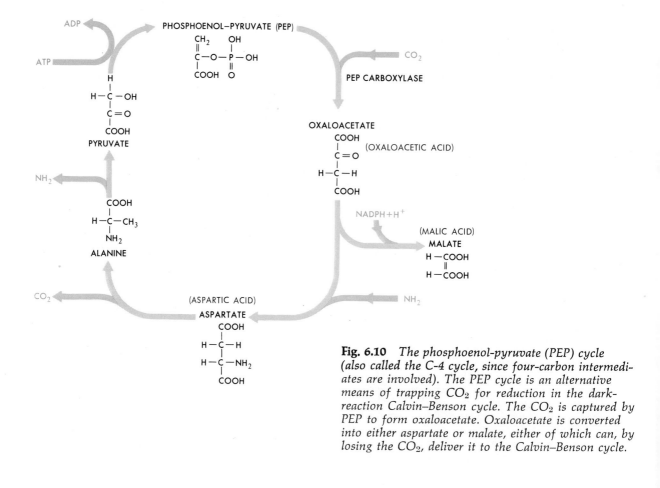

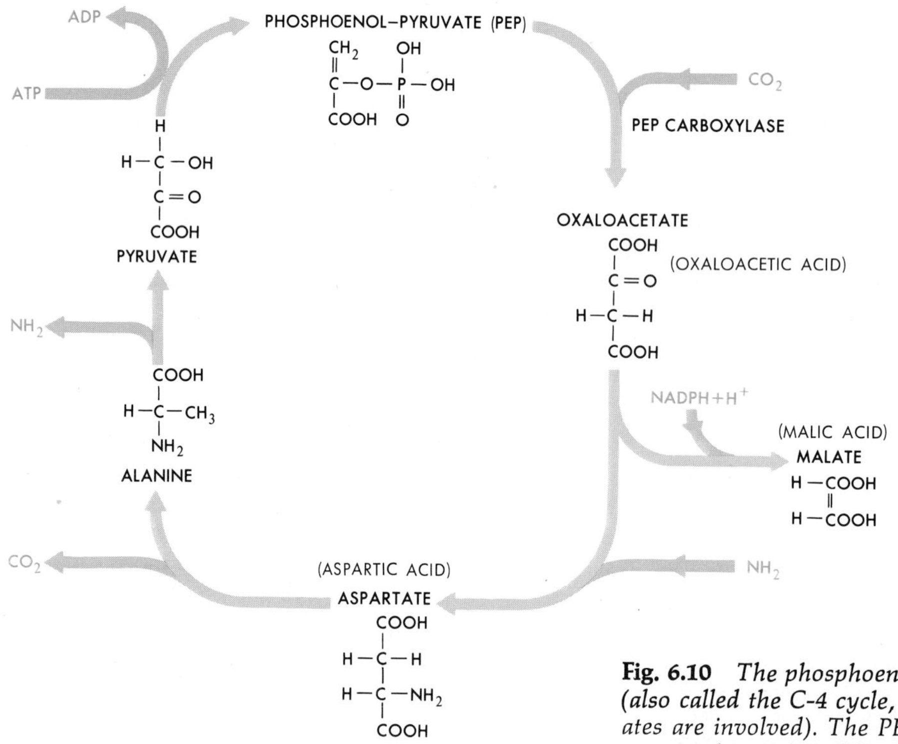

Fig. 6.10 *The phosphoenol-pyruvate (PEP) cycle (also called the C-4 cycle, since four-carbon intermediates are involved). The PEP cycle is an alternative means of trapping CO_2 for reduction in the dark-reaction Calvin–Benson cycle. The CO_2 is captured by PEP to form oxaloacetate. Oxaloacetate is converted into either aspartate or malate, either of which can, by losing the CO_2, deliver it to the Calvin–Benson cycle.*

In the Calvin–Benson cycle, for every three molecules of CO_2 that combine with three molecules of RuDP, six molecules of PGAL are produced. PGAL is the primary product of the dark reaction. One PGAL molecule is funneled off: the other five are reformed into three molecules of RuDP so that the cycle can begin again.

---

patula, which grows in less dry environments and has only the C-3 (Calvin–Benson) cycle for CO_2 fixation; (c) the leaf of *Atriplex rosea,* a close relative, which grows in very hot places and utilizes both the C-3 and C-4 systems.

Here emerges the second interesting feature of the C-4 system. The enzymes for the PEP cycle are localized in the cells of the mesophyll, a layer of cells just beneath the epidermis of the leaves of most plants (see Fig. 6.11a and c). The mesophyll cells are close to air spaces, the openings to which are regulated by the stomata. Two intermediates in the PEP cycle, malate and aspartate, are easily transportable from the mesophyll into cells of a deeper layer in the leaf, the bundle-sheath cells, which contain enzymes for the Calvin–Benson cycle. There they are converted into pyruvate by the removal of the CO_2 that had just been added. The pyruvate is then transported back into the mesophyll cells, where it completes its pathway by being converted into PEP, thus rounding out the cycle. Meanwhile, the CO_2 delivered into the bundle-sheath cells enters the Calvin–Benson cycle by the usual path of combining with RuDP. The C-4 system picks up CO_2 at low concentrations and delivers it to where it can be used in the dark reaction.

It should be pointed out that cells of all green plants contain enzymes for the Calvin–Benson (C-3) cycle. Those plants designated C-4 types engage in the PEP cycle *in addition to* the Calvin–Benson cycle. The plants in which the C-4 system is found are precisely those species that thrive in dry, hot, and sunny regions. The C-4 system is a photosynthetic adaptation to hot, dry conditions under which the stomata of the plants are mostly closed. Accompanying the biochemical and physiological differences between C-3 and C-4 types of plants are anatomical differences. C-3 plants do not have the cell layers of the leaf differentiated into mesophyll and bundle sheath as do C-4 plants, but rather tend to have cells more or less resembling the mesophyll types. Biochemical and anatomical characteristics are clearly interrelated here.

Fig. 6.11 *(a and b) Diagram of leaf section (above)* ▶ *and photosynthetic pathway localization (below) of the desert plant* Atriplex patula, *a C-3 plant. This species lacks the enzymes for the PEP cycle; it also lacks cellular differentiation into mesophyll and bundle sheath layers of the leaf. (c and d) Diagram of leaf section (above) and photosynthetic pathway localization (below) in* Atriplex rosea, *a C-4 plant.* A. rosea *has both the C-4 (PEP) and the C-3 (Calvin–Benson) cycles. These are localized in different layers of cells. The PEP system is found in the mesophyll cells, near the outer surface of leaf, and near the stomata; the Calvin–Benson system is localized in the bundle sheath layer, deeper within the substance of the leaf, and surrounding the vascular tissues. The PEP system allows* A. rosea *to pick up and utilize CO_2 molecules even when the latter are in low concentration within the cellular fluids of the mesophyll layer. Thus, even on extremely hot days in the desert,* A. rosea *can carry on a high rate of photosynthesis; it simply uses every available CO_2 molecule, something its close relative.* A. patula *(which lacks the PEP system) cannot do. (Diagrams after Olle Björkman and Joseph Berry, "High efficiency photosynthesis,"* Sci. Am. *229, no. 4, October 1973, p. 80.)* **Fig. 6.11 appears on pages 218–219.**

Figure 6.11

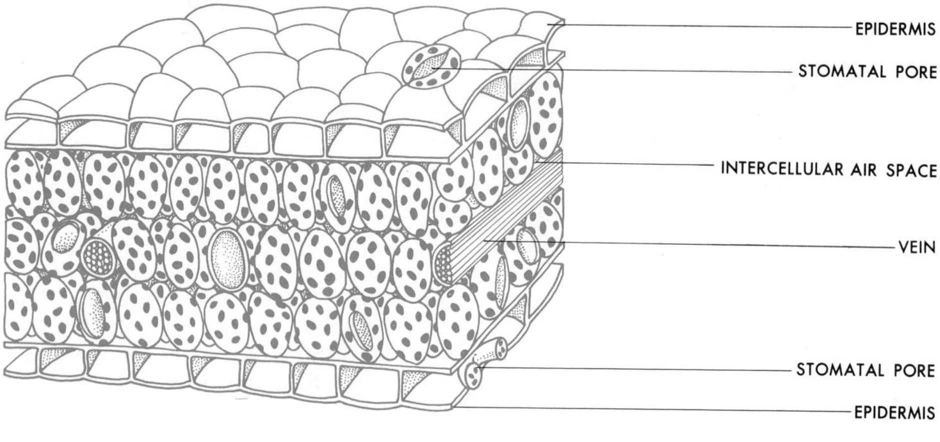

(a)

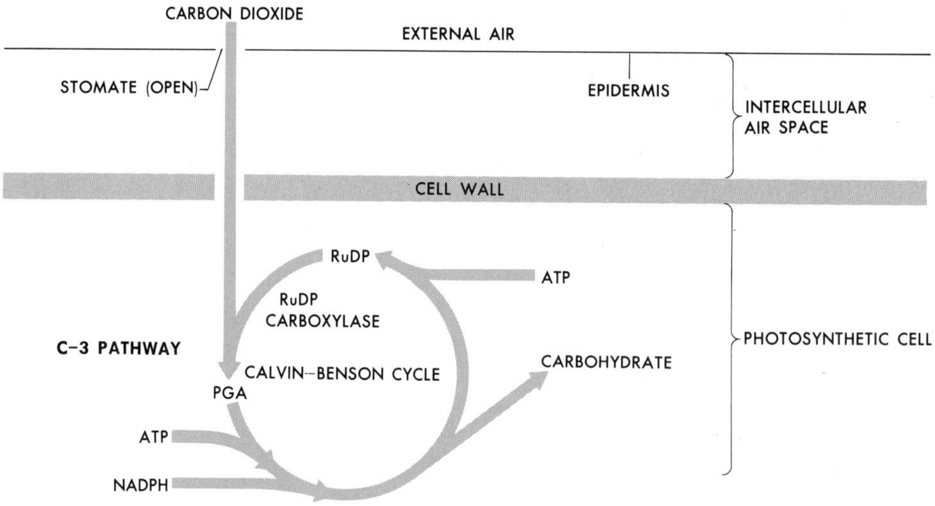

(b)

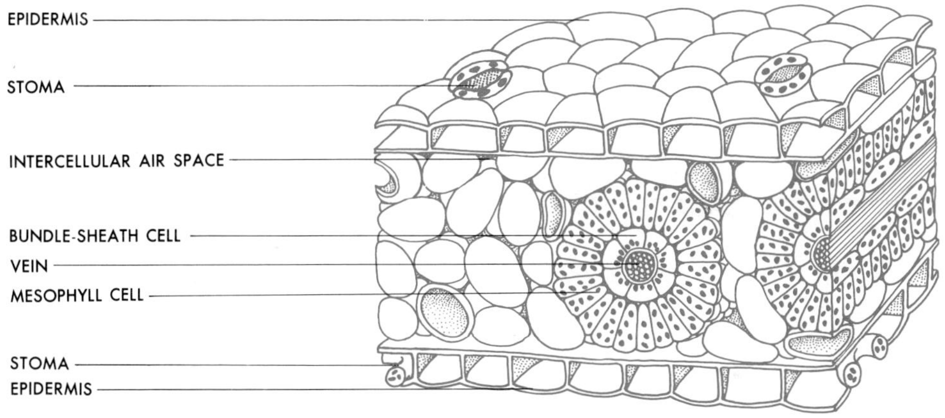
EPIDERMIS
STOMA
INTERCELLULAR AIR SPACE
BUNDLE-SHEATH CELL
VEIN
MESOPHYLL CELL
STOMA
EPIDERMIS

(c)

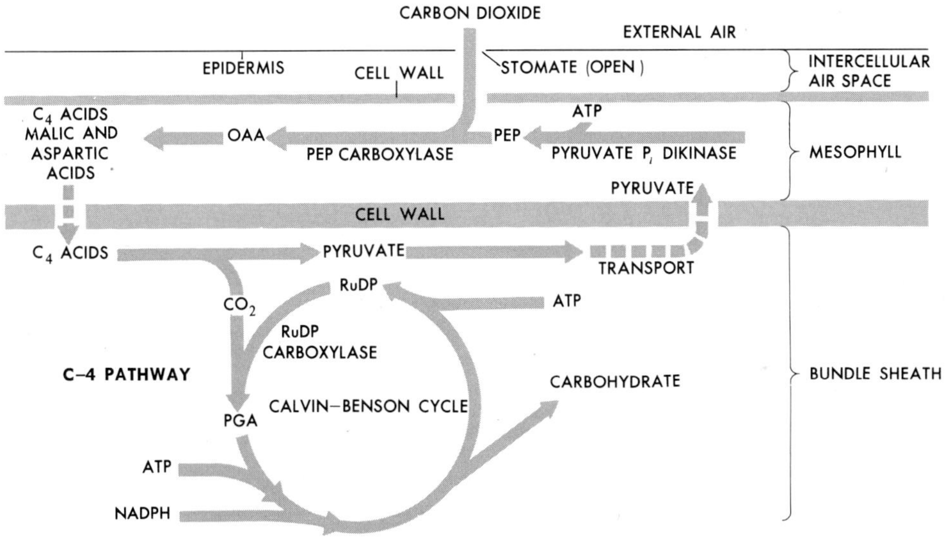
CARBON DIOXIDE
EXTERNAL AIR
EPIDERMIS
CELL WALL
STOMATE (OPEN)
INTERCELLULAR AIR SPACE
C_4 ACIDS MALIC AND ASPARTIC ACIDS
OAA
PEP CARBOXYLASE
ATP
PEP
PYRUVATE P_i DIKINASE
MESOPHYLL
PYRUVATE
CELL WALL
C_4 ACIDS
PYRUVATE
TRANSPORT
RuDP
ATP
CO_2
RuDP CARBOXYLASE
C–4 PATHWAY
CARBOHYDRATE
BUNDLE SHEATH
PGA
CALVIN–BENSON CYCLE
ATP
NADPH

(d)

Figure 6.12

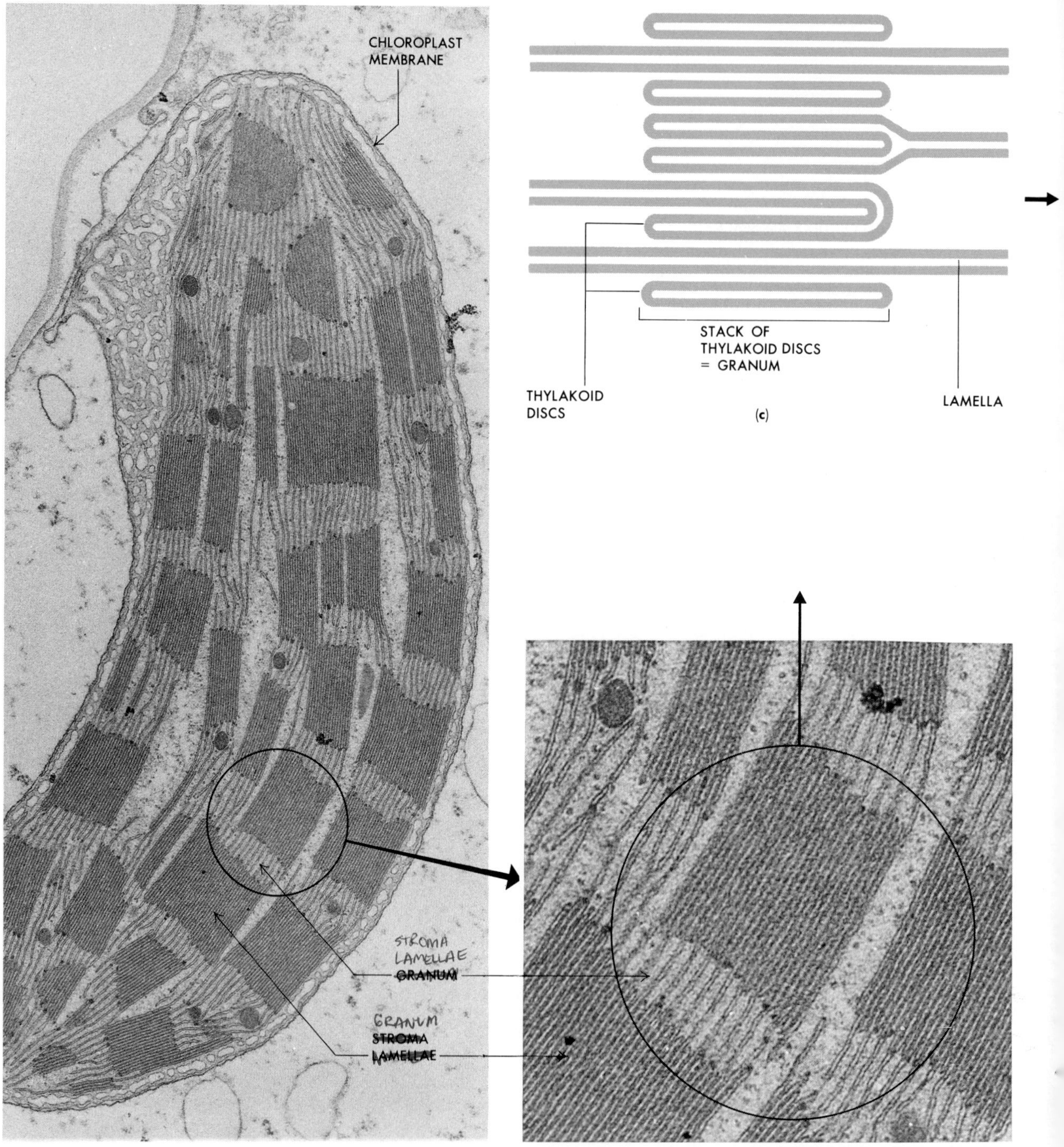
CHLOROPLAST MEMBRANE
STROMA LAMELLAE
GRANUM
GRANUM
STROMA LAMELLAE
STACK OF THYLAKOID DISCS = GRANUM
THYLAKOID DISCS
LAMELLA
(c)

(a)

(b)

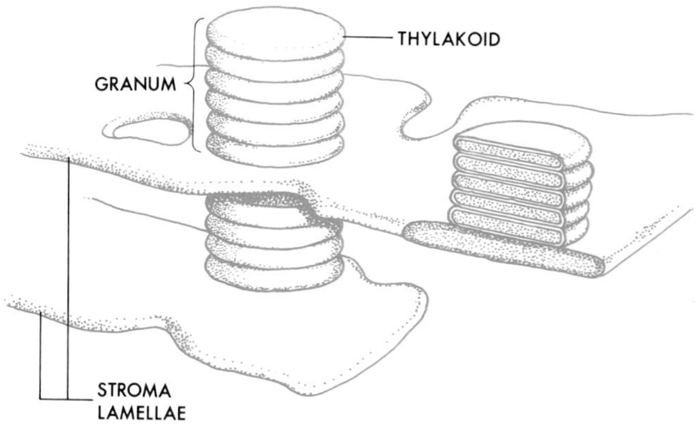
THYLAKOID
GRANUM
STROMA
LAMELLAE

(d)

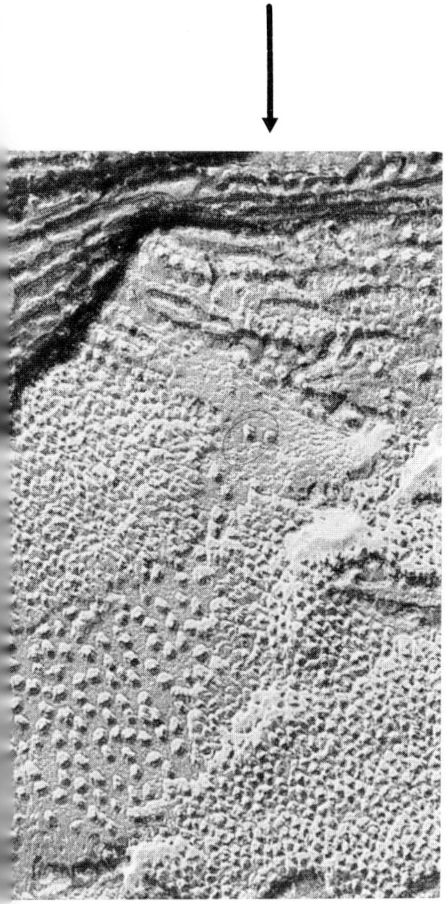

(e)

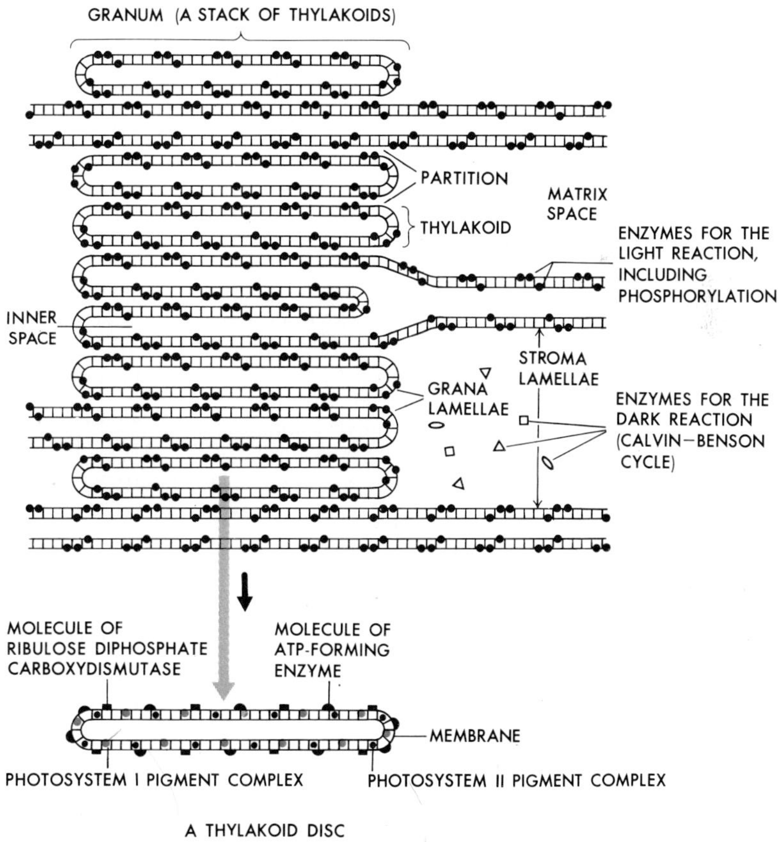
GRANUM (A STACK OF THYLAKOIDS)
PARTITION
MATRIX SPACE
THYLAKOID
ENZYMES FOR THE LIGHT REACTION, INCLUDING PHOSPHORYLATION
INNER SPACE
STROMA LAMELLAE
GRANA LAMELLAE
ENZYMES FOR THE DARK REACTION (CALVIN–BENSON CYCLE)
MOLECULE OF RIBULOSE DIPHOSPHATE CARBOXYDISMUTASE
MOLECULE OF ATP-FORMING ENZYME
MEMBRANE
PHOTOSYSTEM I PIGMENT COMPLEX
PHOTOSYSTEM II PIGMENT COMPLEX
A THYLAKOID DISC

(f)

◀ **Fig. 6.12** *Detailed structure of chloroplasts. (a) Single chloroplast from a leaf of* Froelichia, *showing chloroplast membrane and the internal membrane system composed of lamellae running lengthwise through the liquid matrix, the stroma. Periodically the lamellae are folded into a series of flat discs called thylakoids, which are stacked on top of one another to form grana. (Courtesy L. K. Shumway.) (b) Detail of granum showing stacking of thylakoid discs. (Courtesy L. K. Shumway.) (c) Schematic drawing of interrelationship between lamellae and the thylakoid discs. (After A. L. Lehninger,* Bioenergetics, *2d ed. Menlo Park, Calif.: W. A. Benjamin, 1973, p. 120.) (d) Three-dimensional model of interrelationship between lamellae running through stroma and folded lamellae of the thylakoid discs. (Photo courtesy J. E. Heslop-Harrison, from "Structural features of the chloroplast,"* Science Progress, *54, 1966, 519–541). (e) Electron micrograph of several layers through a granum, showing granules, thought to contain the various pigment complexes and enzymes for the light reaction. Larger granules are the quantosomes, thought to contain pigment complexes. Smaller granules may be various enzyme systems for photophosphorylation and electron transport during the light reaction. (Courtesy Dr. R. B. Park.) (f) A hypothetical scheme for how the localization of various enzymes and pigment complexes could be distributed on the lamellae and within the grana. The enzymes and pigment molecules for the light reaction are bound to the membranes running through the chloroplasts, especially those concentrated in the grana (the thylakoid discs). The enzymes for the dark reaction (the Calvin-Benson cycle) are free-floating in the stroma, between the lamellae. (Upper diagram after A. Trebst, "Energy conservation in photosynthesis electron transport of chloroplasts,"* Annual Review of Plant Physiology *25, 1974, p. 426. Lower diagram from A. L. Lehninger,* Bioenergetics.) **Figure 6.12 appears on pages 220–221.**

6.5 Localization of Photosynthetic Components

Both the photosynthetic pigments involved in the light reaction and the enzymes that catalyze the light and dark reactions are found within plant cell organelles known as chloroplasts (see Fig. 4.23). Furthermore, the molecular machinery for the light reaction is found in different parts of the chloroplast from those in which the dark reaction machinery is found.

As we saw in Chapter 4, the chloroplast is a highly structured organelle made up of an orderly assembly of membranes, or **lamellae.** The liquid matrix that bathes the lamellae is known as **stroma.** Figure 6.12 gives a variety of views of chloroplast structure and the localization of photosynthetic machinery on or within these organelles. Note that the lamellae are periodically folded and flattened into disc-shaped structures piled on top of one another (Fig. 6.12d and f). Each disc is called a **thylakoid disc,** and the pile is known as a **granum.** Grana appear throughout the chloroplast. Lamellae and thylakoid discs are membrane bilayers to which chlorophyll and other pigment molecules are attached. As Fig. 6.12(e) shows, the flat surface of a thylakoid disc appears to contain a number of granules, some larger in size than others. The larger granules are called **quantosomes,** which are associated groups of pigment complexes (pigments for photosystems I and II). The excitation of electrons by light energy takes place in the quantosomes. The smaller granules shown in Fig. 6.12(e) are thought to be various groups of enzyme molecules and electron transport systems associated with the light reaction. Because of their binding to the membranes of the grana, all the various molecules involved in the light reaction lie in close proximity to one another. Electrons removed from P_{700} or P_{684} can thus be picked up readily by the various acceptors, because of the spatial relationships of the molecules bound to chloroplast membranes.

Enzymes for the dark reaction, the Calvin–Benson cycle, are not membrane-bounded, but freely suspended in the liquid of the stroma. Because the whole chloroplast is surrounded by a membrane, these enzymes are kept localized within a very small space in the cell. Thus they are available to complete the job of photosynthesis started by the light reaction. The products of the light reaction—ATP and NADPH + H^+—can be used directly in the reduction of carbon dioxide in the free-floating stroma of the chloroplast.

The pattern of localization in the chloroplast is strikingly similar to that found in mitochondria, where the enzymes for electron transport and phosphorylation are bound to the cristae, while the enzymes for the citric acid cycle are free-floating within the liquid matrix of the mitochondria. Strong evidence suggests that the processes of photosynthetic phosphorylation and oxidative phosphorylation are very similar and may well have evolved from a common biochemical pathway. Such thinking is consistent with an hypothesis proposing that chloroplasts and mitochondria have evolved from a common ancestral organelle (see Supplement 13.2).

It seems likely that this ancestral prototype organelle was a mitochondrion-like structure, involved largely in oxidative phosphorylation. The chloroplast may have evolved from it in the group of organisms leading to what we now call green plants. Certainly, the two organelles have many structural characteristics in common, from their general shape to their membranous inner structure and patterns of enzyme localization. For example, in mitochondria the enzymes for the citric acid cycle are free-floating in the liquid matrix, while those for oxidative phosphorylation are bound to the inner membrane of the cristae. In chloroplasts, the enzymes and other molecules for the Calvin–Benson cycle are free-floating within the stroma, while those (including pigments) for the light reaction are bound to the lamellae and thylakoid discs. It is intriguing to contemplate how two such similar organelles and their associated biochemical functions could have evolved over eons of time.

6.6 The Efficiency of Photosynthesis

The solar energy that strikes the surface of the earth every day is the equivalent of 100,000,000 atomic bombs of the size dropped on Hiroshima. Some of this energy evaporates water in the seas, water which may later fall back on the land and form potential energy sources as it runs downward through streams and rivers on its way back to the sea.

Of much greater significance, however, is the solar energy captured by green plants. The fall of excited electrons in chlorophyll is far more important than all the waterfalls on earth. The chemical energy resulting from carbon dioxide fixation constitutes virtually all of the fuel for living matter. Indeed, the use of light energy in building energy-rich molecules is so important that those organisms capable of using light for that purpose are given a special title, the **autotrophs.** Without the autotrophs' constant conversion of light energy into a chemical form usable by all kinds of organisms, virtually all life on earth would perish. For, directly or indirectly, most nonphotosynthesizing organisms **(heterotrophs)** are dependent upon the autotrophs for the fuels that keep them going. Thus Joseph Priestley's original concept of the need for plant life to "render air fit for breathing again" has been enlarged to include many other needs as well.

Since the photosynthetic process is of utmost importance to all forms of life, including humans, the efficiency of that process must also be a matter of great importance.

There are several ways of looking at the subject. It can be noted, for example, that of the sun's energy reaching the earth (most is wasted by being radiated elsewhere in space), only one part in 2000

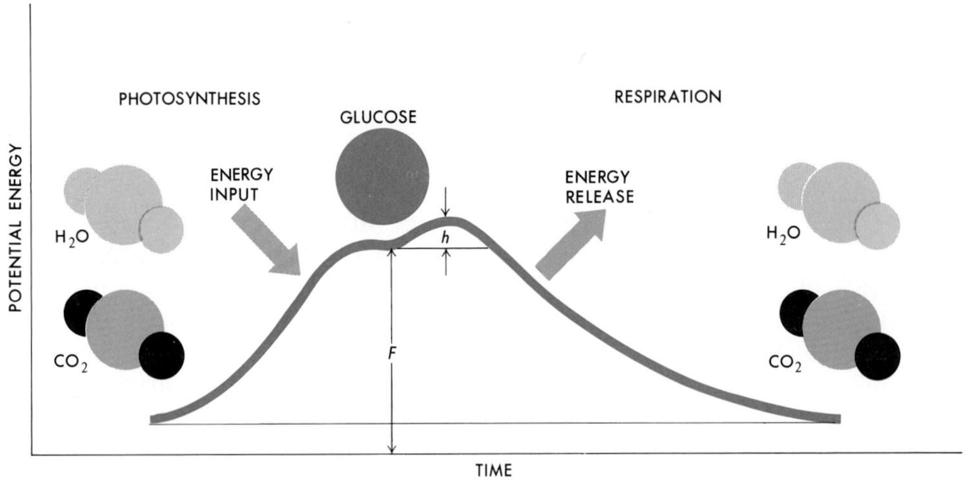

Fig. 6.13 *Photosynthesis and respiration are opposite sides of the "energy hill" in the living world. Photosynthesis is an uphill, endergonic series of reactions, while respiration is a downhill, exergonic series. Both are essential for the continuance of life.*

is actually captured by photosynthesizing plants. This very low efficiency would be far lower were it not for the fact that about 70% of the earth's surface is covered with water. Though we often tend to think of the familiar green land plants whenever we consider photosynthesis, about 90% of all photosynthetic activity is carried on by microscopic plants in the oceans, rivers, and lakes of the world. Indeed, *as far as humans are concerned,* many land plants are actually very inefficient food producers. A great deal of the energy they capture and store is used to build their own complex bodies. A single-called alga, on the other hand, is almost entirely food, and when it is eaten most of its body and stored energy enter the food chain. Compare this with the inefficiency of the apple tree, in which only the fruit is eaten. It must be stressed again, however, that the plant is inefficient *only* so far as we are concerned. For the apple tree, apple production must have been an efficient means of seed dispersal or the structure would not have evolved. Though we have often acted as if it were otherwise, no plant or animal has evolved for the purpose of supplying us with food, efficiently or inefficiently.

It is a basic principle of thermodynamics that whenever energy is transformed from one form to another, some of that energy is "lost"—cannot be made available for useful work. This problem is directly connected to that of photosynthesis and the use of photosynthetic products by all forms of life. Quite obviously (and fortunately) photosynthesis is an immensely profitable venture for the plant. The energy investment costs nothing; the dividends are so great that they can support those forms of life that do not contribute to them (i.e., do not photosynthesize). The Biblical quotation "All flesh is grass" is quite literally true.

Fortunately, once light is absorbed by chlorophyll molecules of the green plant, the synthetic process becomes remarkably efficient. Better than half of the absorbed light energy ends up locked within glucose molecules as potential chemical energy. Any further chemical transformations of this energy are bound to draw on this stored chemical energy, sacrificing more of it. Feeding plant material to animals to fatten them for human consumption is, in terms of energy expenditure, an extremely wasteful process. Only the wealthiest nations can afford it. It is no accident that the poor nations of the world are primarily vegetarian, relying for their nourishment mostly on such crops as rice, corn, or other plants. Such nations cannot afford the considerable energy "leakage" that must inevitably accompany the transformation of plant matter into animal matter before it is eaten.

Photosynthesis and respiration represent two sides of the energy hill that characterizes the metabolic processes within the living world (see Fig. 7.13). Photosynthesis is an uphill, or endergonic, reaction. During photosynthesis, CO_2 is pushed up the energy hill to a higher free-energy level in the form of glucose. Glucose stores free energy in its chemical bonds, which can be released by the oxidative processes of respiration. Respiration is thus a downhill, or exergonic, reaction. During respiration the free energy from glucose is released via gradual breakdown of the molecule through the removal of electrons (oxidation). The end products, CO_2 and H_2O, are the same as the starting reactants of photosynthesis. Since photosynthesis drives the uphill reactions that respiration pushes downhill, respiration could not occur without it. Photosynthesis winds up the world's biological motor, and respiration unwinds it in a controlled way. The energy source from the outside, nonbiological world that makes all this possible is, of course, sunlight.

SUMMARY

1. Photosynthesis is the process by which energy-rich compounds such as glucose are synthesized from low-energy precursors such as CO_2 and H_2O. Photosynthesis is an endergonic reaction driven by the energy of sunlight.
2. An overall input–output equation for photosynthesis is:

   $$6CO_2 + 12H_2O \rightarrow C_6H_{12}O_6 + 6O_2 + 6H_2O$$

   This equation obscures the important intermediate events involved in photosynthesis and indicates only what goes in and what comes out of the process.
3. Photosynthesis is characteristic only of green plants and those containing certain other types of pigments (not necessarily green). The pigments function as means of capturing the sun's light energy in a form the cell can use.
4. Photosynthesis takes place in two interconnected but separate pathways, the light reaction and the dark reaction.
5. The light reaction is the name given to the pathway by which quanta or photons of light energy are used to raise electrons from the two forms of chlorophyll *a*, P_{700} and P_{684}, to higher energy levels. The products of the light reaction, ATP and NADPH + H^+, are the important energy (and hydrogen) sources for driving the dark reaction.
6. Two pigment systems are involved in the light reaction phase of most plant photosynthesis: photosystem I consists of regular chlorophyll *a* and a special form of chlorophyll *a*, P_{700}; photosystem II consists of regular chlorophyll *a*, a special form of chlorophyll *a*, P_{684}, and chlorophyll *b*. The two pigment systems are able to use sunlight to carry out photosynthetic phosphorylation (or photophosphorylation, for short), the use of light energy to produce ATP.
7. Steps in the light reaction are as follows:
   a) Quanta of light strike molecules of regular

chlorophyll *a* and *b*, which act as antennae, funneling their excitation energy into removing electrons from P_{700} and P_{684}. Consider the case of photosystem I first. Two quanta of light raise two electrons from P_{700} to an excited state (to higher energy levels), from which they are picked up by an acceptor molecule, Z. Acceptor Z transfers the electrons to the iron-containing compound ferredoxin. From here they can follow one of two pathways. (1) Cyclic photophosphorylation: the electrons are passed through a group of acceptors (including cytochrome *b*, cytochrome *f*, and plastocyanin) back to P_{700}. During this process two ATP's are generated for each pair of electrons cycled. (2) Noncyclic photophosphorylation: the electrons are passed from ferredoxin to NADP, reducing it to NADPH + H^+. Ultimately, electrons passed from NADPH to reduce CO_2 also generate some ATP in the process.

b) A pair of excited electrons from pigment P_{684} are picked up by the acceptor Z_1, from which they can follow only one course: they are passed through a series of acceptors including plastoquinone, cytochrome b_{599}, cytochrome *f*, and plastocyanin. These electrons are eventually passed to chlorophyll *a*. Every pair of electrons passed from chlorophyll *b* back to chlorophyll *a* generates four ATP's.

c) Removal of electrons from pigments P_{700} and P_{684} leaves a "hole" in the chlorophyll molecules. The presence of this "hole" means the pigments have a tendency to pick up electrons to replace the lost electrons. Chlorophyll *a* gets its electrons either from the cyclic process or by noncyclic flow from chlorophyll *b*. Thus, while chlorophyll *a* does not pass its electrons to chlorophyll *b* (i.e., to photosystem II), chlorophyll *b* passes its electrons to chlorophyll *a*. Chlorophyll *b* replenishes its electrons by splitting H_2O, which releases molecular oxygen.

d) The net products of the light reaction are ATP and NADPH + H^+. These energy-rich substances are used to drive the dark reaction. In addition to supplying energy, NADPH + H^+ supplies hydrogens for the reduction of CO_2 in the dark reaction.

8. The dark reaction consists of the Calvin–Benson cycle, in which a prominent role is played by three-carbon sugars. Therefore it is called the C-3 system.

9. Events of the dark reaction are as follows:

   a) A molecule of CO_2 is picked up by a molecule of the five-carbon ribulose diphosphate (RuDP), making a six-carbon intermediate that is unstable. It breaks down into two three-carbon sugars and phosphoglyceric acid (PGA).

   b) PGA is reduced by hydrogens from NADPH and by energy from ATP to phosphoglyceraldehyde (PGAL). PGAL undergoes a series of transformations (including pathways through four-, six-, and seven-carbon sugar intermediates.

   c) From these transformations, one three-carbon sugar emerges for every three molecules of RuDP that combined with CO_2 in step (a). This three-carbon sugar is actually PGAL. Two PGAL's, for example, can combine to form fructose-1, 6-diphosphate, which is easily convertible into glucose. PGAL is the basic photosynthetic product.

10. The dark reaction is a cycle because the intermediates are regenerated—they show neither an overall increase nor an overall decrease in quantity within the chloroplast—during the process.

11. An alternative pathway for carbon fixation occurs in the cells of certain green plants, es-

pecially those found in hot, arid climates. The C-4 pathway utilizes a number of four-carbon acids as intermediates. The function of this pathway is to pick up CO_2 at low concentrations in the outer layers of the leaf (the (mesophyll) and transport it as a four-carbon intermediate (such as malic acid) to an inner layer of the leaf (the bundle-sheath layer). In the inner layer, the four-carbon intermediate breaks down, releasing the CO_2 it had picked up in the outer layer. This process delivers the CO_2 to the cells where it can be used in the Calvin–Benson cycle. The enzyme systems for the C-3 and C-4 systems in these desert plants are localized. Mesophyll has a high concentration of the enzymes for the C-4 pathway and a low concentration of enzymes for the C-3 pathway. Bundle-sheath cells are just the opposite. The existence of these two systems makes it possible for those desert plants that have the enzymes for the C-4 system to photosynthesize at a high rate even with the stomata completely closed, as would happen at midday when water loss through evaporation would be at its highest.

12. The Calvin–Benson (C-3) cycle has many similarities to the pathways of glycolysis. Both are cycles, both involve numerous intermediates, and both actually incorporate some of the same compounds (PGAL, PGA, pyruvate).

13. Photosynthesis is localized within the chloroplasts of plant cells. Molecules of chlorophyll and the enzyme systems associated with photophosphorylation are bound to the lamellar membranes within the chloroplast. The enzymes associated with the Calvin–Benson cycle are free-floating in the intrachloroplast medium, the stroma.

14. In terms of both general structure and functional organization, chloroplasts and mitochondria have much in common. Both are membrane-bounded organelles with an internal membrane system. Both are concerned with harnessing energy in a useful form for the cell. Both harness their energy from downward transition of electrons (electron transport) coupled with phosphorylation of ADP to produce ATP. Both have the enzymes associated with electron transport and phosphorylation bound to internal membranes, while the enzymes involved in oxidation and reduction of intermediate compounds are free-floating in the spaces between the inner membranes. These similarities have led to some speculation that chloroplasts and mitochondria are evolved from a common ancestral cell organelle.

15. Photosynthesis and respiration are opposite sides of the energy hill in the living world. Photosynthesis generates the high-energy compound glucose from CO_2 and H_2O; respiration breaks glucose down into CO_2 and H_2O. Photosynthesis is overall a reduction process; respiration is overall an oxidation process. Photosynthesis is absolutely essential for respiration to continue. Without photosynthesis there would be no glucose to oxidize, and the atmosphere would soon become exhausted of its oxygen. In turn, respiration supplies some of the CO_2 used by photosynthetic organisms as a raw material for photosynthesis. These two processes, on which all life depends, exist in a balanced state on earth. Preservation of that balance is essential for the continuance of life on earth.

EXERCISES

1. In what way was the origin of the oxygen evolved during photosynthesis eventually determined? Explain the technique involved.

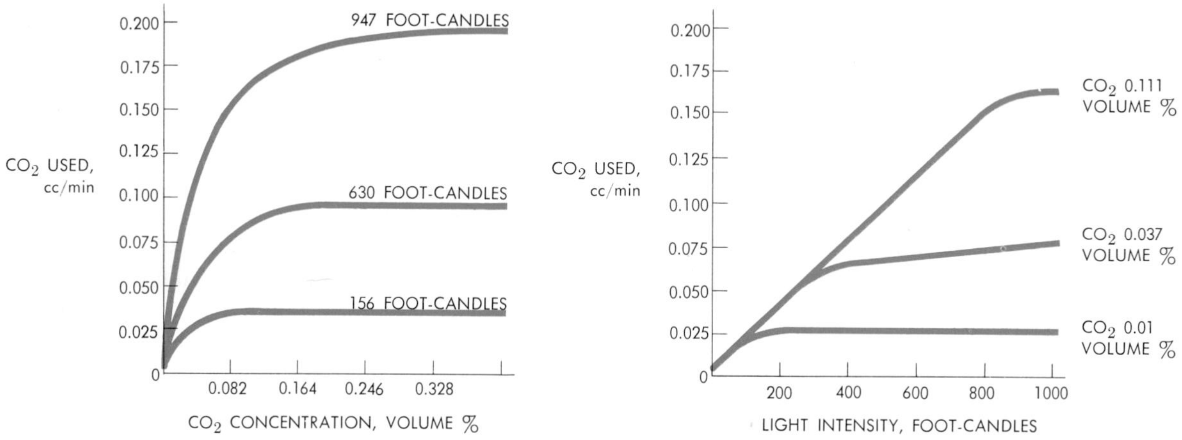

Figure 6.14

2. Evaluate each of the statements below only in terms of the data given in Fig. 6.14. Explain why you agree or disagree with each statement.
   a) Both carbon dioxide concentration and light intensity have upper limits beyond which they will not stimulate photosynthesis.
   b) A concentration of 18% CO_2 would have an inhibiting action on the rate of photosynthesis at *any* light intensity.
   c) When the concentration of CO_2 is 0.111% by volume, a doubling of the light intensity in foot-candles approximately doubles the rate of photosynthesis.
   d) Up to a certain limiting value, a change in either the concentration of CO_2 or the intensity of light produces a change in the rate of photosynthesis.
   e) Carbon dioxide and light energy are each involved in two separate phases of the photosynthetic cycle. Therefore, the rate of utilization of one does not affect the rate of utilization of the other.
3. What does the fact that most leaves are green tell you about the wavelengths of the light used in photosynthesis?

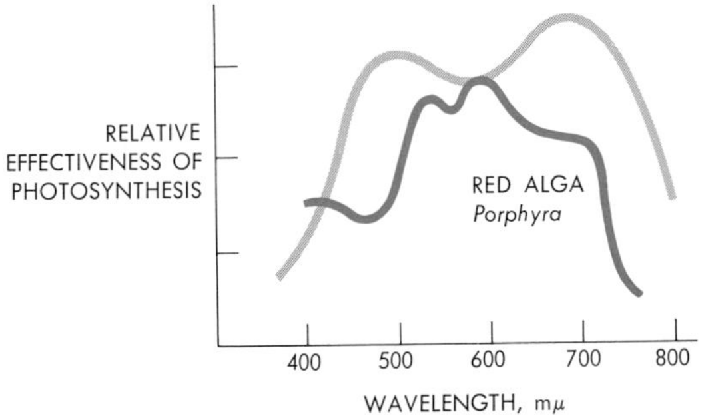

Figure 6.15

4. Observe the action spectra of two organisms (Fig. 6.15). Evaluate each of the following statements in relation to this graph.
   a) The red alga and green plant, both exposed to roughly the same light source, would absorb the same wavelengths of light with the same degree of effectiveness.
   b) Which single wavelength of light would produce about equal photosynthesis in the red algae and the green plant: (i) 450; (ii) 500; (iii) 510; (iv) 585; (v) 600; (vi) 650; (vii) 700?
   c) If a green alga and a red alga were inhabiting the same region of the ocean, in general they (would, would not) compete for the same wavelengths of light energy.
   d) The red alga is more effective in absorbing the total range of solar energy than the green plant.
5. Chlorophyll serves to "capture" light energy for use in the process of photosynthesis. It does this by absorbing the light from the sun (radiant energy) and making it usable by:
   a) losing an electron that is transferred by acceptor molecules, thus producing chemical energy for splitting up water,
   b) gaining an electron that, when it joins the chlorophyll molecule, releases some energy used to build up ATP and to split water,
   c) losing an electron that is transferred by acceptor molecules, thus producing small amounts of energy used to build up ATP and also to split water,
   d) gaining an electron, which is then immediately transferred for the sole purpose of splitting up the water.
6. What is the function of the carotenoid pigments in plants?
7. Distinguish between the light reactions and the dark reactions of photosynthesis.
8. Which of the following substances, supplied to a plant kept in the dark, would

make it possible for the plant to remain alive? a) NADP alone; b) NADP and CO_2; c) PGAL; d) NADH alone; e) NADH and CO_2. Explain your answer.

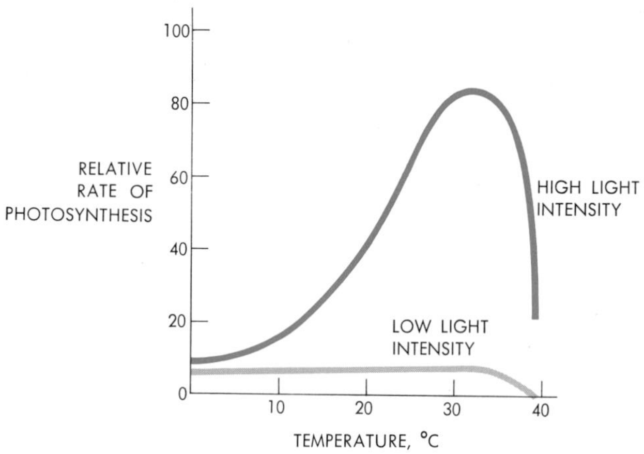

FIG. 9–22

Figure 6.16

9. Observe the graph in Fig. 6.16. Propose an hypothesis to account for the way in which temperature may affect the rate of photosynthesis differently at different light intensities.
10. The process of respiration is said to be the opposite of the process of photosynthesis. In what sense is this statement true? In what ways is it not true?
11. In the past, many textbooks have given the overall equation for photosynthesis as: $6CO_2 + 6H_2O \rightarrow C_6H_{12}O_6 + 6O_2$. What is incorrect about this formulation of the equation? How should this overall equation be written?

SUGGESTED READINGS

Bassham, J. A., "The path of carbon in photosynthesis," *Scientific American*, June 1962, p. 88. Although an older article, this one remains an excellent account of the basic methods involved in unraveling the steps of the dark reaction.

Bassham, J. A., and Melvin Calvin, *The Path of Carbon in Photosynthesis*. Englewood Cliffs, N.J.: Prentice-Hall, 1957. A more detailed account of the work described in summary form in the above *Scientific American* article of the same name.

Björkman, Olle, and Joseph Berry, "High-efficiency photosynthesis," *Scientific American*, October 1973, p. 80. A detailed discussion of the discovery and mechanism of the C-4 pathway in desert plants such as *Atriplex*.

Govindjee, and Rajni Govindjee, "The absorption of light in photosynthesis," *Scientific American*, December 1974, p. 68. A summary of current knowledge of the light reaction. Well written, but technical and highly detailed in certain parts.

Lehninger, A. L., *Bioenergetics,* 2d ed. Menlo Park, Calif.: W. A. Benjamin, 1973. Chapter 6 deals with "Photosynthesis and the Chloroplast." It is an authoritatively but simply written account of the energetics of the light reaction. One of the best summaries of current information available.

Levine, R. P., "The genetic dissection of photosynthesis," *Science* 162 (Nov. 15, 1968), p. 768. A very good (though technical) article dealing with the question of how photosynthesis is controlled genetically.

Levine, R. P., "The mechanism of photosynthesis," *Scientific American,* December 1969, p. 58. An account of both the light and dark reactions, but with emphasis on the light reaction. Older, but more simply presented than the Govindjee article.

SEVEN

CELL REPRODUCTION

7.1 Introduction

Since in most living organisms the cell is a unit of both structure and function, it is not surprising that so much of biological science centers upon it. In no area is this more true than in reproduction, for the reproduction of all cellular organisms must in the final analysis be a cellular process. For those plants or animals that reproduce asexually by fission, cell reproduction is synonymous with reproduction of the individual. For sexually reproducing organisms, the production of reproductive cells is itself the result of a specialized type of cell reproduction. Directly related to reproduction are the biological phenomena of inheritance and embryological development, neither of which can be understood without knowledge of cellular reproduction. In this chapter, then, we shall direct our attention to cell reproduction: first to the mechanics by which it is accomplished, second to some of the questions these mechanics pose, and third to its role in the production of reproductive cells themselves.

7.2 Cell Division: Some Problems of Study

The first scientific papers describing the mechanics of cell division appeared in the late nineteenth century. Many of the observations were imperfect and confusing, however. Not unexpectedly, progress in accurately describing what goes on inside a cell when it divides may be directly equated with improvements in both the tools (e.g., the microscope) and techniques (e.g., micromanipulation) available to the researcher.

Much of the past research on cell reproduction involved the sectioning, staining, and microscopic examination of dividing cells. Unfortunately, sectioned cells do not divide. Furthermore, many of the most useful stains react chemically with certain substances in the cell, changing their nature. Biologists studying cell reproduction have been plagued

Cell division and mitosis are not synonymous terms. Cell division refers to the processes of mitosis and cytokinesis, by which one cell becomes two. Mitosis is one part of that process: the duplication and division of chromosomes in such a way that both daughter cells end up with the same number and kind of chromosomes as the parent cell.

by an "uncertainty principle"—by virtue of the techniques needed to study the phenomenon, the conditions under which it could occur were changed to the point of being incompatible with the phenomenon itself. The problem has been only partially solved by the use of time-lapse photography, which has enabled biologists to telescope the often hours-long process of cell division into a few minutes of observation time.

Although our description of the process will focus on certain distinctive phases through which cells pass during the division process, it is important to keep in mind that cell division is very much a dynamic and continuous process, and that it proceeds steadily from one phase to another.

7.3 Mitosis

Mitosis is a series of changes by which one cell nucleus becomes two. This simple definition shows mitosis to be a nuclear process, primarily concerned with the equal distribution of the chromosomes into the daughter cells. The changes that occur in the cell cytoplasm during mitosis **(cytokinesis)** are not directly a part of the mitotic process (although they may occur simultaneously with it).

Not all cells undergo mitosis. A mammalian red blood cell, with no nucleus, cannot divide. Nor can a highly specialized cell, such as a neuron. On the other hand, some cell types undergo mitosis quite often. In multicellular organisms, these cells are usually found in regions of growth, such as a plant root tip, skin cells, or cells of a developing embryo.

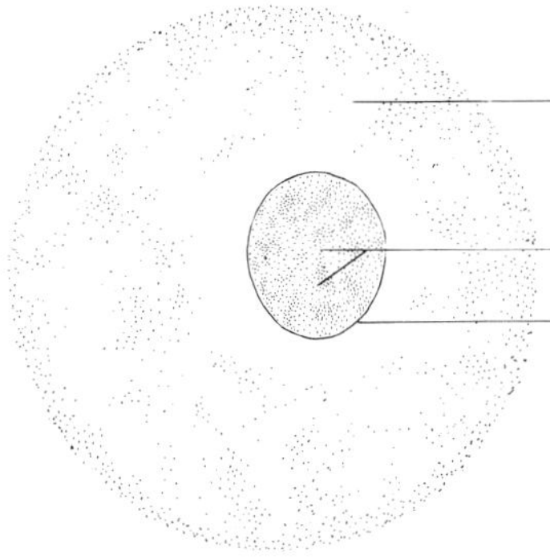

Fig. 7.1 *The nucleus of an amoeba in interphase.*

Of the five main phases associated with mitosis, the longest is **interphase.** As its name implies, interphase includes the period a cell spends between one division and the next, and in this sense some biologists prefer not to consider it a part of the mitotic process. Interphase has often been referred to as a "resting stage." This term is a highly misleading one, however. A cell in interphase is in fact quite active, as we will see.

The cell pictured in Fig. 7.1 is in interphase.* The nucleus and nuclear membrane are clearly distinguishable. No chromosomes are seen, however, since the chromosomal material is so diffused

* For simplicity, and in view of the uncertain role it plays in the process, we have chosen to ignore the actions of the centriole during mitosis. A discussion of some current hypotheses regarding its role will be found in Section 7.4.

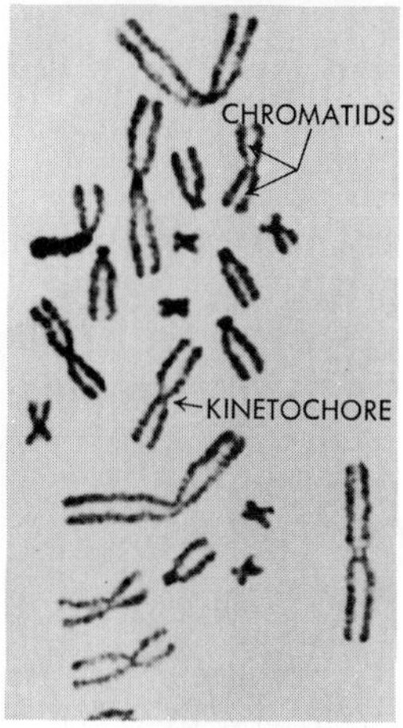

Fig. 7.2 *Replicated chromosomes showing chromatids held together at the kinetochores.*

throughout the nucleus as to be "invisible." It is only when these threads begin to condense into a tight, coillike configuration that the chromosomes become readily visible. When this occurs, the cell enters prophase.

The Mitotic Processes

Prophase begins with the coiling of the chromosomal material (sometimes called chromatin) into compact, recognizable chromosomes. The nuclear membrane also gradually disappears, as do the nucleoli. At high magnification, the chromosomes are revealed to be doubled. There are good reasons, of course, for such a chromosome doubling, or replication. Were this replication *not* to occur, the equal distribution of chromosomes to each daughter cell that takes place during mitosis would result in only half as many chromosomes in each daughter cell as in the parent cell from which it arose. The ultimate result of a series of such divisions, in each of which the chromosome number is reduced by one-half, is obvious.

Still closer examination of the replicated chromosomes reveals that the resulting pair still remain attached to each other by a constricted region, the **kinetochore** or **centromere** (Fig. 7.2). To these chromosome "halves," as long as they remain joined at the kinetochore, the term **chromatid** is applied. When the kinetochores separate later (at anaphase), the result is two separate, full-fledged chromosomes, each with a kinetochore of its own.

Also seen during prophase is the formation of fiberlike processes extending from the poles toward the equatorial plate of the cell. These processes are termed **spindle fibers.** Electron microscope studies have revealed that the spindle fibers are actually hollow microtubules (see (Fig. 7.3 and Section 4.4). The orientation of the spindle fibers establishes the direction in which the chromosomes will later move, as well as the region where the cell will eventually divide in two. The radiating fiberlike processes surrounding each pole in animal cells and most lower plant cells are called **asters** (Fig. 7.4a).

In **metaphase,** the next mitotic phase, the chromosomes line up along the equatorial plate of the cell. In the top photograph in Fig. 7.4(b), the cell has been sectioned as in the prophase photographs, i.e., lengthwise along the polar axis. In the bottom photo, however, the cell was sectioned at right angles to the polar axis. Thus the poles and asters are not seen, and the actual arrangement of the chromosomes along the equatorial plate is revealed to be circular, rather than linear as the center photo, seen alone, might indicate.

Anaphase follows metaphase. Anaphase is characterized by separation of the kinetochores so that the chromosomes (formerly chromatids) are free to separate, the two members of each pair moving to opposite poles. The separating chromo-

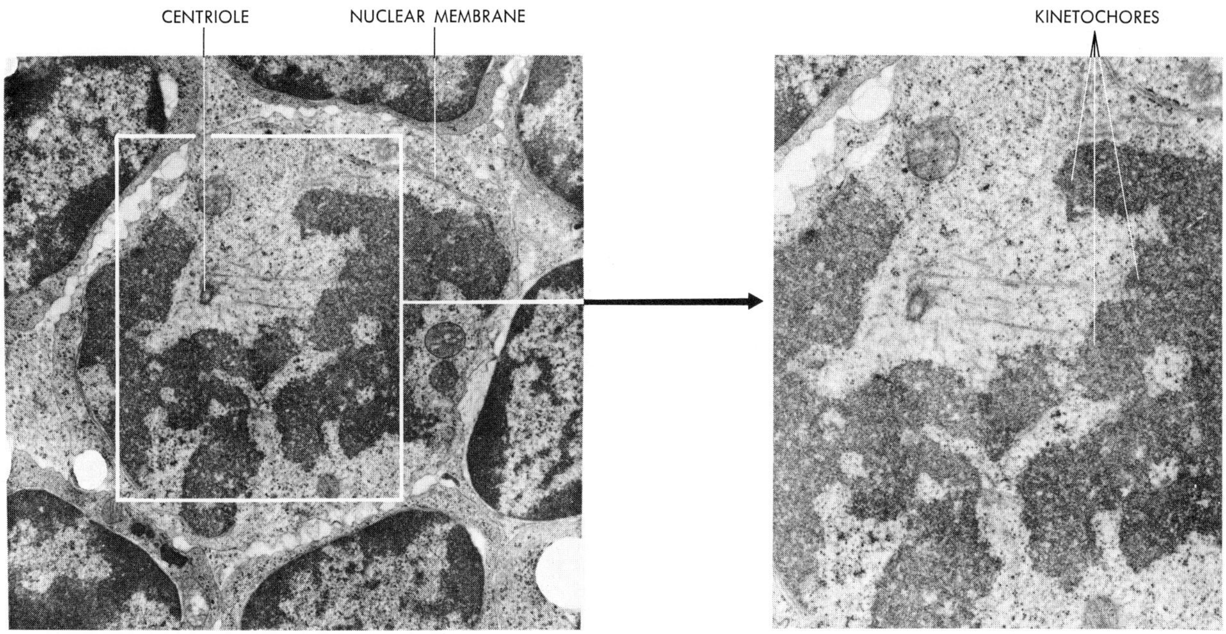

Fig. 7.3 *Electron microscope photograph of rat thymus gland cells in prophase, ×6500. Note that the nuclear membrane is still present and has a duplicated or "stacked" appearance. A centriole can be seen cut in cross section. From the zone around the centriole, spindle fiber microtubules can be seen extending outward. The enlargement at the right (×11,000) shows more clearly that the microtubules lead to the kinetochores of the chromosomes. (Photo courtesy Raymond G. Murray, Assia S. Murray, and Anthony Pizzo, Indiana University.)*

Fig. 7.4a *Stages of the mitotic process. Prophase includes the formation of asters and spindle fibers, the appearance of chromosomes as distinct structures, and the disappearance of the nuclear membrane. At metaphase the chromosomes appear to be aligned on the equatorial plane. During anaphase the chromosomes split longitudinally, resulting in two chromatids that move to opposite poles. The distribution of genetic material is completed in telophase, and the division of the cytoplasm (cytokinesis) takes place. During telophase, chromosomes as observable structures disappear, and the nuclear membrane forms again. The mitotic process accomplishes the distribution of identical amounts and kinds of genetic material to daughter cells.* ► **Fig. 7.4a appears on pages 236–237.**

Figure 7.4a

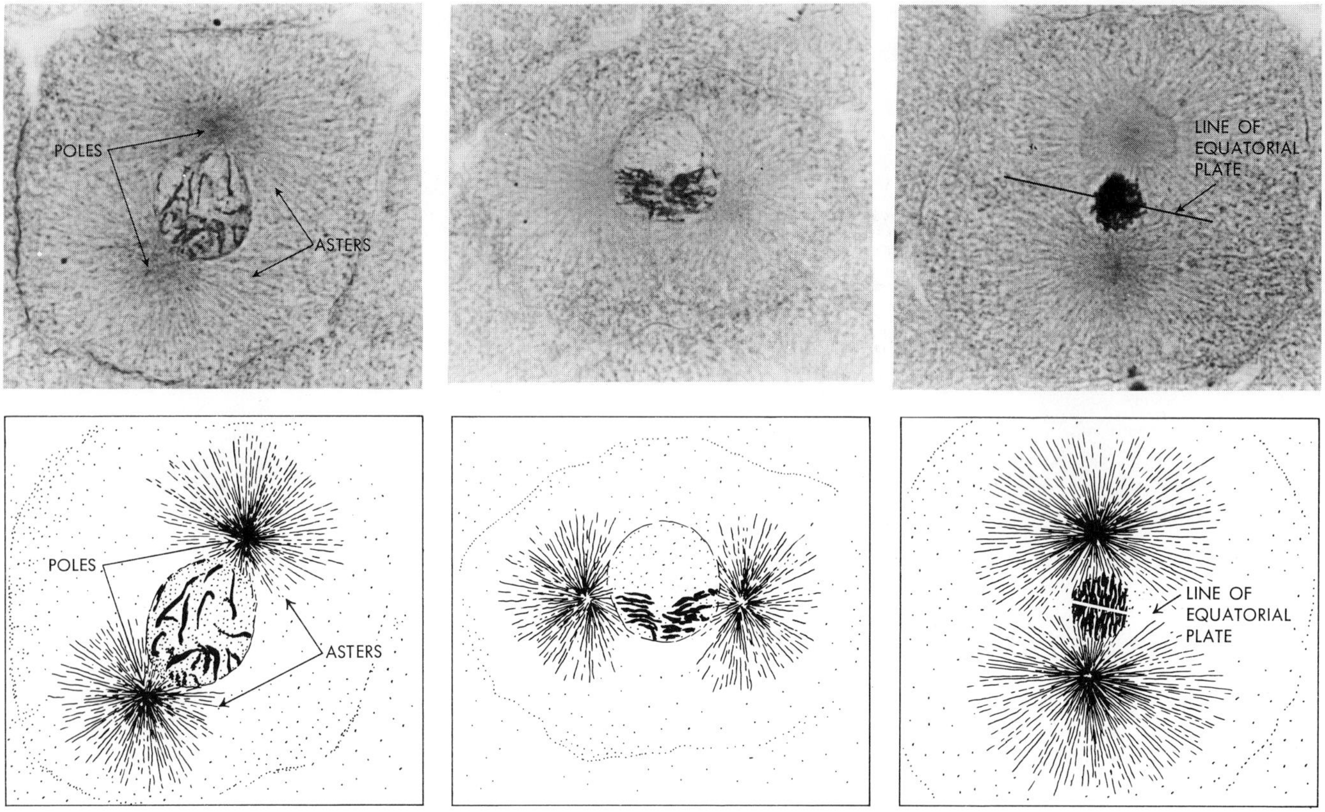

EARLY PROPHASE

MIDDLE PROPHASE

LATE PROPHASE

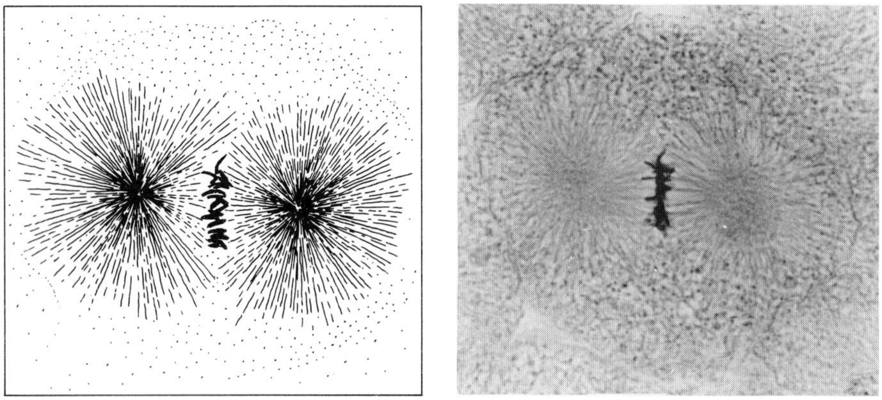

METAPHASE

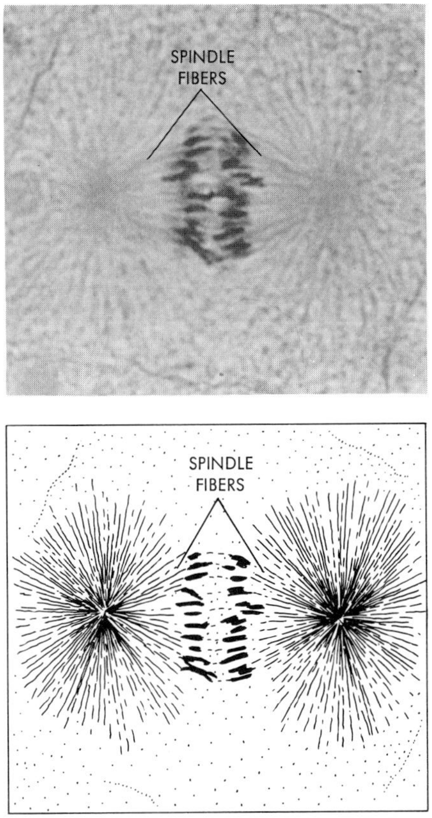

EARLY ANAPHASE

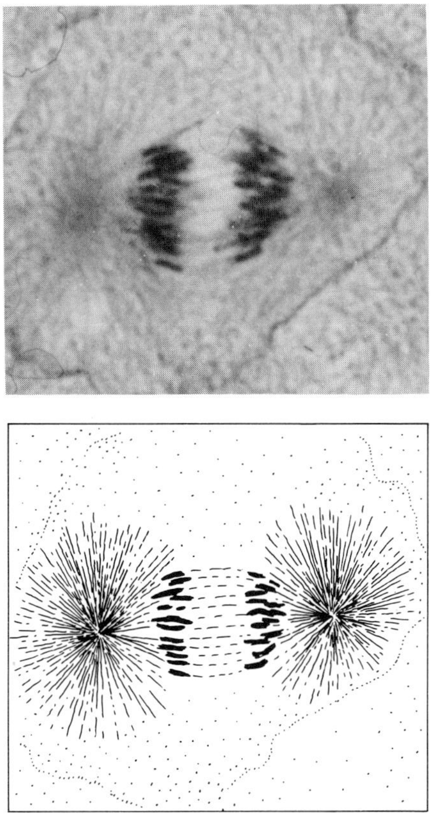

MIDDLE ANAPHASE

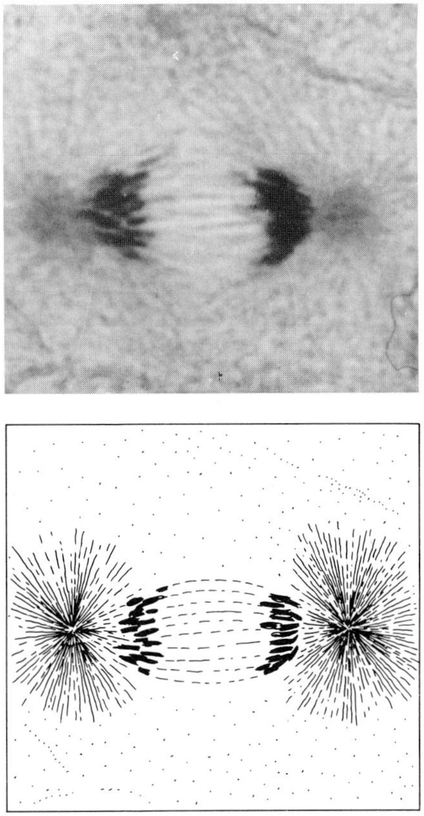

LATE ANAPHASE

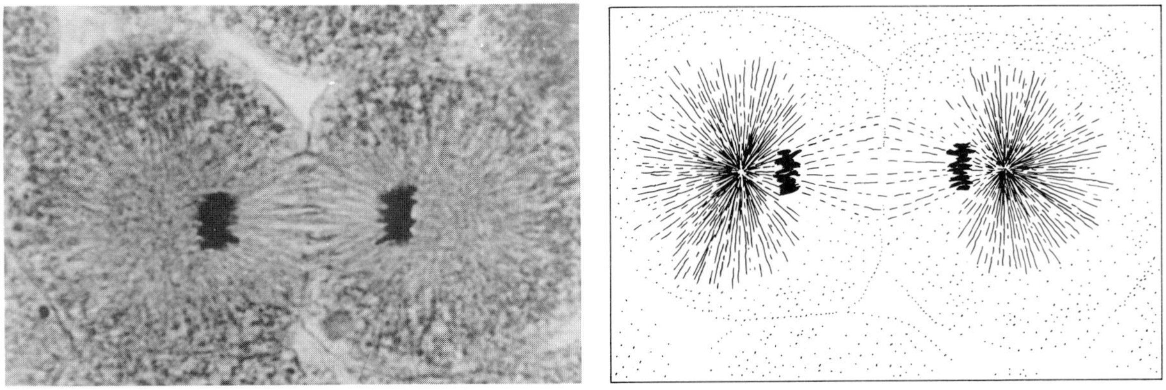

TELOPHASE

Fig. 7.4b *Two views of metaphase. In the top figures the microscope is focused on the plane that runs through the center of both asters (see the sketch at the right). The chromosomes appear as if on a line. In the lower figure the cell has been turned for observation and the microscope focused on the plane of the chromosomes (see sketch). The chromosomes appear to be distributed in a circle. Such microscopic views are only two-dimensional. The reality of three dimensions is best inferred when both views are considered.*

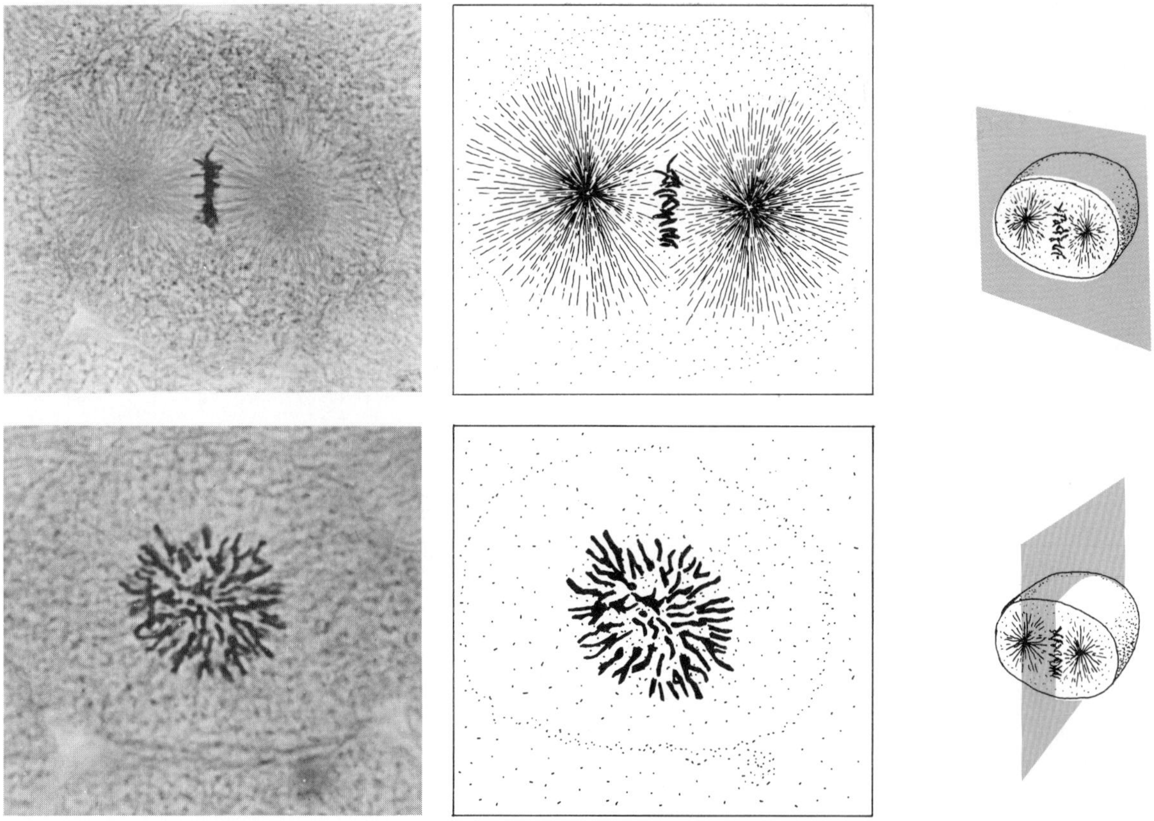

Despite the fact that it is often necessary to present it as a series of discrete steps, mitosis is a dynamic, continuous process. We separate the process into phases with specific names only for the sake of convenience.

somes move along the directions established by the spindle fibers, to which their kinetochores appear to be attached (see Figs. 7.4 and 7.5). The attachment of the chromosome's kinetochore to a spindle fiber is absolutely essential for the anaphase movement to occur. If the kinetochore is lost, the chromatid pair cannot orient on the spindle and the characteristic separation at anaphase fails to occur.

During animal cell **telophase,** the cell undergoes the process of "furrowing," in which it pinches in on all sides until two daughter cells are formed (Fig. 7.4). During late telophase, a nuclear membrane is reformed, the chromosomes "uncoil" and disappear as distinct structures visible under the light microscope, and the cell generally returns to the interphase state.

Differences found between mitosis as it occurs in animal cells (essentially as just described) and the same process as it occurs in plant cells (see Fig. 7.6) are largely due to the anatomical differences between the two kinds of cells. The most notable difference, perhaps, occurs during telophase. Since the plant cell is generally surrounded by a nonliving cellulose cell wall, it cannot pinch inward. Instead, a **cell plate** is formed, at which a new partition of cellulose will be synthesized between the two daughter cells.

The Mitotic Timetable

The five phases of mitosis by no means span equal lengths of time. For example, at one temperature a meristematic tissue cell in the root tip of a pea plant takes approximately 1529 min ($25\frac{1}{2}$ hr) to

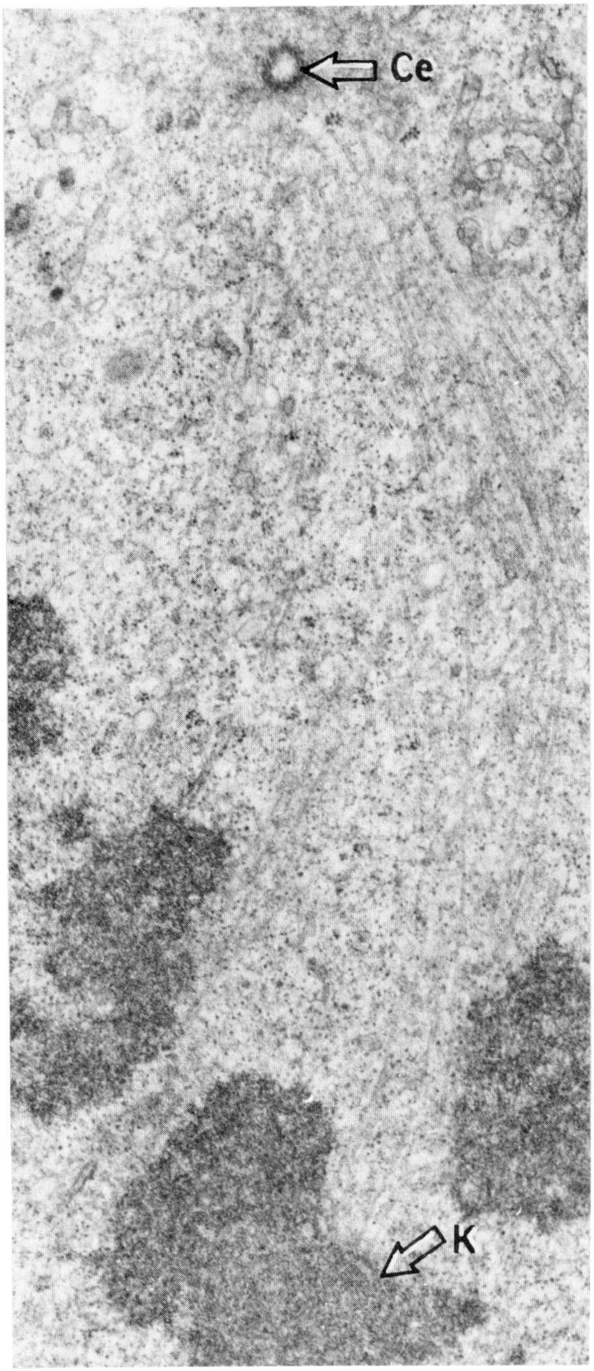

Fig. 7.5 *This electron microscope photograph shows a highly magnified (×15,000) portion of a tissue culture cell at anaphase. Clearly visible are the spindle microtubules extending from the centriole (Ce) to the centromere or kinetochore (K) of a chromosome, here seen as a dense plate separated slightly from the main mass of the chromosome. (From A. Krishan and R. C. Buck,* Journal of Cell Biology *24, 1965. Photo courtesy Dr. R. C. Buck, University of Western Ontario.)*

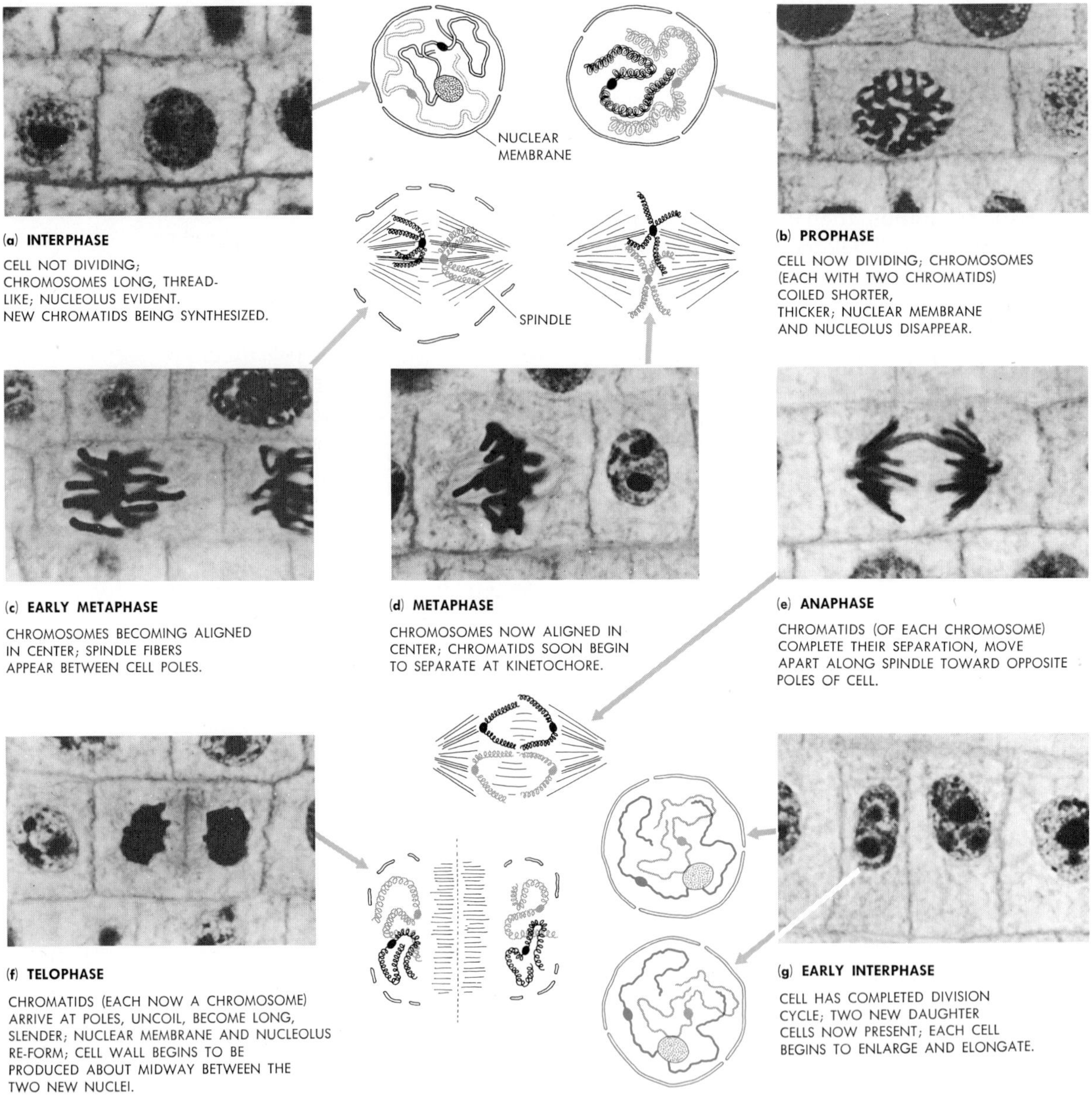

NUCLEAR
MEMBRANE
SPINDLE
(a) INTERPHASE
CELL NOT DIVIDING;
CHROMOSOMES LONG, THREAD-
LIKE; NUCLEOLUS EVIDENT.
NEW CHROMATIDS BEING SYNTHESIZED.
(b) PROPHASE
CELL NOW DIVIDING; CHROMOSOMES
(EACH WITH TWO CHROMATIDS)
COILED SHORTER,
THICKER; NUCLEAR MEMBRANE
AND NUCLEOLUS DISAPPEAR.
(c) EARLY METAPHASE
CHROMOSOMES BECOMING ALIGNED
IN CENTER; SPINDLE FIBERS
APPEAR BETWEEN CELL POLES.
(d) METAPHASE
CHROMOSOMES NOW ALIGNED IN
CENTER; CHROMATIDS SOON BEGIN
TO SEPARATE AT KINETOCHORE.
(e) ANAPHASE
CHROMATIDS (OF EACH CHROMOSOME)
COMPLETE THEIR SEPARATION, MOVE
APART ALONG SPINDLE TOWARD OPPOSITE
POLES OF CELL.
(f) TELOPHASE
CHROMATIDS (EACH NOW A CHROMOSOME)
ARRIVE AT POLES, UNCOIL, BECOME LONG,
SLENDER; NUCLEAR MEMBRANE AND NUCLEOLUS
RE-FORM; CELL WALL BEGINS TO BE
PRODUCED ABOUT MIDWAY BETWEEN THE
TWO NEW NUCLEI.
(g) EARLY INTERPHASE
CELL HAS COMPLETED DIVISION
CYCLE; TWO NEW DAUGHTER
CELLS NOW PRESENT; EACH CELL
BEGINS TO ENLARGE AND ELONGATE.

Fig. 7.6 *Plant cell division is shown here in the apical meristem of an onion* (Allium) *root. Although the onion actually possesses 16 pairs of chromosomes, for simplicity the accompanying drawings show only one pair. Note that the most evident difference between plant and animal cell division occurs at telophase. (Photos courtesy Carolina Biological Supply Company.)*

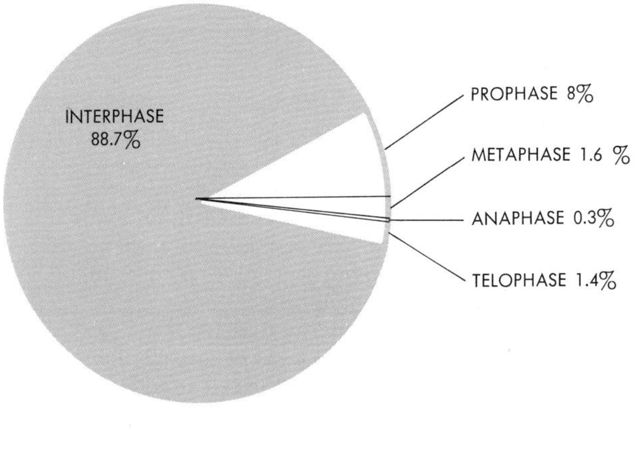

Fig. 7.7 *Diagram showing the percentage of time spent in each phase of mitosis by a meristematic tissue cell in the root tip of a pea plant.*

divide. Of this time, 1356 min are spent in interphase, 126 in prophase, 24 in metaphase, only 5 in anaphase, and 22 in telophase (Fig. 7.7). At a higher temperature, the total length of time spent by a similar cell undergoing mitosis is much shorter (952 min); nevertheless, the lengths of time spent in each phase are in the same proportion as at the lower temperature. A similar difference in phase time is shown by animal cells. A human connective-tissue cell takes 18 hr to undergo mitosis. Yet the time span from the beginning of prophase to the end of telophase is only 45 min. Thus the cell spends more than 17 of the 18 hr in interphase, preparing for division. On the average, however, a cell generally spends about 10% of the cycle time in the kinetic aspects of mitosis, regardless of the temperature.

As might be predicted, cells whose main function is division spend less time in interphase than other cells. Certain cells in a grasshopper embryo, for example, complete mitosis in 208 min (at 38°C). Of this time, 17 min are spent in interphase, 102 in prophase, 13 in metaphase, 9 in anaphase, and 37 in telophase. As these figures demonstrate, the cells undergo the most kinetically active phases of mitosis in a relatively short period of time. This means that a great deal of energy is being expended per unit time by actively dividing cells. As would be predicted, the respiratory rates of such tissues as the meristem of plants or the cells of a growing animal embryo are high compared to those of less actively dividing cells. Note, however, that this statement applies to *populations* of cells, such as found in a tissue. In contrast, it can be shown that with *individual* cells actually undergoing the mitotic process it does *not* apply, since applications of compounds (such as azide) that inhibit respiratory activity fail to stop mitosis once it has been initiated. The situation is somewhat analogous to the use of a playground slide by a child. The entire activity is an energy-requiring one; the child must climb the ladder to the top of the slide. Once the sliding activity is started, however, no further energy input is necessary to complete it. Similarly, while the entire cycle of cell division requires considerable energy input, the most active phases, once initiated, are able to proceed on their own.

At least once in their life span, most cells possess the power to undergo mitosis. Generally, however, they also have some built-in mechanism that periodically (or permanently) makes them stop dividing. Occasionally, a cell arises that seems to have lost this regulatory mechanism. When this cell divides, it passes on its lack of division control to its descendants. These are cancer cells. Such cells often seem to divide in a disorderly, chaotic manner. They may even appear to enter one phase of mitosis without completely finishing the previous one. In carrying out this disorganized form of division, cancerous tissues require large amounts of energy. They thus rob healthy tissues of needed nutrients. Further, by spending so much time dividing, cancer cells fail to perform the regular body functions assigned to them. Left unchecked, they may ultimately cause the afflicted organism's death.

Supplement 7.1

MITOSIS—SOME PROBLEMS

It is obvious that at some time during mitosis the genetic material of the chromosome must replicate. Otherwise, as we pointed out earlier, each daughter cell would receive only half of the genetic material. Indications of gene replication are first detected during prophase, when the doubled chromosomes are visible. It is logical to hypothesize, therefore, that gene replication is initiated either during early prophase or late interphase. Choosing the former possibility, we might reason that if gene replication is initiated during early prophase, then uptake of a radioactively labeled compound unique to gene structure should occur during early prophase. The prediction is not verified. Instead, the radioactive compound never becomes part of the genetic material unless it is administered at a crucial period during interphase. It is evident that this crucial period is the time when gene replication is initiated.

The role of the centriole in cell division poses another problem. In animal cells, the centrioles are involved in considerable activity during mitosis. In middle prophase the centriole (which has already replicated a division in advance) separates into its two parts, which migrate along the nuclear membrane. As they migrate toward the poles, spindle fibers appear between the centrioles, so that when the nuclear membrane breaks down, the chromosomes are imprisoned within the developing spindle. The two centrioles end up at the poles of the spindle, where the fibers extending from pole to pole form the spindle, while those extending into the surrounding cytoplasm form the astral rays. The appearance of the spindle fibers between the centrioles as they migrate apart toward the poles supports a hypothesis proposing that the centrioles play a role in spindle fiber synthesis. In the light of this hypothesis, it is doubly significant that centriole-like structures have been found at the base of both cilia and flagella, themselves fiberlike structures. A serious objection to this hypothesis, however, is the fact that no centrioles have been detected in the cells of higher plants (the primary reason why they have not been mentioned until now), and yet such plant cells manage to form spindle fibers and divide quite successfully. Some investigators, however, are of the opinion that particles equivalent to the centrioles will eventually be found in such plant cells.

What supplies the energy to drive the mitotic machinery? ATP is a likely suspect; there is little reason to assume that the cell would use a less standard means of supplying energy in mitosis. An hypothesis proposing ATP as the main fuel for mitosis predicts the presence in the mitotic apparatus of an active enzyme capable of splitting ATP, and such an enzyme has indeed been found. Although this discovery is definitely consistent with the ATP hypothesis, it by no means provides conclusive support for it.

The movement of the chromosomes at anaphase poses another puzzle. The presence of the kinetochore seems to be essential for this to take place (Diagram A), and the connection of the kinetochores of each chromosome to the microtubules of the spindle has been well established by electron microscope observations, but the cause of chromosome movement along the spindle fibers has yet to be adequately explained. The fact that ordinary muscle fibers use ATP when they contract might suggest the tempting hypothesis that contraction of the spindle fibers may pull the chromosomes from the cell equator to the poles. Certain observations contradict this hypothesis, however. No thickening of the spindle fibers is seen, as might be predicted if they did indeed contract.

Several years ago it was discovered that application of the chemical colchicine blocks the assembly of the spindle fibers. Colchicine does not, however, interfere with the condensation of the chromosomes. As the hypothesis proposing that the spindle fibers are responsible for chromosome separation at anaphase would predict, the chromosomes do not separate. Thus when the nuclear membrane reforms, it encloses

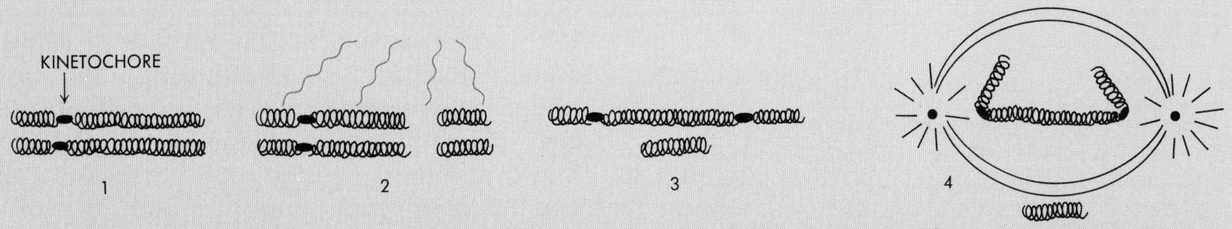

Diagram A *An hypothesis proposing that the presence of the chromosome kinetochore is essential for chromosome movement at anaphase can be tested by this experiment. If exposed to ionizing radiation, normal chromosomes (1) can be broken (2). Broken chromosome ends always rejoin, but they may do so in various ways. The fragments with kinetochores have joined together, as have those without kinetochores (3). The hypothesis predicts that the piece without a kinetochore will not move at anaphase, while the piece with two kinetochores will be stretched toward the opposite poles (4). The hypothesis is supported.*

a double set of chromosomes. Such a nucleus is said to show **polyploidy.** If the cell is removed from the colchicine-containing medium, further mitotic divisions proceed normally. All the daughter cells, however, have a multiple set of chromosomes. Since the chromosomes carry the nuclear genetic material, the properties of the polyploid organism are greatly affected. Occasionally such effects are quite favorable; many of the varieties of flowers sold by plant breeders to gardeners are polyploid. Further, the occasional occurrence of polyploidy in nature has played an important role in the evolution of various plant groups.

If there is still no completely satisfactory hypothesis to explain the phenomenon of chromosome movement at anaphase, neither is there one for the causes of animal cell pinching-in at telophase. An hypothesis proposing that the mitotic apparatus (consisting mostly of the spindle fibers) may play a role seems promising, for the pinching-in at telophase always occurs perpendicular to the axis of the spindle and across its middle section. Experimentally changing the position of the spindle causes a corresponding change in the position of the plane in which the cell divides. Furthermore, if the mitotic apparatus is removed well before cell-body division is due, then this division fails to occur.

However, if the same mitotic-apparatus removal operation is performed *immediately* before division, when the chromosomes are moving toward the poles at anaphase, it has no effect; cell-body division *does* occur. This observation might suggest that the control of cell-body division is transferred from the mitotic apparatus to the chromosomes at anaphase. This hypothesis is also contradicted, however; the Japanese scientists Y. Hiramoto and T. Kubota were able to remove the entire spindle and chromosomes in sea urchin eggs at a stage just prior to cell-body division—yet the division still took place. These results are important, because spindle fibers have been shown conclusively to be tiny hollow tubes or microtubules. Such microtubules are known to be associated with certain biological processes that generate movement within cells.

The Japanese results and those of other workers indicate that the causes of the pinching-in that leads to cell division in telophase may lie in the surface layers of cells, where no microtubules have been found. Acting on this suspicion, Dr. Geoffrey Selman and Margaret Perry of the University of Edinburgh have used the electron microscope to study in detail the ultrastructure of amphibian eggs just at the time when the surface of the egg dips inward to form a groove. They have detected fine filaments, only nine nanometers (approximately one-third of a millionth of an inch) in diameter, each oriented in the direction of the groove. Similar filaments have been found by American biologists in the eggs of jellyfish, polychaete worms, squid, and sea urchins. Dr. D. Szolloski of the University of Washington suggests that these filaments may be anchored to the microvilli that have been detected in the division groove, and that the gradual contraction of the filaments causes the pinching-in of the cell. Unfortunately, it has so far proven too difficult to trace individual filaments from end to end.

Despite the detail in which they have been described, mitosis and cytokinesis are still a very long way from being completely understood. The problems are intellectually stimulating ones, however, and will undoubtedly continue to receive much attention.

7.4 Meiosis

With certain exceptions, the number of chromosomes in the body cells of an organism is constant from one cell to the next. Thus, for example, each body cell of a certain gill fungus has 4 chromosomes, of an elm tree 56, and of a stalk of sugar cane 80. In an earthworm, each body cell has 32 chromosomes, in a bull frog 26, in a chicken 18, and in a human 46. Because of the chromosome duplication that occurs during mitosis, each daughter cell is assured of a full complement of chromosomes after each division is completed.

Every multicellular organism is the result of millions of mitotic divisions, starting with the first division of the fertilized egg. Take the human as an example: the fertilized egg contains 46 chromosomes, of which 23 were contributed by the egg and 23 by the sperm. Thus half of the 46 chromosomes in a person's body cells are duplicate descendants of the mother's (**maternal**) chromosomes and half of them are duplicates of the father's (**paternal**) chromosomes. The 46 chromosomes are thus more accurately described as 23 pairs of **homologous chromosomes.** Each pair carries many inheritance factors that influence many specific traits, such as eye or hair color. One member of the homologous pair carries the maternal factor for the trait, the other the paternal factor. The chromosome numbers given for humans and the other organisms listed earlier are thus double (**diploid** or $2n$) numbers. The gametes of these organisms, however, cannot have a diploid number of chromosomes. If they did, their union at fertilization would result in a tetraploid (or $4n$) number. With each ensuing generation, the chromosome number would increase geometrically. Quite obviously, this cannot and does not occur. Rather, each gamete contains an **haploid** (or n) number of chromosomes. Thus in a dog, diploid number 56, a sperm and an egg each contain 28 chromosomes; in a horse, diploid number 60, a sperm and an egg each contain 30 chromosomes, and so on.

Meiosis is a reduction division: it results in the chromosome number being reduced to one-half that of the diploid number. In eucaryotes, each resulting haploid cell contains one member from each pair of chromosomes found in the parent cell.

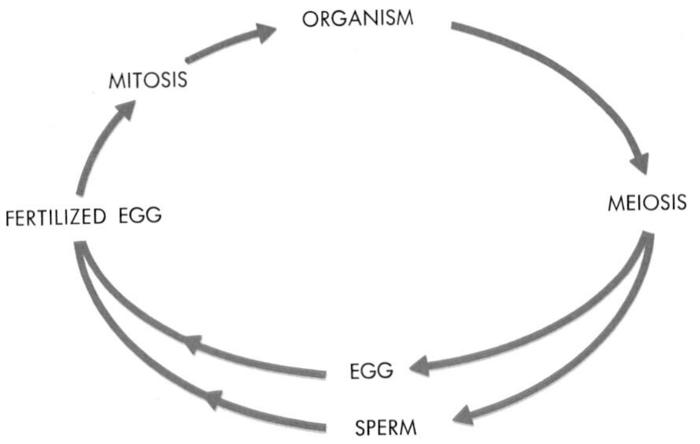

Fig. 7.8 *The life cycle of a sexually reproducing animal in terms of mitosis and meiosis. Plant groups show wide variation in the place of meiosis in their life cycles. All the many cell divisions undergone during embryological development and later growth are mitotic divisions.*

The cellular division process by which the daughter cells receive only the haploid number of chromosomes is known as **meiosis.** In animals this process takes place during the formation of the gametes (**gametogenesis**—called **spermatogenesis** in the male and **oogenesis** in the female), while in most plants meiosis occurs when spores are produced (**sporogenesis**). The life cycle of a sexually reproducing animal is shown in relation to meiosis and mitosis in Fig. 7.8.

Meiosis resembles mitosis in that it, too, occurs concurrently with cell division. Furthermore, meiosis can be divided into the same descriptive phases as mitosis, and much of the mechanics are the same—a spindle apparatus is formed, the chromosomes migrate apart at anaphase, and daughter cells result. Yet meiosis differs significantly from mitosis. Two cell divisions rather than one generally accompany the meiotic process; the actions of the chromosomes during some of the phases are quite different (Fig. 7.9); and finally, as we have just seen, meiosis is a reductional rather than an equational process, yielding a haploid number of chromosomes as the final result. Mitosis, on the other hand, always results in the same number of chromosomes in the daughter cells as in the parent cell. Thus if a haploid cell divides by mitosis, the daughter cells will also be haploid; if a diploid cell divides by mitosis, the daughter cells will be diploid; if a tetraploid cell divides by mitosis, the daughter cells will be tetraploid; and so on.

The first difference between mitosis and meiosis becomes evident at the start. Whereas both mitotic and meiotic prophase are characterized by the first visible evidence of the chromosome doubling initiated during interphase, meiotic prophase is also distinguished by a pairing of the homologous chromosomes. During homolog pairing, the two chromosomes become closely applied to each other along their length, a factor of considerable significance in the events that follow.

The Reduction Division

The process of the pairing of homologous chromosomes during meiosis is known as **synapsis** (Fig. 7.10). During the so-called S phase of interphase, DNA and histone synthesis occurred, so the hereditary material is already duplicated at synapsis. The result is that each homologous chromosome has replicated, forming a four-part structure, the

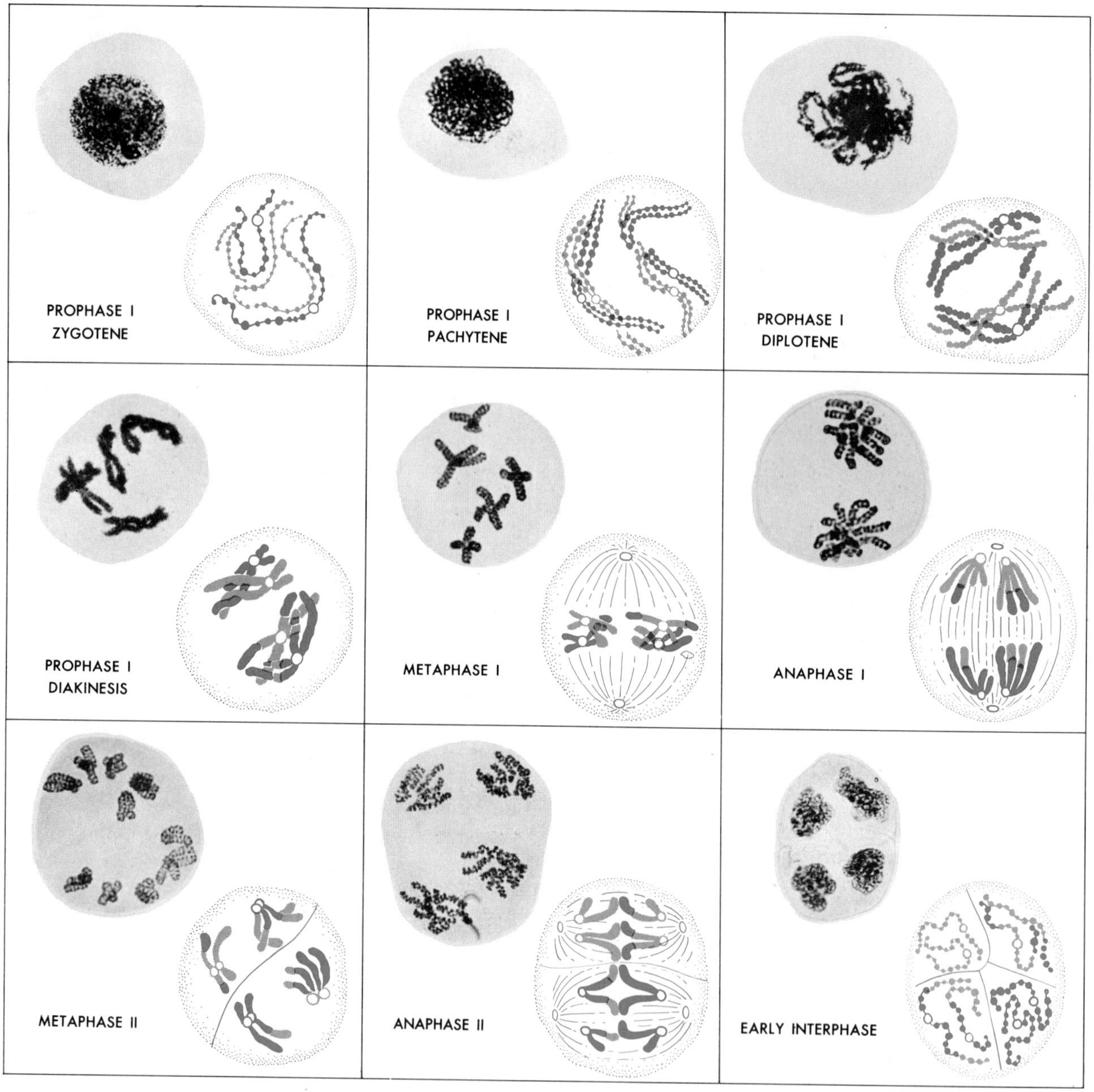

Fig. 7.9 *Stages in meiosis. The terms "zygotene," "pachytene," "diplotene," and "diakinesis" refer to distinctive stages of prophase I. Note the essential chromosome movements: condensing and doubling (prophase), coiling and moving to the cell equator (metaphase), pulling away from the equator toward the poles (anaphase), and finally re-forming into distinct nuclei (telophase). (Photos courtesy Arnold H. Sparrow and the Brookhaven National Laboratory.)*

tetrad (Fig. 7.10b).* Note, however, that the duplicates remain attached at the region of the kinetochore. As long as they remain so attached, each member of the "Siamese twin" is still a chromatid.

During tetrad formation, sections of the chromatids of different pairs often overlap or wrap around each other, forming cytologically observable patterns called **chiasmata** (sing., **chiasma;** see Fig. 7.10c). There is evidence that the formation of chiasmata may be followed by chromosome breakage, *with subsequent exchange of the broken segments between maternal and paternal chromatids.* The process is known as **crossing over.** Since the evidence for it is both genetic and cytological, a complete discussion of the significance of such crossing over will be postponed until the next chapter. Suffice it to say here that its occurrence is of primary importance to the study of both biological inheritance and offspring variation, and that it is a major selective advantage of sexual reproduction.

After tetrad formation, the homologous chromosomes line up at the cell equator in a typical metaphase, with anaphase and telophase following closely (Fig. 7.10d). Note that by virtue of their being joined at the kinetochore, the chromatid pairs must travel together. Thus for any single pair of homologous chromosomes, the maternal chromatids go to one pole while the paternal chromatids go to the other. This does not mean, however, that when two or more pairs of chromosomes are involved the resulting daughter cells will carry only maternal or paternal inheritance factors. If chiasmata formation and crossing over have occurred, as shown here, almost any combination of maternal and paternal inheritance factors is possible.

Because they are joined at the kinetochore, two

* For clarity it is necessary to draw the tetrad in Fig. 7.10 as if spread on a plane. In reality, it is a three-dimensional figure that in end view appears as ⁚⁚

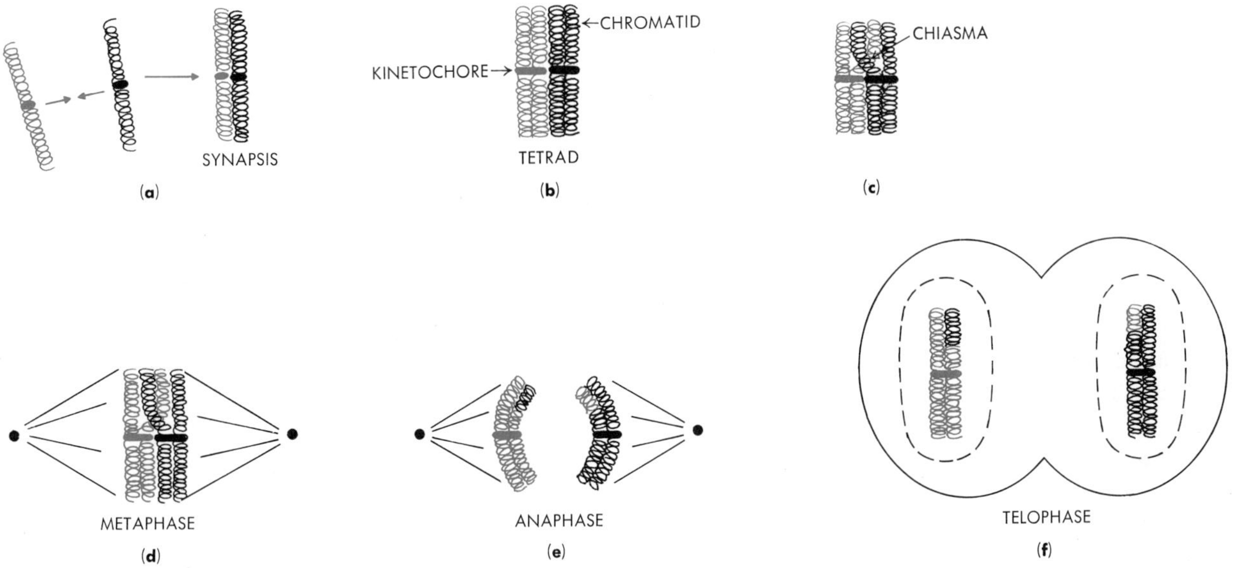

Fig. 7.10 *The process of synapsis and chiasma formation during prophase I.*

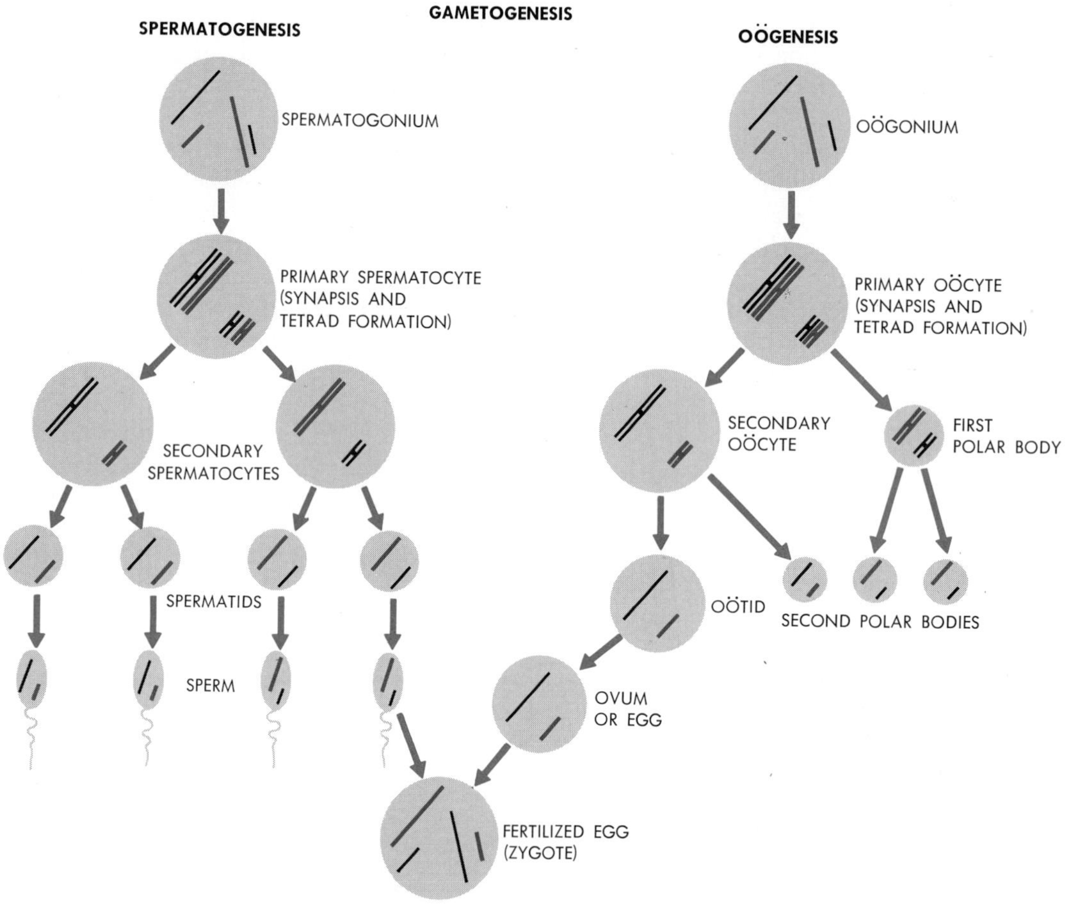

Fig. 7.11 *For each cell that begins gametogenesis, four potential gametes are formed. In oögenesis, however, only one of these matures to form the egg. Which of the four cells becomes the egg is determined entirely by chance. For simplicity, crossing over is not shown.*

chromatids form one chromosome. Thus the division that has occurred is a true **reduction division:** the number of chromosomes in each daughter cell has been reduced by one-half, or from diploid ($2n$) to haploid (n).

A second division follows this reduction division, often without any interphase period for the daughter cells of the first division. In this division, however, the kinetochores divide at metaphase, releasing the chromatids to migrate to the opposite poles at anaphase as full-fledged chromosomes. Note that this occurrence merely retains the haploid number of chromosomes, rather than restoring the diploid number. From each original cell at the beginning of the meiotic process, *four* haploid cells result. In spermatogenesis, these four cells transform into sperm. In oögenesis, the greater mass of cell cytoplasm is given to one cell, the egg (the choice being determined entirely by chance); the rest of the cytoplasm is discarded as part of the polar bodies (Fig. 7.11).

Variations Resulting from Meiosis

Two more important points remain to be stressed. First, it has been expedient to describe synapsis, tetrad formation, and migration of the chromatids and chromosomes to the opposite poles during the meiotic divisions in terms of one pair of homologous chromosomes. As has been seen, however, most organisms have many pairs, and during meiosis each of these pairs undergoes the same activity just described for a single pair. Thus a meiotically dividing cell, particularly one containing large numbers of chromosomes, is a maze of complicated activity.

The fact that mishaps, resulting in unequal distribution of chromosomal material, occur as rarely as they do during meiosis is a testimonial to the intricate near-perfection of living systems. Occasionally, however, offspring are born with deformities that can be traced to a failure of the chromosomes to separate at metaphase. At anaphase both members of the pair may migrate to one pole together and end up in one gamete. The other gamete, of course, receives no chromosome of this pair. This phenomenon, known as **nondisjunction,** was first identified in the fruit fly, *Drosophila melanogaster*. More recently, nondisjunction of chromosome pair 21 has been found to be the cause of Down's syndrome, or Mongolism, in human children.

Since this discovery of the cause of Down's syndrome, several other congenital abnormalities in humans have been traced to nondisjunctional occurrences of other chromosome pairs. **Sex chromosomes** are those chromosomes that determine the sex of an organism. In humans, the sex chromosomes consist of an *X* chromosome and a *Y* chromosome somewhat shorter than the *X*. The diploid cells of every woman contain two homologous *X* chromosomes, while those of a man contain only one *X* chromosome and a *Y* chromosome, which pair at synapsis. Should nondisjunction occur with the sex chromosomes rather than the other chromosomes **(autosomes)** and the resulting gametes participate in formation of the zygote, the resulting individual may have such sex chromosomal combinations as *XXY*, *XYY*, etc.

Nondisjunction is also known in plants. It has been shown to occur in each of the 12 pairs of chromosomes in the Jimson weed (*Datura stramonium*). The individual plant is variously affected, depending on which pair of chromosomes is involved. If chromosome pair A is affected so that the plant has three A chromosomes (instead of two), then the plant is small, its leaves are abnormally narrow, and its fruits are tiny. If chromosome pair J undergoes nondisjunction, the plant has dark puckered leaves similar to those of spinach. In other plant genera, nondisjunction of chromosomes has resulted in the formation of species with fewer chromosomes than their ancestors, for example, in the false dandelion (*Crepis*), the species *C. fuliginosa* ($n = 3$) has most likely been derived from *C. neglecta* or its ancestor ($n = 4$). Through chromosome nondisjunction, numerous new species have apparently been produced in genera such as mustard (*Brassica*) and sedges (*Carex*).

The second detail that must be stressed is the relation of the chromosome pairs to each other during meiosis. At metaphase, all the chromosome pairs are lined up along the cell equator prior to separation at anaphase. For simplicity's sake, let us deal with only two chromosome pairs and assume that no crossing over has occurred. Thus each homolog carries only maternal (M) or paternal (P) genetic factors (Fig. 7.12a). Note that in the figure we have arbitrarily placed the maternal chromosomes of each pair on the left and the paternal on the right. *But there is absolutely no requirement that we do so.* It would have been just as easy to reverse them, placing the paternal chromosomes on

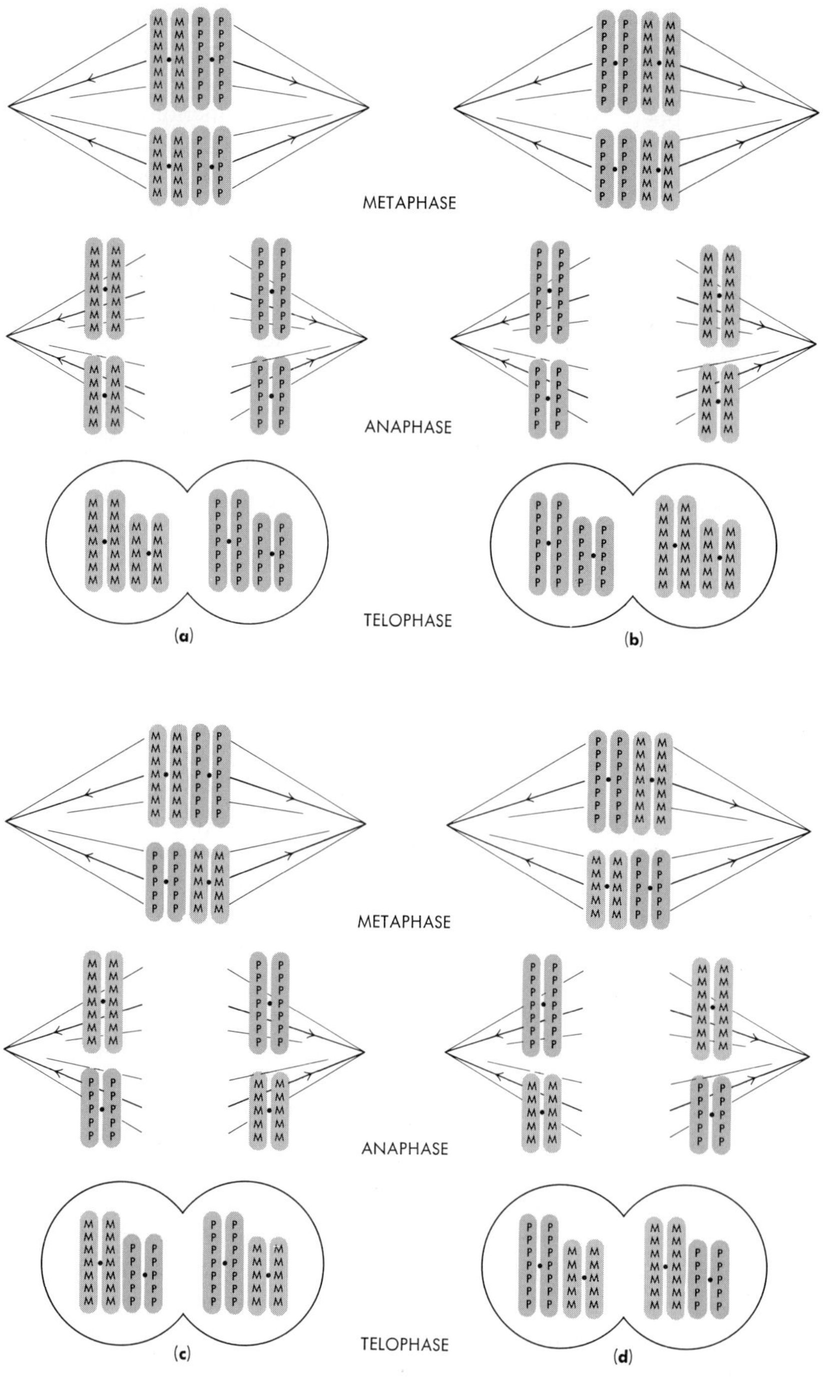

Fig. 7.12 *Schematic representation of chromosome positions during meiosis. Maternal (M) and paternal (P) members of each homologous pair line up at the equator during metaphase I. By the end of the first meiotic division (division I) maternal and paternal homologs have been separated. Each chromosome has duplicated itself, forming two chromatids, by prophase I. Thus by metaphase I, each homologous pair consists of four chromatids collectively known as a tetrad. The several combinations of maternal and paternal chromatids for a cell where 2n = 2 are shown in (a), (b), (c), and (d). All four combinations are possible, although only two will occur for any one cell undergoing division.*

Supplement 7.2

IS THERE A "CRIMINAL CHROMOSOME"?

In 1961 a human male was found showing an *XYY* chromosome complement, a result of nondisjunction of two Y chromosomes (presumably during spermatogenesis in the subject's father). In December 1965, Dr. Patricia Jacobs and her colleagues at Western General Hospital in Edinburgh (Scotland) published a cytogenetic study of male inmates in the hospital's security ward. These inmates all had records of what was considered violent criminal behavior. Of the 197 men surveyed, 7 showed the *XYY* chromosome condition.

There was at the time very little data available on the frequency of *YY* nondisjunctions; but the related *XX* nondisjunctions were known to occur in about 1.3 out of every 1000 births, or about 0.1% of the time. Thus the appearance of *XYY* in 3.5% of the inmates of a penal institution appeared to Dr. Jacobs and others to be a significant departure from the norm. They hypothesized that there might be a cause–effect relationship between presence of the extra Y chromosome and a tendency toward aggressive and violent behavior. As a result of this early work and the publicity it received, many surveys were carried out to screen for *XYY* males. Most of these surveys were done on inmates in various kinds of prisons or mental institutions. More examples of *XYY* males were uncovered. The frequency of *XYY* males in at least some penal institutions appeared to be significantly higher than the frequency estimated for the general population.

Given the apparently higher frequency of the extra Y chromosome in prison populations, geneticists and psychologists put forward the hypothesis that the extra Y chromosome causes males to have a greater tendency toward violent and antisocial behavior. Such behavior, they argued, could lead to various kinds of criminality. The precise means by which an extra Y chromosome could produce a "tendency to violence" was a matter of some speculation. Some workers suggested that presence of the extra Y chromosome might increase the amount of male hormone secreted, thus causing an increased level of general aggressive behavior. Others suggested that the extra Y chromosome might increase the rate of development, especially around puberty, making *XYY* boys grows faster, appear larger than normal for their age, and become hyperactive social misfits. A third possibility suggested was that the extra Y chromosome specifically affected brain development, acting on "violence centers" supposed to exist in areas such as the hypothalamus or amygdala. While no biologist has any idea about the mechanisms that might be involved, by the mid-1960s a number of researchers maintained that there might be a direct link between possession of an *XYY* chromosome complement and a person's chances of becoming a criminal.

The *XYY* story became another example of a theory of biological determinism: a theory attempting to explain a social phenomenon (in this case aggressive behavior and criminality) in strictly biological terms. The opposing hypothesis is that in the vast majority of cases, aggressive and criminal behavior is caused by unfavorable environments (family problems, slum neighborhoods, racial or ethnic prejudice, economic deprivation, etc.).

By the late 1960s the *XYY* story had been widely circulated through leading newspapers and magazines in much of western Europe and the United States. In France and Australia *XYY* defendants in two murder trials were given light sentences (in one case acquitted) on the grounds that their violent behavior was beyond their control, rooted as it supposedly was in the genes. A report appeared in 1968 that Richard Speck, who in 1966 killed 8 nurses in Chicago, was *XYY*. (That the report was later shown to be false got very little publicity). The result of much of this publicity was to gradually convince the public that biologists, particularly geneticists, accepted as valid the hypothesis that an extra Y chromosome caused an increased tendency toward violent and criminal behavior. In some circles the extra Y was referred to as the "criminal chromosome."

One of the chief critics of the theory on scientific grounds is Dr. Digamber S. Borgaonkar, of the Johns Hopkins University School of Medicine. Dr. Borgaonkar made an exhaustive study of most of the *XYY* cases reported, examining both the data and the methods used to obtain the data. His conclusion is that most of the studies were carelessly executed. For instance, he found that the data were often so unreliable as to be virtually meaningless, and that the suggestion of any relationship between an extra *Y* chromosome and criminality was consequently unsupportable.

The criticisms Dr. Borgaonkar has directed against the *XYY* work provide several guidelines to data collection and analysis:

1. One of the important assertions of the *XYY* work is that *XYY* males are more disposed toward violence than *XY* males. Dr. Borgaonkar found from analyzing papers reporting on behavior traits of *XYY* males within penal institutions that at least in these circumstances, *XYY* males were on the whole more cooperative than their *XY* counterparts. Thus the claim that *XYY* males are more aggressive cannot be considered valid *without specifying the environment involved.*

2. Few physiological or psychological traits have been found that actually distinguish *XYY* males from other males. Only height appears to be distinctive (*XYY* males are on the average slightly taller than *XY* males). Other traits—skeletal structure, electroencephalograms (EEG's), electrocardiograms (EKG's), skin traits, etc.—all seem to be average. Hormonal levels are no different for populations of *XYY*'s and *XY*'s. The IQ of *XYY* males appears to be about the mean for inmates of penal institutions. No significant differences in personality traits distinquish *XYY*'s from *XY*'s. In short, by all significant physical or physiological criteria that might affect behavior, *XYY* males rate about the same as *XY* males.

3. Methodologically, the techniques of collecting and analyzing behavioral data about *XYY* males appeared to Borgaonkar, to be very unreliable and nonrigorous.

a) All the studies lacked either a "blind" or "double-blind" procedure. In a "blind" experiment, an investigator interviewing a subject to determine behavioral and personality traits does not know what is being tested for (i.e., that the patient is suspected of displaying violent behavior); the patient may know, however, the purpose of the study. In a "double-blind" experiment, neither investigator nor subject would know what relationships were being sought. "Blind" and "double-blind" procedures help ensure that neither investigators nor subjects will find more of what they are looking for than is actually there.

b) Virtually none of the studies of *XYY* males had been conducted with matched control groups against which data on the behavior of the *XYY* subjects could be compared. That is, in testing the hypothesis that an extra *Y* chromosome is a significant cause for criminal behavior, it is necessary to eliminate other variables (such as poor socioeconomic status, bad family life, etc.) that may also have profound effects in molding personality. Most of the studies compared the behavior of *XYY* males to control groups of randomly-chosen *XY* males not matched for social class, family background, economic status, and the like. Thus, several variables are introduced simultaneously: the presence of different chromosome complements, different social class and various family backgrounds. When two or more variables are present it is impossible to say which may be the more important cause for a particular phenomenon.

c) Researchers placed much reliance for descriptions of violent or criminal behavior on sources such as police records, legal documents, or records from correctional institutions. Not only are these descriptions likely to be highly variable; they are also subjectively biased in very specific ways. For example, there is no standard definition of what constitutes "violent" behavior. Is swearing at a prison authority or police officer evidence of a tendency toward violence? Or, does physical violence have to be involved? Moreover, police and prison administrators are likely to classify as violent any behavior openly disrespectful of or hostile to their own authority. Yet mere resistance to authority is not by itself adequate indication of a propensity to violence and criminality.

d) There was an element of selection involved in the subjects who were investigated in most of the studies. Only a small fraction of violent behavior actually comes to the attention of authorities and is recorded. Much more needs to be known, Dr. Borgaonkar argues, about the kinds of violent behavior displayed by people who do not come to official notice. Is it the same, less, or more than that displayed by those who are actually caught, convicted, and placed in penal institutions to become readily available objects for study? To claim that *XYY* males are more violent than *XY* males requires knowing what kind of violence *XY* males in the general population perpetrate. In fact, Dr. Borgaonkar reports that penal institution records indicate that *XYY* males have committed no more violent crimes than the *XY* males in prison with them. In the absence of data about the differences between the kinds of crimes that arouse official attention and those that do not, it could be hypothesized that *XYY* males are simply more open and honest than *XY* males and thus get caught more readily. In other words, if there is any genetic basis to the argument at all, it is that perhaps the extra *Y* chromosome is an "honesty-determining," rather than a "criminal-determining" chromosome!

In concluding his study,* Dr. Borgaonkar points out that some behavioral and developmental conditions (such as Down's syndrome, or "Mongolism") are definitely hereditary in nature. He does not deny the role of heredity in determining, in a general way, some broad patterns of personality development. On the other hand, he emphasizes that behavioral problems once thought to be largely hereditary (certain kinds of epilepsy or abnormalities in EEG) have been strongly linked to social class and family stability. He cautions against assuming a genetic cause for something as specific as behavior, when the obvious environmental influences that can affect such behavior have been largely ignored. He writes:

Inadequate understanding of the phenomena and the premature conclusions about the XYY phenotype, which have been reported with distressing frequency, have produced remarkably simplistic views of the interactions between XYY genotypes and the almost infinitely varied environments with which they interact. We should always keep in mind that even the demonstration of a genetic contribution to poor impulse control warrants only the conclusion that in certain environments some persons with particular genotypes will respond by developing certain behavioral problems more frequently than others. However, this does not preclude the possibility that in some other environments persons with the very same genotypes (i.e., XYY) may well manifest socially adaptive behaviors.

The *XYY* case has raised other questions about research on human subjects, especially where negative aspects of an individual's makeup are the main focus. The issue came to light dramatically in 1974 and 1975 at the Harvard University Medical School in Boston. A large project screening for *XYY* babies had been in progress at Boston Hospital for Women (formerly Boston Lying-In Hospital) from 1968 to 1975. The researchers heading the project, psychiatrist Stanley Walzer and geneticist Park Gerald, both

* From D. S. Borgaonkar and Saleem A. Shah, "The *XYY* chromosome male—or syndrome?" *Progress in Medical Genetics* X, 10 (1974), pp. 135–222.

from the Harvard Medical School, wanted to identify *XYY* males born in the hospital and follow their personality and behavioral development through adulthood. Walzer and Gerald explained that the purpose of their study was to identify *XYY* genotypes early in a child's life, so that psychiatric counseling could be provided to help overcome personality problems if and when they arose. Their research was funded by the Crime and Delinquency Division of the National Institutes of Mental Health.

Dr. Jonathan Beckwith, also of the Harvard Medical School, and Dr. Jonathan King of Massachusetts Institute of Technology mounted an extensive campaign, beginning in 1974, to have the *XYY* project closed down. They argued that parents participating in the project (that is, who had agreed to have their children studied) were not adequately informed of what the project was about. They claimed that parents did not understand the stigma that might be attached to their child, even if he appeared perfectly normal in his behavior, if he were known as an *XYY* type. (A well-known American geneticist appalled his colleagues at a professional meeting recently by claiming that he "wouldn't invite an *XYY* home to dinner.")

Given the misinformation the public has received in considerable doses about "criminal chromosomes," Beckwith claims that Walzer and Gerald did not take adequate steps to inform the parents of how participation in the project might affect their child's future. Moreover, Beckwith and King argue that given what most people now think about the *XYY* genotype, the hypothesized relationship between an extra *Y* chromosome and aggressive behavior could become a self-fulfilling prophecy. In other words, parents who know their children are *XYY* will treat them differently, perhaps pushing them almost unconsciously toward violent behavior. The medical school's Human Studies Committee found that Walzer and Gerald's work did comply with their requirements that (1) informed consent be properly obtained, (2) the patients' rights be protected, and (3) the benefits of participating in the study outweigh the risks.

Beckwith and King do not agree. They argued that the committee was composed of established doctors who had a stake in protecting themselves and their colleagues from challenges to the fundamental nature of the research. Beckwith and King have aimed their criticism of the Boston project largely at the moral and ethical implications of the *XYY* research. However, they would not necessarily have done so if the scientific basis of the work were not also faulty. This is an important point. Of course the rights of patients should be protected in all such research projects, no matter how valid the scientific basis of a piece of research. But when the research methods and the data are themselves questionable, the harm can be considerably greater. Beckwith and King argued that the patient's rights are considerably more in jeopardy when the conclusions from supposedly scientific work are erroneous. To be stigmatized for life is bad under any circumstances; to be stigmatized erroneously is even worse.

As a result of the pressure brought to bear on the project by Beckwith, King, and others, the *XYY* project in Boston was discontinued in the spring of 1975. Some people think Walzer stopped screening *XYY*'s because he finally realized that the risks outweighed the benefits. Walzer denies this. "I hope no one thinks I don't still believe in my research," he declares. "I do. But this whole thing has been a terrible strain. My family has been threatened. I've been made to feel like a dirty person. I was just too emotionally tired to go on." Walzer agrees that talk of a "criminal chromosome" is nonsense, but he still thinks there is enough evidence of certain learning difficulties in *XYY* children to justify an early identification leading to corrective therapy. King and Beckwith, on the other hand, claim that the potential harm to individual people is far greater than the potential good, and that under such conditions stopping the screening project is justifiable.

In the first anaphase of meiosis, the direction of separation of maternal and paternal homologous chromosomes of any one pair is totally independent of the direction of separation of those in any other pair.

---

the left and the maternal on the right (Fig. 7.12b). Two other combinations are also possible (Fig. 7.12c and d). The point here (and it is a *very* important one) is that the *direction of migration taken by one member of any chromosome pair in no way influences the direction of migration taken by a member of any other chromosome pair.* In other words, the members of individual chromosome pairs assort into their respective gametes *at random. Thus meiosis results in variability in assortments of chromosomes.* This is a factor of vast biological importance.

SUMMARY

1. Individual cells reproduce by cell division. Cell division consists of several processes: (a) mitosis, the replication and division of the chromosomes among daughter cells; and (b) cytokinesis, the splitting of the cell into two daughter cells.
2. Growth in knowledge about the mechanism of mitosis has paralleled the development of tools and techniques allowing experimental study of actively dividing cells.
3. Although it is convenient to describe it in terms of a number of specific, discrete stages, the entire process of cell division is a dynamic and active one. The stages are:
   a) Prophase: The duplicated chromatin strands begin to condense into chromosomes.
   b) Metaphase: The chromosomes assemble at the equatorial plate, each chromosome attached by its kinetochore (centromere) to a spindle fiber, in turn connected to the centrioles at either end of the cell. (The cells of higher plants lack centrioles, yet mitosis occurs just the same. In this case the spindle fibers appear to be attached to some other structure at the polar region of the cell.)
   c) Anaphase: The spindle fibers appear to be the vehicles for pulling the chromosomes away from the equator toward the poles. This separation occurs in such a way that each daughter cell gets a duplicate set (the diploid number) of chromosomes. The daughter cells are now genetically identical to each other and to the parent from which they are derived.
   d) Telophase: The chromosomes group around the poles of the cell, and the nuclear membrane begins to form.
4. Following telophase the process of cytokinesis goes to completion. The old cell, now with two identical nuclei arranged at opposite ends, begins to pinch in at the equator (or, in plants, build a cell plate and new membrane), creating two new cells.
5. Among the unanswered questions about mitosis are: (a) What initiates the process of cell division in the first place? (2) What forces are responsible for moving the chromosomes about during early metaphase, and especially at anaphase? (c) What controls the rate of or cessation of mitosis? What mechanism is responsible for

maintenance of continual mitosis in some tissues (bone marrow, gamete-producing tissues, epithelium, etc.), and for the lack of mitosis in others (nerve cells, muscle cells)?

6. Meiosis is similar in many ways to mitosis. However, it is a reduction not an equational division, producing daughter cells with one-half the number of chromosomes as the original parental cell. Meiosis converts a diploid parental cell into haploid daughter cells.
7. Meiosis occurs in two series of divisions:
   a) Prophase I: Duplicated chromatin condenses into rod-shaped chromosomes; each chromosome is duplicated, the two duplicate bodies being held together at the kinetochore. (Each duplicate member of such a pair is called a chromatid. Two duplicated chromatids, bound together, make up a chromosome.) Homologous chromosomes now come together and undergo synapsis. Since four chromatids are involved, the group is called a tetrad. Crossing over (chiasma) can also occur between homologous strands of the tetrad. In chiasma, equivalent sections of homologous strands may interchange.
   b) Metaphase I: Tetrads line up at the equatorial plate; kinetochores are attached by means of spindle fibers to the polar regions, often to the centriole.
   c) Anaphase I: Homologous chromosomes separate from one another, each chromosome still consisting of two chromatids. There is no regularity to the separation of maternal and paternal chromosomes. Each pair of homologous chromosomes separates from the equator without any regard to the direction in which other pairs are separating. Thus the newly formed nucleus might contain all maternal, all paternal, or any possible combination of chromosomes. Normally, however, just as in mitosis, the two members of a homologous pair go to opposite ends of the dividing cell.
   d) Telophase I: The new nuclei are surrounded by a nuclear membrane, and cytokinesis proceeds to completion.
   e) Prophase, metaphase, anaphase, and telophase II: The duplicated chromatid strands are now separated from one another and go into separate cells.
8. Meiosis consists of one chromosome duplication and two cell divisions. The net result is four haploid cells. In spermatogenesis, four sperm cells are produced; in oögenesis only one haploid product actually becomes a functional egg.
9. Two observations are of primary importance with respect to the role meiosis plays in the process of heredity:
   a) During synapsis and tetrad formation, homologous chromosome pairs may break and exchange corresponding segments with one another. This creates the possibility of new genetic combinations.
   b) During separation at anaphase, the direction of migration toward a pole taken by a member of one homologous chromosome pair in no way affects the direction taken by a member of another pair. During metaphase and anaphase I, then, the chromosomes assort at random.
10. Meiosis occurs in the process of gametogenesis. It is the principal process by which haploid eggs and sperm are produced from the diploid germ cells, spermatogonia and oögonia.
11. Replication is a property of some cell organelles as well as of whole cells. Mitochondria and chloroplasts both have the ability to replicate; their replication process is governed by DNA contained within the organelles. The details of this process are less clearly understood than is mitosis in whole cells.

EXERCISES

1. What is the basic accomplishment of mitosis as far as the genetic material is concerned?
2. Why is duplication of the chromosomes an indispensable step in mitosis?
3. How would the respiratory rate of cancer cells compare with the respiratory rate of cells in a mature, nongrowing tissue?
4. It has been hypothesized that the spindle fibers represent a magnetic field within the cell. Suggest an experiment to test this hypothesis.
5. What experimental evidence suggests DNA replication occurs during interphase?
6. Explain why meiosis is a necessary part of the life cycle of any sexually reproducing organism.
7. In gametogenesis, how many sperm can arise from 100 spermatogonia? From 100 primary spermatocytes? From 100 secondary spermatocytes? From 100 spermatids? How many eggs can arise from 100 oögonia? From 100 primary oöcytes? From 100 secondary oöcytes? From 100 oötids? (Use Fig. 7.11 in determining your answers.)

SUGGESTED READINGS

Kimball, John, *Cell Biology*. Reading, Mass.: Addison-Wesley, 2d ed., 1978. A simple account of cell division.

Mazia, Daniel, "How Cells Divide," *Scientific American* reprint #93 (September 1961). Good discussion of cell division and some of the investigational techniques and problems.

Swanson, Carl P., *The Cell*, 3rd ed. Englewood Cliffs, N.J.: Prentice-Hall, 1969, Chapter 5. This paperback book discusses in detail the several stages of mitosis, as well as the five stages recognizable in prophase of meiosis.

EIGHT

GENETICS I: FROM MATH TO MENDEL

8.1 Introduction

Genetics, the study of biological inheritance, is one of biology's youngest branches—and possibly its most important. Practical applications of knowledge attained through genetic research have had considerable impact in many areas; the improvement of cultivated plant and animal stocks is but one example. On a less technological and more scientific plane, genetics has provided considerable support for the theory of evolution by natural selection. Darwin was without the benefit of a satisfactory theory of inheritance, and many of his critics persistently attacked this weakness.

Since genetics deals with inherited characteristics, it must also deal with the cellular, organismal, and environmental factors influencing these characteristics. Further, genetics has found it fruitful to approach its subject matter on several fronts, from the molecular to the population level of investigation. Genetics has indeed come to occupy a central position among the life sciences. In this and the next chapter, we will arrive at an understanding of the major concepts on which modern genetics is now firmly based.

8.2 A Mathematical Basis

It will be helpful to turn from genetics (and even biology) for a moment, and direct our attention to certain things that are determined completely by chance. The use of the word "certain" is not entirely accidental. *There are few things that are more "certain" than those based on the entirely uncertain.* Put another way, *there are few things more predictable than those based on the completely unpredictable.*

The Product Principle of Probability

In the tossing of coins, the laws of chance are simply and directly involved. When a coin is tossed into the air, the chances that it will fall heads (or tails) are even, or 50–50 (which can be expressed as $\frac{1}{2}$). Suppose that two coins are tossed simultaneously. What *now* would be the chances that both of them would fall heads (or both tails)? The answer is given by the **product principle of probability.** This principle states that *the chance that two or more independent events will occur simultaneously is given by the product of the chances that each of the events will occur individually.* In the example of the simultaneous flipping of two coins, each has a 50–50 chance of coming down heads (or tails). In other words, the probability that either coin will come down heads (or tails) is $\frac{1}{2}$. Therefore, the probability that *both* of them will come down heads (or tails) is the product of $\frac{1}{2}$ times $\frac{1}{2}$ or $\frac{1}{4}$. The probability that one coin will fall heads while the other comes down tails is $\frac{1}{4}$, and the probability that the other coin will come down heads while the first comes down tails is also $\frac{1}{4}$. Thus the total probability for a head and tail combination is $\frac{1}{4}$ plus $\frac{1}{4}$, or $\frac{1}{2}$.

Tossing coins produces the following chance distribution of results:

THREE COINS

DISTRIBUTION:	HHH	HHT	HTT	TTT
PROBABILITY:	$\frac{1}{8}$	$\frac{3}{8}$	$\frac{3}{8}$	$\frac{1}{8}$

FOUR COINS

DISTRIBUTION:	HHHH	HHHT	HHTT	HTTT	TTTT
PROBABILITY:	$\frac{1}{16}$	$\frac{4}{16}$	$\frac{6}{16}$	$\frac{4}{16}$	$\frac{1}{16}$

It would be possible to go on in a similar manner computing the chances of various head–tail distributions among the simultaneous tossings of as many coins as desired. It is obvious, however, that for a problem involving a large number of coins, the arithmetic would get a little tedious.

There is a very simple and convenient algebraic method of computing the probability of occurrence of any combination of heads and tails in any given number of coins. It consists of expanding a binomial $(a + b)$ to the nth power, where a and b are simply symbols for the two possible results (in this case, heads and tails), and n represents the number of units participating in the event (in this case, the number of coins involved). In other words, any coin problem of this type is solvable by expanding $(a + b)^n$.

If the binomial expansion works for coins, it should work for anything that involves two chance possibilities. Since a and b are merely symbols, they can be maneuvered at will. Consider, for example, the sex distribution of human infants. The chances that a child will be male (or female) are approximately 50–50 or $\frac{1}{2}$. What, then, are the chances that in a family of four children all the children will be girls? We will let a represent the chances for boys, which are 50–50, or $\frac{1}{2}$, and b the chances for girls, which are also $\frac{1}{2}$. Using $(a + b)^n$, with $n = 4$:

$$(a + b)^4 = a^4 + 4a^3b + 6a^2b^2 + 4ab^3 + b^4.$$

Since b represents the chances for girls, we want the term in the expansion that gives us 4 b's. Clearly this is b^4. Substituting, we get

$$b^4 = (\tfrac{1}{2})^4 = \tfrac{1}{16}.$$

In other words, the chances of getting all girls in a family of four children are the same as the chances of each child being a girl multiplied by each other child's chance of being a girl, or $\frac{1}{2} \times \frac{1}{2} \times \frac{1}{2} \times \frac{1}{2} = \frac{1}{16}$.

8.3 Experimental Genetics: Mendel's Work

The laws of chance and probability were well known in certain of the natural sciences during the latter part of the nineteenth century. They had been applied to such areas of physics as statistical mechanics and dynamics. However, virtually no application had been made to any areas of biology.

Fig. 8.1 *Gregor Johann Mendel (Photo courtesy Burndy Library, Norwalk, Conn.)*

It was in just such an application that Gregor Mendel (Fig. 8.1), an Augustinian monk in Brünn, Austria (now Brno, Czechoslovakia), made his outstanding contributions to the study of heredity.

Mendel published his important work on heredity in peas in 1865. Despite, or perhaps because of, its innovative nature, Mendel's work made virtually no impression on anyone for the next 35 years. Only in 1900 was Mendel's work recognized for the pioneering effort it had been. In that year, sometimes called the birthyear of modern genetics, three investigators more or less independently rediscovered Mendel's original paper and publicized it to the biological world.

The Right Person, the Right Time, the Right Place

The study of heredity did not originate with Mendel. From ancient times, questions about how parents influence the appearance of their offspring had attracted considerable attention and speculation. Especially speculation. Every sort of hypothesis imaginable had been proposed to account for the various patterns or lack of patterns of heredity that people thought they observed in their families, their pets, their livestock, their crops. No scheme had met with any widespread acceptance. With the publication of Darwin's *Origin of Species* in 1859, the issue of how heredity is controlled became even more important. Darwin's theory of natural selection depended upon the fact that certain characteristics of organisms showed slight variations that could be inherited. Darwin assumed that this happened, though he had very little evidence on which to do so. Many of Darwin's critics as well as his supporters sought to determine whether this assumption was true. Their attempts usually involved experimental breeding of organisms. Most of the experiments were carried out with small numbers of organisms, and often there was no systematic method devised for making observations and keeping records of the offspring.

Mendel was familiar with many of the animal and plant breeders, beekeepers, and other agricultural investigators in the rich farming area where he lived. Many agricultural workers were also interested in finding patterns of heredity, so that they could more successfully breed the kinds of crops and animals they wanted. Mendel also read Darwin's *Origin of Species* and was intrigued by the problems of heredity transmission it raised. These two sets of influences—Darwin's theory of natural selection, and the needs of agricultural breeders—appear to have stimulated Mendel to actually undertake some breeding experiments of his own in about 1854 or 1855.

Mendel was an amateur biologist but he was not unfamiliar with the natural science of his day. In Brünn he served as a teacher of science in the monastery school. To prepare himself for teaching, he had attended the University of Vienna to study physics, zoology, and mathematics. He worked there with some of the most advanced physicists of the day, including Christian Johann Doppler (discoverer of the Doppler effect, an observed shift toward the red end of the spectrum in light coming from a source moving away from the observer). Although Mendel became conversant with mathematics and natural science, he was never able to pass the examination to qualify for a high school teaching certificate. Particularly significant was the fact that he passed the part of the exam dealing with the more mathematical physical sciences but was unable to pass the natural history (biology) portion.

Most of Mendel's important experiments were done between 1856 and 1863. In 1865 he presented his paper reporting results obtained from thousands of tedious breeding experiments. The report was read aloud at a meeting of the Brünn Society for the Study of Natural Science and was later published in the transactions of that society. At the meeting Mendel received polite attention. However, his application of mathematics to the ratios of plant offspring seemed too much for the audience to take. Attention wandered. It is recorded in the minutes that "there were neither questions nor discussions." The fact that scientific history was made that night remained unnoticed for almost 35 years.

Probability and the Pea

Mendel performed breeding experiments on several plants and animals. He is best known, however, for his work with the garden pea, a plant that is normally self-pollinating and hence inbred in the natural state. By close anatomical examination of these plants he was able to distinguish several distinct and inherited characteristics. Among these were the color and shape of the seeds, the positioning of the flowers on the stems (some at the end, others on the side) and the height of the plants (whether they were short or tall). In describing Mendel's experimentation and his conclusions we will, for simplicity, deal with the characteristic of height.

Mendel crossed true-breeding tall plants with true-breeding short ones. Several possibilities were available. The plants might be all tall, all short, or a mixture of talls and shorts, or they might even be of an intermediate height. Figure 8.2 represents Mendel's results; all of the F_1 (first filial, or offspring, generation) plants were tall.

Mendel found himself faced with many questions. Had the shortness characteristic been destroyed, or was it merely hidden? Were the F_1 tall plants genetically similar to their tall parents; that is, would they breed true?

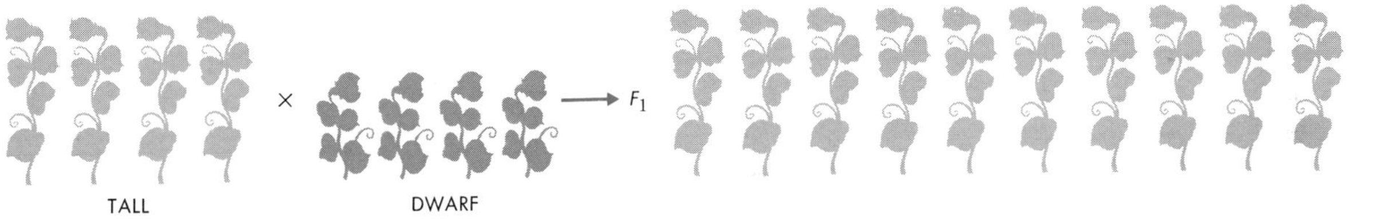

Fig. 8.2 *A representation of Mendel's first cross of garden peas, which resulted in an F_1 of all tall plants.*

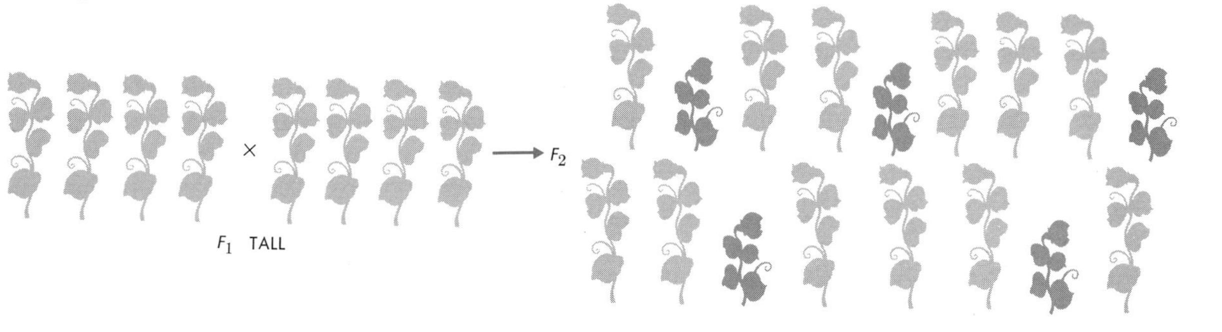

Fig. 8.3 *A representation of Mendel's second cross of garden peas, which resulted in an F_2 of 787 tall plants to 277 short plants.*

Mendel then crossed two tall plants from the F_1 generation. Figure 8.3 represents the results: out of 1064 F_2 plants, 787 were tall and 277 short.

Mendel explained the results of his first cross by coining two terms still used in present-day genetics. Since the shortness factor was completely hidden by the tallness factor, he postulated the tallness characteristic to be **dominant** over the shortness, which he termed **recessive.**

It was in his interpretation of the second cross, involving the F_1 tall plants, that Mendel's real genius was asserted. Long before he began his work, many people had noticed numbers like those just cited in their experimental crosses. Few, however, had bothered to look at the significance of these numbers. Fortunately Mendel was an extremely talented mathematician, and it was this ability that enabled him to discover the prize that had so long eluded earlier naturalists.

Mendel noticed that 787 tall plants to 277 short ones constituted a ratio of 2.84:1. In similar crosses involving other characteristics of the pea plant, he obtained similar results; a dominant characteristic appeared in the F_1 generation, with ratios of 3.15:1, 2.96:1, 2.82:1, 3.14:1, 2.95:1, and 3.01:1 in the F_2 generations.

The suggestion that a ratio of three to one was predictable in the F_2 of such crosses was too strong to be overlooked. To Mendel, this discovery was very exciting. He recognized the ratio as being indicative of the operation of the laws of chance and probability. He therefore hypothesized as follows: Suppose that in true-breeding tall plants there is not just one factor influencing height, but *two.* Suppose further that two other factors are present in short plants. If only *one* of these factors gets into a sperm cell and only *one* into an egg cell, then the uniting of these cells will produce a plant with two factors, one from the tall plant and one from the short. Since the factor for tallness is dominant over the factor for shortness, the hybrid in the F_1 will be tall. If we let T stand for the tallness factor and t for the shortness factor, the first cross can be diagrammed as follows.

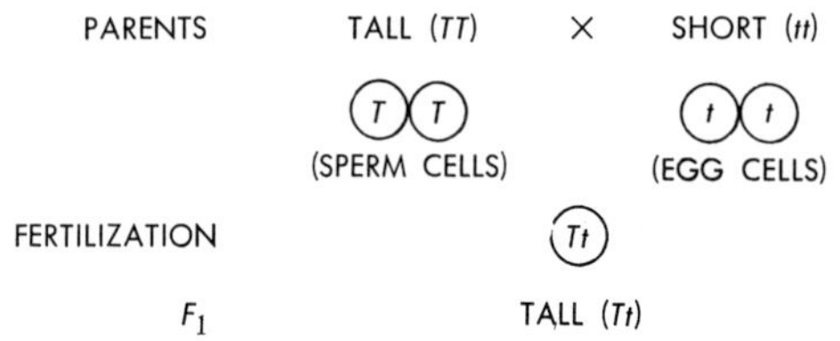

The Law of Segregation

Recall that Mendel performed his experiments around 1860; such things as chromosomes were unknown to him. But notice that homologous chromosomes behave suspiciously like Mendel's hypothesized factors; they occur in pairs and separate from each other during gamete formation (meiosis), one going to each gamete. Fertilization brings two homologous chromosomes together again, one from each parent.

Thus, by hypothesizing that only one factor for an inherited characteristic gets into each gamete, Mendel recognized the necessity of meiosis, though of course he knew nothing about the process. He also recognized that the two factors paired again during fertilization when he noted the restoration of the pairing in the F_1 hybrid. In other words, Mendel recognized that the two factors influencing height in pea plants must separate or **segregate** from each other in the production of the germ cells and then be reunited at fertilization. This concept is known as Mendel's first law, the **law of segregation,** which states that *the factors for a pair of characteristics are segregated.*

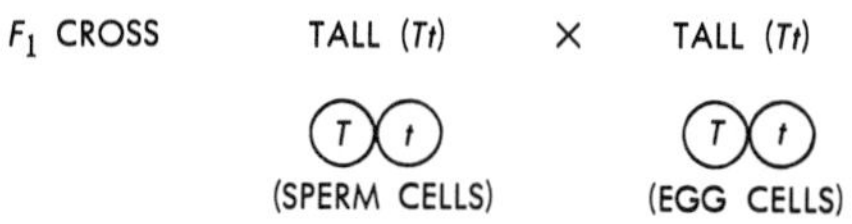

Here each parent plant can produce *two* kinds of gametes instead of only one. Half of the gametes produced will carry the tallness factor T. The other half will carry the shortness factor t. Mendel saw that if two types of gametes were produced in equal numbers by both plants, his three-to-one ratio

Mendel's first law (the law of segregation) states that in any single gamete-forming cell of a parent, the two factors for any trait segregate so that each ends up in a different gamete.

could be explained on the basis of chance and probability. Since any type of sperm cell has an equal chance of fertilizing any type of egg, there are four possible fusions that can take place.

1. A sperm cell carrying the *T* factor may fuse with an egg carrying the *T* factor, yielding *TT*.
2. A sperm cell carrying the *T* factor may fuse with an egg carrying the *t* factor, yielding *Tt*.
3. A sperm cell carrying the *t* factor may fuse with an egg carrying the *T* factor, yielding *Tt*.
4. A sperm cell carrying the *t* factor may fuse with an egg carrying the *t* factor, yielding *tt*.

Since the F_1 generation showed that the *T* factor is dominant over the recessive *t* factor, the first three fertilization possibilities will result in tall plants. Only the fourth produces a short plant.

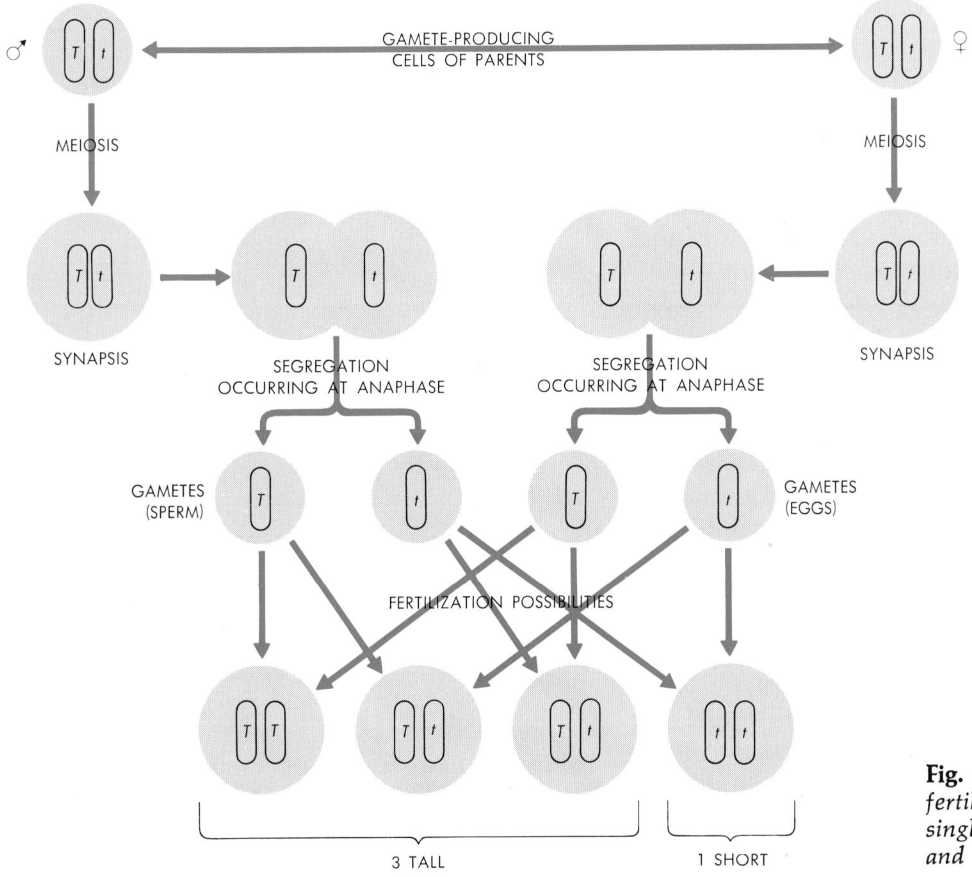

Fig. 8.4 *Gametogenesis and fertilization possibilities for a single pair of alleles, T (tall) and t (short).*

Using present-day knowledge of meiosis, we can represent what happened in Mendel's experimental crossbreeding by means of a highly simplified diagram (Fig. 8.4).

8.4 Application of Mendel's Hypothesis to Animals

As powerful as it is, Mendel's hypothesis would be of considerably less interest if it applied only to peas. Much experimental work in the first three decades of the twentieth century showed that Mendel's basic ideas applied to a wide variety of animals as well as plants. Consider several examples of simple Mendelian principles in animals.

Mice are often used in genetic studies. They reproduce rapidly, are relatively inexpensive, and are easy to care for in the laboratory. One unusual inherited characteristic in mice results in a complete lack of body hair, coupled with heavy folds or wrinkles in the skin (see Color Plate I). For rather obvious reasons, the condition is known as rhino. Since it is inherited and easy to distinguish, the rhino condition would seem to be a good choice for use in experiments designed to shed light on how inherited characteristics are passed from generation to generation.

Further considerations rule against this choice, however. The rhino condition couples together at least two inherited conditions, hairlessness and wrinkled skin. Further, the skin itself is a complex organ, composed of several different kinds of tissues (glandular, muscular, connective, etc.). It seems evident that there must be many inherited factors involved in the rhino condition. Finally, rhino mice reproduce very poorly, if at all; the very physical deformity that makes them rhino interferes with their mating capability. It is reasonable, therefore, to begin an attempt to explain the transmission of inherited characteristics by working with characteristics of a simpler nature.

Coat color in mice is such a characteristic. Differences in coat color are easy to distinguish, and the required matings can be accomplished without difficulty. Further, differences in coat color are due to interacting factors far less complex than those that cause the rhino condition. In general, the coats of different-colored mice are the same; the only difference is in the pigments deposited in the hairs composing these coats.

A certain well-established strain of mice possesses a solid black coat of fur.* Another strain has a solid brown coat (see Color Plate I). Since both are pure strains, crosses between black mice always yield black offspring and crosses between brown mice always yield brown offspring. It is reasonable to wonder, therefore, what would be the color of the offspring resulting from a cross between brown mice and black. Here are a few of the possibilities:

1. An intermediate color between black and brown.
2. Some black mice and some brown.
3. Spotted black and brown mice.
4. A color entirely different from black and brown.
5. All black mice.
6. All brown mice.

Several matings are made. In some, the male is black and the female brown, while in others the reverse is true. This eliminates sex as a variable in the experiment, while the large number of offspring produced by many matings allows more accurate generalizations.

Figure 8.5 shows that prediction 5 is the cor-

* Mice that have been mated brother-to-sister for at least twenty generations are considered an established inbred strain.

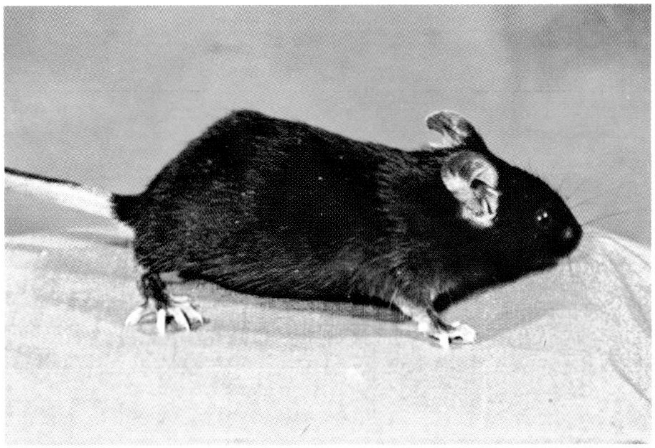

BLACK

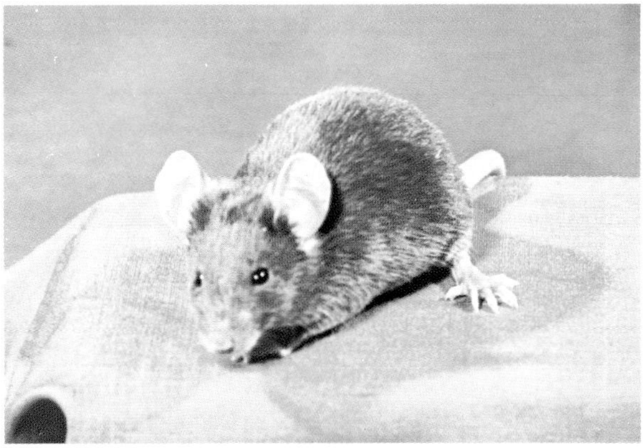

BROWN

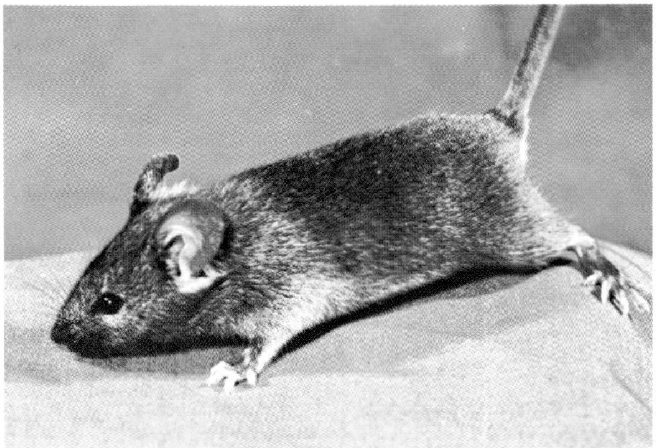

BLACK AGOUTI

BROWN AGOUTI

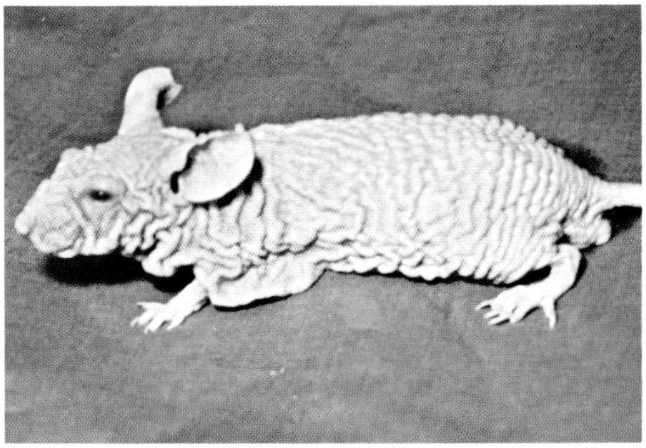

RHINO

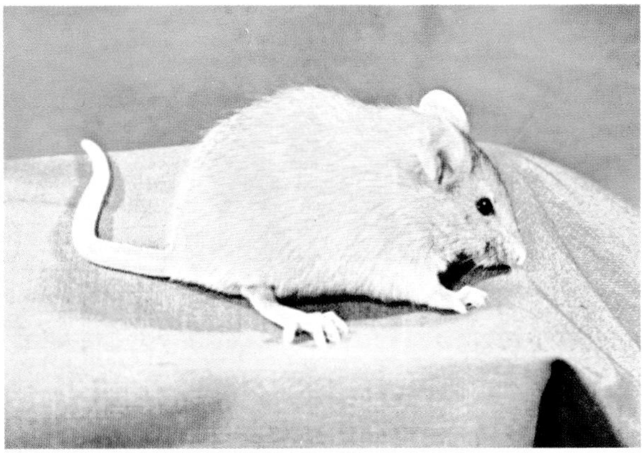

YELLOW

Plate I *Coat characteristics inherited in mice. (Courtesy Roscoe B. Jackson Memorial Laboratory.)*

(a)

(b)

(c)

Plate II *Three stages of succession in a freshwater pond. (a) Pond stage, with dense vegetation around the edges. (b) Pond fills in to become a bog, where cattails and other shallow-water species obtain a hold. (c) Land stage, characterized by the beginnings of forestation.*

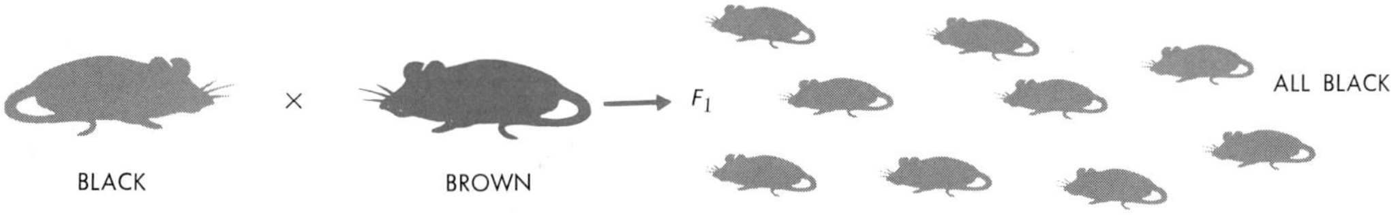

Fig. 8.5 *A representation of the first cross of black mice with brown mice, yielding an F_1 of 992 mice, all black.*

rect one. All 992 mice obtained (the F_1 generation) are solid black and indistinguishable in appearance from their black parent.

This result raises several questions. First, what has happened to the brown color? Has it been completely destroyed by the black color? Or is it still present but hidden? Second, does the fact that the F_1 black mice *appear* identical to their black parent mean that they are also genetically identical? Granting that crosses between two inbred-strain parents always produce black offspring, is the same true for the F_1 black mice? In other words, if they are crossbred, will they too produce only black mice?

To answer these questions, a second set of breeding experiments must be performed. The same procedure is followed as before, but this time the F_1 mice, all of which are black, are crossed with each other.

Figure 8.6 reveals the results of this cross: of the 1278 progeny, 961 are black and 317 brown.

The results answer the questions asked prior to the second crossbreeding. First, quite obviously, the brown color was not destroyed by the black. Though not in evidence, it must have been present all the time. The brown mice obtained in the F_2 are just as brown as the original brown parents, and when these F_2 browns are mated, their offspring are all brown. It can be seen, therefore, that the factor for brown coat color must be both independent of, and unaffected by, the black color.

Second, although the F_1 blacks are indistinguishable from their black parents in appearance,

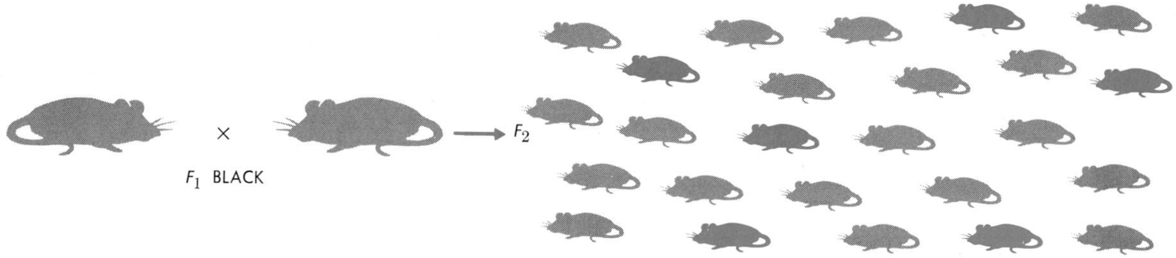

Fig. 8.6 *A representation of the second cross of mice, yielding an F_1 of 961 black mice to 317 brown mice.*

they still differ genetically. They fail to breed true, i.e., to produce only black mice.

Third, while the brown-coat factor is not affected or changed by the black-coat factor, it is still completely hidden in the F_1 generation. This leads us to consider the possibility that the black factor may in some way be "stronger" than the brown factor.

Mendel's hypothesis provides a ready explanation for the results of a cross between two black F_1 mice. Note that 961 black mice to 317 brown mice closely approximates a three-to-one ratio. Therefore, it can be hypothesized that the F_1 male parent produced two kinds of sperm, half carrying a *B* factor for black, the other half carrying a *b* factor for brown. The female produced two kinds of eggs, half carrying a *B* factor, the other half carrying a *b* factor. Given complete dominance of black over brown, and the fertilization possibilities governed by the laws of chance and probability discussed in Section 8.2, a three-to-one ratio is readily accounted for.

As has been constantly stressed throughout this book, a good hypothesis must not only explain the observed, it must also act as a basis for accurate predictions. Thus, we can reason that if in mice, factors for the black and brown coat colors segregate into the gametes and recombine at fertilization according to the laws of chance and probability, and if black is completely dominant over brown, then crosses between F_1 black mice and brown mice should result in offspring of which one-half are black and one-half are brown.

An F_1 black mouse, being the result of crossing a pure black (*BB*) with a pure brown (*bb*) mouse, must carry both coat-color factors, *B* and *b*. The dominance of black over brown makes the animal black. Half the gametes it will produce will carry the *B* factor and half will carry the *b* factor. The brown mouse, on the other hand, can produce only gametes that carry the *b* factor. Since these gametes have an equal chance of fusing with either of the other gametes, a 50–50 distribution of black to brown mice is the prediction. Of the 833 offspring of such a cross, 412 were black and 421 were brown. Since the slight deviation in the predicted ratio is not statistically significant, Mendel's hypothesis is supported. Moreover, it gains stature by this experiment because his generalization concerning inheritance in a species of plant has been successfully extrapolated to a species of animal. Indeed, it will be seen that the principles of Mendelian genetics play a role in the inheritance of many different species of living organisms. For, as has probably been obvious to the reader for some time, what Mendel thought of as "factors" are the hereditary units, the genes.

Incomplete Dominance

Not all characteristics are inherited in as simple a manner as the tallness and shortness in peas or the black and brown coat colors in mice. Indeed, simple inheritance in which only one pair of genes is involved is very much the exception rather than the rule. Nor is it required for one characteristic to be completely dominant over the other. Quite often the combination of different genes tends to produce varying degrees of partial or **incomplete dominance,** with the latter usually resulting in a blending of the two genes to produce a different appearance.

An example of incomplete dominance is seen in the breeding of certain types of cattle. If a red animal is crossed with a white one, an intermediate-colored animal, a roan, is produced. No other color ever appears; however, crosses between two roans yield a ratio of one red to two roans to one white. Once again, Mendelian genetics provides a ready explanation for the results (Fig. 8.7).

Note that if red coat had been dominant over white (or vice versa), a 3:1 F_2 ratio would result. However, since the genes modify each other to produce an intermediate color, a 1:2:1 ratio is obtained.

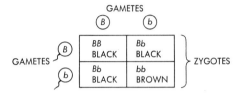

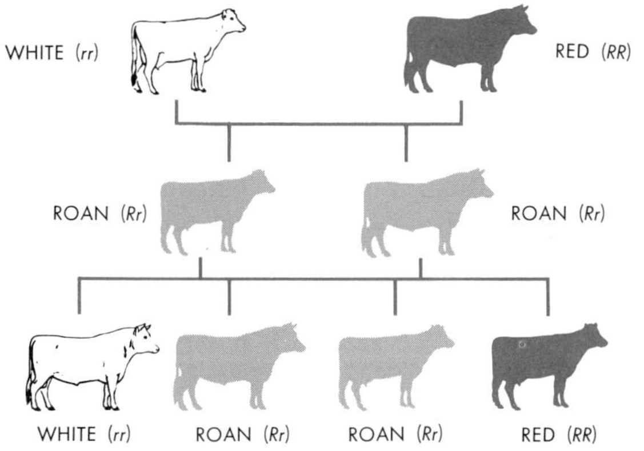

Fig. 8.7 *Pattern of inheritance of coat color in cattle. White and red parents produce roan F_1's, two roans interbred produce a 1:2:1 F_2 ratio of white: roan:red. The R and r alleles show incomplete dominance.*

Further, in this case the F_1 animals are easily distinguishable from their parents.

Note also that this 1:2:1 ratio is the same obtained in flipping two pennies simultaneously. The number of times we obtain both heads, a head and tail combination, or both tails is 1:2:1, respectively. Further, the same mathematics discussed in Section 8.2 can be used to predict such things as the frequency of occurrence of inherited characteristics in the offspring of living organisms.

Prior to Mendel it was widely believed that inherited characteristics represented a blend of traits shown by the parents. Undoubtedly, mulatto children produced by marriages between blacks and whites lent considerable support to this idea. It might be noted in passing that an hypothesis proposing blending of inherited traits is just as satisfactory as Mendel's particulate hypothesis in explaining the results of the cross between red and white cattle. The F_2 generation, however, yields results that contradict the blending hypothesis and support Mendel's particulate one.

Usually it is easier to determine the possible combinations that can occur at fertilization by arranging the gametes in a matrix, or Punnett square:

GAMETES	(B)	(b)	
(B)	*BB* BLACK	*Bb* BLACK	ZYGOTES
(b)	*Bb* BLACK	*bb* BROWN	

The Punnett square prevents our overlooking a possible gamete combination, easily done in complex crosses involving more than one pair of genes. However, do not forget that the external symbols represent gametes (sperms and eggs) and the internal ones zygotes of the potential organism.

Some Genetic Terminology

A brief look at some genetic terminology will facilitate understanding of the material to follow. Genes that carry contrasting inheritance factors are called **alleles.** Alleles line up opposite each other during the synapsis of homologous chromosomes (Section 7.4). Genes *T* and *t* in Mendel's tall and dwarf pea plants are alleles. The genes *B* and *b* in the black and brown mice are also alleles. In the case of the pea, alleles *T* and *t* carry the contrasting inheritance factors for height; in the mice, alleles *B* and *b* carry the contrasting factors for coat color.

The term **homozygous** is used to designate an individual in which a pair of genes are identical; if the genes are different, the individual is **heterozygous.** For example, both the tall and the dwarf parent pea plants used by Mendel were homozygous, since they both contained like genes for height (*TT* and *tt* respectively). However, two-thirds of the F_2

tall plants and all of the F_1 plants were heterozygous (or hybrid), since they contained genes that were different (*Tt*). In the experiment with the mice, the parent black mice used in the first cross were homozygous (*BB*), as were the brown parents (*bb*). The F_1 blacks and two-thirds of the F_2 blacks were heterozygous (*Bb*). Of course, an individual may be heterozygous for some pairs of genes but homozygous for other pairs.

A distinction must be made between the *appearance* of an organism and the inheritance factors or genes that it will pass along to its descendants, for, as we have already seen, the one may not reveal the other. The appearance registered by means of our senses is called the **phenotype.** The phenotype of the plant *TT* is tallness. The phenotype of the plant *Tt* is also tallness. The plant with *tt* has a phenotype of shortness. Similarly, mice *BB* and *Bb* are phenotypically black, while mice *bb* are phenotypically brown.

On the other hand, the code or classification given to an organism on the basis of data from breeding experiments is its **genotype.** While phenotype is a classification according to appearance, genotype is a classification according to genetic makeup. For example, the symbols *TT* represent the genotype for a homozygous tall pea plant; the symbols *Tt* represent the genotype of a heterozygous tall pea plant. The genotype of a dwarf plant is *tt*. Likewise, in the brown and black mice, there are only three possible genotypes for coat color: *BB*, *Bb*, and *bb*.

A distinction must be made between genotype and phenotype in recording ratios given by an experimental genetic cross. In the cross between the heterozygous F_1 black mice, our phenotypic ratio was 3:1. But the genotypic ratio was 1 homozygous black, 2 heterozygous black, and 1 homozygous brown—or 1:2:1. In the cross between red and white cattle, yielding an intermediate phenotype of roan, the cross between the F_1 roans produced offspring in which the phenotypic and genotypic ratios were identical.

8.5 Two Pairs of Genes

Thus far, we have dealt with the principles of simple Mendelian inheritance in crosses involving a single pair of genes. Relatively few inherited characteristics, however, are influenced by only one pair of genes. For example, the genes concerned with the formation and proper functioning of the pituitary gland are not directly involved in the growth of the body. Yet should they fail to carry out their assigned job, dwarfism results. It is often difficult to say that a certain characteristic is carried by a particular gene or genes. It is far more accurate to view it as the expression of the interaction of several groups of genes.

In addition to the black and brown coloration in mice, there is another factor that influences coat color. This factor is known as the **agouti** condition. When the agouti condition is combined with black, brown, or some other coat color, it produces a characteristic muted appearance (see Color Plate I). If we examine the individual hairs of the coats of agouti mice, the reason for this appearance becomes clear. The hairs of a completely black or brown mouse are solidly colored along their entire length. The hairs of an agouti animal have a distinct band across them, near the end (Fig. 8.8).

As in the case of black or brown coat color, the inheritance of the agouti condition is dependent on one pair of genes and is dominant. Therefore, crosses between homozygous agouti and nonagouti

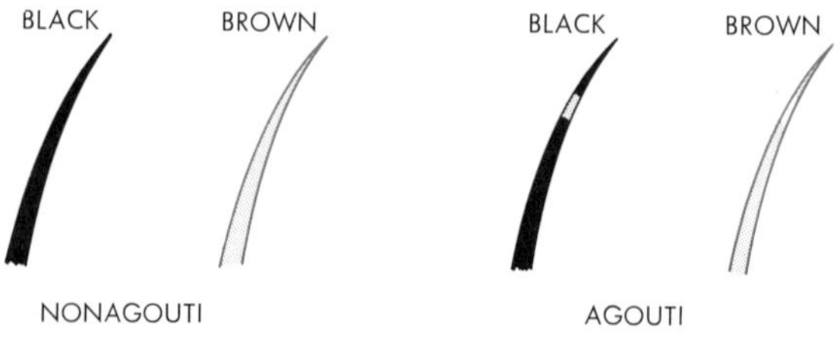

Figure 8.8

mice yield an F_1 that are all agouti. Crosses of the F_1 agouti animals yield a ratio of three agoutis to one nonagouti.

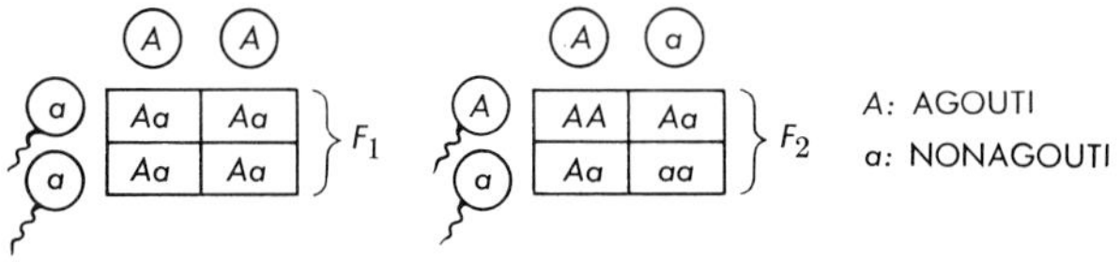

Such F_2 3:1 ratios of black to brown or agouti to nonagouti are obtained when the characteristics are considered *separately*. Suppose, however, that they are considered *together*. In particular, consider a cross involving black agouti mice from a pure inbred strain (genotype *AABB*) with brown nonagouti mice (genotype *aabb*). Black, of course, is dominant over brown, as is agouti over nonagouti. It might be predicted, therefore, that the F_1 mice will show both dominant characteristics—that they will all be black agoutis. This is precisely what occurs. In a series of experiments, many matings of black agoutis with brown nonagoutis produced 1624 F_1 animals, all black agoutis.

Were these F_1 black agoutis genotypically the same as their dominant parent? If so, then crosses between F_1 mice should produce only black agouti mice. Previous experience, however, indicates that these F_1 individuals will not breed true, despite the fact that their appearance is indistinguishable from that of their black agouti parents. Past experience also leads us to expect the three-to-one ratio. *But three whats to one what?* Three black agoutis to one brown nonagouti? Or three black nonagoutis to one brown agouti? Is there any justification for assuming that the black color always appears linked to the agouti condition and the brown to the nonagouti? If not, then it must be admitted that we can obtain *four* possible kinds of mice instead of just two, namely black agoutis, black nonagoutis, brown agoutis, and brown nonagoutis.

Many such F_1 crosses were made, and a total of 1625 baby mice were obtained. Of these, 909 were black agoutis, 304 were black nonagoutis, 299 were brown agoutis, and 103 were brown nonagoutis, a ratio of 9:3:3:1.

The Law of Independent Assortment

Another examination of Mendel's work provides an hypothesis to explain this 9:3:3:1 ratio. Mendel had noticed that one inherited difference in the garden pea was seed color; some plants had yellow seeds, others green. Crosses of yellow-seed producers with green-seed producers yielded only plants that produced yellow seeds. Mendel hypothesized the dominance of yellow over green, combined with segregation of the factors for these seed colors, and saw that a three-to-one ratio of yellows to greens should result from crosses of the F_1 yellows. Experimental results supported this hypothesis.

However, Mendel also noticed that the pea seeds differed in still another characteristic. Some were round and smooth, while others were wrinkled or shriveled in appearance. Mendel found that crosses between plants that produced round seeds and plants that produced wrinkled seeds gave an F_1 of plants that produced round seeds; thus roundness is carried by the dominant gene. Crosses between plants of the F_1 generation produced a 3:1 ratio of round seeds to wrinkled ones.

Mendel then considered the factors of seed color and seed shape together. He noticed that crosses between two pure varieties of pea plants—one showing the dominant characteristics of round yellow seeds, the other the recessive characteristics of wrinkled green seeds—yielded an F_1 phenotypically identical to the dominant parent; that is, the F_1 plants all produced round yellow seeds. However, crosses of these F_1 plants yielded a ratio of nine plants that produced round yellow seeds, to three that produced round green seeds, to three that produced wrinkled yellow seeds, to one that produced wrinkled green seeds. Notice that the ratios obtained by Mendel for the peas were the same as those obtained with the mice in the experiment mentioned above. In interpreting his results, Mendel turned his attention to gamete production. Of course, Mendel knew nothing of meiosis, but it will be helpful for us to consider his hypothesis in the light of modern-day knowledge of the process.

Recall that when Mendel was working with just one characteristic, such as tallness/dwarfness in pea plants, he hypothesized the existence of one pair of "factors" (genes) that segregate at gamete formation. When he began dealing with two characteristics, Mendel did not hesitate to hypothesize the existence of *two* pairs of genes, each pair of which was concerned with the inheritance of one of the two contrasting traits (round-wrinkled and yellow-green). The members of each of these pairs also segregate during gametogenesis, with each seed-color gene going to a different gamete. The same is true for the genes for seed shape; these also go into separate gametes. But, Mendel reasoned, if the segregation of the genes for seed color is *entirely independent* of the segregation of the genes for seed shape, and occurs *completely at random*, then once again the mathematics of chance and probability will apply. This is known as Mendel's second law, the **law of independent assortment.**

Mendel's first cross can be diagrammed:

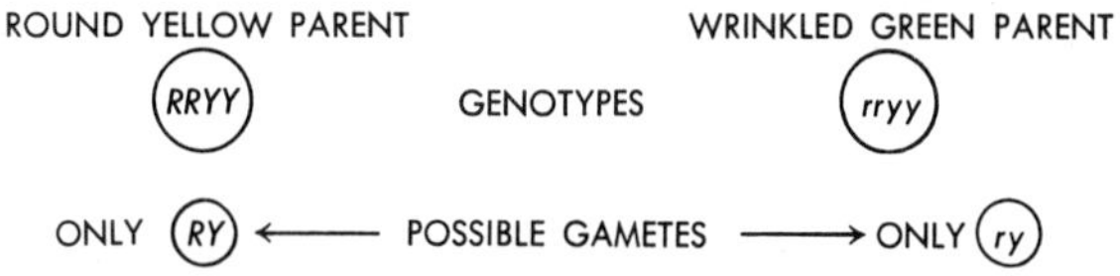

Here, R = round, r = wrinkled, Y = yellow, and y = green. Since each parent can produce only one type of gamete, only one fertilization combination is possible:

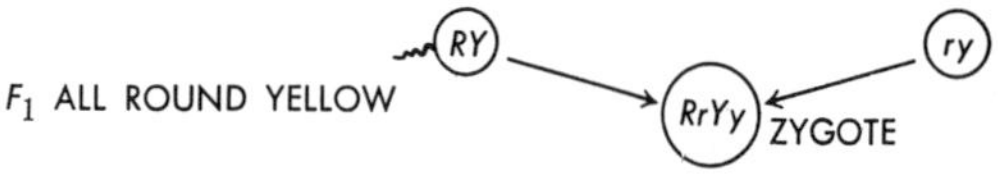

Note that Mendel's F_1 plant (producing round yellow seeds) differs genotypically from the parent that produces round yellow seeds. All the F_1 plants must be doubly heterozygous, or dihybrid. Thus the F_1 plants produce gametes that are unlike those of their homozygous parents. Mendel stressed completely random distribution of the allelic genes and completely independent assortment of the nonallelic genes. With this in mind, it becomes evident that each F_1 plant can produce four types of gametes:

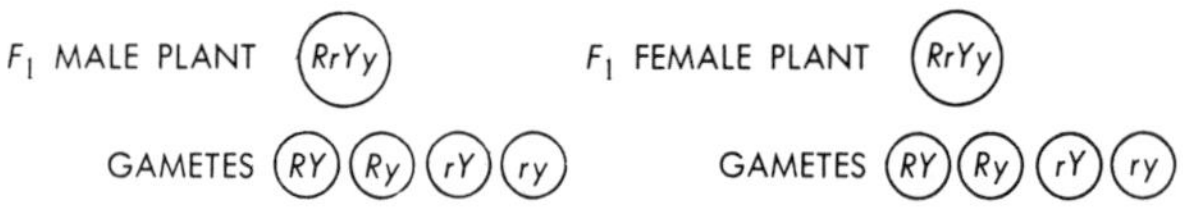

There are now several fertilization possibilities, each determined entirely by chance. With just one pair of genes involved, only two types of gametes could be produced by the F_1 hybrids. There were therefore 2×2, or 4, fertilization possibilities. With complete dominance, a three-to-one ratio is obtained. In this case, however, each F_1 individual produces *four* different types of gametes. The fertilization possibilities are thus 4×4, or 16. The three-to-one ratio in the monohybrid cross yields the total number of fertilization possibilities with one pair of genes, for $3 + 1 = 4$. The same holds true for the ratio obtained when two pairs of genes are involved, for $9 + 3 + 3 + 1 = 16$.

This F_1 dihybrid cross can be diagrammed by use of a Punnett square, as shown in Fig. 8.9. A count of the phenotypes reveals the experimentally obtained 9:3:3:1 ratio.

Experimental and Mathematical Evidence

Mendel recognized that his hypothesis had to have predictive value if it was to be accepted by his fellow scientists. Accordingly he crossed plants of known genotypes. Before obtaining his experimental results, he determined theoretically the types of gametes that each would produce and calculated the fertilization possibilities. Mendel then recorded his predictions of the types of plants and the proportion of each that these crosses would be

---

Mendel's second law (the law of independent assortment) states that the individual members of pairs of alleles segregate independently of one another during pollen or egg cell formation.

MALE GAMETES OF F_1 GENERATION

FEMALE GAMETES OF F_1 GENERATION

	RY	*Ry*	*rY*	*ry*
RY	*RRYY* ROUND YELLOW	*RRYy* ROUND YELLOW	*RrYY* ROUND YELLOW	*RrYy* ROUND YELLOW
Ry	*RRYy* ROUND YELLOW	*RRyy* ROUND GREEN	*RrYy* ROUND YELLOW	*Rryy* ROUND GREEN
rY	*RrYY* ROUND YELLOW	*RrYy* ROUND YELLOW	*rrYY* WRINKLED YELLOW	*rrYy* WRINKLED YELLOW
ry	*RrYy* ROUND YELLOW	*Rryy* ROUND GREEN	*rrYy* WRINKLED YELLOW	*rryy* WRINKLED GREEN

F_1 GENOTYPES AND PHENOTYPES

Fig. 8.9 *Punnett square for the cross between two pea plants, each heterozygous for two traits: seed color and seed shape. In such a cross, Mendel found that the offspring showed a 9:3:3:1 phenotype ratio.*

expected to produce. Two such crosses were as follows:

1. *Genotype RrYy with RRYY.* Mendel hypothesized that in terms of these genetic factors, the dihybrid would produce four kinds of gametes, *RY*, *Ry*, *rY*, and *ry*. The other would produce only *RY* gametes. Thus no matter which fertilization combination occurred, the seeds of the resulting plants would all be round and yellow. *Results:* 192 plants, all of which produced round yellow seeds.

2. *Genotype RrYy with rryy.* Mendel hypothesized that four kinds of gametes would be produced by the dihybrid (*RY*, *Ry*, *rY*, and *ry*) and one kind (*ry*) by the other plant. By considering all the possible fertilization combinations, Mendel predicted four different kinds of plants in equal numbers. *Results:* 55 with round yellow seeds, 51 with round green seeds, 49 with wrinkled yellow seeds, and 53 with wrinkled green seeds.

It would take a professional Doubting Thomas to question the validity of Mendel's hypothesis in the light of the extraordinary predictive abilities demonstrated by these crosses.

Lest the reader lose sight of the importance of the part played by the product principle of

Supplement 8.1

SOME HINTS ABOUT SOLVING GENETICS PROBLEMS

The study of genetics has been the nemesis of many biology students because of difficulties they encounter in the solution of problems. On the other hand, some students who may have been doing quite poorly in a biology course suddenly blossom in the study of genetics. It is difficult to see why these differences occur; possibly it is because genetics problems, by their mathematical nature, demand very precise reasoning. Nevertheless, almost all genetics problems encountered in introductory biology can be attacked in a simple and straightforward manner.

The following steps will help in the solution of genetics problems. They are applicable to the problems in both Chapters 8 and 9.

1. Determine the type of inheritance dealt with in the problem. Does it show complete dominance, incomplete dominance, or some other inheritance feature? Are there one, two, or more pairs of genes involved? Usually this information is given in the problem. If not, it can be deduced from the phenotypic ratios of the offspring.

2. Determine the genotypes of the individuals involved.

3. Determine the types of gametes each parent can produce. Arrange them into a Punnett square and fill in the possible progeny genotypes.

4. Count the resulting phenotypes and express them as a ratio. Often this is as far as the problem will require you to go.

5. If chance predictions are involved, expand the binomial $(a + b)^n$ to the proper nth power. Substitute the fractions given by the Punnett square in step 4 and solve.

Sample Problem: In humans, the ability to taste the bitter chemical phenylthiocarbamide (PTC) is due to a dominant gene T; inability to taste it is due to the recessive allele t. A man who can taste PTC, but whose father could not, marries a woman who can also taste PTC, but whose mother could not.

a) What proportion of their children will probably have the ability to taste PTC?

b) If they have five children, what are the chances that four will be tasters and one a nontaster?

Step 1. We determine that this is a case of simple dominance and that there is only one pair of genes involved.

Step 2. We determine that the man's genotype must be Tt (since he is a taster, he must have at least one gene T, but since his father was a nontaster (tt), and he is the product of his father's sperm, he must also have a t gene.) Likewise, the woman must also be Tt, since her mother was a nontaster (tt) and she is a product of her mother's egg.

Step 3. We determine that the man and woman can each produce two types of gametes, carrying either a T or t gene. We next put them in a Punnett square, as follows:

	(T)	(t)
(T)	TT	Tt
(t)	Tt	tt

Step 4. We count up the phenotypes. In this case there are two, tasters and nontasters. The phenotypic ratio in this case is 3:1. This gives the answer to part (a); $\frac{3}{4}$ are tasters.

Step 5. We expand $(a + b)^n$ with $n = 5$ as follows:

$$a^5 + 5a^4b + 10a^3b^2 + 10a^2b^3 + 5ab^4 + b^5.$$

Letting a = tasters and b = nontasters, we choose the proper term in the expansion that gives us four tasters and one nontaster: four a's to one b. Clearly this is the term $5a^4b$. Substituting the Punnett square fractions of $\frac{3}{4}$ tasters to $\frac{1}{4}$ nontasters in this term, we obtain $5a^4b = 5(\frac{3}{4})^4(\frac{1}{4}) = 5(\frac{81}{256})(\frac{1}{4}) = 5(\frac{81}{1024}) = \frac{405}{1024}$, or 405 of 1024 chances. This is the answer to part (b).

probability, it is worthwhile pointing out that the 9:3:3:1 ratio could have been predicted in yet another way. A cross involving two individuals with a single pair of allelic genes (*Aa*), in which gene *A* shows complete dominance, yields a 3:1 ratio. The same ratio is obtained from a cross involving two individuals of genotype *Bb* with *B* completely dominant. By considering both gene pairs together in one cross (*AaBb* × *AaBb*), we simply increase the number of fertilization possibilities from four (the sum of 3:1) to 16 (the sum of 9:3:3:1). If we assume, as did Mendel, that the fertilization possibilities are entirely governed by the laws of chance and probability, the product principle of probability should apply directly:

$$\begin{array}{r} 3{:}1 \\ \times\ 3{:}1 \\ \hline 9{:}3{:}3{:}1 \end{array}$$

And so it does.

Mendel, by incorporating the mathematical laws of chance and probability into the science of genetics, provided a solid foundation upon which all later investigations in genetics could be based. Once it had been determined how genes interact with each other (whether they demonstrate dominance, incomplete dominance, etc.), reliable predictions could be made concerning the phenotypic and genotypic ratios of the progeny.

SUMMARY

1. Breeding experiments in genetics are based on the laws of probability. Even though any one event in itself is unpredictable, given enough events the overall outcome can be predictable.
2. The product principle of probability states that the chance that two or more independent events will occur simultaneously is given by the product of the chances that each of the events will occur individually. In flipping two pennies, for example, the chance of either by itself turning up heads is $\frac{1}{2}$ (1 out of 2, or 50–50). The chance that *both* will turn up heads simultaneously is given as $\frac{1}{2} \times \frac{1}{2}$ or $\frac{1}{4}$.
3. Mendel's original observations on peas suggested that the ratios of different types of offspring from any cross followed the laws of probability. Mendel assumed that every parent organism contains, in its germ cells, two "factors" for every observable trait. One factor derived from that organism's male parent, the other from its female parent. Mendel claimed the factors could be either dominant or recessive with respect to one another. A dominant trait is one that completely masks its recessive counterpart (complete dominance). A recessive trait is one masked by a dominant counterpart. Dominant and recessive are relative terms; that is, one factor is always dominant with respect to some other factor. In incomplete dominance, the two traits are phenotypically blended.
4. Mendel's first law, the law of segregation, states that in the process of forming gametes from a parental germ cell, the two factors for any trait always segregate and are distributed to two different gametes. Mendel's second law, the law of independent assortment, states that the segregation of two or more pairs of factors occurs independently—that is, randomly.
5. Differences in appearance and genetic makeup have led to geneticists' making a distinction between genotype and phenotype. Phenotype is the appearance of the organism for any given trait or combination of traits, while genotype refers to the actual genetic makeup of the individual with respect to any trait or traits. Thus, individuals homozygous and heterozygous for a trait showing complete dominance will have the dominant phenotype. Their genotypes will be different, however, since one is homozygous, the other heterozygous.

6. The factors for any given trait can exist in any organism in one of several combinations: homozygous dominant, homozygous recessive, or heterozygous. Homozygous means that both factors are alike, whether both are dominant or both recessive. Heterozygous means that the factors are different. In complete dominance, homozygous dominants and heterozygotes look alike; only the homozygous recessives appear different. In incomplete dominance, the homozygous dominants show the dominant phenotype, while the heterozygotes appear as an intermediate phenotype between the dominant and recessive condition. Homozygous recessives show the pure recessive trait. The terms homozygous and heterozygous thus refer to the genetic makeup, or genotype, of the individual.
7. In determining offspring ratios from any given cross, both phenotype and genotype can be used. It is very important to indicate which ratio is meant. Usually ratios are determined by looking at the phenotype only. Mendel's ratio of 3:1 in crosses between two heterozygous tall pea plants is a phenotype ratio. The genotype ratio is 1:2:1 (1 homozygous tall; 2 heterozygous talls; and 1 homozygous recessive).

EXERCISES

1. A spotted rabbit and a solid-colored rabbit were crossed. They produced all spotted offspring. When these F_1 rabbits were crossed, the F_2 consisted of 32 spotted and 10 solid-colored rabbits. Which characteristic is determined by a dominant gene?
2. What proportion of the F_2 spotted rabbits in Exercise 1 would be heterozygous? Homozygous? How many of the F_2 solid-colored rabbits would be expected to be homozygous?
3. What method would most easily tell which of the spotted rabbits in the above problem were heterozygous and which homozygous? Is there any other method? If so, what is it?
4. If we cross a heterozygous tall pea plant with a dwarf pea plant, what proportion of the F_1 will we expect to be tall?
5. Suppose we cross one dwarf pea plant with another. What proportion of the F_1 will be expected to be dwarf?
6. A brown mouse is crossed with a heterozygous black mouse. If the mother has a litter of four, what are the chances that all of them will be brown?
7. What are the probable genotypes of the parents in a cross that gives a 3:1 ratio? A 1:2:1 ratio? A 1:1 ratio? (Use the symbols B and b).
8. A cross between two mice produced an F_1 with a ratio of one brown to one black. What are the probable genotypes of the parents? Of the progeny?
9. In horses, black is due to a dominant gene B, chestnut to its recessive allele b. The trotting gait is due to a dominant gene T, pacing to its recessive allele t. A homozygous black trotter is crossed with a chestnut pacer. What sort of foals will result in several crosses of this sort so far as coat color and gait are concerned?
10. If two of the F_1 animals in Exercise 9 are crossed, what kinds of foals can be produced? In what proportion will each phenotype be expected to appear?

11. In poultry, black color is due to a dominant gene *E* and red color to its recessive allele *e*. Crested head is due to a dominant gene *C*, plain head to its recessive allele *c*. A male bird, red and crested, is crossed with a black, plain female. They produce many offspring, half of which are black and crested, the other half red and crested. What would you infer about the genotypes of the parents?

12. A mating is made between two black, crested birds. The F_1 contains 13 offspring in the following proportions: 7 black, crested; 3 red, crested; 2 black, plain; 1 red, plain. What are the probable genotypes of the parents?

13. In humans, aniridia, a type of blindness, is due to a dominant gene. Migraine, a headache condition, is the result of a different dominant gene. A man with aniridia, whose mother is not blind, marries a woman who suffers with migraine but whose father does not. In what proportion of their children would *both* of these conditions be expected to occur?

14. Suppose that a woman who is not blind, but whose parents both suffer from aniridia, goes to a genetic expert for advice. She suffers from migraine, which her father also has. She wants to know what the chances are that her children will have aniridia or migraine. What would the geneticist tell her?

15. In cocker spaniels, black coat color is due to a dominant gene *B*, reddish-tan coat to its recessive allele *b*. Solid coat color is determined by a dominant gene *S*, while white spotting is determined by the recessive allele *s*. A black and white female was mated to a reddish-tan male. The litter contained five puppies: one black, one reddish-tan, one black and white, and two reddish-tan and white. What are the genotypes of the parents?

16. In summer squash, colorless fruit is due to a dominant gene *W*; colored fruit is due to its recessive allele *w*. Disc-shaped fruit is determined by a dominant gene *S*, sphere-shaped fruit by its recessive allele *s*. How many different genotypes may squash plants have in regard to color and shape of fruit? How many categories of phenotypes could be expected from their genotypes? How many different homozygous genotypes are possible? What phenotypic ratio would you expect from a cross between two heterozygous plants?

SUGGESTED READINGS

Levinson, Horace, *Chance, Luck and Statistics.* New York: Dover Publications, 1963. A well-written book discussing probability in relation to many types of games. This book is highly recommended to any student who is interested in the theory of probability.

Srb, A. M., R. D. Owen and R. S. Edgar, *General Genetics*, 2d ed. San Francisco, W. H. Freeman, 1965. An up-to-date account of both classical (Mendelian) and modern (molecular) genetics, aimed at upper-level genetics students.

Strickberger, M. W., *Genetics*, 2d ed. New York: Macmillan, 1976. A very fine text, more detailed than Srb, but readable. Contains useful problems.

NINE

GENETICS II: GENES AND CHROMOSOMES

9.1 Introduction

A good hypothesis must explain the phenomenon to which it pertains. Mendel's hypothesis, involving the segregation and random assortment of genetic "factors," pertained to the inheritance of contrasting characteristics in peas (tall and short plants, round and wrinkled seeds, etc.) His hypothesis did indeed account for the results very nicely; it also allowed him to predict accurately the ratios of the phenotypes of offspring resulting from crosses between parent plants of known genotypes.

However, a really good hypothesis about inheritance must be applicable to more than just one kind of organism and more than just a few simple inherited characteristics. Mendel's hypothesis meets the first of these challenges, since it accounts for the inheritance of coat variations in mice. But this is still a simple inheritance situation, involving the activity of only a few gene pairs. In this and the next chapter, we focus attention on the applicability of Mendel's hypothesis to the inheritance of more complex traits. And, more important, we shall tackle the two most fundamental questions of modern genetics—What *is* a gene, and how does it work?

9.2 Linkage

To account for the phenotypic ratios he observed in his pea plants, Mendel proposed a purely intellectual model based on segregating and randomly assorting factors. Later the behavior of chromosomes during meiosis was seen to parallel the behavior of Mendel's hypothesized factors, and his intellectual model seemed to attain reality.

The close parallel between the behavior of Mendel's factors (genes) and the behavior of the chromosomes during meiosis might lead to a tentative hypothesis that a gene and a chromosome are one and the same thing. We can reason that if a chromosome is the same thing as a gene, then chro-

Mendel never observed chromosomes, nor did he know anything about the events of mitosis and meiosis. His hypothesis was based solely on observed phenotypic ratios among the offspring from specific breeding experiments.

mosomes should demonstrate segregation during gametogenesis; furthermore, we can also predict that the members of separate pairs of homologous chromosomes should show random and independent assortment into the prospective gametes. Both of these predictions are verified.

Other observations, however, seem inconsistent with this hypothesis. Most obvious is the chromosome number. A red fox, for example, has only 34 chromosomes. It is difficult to imagine that all its inherited characteristics could be controlled by only 17 pairs of genes. Further, the arctic fox has 52 chromosomes. Such a wide variation in the number of "chromosome-genes" would not be predicted between these closely related forms. Finally, there seems to be no consistent principle underlying the variation in chromosome number between organisms of widely differing evolutionary status; this fact becomes extremely difficult to explain if we are to accept the hypothesis of oneness between chromosome and gene. For example, it is difficult to see why a one-celled radiolarian should have 1600 "chromosome-genes," while a crayfish has 200, and a human only 46.

There is one very simple and obvious way out of this dilemma. The hypothesis that a chromosome and a gene are one and the same thing can be modified to propose that *a chromosome represents several genes, and that these genes are located in a linear order along it.*

The first portion of this hypothesis, that a chromosome represents several genes, nicely overcomes the objections just raised to the one chromosome-one gene hypothesis. The second portion, dealing with the placing of genes on the chromosomes, is quite another matter. Such an hypothesis necessarily leads to certain predictions: if there are several genes located on a chromosome, then certain characteristics should tend to be inherited together. The reason for this prediction becomes apparent if the chromosome is visualized as a string of beads, each bead representing one gene. Since these genes are joined (or **linked**) together, this hypothetical model predicts that *wherever one gene on a chromosome goes, so must all the other genes on that chromosome.*

In seeking support for this hypothesis, it is reasonable to ask whether any cases have been definitely established in which one genetic characteristic always appears with another. It is easy to feel intuitively that there are such cases. In humans, for example, we generally associate the occurrence of freckles with sandy or reddish hair. It might be proposed, therefore, that the genes influencing freckles and those influencing red hair are linked—that they are on the same chromosome.* However, humans have many chromosomes and cannot be bred experimentally, so such a linkage is difficult to establish.

It is necessary, therefore, to turn to an organism that has fewer chromosomes and adapts more easily to controlled breeding. The tomato plant, with 12 chromosome pairs, is such an organism. In tomatoes, tall growth habit is the result of a dominant gene D; dwarf growth habit is the result of a recessive allele d. Smooth epidermis is due to a dominant gene P; pubescent (hairy) epidermis is

* It is also possible, of course, that freckles and red hair may have the same underlying genetic cause, in which case they would be an example of **pleiomorphism.**

due to a recessive allele *p*. As would be predicted, crosses between homozygous tall smooth plants (*DDPP*) and dwarf pubescent ones (*ddpp*) yield an F_1 of all tall smooth plants (genotype *DdPp*). Thus far, then, Mendel's first and second laws regarding segregation and random assortment are supported.

Assume now that a cross is made between a tall smooth F_1 tomato plant and a dwarf pubescent tomato plant. The F_1, a tall smooth plant with genotype *DdPp*, should produce four types of gametes, *DP*, *Dp*, *dP*, and *dp*. The dwarf pubescent plant could produce only one type of gamete, *dp*. The fertilization possibilities are thus:

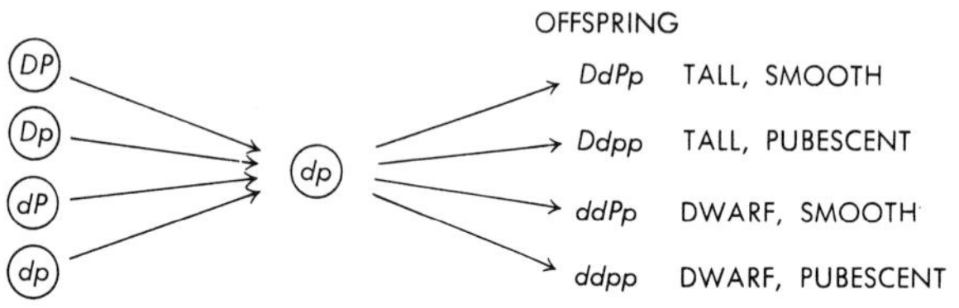

which yields a ratio of one tall smooth, to one tall pubescent, to one dwarf smooth, to one dwarf pubescent.

Such a cross has been made, yielding 112 plants. Of these, 54 were tall smooth plants and 58 were dwarf pubescent plants, a ratio of approximately 1:1.

Clearly Mendel's second law is contradicted by these results; random assortment of the factors for height and skin texture cannot have occurred. Instead, wherever the factor for tallness went, the smooth-skin factor must have followed. Likewise, wherever the factor for dwarfness went, the pubescent factor went also. Such a result, and the 1:1 ratio of tall smooth to dwarf pubescent plants, would be predicted by an hypothesis that the genes for tallness and smooth skin were located on the same chromosome—that they were linked. According to this hypothesis, the genes for dwarfness and pubescent skin must also be linked.

The experimental results just cited are matched by similar results obtained with certain inherited traits in other plants and animals. Such results strongly support the hypothesis of gene linkage and force us to impose a strong qualification on Mendel's second law: *Genes assort at random if and only if they are located on separate chromosomes.* Thus in many organisms, particularly those with few chromosomes, gene linkage is the rule rather than the exception.

Of course, it may be that the genes for tallness and smooth skin are not only on the same chromosome but are actually the same gene. The same might be true for dwarfness and pubescent skin. As we will see in the next section, however, these hypotheses yield predictions that are sharply contradicted.

9.3 The Fruit Fly Era

In the period between 1910 and 1940, geneticists turned to the fruit fly, *Drosophila melanogaster*, for an intensive study of the chromosomal basis of inheritance. This small insect, often seen around decaying fruit, has a short life cycle (approximately two weeks) and is easily raised in the laboratory. *Drosophila* also has a low chromosome number (four pairs), with each pair easily distinguishable from the others. Finally, and most important, *Drosophila* shows hundreds of inherited variant characteristics.

When certain traits appear to be inherited together a large part of the time, they are said to be linked. Linkage provides an exception to Mendel's second law (the law of independent assortment).

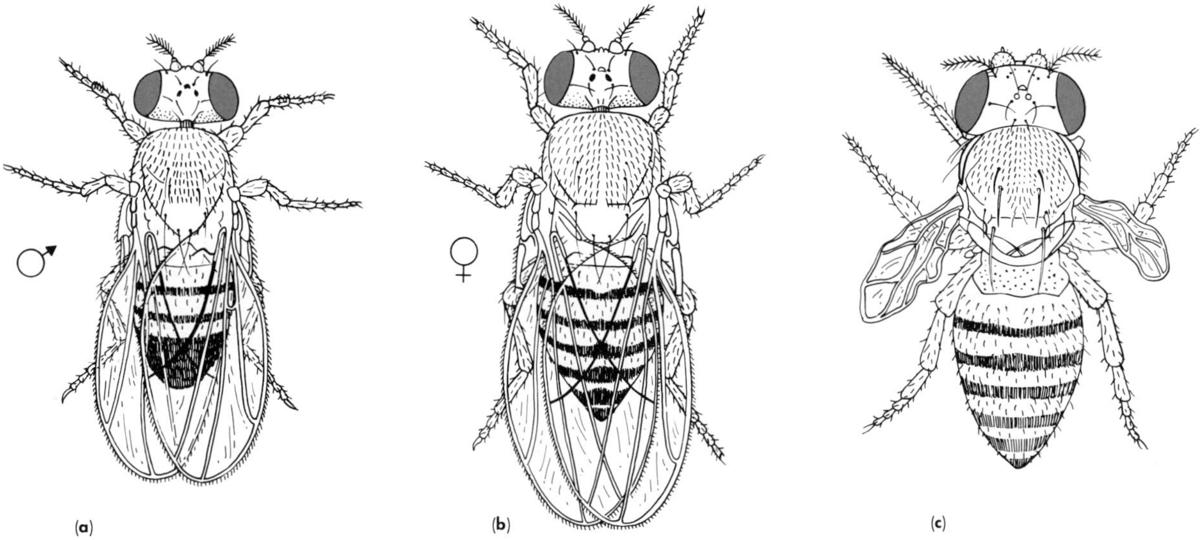

Fig. 9.1 *The fruit fly,* Drosophila melanogaster. *Normal (wild-type) male (a) and female (b) show the wild-type characteristics of straight wings and red eyes. A vestigial-wing mutant is shown in (c); the much shorter wings are nonfunctional. The vestigial mutation is recessive to wild-type normal wings and is located on the second chromosome pair.*

One such variant characteristic is vestigial wings. Flies with this characteristic have only stumps where their wings should be (Fig. 9.1). The condition is caused by the presence of a single pair of recessive genes, for crosses between vestigials yield only vestigials, while crosses of purebred winged flies with vestigials yield all winged. The F_1 intercross yields the familiar 3:1 ratio of dominant to recessive phenotypes.

A major center for fruit fly research was the Columbia University laboratory of Thomas Hunt Morgan (1866–1945) and his associates. Their work, genetically speaking, made *Drosophila melanogaster* the most thoroughly understood multicellular organism in the world. A part of Morgan's work focused on the inheritance of eye color in *Drosophila*. The regular, or "wild type" insect has red eyes. One day a white-eyed male fly appeared in one culture bottle. It was crossed with a red-eyed female and the resulting progeny were all red-eyed. The F_1 intercross yielded a ratio of three red-eyed flies to one white-eyed fly. It was noticed, however, that *all the white-eyed flies were male.*

Since the location of the genes on the chromosomes had already been hypothesized, it was reasonable to look for differences in the chromosomes of male and female fruit flies to explain the connection of the white-eyed condition to maleness. Morgan was already familiar with such a difference. In the male fruit fly, one pair of chromosomes is markedly different from the other three pairs. This pair consists of one normal-appearing chromosome, the X chromosome, and one short, bent chromosome, the Y chromosome. In the female, two matching X chromosomes are found.*

Morgan saw immediately that the occurrence of white-eyed males in the F_2 generation could be explained by postulating the white-eye gene to be recessive and located on the X chromosome. The Y chromosome, being shorter, has no locus for the gene. Thus the mere occurrence of the white-eye gene would be enough to cause white eyes in the male, since no dominant gene for red eyes would be present to override it:

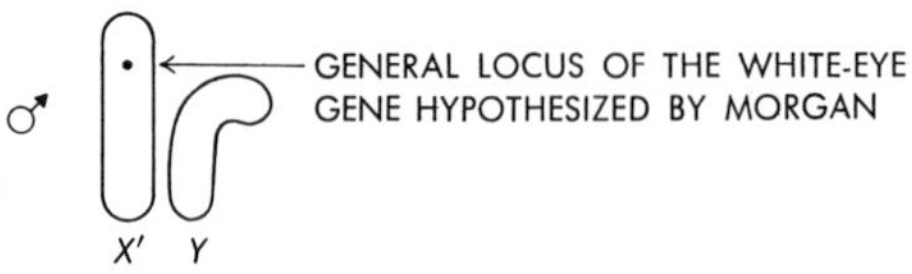

[The prime (′) symbol on the X indicates the presence of the recessive white-eye gene.] Morgan called this condition **sex-linkage,** or **sex-linked inheritance,** to indicated a condition determined by genes on the X chromosome.

In the female, with two X chromosomes, white eyes would occur only rarely. Even if the white-eye gene were present on one chromosome, it would be masked by the dominant red-eye allele on the other, and this female would still be red-eyed:

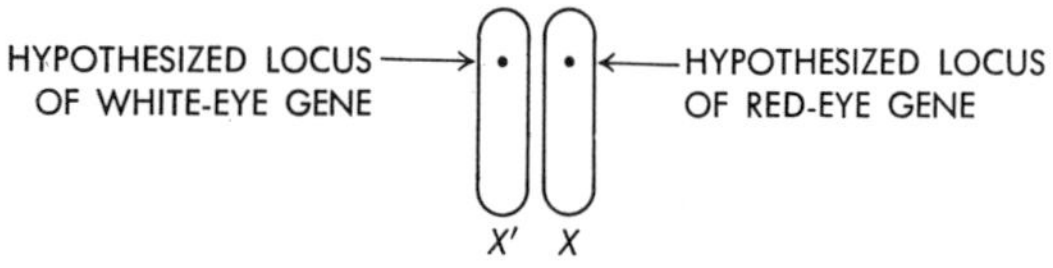

Only occasionally, when a female contained the white-eye mutation on *both* X chromosomes, would it be visible. Morgan thus pictured the first cross of the white-eyed male with the red-eyed female as follows:

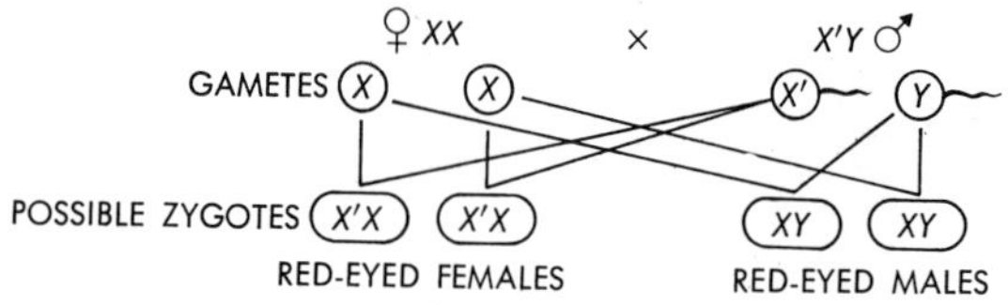

Crossing the F_1 flies resulted in a ratio of three red-eyed flies to one white-eyed fly, with white-eye appearing only in the male flies.

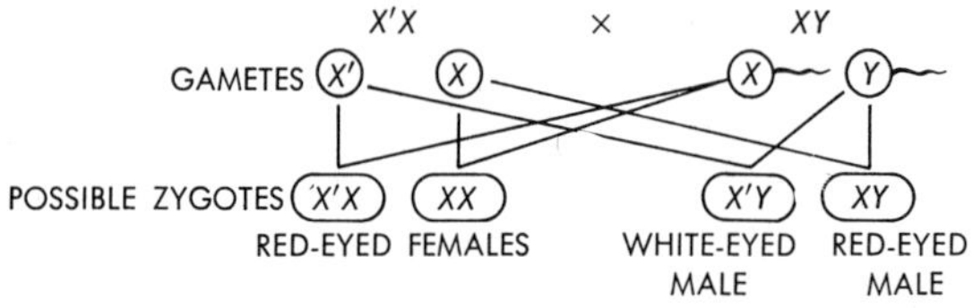

Thus Morgan's hypothesis, placing the locus of the recessive white-eye gene on the X chromosome, nicely accounted for his experimental results. But it also predicted the possibility of obtaining white-eyed females. If the locus of the recessive white-eye gene is on the X chromosome, then crosses between white-eyed males ($X'Y$) and red-eyed females whose fathers were white-eyed ($X'X$) should produce a 1:1 ratio of red-eyed to white-eyed flies. Of the white-eyed flies, half should be females. In terms of the Punnett square, it can be represented:

	X′	X
X′	X′X′ WHITE-EYED FEMALE	X′X RED-EYED FEMALE
Y	X′Y WHITE-EYED MALE	XY RED-EYED MALE

The prediction is verified; the hypothesis is supported.

In humans, both red–green color-blindness and hemophilia (a condition of the blood that prevents

* At the time of Morgan's work on the white-eye condition, the Y chromosome had not been detected in *Drosophila.* Morgan therefore assumed the X chromosome had no homolog in the male. In terms of the hypothesis he proposed, the presence or absence of a Y chromosome makes little difference, so we shall deal with his hypothesis as if he had been familiar with the Y chromosome in this insect.

it from clotting properly) are sex-linked conditions. Studies of family histories show that the conditions are recessive, are rare in females, and are transmitted to an affected man's grandchildren through his daughters. Morgan's hypothesis, devised solely to explain the inheritance of the white-eyed condition in *Drosophila*, also accounts for the inheritance of red–green color-blindness and hemophilia in humans. His work is but one of many examples of basic research that result in unforeseen benefits to the human species.

9.4 Broken Links and the Gene–Chromosome Theory

If the same cross between a tall smooth F_1 tomato plant and a dwarf pubescent plant is carried out many times, some tall pubescent and dwarf smooth plants *do* appear. In a total of 402 progeny, a typical phenotype breakdown is as follows: 198 tall smooth, 8 tall pubescent, 6 dwarf smooth, 190 dwarf pubescent. The appearance of these recombinants contradicts the hypothesis (mentioned at the end of the last section) proposing that "tall smooth" and "dwarf pubescent" may each be due to one rather than two genes. Thus there seems no way to explain the appearance of these few tall pubescent and dwarf smooth recombinant plants other than to assume that occasionally *linkage can be broken*. By itself, however, this assumption is not enough. We must go further and hypothesize not only that linkage can be broken (that chromosomes break), but that *the broken pieces must actually be exchanged between homologous chromosome pairs*. This hypothesis, posed to explain the occasional occurrence of phenotypes that an unmodified linkage hypothesis would predict to be impossible, leads us to examine the behavior of the homologous chromosome pairs themselves. Do they ever assume positions that might be conducive to an exchange of portions of their length?

The reader may recall from Chapter 7 that the answer to this question is yes. Prior to the first meiotic division, during tetrad formation, homologous chromosomes often form chiasmata. What might have happened with the tomato plants can be visualized as follows. Most of the time, the following occurs:

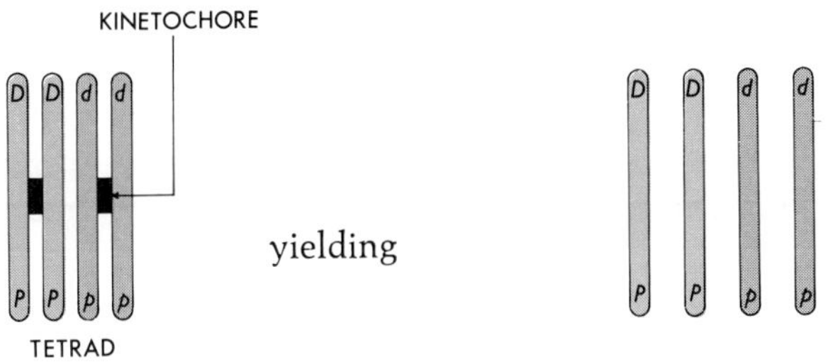

Therefore, only *DP* and *dp* gametes are produced. Occasionally, however, the following occurs:

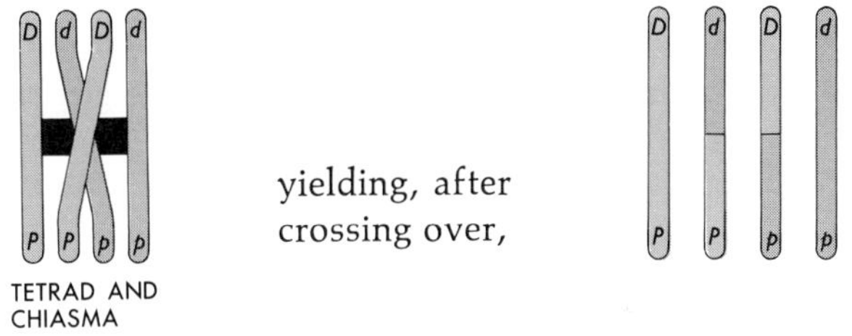

Therefore a few *Dp* and *dP* gametes are produced, as well as the expected *DP* and *dp* gametes.

The chiasmata formed by homologous chromosomes within living cells can be observed under the microscope. Once again, a hypothetical model constructed to explain observed genetic ratios has physical reality in actual chromosome behavior.

The idea has now been well established for eucaryotic (and many procaryotic) organisms that chromosomes consist of linear arrays of genes with a specific order characteristic for each chromosome pair. The lines of evidence contributing to this concept are both direct and indirect. Among the indirect lines is the fact that chromosomes for many different organisms have been mapped. The process of **mapping** takes advantage of crossing over and exchange of parts between homologous chromosomes during meiosis. The mapping procedure gives clear results: gene locations can be determined with great accuracy. That mapping works is only indirect evidence, however, for the linear arrangement of genes. Mapping does not "prove" linear order; it is only logically consistent with it. In fact, without linearity, the procedure for mapping could not work. The mapping process is a logical consequence of linearity.

Direct evidence for the existence of genes on chromosomes, and for a linear order, comes from the field of cytogenetics. The genetic maps described above provide only relative distances of genes apart—given in arbitrary units known as **map units.** Genetic maps say nothing about the location of any single gene on any particular part of a chromosome. However, it is possible to determine the correspondence between specific genes and specific chromosomal regions by taking advantage of what are called **chromosomal aberrations.** Chromosomal aberrations are structural changes in chromosomes that affect synaptic patterns and phenotypic ratios. By correlating certain chromosomal changes with particular alterations in expected ratios, direct correspondence of gene with chromosome segment can be made.

9.5 Chromosome Mapping in Eucaryotes

The mapping of gene loci is not as difficult as it might seem. Indeed, the process of mapping takes advantage of crossing over during meiosis. If genes are located in a consistent linear order along a chromosome, then the amount of crossing over that occurs between any two genes should be in direct proportion to the distance between them. We detect crossing over when it leads to exchange of chromosome parts and consequently to recombination of traits in the offspring. The more exchange between any two linked genes in the gamete-producing cells of the parents, the greater the proportion of offspring showing the crossover phenotype. This percentage gives a direct measure of the amount of chromosomal crossing over. This, in turn, gives a direct measure of the distance between the genes involved. The distance is of course a *relative* one—formed by translating percentage of crossover numbers into "map units."

A New System of Symbols

Thus far we have been using the classical Mendelian symbols for genes (capital letter for the dominant form, small letter for the recessive). Twentieth-century geneticists have gradually evolved another, shorthand system, which is now in common use. The wild-type form of any characteristic is usually symbolized by +; the mutant form by a letter, such as a, b, c, w, B, etc. The letter is usually the first letter of the word describing the mutation (such as w for white eye, m for miniature wing), or B for bar eye). If the mutant form is recessive to the wild type, a small letter is used; if the mutant is dominant over the wild-type, a capital letter is used.

Characteristically, these symbols are used to indicate either phenotype or genotype. If the symbols are used alone (abc, or mwf, or +++), then they refer to phenotype only. If, however, they are on one side of a straight line (for haploid cells or organisms) on both sides of a straight line (for diploid cells or organisms), they indicate genotype; for example:

$\underline{\text{a b c}}$	$\dfrac{\text{a b c}}{+\ +\ +}$
HAPLOID	DIPLOID

The line represents the chromosome on which the genes are linked. For diploid organisms, the symbols on each side of the line represent the particular combination of genes (for whatever traits are being observed) located on one member of the chromosome pair; the symbols on the other side of the line represent the array of genes on the homologous chromosome. In the case of the above diploid organism, for the three traits symbolized by a, b, and c, one chromosome has all three mutant genes, while the other has all three wild-type genes. Of course, it is not necessary for all the mutant genes to be found on one chromosome and all the wild-type genes on the other. We could have any of the following combinations for diploid organisms where three pairs of genes are involved:

$$\frac{a\ b\ +}{+\ +\ c},\ \frac{a\ +\ c}{+\ b\ +},\ \frac{+\ +\ c}{a\ b\ +}.$$

Each combination represents a different genotype, though all have the same phenotype.

Finding the Frequency of Crossing Over

To understand how a genetic map is constructed, consider the following hypothetical example using three genes, a, b, and c, located on one pair of chromosomes in the fruit fly, *Drosophila melanogaster*. The original cross is made between two flies, one of which is homozygous wild-type for all three traits; phenotypically it would be symbolized as + + +, genotypically as

$$\frac{+\ +\ +}{+\ +\ +}.$$

The other fly is homozygous recessive for all three traits and would be written phenotypically as abc; genotypically it would be written as

$$\frac{a\ b\ c}{a\ b\ c}.$$

The order of the genes is not necessarily that shown here (abc); it could be bac or bca. The ordering of the genes is the unknown. We must determine that before we can begin to calculate map distances.

The original cross gives an F_1, as shown:

ORIGINAL CROSS $\quad \dfrac{+\ +\ +}{+\ +\ +} \times \dfrac{a\ b\ c}{a\ b\ c}$

$F_1 \quad \dfrac{+\ +\ +}{a\ b\ c}$ and $\dfrac{+\ +\ +}{a\ b\ c}$

The function of the first cross is simply to get a number of heterozygotes in whose germ cells crossing over can be detected. (Crossing over is assumed to take place in homozygotes, too. However, since the exchange is between identical parts, no phenotypic change can be observed). If F_1's from the above cross are now mated with homozygous recessive forms $\left(\dfrac{a\ b\ c}{a\ b\ c}\right)$, any crossovers will show up as recombinations—that is, as any of the following phenotypes: ab+, ++c, a+b, or +b+. The proportion of these various recombinations, as compared to the number of nonrecombinant types, provides a measure of the *frequency* of crossing over. The data in Table 9.1 give the number of flies showing each phenotype with regard to the three genes. The first aspect of the data we can look at is the category called "Double Crossover," *determined as the smallest phenotype category for any cross (where thousands of offspring are produced).*

Finding Gene Sequence

The double crossover class can be used to indicate the *sequence* of genes on the chromosome. If the order of genes is that shown above (i.e., abc), a double crossover would involve two single crossovers—one between a and b and the other between

Table 9.1

Phenotype:	+++	abc	a++	+bc	ab+	++c	a+c	+b+
Totals	436	741	107	122	313	236	390	382
	1177		229		549		772	
	Parental types		Double crossovers		Single crossovers			

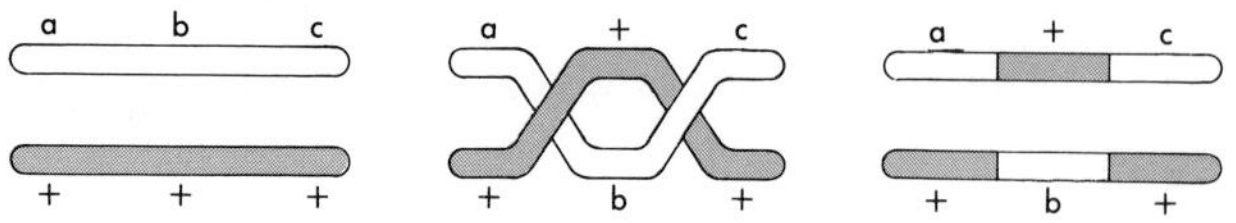

Fig. 9.2 *The recombination results of double crossing over.*

b and c—as shown below in Fig. 9.2. Thus we would expect the double crossover classes to show the phenotype combinations of a, +, and c on the one hand, and +, b, and + on the other. However, as the data shown in Table 9.1 indicate, this is not the double crossover category, hence the order is not abc. Only one particular order of any three genes can yield any given double crossover combination. By trial and error, we can determine relatively easily that the order bac is the only one that could give double crossover categories of a++ and +bc, as shown below in Fig. 9.3.

Once the order of genes is determined, the relative distances apart can be calculated. Crossovers between b and a are found in the last two columns of Table 9.1. (Note the reasoning involved here: If bac is the correct sequence, then a crossover between b and a would yield phenotype recombinations of b++ and +ac; these are indeed the combinations found in the last two columns). The number of crossovers between b and a is determined first by adding together the values for each group (390 + 382) to give the total of 772. But this does not represent quite all of the single crossover occurrences between genes b and a. Every double crossover involves two single crossovers. Thus, to be accurate, the number of double crossovers must also be included when considering the total number of crossovers. To 772, then, we should add 229, since in each of these latter cases a single crossover between b and a also occurred. The total of 1001 (772 + 229) is then divided by the total number of F_2 counted (for all 8 categories). Since this was 2727 (1177 + 229 + 549 + 772), the fraction 1001/2727 = 0.36 or 36% crossover between b and a. Thus b and a are 36 "map units" apart on the chromosome. The crossover percent between a and c is found in the fifth and sixth columns of Table 9.1. This is calculated in the same way as before; thus, 549 + 229 = 778: 778/2727 = 0.28, 28%, or 28 map units. The appropriate genetic map would be as shown in Fig. 9.4.

Validity and Utility in a Concept

The concept of genetic mapping is a highly imaginative one, and it is based on a number of assumptions. One of these is the linearity of gene arrangement along the chromosome. Another is that the chances of breakage are about equal anywhere along the length of the chromosome. If some sections tended to break more easily than others, the

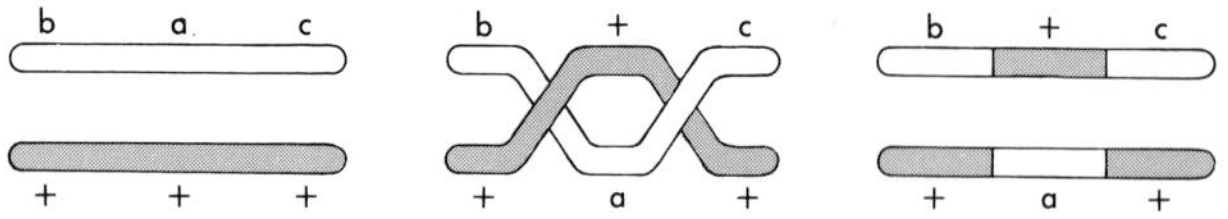

Fig. 9.3 *Recombination possibilities from double crossing over if gene order is bac.*

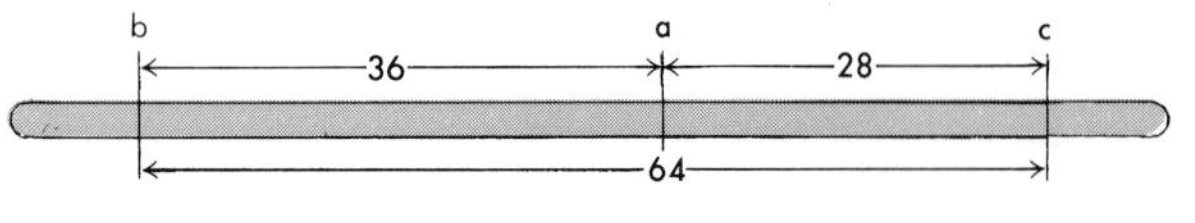

Figure 9.4

Supplement 9.1

CHROMOSOMES: THE MATERIAL BASIS OF MENDEL'S GENES

The fact that genes could ultimately be shown to exist as specific **loci** (singular, "locus") or points, at distinct positions on a given chromosome, is due largely to two lines of investigation. One involves the cytological study of chromosomal aberrations, the other the cytological study of the fine structure of individual chromosomes. Both contribute direct evidence in support of the idea that genes are physically a part of the chromosome structure.

As we have seen in earlier chapters, chromosomes are small, though visible, structures within the cell nucleus. Observations of chromosomes from most cells in the body of any eucaryotic organism indicate that each pair of chromosomes has its own characteristic size, shape, and banding pattern. But the chromosomes are generally too small to allow the cytologist to make a detailed study of the bands. The so-called "fine structure" of the chromosomes remained largely a mystery until the late 1920s.

At that time several workers, most notably Theophilus S. Painter (1889–1969), then at the University of Texas, found that the fruit fly, *Drosophila melanogaster*, offered a rich source not only for breeding experiments but also for chromosome studies. Painter and others found that the salivary gland cells of older *Drosophila* larvae (see Diagram A) contained extremely large chromosomes. Their increased size is due to the fact that although the chromosome replicates numerous times, the replicated strands do not separate. Thus the "giant" salivary gland chromosomes are in reality several hundred duplicate strands bound together. This means that the banding structures are magnified a hundred or more times even before the cytologist turns a microscope on them! The discovery of salivary gland chromosomes greatly aided the study of the fine structure of chromosomes. In particular, it made possible the detailed analysis of structural changes known collectively by the term "chromosomal aberrations."

Chromosomal aberrations include deletions, duplications, translocations, and inversions. These various aberrations can occur from time to time during meiosis in the gamete-producing cells of adult organisms. The aberrant chromosome is passed on to the next generation through the egg or sperm. Depending on the nature of the homologous chromosome with which it pairs, the aberrant chromosome may or may not affect the phenotype of the offspring inheriting it. Through analyzing the relationship between phenotypic variations and specific chromosomal aberrations, further support has been found for the idea that

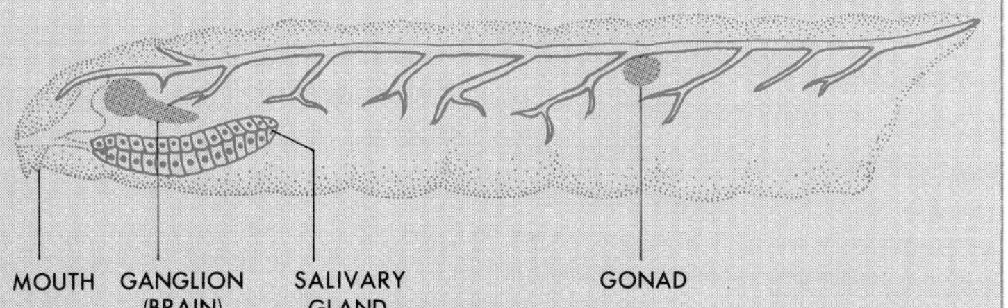

Diagram A *Lateral (side) view of a* Drosophila *larva, showing position of salivary glands, in whose cells the "giant" chromosomes are located. A photograph of a chromosome from the salivary gland, showing considerable fine structure, appears in Fig. 4.29.*

chromosomes are the physical bearers of genes. Let us consider two aberrations as an example of how such studies support the chromosome theory. Our two examples will be deletions and inversions.

Deletions

Occasionally a broken piece of chromosome fails to become attached to another chromosome and is simply lost. Such occurrences, called **deletions,** offer a unique chance to discover gene loci. For example, if the absence of a particular chromosome segment is always accompanied by a certain phenotype deficiency, it is reasonable to suppose that the gene normally responsible for preventing this deficiency is located on the missing piece.

A deletion may also produce evidence for the linearity of genes on the chromosomes. Assuming linearity of genes on homologous chromosomes, it follows that genes influencing the same trait will pair off opposite each other during synapsis

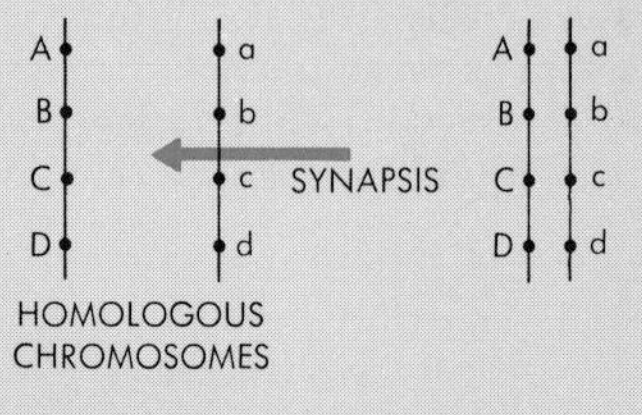

Diagram B

prior to the first meiotic division (see Diagram B). Suppose that gene B is deleted from one of the homologs. In order for similar genes to pair, an unusual synapsis figure must occur, such as shown in Diagram C. Such synaptic figures have been observed in chromosomes after a segment of one chromosome has been deleted, a fact that strongly supports the hypothesis that genes are

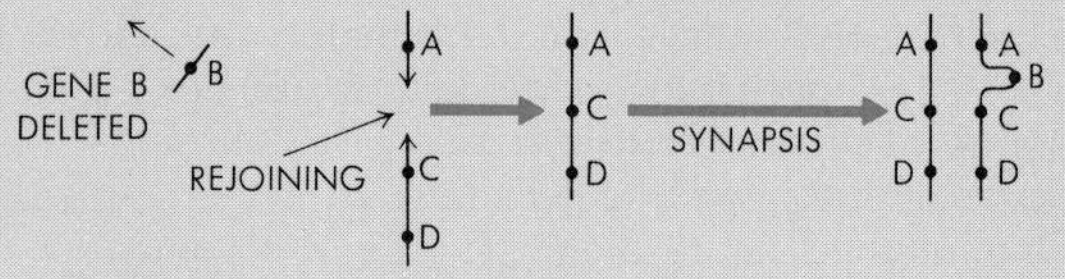

Diagram C

arranged in a linear fashion on the chromosomes. Moreover, in the above case the organism shown in Diagram C will phenotypically display the mutant recessive *b* trait, due to the deletion of the *B* gene from the homologous chromosome. Thus the observation of a phenotypic change in the offspring can be correlated with a change in a specific region of a chromosome. This is strong evidence that genes are located on chromosomes.

Inversions

Occasionally an entire midsegment of a chromosome may break and rejoin in a completely reversed position, as shown in Diagram D. Such

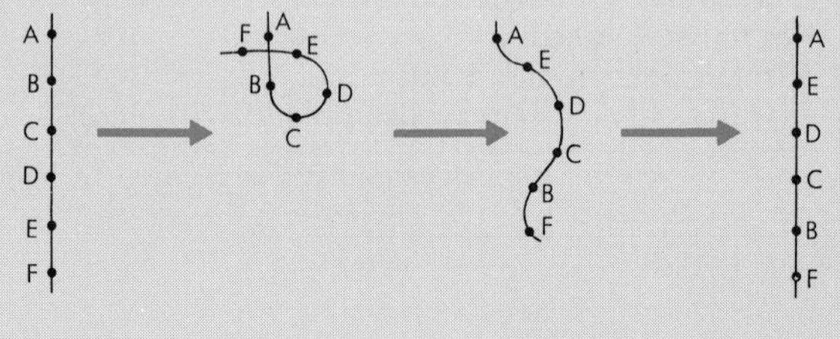

Diagram D

an occurrence is known as an *inversion.* Since it is extremely unlikely that the homolog of an inverted chromosome would undergo a similar inversion, we can predict that the chromosomes will have to be greatly contorted before proper synapsis can take place. Using line model chromosomes, one of which has the inversion shown in

Diagram D, at least two theoretically possible synaptic figures may be drawn as shown in Diagram E. Such postinversion synaptic figures have actually been observed in corn.

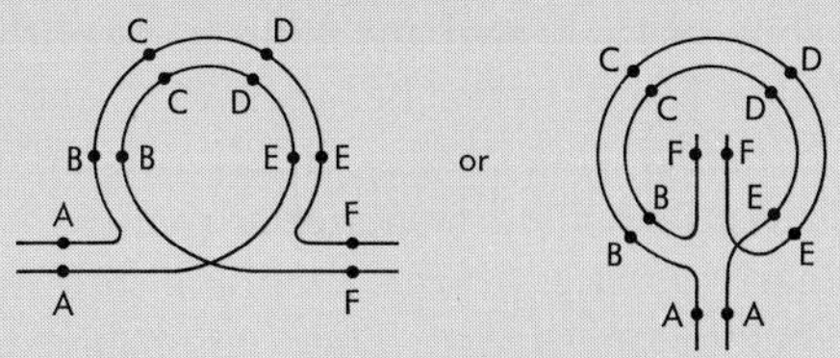

Diagram E

The existence of inversion suports the idea of a linear arrangement of genes on chromosomes in two ways. First, since chromosomes tend to synapse with corresponding genes across from one another, the unusual synaptic patterns of postinversion chromosomes are in agreement with predictions. If the genes were not parts of chromosomes and if they were not linearly arranged, such synaptic contortions in inverted chromosomes would not have to occur. Second, inversions seldom show crossing over within the inverted region of the chromosome. This means that recombination for the traits contained within the inverted region is effectively suppressed. When inversion strains of the fruit fly are bred, for example, fewer recombinants are indeed found for those very traits contained within the inverted region of the chromosome. Again, support for the idea of genes as physical parts of chromosomes arranged in a linear fashion comes from a correlation between genetic and cytological data.

Diagram F *Detailed comparison between the genetic map (above) and a corresponding section of the salivary gland chromosome of* Drosophila *(below). The genetic map is based on analysis of cross-over frequencies from breeding studies. The loci determined in this manner are relative to one another and do not have any distinct physical location. The chromosome map is based on cytological analysis of giant salivary gland chromosomes; rearrangements of banded sections are compared to observed phenotypic changes in particular traits. In this way the corresponding gene locations between the genetic and cytological maps are determined. The letters on the genetic map (above) symbolize specific genes. The letters below the chromosome map show the physical location of the gene on the chromosome.*

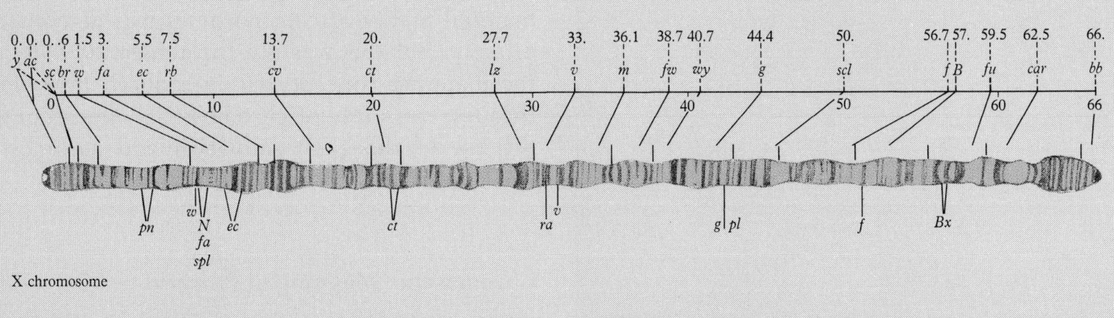

How has all this been aided by discovery of the salivary gland chromosomes? The availability of large, easily observable chromosomes from an organism about which so much breeding data existed made possible the detailed correlation between phenotypic and chromosomal variation. Without the giant salivary gland chromosomes, it would have been impossible to observe that inversions had actually taken place in the chromosome. Detecting an inversion cytologically requires being able to see the bands in enough detail to actually recognize when they are out of place. Such detailed observations are possible only on the giant salivary gland chromosome. By making such detailed comparisons between breeding data (crossover frequencies) and cytological study of giant chromosomes, a detailed correspondence between genetic and cytological maps is possible (see Diagram F).

fact would tend to alter distance calculations. These assumptions were explicitly recognized by some of the early workers who pioneered in the mapping techniques. The model worked—it gave consistent values for map distances. But there was no direct evidence to demonstrate that the model actually corresponded to real events at the chromosomal level. Many critics of the early theories of gene mapping complained that the process was purely hypothetical, since it could not be shown that genetic maps necessarily had anything to do with real chromosomes. Many critics between 1912 and 1920 felt that a model was useless in science unless it could be directly related to material reality.

The early critics of the mapping procedures were correct in one sense. They emphasized that such a concept was of no value unless it could ultimately be shown to represent reality. However, the critics missed a main point about the use of models in science (or any field of thought, for that matter). Models can be valuable in suggesting further lines of work even if the material reality of the model cannot be shown at the moment. The model of the chromosome as a linear array of genes led to the investigation of many problems in classical genetics: the interaction of genes, the effect of position of genes on their expression, the concept of sex linkage, and many others. Thus the model served a distinct purpose in the history of our ideas about gene structure and function. Ultimately, of course, its lasting value came from the fact that the assumed relationship between the model and the material reality of chromosomes was demonstrated directly. But this was not for almost 20 years after the mapping concept was introduced into biology through the study of chromosomal aberrations and the use of the giant salivary gland chromosomes of *Drosophila*.

9.6 Chromosome Mapping in Procaryotes (Bacteria)

In studying the genetic makeup of procaryotic cells, some different techniques are employed from those used with eucaryotic organisms such as *Drosophila*. Let us consider the bacteria.

Bacteria

Joshua Lederberg and E. L. Tatum devised a procedure for mapping the bacterial chromosome in the 1940s. They took advantage of the fact that most bacteria "mate." That is, two bacteria come to lie close together (conjugate), and form a bridge between them. Some, but usually not all, of the genetic material from one bacterium (the donor strain) passes over the bridge into the recipient cell. What Lederberg and Tatum noted was that the donor chromosome always passes across the bridge at the same rate of speed with the same end going first. Since the rate of passage appeared to be uniform, they reasoned that by interrupting the process at various points, they could produce a recipient cell with various lengths of donor DNA. By systematically allowing only certain lengths of chromosome to pass into the recipient, the sequence of genes could be mapped.

Lederberg and Tatum used as a recipient a strain of bacteria that had a number of mutations along its chromosome. These mutations were metabolic in nature; that is, their phenotype was expressed by being unable to grow without the addition of one or another nutrient ("raw material") to their medium. Thus a recipient strain might *need* thymine, aspartic acid, and a variety of other nutrients (symbolized in each case by −); the donor strain would be wild type for all of these traits (symbolized by +). The particular mating can be schematically shown as in Fig. 9.5. Conjugation takes up to an hour or more in different strains. Hence the process can be interrupted easily at any point. Interruption is usually accomplished by placing the bacterial culture into a blender and churning it for a few minutes. This separates most of the conjugating pairs, leaving the recipient cell with some portion of the donor chromosome, in addition to its own complete chromosome, inside.

Suppose that interruption occurs after genes A^+ and B^+ have passed over from donor to recipient cell. The recipient will now contain two copies of genes A and B (represented A^+A^- and B^+B^-). The recipient cell will be able to grow on a medium lacking substances A and B, but it must be supplied with substances C, D, and E. (F is a symbol of the fertility or sex factor; the two mating strains, F^+ and F^-, are thus equivalent to donor and recipient, respectively.) The time period required to get a recipient strain that could grow without A and B is a measure of the amount of chromosome that must have passed through the conjugation tube. Time can thus be translated into distance measuring the spacing of genes along the bacterial chromosome. Allowing the transfer process to proceed a little longer produces a recipient strain that can grow without substance C. This process is carried out hundreds of times, in each case allowing the conjugation to proceed a little longer than the time before. In this way, the sequence of genes along the bacterial chromosome can be mapped.

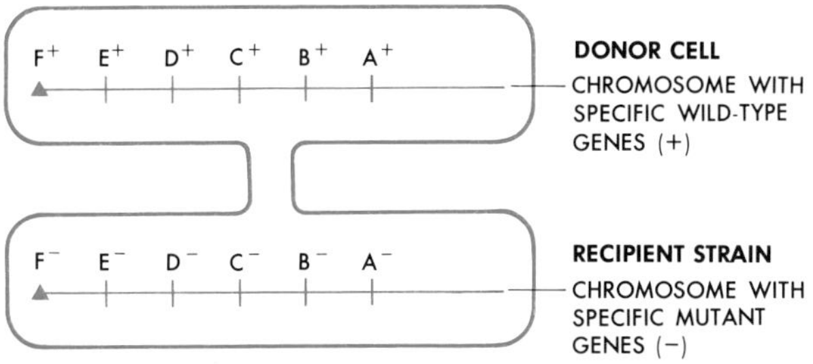

Fig. 9.5 *Conjugating bacteria. The top cell is the donor—its chromosome is passed across the bridge to the lower, recipient cell.*

From such studies a chromosome map of bacteria such as *E. coli* (see Fig. 9.6) can be constructed. For *E. coli*, the most thoroughly studied bacterium, the chromosome exists most of the time (except during conjugation, for example) as a circle. In principle, however, it preserves the linear arrangement of genes characteristic of rod-shaped chromosomes.

Fig. 9.6 *Recently revised chromosome map of the bacterium* Escherichia coli., *strain* K_{12}. *Outer circle with letters and abbreviations shows the relative position of genes coding for specific traits (often specific, known enzymes). Inner circle shows the position of the sex factor, or F factor, for a number of different strains that have been studied.*

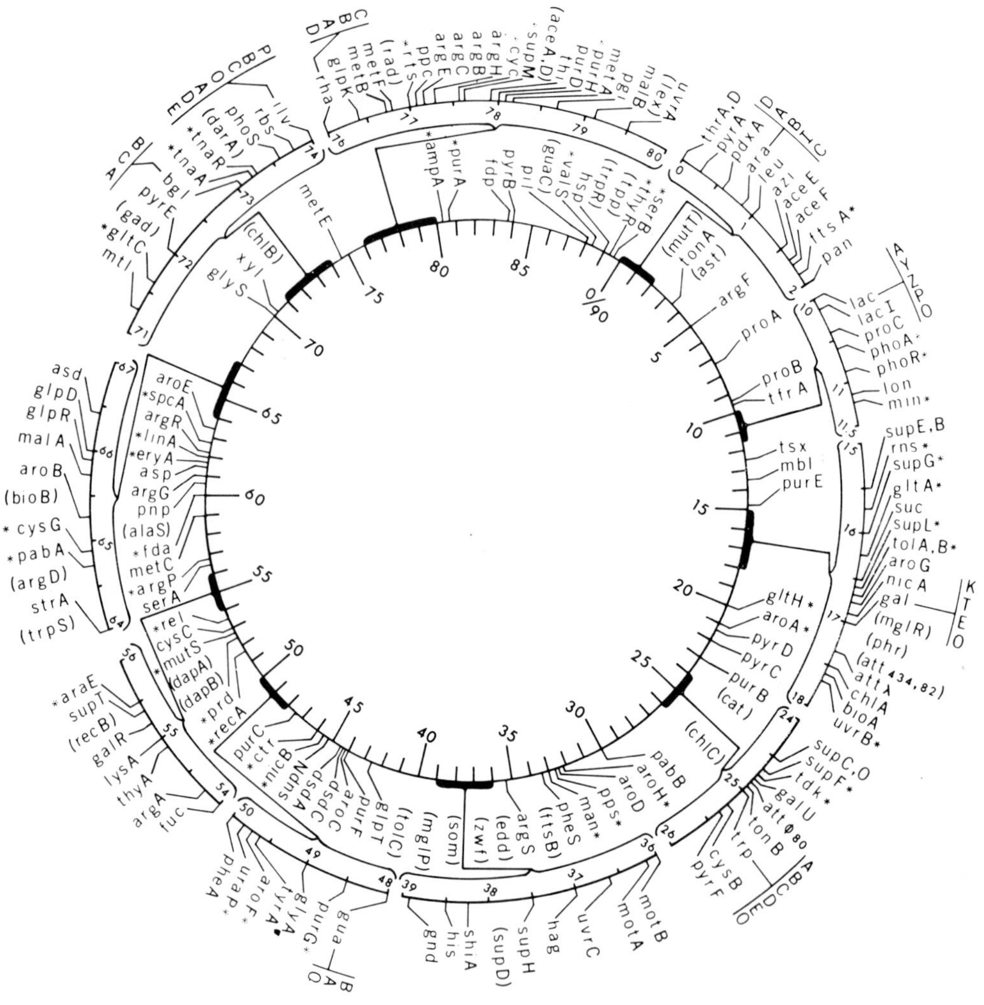

9.7 Epistasis and Multiple Alleles

Certain breeds of poultry have feathers on their shanks, while other breeds do not. Crosses between homozygous feathered and unfeathered birds yield all feathered offspring. An hypothesis that one pair of genes influences feathering is untenable, however, for crosses between F_1 feathered birds do not yield the predicted 3:1 ratio of feathered to unfeathered birds. Instead a 15:1 ratio is obtained. The expression of the obtained ratio in sixteenths rather than in fourths makes it evident that two pairs of genes, rather than one, must be involved. We have already dealt with cases of complete dominance involving two pairs of unlinked genes, and in those cases the ratio was 9:3:3:1. An hypothesis to account for the deviant 15:1 ratio must therefore be proposed.

One such hypothesis follows directly from the assumption that Mendel's law of independent assortment holds just as well in this case as in one yielding a 9:3:3:1 ratio, but that a different sort of gene action is in effect. The crosses between the feathered and unfeathered birds can be diagrammed as shown in Fig. 9.7. It will be seen that the 15:1 ratio can be accounted for by assuming that feathering will occur even if just one dominant gene from either pair is present. In these chickens, unfeathered shanks will result only from the complete absence of dominant genes. This hypothesis is supported by crosses between unfeathered birds and F_1 dihybrids. A 3:1 ratio of feathered to unfeathered birds would be predicted, and this is the ratio obtained.

In dogs, a cross between certain homozygous brown animals and homozygous white animals yields all white puppies. Many crosses of F_1 white animals yield an offspring ratio of 12 whites to 3 blacks to 1 brown.

Again a modification of the 9:3:3:1 ratio has occurred and an explanatory hypothesis is needed. The most satisfactory hypothesis is one that makes the following three assumptions.

1. Two pairs of randomly assorted genes are in-

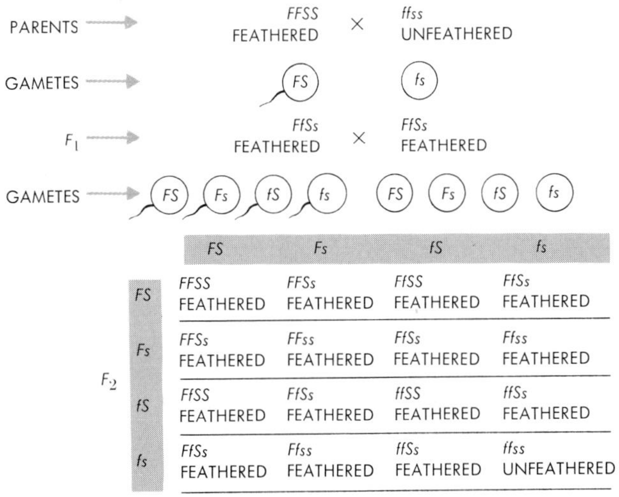

Fig. 9.7 *Punnett square for the cross between a feathered and an unfeathered variety of bird. The phenotype of the F_2 offspring is given under each genotype in the Punnett square. Note that the phenotype ratio among the F_2 is not the expected 9:3:3:1, but 15:1.*

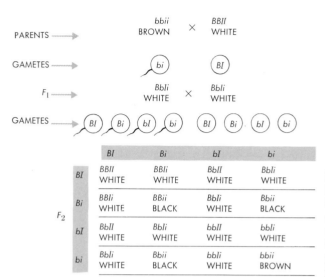

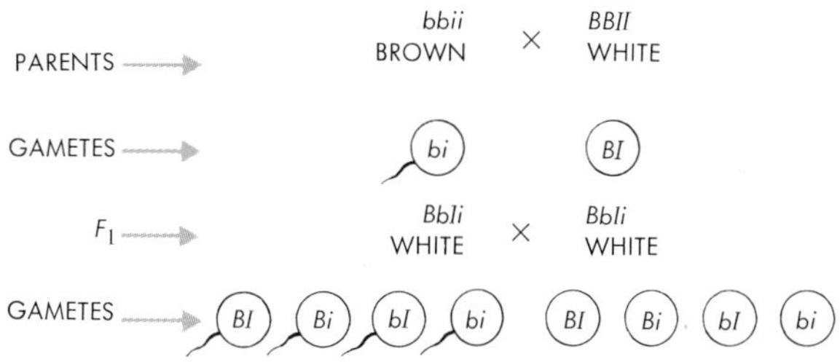

F_2

	BI	Bi	bI	bi
BI	BBII WHITE	BBIi WHITE	BbII WHITE	BbIi WHITE
Bi	BBIi WHITE	BBii BLACK	BbIi WHITE	Bbii BLACK
bI	BbII WHITE	BbIi WHITE	bbII WHITE	bbIi WHITE
bi	BbIi WHITE	Bbii BLACK	bbIi WHITE	bbii BROWN

Fig. 9.8 *Punnett square for the cross between a brown and a white dog. The F_1 is all white, but the F_2 shows a modification of the 9:3:3:1 ratio: there are 12 whites to 3 blacks to 1 brown. Hypothesizing that one of the two pairs of alleles involved in this trait is a color-determining gene, I, whose presence in the dominant form inhibits any color formation accounts for the 12:3:1 ratio.*

volved; i.e., the genes are not linked and Mendel's second law of random assortment is in effect.

2. One of these gene pairs influences coat color to the extent that it causes the coat to be black or brown, with the former being dominant.

3. The second gene pair is one that, when the dominant form is present in either the homozygous or heterozygous state, *inhibits* the production of any pigment at all, i.e., the animal is white.

The cross depicting this hypothesis is diagrammed in Fig. 9.8. This hypothesis, proposing a pigment-inhibiting gene, can be tested by crossing brown dogs with F_1 whites. The predicted 12:1 ratio of black to white to brown is obtained.

Feathered shanks in poultry and coat color in dogs are but two of many examples of modifications of two-pair ratios. Such examples illustrate two important points. First, what seem to be contradictions of the ratios predicted by Mendelian genetics are often not contradictions at all. Indeed, close analysis reveals that these results provide still further evidence in support of Mendelian genetic principles. Second, and more important, *gene pairs often work together to produce their phenotypic effects, the action of one gene pair often profoundly influencing the action of others.* A situation such as that shown in the two examples described above is known as **epistasis.**

Thus far, only genes that exist in pairs have been considered. Many genes, however, have more than one allele. Besides a dominant and a recessive gene, there may be one or more intermediate genes at the same locus forming a series. Genes of this sort are called **multiple alleles.** In rabbits, for example, there is a gene *C* that causes a colored coat; the genotype *cc* results in albino. But two other genes at the same locus are c^h and c^{ch}. Genotype c^hc^h causes the "Himalayan" pattern, in which the body is white but the extremities (the tips of nose, ears, tail, and legs) are colored. Genotype $c^{ch}c^{ch}$ results in a "Chinchilla" pattern, a light gray color over the entire body. In this case, the genes can be arranged in a series, *C*, c^{ch}, c^h, *c*, with each gene

The existence of epistasis and multiple alleles indicates that genes do not function as isolated units, but work together. No gene affects only a single phenotypic character, but may affect all other characters of the organism to one degree or another. Similarly, no phenotypic character is produced by a single gene, but by an interaction of many of the genes in the organism.

dominant to the genes that follow it and recessive to those that precede it. In other examples of multiple alleles, the genes may be incompletely dominant, so that the heterozygote shows a phenotype intermediate between those of its parents.

The foregoing examples of epistasis and multiple alleles indicate a very important point that emerges from a study of classical genetics. *Genes interact with one another.* They are not discrete, isolated units acting totally independently of one another. This suggests that the organism's entire genome (the sum total of all the genes carried by the organism) is composed of interacting units, the nature of any one affecting to some degree the expression of all the others. The question then becomes: By what means does interaction between genes take place? Such a question leads to the basic topic of the next chapter: how genetic information coded in DNA is translated into adult phenotype.

9.8 Extrachromosomal Inheritance

The reader may recall from Chapter 7 that both mitochondria and chloroplasts are capable of self-replication and possess their own DNA- and protein-synthesizing equipment. This alone shows that not all of the cell's genetic material is located within the chromosomes. As long ago as 1908, the German botanist Carl Correns, noted as one of the three biologists who in 1900 "rediscovered" Mendel's work, described the first example of a nonchromosomal gene. The explosive growth of chromosomal genetics soon eclipsed Correns' study of nonchromosomal inheritance in a number of plant species. In the 1950s, however, Dr. Ruth Sager of Columbia University and her coworkers showed that certain inherited characters in the green alga *Chlamydomonas* were due to nonchromosomal genes. A nonchromosomal gene is identified by its failure to follow the Mendelian pattern of inheritance, in which the genes from the male and female contribute equally to the genetic constitution of the progeny. Instead, the nonchromosomal genes from the female parent are usually transmitted during meiosis to all the progeny, while those from the male are lost. Furthermore, the data indicate that nonchromosomal genes in *Chlamydomonas* are distributed by a highly oriented mechanism, rather than at random.

The existence of some extrachromosomal inheritance has now been conclusively demonstrated in a large number of organisms. It is interesting to speculate on the selective value of nonchromosomal inheritance in terms of evolution. It has been shown that, unlike chromosomal replication, the replication of organelles such as mitochondria and chloroplasts is somewhat independent of cell division, and the same may be true of other cytoplasmic systems. It is not difficult to imagine situations in which the ability to increase the numbers of mitochondria or chloroplasts in a cell, independent of the division of that cell, would be of considerable adaptive value. Thus nonchromosomal genetic systems may help by providing flexibility to the manner in which an organism responds to a changing environment.

9.9 Human Heredity

Few areas of modern biology have commanded as much attention over the years as human genetics. In most ways human genetics is simply a branch of the more general study of all genetics. The principles of heredity discovered in experiments with *Drosophila, Escherichia coli,* corn, and bacteriophage are basically the same as those for other species, including human beings. Although every species has its own particular patterns of heredity, the fundamental concepts of genetics at the population, chromosome, gene, and biochemical level appear to apply throughout the living world. Human beings are no exception.

While there do not appear to be fundamental differences in principles of heredity between the human and other species, there are some distinct differences in the methods of study. Human beings cannot be mated at will like experimental animals; hence rigorous tests of hypotheses about the genotype of individuals, or patterns of inheritance, are not possible in the ways that are customary with other species. Furthermore, human beings produce so few offspring each generation that statistical analysis of ratios of progeny is subject to enormous error. These problems do not mean, of course, that it is impossible to learn anything about human heredity. They simply mean that we cannot rely upon all the same methods used for other species, and that we must devise other methods that can be directly applied to human beings. Some of these methods will be discussed in Section 9.10.

The study of human genetics, like that of other species, is largely the study of *variation* within family lines and larger populations. Without the presence of phenotypic variation, it is impossible to study the pattern of inheritance of any trait and thus impossible to determine its genotypic base. Only from variations in the pattern of transmission and from the frequency of occurrence (or phenotypic expression) can the nature of the genetic process in human beings and other organisms be understood. Thus the study of human genetics is largely the study of patterns of variation from the so-called normal phenotype (what we might call "wild type" in other species) within a population or family line.

For several reasons the study of human genetics has become an important aspect of modern biology. Medically, knowledge about the hereditary nature of certain diseases (for example, such metabolic deficiencies as the inability to break down certain substances, forms of muscular dystrophy, and diabetes) can advance the search for appropriate cures. Knowing that a disease is the result of a genetic change suggests that an absent or defective enzyme may be involved. In some cases, medical investigators have already found ways in which they might supply the missing enzyme to deficient individuals. Although such practice is still largely in the testing phase, there is hope that all hereditary diseases will not remain "incurable." In addition, knowledge about the presence of harmful recessive genes in a family line can be useful in what is called genetic counseling. If it is found that a serious and incurable genetic disease has a high risk of occurrence in certain family lines, the parents may want to avoid having more (or any) children. Although it must be used cautiously, genetic counseling can help avoid a great deal of misery for parents and children where serious, debilitating, or fatal genetic diseases are involved.

Certain dangers, as well as benefits, attend the recent increase of interest in human genetics. Chief among them is oversimplification. Genetic hypotheses, especially about human behavioral or personality traits, are highly attractive because of their apparent simplicity. Recently such behavioral traits as performance on I.Q. tests, alcoholism, spite, industriousness, laziness, homosexuality, and bad temper have all been claimed to be determined to a significant extent by genes. Needless to say, no rigorous evidence whatsoever exists for such claims. Because the traits involved are behavioral, simplistic genetic explanations have often led to simplistic treatments, sometimes with great harm to the individuals or groups involved.

9.10 Detecting and Studying Human Genetic Traits

Because we cannot mate human beings at will and thus obtain data from specific experimental crosses, it is necessary for investigators in human genetics to use other methods for (a) detecting genetically determined traits and (b) analyzing their patterns of inheritance through a series of generations. Consider first the problem of detection.

Detecting Genetic Traits

Inspection of gross anatomy. There is a large genetic component in determining the final adult phenotype of all physical traits. The easiest and time-honored method of detecting such traits is by observation of the adult phenotype. An organism with blue eyes has one genotype, while an organism with brown eyes has another. Inspection of the phenotype does not automatically determine what the genotype is, of course, or indicate its pattern of inheritance. For example, since brown eyes is dominant to blue, a brown-eyed individual may be either homozygous dominant or heterozygous. A blue-eyed individual, however, is almost certainly a homozygous recessive.

Biochemical differences between genotypes. Since genes determine the structure of the molecules within the cells of a human being, it is possible to detect genotypic differences by differences in molecular structure. The molecules most frequently studied for this purpose are proteins, though the methods can be used for lipids and carbohydrates as well.

Such methods are valuable ways of detecting phenotypic variations at a lower level of organization than that of gross anatomy or physiology. The protein molecule becomes a kind of phenotype, albeit one we can only observe indirectly. The effect of a changed protein phenotype sometimes produces gross effects of a more directly observable and dramatic sort. As a result of employing biochemical methods, it has been discovered that a great deal more variation among any one type of protein exists in a human (or other animal and plant) population than was previously thought. Biochemical methods also can be used to distinguish between homozygous dominant and heterozygous individuals. Quite often a heterozygous individual will show two forms of a given protein (one produced by the dominant gene, the other by the recessive), even though at the level of gross phenotype there will be no observable difference in anatomical or physiological characteristics from a homozygous recessive. Being able to detect heterozygotes directly, rather than having to rely upon observing the effects in their offspring, is of obvious advantage in organisms where planned matings are not practical, reproductive rates are slow, and the number of offspring is small.

Determining Patterns of Inheritance

Once a given trait is detected, its pattern of inheritance can be traced by analysis of what are called family trees, or pedigree charts. Pedigrees trace the phenotypic appearance of a trait through as many generations of a family line as possible. A fairly extensive pedigree appears in Fig. 9.9. This pedigree traces the phenotypic trait known as brachydactyly (see Fig. 9.10) through seven generations.

If pedigree charts contain enough entries, it is often possible to determine a great deal of genetic information from them. For instance, one can tell whether a trait is inherited or not. If the trait appears irregularly, with no particular pattern, it is not likely to be genetically based. On the other hand, if it appears frequently, with some regular pattern of occurrence over several generations, it might be considered genetic. Pedigree charts can also indicate whether a genetic condition is dom-

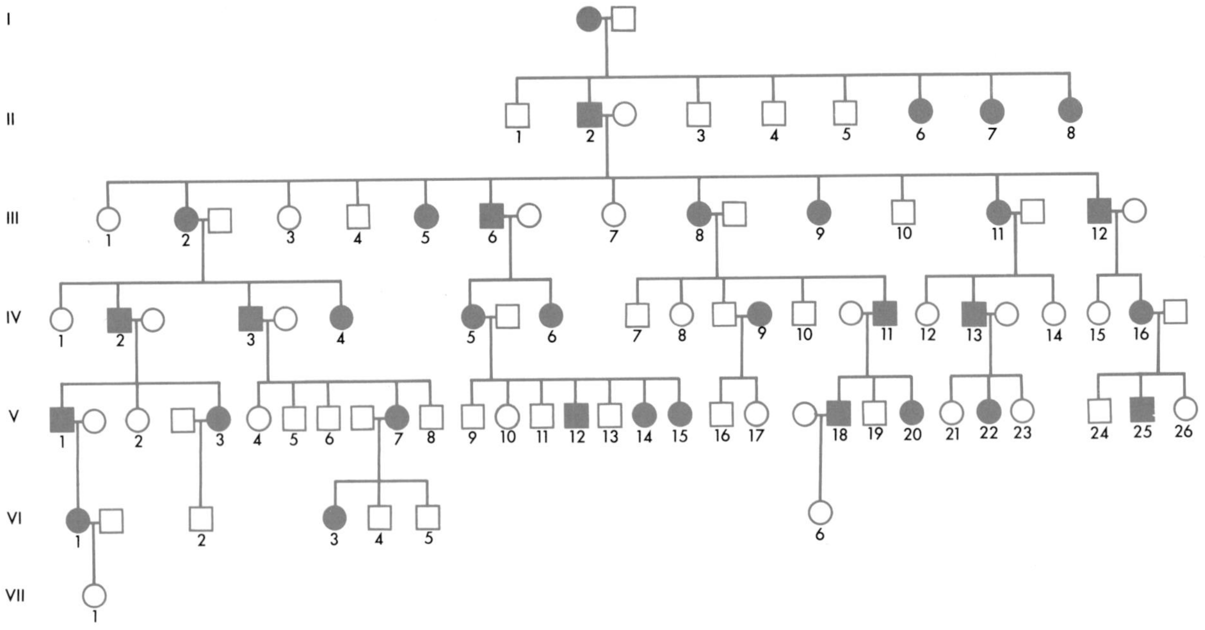

Fig. 9.9 *Partial family tree or pedigree for the dominant autosomal trait, brachydactyly (for illustration of phenotype, see Fig. 9.10). Symbols characteristically used for constructing pedigree charts are as follows: squares represent males, circles females; figures representing individuals affected by the condition phenotypically are solid, while unaffected phenotypes are represented by an open figure. Parents are linked by a horizontal line between them, and siblings appear beneath a horizontal line, from left to right in chronological order of birth. Generations are indicated by the Roman numerals at the left. Numbers underneath offspring are for identification of each child within a given generation. Other symbols are used for more special cases, such as matings between close relatives, the appearance of monozygotic and dizygotic twins, or the presence of multiple traits in the same individuals. For simplicity's sake, not all the offspring of each member of a generation are shown. (Modified from Victor McKusick,* Human Genetics, *2d ed. Englewood Cliffs, N.J.: Prentice-Hall, 1969, p. 48.)*

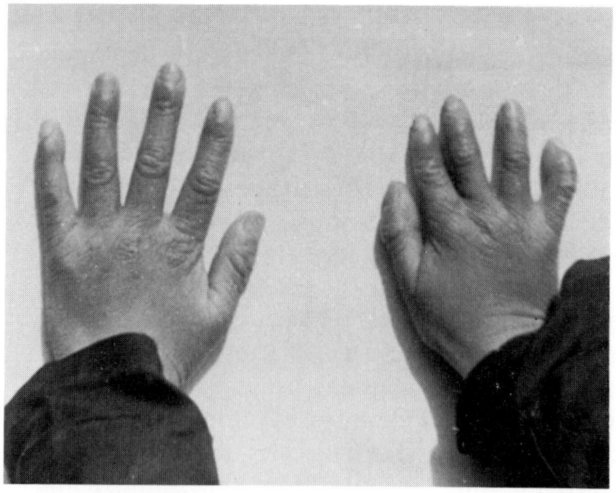

Fig. 9.10 *Brachydactylous (right) and normal hands. (From Komai,* Journal of Heredity **44,** *84, 1953.)*

Human genetics is similar to that of other multicellular organisms. For example, humans show dominance and recessiveness, sex-linkage, point and chromosomal mutations, simple and complex Mendelian inheritance, and the direct effect of gene mutations on protein structure and function.

inant, recessive, due to a single pair of genes or to several pairs (polygenic), sex-linked, and even whether it may be a new mutation. Note that the pedigree chart for brachydactyly records that in the first generation the female parent was known to have had the condition. Of her eight offspring, four showed brachydactyly. Male son number 2 in the second generation displayed the trait, and of his twelve children, seven (over 50%) showed the condition. Following affected individuals down through the sixth or seventh generation indicates that (1) the condition appears regularly enough to probably be genetic; (2) the condition is probably due to a single dominant gene, since it appears in about 50% of the offspring of parents in which one member was normal and the other affected; and (3) the condition is not sex-linked (it appears about as frequently in males as females). Pedigree charts that contain many fewer entries often do not yield so much information because they are subject to sampling error. In other words, there are simply too few cases to suggest any statistically valid conclusions.

9.11 Simple Mendelian Inheritance in Human Beings

Blood Groups

Among the human genetic conditions first found to follow a Mendelian pattern were the different blood groups. In the nineteenth century doctors knew that it was impossible to transfuse blood from certain individuals to other individuals without producing a massive and often fatal reaction in the recipient. In other cases, however, transfusions produced no adverse reaction.

Studies to determine the differences in these cases led to the discovery by Karl Landsteiner in 1900 of blood groups among human populations. As a result of such studies, all human beings are now classified into four basic groups, A, B, AB, or O. It is possible to transfuse blood from an individual of one group into another individual of the same group. But successful transfusion from individuals of one group to those in another group can only occur in certain specified ways. We now know that differences among the blood groups are due to chemical differences in substances located on the red blood cell surface, as well as on the surface of other cells in the individual's body. The chemical differences result from the presence or absence of certain enzymes in the cells of individuals of each blood type. Thus it is possible to confirm biochemically what was deduced from pedigree studies over 70 years ago: blood groups are determined genetically.

The genetics of the ABO blood group system is relatively simple. There are three alleles in the system, characteristically symbolized A, B, and O. As we saw in Chapter 8, with three alleles there can be six possible genotypes: AA, AO, BB, BO, OO, and AB. Since O is recessive to A and B, only four phenotypes are actually observed: the four blood groups A, B, AB, and O. Using the basic patterns of Mendelian genetics, it is possible to determine from blood group phenotypes of both parents and offspring what their genotypes are. The relationships between blood group phenotype and genotype are shown in Table 9.2.

Table 9.2

Genotype	Phenotype
AA AO	A
BB BO	B
OO	O
AB	AB

Blood group analysis, including many more blood groups than the common ABO, is often a means of determining paternity in cases where identification of the father of a child is disputed. Genetic evidence of blood groups is thus frequently introduced in court in paternity suits. However, blood tests can never demonstrate conclusively that a given man *is* a child's father—only whether it is genetically *possible* for him to be. On the other hand, it is possible to demonstrate with greater certainty that a man is probably not a child's father.

Consider the following example. A woman with blood type O has a child with blood type O. She claims the father was a man who turns out to have blood type AB. It can be demonstrated that this man is very likely *not* the father. The mother's genotype is obviously OO, since that is the only genotype that shows type-O blood. The disputed father's genotype is obviously AB phenotype. The offspring of such a cross would thus be likely to be AO or BO, as follows:

PARENTAL GENOTYPES	AB		OO	
GAMETES	(A) (B)		(O)	
OFFSPRING GENOTYPES		AO		BO
POSSIBLE PHENOTYPES		A		B

No type-O offspring could result from the above cross unless by mutation. The latter is possible, but would not be expected. Thus barring mutation, the suspected father's paternity can be rejected. If the suspected father had also been type-O, however, he could be the child's father. Such evidence is only *consistent* with the hypothesis proposing that he is the father; it does not establish that hypothesis with certainty. Another man with type-O blood could be the father instead. The problem would be more complex if the disputed father's blood type had been phenotypically A or B. Since both types have two genotypic forms, AA or AO and BB or BO, respectively, an AO or a BO man could, with an OO woman, father a child with OO blood. Here, another variable would be introduced, the disputed father's genotype, which cannot be readily determined from his phenotype.

The frequency of the various blood group phenotypes is well known, and turns out to be relatively stable for most human populations. For instance, among the Caucasian (white) population, about 45% of the population is type A (of that, 37% is AO, while about 8% is AA), 8% is type B (of that 7.6% is BO and 0.4% BB), 44% is type O, and 3% is type AB. In Negro (black) populations the percentage of type-B individuals is slightly higher. Among American Indians there are virtually no type-B individuals present. The significance of such differences is not clear. As far as can be determined, they have little to do directly with survival, at least under present environmental conditions.

Two other blood types inherited independently are the M-N factors and the series of Rh alleles, the latter obtaining their name from the fact that they were first identified in the blood of rhesus monkeys. The Rh factor is another agglutinogen, and its presence or absence in the blood indicates whether the individual is Rh-positive or Rh-negative. The Rh factor may cause difficulties if an Rh-negative woman marries an Rh-positive man who is homozygous for the condition. Since the presence of the Rh factor (Rh-positive) is dominant, and the growing fetus develops its own blood system, the woman will carry an Rh-positive child. If, as often happens, blood from the fetus contacts the

Supplement 9.2

BLOOD GROUP DIFFERENCES AND BLOOD TRANSFUSION

Tests to determine which blood group an individual belongs to are quite simple. Required for the test are two drops of the individual's blood and two reagents called anti-A and anti-B serums. These substances contain antibodies that have been generated in the body of an organism such as a rabbit into which red blood cells of type A or B have been injected. Type-A blood has A substances on its red blood cell surface, while type-B blood has B substances on the cell surfaces. The rabbit's immunological system has produced antibodies to the specific substances on the red blood cell surface, and thus the red cell substance acts as an antigen to the rabbit's immunological system. Anti-A reagent contains antibodies that specifically react with A-substances on red blood cell surfaces, while anti-B reagent contains antibodies that specifically react with B-substances on the red blood cell surface. When an antibody for A-substance comes in contact with a blood sample containing type-A red blood cells, the antibody binds to the A-substance on the red blood cell surface. With appropriate differences for the substances involved, this causes the cells to stick together, a process known as **agglutination.** The same happens for B-substance.

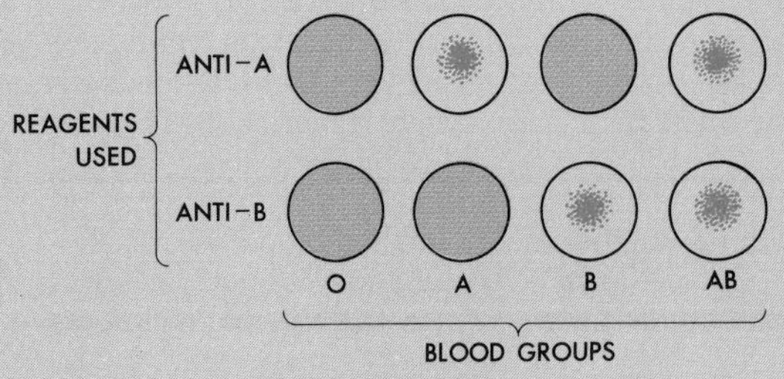

Diagram A

To type blood, one drop of donor blood is mixed with anti-A reagent, the other with anti-B. After a few minutes, one of two reactions will be observed in each drop: (1) the blood will remain whole and unclumped, or (2) the blood will collect in clumps or agglutinate (see Diagram A). The former possibility is called a **negative reaction** (no agglutination), while the latter is called a **positive reaction.** As Diagram A shows, in a test of the blood of any one individual, the reaction can be positive or negative for each reagent; hence there are four possibilities altogether. The four possibilities correspond to the four major groups, into one of which any human's blood will fall. A person with a negative test with both reagents A and B belongs to blood group O; a person with a negative reaction to reagent B, but positive reaction to reagent A, belongs to group A; a person negative for reagent A but positive for B belongs to the B group; and a person with positive reactions to both reagents A and B belongs to group AB. It is necessary to observe the reaction in both drops of blood sample in order to classify any individual into a blood group.

Blood group substances A and B are present on the cell surface not only of humans but of other organisms as well, including bacteria. Hence it is not surprising to find that from birth, the immune system of human beings contains antibodies against substances A and B. The antibodies are part of the individual's ready-response immune system.

Usually individuals do not contain or produce antibodies against their own specific body's molecules. (When the body does develop immunity to some of its own molecules, a severe condition known as the autoimmune disease develops. Though rare, the disease is usually fatal.) With regard to blood group substances, it has been found that an individual's blood plasma or serum contains antibodies against the A or B

Table 1
Antigens and Antibodies Contained in the Blood of ABO Blood Groups

If the red cells contain the antigen	Then the serum contains the antibodies
A	anti-B
B	anti-A
AB	none
O	anti-A and anti-B

substances that he or she lacks on the surface of red blood cells at birth. The antigens and antibodies found in the blood of individuals of the different blood groups are listed in Table 1. From this list the possible patterns of blood transfusion can be determined.

One factor that has a considerable influence on the direction of successful transfusions has been neglected. Transfusions involve transferring *both* red blood cells and blood plasma from donor to recipient. Any transfusion thus involves passage of some donor plasma containing the donor's already-existing antibodies of whatever blood type he or she represents. We might expect that this plasma could, in certain cases, cause the recipient's cells to agglutinate. However, in comparison to the recipient's total blood volume, the amount of plasma transferred is very small. Consequently, there will be relatively few antibodies transferred that will agglutinate the recipient's red blood cells. Some small amount of agglutination will occur, but no serious harm will be done. So far as the donor is concerned, the important transfusion question is what antigen the donor red blood cells contain; as far as the recipient is concerned, the question is what antibodies the recipient's plasma contains.

Individuals of type A can donate to other individuals of type A or type AB, since these recipients do not contain anti-A antibodies in their serum. Type-B individuals can donate to type-B or type-AB recipients, since neither contains the anti-B antibodies in the blood plasma. Individuals of type AB can donate only to other AB's, since the serum of all other recipients would contain one or the other or both A and B antibodies. Type-O individuals, however, can donate to themselves or *any* of the other groups, since type-O red blood cells contain neither A nor B antigens. For this reason, type-O individuals are called **universal donors.** Their blood can be used for transfusion into anyone else with no risk as far as blood group complications are concerned.

Now let us look at the recipient of blood. Individuals of blood group A can receive blood from individuals of type A or O, but not from B or AB. Type-A blood serum contains anti-B antibodies and, since both red blood cell types contain B-substance, would agglutinate the newly transfused red blood cells from B or AB donors. In a similar fashion, type-B individuals can receive blood from other B's or from O, but not from A or AB. Type-O individuals cannot receive transfusions from any other individuals than O's. People with type-O blood have both A and B antibodies in their plasma and would agglutinate the transfused red blood cells from A, B, or AB donors, all of whom have either A or B antigens on their red blood cells. Type-AB individuals, on the other hand, can receive transfusions from all blood group types, including their own. AB individuals have no antibodies in their plasma and will not agglutinate donor red blood cells from A, B, or AB types. For this reason, persons with blood type AB are known as **universal recipients.**

mother's blood, her white blood cells will be stimulated to produce antibodies to the Rh factor. Usually more than nine months are needed for this antibody level to build up high enough to cause trouble. In later pregnancies, however, some of these antibodies may pass into the fetus' bloodstream and cause agglutination. The resulting condition, *erythroblastosis fetalis,* may be severe enough to cause death before birth, but more often the baby dies after birth.* Familiarity with the parental blood types and knowledge of their genetic basis now generally enables afflicted infants to be saved through massive blood transfusions given at birth.

The fact that the Rh factor was discovered in rhesus monkeys before it was known in humans points up the fact that the blood groups have been helpful in establishing evolutionary relationships as well. Although many mammals and birds have substances similar to agglutinogens A and B in their blood, only in apes such as the orangutan, gorilla, and chimpanzee are such blood types as O, A, B, and AB known. The M and N substances of the chimpanzee are most like those of humans, a fact that can be added to much other evidence suggesting that chimpanzees are the anthropoid apes most closely related to humans.

Genetically speaking, the human species is enormously heterogeneous. It is sometimes said that the human species is **polygenic** (many genes per trait), or **polymorphic** (many phenotypic forms). This simply means that there is a great deal of variation for any given trait, blood group or otherwise. For example, over thirty different blood groups have been identified. While the frequencies of these groups vary from one subpopulation, or group, to another, almost every population shows a large number of the different groups in at least some proportion.

* Interestingly enough, the frequency of *erythroblastosis fetalis* is much less than would be predicted from the known frequency of marriages where the male is Rh+ and the female Rh−. There is now some evidence that the ABO blood group system is possibly associated with minimizing the effects of the fetus's Rh+ blood on the immunizing system of the mother.

Sex-Linkage in Human Beings

Sex-linkage has been determined for a number of traits in human beings. The most common of these are red-green color-blindness and hemophilia. Figure 9.11 shows the typical sex-linked inheritance pattern for color-blindness. The pattern is identical for other traits such as hemophilia. If a woman heterozygous for color-blindness marries a normal man, one-half of the sons among their offspring would be expected to show color-blindness, and one-half of the daughters would be phenotypically normal, though carriers of the gene for color-blindness. As in *Drosophila,* sex-linkage in humans follows the pattern in which affected male children always inherit the sex-linked gene from their mothers. Affected female children only show the condition phenotypically when they inherit one sex-linked gene from each parent. For this to happen, the mother must be a carrier and the father must show the sex-linked trait phenotypically. The gene or genes for color-blindness are recessive to the normal allele. Although the inheritance pattern appears to be by-and-large Mendelian, there are many variant phenotypes of each condition. For example, not all color-blind people are equally color-blind, nor are all cases of hemophilia as severe as others. This has led geneticists to recognize that both conditions are due to the interaction of a number of different genes, most of which are on the *X*-chromosome but not at the same locus. Like most human genetic conditions, color-blindness and hemophilia are examples of multiple-factor inheritance.

Numerous conditions are now known to be determined by genes on the *X*-chromosome in human

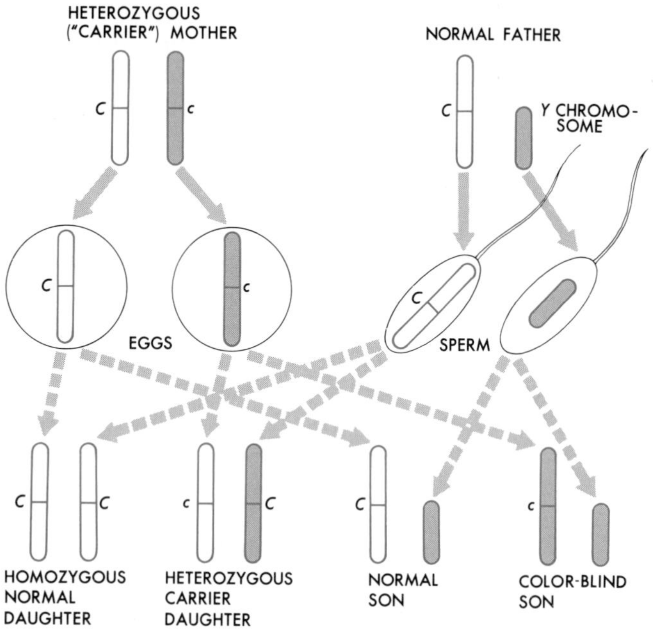

Fig. 9.11 *Diagram of inheritance for a typical sex-linked characteristic, red–green color-blindness, in human beings. The trait is carried on the X-chromosome, so that in a marriage between a carrier mother and a normal father, one-half the children will possess the gene for color-blindness. Of these, only half the males will be expected to show the condition. The heterozygous daughters will not show the trait because it is masked by the dominant allele on the other X-chromosome. Half the sons will be expected to be normal. Only the sons who inherit the single X-chromosome bearing the recessive gene for color-blindness will show the condition. The same inheritance pattern applies for hemophilia.*

beings. These include a certain form of albinism, production (or lack of production) of the enzyme glucose-6-phosphate dehydrogenase, and *ichthyosis gravior,* a drying out and scabbing of the skin. In addition, however, a few genes have been tentatively localized on the Y-chromosome, including genes responsible for determining development of maleness. Why this chromosome is so lacking in identifiable genes is not understood.

9.12 Variations in Autosome Number: Effects on Heredity

The most dramatic genetic variations in human beings are those which involve the addition or loss of whole chromosomes. Examples of such conditions are Down's syndrome (Mongolism) and various abnormalities in the number of sex chromosomes.

Down's Syndrome

In 1866 Dr. John Langdon Down (1828–1896), a London physician, described a series of traits that always appeared to be associated with a form of mental retardation. The condition he described became known as Down's syndrome, Mongolism, or Mongolian idiocy. The latter terms were incorrectly applied by Down, since the eyes of affected children superficially resemble the skin folds around the eyes of Orientals (see Fig. 9.12).

The physical traits Down associated with "Mongolism" included a prominent forehead, flattened nasal bridge, habitually open mouth, projecting lower lip, skin fold at the inner corner of the eyes, short and broad neck, rough and dry skin, abnormally shaped and aligned teeth, wide gap between the first and second toes, and a broad crease across the palm (called the "simian line"). Children with Down's syndrome are mentally retarded, be-

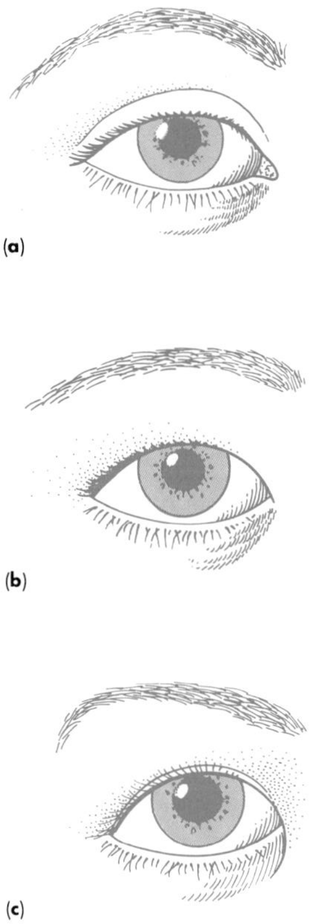

Fig. 9.12 *Differences in the appearance of (a) the Caucasian eye, (b) the Oriental eye, and (c) the eye of an individual afflicted with Down's syndrome (trisomy-21, or "Mongoloid idiocy"). The Down's syndrome eye is quite different from the Oriental eye, and Down's syndrome is not related to Oriental ancestry. Hence "Mongolism," the name originally given to the condition, is really a misnomer. (After E. Peter Volpe,* Human Heredity and Birth Defects. *(New York: Pegasus, a Division of Bobbs-Merrill, 1971, p. 72.)*

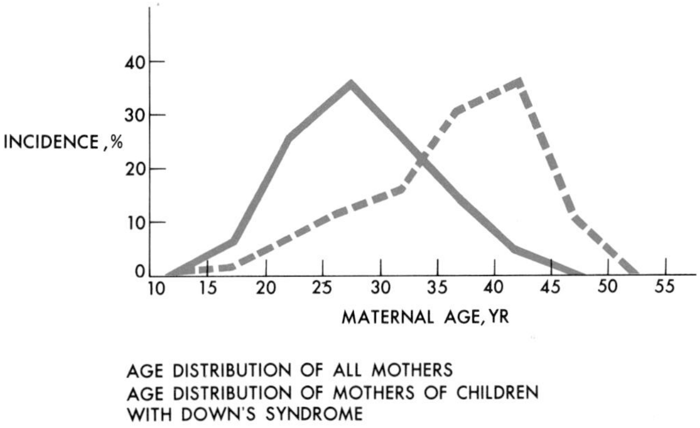

Fig. 9.13 *Age distribution of mothers of Down's syndrome patients compared with that of all mothers. Vertical axis of graph shows the relative percent of mothers in each age group. (Based on data from Lionel S. Penrose; from Victor McKusick,* Human Genetics, *2d ed. Englewood Cliffs, N.J.: Prentice-Hall, 1969, p. 23.)*

ing technically classified as imbeciles. Affected children are slow learners but often enjoy music and crafts and are sometimes able to learn to read and write. However, most Down's syndrome patients are unable to care for themselves, and the trend has been to institutionalize them. Recent evidence suggests, however, that home care for the child produces greater development of mental and physical abilities than the kind of institutional care to which most afflicted children have been exposed.

Because many children with Down's syndrome also have congenital defects of the heart or other organs, their chances of survival are severely reduced. A number of years ago the geneticist Lionel Penrose estimated that the life expectancy of children with Down's syndrome was about 12 years. Sixty percent of all children afflicted with Down's syndrome do not survive beyond 10 years of age. Those who do survive into adulthood generally do not reproduce. Males with Down's syndrome are sterile. Females are not usually sterile; as of 1970, 13 cases were known where females with Down's syndrome had children. About half of the recorded offspring have shown Down's syndrome, and the other half have been normal.

What is the genetic cause for Down's syn-

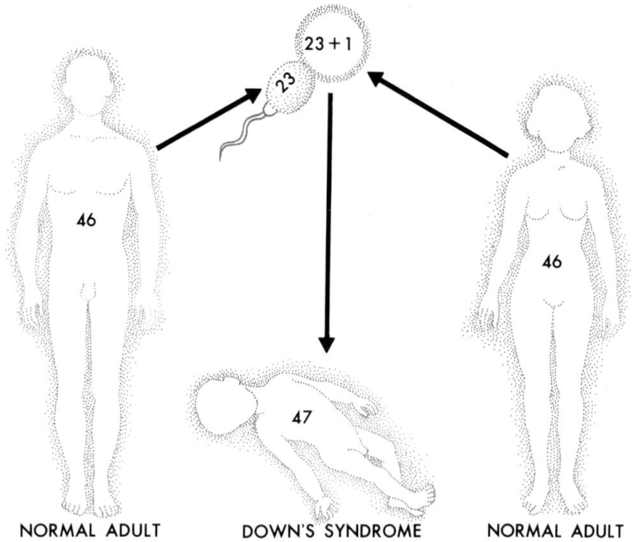

Fig. 9.14 *Chromosomal formation of individual afflicted with Down's syndrome born to two normal parents. Both parents have the normal chromosome complement of 46. Nondisjunction of the two members of pair 21 during oögenesis leads to production of an egg with 23 + 1 chromosomes; the extra chromosome is a member of pair number 21, which is thus represented by both homologs from the mother. (After E. Peter Volpe,* Human Heredity and Birth Defects. *New York: Pegasus, a Division of Bobbs-Merrill, 1971, p. 72.)*

drome? The earliest clinicians describing the syndrome noted that it occurred predominantly (though not always) in children of older mothers. Statistical data suggested that above age 35, a woman's chances of bearing a child with Down's syndrome increase eight-fold: The incidence is 1 per 1500 for women under 30, 1 per 750 for women between 30 and 34, 1 per 600 for women 35–39, and 1 per 300 in women aged 40–45 (see Fig. 9.13). Through the first half of the twentieth century it was not clear whether Down's syndrome was caused at the genetic level or could be attributed to some abnormality of development associated with an aging mother. Since affected patients usually do not have children, it was impossible to follow the trait through several generations. In 1959 however, three French scientists at the University of Paris provided cytological evidence that Down's syndrome was associated with a chromosomal abnormality in which one of the autosomes occurred in triplicate. Chromosome 21 is the pair found to exist in triplicate in patients with Down's syndrome. Studies in the past two decades have revealed that nondisjunction is responsible for the failure of the two homologous chromosome 21's to separate at meiosis during either oögenesis, as shown in Fig. 9.14, or sometimes during spermatogenesis. Cytological examination of many Down's syndrome patients has shown this pattern to exist in the vast majority of cases.

9.13 Sex Determination

Sex determination in human beings sheds further light on the effect of extra (or fewer) chromosomes. For many years, the model of sex determination in *Drosophila* was held to apply to all other animals with a similar pattern of sex chromosomes (that is, in which males were *XY* and females *XX*). Through the work of Calvin Bridges in T. H. Morgan's laboratory, and others, it became clear that in the fruit fly the *X*-chromosome was the sex determiner. One *X* produced a male and two *X*'s a female. The *Y* chromosome was present but did not appear to affect the determination of sex; though usually sterile, flies without a *Y* chromosome were still phenotypic males. While the *X*-chromosome in *Drosophila* contains many non-sex-determining genes such as that for eye color, it was also thought to contain some genes influencing the direction in which sexual development occurs in embryo flies. In *Drosophila,* however, sex determination is not a matter of specific genes on any one chromosome. The factor that seems to trigger a given embryo to develop in the direction of maleness or femaleness is the *ratio* of *X*-chromosomes to the number of members in each autosomal pair. A ratio of one *X* to two members of each autosomal pair produces maleness. A ratio of two *X*'s to two members of each autosomal pair produces femaleness. By "produces" it is understood that the ratio of *X*'s to number of members per autosomal pair *triggers* the expression of particular genes governing the development of male or female traits, on whatever chromosomes those genes may be located. It is not known how this ratio of *X*'s to autosomes affects the expression of other genes. But for *Drosophila* a chromosomal ratio is the primary determiner for the development of a specific sexual phenotype.

In human beings, however, the situation is different. The *Y*-chromosome is extremely active in sex determination. It appears to contain potent male-determining genes. Among humans, an individual carrying a *Y*-chromosome is male, even if he also has one, two, or three *X*-chromosomes! Conversely, an individual lacking a *Y*-chromosome is a female, whether she has one, two, or three *X*'s.

Variations in Sex Chromosomes

A range of variations in sex-chromosome number for human beings is shown in Table 9.3. All of these result from some form of nondisjunction. Other than the normal phenotype, only two of the types shown in the table appear to be phenotypically normal though they possess an unusual chromosome complement: *XYY* males (see Supplement 7.2) and the triple-*X* females. Both have normal sexual development and both are fertile. Most of the others show some variation in sex development and are infertile, or are at least of doubtful fertility.

Women with Turner's syndrome (*XO*) have rudimentary ovaries, or in some cases none at all;

Table 9.3
Chromosomal Complements in Sex Determination in Human Beings

Descriptive name of phenotype	Sex phenotype	Fertility	Sex-chromosome complements
Normal male	Male	+	XY
Normal female	Female	+	XX
Turner's syndrome	Female	−	XO
Klinefelter's syndrome	Male	−	XXY
XYY syndrome	Male	+	*XYY*
Triple *X* syndrome	Female	±	*XXX*
Triple *X*-*Y* syndrome	Male	−	*XXXY*
Tetra *X* syndrome	Female	?	*XXXX*
Tetra *X*-*Y* syndrome	Male	−	*XXXXY*
Penta *X* syndrome	Female	?	*XXXXX*

Adapted from V. McKusick, *Human Genetics.* Englewood Cliffs, N.J.: Prentice-Hall, 1969, p. 19.

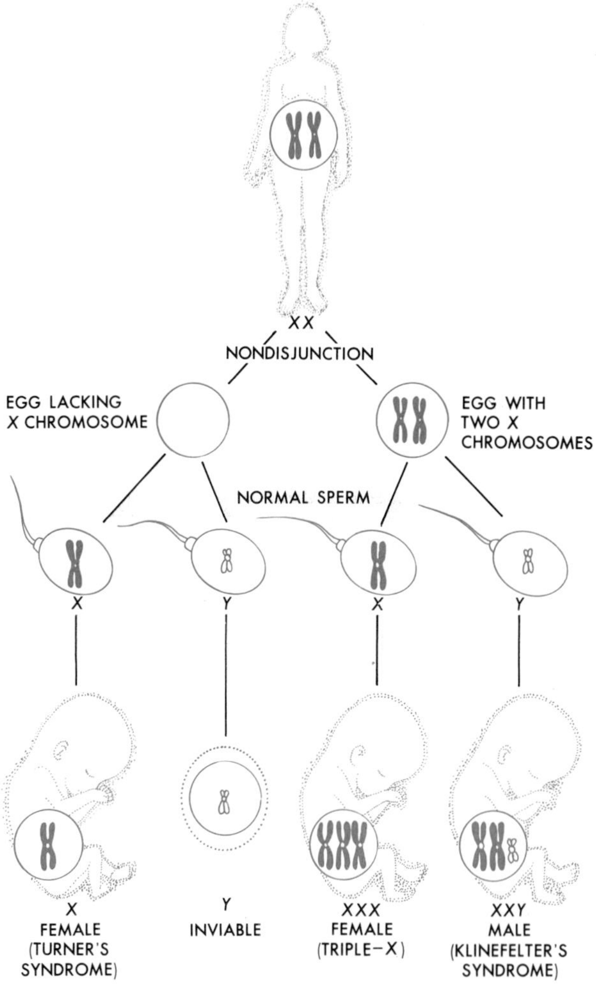

Fig. 9.15 *Nondisjunction in the X chromosomes of a human female can result in an egg cell with two X-chromosomes and an egg cell with no X-chromosome (designated 0). Even when fertilized by normal sperm, such eggs always give rise to defective offspring.*

they have undeveloped breasts, often show "webbing" or folding of skin around the neck, are unusually short, and generally show subnormal intelligence. Men with Klinefelter's syndrome (*XXY*) are outwardly male (they have male genitalia), but their testes are undeveloped and their breasts enlarged; they have longer limbs than average and sparse body hair. Many are mentally defective, and almost all reported cases have been sterile. Both Klinefelter's and Turner's syndromes are produced by the nondisjunction mechanism shown in Fig. 9.15. Although individuals with Klinefelter's syndrome have two *X*'s, the presence of the *Y* chromosome determines their generally masculine appearance. Although individuals with Turner's syndrome have only one *X*, the lack of a Y-chromosome determines a generally female appearance.

As Table 9.3 shows, still other abnormalities of sex-chromosome constitution are found in human beings. A few triple-*X* females have been observed. They are produced by the same type of nondisjunction processes that give rise to Turner's or Klinefelter's syndrome. Although phenotypically female, many show various degrees of mental retardation and are infertile. However, some triple-*X* females show no signs of retardation and *are* fertile. More surprisingly, children born to fertile triple-*X* females have been either normal males or females. This is unexpected, since we would predict that a triple-*X* mother could produce two kinds of eggs: an *XX*-bearing egg, and an *X*-bearing egg. Accordingly, half of her daughters would be expected to have the triple-*X*-condition, and half her sons would be *XXY* (Klinefelter's syndrome). Geneticists are uncertain as to why the latter abnormal offspring have not turned up. It could be chance, but it may also be that *XX*-bearing eggs are inviable. If this is true, however, we are left without a mechanism for producing the triple-*X* female in the first place and also without a mechanism for producing *XXY* males (Klinefelter's), both of which require a nondisjunction egg with two *X*-chromosomes in the parental generation. The small number of triple-X females observed appears to result from a high abortion rate among triple-X fetuses.

Other sex-chromosome constitutions also occur, including tetra-X, tetra-XY, and penta-X. These are relatively rare and result from several generations of compounded nondisjunctions of the X-chromosome. Fertility in these individuals is extremely low.

Sex determination in human beings and other mammals thus differs considerably from the process in *Drosophila*. Here is another example of the fact that it is not always possible to generalize a mechanism in one species to all species, even when the same elements (such as X- and Y-chromosomes) are involved. Indeed, determination of sex in human beings parallels far more closely the process in the plant *Melandrium* than that in the fruit fly.

Through evolutionary history, several different methods of achieving two separate sexes (**sexual dimorphism**) have evolved. The existence of such variations in the chromosomal basis of sex determination leaves modern biologists still very much in the dark about what actually determines the development of an individual into a male or a female. More will be said about the developmental aspects of this process in Chapter 11.

The Lyon Hypothesis

Abnormalities in the number of X-chromosomes appear to have less severe effects on overall phenotype than similar abnormalities of the autosomes. Triple-X females are often retarded, as are individuals with Klinefelter's or Turner's syndromes, but they are not generally so retarded as individuals with Down's syndrome. Other physical abnormalities associated with different doses of sex chromosomes are also less severe than Down's syndrome.

Why should this difference exist? One hypothesis suggests that since it is normal in a human population for the X-chromosome to exist in two different doses (a double dose for normal females, and a single dose for normal males), the sex chromosomes have evolved over the generations a "dosage-compensation mechanism."

The first clue as to what this mechanism might be came in 1949. At that time M. L. Barr made the seemingly minor observation that most nondividing nerve cells of female cats contain a dark-staining, small body in the nucleus. This small body is never present in the nerve cells of male cats. Called the Barr body, this small object was found to exist in the cells of females of many other mammalian species, including human beings. The fact that the Barr body was lacking in women with Turner's syndrome, who are XO, and present in men with Klinefelter's syndrome, who are XXY, suggested that it might be sex chromatin. But where did this sex chromatin come from, and why did it appear as an extra body in the cell nucleus of females only?

A second clue came from observations of the variegated coat color patterns typical of many mammals. It was known, for example, that the condition known as tortoise shell in cats is almost always found only in females. A similar condition is known in mice. These observations suggested to observers that cells in different parts of the animal's body were genetically different, like a genetic mosaic.

In 1961 British investigator Mary Lyon proposed an hypothesis to account for the following observations:

1. The presence of the Barr body in the cells of female, but not male, mammals.
2. The apparent compensatory mechanism that mammalian cells have for dealing with either one or two X-chromosomes.
3. The appearance of coat color variegation only in female mammals.

Lyon proposed that relatively early in the development of females, one X-chromosome in each cell is inactivated. This means that DNA on the inactive

X-chromosome is somehow permanently prevented from transcribing *m*RNA and producing protein. The inactivation is passed on to all replicates of that chromosome. Either the maternally or paternally derived *X*-chromosome can be inactivated, probably at random. This means that in about half the body cells of a female, the maternally derived *X*-chromosome is transcribing *m*RNA, while in the other half it is the paternally derived *X*-chromosome. Such random inactivation would account for the variegated, patchwork coat patterns. It also suggests that the Barr body may be the inactive chromatin (that is, DNA) from the one *X*-chromosome. No one is certain how inactivation is brought about.

The Lyon hypothesis can be tested in an elegant but simple way. A sex-linked mutation on the *X*-chromosome has been discovered that results in a deficiency (lack) of the enzyme glucose-6-phosphate dehydrogenase, the enzyme responsible for one step in the biochemical pathway of glucose oxidation. Women heterozygous for the mutation would be expected to have a normal allele for the enzyme on one *X*-chromosome and the mutant form on the other. If the Lyon hypothesis is correct, then heterozygous women should display two populations of cells: one with a normal amount of enzyme activity, and one with very little or no activity. If the Lyon hypothesis is incorrect, then heterozygous women should have only one population of cells, all with the same relatively high level of enzyme activity. Studies of sample tissues from heterozygous females indicate that there are, indeed, two populations of cells: one with high and the other with extremely low glucose-6-phosphate dehydrogenase activity. The evidence supports the Lyon hypothesis.

The Lyon hypothesis provides a way of understanding the gross phenotypic abnormalities that result from extra autosomes, compared to the relatively small abnormalities that result from extra sex chromosomes. Since the presence of one or two *X*-chromosomes is a regular occurrence, a mechanism for dosage compensation has obviously been selected for and has evolved. But the presence of extra autosomes is a very unusual occurrence, and a dosage-compensation mechanism has not evolved. Thus, extra autosomes yield complexities of cell metabolism that alter or completely prevent normal development.

9.14 Human Biochemical Genetics

In testing the Lyon hypothesis, it is assumed that genes produce proteins in human beings in the same way as in other organisms. Back in the early part of the twentieth century, the English physician Archibald Garrod discovered two examples of what he called "inborn errors of metabolism." Both were diseases associated with metabolism of the amino acids tyrosine and phenylalanine (see Fig. 9.16). One disease, called alkaptonuria, will be useful in demonstrating the relationship between genes and enzymes in human beings. The disease is caused by a defective enzyme, homogentistic acid oxidase, which is involved in the metabolism of homogentistic acid (see Fig. 9.16). Because the enzyme is defective and cannot catalyze the oxidation of homogentistic acid, the latter accumulates in the body tissues and is excreted in the urine. Upon exposure to light at urination, homogentistic acid turns black, readily calling attention to the existence of the disease. In the body, homogentistic acid molecules bind to the protein collagen of connective tissues and can cause severe arthritis in the joints.

With the help of geneticist William Bateson, Garrod concluded that alkaptonuria was inherited in a Mendelian fashion. Alkaptonuria is caused by a recessive gene which we can infer must produce a defective enzyme. Heterozygotes do not show the gross clinical manifestations of the disease. A double dose of the enzyme is apparently no more effective than a single dose in metabolizing homogentistic acid. Only homozygous recessives show the overall phenotypic traits of alkaptonuria.

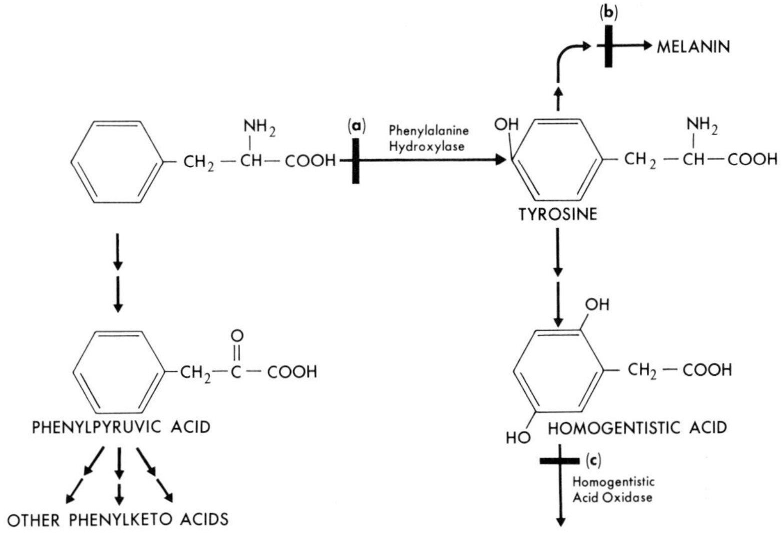

Fig. 9.16 *Outline of the pathways involved in tyrosine and phenylalanine metabolism in human beings. The heavy black lines indicate a block in the metabolic pathway due to a defective enzyme. Substances synthesized prior to the block in the pathway accumulate. A block at (a) produces phenylketonuria, a disease characterized by accumulation of phenylpyruvic acid in the body. A block at (b) produces albinism, since tyrosine cannot be converted into the pigment melanin. A block at (c) produces alkaptonuria, a genetic disease characterized by accumulation of homogentistic acid in the body.*

Both phenylketonuria and albinism result from similar enzyme deficiencies in the same amino acid pathways. As with alkaptonuria, the inheritance pattern appears to be strictly Mendelian. The fact that a Mendelian inheritance pattern is associated with measurable defects in specific enzymes is good evidence that genes in humans, as in other organisms, function by coding for specific proteins. Recent evidence also suggests that a defect in a single enzyme can influence many different traits. For example, in 1970 Victor McKusick of Johns Hopkins University School of Medicine estimated that over 100 known disorders in human beings are the result of a defect in single enzymes.

SUMMARY

1. The chromosome theory of inheritance, as developed for eucaryotic organisms, claims that individual genes exist as discrete units on the chromosomes in the cell nucleus. Genes are thus linked together by existing side-by-side on the same chromosome. In all eucaryotic organisms studied to date, the number of linkage groups (groups of traits inherited together) corresponds closely to the number of chromosome groups.
2. Linkage of genes on the same chromosome is sometimes broken by crossing over between homologous chromosomes, a process that occurs occasionally during synapsis of meiosis I in producing gamete cells. Breakage is usually followed by rejoining of the broken pieces. Sometimes this occurs so that the end of one member of the homologous pair is rejoined to the opposite end of the other member. This process is called recombination, since it produces a new combination of alleles.
3. The existence of linkage, crossing over, and recombination has been used to make genetic maps of genes on chromosomes. Genetic map

distances are given in arbitrary units (map units) that are purely relative. The genetic map itself has no necessary correspondence to the physical chromosome itself. Devising a method to show that genetic maps correspond directly to physical regions of a given chromosome pair required the development of cytological techniques.

By studying changes in fine structure of chromosomes resulting from aberrations such as deletions, inversions, or translocations, geneticists have been able to correlate physical changes in small chromosome regions with alterations in phenotype for certain traits. Such correlations suggested that the physical region observed on the chromosome contained the gene or genes for the traits that were phenotypically altered.

4. Morgan and others showed that sex linkage occurred in *Drosophila*. Sex linkage is a phenomenon by which certain traits appear most frequently (but usually not exclusively) in members of one sex. For organisms like *Drosophila* and human beings (where males are *XY* and females *XX*), sex-linked traits are phenotypically expressed most frequently in males. Sex-linked traits are found as genes on the *X* chromosome. Because most mutants are recessive, they will show up only in males, where the *Y* chromosome has no corresponding allele. In females, the recessive sex-linked traits will not show up very frequently, since most of the times the homologous *X* chromosome will contain a dominant allele. Only when the daughter receives two mutant alleles, one from her father, the other from her mother, will she show the sex-linked recessive trait. White-eye is a recessive, sex-linked trait in *Drosophila;* red–green colorblindness and hemophilia are two sex-linked traits in human beings. A few traits are determined by genes on the *Y* chromosome in many kinds of organisms. These would be sex-linked traits also, and they would appear exclusively in males.
5. Genes within a genome interact in a variety of ways. Sometimes two pairs of alleles interact in a manner called epistasis. Epistasis occurs when the action of one gene masks the action of another or, alternatively, when a certain combination of dominant or recessive genes produces an effect that is unexpected on the basis of the action of other combinations of the same alleles.
6. Sometimes a single gene may have more than one effect, a condition known as pleitropy.
7. Multiple alleles is a condition in which more than two alleles are known to exist at the same locus—such as the many different eye-color alleles in *Drosophila.*
8. While most genes in eucaryotic organisms are contained within the nucleus as part of the chromosomes, some traits are determined by extrachromosomal genes. Such genes do not follow Mendelian inheritance patterns; extrachromosomal (or cytoplasmic genes) are contributed to an offspring solely by the mother, through the egg cytoplasm.
9. The human species shows enormous amounts of variation in every trait known.
10. Human beings show classical Mendelian inheritance for a number of traits, such as red–green color blindness (sex-linked), the ABO blood group alleles, and brachydactyly. As in all higher animals and plants, many other human traits involve the interaction of genes and do not follow simple Mendelian patterns.
11. An increase or a decrease in the number of chromosomes of a given pair due to nondisjunction is responsible for conditions such as Down's, Klinefelter's, and Turner's syndrome. All of these are examples of genetic changes resulting from an increase (Down's and Kline-

felter's) or decrease (Turner's) in chromosome number.

12. In human beings sex is determined largely by the presence or absence of a Y-chromosome. Presence of a *Y* determines a male, regardless of whether one or more *X*'s are present. Lack of a *Y*, regardless of how many *X*'s are present, determines a female. This situation is the reverse of that found in *Drosophila,* though in the insect, as in humans, a normal male is *XY* and a normal female *XX*. In *Drosophila,* one *X* determines a male; two or more *X*'s a female. How *X*- and Y-chromosomes actually affect the development of sex in organisms is not yet clear.

EXERCISES

1. The male clover butterfly is always yellow, but the female clover butterfly can be either white or yellow. White color is due to a dominant gene *W*; yellow is due to its recessive allele *w*. What is the genotype of a yellow female? Propose an hypothesis to suggest the reason why a male of the same genotype does not show the same characteristic. Plan an experiment to test your hypothesis.
2. In rats, pigmentation is determined by a dominant gene *P*, albinism by its recessive allele *p*. Black coat color is due to a dominant gene *B*, whitish color to its recessive allele *b*. For the black gene to express itself, however, the gene *P* must be present. If a rat with the characteristics *PPBB* is crossed with another whose characteristics are *ppbb*, what will be the phenotypes and genotypes of the first generation?
3. Among the children of a man and woman who are both heterozygous for albinism, the distribution of their offspring will be expected to be three with normal pigmentation to one albino. However, studies of a number of families in which albinism is known reveal an interesting situation: Albinos make up considerably more than 25% of the offspring of these families. Furthermore, the fewer the children in each family, the greater the proportion of albinos among them. Geneticists, however, maintain that these results still agree perfectly with Mendelian principles. Offer an explanation for this discrepancy.
4. Explain why so much is known about the means by which *Drosophila melanogaster* and *Escherichia coli* inherit their characteristics.
5. In mice, yellow coat color is known (see Color Plate I). Crosses between black-coated and yellow-coated mice yield a 1:1 ratio of black- to yellow-coated mice. These same blacks, mated to each other, produce only black-coated young. These results are consistent with an hypothesis proposing a genotype of *BB* for the black mice and *Bb* for the yellow mice. However, this hypothesis also predicts that crosses between yellow mice should yield a ratio of one black to two yellow to one of whatever color results from genotype *bb*. This ratio, however, is not obtained. Instead, the result is a 1:2 ratio of black- to yellow-coated mice. Further investigation salvages the hypothesis, however. Dissection of the uteruses of the yellow females yielding such 1:2 ratios shows that one-fourth of the young are arrested at an early stage of their embryological development and die before birth. Evidently, then, the *bb* genotype represents a lethal combination, and the organism

cannot survive. Suggest how, in terms of the modern concept of gene action, a lethal gene combination might act to produce its effects.

6. The fruit fly, *Drosophila melanogaster*, has four pairs of chromosomes. The loci of several genes have been determined on each chromosome. Suppose you discover a new inherited trait in *Drosophila melanogaster*.
   a) How would you determine that the new trait is due to a single pair of genes?
   b) How would you determine on which of the chromosomes this new gene is located?
   c) After determining which chromosome the gene is on, how would you go about determining its specific locus relative to known genes X, Y, and Z?
7. In cats, short hair is dominant over long hair. A long-haired male is mated to a short-haired female whose father had long hair. They produce a litter of ten kittens.
   a) What proportion of their kittens would you predict to have long hair?
   b) Give *two* hypotheses that might account for all ten kittens being long-haired.
   c) Devise an experiment that would distinguish between the two hypotheses given in (b).
8. Radishes have either a long, oval, or round shape. Crosses between radishes of various phenotypes yield the following results:

Parental phenotypes	F_1 phenotypes
Long × Oval	52 long: 48 oval
Long × Round	99 oval
Oval × Round	51 oval: 50 round
Oval × Oval	24 long: 53 oval: 27 round

Answer the following questions:

   a) How many pairs of alleles are involved in determining body shape? How can you tell?
   b) What are the genotypes of each body shape discussed here? Be sure to specify the meaning of the symbols you use.
   c) Is dominance involved or not? If so, what kind of dominance? If not, what determines the various phenotypic results in each cross shown?
9. In corn the following allelic pairs have been identified in chromosome 3:
   1. + and b — Plant color booster (+) versus noncolor (i.e., the + allele produces a dark color in the kernels, and the b allele a light color).
   2. + and lg — Presence of ligules (+) on the leaf versus absence of ligules (lg). Ligules are membranous extensions of the tip of the leaf.
   3. + and v — Green plant color (+) versus virescent (v) color (i.e., a light green color of the leaf).

   In the above case, the wild-type allele is dominant over the mutant condition (symbolized by the small letter). Thus a genotype of +b would be the same as *Bb*,

++ the same as *BB*, etc. A test cross involving triple recessives and F_1 plants heterozygous for the three alleles would be expressed as:

$$\frac{+\,+\,+}{b\ lg\ v} \times \frac{b\ lg\ v}{b\ lg\ v}$$

The following phenotypes were obtained from this cross:

Phenotype	Number Out of Total Progeny
b lg v	305
b + +	128
b lg +	18
+ lg +	74
b + v	66
+ + v	22
+ lg v	112
+ + +	275
Total offspring	1000

On the basis of these data, answer the following questions:

a) Determine the sequence of genes on chromosome 3 and the map units apart.

b) Why is it important to look closely at the double crossover values in such crosses in order to determine actual sequence of genes?

c) How does the process of mapping support the idea that genes are located in a linear fashion on chromosomes?

d) What general assumptions about the phenomenon of chromosome breakage is it necessary to make in order to use crossover values as an indication of distance of genes apart?

10. If a healthy man marries a woman who is a hemophiliac, is it possible for them to have normal children? What could you say about the sex of the normal offspring resulting from this marriage?

11. If a color-blind man married a normal, healthy woman whose father was color-blind, could they have a son with normal vision? Could they have a color-blind daughter?

12. A man and woman both have normal vision, but their first child is color-blind. What are the chances that their next child will be a color-blind daughter? What are the genotypes of the parents of the man and of the woman?

13. In humans, migraine (a type of headache) is due to a dominant gene. A woman who has normal vision and does not suffer from migraine takes her daughter to a doctor for a checkup. During the examination the doctor finds his patient to be suffering from both migraine and color-blindness. What can he immediately infer about her father?

14. What proportion of human progeny receive an *X* chromosome from the father? What proportion receive one from the mother? What proportion receive an *X* chromosome from the mother and a *Y* chromosome from the father?

15. Assume that both Mrs. Baker and Mrs. Allen had babies the same day in the same hospital. After arriving home, Mrs. Baker began to suspect that her child had been accidentally switched with the Allen baby. By assigning hypothetical blood types to all four parents and the two babies, design a combination that will prove Mrs. Baker's suspicions are correct.

SUGGESTED READINGS

General Genetics

Levine, Paul, and Ursula Goodenough, *Genetics.* New York: Holt, Rinehart and Winston, 1974. A thorough introductory text that can be used as reference for individual topics raised in this chapter.

The Chromosome Theory

Moore, John A., *Heredity and Development.* New York: Oxford University Press, 1963. A semihistorical approach to understanding the problems of classical chromosome theory. Approaches the experiments of Morgan, Sturtevant, Bridges, and others in terms of the questions the investigators were asking and the data they obtained.

Whitehouse, H. L. K., *Towards an Understanding of the Mechanism of Heredity,* 2d ed. New York: St. Martins Press, 1969. A detailed analysis of many of the classical experiments that established the chromosome theory.

Human Genetics

Bodmer, W. F., and L. L. Cavalli-Sforza, *Genetics, Evolution and Man.* San Francisco, W. H. Freeman, 1976. An excellent introduction to human genetics from an evolutionary perspective.

Cavalli-Sforza, L. L., *Elements of Human Genetics,* 2d ed. Menlo Park, Calif.: Benjamin, 1977. A clear, brief introduction to the principles of human genetics, with emphasis on their applications in many areas of social and ethical concern.

McKusick, Victor, *Human Genetics.* Englewood Cliffs, N.J.: Prentice-Hall, 1969. A more detailed introduction to many aspects of human genetics. Well written and clear. A thorough introduction.

Stern, Curt, *Principles of Human Genetics,* 3d ed. San Francisco: W. H. Freeman, 1973. The most detailed and comprehensive textbook now on the market for human genetics. A useful reference, but many parts require a good genetics background.

TEN

GENETICS III: THE MOLECULAR BIOLOGY OF THE GENE

10.1 Introduction

The preceding chapter focused largely on the physical nature of heredity: how genetic factors are transmitted from one generation to another and how genes are arranged on chromosomes. The other side of the story of heredity is the question of how genes *function*. This has commonly been referred to as the problem of gene **translation.** It involves studying the molecular nature of genes and their relationship to protein structure. It also involves studying the relationship between genes and biochemical pathways within the cell. In the last analysis, genes can be thought of as molecules containing coded information that is translated into specific protein structure. As enzymes, some of these proteins control biochemical pathways and ultimately determine the organism's phenotype.

10.2 In Search of the Gene

Thus far we have dealt with Mendel's "factors," the genes, in terms of an hypothesized model that pictures the chromosome–gene relationship as analogous to the one existing between a necklace and its individual beads. This model depicts the chromosome as little more than a physical entity formed by a series of connected genes. The model is a completely satisfactory one in terms of its ability to account for random assortment, linkage, etc. And if we make the chromosome–gene necklace the snap-in kind, in which individual beads or groups of beads can be exchanged one for another, the phenomena of crossing over, chromosomal inversions, deletions, etc., are satisfactorily pictured (though not, of course, either accounted for or explained).

Modern genetics has zeroed in on the gene itself. What is the gene composed of? What is its structure? How does it work? In answering these questions, the string-of-beads model is of little or no use. Indeed, close examination of chromosome structure casts serious doubt on whether the model is at all correct! Genes, for example, might be expected to have a rather definite chemical structure. Thus if the chromosome is simply a string of genes, it, too, should reflect gene structure. Yet chemical and physical analysis of chromosome structure does not bear out predictions based on the string-of-beads model. Instead, chromosomes are found to be composed of varying amounts of DNA and RNA, a protein of low molecular weight (histone), a more complex protein, and some lipid substances. *Which one or more of the substances found in chromosomes represents the primary genetic material?*

Griffith's Seminal Observations. In 1927 an experiment bearing indirectly on this question was performed by F. Griffith. Griffith worked with two strains of pneumonia bacteria. Strain S forms smooth colonies on a bacterial agar plate and each cell is encapsulated (i.e., surrounded by a carbohydrate wall). Strain R forms rough colonies, and R cells are unencapsulated (i.e., not surrounded by a carbohydrate wall).

More important to Griffith's experiment, however, was the fact that when injected into mice, strain S causes death—and this death-causing ability is inherited. The injection of strain R cells into mice, on the other hand, does not cause death. The lack of death-causing ability is also an inherited feature of these strain R bacteria.

Griffith found that when strain S bacteria were heated to 60°C they were killed. Injection of such dead strain S bacteria was no longer fatal to mice. However, if the dead strain S bacteria were injected along with some living strain R bacteria, some of the mice receiving such injections died. The blood of these mice was always infected with living harmful bacteria *bearing capsules.*

The interpretation of these experimental results seems clear: something must have passed from the dead strain S bacteria into the living strain R to transform them from harmless, unencapsulated cells to a harmful, encapsulated kind. Further, the fact that these transformed bacterial cells passed on their new characteristics to their descendants indicated it was the genetic material of the dead strain S cells with which they were mixed before injection that must have entered the strain R cells and transformed them.

Isolating the "Active Fraction." The next step was a logical one—isolate the genetic substance responsible for Griffith's bacterial transformation and chemically identify it. This task proved not to be an easy one. From the fact that the main bulk of chromosomal material is protein, it was generally thought that the genes, too, must be proteins. Yet none of these isolated proteins caused bacterial transformation. Finally, in 1945, Avery, MacLeod, and McCarty, who were working with pneumococcus bacteria, isolated a "biologically active fraction" from encapsulated strain S bacteria. Under appropriate culture conditions, it was shown that this active fraction could genetically transform the R strain to the S strain, a strain genetically different from the former. By various chemical and physical techniques, this active fraction was found to be composed primarily of DNA. Avery and his coworkers concluded their historic paper with the words:

The evidence presented here supports the belief that a nucleic acid of the de[s]oxyribose type is the fundamental unit of the transforming principle of Pneumococcus Type III.

It seemed, then, that DNA was the primary genetic material. By some still-undetermined process, one strain of bacteria had absorbed the DNA of another strain and had thereby acquired characteristics of the donor strain. Further, and most important, this transformation of the recipient strain of cells was passed on to their descendants.

There were, however, some objections to acceptance of DNA as the primary genetic material on the basis of these experiments. These objections were based on the fact that the transforming DNA fraction was only about 95% pure. It was reasoned that the 5% impurities (mostly protein) could be the genetically active fraction, rather than the DNA. This argument was later somewhat weakened when it was shown that highly purified DNA, containing negligible amounts of other compounds and less than 0.02% protein, was still capable of causing transformation.

The crucial experiment, however, utilized the enzyme deoxyribonuclease. This enzyme breaks down DNA, destroying its potency, but leaves other compounds, such as RNA and proteins, unaffected. It could be reasoned that if DNA is the primary genetic material causing bacterial transformation, then addition of the enzyme deoxyribonuclease to the transforming mixture before it is exposed to the recipient bacteria should destroy its ability to cause transformation. On the other hand, if the genetic activity of the transforming mixture is *not* due to the DNA, but rather to the protein or other impurities it contains, then exposure of the enzyme deoxyribonuclease to the transforming mixture should have no effect on its ability to cause transformation. The results of this experiment clearly supported the first hypothesis. No bacterial transformation was caused by mixtures exposed to deoxyribonuclease.

Further Evidence. The bacteriophage viruses discussed in Chapter 4 (Fig. 4.37) provide still further insight into the possibility that DNA is the primary genetic material. These viruses become attached to the surfaces of bacteria as shown in Fig. 10.1. After about 20 min the bacteria burst, each cell releasing about 100 complete new T2 viruses. It is clear that some substance (or substances) must pass from the infecting virus into the bacterial cell and cause the formation of new viruses. This substance, therefore, must contain the genes of the virus.

Chemical analysis of the T2 virus reveals it to be composed only of DNA and protein:

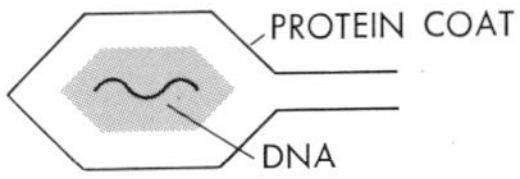

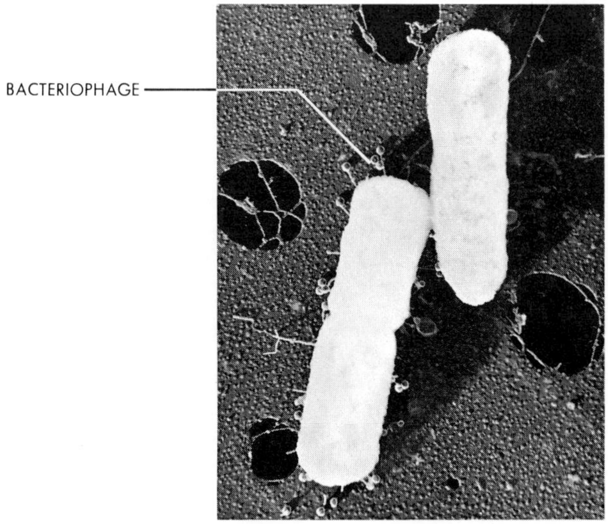

Fig. 10.1 *The bacterial cell at the left shows several bacteriophages attached to it. Some bacteria, however, are immune to them. Luria and Delbrück performed an experiment to determine whether the mutation causing immunity was a random one or was due to exposure to the phages.*

DNA contains phosphorus; protein does not. Conversely, the viral protein contains sulfur; DNA does not. To investigators A. D. Hershey and Martha Chase this fact suggested a crucial experiment. By tagging the DNA with radioactive phosphorus or the viral protein with radioactive sulfur, they could tell which part of the virus entered the bacterial cell and participated in the formation of new viruses. The results showed that almost all of the DNA entered the bacterial cell on infection. In contrast, only 3% of the protein did so. The conclusion to these experiments on bacterial transformation and viruses seems inescapable: Genes, at least those of the pneumococci bacteria and T2 viruses, are made of DNA.

Note, however, that we have not yet generalized beyond the organisms used in these experiments. The crucial question is: Are *all* genes made of DNA? The answer is no, for there are some viruses that do not contain DNA. In such viruses, RNA has been shown to be the primary genetic material. But these viruses are certainly exceptions, comprising only a tiny fraction of the vast spectrum of life. In terms of studies done to date, it seems clear that DNA is the molecule of heredity for virtually all species of plants and animals. DNA appears to be the basis of most hereditary processes in the living world.

10.3 Gene Structure

Granting that genes are composed of DNA, it is reasonable to inquire about the molecular structure of DNA itself. Determining this structure necessarily entails two steps:

1. The molecular subunits of which DNA is composed must be identified.
2. The way in which these parts are fitted together to form the entire DNA molecule must be determined hypothetically.

Step 2 is subject to two limitations. First, the hypothesized model of DNA structure must explain how genes are able to form copies of themselves. Most certainly, if genes are composed of DNA, then DNA itself must be a self-replicating molecule. Second, the hypothesized model of DNA structure must also account for the way in which genes are able to carry out their functions leading to the expression of phenotypes they control.

Components of DNA

Step 1 involves a chemical analysis of pure DNA. Such an analysis reveals DNA to be composed of just three different chemical substances.

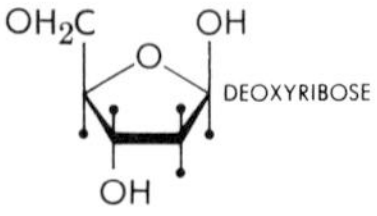

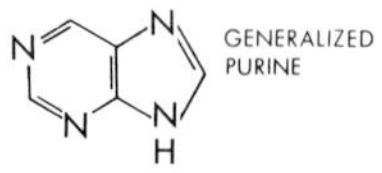

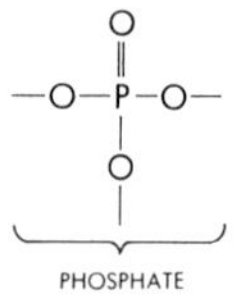

1. A five-carbon (pentose) sugar, **deoxyribose.**

2. **Phosphates.** Each phosphate group is composed of an atom of phosphorus surrounded by four atoms of oxygen and two of hydrogen (the latter are not shown).

3. Usually four nitrogenous (nitrogen-containing) bases—**adenine, guanine, cytosine,** and **thymine.** Adenine and guanine are **purines.** Purine molecules are double-ring structures. Cytosine and thymine are **pyrimidines.** Pyrimidine molecules consist of a single ring of atoms.

Note that the pyrimidine bases are smaller than the purines. The significance of this size difference will be seen shortly.

DNA is the molecule of heredity in nearly all organisms, from bacteria and viruses to human beings. Only in certain types of virus, such as tobacco mosaic virus or polio virus, is the hereditary material different. There it is a form of RNA.

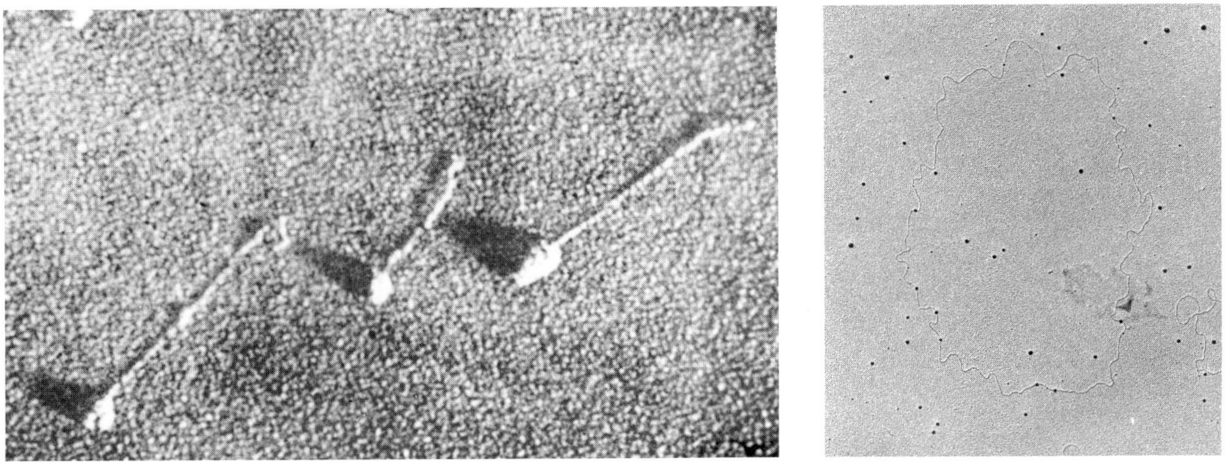

(a) (b)

Fig. 10.2 *Under the electron microscope, DNA appears long and threadlike. (a) DNA is seen extruding from influenza viruses. (b) The circular DNA strand of the lambda viruses that infect* E. coli. *Each such DNA strand is calculated to contain approximately 50,000 nucleotide pairs. (Lambda virus micrograph courtesy Lorne A. McHattie and Vernon C. Bode, Harvard University Medical School.)*

Structure of DNA

With these substances identified, many biologists turned their attention to Step 2—devising a satisfactory hypothetical model for the structure of DNA. Among these biologists were J. D. Watson and F. H. C. Crick. Using data obtained from many different experiments, Watson and Crick proposed a model of DNA structure that has proved highly successful both in its ability to account for gene replication and function and in the accuracy of the predictions that can be derived from it.

First, the physical appearance of purified DNA suggests that the molecule is long and stringlike. When extracted and precipitated in a cold solution of ethanol sodium chloride, DNA rather resembles strands of a spider web or spun glass. Electron microscope photographs confirm the threadlike nature of DNA (Fig. 10.2). Such photographs show each DNA molecule to be quite long but only about 20 A wide. This latter figure is an important one, for it reveals that the DNA molecule can only be about 10 or 12 atoms across.

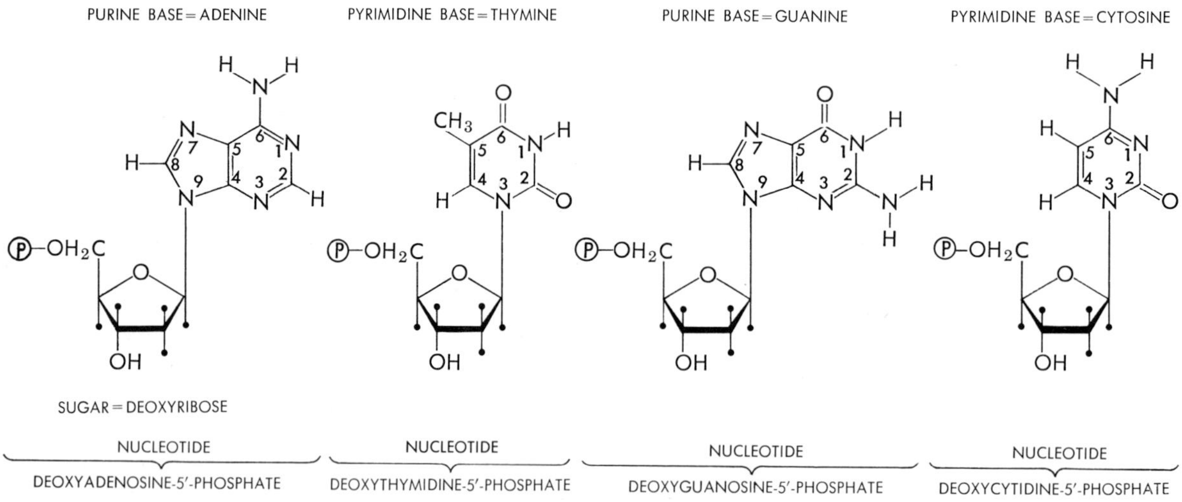

Fig. 10.3 *The four nucleotide bases that form the building blocks of deoxyribonucleic acid, DNA. A nucleotide consists of a purine or pyrimidine base bonded to a sugar and a phosphate group. The sugar–phosphate groups are identical for all nucleotides in DNA. Only the bases differ from one nucleotide to another.*

A second important consideration in determining DNA structure is that one must know which parts are capable of being chemically united; one does not attempt to do a jigsaw puzzle by forcing together pieces that obviously do not fit. It can be shown that the molecular subunits of DNA are joined together into larger subunits called **nucleotides.** Each DNA nucleotide consists of one of the nitrogenous bases, one molecule of deoxyribose, and a phosphate group. Since four nitrogenous bases are generally found in DNA, there are four different nucleotides (Fig. 10.3).

It can be shown that these nucleotides will join together to form long, polynucleotide strands (Fig. 10.4). This fact suggests that the same arrangement might be found in the threadlike DNA molecule. But the distance across one polynucleotide strand is only about 10 A, and DNA, as already mentioned, is generally about 20 Å wide. This suggests that DNA might be composed of paired polynucleotides lying side by side in chemical union.

Determination of the way in which these two strands might be oriented is based largely on information supplied by the physical chemist. First, the molecular structure and configuration of the nitrogenous bases shows them to be capable of forming weak hydrogen bonds with each other. Second, the nitrogenous bases are repelled by water (they are **hydrophobic**), while the sugar–phosphate portions of each nucleotide readily form bonds to water molecules (they are **hydrophilic**). These facts favor an arrangement in which the nitrogenous bases face the interior of the DNA molecule (from which the water of the surrounding cellular environment would be essentially excluded) with the sugar-phosphate units on the outside. This arrangement can be tentatively represented as shown in Fig. 10.5(a).

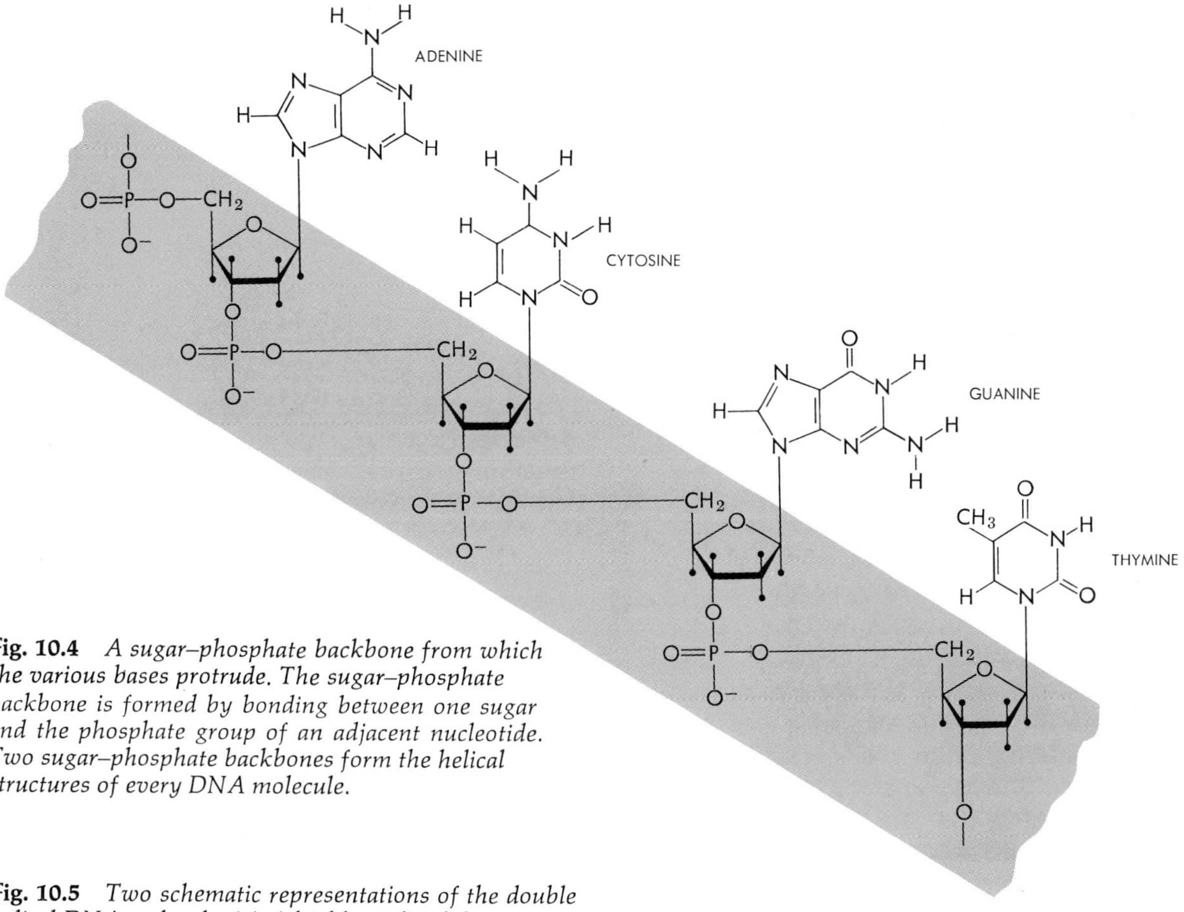

Fig. 10.4 *A sugar–phosphate backbone from which the various bases protrude. The sugar–phosphate backbone is formed by bonding between one sugar and the phosphate group of an adjacent nucleotide. Two sugar–phosphate backbones form the helical structures of every DNA molecule.*

Fig. 10.5 *Two schematic representations of the double helical DNA molecule. (a) A highly stylized diagram of the sugar–phosphate backbone and the nitrogenous bases joined across the center by hydrogen bonds. (b) A more three-dimensional representation, showing the coil of the helix and the position of the bases with respect to the sugar–phosphate backbone.*

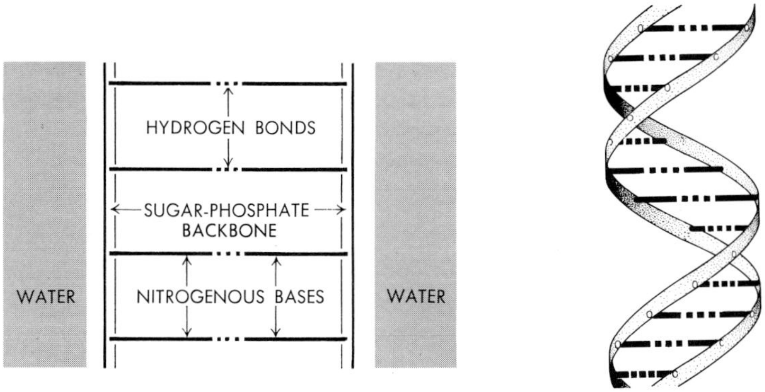

(a) (b)

All the processes of life—e.g., development of the egg into a fully differentiated adult, the chemical functioning of a cell, or the transformation of a normal cell into a cancer cell—can be viewed in terms of the functioning of DNA molecules.

The next clue to DNA structure was provided by X-ray diffraction.* Data obtained from the use of this technique on crystallized DNA showed the model to have a helical rather than a planar structure, with one complete twist of the helix occurring every 34 A, or about every 10 base pairs. The previous representation of DNA structure in Fig. 10.5(a) must therefore be modified to convey its helical nature (Fig. 10.5b).

Obviously, a most desirable way of confirming the hypothesized helical structure of DNA would be to take a "picture" of it. This feat was accomplished in 1970 (Fig. 10.6), and the intellectual creation of Watson and Crick given still more concrete reality.

* X-ray diffraction is a technique that involves passing X-rays through crystallized DNA and determining how they are deflected by the structure through which they pass. X-ray diffraction is *somewhat* analogous to determining the three-dimensional shape of an object by the shadow it casts, but the process is considerably more complex than this analogy indicates.

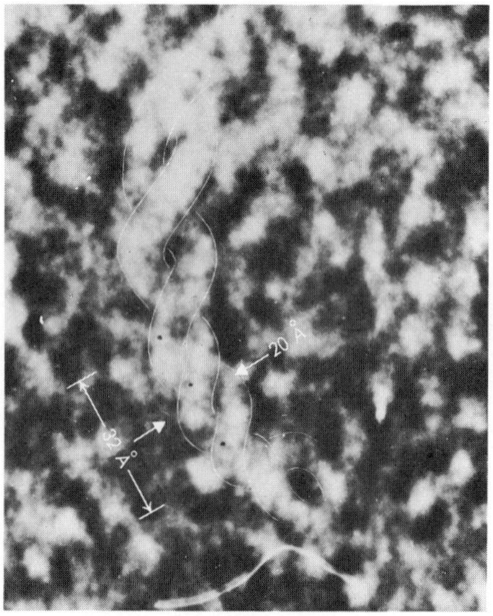

(a)

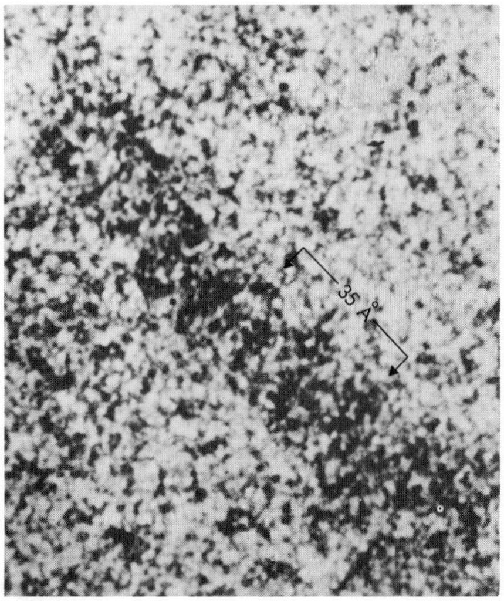

(b)

Fig. 10.6 *These two micrographs dramatically confirm the helical structure of DNA. (a) A short length of one DNA molecule from chromosomes of the pea. The double-stranded helix is plainly discernible. For additional clarity, the artist has added white hairlines around the edge of the helix. (b) A portion of a filament of calf thymus DNA. The double helical substructure has the predicted diameter of 20 angstroms, with a period of approximately 35 angstroms for one complete turn. [(a) courtesy Jack Griffith, California Institute of Technology; (b) courtesy F. P. Ottensmeyer, University of Toronto.]*

The Base Pairs

Another problem remains, however. Granted that the helical strands of DNA polynucleotides are held together by the hydrogen bonds that form between them, it must still be determined which bases pair with which. Do they, for example, pair purine to purine, pyrimidine to pyrimidine, or purine to pyrimidine?

The answer is partly given by our knowledge of the physical nature of hydrogen bonds and the conditions necessary for their formation. Besides calling for the presence of a covalently bonded, positively charged hydrogen atom and a covalently bonded, negatively charged acceptor atom, hydrogen bonds can only form over certain critical interatomic distances. Recall that the purine bases are double-ring structures, larger than the single-ring pyrimidines. If two purine bases formed hydrogen bonds between them, any two pyrimidines in the double strand of DNA would be held too far apart for the formation of their own hydrogen bonds. This can be represented diagrammatically as shown in Fig. 10.7(a). Conversely, if the pyrimidine bases were at the proper distance for hydrogen-bond formation, the purine bases would overlap (Fig. 10.7(b). (For simplicity, the helical shape of DNA is ignored in these diagrams.) It might be thought that the sugar–phosphate strands could bend sufficiently to establish the proper distances for hydrogen bonding both between the pyrimidine bases and between the purine bases. However, physical and chemical considerations reduce this possibility almost to the vanishing point. One might hypothesize that infolding or puckering along the sides of a DNA molecule would allow the proper distances to be established, but electron micrographs fail to show any such puckering.

The alternative hypothesis, which eliminates these difficulties, is clear. Purine-to-pyrimidine bonding must occur (Fig. 15.8c). However, there are two purines involved, adenine and guanine, and two pyrimidines, cytosine and thymine. A new question therefore arises: Granting a purine-to-pyrimidine base pairing, which of the two purine bases pairs with which pyrimidine? Again, the an-

Fig. 10.7 *Schematic diagram showing that base pairing across the double helix of DNA can only occur between a purine and a pyrimidine, not between two purines (overlap) or two pyrimidines (gap). Purine–pyrimidine base pairing is consistent with the observation that all along its length the DNA molecule has a constant diameter.*

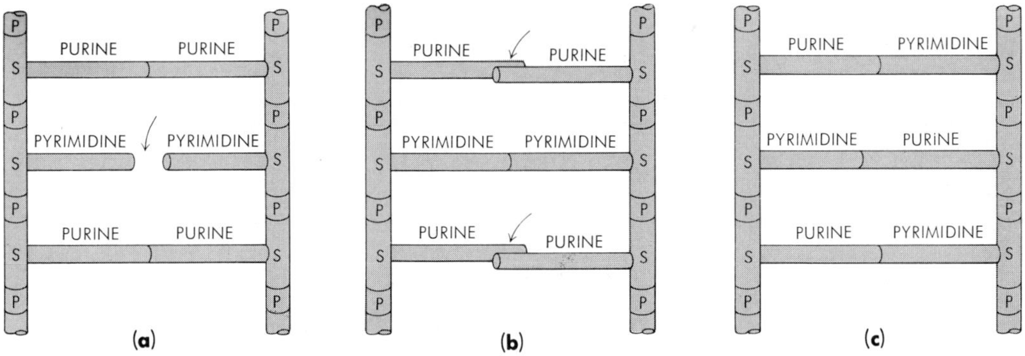

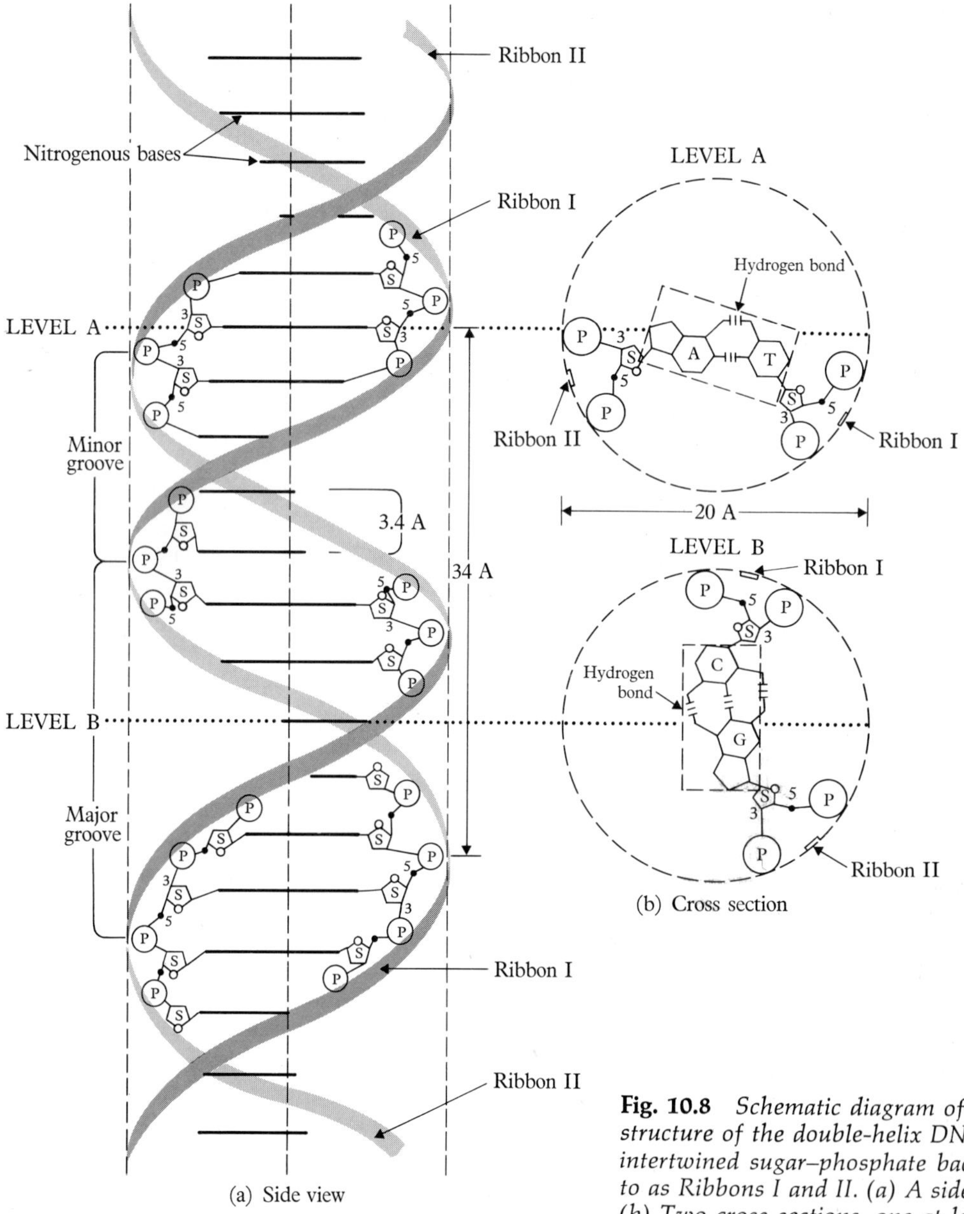

Fig. 10.8 *Schematic diagram of the three-dimensional structure of the double-helix DNA molecule. The two intertwined sugar–phosphate backbones are referred to as Ribbons I and II. (a) A side view of the molecule. (b) Two cross sections, one at level A, the other at level B, as indicated on the side view. This diagram has been prepared to represent actual molecular dimensions and thus to give an indication of how the molecule fits together in space. [From Dr. William Etkin,* BioScience *23 (November 1973), p. 653, with permission of the author and the American Institute of Biological Sciences.]*

Table 10.1
Adenine-to-Thymine and Guanine-to-Cytosine Ratio in DNA

Source of DNA Template	Adenine	Thymine	Guanine	Cytosine
Bovine thymus	28.2	27.8	21.5	21.2
Bovine spleen	27.9	27.3	22.7	20.8
Bovine Sperm	28.7	27.2	22.2	20.7
Rat bone marrow	28.6	28.4	21.4	20.4
Herring testes	27.9	28.2	19.5	21.5
Paracentrotus lividus	32.8	32.1	17.7	17.3
Wheat germ	27.3	27.1	22.7	22.8
Yeast	31.3	32.9	18.7	17.1
Escherichia coli	26.0	23.9	24.9	25.2
Mb. tuberculosis	15.1	14.6	34.9	35.4
Rickettsia prowazeki	35.7	31.8	17.1	15.4
Micrococcus lysodeiticus	15.0	15.0	35.0	35.0
Aerobacter acrogenes	22.0	22.0	28.0	28.0
Phage T2	32.0	32.0	18.0	18.0

swer is provided by the factors leading to the formation of hydrogen bonds between the nitrogenous bases. The only base-pairing combinations that allow the formation of the hydrogen bonds found in DNA are those of adenine–thymine and guanine–cytosine.

It is interesting to note that the limitation of purine–pyrimidine base pairing to adenine–thymine and guanine–cytosine combinations could have been deduced from other data. Careful chemical analyses of DNA had been made earlier to determine just how much of each kind of base is present in the DNA of a particular species. The results of some of these analyses are presented in Table 10.1. Note that in any one species, the amount of adenine equals or closely approximates that of thymine, while the amount of guanine closely approximates or equals that of cytosine, the deviations in each ratio lying well within the range to be expected from errors in experimental measurements. These results would be predicted by the Watson and Crick model, which specifies an adenine–thymine and guanine–cytosine base pairing within the DNA molecule.

Given these details of structure, an accurate model of the DNA molecule can now be drawn (see Fig. 10.8). This diagram clearly shows the double helical nature, the purine-to-pyrimidine bonding, and the role of the sugar-phosphate groups in establishing the helical "backbone."

10.4 DNA Replication: The Modern Picture

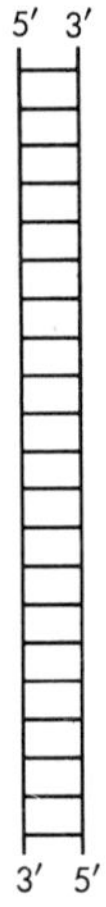

In 1956 a team of scientists headed by Dr. Arthur Kornberg, then at Washington University in St. Louis, Missouri, succeeded in isolating an enzyme capable of synthesizing DNA in the test tube. Kornberg named this enzyme DNA polymerase. In the course of their work, Kornberg and his St. Louis team also confirmed another prediction of the Watson–Crick structure: that the two strands of DNA in the double helix lie "head to tail." In chemical terms this meant that the end called 3′ of one strand lay opposite the 5′ end of the other (see the diagram). The importance of this fact will become apparent shortly. Late in 1967 at Stanford University, Kornberg achieved another major suc-

Supplement 10.1

DNA REPLICATION: HOW DO WE KNOW?

The Watson–Crick model of DNA must account for the ability of the molecule both to direct cell activity (that is, gene function) and to replicate itself (reproduce). The beauty of the model is that it suggests ways in which both processes could occur. Subsequent experimentation has provided considerable understanding of both processes. Let us consider here the second question: how the DNA molecule makes a perfect copy of itself. Work on this problem began almost immediately after the publication of Watson and Crick's original paper in 1953.

It is quite easy to imagine that the process of replication involves an unwinding and separation of the two DNA polynucleotide strands, with "unzipping" occurring through the breakage of the hydrogen bonds of each base pair. This mental picture is supported by the fact that the amount of energy needed to separate the DNA strands is equivalent to the amount of energy needed to break hydrogen bonds. Once separated from their partners, the unpaired nucleotides on each strand would attract their complementary nucleotides from the surrounding medium. Thus each unpaired polynucleotide strand would build a strand complementary in structure to itself. The result would be two DNA molecules, each an exact replica of the original (Diagram A).

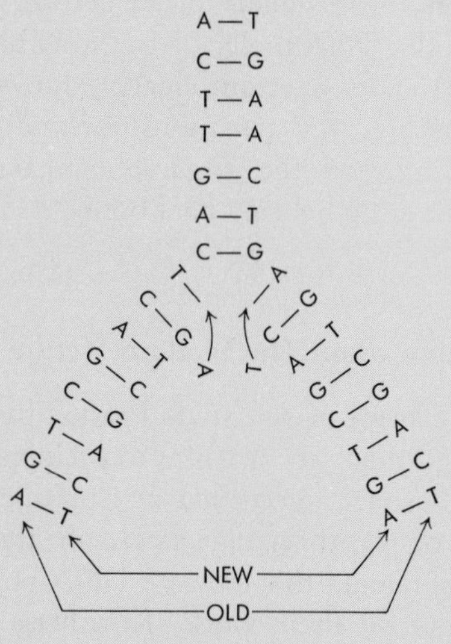

Diagram A *A highly diagrammatic representation of base pairing and DNA replication as hypothesized by Watson and Crick. The four bases—adenine, thymine, guanine, and cytosine—are represented by the letters A, T, G, and C. The total length of the segment of molecule represented here would be approximately 50 Å (0.005 μ).*

One immediate prediction of the Watson–Crick hypothesis suggested by the diagram is that DNA in the process of replication should have the form of the letter Y. Dr. John Cairns of the Cold Spring Harbor Laboratory in New York has verified this prediction.

An elegant experimental test of the means of DNA replication suggested by Watson and Crick was performed in 1958 by Matthew Meselson and F. W. Stahl, then at the California Institute of Technology. *Escherichia coli* bacteria were grown in a medium containing glucose, mineral salts without nitrogen, and ammonium chloride (NH_4Cl) in which almost all of the nitrogen atoms were the heavy isotope, N^{15}. The bacteria were allowed to grow in this medium until fourteen generations of bacteria had arisen from those used to inoculate the culture. Hence virtually all of the nitrogen in the DNA of these bacteria was heavy nitrogen.

The key to the Meselson–Stahl experiment lay in the experimental technique. When a cesium chloride solution is spun in an ultracentrifuge up to about 60,000 times the force of gravity, the molecules of cesium chloride begin to sediment. However, due to their relatively low molecular weight, they never sediment completely. Instead, the result is a gradation of low-to-high solution density, from the top of the centrifuge tube to the bottom. Any foreign molecule centrifuged in such a cesium chloride gradient will come to rest at a level at which its density equals that of the surrounding solution. Thus DNA extracted from *E. coli* bacteria grown in a medium containing the regular light isotope of nitrogen, N^{14}, will form a band at a higher point in the centrifuge tube than DNA from *E. coli* bacteria grown in a medium containing N^{15}.

After centrifuging, Meselson and Stahl removed the experimental bacteria from the heavy nitrogen medium and allowed them to undergo just one more generation, one more DNA replication. Note that as a result of this step, heavy DNA molecules containing N^{15} replicated in a medium in which only light nucleotides containing N^{14} were available. According to the hypothesis of Watson and Crick, the result of such replication should be "hybrid" DNA molecules, each molecule containing one heavy strand and one light strand. After centrifugation, these hybrid DNA molecules should appear as a new band in the cesium chloride density gradient. This new band should lie in the region between the completely light DNA band formed from bacteria grown in a medium containing N^{14}, and the heavy DNA band formed from bacteria grown in a medium containing N^{15}. This prediction is verified by the experimental results.

The test can be carried further by allowing the N^{15}-labeled bacteria to replicate for *two* rather than one generation in a medium containing N^{14}. Here one would predict two DNA bands in the cesium chloride density gradient tube. Half of the DNA should be found in the area occupied by light DNA. The other half should be in the area occupied by the "hybrid" DNA. Again, the results obtained by Meselson and Stahl verified these predictions and supported the hypothesis of Watson and Crick (see Diagram B).

E. coli, of course, is a bacterium, a fairly primitive organism. What about higher organisms, in which the DNA is located in distinct chromosomes? We have only working hypotheses to explain just how DNA is arranged within the chromosome. However, despite this uncertainty, an experiment performed by J. A. Taylor of Columbia University sheds some light on the problem of extrapolation from the replication of DNA in *E. coli* to the replication of DNA in higher organisms.

Taylor worked with plant root tips, in which the cells constantly undergo mitosis. He immersed these root tips in a solution containing the nucleoside thymidine, which had been labeled with radioactive hydrogen (tritium, H^3). The root tips were left in this solution long enough for many of their cells to double their DNA content, but not long enough for it to be doubled again. Any DNA molecules formed during this time would incorporate the radioactive thymidine into their structure and thus themselves become radioactive (see Diagram C).

As soon as the pair of daughter chromosomes became visible, they were tested by autoradiography. The hypothesis of Watson and Crick predicts that the members of each pair of chromosomes should contain radioactive DNA and that they should contain it in equal amounts. This prediction was verified by a count of the black dots found on the photographic film where radiation had fogged it.

If one allows cells containing radioactive daughter chromosomes to undergo another cycle of duplication in a solution containing no radioactivity, the result is a completely different prediction, which provides yet another test of the Watson–Crick hypothesis. Here, autoradiography should reveal one member of each pair of daughter chromosomes to be radioactive and the other nonradioactive. This prediction is verified. Because of the many complexities involved in experimentation of this sort, which introduce the possibility of uncontrollable variables, Taylor's observations have been questioned by other investigators. However, the reported results are consistent with the Watson–Crick hypothesis, and similar experiments on human chromosomes in tissue culture yield the same results.

From the Meselson–Stahl and Taylor experiments, it is apparent that: (1) DNA does replicate itself, by one strand acting as a template on which a second whole strand is synthesized, and (2) this process appears to be general, from the DNA of bacteria to that of eucaryotic organisms such as plants or human beings. Many of the details of the process are still unclear (see Section 10.4); its broad outline, however, is now well established.

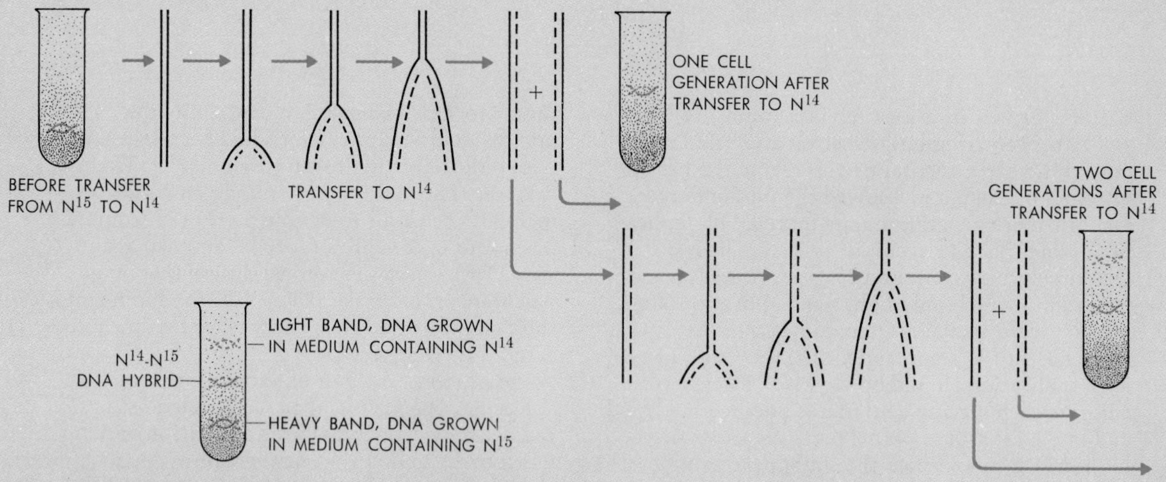

Diagram B *The Meselson–Stahl experiment is based on a density gradient caused by ultracentrifugation for 48 hr in a cesium chloride solution. When DNA molecules are spun in such a solution, they separate into regions of the centrifuge tube at which their density equals that of the CsCl solution.*

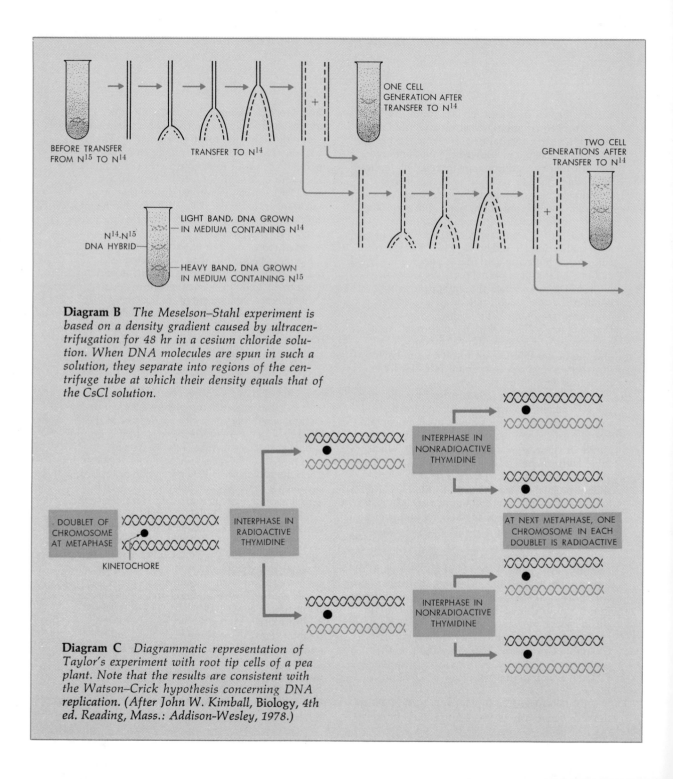

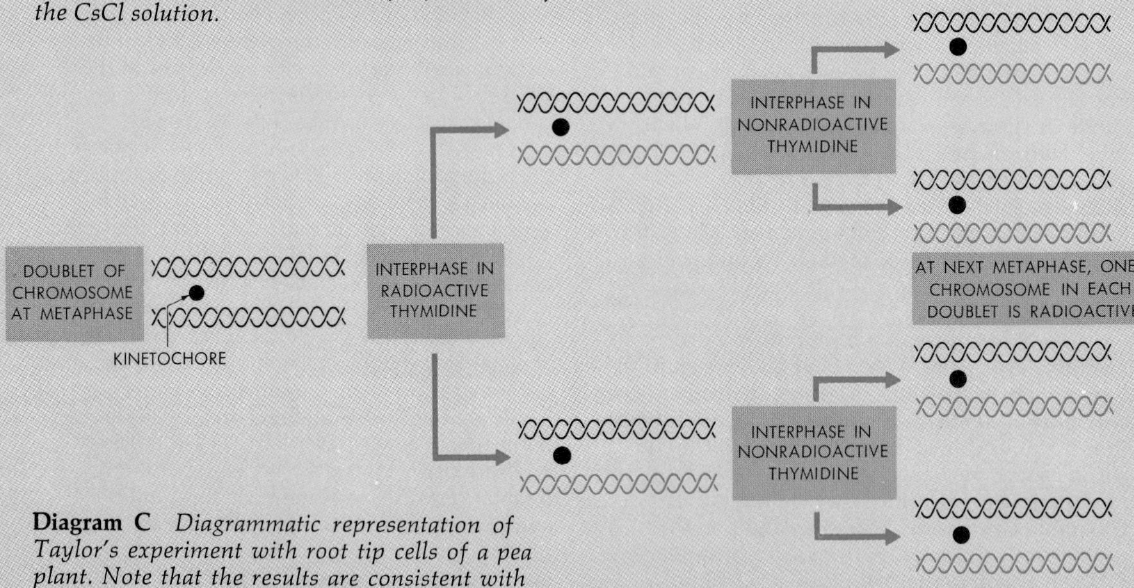

Diagram C *Diagrammatic representation of Taylor's experiment with root tip cells of a pea plant. Note that the results are consistent with the Watson–Crick hypothesis concerning DNA replication. (After John W. Kimball,* Biology, *4th ed. Reading, Mass.: Addison-Wesley, 1978.)*

cess when he used DNA polymerase to synthesize a perfect copy of bacteriophage DNA.

The Kornberg enzyme (as it is called) is one of the largest single-chain globular proteins known—it has a molecular weight of 109,000. Only one site on the enzyme can interact with free nucleotide triphosphates. It is hypothesized that once the DNA polymerase encounters a given base in the template strand, it can catalyze formation of the ester bond (in the sugar-phosphate backbone) in the daughter strand *only* when the correct, complementary base (nucleotide) is present. Presumably only the complementary base will have the correct geometry to align properly with the elongating daughter strand and thus be in the correct position to form an ester bond. For example, if a pyrimidine is called for, a purine might stick out too far to align properly. Once an ester bond is formed, the enzyme is presumably free to move on to the next base in the template.

However, the bacteriophage DNA synthesized by Kornberg forms a closed circle (similar to that shown in Fig. 10.9) and, more important, consists of a single rather than a double strand. Most DNA is double-stranded. Like most enzymes, DNA polymerase is quite specific in its action, not only as to substrate but also as to how it acts on this substrate. In particular, Kornberg found that *DNA polymerase could synthesize a strand of DNA only in a manner causing the new polynucleotide chain to grow in the 5′-to-3′ direction.*

Kornberg's DNA polymerase presented geneticists with an immediate dilemma, however. DNA polymerase can synthesize a strand of DNA only in one direction: starting at the 5′ end of the template strand, and moving toward the 3′ end. Since most DNA is double-stranded, the question became, how does synthesis of new DNA take place from a parental molecule? Does only one of the two parental strands get copied? Obviously not, since in DNA replication it is known that *both* parental strands are replicated. Are there two different DNA polymerases, one which copies one of the parental strands in the 3′-5′ direction, and another which copies the other parental strand in the 5′-3′ direction? A long and arduous search by biochemists and molecular geneticists has revealed no such reverse DNA polymerase. All DNA polymerases studied, from whatever organism, appear to catalyze synthesis only in the 5′-3′ direction.

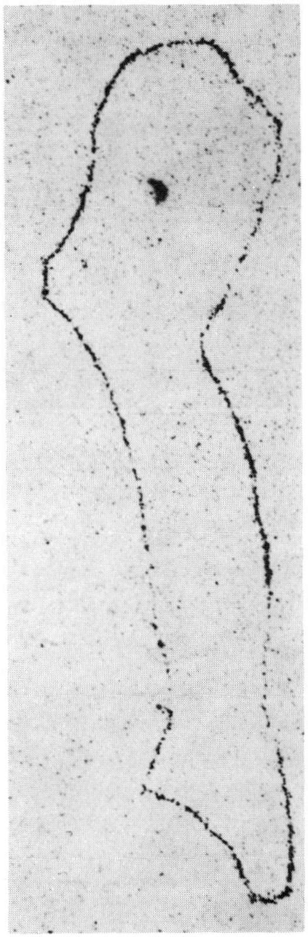

Fig. 10.9 *Shadowed electron micrograph of circular DNA of a bacteriophage, λ, similar to that found in the phage φX-174 used by Kornberg. λ DNA is double-stranded, while φX-174 DNA is single-stranded. The length of the circular strand is about 16 μ.*

Kornberg hypothesized that as the parental double helix unwinds, DNA polymerase synthesizes short groups of nucleotides on both parental strands. In this way, the enzyme could always join nucleotides onto the 3′ (OH group) end of another nucleotide. Apparently both parental strands are copied in very short segments, rather than growing as single, long chains. But how, then, are whole, intact daughter strands produced? The answer comes from the recent work of Reiji Okazaki and his col-

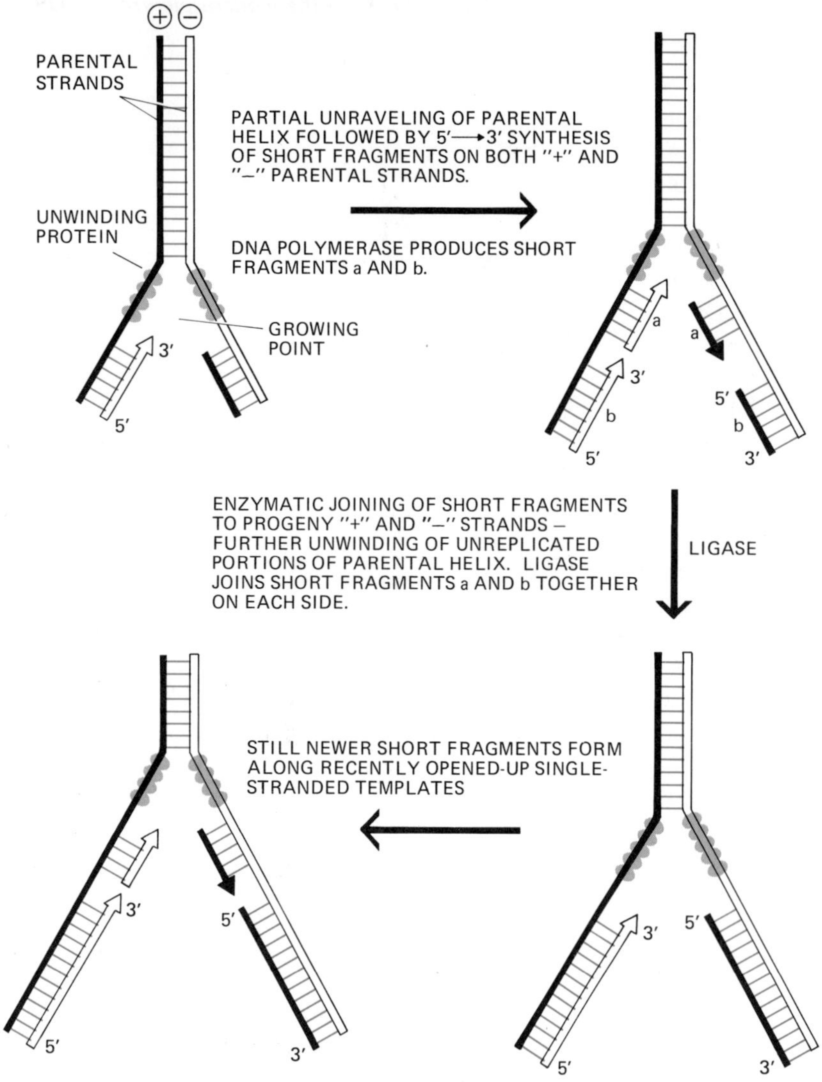
+
−
PARENTAL STRANDS
UNWINDING PROTEIN
GROWING POINT
3′
5′
PARTIAL UNRAVELING OF PARENTAL HELIX FOLLOWED BY 5′⟶3′ SYNTHESIS OF SHORT FRAGMENTS ON BOTH "+" AND "−" PARENTAL STRANDS.
DNA POLYMERASE PRODUCES SHORT FRAGMENTS a AND b.
a
b
ENZYMATIC JOINING OF SHORT FRAGMENTS TO PROGENY "+" AND "−" STRANDS – FURTHER UNWINDING OF UNREPLICATED PORTIONS OF PARENTAL HELIX. LIGASE JOINS SHORT FRAGMENTS a AND b TOGETHER ON EACH SIDE.
LIGASE
STILL NEWER SHORT FRAGMENTS FORM ALONG RECENTLY OPENED-UP SINGLE-STRANDED TEMPLATES

Fig. 10.10 *Okasaki's modification of Arthur Kornberg's original hypothesis concerning DNA replication. As Kornberg originally hypothesized, a single enzyme, DNA polymerase, is involved in synthisizing new DNA by moving from the 5′ to the 3′ end of an open, or single, strand of parental DNA. The dilemma for Kornberg was how to account for two-directionality of synthesis (since both strands of the double helix are copies, yet both run in opposite directions) with the same enzyme. Okasaki has recently demonstrated that Kornberg's enzyme does guide synthesis on both strands, in opposite directions, but for only very short segments. Another enzyme, known as ligase, joins these short strands together. In the diagram to the left, the two strands of the parental DNA are labelled (+) or (−) and shown in black and white, respectively. (Adapted from J. D. Watson,* Molecular Biology of the Gene, *2d ed. Menlo Park, Calif.: W. A. Benjamin, 1970, p. 286.)*

leagues at Nahoyo University in Japan. Exposing cells very briefly to radioactive nucleotides so that only the most recently synthesized DNA was labelled, Okasaki found that the newest DNA existed in extremely short fragments (of only a few nucleotides), held to the parental strand by hydrogen bonds. Soon thereafter, Okasaki found, these short fragments are joined together by a second enzyme, which is now called **ligase,** or the "joining enzyme." Thus, DNA polymerases joins nucleotides together to produce short fragments, with ligase catalyzing the joining together of these fragments to make a complete strand of DNA. Thus, it appears that while the two parental strands of any single DNA molecule grow in the same direction in the overall sense (see Fig. 10.10), they grow in opposite directions for very brief periods of time.

In summary, then, though somewhat more complex than the older model, the Kornberg model of DNA replication is plausible and, more important, fits the facts. Despite verification of predictions stemming from the Kornberg hypothetical model of DNA replication, one major objection remains: *no one has ever demonstrated that DNA polymerase can synthesize a double helix of DNA in the test tube.* Recall that Kornberg achieved his DNA-synthesis feat with a bacteriophage DNA that is *single*-stranded.

Primarily because of this failure to synthesize double-helix DNA with DNA polymerase, doubts began to be expressed whether DNA polymerase was the primary DNA-replicating enzyme at all! The more the enzyme was studied, the more apparent it became that it constituted a perfect DNA *repair,* rather than replicating, enzyme. (DNA repair is a universal and vital function, involving "snipping out" wrongly matched sections of DNA and replacing them with sections having the correct base sequences.) The case against DNA polymerase as the DNA-replicating enzyme was greatly strengthened when, after laboriously screening almost 3500 mutants of *E. coli,* Drs. John Cairns and Paula de Lucia at Cold Spring Harbor found one that had no active form of DNA polymerase, yet could still carry out DNA replication. Such a finding suggested there must be a second kind of enzyme that could catalyze DNA synthesis. Furthermore, the new mutant was far more sensitive to ultraviolet light than normal *E. coli.* Since ultraviolet radiation is known to cause mutations, and since the mutant *E. coli* seemed less capable of repairing these mistakes, this result provides strong support for the hypothesis proposing a DNA-repair rather than DNA-replicating role for DNA polymerase. The postulated existence of a second enzyme, known as DNA polymerase III, was confirmed in 1968. Its molecular structure is poorly understood. Like Kornberg's original enzyme, now called DNA polymerase I, polymerase III synthesizes in the 5′-to-3′ direction. Unlike polymerase I, however, polymerase III can *initiate* DNA synthesis using RNA primer. Polymerase I can only continue synthesis of a strand that is already going. Again, it appears that Kornberg's original enyzme may serve primarily a repair function.

10.5 Protein Synthesis

Differences in the DNA of different organisms can be traced to differences in the order of base pairs along the DNA molecule. Similarly, differences between proteins can be traced primarily to differences in the kinds and order of amino acids along the polypeptide chains. The conclusion seems clear and inescapable: *the order of base pairs along DNA molecules must somehow control the kind and order of amino acids found in the proteins of an organism.* DNA, then, carries a **genetic code**—the blueprint establishing the kinds of proteins synthesized by the cellular machinery that make an individual organism unique.

On the basis of the results of many different kinds of experiments performed in laboratories all over the world, a model system has been hypothesized to explain the sequence of events comprising protein synthesis, from gene to final product. In order to fully understand this model system, we must direct our attention toward ribonucleic acid, or RNA. Chemically, RNA molecules are very similar to those of DNA; they consist of sugar and phosphate units connected to four different kinds

In molecular terms, the genotype of an organism is the information coded in the base sequence of DNA; the phenotype is the collection of different kinds of protein molecules that DNA produces in the cell.

Supplement 10.2

GENES PRODUCE PROTEINS: HOW DO WE KNOW?

There is a condition in humans known as phenylketonuria, the most prevalent symptoms of which are a severe mental deficiency and the execretion in the urine of about 1 g of phenylpyruvic acid per day. Phenylpyruvic acid is derived from phenylalanine (an essential amino acid in the human diet) by replacement of the amino (NH_2) group by an oxygen atom.

In normal persons, phenylpyruvic acid is converted to *p*-hydroxyphenylpyruvic acid, which in turn is converted into another compound, and so on. Eventually the reaction series ends in the production of carbon dioxide and water. Each step in the series is catalyzed by a specific enzyme. A representation of this reaction series can be given as follows:

$$\text{PHENYLALANINE} \xrightarrow{\text{enzyme 1}} \text{PHENYLPYRUVIC ACID} \xrightarrow{\text{enzyme 2}} p\text{-HYDROXYPHENYLPYRUVIC ACID} \xrightarrow{\text{enzyme 3}} \cdots \xrightarrow{\text{enzyme 4}} \cdots \xrightarrow{\text{enzyme } n} \cdots CO_2 + H_2O.$$

Examination of the pedigrees of human families wherein phenylketonuria occurs suggests that the condition is recessive and due to a single pair of genes.

Early in this century the Oxford physician–biochemist Sir Archibald E. Garrod proposed that persons with conditions similar to phenylketonuria differ from normal persons only in that they lack the enzyme that catalyzes a certain vital reaction in a series. In the case of phenylketonuria, this would be the enzyme catalyzing the reaction that converts phenylpyruvic acid to *p*-hydroxylphenylpyruvic acid. The hypothesis selects this enzyme over the others involved in the reaction series from phenylalanine to carbon dioxide and water because only the blockage of the action of this enzyme could account for the accumulation of phenylpyruvic acid. Phenylpyruvic acid, unable to cross the energy barrier of the reaction without the aid of the missing enzyme, increases in amount and concentration within the body. This increase damages the sensitive brain tissue, leading to symptoms of mental deficiency. With still further accumulation, the compound spills over into the urine for excretion.

In brief, then, this hypothesis proposes that what is inherited is not phenylketonuria, but the failure of the body to produce a certain enzyme, and that this enzyme deficiency results in phenylketonuria.

It is easy to extend this hypothesis to suggest that genes carry out their action by directing the formation of enzymes (all enzymes, of course, are proteins), and that these enzymes, in turn, catalyze one step of a particular reaction series, such as the series leading to the production of black pigment in the hair of a mouse.

Here is an hypothesis with obvious potential for explaining a great deal about gene action. Considerable difficulty was encountered in testing it, however, a major limitation being the absence of a suitable experimental organism. The fruit fly, invaluable for examining the various ways in which phenotypic variations are inherited and in furthering the gene-on-chromosome hypothesis, is unsatisfactory here. Between the action of the genes and the formation of a vestigial wing is a broad area of very complicated chemistry, with many reaction pathways available. In the late 1930s large numbers of investigators, attracted by the promising outlook of biochemical genetics, turned their attention to other experimental organisms. Among these were *E. coli* (about which more genetic knowledge has been accumulated than about any other form of life), and the red bread mold, *Neurospora crassa*. *Neurospora*, in particular, possesses several advantages as an experimental organism. First, it has an extremely short life cycle. Second, it has a low number of chromosomes (seven), on which the genes can readily be mapped. Third, its life cycle, though complex, is short, with the majority of the cycle being spent in the haploid condition. This means that any recessive gene present is immediately detectable in the phenotype, there being no dominant allele to mask it. Finally, the reproductive apparatus of *Neurospora* can be neatly dissected to isolate individual reproductive cells. This procedure in effect isolates a single complete set of genes.

Molds are generally thought of as being rather simple and primitive forms of life. In terms of their status on the evolutionary tree, such a concept is not without validity. Biochemically speaking, however, molds are far from simple. Indeed, they retain the ability to synthesize compounds necessary for life which humans cannot synthesize and must include in their diet. Thus *Neurospora* can be grown in a laboratory on a medium containing only cane sugar, inorganic salts, and one vitamin (biotin). Such a medium is called a **minimal medium,** for it contains *only* those substances *Neurospora* needs but cannot synthesize for itself. *Neurospora* possesses, as an inherited characteristic, the ability to synthesize all the other vitamins and amino acids essential for life.

Biologists George Beadle and Edward L. Tatum devised a way to use *Neurospora* to test the hypothesis that genes act by directing the formation of specific enzymes. First they exposed spores of *Neurospora* to X-rays or ultraviolet light. It was found that some of the molds produced by such irradiated spores could no longer grow on a minimal medium. Evidently the experimental molds had lost the ability to produce certain essential substances. By adding vitamins and amino acids to the minimal medium one by one, investigators could determine which substance or substances a mutant mold had lost the ability to synthesize.

A specific example may help to make this important concept clear. Three strains of mutant molds that could no longer live on the minimal medium were isolated. The strains differed from each other in their requirements for additives to the minimal medium. Strain A would grow only if the amino acid arginine were added; strain B would grow if either arginine or citrulline were added; strain C, on the other hand, would grow with the addition of arginine, citrulline, or ornithine. The explanation for these different requirements was found by examining the following reaction series.

It had been determined that there is a single reaction series leading to the formation of arginine. Each step is catalyzed by a specific enzyme:

$$\text{PRECURSOR } X \xrightarrow{\text{enzyme 1}} \text{ORNITHINE} \xrightarrow{\text{enzyme 2}} \text{CITRULLINE} \xrightarrow{\text{enzyme 3}} \text{ARGININE.}$$

On the basis of the gene–enzyme hypothesis, the gene responsible for the synthesis of enzyme 3 must have been mutated in mutant strain A to a form incapable of guiding production of the enzyme. The hypothesis proposing such a mutation predicted that adding any precursor compound, such as ornithine or citrulline, would not enable strain A to grow on the medium, since the essential reaction converting citrulline to arginine would be blocked by the absence of enzyme 3. This prediction was verified. Similar reasoning led to the hypothesis that in mutant strain B, the gene responsible for producing enzyme 2 had been mutated. The predictions of this hypothesis were consistent with the observation that the addition of either arginine or citrulline (but *not* ornithine) enabled strain B to grow. It was easy to see that mutant mold strain C would be able to grow with the addition of ornithine or any of the compounds coming after ornithine in the reaction series. This prediction was also verified by the experimental results: the gene responsible for producing enzyme 1 must have been mutated.

Two important steps had then to be taken to support conclusively the gene–enzyme hypothesis. First it had to be shown that any given deficiency in a mutant strain was due to a single gene. This was done by showing that the ratios of defective to normal molds resulting from experimental crosses were those predicted by Mendelian genetics for one-gene characteristics. Next it had to be shown that the enzyme in question was indeed absent from molds showing the defect. Thus, for example, a biochemical analysis of mutant mold strain A must reveal that no enzyme 3 is present. Similar analysis of strains B and C must show an absence of enzymes 2 and 1, respectively. In each case, the predictions of the gene–enzyme hypothesis were verified; the appropriate enzymes were missing.

Beadle and Tatum shared a 1958 Nobel prize for their work with *Neurospora.* In a speech delivered at the award ceremony, Beadle said:

> *In this long, roundabout way, first in* Drosophila *and then in* Neurospora, *we had rediscovered what Garrod had seen so clearly so many years before. By now we knew of his work and were aware that we had added little if anything new in principle. We were working with a more favorable organism and were able to produce, almost at will, inborn errors of metabolism for almost any chemical reaction whose product we could supply through the medium. Thus, we were able to demonstrate that what Garrod had shown for a few genes and a few chemical reactions in man was true for many genes and many reactions in* Neurospora.

Garrod, however, had linked deficient or missing enzymes in humans with specific *abnormal* conditions such as phenylketonuria or alcaptonuria (symptomized by a marked discloration of the urine) only by the indirect evidence provided by Mendelian ratios. The direct connection established by Beadle and Tatum in *Neurospora* still left open a wide extrapolation gap from mold to human. In 1970, however, Drs. A. Kobata and V. Ginsburg of the National Institutes of Health in Bethesda, Maryland, showed that the enzyme responsible for the synthesis of the A antigen in individuals of blood types A and AB is not present in persons with O or B blood. Thus, nearly 70 years after Garrod and 30 years after Beadle and Tatum, direct experimental evidence for the one gene–one enzyme hypothesis was obtained in normal human beings.

Despite this recent triumph, it was the Beadle and Tatum experiment that provided conclusive support for a one gene–one enzyme hypothesis. It later became evident that certain genes control the formation of certain proteins that are not enzymes. The protein collagen, for example, is a gene-coded, nonenzyme structural protein that is the main constituent of connective tissue. Collagen accounts for approximately one-third of the protein in the human body. Thus the one gene–one enzyme hypothesis was modified to the one gene–one protein, or the one gene–one polypeptide, hypothesis.

of nitrogeneous bases. In RNA, however, thymine does not occur, its place being taken by **uracil.** Like DNA, RNA can store genetic information in its base sequence. Like the circular DNA of Kornberg's bacteriophage, RNA molecules are single-stranded. Finally, the sugar in RNA is ribose rather than the deoxyribose of DNA.

Ribonucleic Acid (RNA)

Present ideas concerning the molecular configuration of RNA molecules are much less clear than the Watson–Crick model of DNA. There are good reasons for this. RNA is far more difficult to obtain in pure crystalline form than DNA. Hence RNA is less easily studied by X-ray diffraction techniques. As a result, the type of information this technique can provide has been lacking for RNA, at least until fairly recently. In addition, RNA occurs in at least three forms. Each of these forms has a different structure and function. The three forms of RNA recognized today are *transfer* RNA, *ribosomal* RNA, and *messenger* RNA. Structural differences between these forms are not due primarily to differences in the nucleotides involved, but result from differences in molecular weight and configuration.

Transfer RNA. Transfer RNA (*t*RNA), often called soluble RNA, is the smallest type of RNA. Each transfer RNA molecule contains only 70 to 80 nucleotides. Transfer RNA is the only type of RNA for which a fairly definite molecular structure has been determined.

Transfer RNA picks up individual amino acids in the cell and carries them to the sites of protein synthesis. Since each RNA molecule will pick up only one type of amino acid, there are many molecular variations of transfer RNA, one for each type of amino acid. Each has a slightly different sequence of bases. This enables each transfer RNA molecule to unite with one specific type of amino acid.

In December 1964, the precise sequence of the 77 nucleotides of the transfer RNA coding for the amino acid alanine was worked out by a team of Cornell University scientists headed by Dr. Robert Holley. Holley's work, published in 1965, gave precise data only on the sequence of nucleotides in one transfer RNA molecule. At that time there was little evidence about the three-dimensional structure and thus about how the molecule could actually work. In recent years, however, X-ray diffraction studies and even electron microscopy have revealed that the *t*RNA molecule has generally constant shape, although each of the twenty different types (one specifically for each amino acid) have recognizable differences. The generalized shape is shown in the three-dimensional sketch in Fig. 10.11. The molecule tends to assume this shape as the thermodynamically most stable configuration, with hydrogen bonds forming across the strands between complementary bases. The loop containing the anticodon is where the *t*RNA molecule attaches to the messenger RNA; the anticodon is a sequence of three bases (triplet) complementary to a specific triplet on the *m*RNA (see page 632). The 3′ (—OH) end, or amino acid arm, of the *t*RNA molecule is where the specific amino acid attaches. The distance from the anticodon loop to the amino acid arm appears to be constant for all *t*RNA molecules, ensuring that all amino acids will be brought adjacent to each other when the *t*RNA attaches to the messenger. Schematic representations of a few types of *t*RNA molecules are shown in Fig. 10.12.

Ribosomal RNA. Ribosomal RNA (*r*RNA) has a relatively high molecular weight. As pointed out in Chapter 4, it is contained in ribosomes, located in the cytoplasm of the cell. Ribosomes are composed of two thirds RNA and one third protein. The ribosomes function as centers of protein synthesis. It is to the ribosomes that the transfer RNA molecules carry their amino acids. During this process, some interaction occurs between transfer and ribosomal RNA.

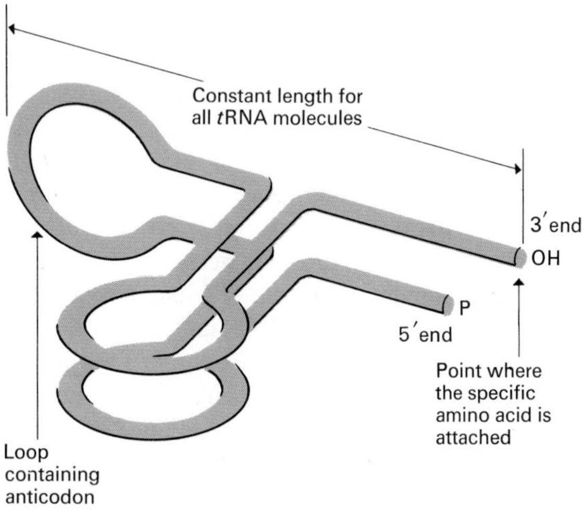

(a)

(b)

Fig. 10.11 *(a) One of the possible three-dimensional models for the structure of transfer RNA. This generalized picture applies to all tRNA molecules; though there are many differences between the twenty-odd types, they may all possess at least as much in common as the pattern of folded loops and the constant distance from the anticodon to the amino acid attachment site. (Adapted from Loewy and Siekevitz,* Cell Structure and Function, *2d ed. New York: Holt, Rinehart, 1969, p. 168.) (b) The actual physical structure of the tRNA molecule is probably not so neat looking as that shown in (a). A representation of one type of tRNA molecule, based on X-ray diffraction data, shows that it contains a number of twists and loops, folding back and base-pairing on itself in many places. For purposes of illustrating specific kinds of tRNA, however, the more schematic model shown in (a) will be used (for example, see Fig. 10.12).*

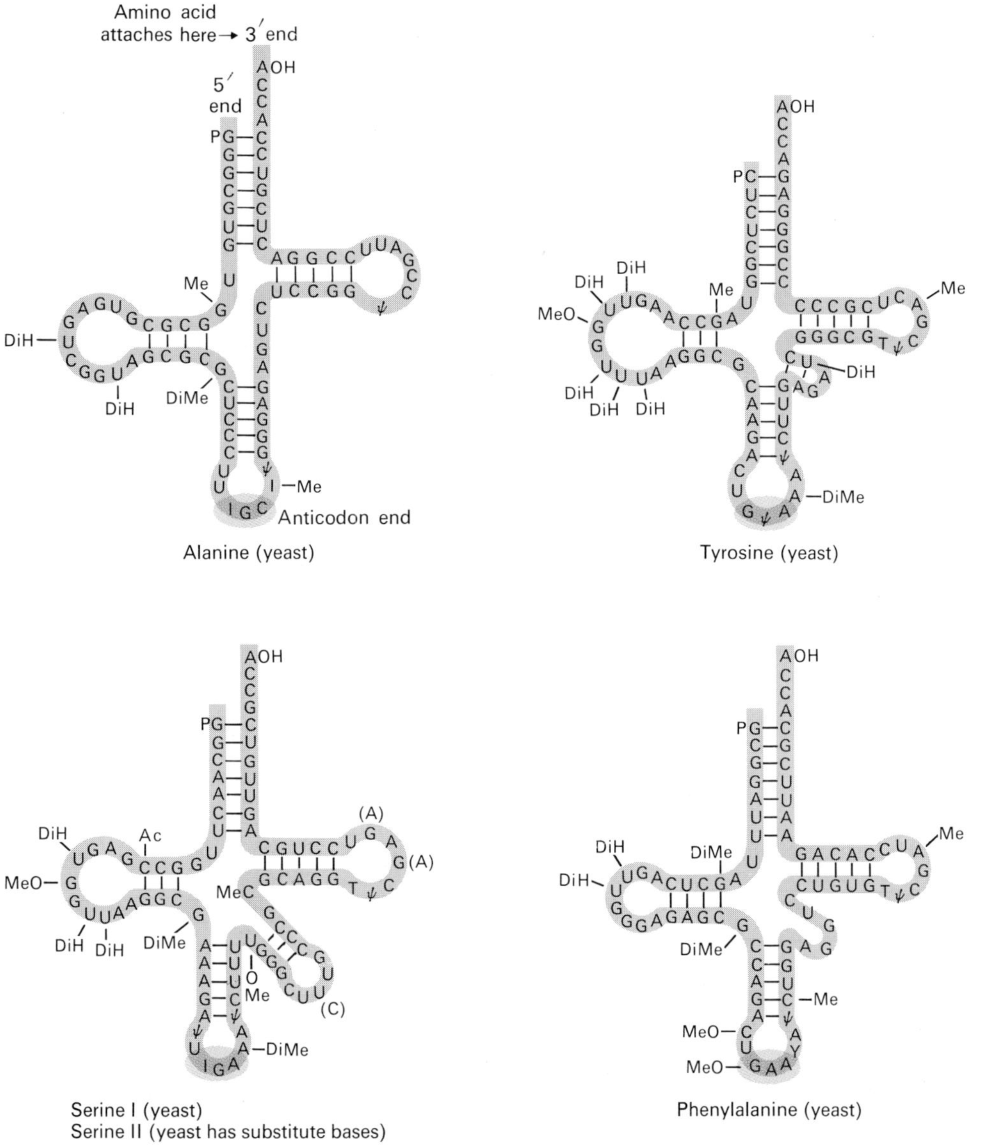

Fig. 10.12 *Several representations of the "cloverleaf" pattern of tRNA, showing possible variations in the structure of specific types. Each type of tRNA attaches to one type of amino acid at the OH arm shown at the top of each molecule. Some slightly modified bases are incorporated into tRNA. These are indicated as DiH-U (dihydroxy-uridine), Me-G (methyl-guanidine), MeO-G (methoxyguanidine), I (inosine), and ψ (a form similar to uridine). P = phosphate group attached to the 5′ end of the molecule (after Holley). [Struther Arnott, "The structure of transfer RNA,"* Prog. in Biophys. and Mol. Biol., *22 (1971): 181–213; the various* tRNA*'s are diagrammed on pp. 183–185.]*

No satisfactory picture has yet been developed for the molecular configuration of ribosomal RNA. For at least part of the length of the molecule, the structure appears to be that of a double helix. The rest of the molecule has an unknown shape. Working with this form of RNA presents special problems of technique. Attempts to crystallize whole ribosomes have met with only partial success. However, today neutron-scattering techniques have been used with considerable success to characterize with greater accuracy than previously the molecular configuration of ribosomal RNA.

Messenger RNA. Messenger RNA (*m*RNA) carries the genetic code from DNA to the ribosomes, where protein synthesis occurs. As its name implies, messenger RNA carries a message. This message is the genetic "blueprint," or building plan for inherited variation. Messenger RNA transmits the plan in the sequence of its own bases, forming a pattern complementary to that of the DNA that formed it. In the ribosome, the coded message that messenger RNA carries is translated into a specific amino acid sequence. Messenger RNA thus acts as an intermediary between DNA and protein.

Fig. 10.13 *Generalized schemes depicting protein synthesis. (a) Messenger RNA is synthesized from a single strand of DNA. The DNA unwinds in an enzymatically controlled reaction and one strand serves as the template from which* mRNA *is synthesized by base pairing. The resulting messenger thus contains a sequence of bases that can be read as a linear series of triplets (each called a "codon"). Which strand of the DNA molecule served as the template for the section of* mRNA *shown? (b) The sequence of steps involved in peptide formation at the ribosome. For the sake of convenience, only a single ribosome is shown here, though several ribosomes are usually attached to any one messenger molecule. The sequence illustrated here (from 1 to 5) shows the steps involved in the addition of one amino acid, and the termination reaction by which the peptide chain is hydrolyzed away from the ribosome–tRNA complex.*

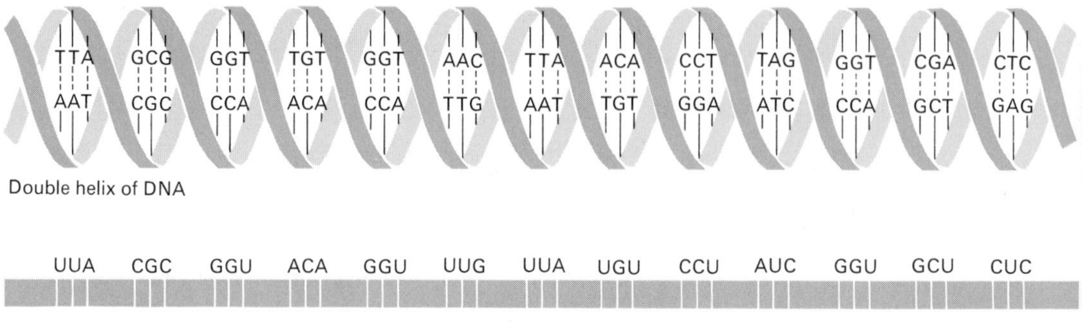

(a)

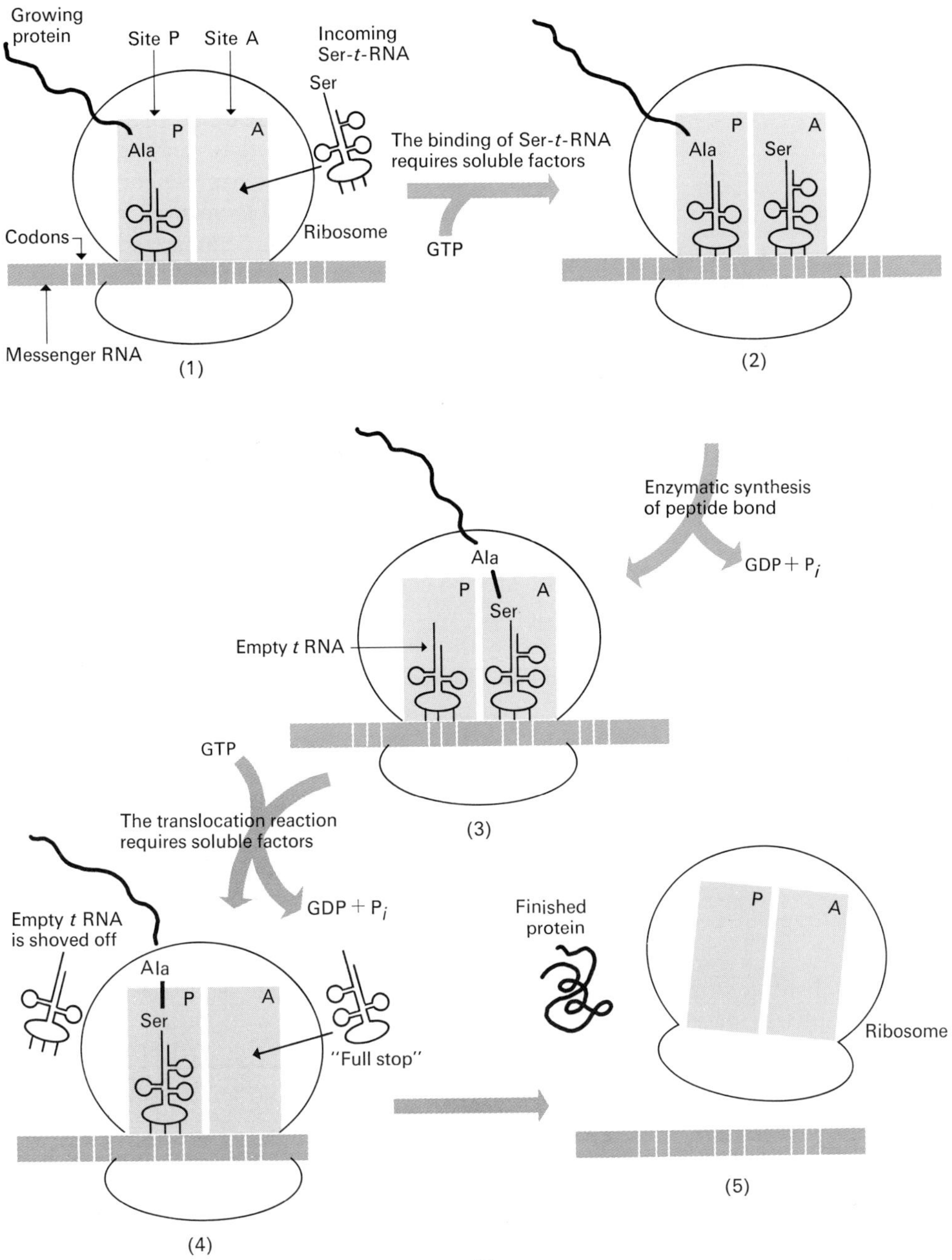
Growing protein
Site P
Site A
Incoming Ser-t-RNA
Ser
P
A
Ala
The binding of Ser-t-RNA requires soluble factors
Codons
Ribosome
GTP
Messenger RNA
(1)
(2)
Enzymatic synthesis of peptide bond
GDP + P_i
Ala
Ser
Empty t RNA
(3)
GTP
The translocation reaction requires soluble factors
GDP + P_i
Empty t RNA is shoved off
"Full stop"
Finished protein
Ribosome
(4)
(5)

(**b**)

The Steps in Protein Synthesis

The sequence of events involved in producing a specific peptide chain from a genetic code on DNA can be represented as follows:

DNA → MESSENGER RNA → RIBOSOMAL RNA → PROTEIN
↑
TRANSFER RNA
+
SPECIFIC AMINO ACID

The completed protein may be an enzyme. This enzyme acts to control one specific reaction or set of reactions in the cell. Since all enzymes are thought to be produced in this way, the genetic code of DNA ultimately controls the entire metabolic activity of the cell. The experimentally established details of protein synthesis by nucleic acids are described below and illustrated diagrammatically in Fig. 10.13.

By base pairing, DNA forms a molecule of messenger RNA. The nucleotide sequence of this newly formed RNA molecule will be complementary to that of DNA. In other words, if the sequence of bases in one strand of the DNA molecule is

A T C C G T G G G A

then the complementary sequence of bases in the new messenger RNA is

U A G G C A C C C U

(Recall that RNA molecules substitute the base uracil for thymine.) The completed messenger molecule passes out of the nucleus into the cell cytoplasm where it comes in contact with ribosomes, the organelles in which the genetic code is "translated" into protein structure.

Meanwhile, another series of reactions is taking place in the cytoplasm. Before the amino acids can be joined together into a polypeptide chain, each amino acid must be activated by complexing with an energy-rich compound such as ATP (catalyzed by an "activating enzyme"). The activated amino acid is at the same time joined to the *t*RNA molecule specific for that amino acid. The bond formed between the amino acid and the *t*RNA contains slightly more energy than that required for the eventual synthesis of the peptide bond between amino acids. The reaction between an amino acid and the appropriate *t*RNA is summarized as:

AA (e.g., alanine) + ATP + *t*RNA (specific for alanine) —"Activating enzyme"→ AA . . . *t*RNA + AMP + 2 INORGANIC PHOSPHATES

The amino acid is attached to the long free arm (amino acid arm) of the *t*RNA molecule as shown in Fig. 10.11. The specificity of the enzyme catalyzing the above reaction assures that alanine and only alanine is linked to the *t*RNA bearing the anticodon for this amino acid. There is a different enzyme with a different specificity for each amino acid and its appropriate *t*RNA.

It is the nature of the *t*RNA molecule, not of the attached amino acid, that now determines where the amino acid is to go in a peptide chain. Experiments have been performed, for example, in which the amino acid cysteine is converted chemically to alanine while attached to the specific cysteine *t*RNA molecule. The modified amino acid is incorporated into the protein as if it were cysteine, indicating that the specific structure of the amino acid does not determine the sequence of units in protein synthesis.

The activated amino acid–*t*RNA complex meets the messenger RNA (*m*RNA) molecule at the ribosome. Ribosomes are the protein-synthesizing machinery of the cell. During protein synthesis, ribosomes are attached to *m*RNA and the AA–*t*RNA complex by weak bonds. There are three specific sites on the ribosome. One of these will attach to a portion of the *m*RNA, while a molecule of AA–*t*RNA will associate with each of the other sites (sites P and A). Just which amino acid-bearing *t*RNA associates with the ribosome at any given moment is determined by the triplet code of the *m*RNA interacting with the anticodon of the *t*RNA. Figure 10.13(b) makes the relationship between these components clear.

The Role of the Ribosome

Ribosomes perform many chemical functions during protein synthesis; one of these is to bring amino acids into the proper orientation so that covalent bonds (the peptide linkage) can be formed between them. The description that follows presents the major events occurring on the ribosome to accomplish the synthesis of a specific sequence of amino acids.

1. The first step involves the association of two AA–*t*RNA molecules with the *m*RNA and ribosome complex. Specific soluble factors must be present in the cytoplasm to accomplish this binding by weak forces (see Fig. 10.13b).

2. An enzyme on the ribosome catalyzes the formation of a peptide linkage between the two amino acids. One molecule of *t*RNA now has two amino acids attached to it. The other has none; it is "empty."

3. The *m*RNA moves one step across the ribosome. In the process, the "empty" *t*RNA is shoved off the ribosome, and the other *t*RNA bearing the two amino acids is moved with the *m*RNA to occupy the site (site P) vacated by the "empty" *t*RNA. This translocation reaction requires certain soluble factors (e.g., G factor) and an energy input from the hydrolysis of a molecule of guanine triphosphate (GTP) to guanine diphosphate (GDP) and P_i.

4. As Fig. 10.13(b) indicates, there is still an empty site (site A) on the ribosome. The amino acid-bearing *t*RNA specified by the triplet code newly aligned with this site will bind to the ribosome, and the events outlined above will be repeated. This accomplishes the addition of another amino acid to the peptide chain.

The growing peptide chain remains associated with the ribosome through each successively added amino acid. Always attached to a *t*RNA molecule, it is passed back and forth from site A to site P, P to A, A to P, and so on as the *m*RNA moves across the ribosome. Eventually the peptide chain will be terminated and thus become free from the ribosome.

The Final Product

Termination is not a random process, but turns out to be highly controlled by special "full-stop" or termination codons. Recent studies have shown that certain triplets or codons (called "nonsense codons") in the *m*RNA chain automatically bring about termination of the peptide chain at that point (see Fig. 10.13, step 4). The nonsense codons are UAA, UAG, and UGA. When the ribosome reaches a nonsense codon, the bond between the final amino acid and the *t*RNA molecule to which it is attached is hydrolyzed. This reaction is mediated by a protein "release factor," which may act as an enzyme or in some other capacity not clearly understood at present. Thus the final amino acid is released from its *t*RNA molecule without forming a peptide bond with another amino acid, as is usually the case. The exact chemistry of termination is still being actively investigated. It appears to be built into the genetic message of the DNA as accurately as the position of each amino acid.

Alexander Rich and his coworkers have found that the long molecule of *m*RNA may have more than one ribosome associated with it. Electron microscopy can reveal several ribosomes spaced at intervals along the length of *m*RNA. Such a cluster of ribosomes held together by *m*RNA is called a

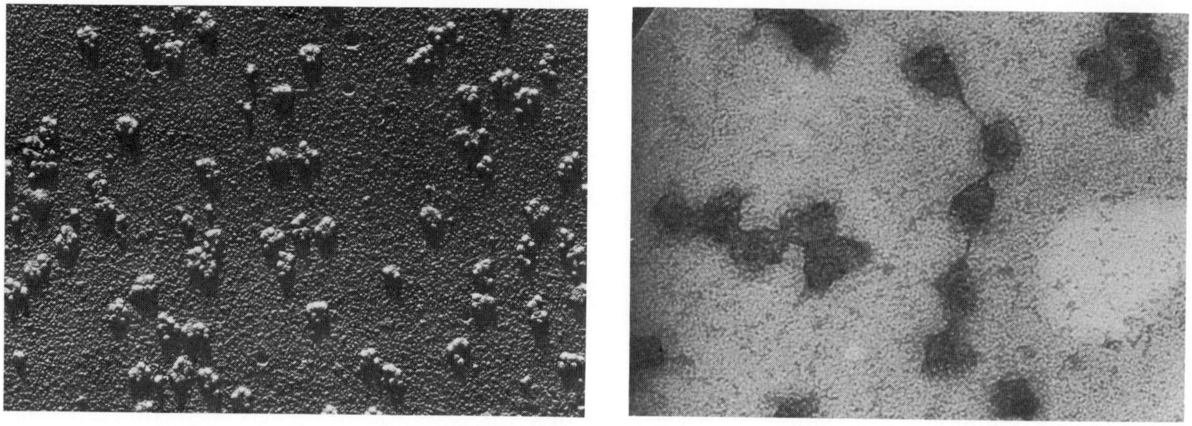

(a) (b)

Fig. 10.14 *(a) An electron micrograph of clusters of ribosomes, called polysomes. (a) Polysomes engaged in the synthesis of hemoglobin polypeptides. The strand connecting the five ribosomes in the center is believed to be a molecule of messenger RNA. (Photos courtesy Dr. Alexander Rich.)*

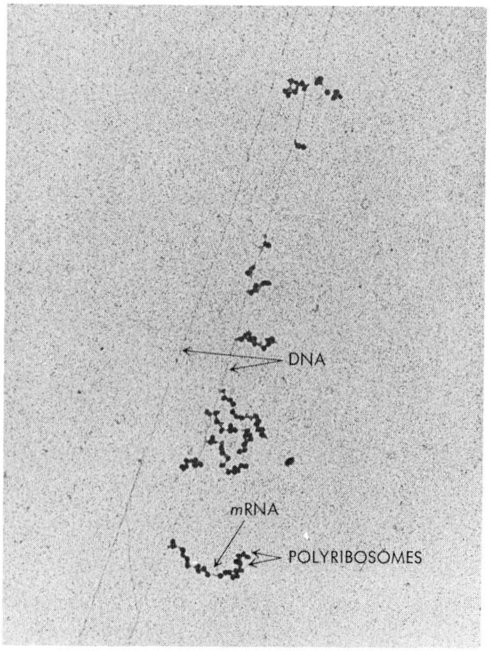

Fig. 10.15 *This electron micrograph, magnified 36,000 times, gives dramatic support to the hypothesized roles of DNA, RNA, and the ribosomes in bacterial protein synthesis. Shown is a portion of the single-stranded, circular chromosome of the bacterium* Escherichia coli. *Most of the chromosome is inactive, while the active portions show* mRNA *transcribed by the chromosomal DNA attached to and interconnecting the ribosomes into polyribosomes. This type of close contact between DNA, RNA, and ribosomes had earlier been predicted in procaryotic cells, in which there is no organized nucleus. (Photo courtesy O. L. Miller, Jr., Oak Ridge National Laboratory; and Barbara A. Hamkalo and C. A. Thomas, Jr., Harvard University Medical School.)*

polysome. (See Fig. 10.14.) Each ribosome of a polysome is involved in protein synthesis. The quantity of growing protein associated with any given ribosome will depend on how far that ribosome has traveled from the starting end of the *m*RNA. Thus it is clear that each molecule of *m*RNA may serve to generate a number of identical proteins from its coded message. Dramatic evidence in support of the hypothesized roles of DNA, *m*RNA, and ribosomes in protein synthesis has been obtained through electron microscopy (see Fig. 10.15).

The events of DNA replication and protein and RNA synthesis all depend upon the formation of specific, though weak, chemical interactions among the nucleic acids involved. Chief among these weak interactions is hydrogen bonding between specific pairs of complementary bases. These are highly specific, ensuring the accuracy of the processes. Their weak character ensures that the chemical associations formed in determining a sequence will be temporary. In the case of protein synthesis, the weak bonds ensure that the newly created substance can be easily detached from the messenger RNA and the ribosome.

The end product of the above sequence of reactions is a completed protein molecule. In many cases, however, a protein can acquire all of its functional properties (e.g., enzymatic) only when it is closely bound to other proteins. The fully functional protein is then said to be made up of **protein subunits.** Such a protein may be an enzyme, or it may be a structural or transport protein such as collagen or hemoglobin. The primary structure of all proteins is determined at the ribosome by the mechanism outlined above. The coiling of a single peptide chain into the alpha helix is determined by intramolecular forces associated with the elements of the peptide linkage. The combination of various coiled proteins into a fully functional protein molecule is determined by the number and location of each amino acid and the interactions that are possible between their side chains (secondary, tertiary, and quarternary structure). Because these intricate conformations of proteins are dependent upon the specification of the amino acid sequence, we can say that all information about the cell's structure and function resides in the sequence of amino acids in the proteins, and thus in the bases in DNA.

The reactions described above have been studied in greatest detail in bacteria. To a large extent, a similar sequence of steps is thought to occur in the cells of higher organisms (animal and plant).

10.6 The Genetic Code

We have seen that the sequence of bases on ribosomal RNA, originally obtained via messenger RNA from DNA, determines the sequence of transfer RNA molecules along the ribosomal RNA molecule. This transfer RNA sequence, in turn, establishes the amino acid sequence in the peptide chain being constructed. The problem of the genetic code, however, is even more specific. The question it poses has three parts. First, how many nitrogenous bases are involved in selection of the amino acid-carrying transfer RNA molecules? Second, which of the four bases are involved? Third, in what order are these bases arranged?

The first question was attacked by simple arithmetic. It was obvious that more than one base was involved. With only one base playing a role, only four amino acids could be selected. After all, there are only four kinds of bases in RNA.

Since the code on a segment of mRNA is complementary to a corresponding segment of DNA, the mRNA code is called an anticodon, or "nodoc" for short.

Supplement 10.3

DNA AND PROTEIN SYNTHESIS IN EUCARYOTIC ORGANISMS

The entire protein-synthesizing mechanism described in the text is a good example of a scientific model. Evidence for many steps in the process is conclusive. Yet evidence for the whole story is only indirect—based on the probability that each of the specific steps is true. This model has been developed largely in relation to a single type of organism, bacteria. Its usefulness as a biological model lies in its generality. Does it apply only to bacteria or to other organisms as well? The problem is, of course, that study of the molecular events in protein synthesis is difficult with eucaryotic organisms. Their chromosomes are much more complex, being composed of not only DNA but histones (proteins) in a way that it is not clearly understood. Eucaryotes have two copies of every gene, so it is more difficult to isolate the activity of any one gene. And, of course, it is much more difficult to study eucaryotic cells in mass culture—they take longer to grow and require much more specialized culture conditions than most bacterial cells. Despite this, however, some evidence consistent with the model developed in bacteria is now available for eucaryotic cells.

In our discussion of the steps in protein synthesis, it was proposed that in the early stages DNA synthesizes an *m*RNA molecule with a complementary sequence of bases. This hypothesis can be tested by exposing chromosomes of the midge, *Chironomus tentans*, to a tritium-labeled base, uracil. Uracil, of course, would be built into any *m*RNA synthesized along the chromosome, but not into DNA (which contains thymine rather than uracil). We can reason that if *m*RNA is synthesized by DNA along the chromosome, then tritium-labeled uracil should show up in the chromosome at those places at which genes are active. Thus, we would expect that autoradiographic film will show black areas over certain regions of the chromosome (see Diagram A). Almost certainly, these dark areas represent points at which *m*RNA is being synthesized. Thus the hypothesis is supported.

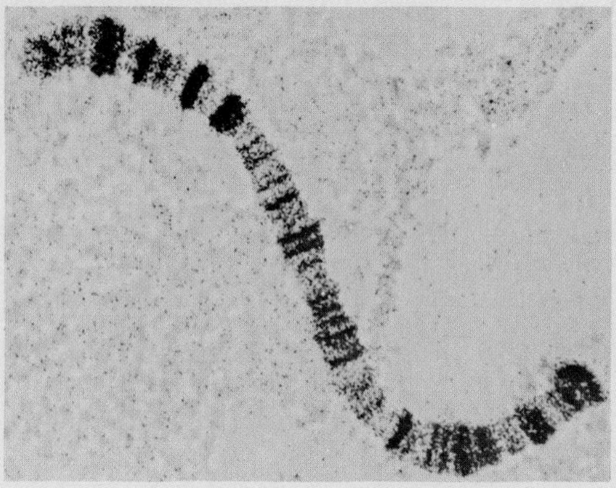

Diagram A *Evidence obtained by autoradiography of the synthesis of* mRNA *at certain regions of the chromosome. (This chromosome is from the larva of a midge,* Chironomus tentans.*) As the animal develops, the regions along the chromosome in which* mRNA *is being synthesized change, suggesting that different sets of genes are called into action at different times. (Photo courtesy Dr. C. Pelling, Max Planck Institute for Biology, Tübingen.)*

Autoradiographic evidence has also been found that DNA remains within the nucleus while *m*RNA passes into the cytoplasm, where it can be conceived of as transmitting the genetic code to the ribosomes (see Diagrams B and C).

In 1969, Drs. O. L. Miller, Jr., and Barbara R. Beatty of the Oak Ridge National Laboratory in Oak Ridge, Tennessee, obtained photographic evidence of the predicted type of close contact between DNA, RNA, and ribosomes (see Fig. 10.15), although the system being studied was in the bacterium *E. coli*, not in a eucaryotic cell. All in all, evidence suggests that the model of DNA control of protein synthesis that is now accepted may apply, at least in broad outline, to eucaryotic as well as to procaryotic cells.

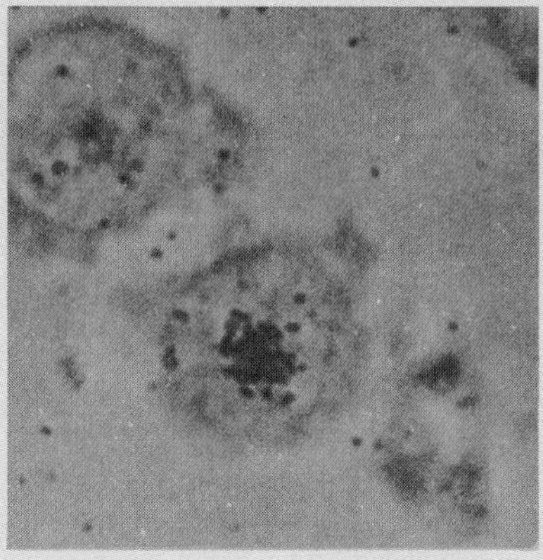

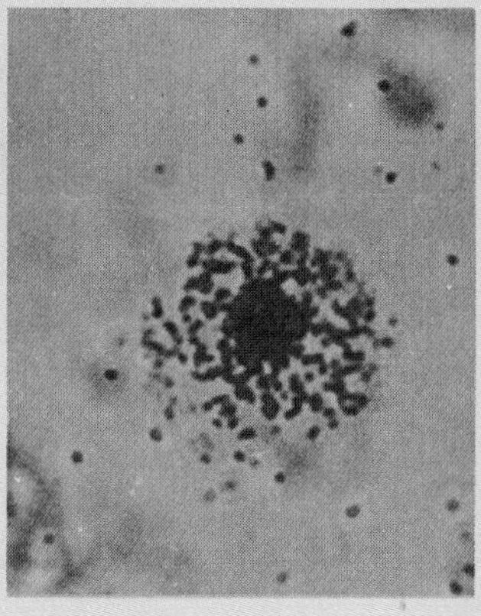

▲

Diagram B *Evidence that DNA remains within the nucleus while mRNA is found in both nucleus and cytoplasm is provided by autoradiography. The moth* (Cecropia) *blood cell at above left has been exposed to isotope-labeled thymidine. Each dark spot represents the emission of electrons from the isotope, tritium (H^3). Since thymidine occurs in DNA but not in RNA, the autoradiograph shows the radioactivity limited to the nucleus. The photo at above right shows the same kind of cell, this time exposed to tritium-labeled uridine, which occurs in RNA but not in DNA. Note the distribution of radioactivity in both nucleus (where* mRNA *is presumed to be synthesized by DNA) and cytoplasm (where the* mRNA *is presumably associated with the ribosomes during polypeptide synthesis). (Photos courtesy Dr. Spencer Berry, Wesleyan University.)*

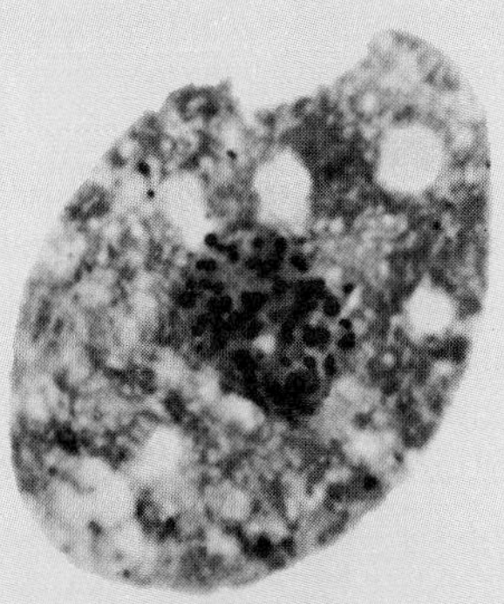

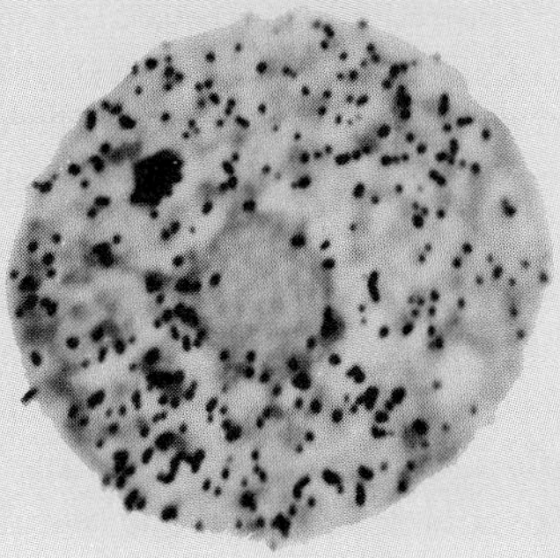

◀ **Diagram C** *The* Tetrahymena *cell at above left has been grown in a medium containing isotope-labeled cytidine. Note the concentration of dark spots within the nucleus. Below it is an autoradiograph of a cell similarly exposed to tritium-labeled cytidine, but then allowed to grow further in a medium free of labeled cytidine. Note that the pattern of dark spots supports an hypothesis proposing that nuclear-synthesized RNA passes into the cytoplasm and, it is assumed, transmits the genetic code to the ribosomes. (Photos courtesy David M. Prescott, University of Colorado Medical School. Reproduced by permission, from J. D. Watson,* Molecular Biology of the Gene. *Menlo Park, Calif.: W. A. Benjamin, 1965.)*

Table 10.2
Possible Genetic-Code Letter Combinations as a Function of the Length of the Code Word (A = adenine, G = guanine, U = uracil, C = cytosine)
(After M. Nirenberg, 1963)

Singlet code (4 words)	Double code (16 words)				Triplet code (64 words)			
A	AA	AG	AC	AU	AAA	AAG	AAC	AAU
G	GA	GG	GC	GU	AGA	AGG	AGC	AGU
C	CA	CG	CC	CU	ACA	ACG	ACC	ACU
U	UA	UG	UC	UU	AUA	AUG	AUC	AUU
					GAA	GAG	GAC	GAU
					GGA	GGG	GGC	GGU
					GCA	GCG	GCC	GCU
					GUA	GUG	GUC	GUU
					CAA	CAG	CAC	CAU
					CGA	CGG	CGC	CGU
					CCA	CCG	CCC	CCU
					CUA	CUG	CUC	CUU
					UAA	UAG	UAC	UAU
					UGA	UGG	UGC	UGU
					UCA	UCG	UCC	UCU
					UUA	UUG	UUC	UUU

Could there be two bases involved? Again, with only two bases, the possible arrangements are 4^2, or 16. This number is not large enough to allow for the selection of the 20 amino acids known to be used in protein synthesis.

The minimum number of bases that could be involved in amino acid selection seemed to be 3. This number gave a possibility for the selection of 4^3, or 64 amino acids, which is more than enough. Higher numbers than 3 were possible, of course, but seemed unnecessary and thus, it was assumed, less likely. The relationship between the number of bases involved in the code and the number of words that can be coded for is given in Table 10.2.

Evidence for the Triplet Code

Work on the genetic code proceeded, therefore, on the assumption that the code was a triplet involving only three bases. But the hypothesis was based largely on what appeared to be logical. What actual evidence exists for the triplet code hypothesis? One of the most direct tests rests on a refined technique that allows the geneticist to insert one or more bases into a synthetic *m*RNA chain. The *m*RNA is then "fed" activated amino acids (phosphorylated, in the form of amino acid-AMP) and the activating enzymes specific for each type of amino acid. The system can synthesize proteins in a test tube. The resulting protein then reflects the significance (in terms of amino acid sequence) of changes in one or more bases in the DNA code. To understand this experiment, keep in mind the fact that *m*RNA carries the code without any punctuation marks or spaces—the code is read by the ribosome as one sequence of triplets after another. The ribosome would thus move along the *m*RNA molecule (or vice versa) three bases at a time (on the hypothesis that the code is triplet). Thus a mistake in the code caused by adding or deleting one base would be multiplied throughout the remainder of the sequence. It would be read as "nonsense" and would produce an incomplete protein.

An analogy helps illustrate how the production of nonsense codes can be used to indicate whether the code is a triplet or not. Consider the following sentence made up of only three-letter words (a triplet code):

THE MAN SAW TOM BUY THE TOY

This sentence is written with punctuation (in this case spaces) between the triplet codes. On *m*RNA, however, the sentence would be written:

THEMANSAWTOMBUYTHETOY

Even though this looks like nonsense at first glance, it is apparent that the "message" can be read by moving along the line three letters at a time:

THEMANSAWTOMBUYTHETOY

However, if we add another letter in the line somewhere near the beginning, it is obvious that this will throw off the sense of the sentence from that point on—that is, as long as we continue to read three letters at a time:

THEMANSAWQTOMBUYTHETOY

THE MAN SAW QTO MBU YTH ETO Y

After the third triplet, the message becomes nonsense. Such experiments, in which one base was added to an already-existing code, have produced nonsense messages. That is, protein synthesis stops beyond the point where the message becomes garbled (at Q in our sample). Now, it can be reasoned that if the code is a triplet, then addition of three bases (in all) ought to restore sense to the message. Using special techniques to invent a new base at various specific points in viral DNA (of the strain known as T4 bacteriophage), F. H. C. Crick, Sidney Brenner, L. Barnett and R. Watts-Tobin set about to test this prediction. They used an acridine dye known as proflavin, which produces mutations by causing the addition or deletion of a nucleotide in DNA. They developed a series of "addition" mutants of the rII region of bacteriophage T4. With one or two bases added to the code, the DNA segment from that point on is read as "nonsense." But Crick and his co-workers reasoned that the insertion of three bases should restore the "sense" of the message on the DNA beyond the mutant region. They set about to test their hypothesis. They combined two addition mutants together (by infecting the same culture of *E. coli* with both) and observed that no wild-type phage plaques appeared. When they combined *three* such mutants, however, they obtained some wild-type plaques. These results dramatically confirmed the triplet nature of the genetic code.

This experiment is a genetic one: it makes changes in the genetic code at the DNA level and then looks for some phenotypic expression of that change. It is also possible to test the triplet code hypothesis biochemically. Marshall Nirenberg and

---

Messenger RNA molecules are like a sequence of letters in a sentence with no punctuation marks. The only signals that approximate punctuation in the mRNA are start–stop indicators. Any given mRNA molecule may have several start and stop indicators. The region between the start and stop signals codes for a continuous polypeptide or protein chain. The region of DNA corresponding to the mRNA segment between start–stop signals is called a gene.

---

some of his coworkers have taken advantage of the fact that in cell-free extracts of *E. coli* (extracts in which components of cells have been released from being bound within the cell membrane), *m*RNA and *t*RNA bind to ribosomes. The researchers found that the amount and kind of *t*RNA (carrying specific amino acids) bound to ribosomes depended on the length and sequence of the *m*RNA added to the system. If *m*RNA consisting of only two nucleotides was added, no *t*RNA was bound. If *m*RNA three nucleotides long was added, however, *t*RNA and ribosomes would bind to it. Furthermore, if different types of *m*RNA three nucleotides long were added, *t*RNA's carrying *particular* amino acids were bound (for example, the sequence UUU caused the *t*RNA carrying phenylalanine to bind). This approach allowed Nirenberg and others to determine systematically which sequences of three nucleotides corresponded to each specific amino acid. These sequences are now called triplets or **codons.**

All the possible codons have now been associated with specific amino acids or a role in polypeptide synthesis. Table 10.3 gives the triplet of base pairs for each of the 20 most common amino acids. To select the proper triplets or codons for any amino acid, simply read in order the letters appearing to the left, above, and to the right of it. Thus the codons for glycine (gly) are GGU, GGC, GGA, and GGG, while those for lysine (lys) are AAA and AAG. Some of the codons assigned here are less certain than others, and those suspected of being connected with the beginning of polypeptide chain synthesis are not included. The two codons labeled

Table 10.3
The Genetic Code as It Would Appear in a Molecule of *m*RNA

1st	2nd				3rd
	U	C	A	G	
U	phe	ser	tyr	cys	U
	phe	ser	tyr	cys	C
	leu	ser	NONSENSE (OCHRE)	STOP	A
	leu	ser	NONSENSE (AMBER)	tryp	G
C	leu	pro	his	arg	U
	leu	pro	his	arg	C
	leu	pro	gluN	arg	A
	leu	pro	gluN	arg	G
A	ileu	thr	aspN	ser	U
	ileu	thr	aspN	ser	C
	ileu	thr	lys	arg	A
	met	thr	lys	arg	G
G	val	ala	asp	gly	U
	val	ala	asp	gly	C
	val	ala	glu	gly	A
	val	ala	glu	gly	G

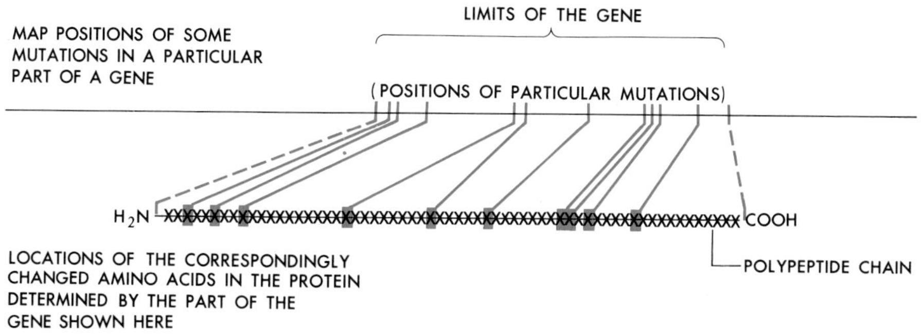

Fig. 10.16 *Diagrammatic representation of the colinearity of a gene and the polypeptide chain for which it codes. Specific regions of the genetic map are known to code for a certain segment of the polypeptide chain; the arrangement of these identified areas along the DNA corresponds to the sequence of amino acids being coded for in the corresponding sections of the protein.*

nonsense, ochre and amber (UAA and UAG), are related to no amino acid and are hypothesized to represent "punctuation" in the chain. The codon UGA is hypothesized to terminate polypeptide chain synthesis.

It is clear from Table 10.3 that the genetic code must be **degenerate;** i.e., more than one codon can select for one of the amino acids. More significant, however, is the fact that all the experimental evidence to date points toward the universality of the genetic code throughout the living world. Experiments on tissue extracts show that the triplets that select particular amino acids in mammals are identical to those that select the same amino acids in *E. coli.* Through genetics, we gain even more insight into the way all living systems share processes in common.

One result of working out the precise nature of the genetic code is that biologists now know genes and proteins to be **colinear.** Colinearity means that the sequence of amino acids in the completed polypeptide follows the same order as the sequence of triplets (codons) in the DNA molecule coding for that polypeptide. Thus a single base-pair change at one end of a DNA molecule can result in the change of a single amino acid at one end of the corresponding protein; similarly, a single mutation at the other end of the DNA molecule results in the change of a single amino acid at the other end of the protein. Furthermore, the order of mutations all along the DNA corresponds to the order of the amino acids along the polypeptide. In other words, the genetic map within the limits of a single "gene" (so far as it has been determined) corresponds point-by-point with the amino acid sequence of the polypeptide (see Fig. 10.16).

Mutations

Another result of working out the genetic code is that biologists now have a much clearer idea of the molecular basis of mutation. Recall that we have discussed mutations in the last two chapters largely in terms of phenotypic changes. Not only that, but the phenotypic changes have usually been gross, visible changes—from red to white eyes, from normal to vestigial wing, etc. To the extent that we have discussed the genotypic basis of mutations, we have simply stated that genes undergo a change, whose existence we infer through observations on the phenotype. But what are mutations? What do they mean on the genetic level? And what do they mean on the biochemical level—that is, on the molecular phenotype within individual cells?

Geneticists now define mutations as changes in the base pairs on DNA. More precisely, **mutations** are changes in one or more base pairs that can have one of three effects: (1) They can make the code so nonsensical that no messenger and hence no protein is synthesized at all. (2) They can cause the coding of messenger that contains a more limited amount of "nonsense" so that protein is produced but is nonfunctional. (3) They can cause the coding of protein that has an altered function; that is, the protein (for example an enzyme) can function either less or more effectively in its particular biochemical role than the nonmutated (normal) form. The degree to which the functioning of the protein is altered depends of course, on the location of the mutation—whether it is in a "sensitive" or an "insensitive" region of the protein.

The latter are known as structural gene mutations. Such a mutation occurs in a gene that codes for a protein that enters directly into a metabolic reaction related to the phenotype. There is a second kind of gene, known as a regulatory gene, whose function is to code for a protein that regulates the expression of other genes in the genome. We will discuss the nature and role of regulatory genes in more detail in Chapter 11. For the moment, it is sufficient to say that regulatory genes can also show mutations that can alter the phenotype of the organism in a variety of ways.

Mutations are like typographical errors in the genetic code. These errors get translated into further typographical errors in the protein itself. What causes mutations? How do base-pair substitutions take place? What determines the exact nature of any particular kind of base-pair substitution? The answer to these questions is not completely understood by geneticists today, but some answers are available. To understand these answers, it will be necessary to digress momentarily to discuss the kinds of processes known to induce mutations artificially. For it is through these methods that mutations are most frequently studied.

Mutations can be induced by exposing organisms to various physical and chemical processes: to mutagenic chemicals such as nitrous acid or hydroxylamine or to physical factors such as high-energy radiation. Nitrous acid deaminates nucleotides, removing amino groups and substituting in their place a keto group (=CO), as follows:

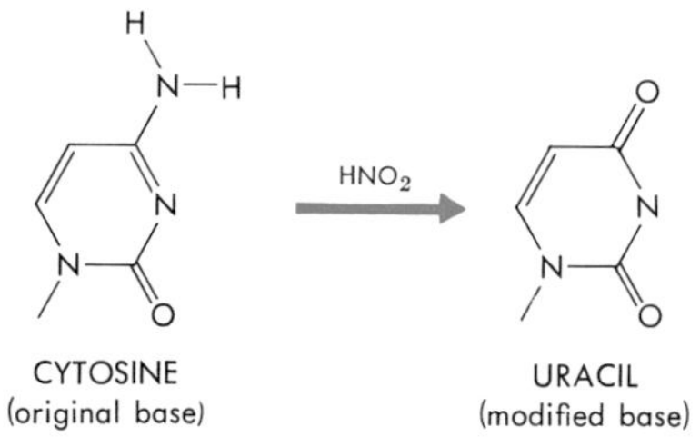

Without going into great detail, suffice it to say that, as in the example above, cytosine can be converted into a modified form that is capable of binding not with guanine, but with adenine. The result of exposure of organisms to mutagenic chemicals, then, is to alter base structure in such a way that new pairing partners get substituted into a DNA molecule. A given triplet sequence is changed, and consequently the genetic message is modified.

Ultraviolet light, X-rays, or other sources of high-energy radiation can cause other kinds of modification of existing bases in DNA. Ultraviolet (UV) light is known to cause adjacent thymines to undergo configurational shifts (a so-called **tautomeric shift**) so that they form a thymine dimer (see Fig. 10.17). This dimer sometimes causes the daughter strand to have a gap across from the dimer. The cell's DNA-repair enzymes often make mistakes in trying to fill in this gap. That is, the repair enzyme cannot adequately resolve the fact

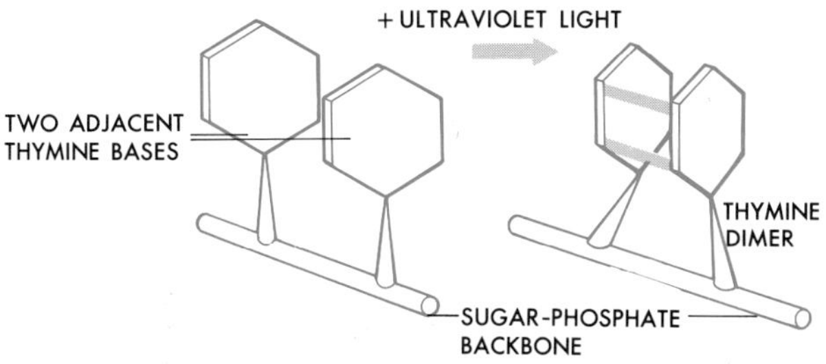

Fig. 10.17 *Formation of thymine dimer by configurational shift of two adjacent thymine bases. The shift results from electron rearrangements in the orbitals of the bases; the rearrangement is due to high-frequency, ultraviolet light.*

that pairing to the dimer is impossible. Hence the enzyme may cause a different kind of base to be substituted in the daughter strand. This, of course, changes the code for the daughter strand.

One very important feature of mutations is that they are random. That is, mutations occur (mistakes are made) more or less randomly along a DNA strand. The occurrence of a given mutation, such as substitution of bases in one part of the DNA rather than in another, is not made in direct response to any environmental factor. This may sound contradictory to the above discussion of mutagenic chemicals and radiation, which are, after all, environmental factors that cause mutations. There is no contradiction, however. Mutagenic factors such as chemicals or radiation increase the general level of all mutations. Usually they do not selectively favor any one mutation over another. They increase the propensity for mistakes to be made, but they do not selectively determine which mistakes are made. As far back as 1927, geneticist H. J. Muller demonstrated that X-radiation directly causes an increase in the random mutation rate. Through elegant experiments he showed that the rate of general mutation in *Drosophila* is proportional to the amount of X-radiation to which the insects were exposed. More evidence that gene mutations are indeed spontaneous, rather than specifically elicited by one or another environmental agent, came from the experiments of Luria and Delbrück (Supplement 10.4).

10.7 Genes, Cistrons, and Operons

What, then, is a gene? We have come a long way from Mendel's genetic "factors." Throughout the development of genetics, many satisfactory gene models have been proposed that, on closer analysis, later proved unsatisfactory (recall the string-of-beads model of chromosome–gene relationships). In molecular genetics, however, breakdown of the DNA molecule destroys the gene. Therefore, while much has been learned from a submolecular level of genetic analysis, any meaningful definitions of a gene must deal with the intact working entity represented by the gene concept.

Such a definition is now possible. The relationship of DNA to the protein whose structure it specifies can be represented as follows (the arrow encircling DNA shows that it is a template for its own replication):

$$\circlearrowright \text{DNA} \xrightarrow{\text{(transcription)}} \text{RNA} \xrightarrow{\text{(translation)}} \text{PROTEIN (POLYPEPTIDE)}$$
(replication)

This representation, often simplified verbally to "DNA, RNA, protein," has been referred to as biology's "central dogma."

A Significant Discovery

In 1965 some interesting results were reported by biochemist Dr. Howard M. Temin, now at the Uni-

Supplement 10.4

ARE MUTATIONS RANDOM? HOW DO WE KNOW?

Biologists have long debated the question of whether genetic mutations occur largely by chance, or whether specific mutations are called forth, over others, by certain environmental agents or conditions. In 1943, using bacteria as their experimental organisms, Salvador Luria and Max Delbrück set out to test the hypothesis that such mutations occur at random.

If a culture of bacteria is allowed to grow for several days and then exposed to bacteriophage, most of the cells are killed. A few may survive, however. These survivors are "resistant" varieties. Their resistance to bacteriophage is passed on to their offspring and is, therefore, a genetic trait. It is the result of a mutation. When does the mutation occur? Two working hypotheses can be formulated:

1. The resistance to phages arises in the bacteria by spontaneous mutation. Such mutant bacteria will appear whether or not the bacterial culture is exposed to phage. In the absence of the selecting agent (the phage), resistant cells are simply not detected among the greater masses of nonmutant bacterial cells. When phage is introduced, however, only the mutant forms survive and reproduce.

2. The resistance is caused to appear in some of the bacteria as a result of their contact with the phage. The bacteria that respond to this change in environment (the introduction of the phage) by mutating survive. The bacteria that do not so respond are destroyed.

In other words, the second hypothesis holds that the presence of the phage is the causal agent for mutation and that mutants for phage resistance do not appear until the bacteria come in contact with the phage. By contrast, the first hypothesis holds that the mutants are present all along but are simply not detectable until the phage is introduced.

To test the first hypothesis, Luria and Delbrück set up a number of bacterial cultures of the same species. Each culture was grown from a small group of bacterial cells. All the bacterial cultures were simultaneously exposed to phage, and the number of resistant cells, or survivors, were counted.* The experimenters reasoned that if mutations occur spontaneously, then the number of resistant cells in the various culture dishes should be quite different. If, for example, the mutation occurs early, when the growing culture contains few cells, the mutant cell will multiply and leave a large number of offspring bearing the mutation. By the time the phages are introduced into the culture, many resistant bacteria may be present. Conversely, if the mutation takes place just before the introduction of phages, only a few resistant cells will be present. The laws of chance predict that there would be considerable variation in the number of surviving cells per culture dish.

On the basis of the second hypothesis, Luria and Delbrück reasoned that if mutations occur in response to the presence of phages, then the number of resistant cells per culture dish should be quite uniform. Since each culture dish contains roughly the same number of bacteria, and the amount of phage introduced in each case is the same, the second hypothesis predicts that the number of mutants should be roughly the same from one culture to the next.

The results of this experiment (Diagram A) show that the variation in number of surviving cells per culture dish is quite large. Some cultures have only two or three surviving colonies, while

* The number of survivor cells can be counted quite easily by allowing the culture to incubate for a few days. Each surviving cell will reproduce to form a colony that can be detected by simply examining the culture dish.

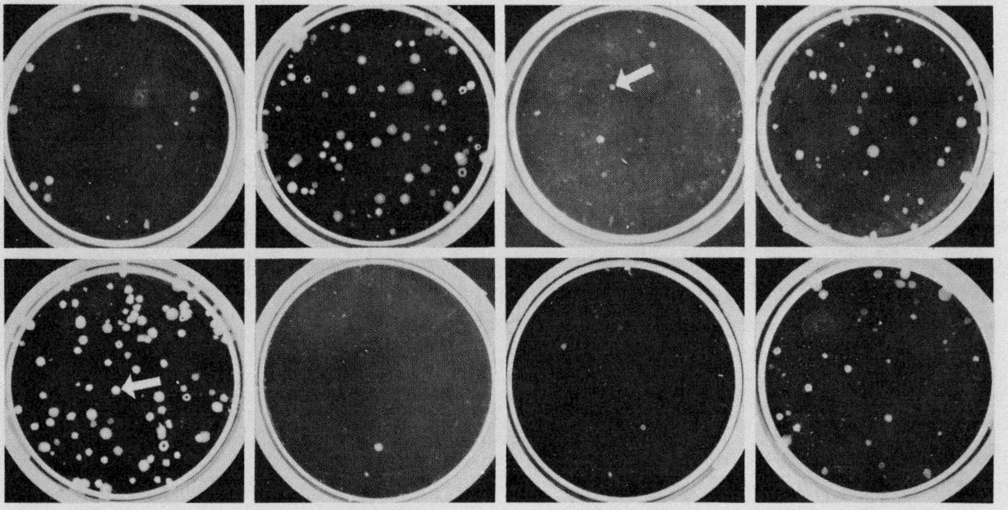

Diagram A *The random variation in bacterial colony numbers obtained in the Luria and Delbrück experiments clearly supports the hypothesis proposing that mutations occur spontaneously and at random. The arrows point to one colony.*

others have twelve, fifteen, or more. The experimental results bear out the prediction of the first hypothesis, which can be said to be supported. The results contradict the prediction of the second hypothesis, which, barring experimental error, can be said to be rejected. Mutations *do* occur at random and are entirely independent of the environmental changes that may give them selective value. Note that the Luria and Delbrück experiment tests both hypotheses simultaneously. Since the predictions made by the two hypotheses are contradictory, the results can support only one of them. Such an experiment is called a **crucial** experiment.

Mutations are randomly occurring events. Environmental factors such as high-energy radiation or mutagenic chemicals cause an increase in the overall mutation rate. Mutations do not occur, however, in response to more general environmental conditions such as temperature, food, or specific "needs" of the organism.

versity of Wisconsin. As pointed out earlier, some viruses (e.g., the tobacco mosaic virus) contain RNA but no DNA. Dr. Temin noted that when such an RNA-only virus invades a cell, strands of DNA complementary to the viral RNA can be found. The implications of this discovery were rather profound: *it suggested that these DNA fragments had been synthesized with the viral RNA as the template,* instead of vice versa. At the time, however, not much attention was paid to Temin's results.

In 1970 Temin and his colleague Dr. Satoshi Mizutani reported that they had found RNA-dependent DNA polymerase that would synthesize DNA using the viral RNA found inside cells infected with an RNA virus. Within days, similar results were reported by David Baltimore of the Cold Spring Harbor laboratories. Still more convincing evidence followed from Dr. Sol Spiegelman's laboratory at the Columbia University Institute for Cancer Research. The four nitrogenous bases of DNA—adenine, guanine, cytosine, and thymine—were labeled with tritium and mixed with viral RNA. The labeled bases soon showed up in intact DNA. Thus the evidence seemed conclusive: viral RNA could synthesize DNA.

Not satisfied, Spiegelman went a step further, reasoning that if the viral RNA had served as a template for DNA synthesis, then the RNA should be complementary to one strand of the DNA and would form a double-stranded hybrid with it. Spiegelman mixed viral RNA and its hypothesized DNA product and spun the mixture in an ultracentrifuge for three days. Since RNA and DNA have different molecular weights, they will form separate layers or fractions in the centrifuge tubes. Upon inspection, these two layers were found—but so too was a third, intermediate layer. Undoubtedly this third layer, lying between the other two, was the RNA–DNA hybrid.

Spiegelman has tested twelve RNA-only viruses for their ability to synthesize DNA. Eight of them can do so; four cannot. Of considerable interest is the fact that the eight that can synthesize DNA from RNA cause tumors in animals, while the four that cannot synthesize DNA from RNA do not cause tumors. With the growing evidence for the role of viruses in at least some forms of cancer, the possibility exists that identification of the enzyme responsible for RNA-directed DNA synthesis might enable the process, and thus perhaps the cancer, to be arrested. Since the transfer of genetic information from DNA to RNA can be blocked by an antibiotic that knocks out the crucial enzyme, this reverse blockage possibility does not seem too remote.

It thus seems that the representation of biology's "central dogma" must be modified to include the possibility of reverse transcription of DNA by RNA:

$$\underset{\text{(replication)}}{\circlearrowleft \text{DNA}} \overset{\text{(transcription)}}{\rightleftarrows} \text{RNA} \xrightarrow{\text{(translation)}} \text{PROTEIN (POLYPEPTIDE)}$$

To date, there is no evidence that proteins might "code" for RNA. It should not take too much reflection to realize that the implications would be profound if this were ever shown to be the case. However, knowing that proteins are an end product of gene action enables us to work backward to pinpoint the gene. Consider a protein containing 500 amino acids. For the selection of each of these, a triplet of three bases is required. Thus, for this protein, *the gene is a portion of the DNA molecule containing 1500 base pairs.* On the basis of average molecular weights for amino acids, it can be further predicted that this gene will have a molecular weight of approximately 10^6. For other proteins, depending on their size, the gene is correspondingly larger or smaller. It is both interesting and intellectually satisfying to note that this estimate of gene size agrees well with calculations made on the basis of more macroscopic investigation, such as data obtained from genetic mapping experiments.

A more direct means of reaching the same conclusion concerning gene size is simply to divide the number of nucleotides in a chromosome by the number of genes located along it. (For example, there are over 70 known genes on the chromosome of the bacterial virus T4). The resulting figure, of course, represents an average number of nucleotides per gene (though there is no reason to assume that all the genes are necessarily the same size). Knowing the average size of the nucleotides allows calculations concerning the size of a gene.

But size is not a very useful way of describing a gene. We want to know what a gene does, not just how big or small it is. As a result of work on the relations between DNA, RNA, and proteins, most molecular geneticists now consider a gene to be a unit of DNA that codes for a complete polypeptide chain. Thus a gene is a functional unit—a unit that guides the production of a biologically significant entity. Today the term **cistron** is sometimes used to refer to the structural and functional unit we have been calling a gene. The term was invented to try to avoid some of the misinformation associated with the more classical term "gene." That older gene concept was much more a structural than a functional concept. The modern concept tries to include both structural (number of nucleotides) and functional (production of specific polypeptides) aspects. We continue to use the term "gene," but in its newer rather than its older sense.

An Enlarged View

With this new concept of the gene, we can take a fresh and meaningful look at some phenomena associated with it. It is possible, for example, to hypothesize that crossing over involves the breakage and rejoining of intact DNA molecules, and there is solid experimental evidence to support this hypothesis. Mutations can be viewed as either a change in the order of base pairs along the DNA molecule or a change in the kinds of bases that occur there (Fig. 10.18). Either change will result in a changed sequence of amino acids in the protein synthesized or a substitution of one amino acid for another. Sickle-cell anemia, to be discussed in Section 14.6, results from the substitution of valine for glutamic acid in but 1 of 200 amino acids in hemoglobin. A very slight change can have far-reaching results for the organism.

The new gene concept also enables us to picture ways in which chemical agents, such as nitrous acid, can cause gene mutations. Nitrous acid is capable of converting one nitrogenous base into another. This results in the selection of an amino acid different from the one that would normally have been selected.

Even Mendel's concept of a recessive "factor" takes on new meaning in light of the present gene concept. The recessiveness of a gene can often be viewed as its failure to produce any functional protein at all. The dominant gene, however, produces enough functional protein to hide the recessive gene's failure. There are probably enough good enzyme molecules to catalyze the metabolic reaction, even though the total number may be reduced. Or possibly there is a control mechanism that increases the output of *m*RNA by the dominant gene to compensate for the inactivity of its recessive allele. Without this compensatory increase, an intermediate phenotype may result.

The term "gene" is still used by most geneticists today, but its definition has been refined. A gene consists of one or more cistrons, units of DNA that code for a functional polypeptide. Groups of genes working together in a related way (especially involving control mechanisms) are called operons.

Usually DNA codes for RNA, which in turn codes for the synthesis of protein. In RNA-containing viruses, however, viral RNA sometimes codes for DNA, which in turn codes for mRNA, which codes for protein. The direction of coding is not universally from DNA to RNA.

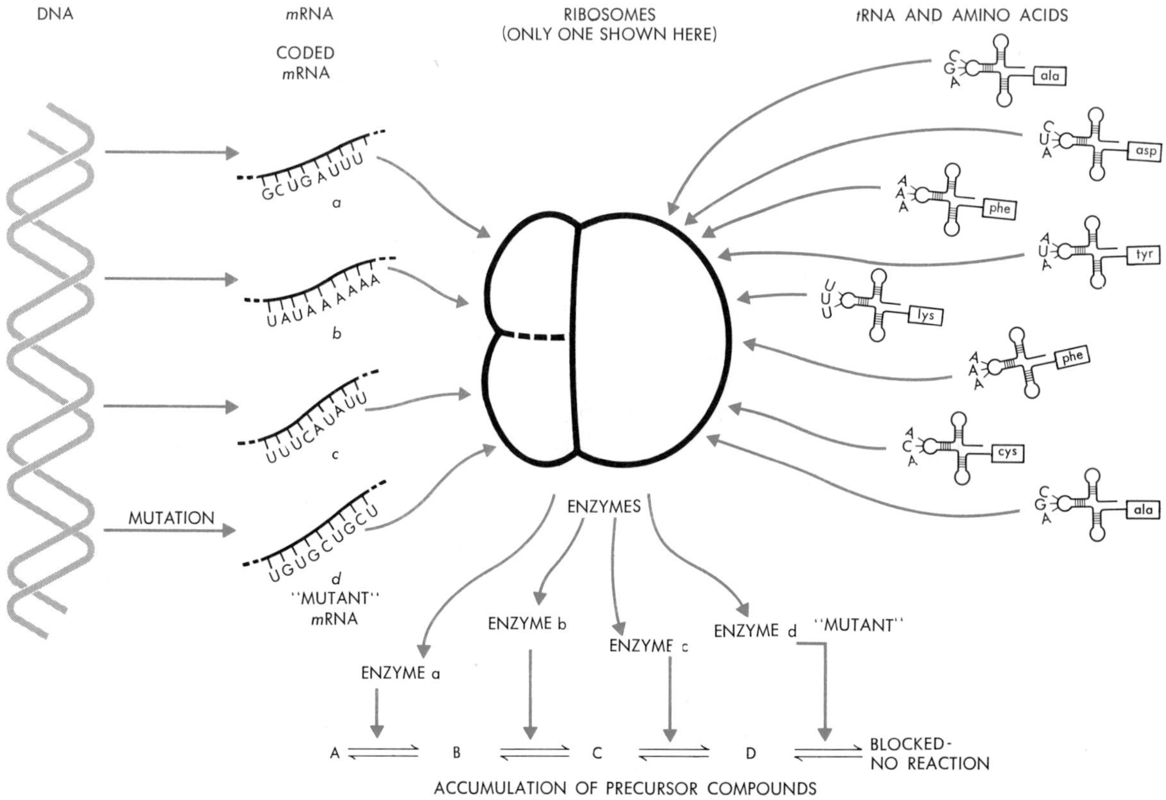

Fig. 10.18 *The Watson–Crick hypothesis that a mutation may be due to a change in the kind of order of base pairs along the DNA molecule provides a useful way to view gene mutations and their effects on the phenotype. Compound E cannot form in the above reaction series because the enzyme catalyzing its production from compound D is either absent or defective.*

SUMMARY

1. In nearly all organisms, deoxyribonucleic acid (DNA) is the molecular substance of the gene. Only certain viruses differ in this respect. Some of these, such as the tobacco mosaic virus and the polio virus, have ribonucleic acid (RNA) as their genetic material.

2. The DNA molecule consists of two spiral backbones wound around each other as a double helix. The backbone of the spirals is composed of sugar-phosphate linkages. Joined to each sugar-phosphate group is one of four nitrogenous bases: adenine and guanine (double-ring molecules known as purines), and cytosine and thymine (single-ring molecules known as pyrimidines). The double strands are held together by hydrogen-bonding between bases; such bonds form across the molecule by a purine bonding only with a pyrimidine and vice versa. The bonding pattern is always adenine–thymine (A–T) and guanine–cytosine (G–C). The two complementary strands also lie head-to-tail with regard to the chemical groups. The 3′ phosphate is exposed on one end of one strand; lying across from it is the 5′ phosphate group of the complementary strand. The reverse is true at the other end of the molecule. This model for DNA structure was proposed in 1953 by James D. Watson and Francis H. C. Crick.

   a) DNA replicates by separation of the two helical strands and the enzymatic catalysis of new strand synthesis from each separated strand. Each old strand acts as a template for the formation of the new strand. Each new molecule is thus composed of one old strand against which a new, complementary strand has been built.

   b) Several enzymes have been identified as contributing to the replication process (Kornberg enzymes, DNA polymerase I and III). It is now thought that DNA polymerase I serves primarily a repair function, whereas DNA polymerase III may be the most active enzyme in synthesis of new DNA from old.

3. DNA not only replicates itself but also guides the synthesis of proteins. It does this through the synthesis of messenger RNA (*m*RNA) directly from one strand (in a process called transcription). RNA contains the purine uracil in place of thymine; thus wherever an adenine appears in the DNA template, a uracil appears in the *m*RNA. It is *m*RNA that carries genetic information from the nucleus to the cytoplasm. The "message" on *m*RNA is translated into a specific sequence of amino acids in a newly synthesized polypeptide. The translation process occurs at the ribosome level. A ribosome attaches to the *m*RNA molecule and moves slowly along the "messenger." This process involves two other types of RNA: transfer RNA (*t*RNA) and ribosomal RNA (*r*RNA). There are as many different types of *t*RNA molecules as there are different types of amino acids (roughly 20). Each type of *t*RNA attaches to its particular type of amino acid, a process that requires the expenditure of energy from ATP and involves an "activating enzyme." The activated amino acid and *t*RNA complex is carried to a ribosome, where it is incorporated into a protein. This incorporation is determined by the genetic code contained in the *m*RNA molecule (copied in turn from the DNA).

   a) The genetic code is a triplet code. Information is stored in sequences of three bases each. The code is contained in the linear sequence of bases along the molecule's length. Each three bases on DNA are called a codon. Each codon determines a complementary sequence on *m*RNA; this sequence is called the anticodon. Each anticodon binds to a specific, complementary portion of the *t*RNA molecule.

b) The final stages of protein synthesis are as follows. Ribosome moves along the *m*RNA molecule three bases (one anticodon) at a time. As it stops at each triplet, the appropriate *t*RNA molecule containing its activated amino acid binds to the triplet by base-pairing. This holds the amino acid in place, and it is joined to the amino acid next to it in the growing polypeptide chain. Joining involves formation of a peptide bond through hydrolysis of the activated phosphate group. Enough energy is provided to drive the uphill reaction of forming the peptide bond. The polypeptide chain grows in this way, one amino acid at a time.

4. Genes and proteins are colinear. The sequence of amino acids in a protein chain corresponds to the sequence of codons for those specific amino acids in the DNA molecule.
5. Mutations are typographical errors in the linear sequence of bases in a DNA molecule. If one base is substituted for another, the sense of the genetic message can be altered, at least for one portion of the resulting protein chain. If a base is added or deleted, the entire message beyond that point in the DNA molecule will become "nonsense," and thus result in a nonfunctional or barely functional protein.

   a) Mutations can be induced by agents such as ultraviolet light or certain chemicals. These chemicals or high-energy radiation can cause substitution of one base for another, binding of two neighboring bases, or sometimes a deletion.

   b) Mutations occur at random. Environmental agents such as high-energy radiation or certain chemicals increase the overall mutation rate, but they do not increase the likelihood of one mutation over another.

6. The term cistron has been used to designate that section of DNA that codes for a functional protein (or polypeptide). The classical term "gene" now refers to one or more cistrons, whose products form a complete protein molecule (such as an enzyme).

EXERCISES

1. Describe the major conclusions drawn from the experimental work on DNA performed by: (a) F. Griffith; (b) Avery, McCarty, and MacLeod; (c) Hershey and Chase; (d) Watson and Crick.
2. Explain the graph lines in Fig. 10.19 in terms of the experiment concluded by

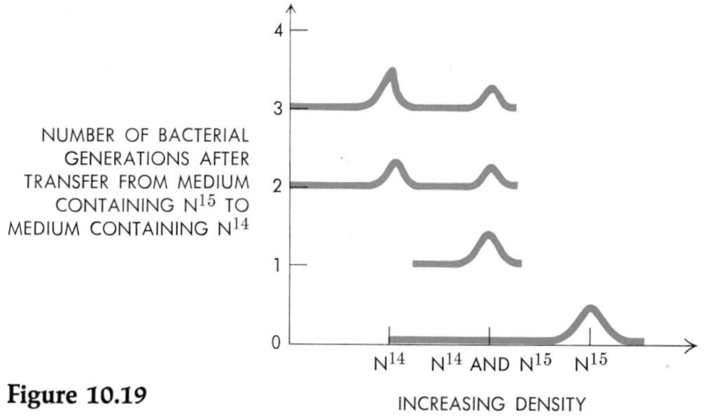

Figure 10.19

Meselson and Stahl to test the Watson–Crick hypothesis of DNA replication. What sort of line would be found after four generations of bacteria?

3. Two identical cultures of HeLa cells (a human cancer cell line that has been cultured in the laboratory for many years) were infected with equal numbers of polio virus. Radioactive uridine was added to both cultures, while actinomycin D was added to culture Number 2. Actinomycin D inhibits DNA-dependent RNA synthesis. Uridine is a form of uracil and is incorporated into RNA. After a 2-hr incubation at 37°C, two measurements were made on each culture: (1) the amount of radioactivity incorporated into the RNA, and (2) the number of new virus particles that had been produced. The following data were obtained:

Culture number	Actinomycin added	Amount of radioactivity in RNA, counts/min	Number of new virus particles, millions
1	no	530	111
2	yes	23	102

   a) Assuming that all the cells were infected with virus, how would you account for these results? Explain your reasoning specifically in relation to the data.
   b) How does your answer relate to the "central dogma" concept of molecular biology (DNA → RNA → Protein)?

4. How has molecular genetics thrown light on the following classical problems, and how can we explain each of these phenomena in molecular and biochemical terms?
   a) dominance and recessiveness
   b) incomplete dominance
   c) mutation
   d) epistasis
5. How do we know that the genetic code is a triplet code?

SUGGESTED READINGS

General Genetics

Levine, Paul, and Ursula Goodenough, *Genetics.* New York: Holt, Rinehart and Winston, 1974. A modern and thorough introduction to general genetics. Requires some background but most explanations are well presented.

Biochemical and Molecular Genetics

Crick, F. H. C., "The Genetic Code," *Scientific American,* October 1962, p. 66. Discusses how the triplet code of DNA was discovered and shows how each amino acid in a protein is coded by a specific triplet. A good survey article.

Croce, Carlo M., and Hiliary Koprowski, "The Genetics of Human Cancer." *Scientific American,* February 1978, p. 117. A very enlightening discussion of the relationship between cancer and chromosomal complement.

Deering, R. A., "Ultraviolet Radiation and Nucleic Acid," *Scientific American,* December 1962, p. 135. This article shows how the effects of ultraviolet radiation on living organisms can be traced to changes in nucleic acids, especially DNA.

Fiddes, John C., "The Nucleotide Sequence of a Viral DNA." *Scientific American,* December 1977, p. 54.

Describes how one complete sequence for ϕX-174 DNA was worked out.

Hurwitz, Gerard, "Messenger RNA," *Scientific American,* February 1962, p. 41. This article discusses the basic information known about structure and function of messenger RNA. Also discusses the general mechanism of protein synthesis.

Medawar, P. B., and J. S. Medawar, *The Life Science: Current Ideas of Biology.* New York: Harper & Row, 1977. A highly readable essay on the frontiers of biological science by two world-famous biologists.

Nirenberg, Marshall W., "The Genetic Code: II," *Scientific American,* March 1963, p. 80. This article is a sequel to the one by Crick. It goes into more detail on the actual mechanism of protein synthesis.

Rich, Alexander, "Polyribosomes," *Scientific American,* December 1963, p. 44. This article reviews the recent findings that show that ribosomes line up along a molecule of messenger RNA during protein synthesis.

Sayre, Anne, *Rosalind Franklin and DNA.* New York: Norton, 1975. Written partly as a response to J. D. Watson's *The Double Helix* (which portrays Rosalind Franklin as a paranoid, secretive, and volatile woman), Sayre's book is a sensitive treatment not only of many aspects of the working out of the model of DNA, but of the problems women encounter in science the world over.

Sinsheimer, Robert, "Single-Stranded DNA," *Scientific American,* July 1962, p. 109. The discovery of single-stranded DNA in a bacteriophage that parasitizes *E. coli* has led to new studies on the structure of DNA and RNA. A good article for those interested in further detail about biochemistry of the DNA molecule.

Watson, J. D., *The Double Helix.* New York: Atheneum, 1968. A fascinating and lively account of the discovery of the three-dimensional structure of the DNA molecule.

———, *Molecular Biology of the Gene,* 3d ed. Menlo Park, Calif.: W. A. Benjamin, 1976. Excellent coverage of molecular genetics, up-to-date, and clearly explained.

ELEVEN

DEVELOPMENTAL BIOLOGY

11.1 Introduction

Through the processes of transcription and translation, genes give rise to proteins. Proteins act either as structural elements or as enzymes for the synthesis and degradation of other molecules. As enzymes, proteins make possible a multitude of biochemical pathways. Those pathways generate cell materials, which makes possible cell growth and differentiation. Such growth and differentiation produce the overall phenotype of the organism. Thus growth and differentiation are from the start the products of gene action. The study of growth and differentiation and the factors that affect and control them comprises the field of **developmental biology.** Developmental biology is concerned with the *changes* organisms undergo in some regular way throughout their life cycle.

Developmental biology is much more than the study of growth and development of young embryos into young organisms. It is concerned with every aspect of regular change that characterizes a particular species of animals or plants. It is concerned with seed germination, with repair of wounds and tissue replacement, with regeneration, with immunology, with the menstrual cycle, with aging and death. Nothing concerning the regular changes that organisms experience falls outside the sphere of developmental biology. Though it is often equated with embryology (the study of the growth and differentiation of embryos), developmental biology is a much broader field.

11.2 Embryonic Development: An Overview

The development of embryos consists of two different but interrelated processes: growth and differentiation. **Growth** involves increase in the size of the organism, generally by increase in the number of cells (and, to a limited degree, increase in the size of individual cells). The biological issues of interest in the study of growth concern the *regu-*

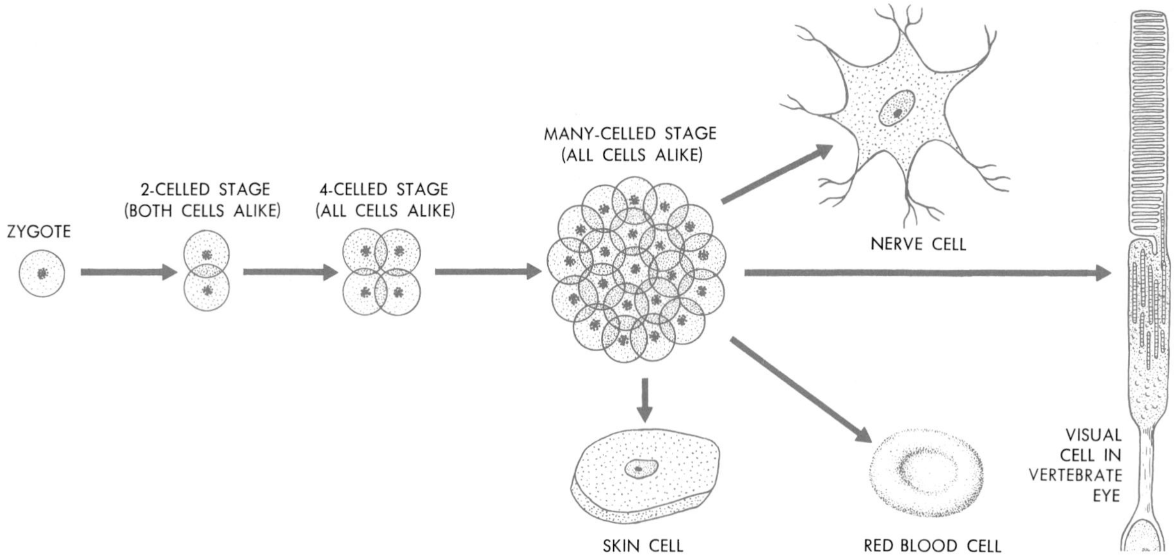

Fig. 11.1 *Schematic representation of differentiation, the central phenomenon in embryonic development. Differentiation is the process by which initially identical cells become structurally and functionally different.*

lation of growth activities. What initiates growth in one or several parts of the organism? And what controls the *rate* of growth? How is growth stopped at certain points? Obviously, the control of growth processes is programmed into the genome of most organisms. All members of a given species have certain growth limits, an indication that control processes are operating.

Differentiation is the process by which like cells become specialized. All cells of an adult multicelled organism start from a single fertilized egg, the zygote. The zygote undergoes numerous cell divisions by mitosis, leading to a group of cells, arranged in a ball-like form (see Fig. 11.1). At this stage all the cells in the "ball" look quite similar, though some are slightly larger or smaller than others. But then, as cell division continues, certain cells begin to take on different sizes, shapes, and biochemical properties. Some cells become embryonic nerve cells, others visual cells, others blood cells, and still others skin or muscle cells (see Fig. 11.1). The cells begin to take on *specialized* structures and functions. This process of differentiation into specialized cell types is of fundamental interest to developmental biologists. Indeed the problem of differentiation has been one of the key issues in developmental biology since Aristotle.

Cells that have undergone differentiation generally lack the ability to live independently of one another. A nerve cell cannot live outside the organism, except in highly controlled laboratory conditions that recreate the conditions which prevail within the organism. While even the most highly specialized cells carry out many, if not all, of the major life activities, their differentiation has made them dependent upon a collective life. The advantage gained, of course, is that the complex organism is much better adapted to meet a variety of external changes. A one-celled organism is completely self-sufficient, but it must live in a relatively narrow range of environments, compared to an oak tree or a human being. While all organisms are adapted to the environments in which they live, generally speaking multicelled organisms with specialized cells can tolerate a far wider range of environmental conditions. Clearly the process of differentiation has enormous adaptive value.

But how does differentiation occur? It is thought that all cells of a multicelled organism are genetically alike; that is, they contain all the same genetic information. Yet chemical analysis reveals that nerve cells contain different proteins from muscle or skin cells. How does this come about? What process is involved in triggering some cells to become nerve and others muscle or skin? What is happening on the level of molecules and genes? Conversely, what is it that causes some cells to lose their special properties, as in the conversion of certain cells into cancer cells? Cancer cells are essentially descendants of once-specialized cells that lost their special functions, and hence *de-differentiated.* Cancer cells resemble cells of embryonic stages, losing, among other things, the controls on their rate of growth. Even the problem of cancer is fundamentally one of development.

To understand embryonic development in a more general way, consider the basic pattern by which it occurs. Embryos grow from fertilized eggs. They require a source of energy for cell division and specialization—that is, for growth and differentiation. The first food an organism receives in its life is often **yolk.** Yolk is a mixture of fatty compounds and proteins. The fatty compounds produce a rich source of energy and the proteins supply building materials for growth.

Almost everyone is familiar with the yolk of a chicken's egg. As in all birds' eggs, the amount of yolk is quite large. The eggs of reptiles such as snakes, alligators, turtles, and lizards also contain a great deal of yolk. Eggs of such animals as frogs and insects contain yolk, too, though far less than the eggs of reptiles and birds. The eggs of mammals, such as the cat, dog, rabbit, or human, contain very little yolk.

At first glance, this last statement is surprising. It seems that it would take far more energy and raw material to build a human being than a baby chick, and of course it does. Why, then, does a human egg contain less yolk than the egg of a chicken?

The answer is quite simple. It lies in the manner in which each embryo develops. Consider first eggs with large amounts of yolk. Birds and reptiles lay their eggs with very young embryos already developing inside (fertilization, of course, has to take place *before* the shell is added to the egg). Until the egg hatches, the embryo has no source of food other than that provided by the yolk. Therefore, there must be enough yolk to supply energy and raw materials for complete development from fertilized egg to young animal.

Consider next the egg with an intermediate amount of yolk. This type of egg is produced by animals such as the frog and most insects. Like the chicken, these are complex animals, and it takes a great deal of energy to carry them from the fertilized egg to adulthood. However, the yolk of these eggs supplies just enough energy and building materials to carry development part of the way. By means of this energy, an intermediate, free-living stage is produced. This intermediate form is the **larva.** Larvae are capable of feeding, and thus storing up more food. The larva of the frog is the tadpole. Unlike the adult, the tadpole is vegetarian, and feeds on water plants. In addition, it gets energy and raw materials by gradually absorbing its tail. The moth larva, or caterpillar, is an example of an insect larva.* The caterpillar is notorious for its appetite. By feeding it stores up energy to be used for the final change into adult form. After a few

* The grubs of beetles and the maggots of flies are two other examples.

weeks, the caterpillar spins a cocoon and changes into a **pupa.** Movement is at a minimum in the pupal stage; thus almost all the energy released from the food the larva ate can be used in the change to adulthood. This type of development is called **metamorphosis.** Complete or partial metamorphosis is typical of organisms that produce eggs with intermediate amounts of yolk.

Most mammalian eggs have very little yolk. Nor is there any mammalian larval stage. However, the small amount of yolk present is enough to get the fertilized egg through its early cell divisions. These divisions occur as the egg passes down the oviduct. When it enters the uterus, the egg becomes implanted in the uterine wall. Until birth, energy and raw materials come from the mother's bloodstream. In a very real sense, the developing embryo is a parasite.

Animals whose developing young are separate from the mother, and who derive their nourishment entirely from the egg yolk, are said to be **oviparous.** Fishes, amphibians, reptiles, and birds are nearly all oviparous. In some cases the young derive their nourishment from the yolk, but still develop within the mother's body. Such animals are said to be **ovoviviparous.** Two examples are the garter snake and the dogfish, or sand shark. In cases in which the embryo derives almost all its food from the mother and develops within her body, the term **viviparous** is used. With a few exceptions, mammals are viviparous.

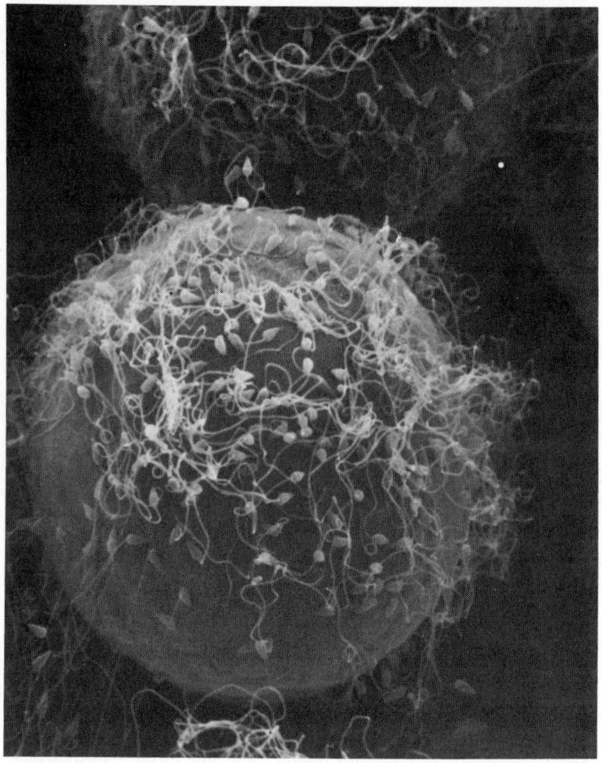

Fig. 11.2 *Scanning electron micrograph of the surface of a sea urchin egg, covered with sperm. Many sperm swarm over each individual egg, but only one sperm usually penetrates the membrane. After penetration, the egg is activated, a process that produces changes in the cell surface preventing other sperm from entering. (Photo courtesy Dr. Mia J. Tegner, Scripps Institution of Oceanography.)*

11.3 Fertilization

Embryonic development for all multicelled animals and plants generally begins with fertilization. There are actually two processes involved in fertilization: activating the egg and uniting of egg and sperm nuclei. As we learned in Chapter 8, which sperm actually fertilizes any given egg is largely a matter of chance. Generally speaking, large numbers of sperm are attracted to each individual egg, thousands of sperm sometimes congregating on or near the surface of a single egg (see Fig. 11.2). As we shall see in later chapters, different species of animals and plants have different ways to ensure that sperm find eggs. The process is least efficient where sperm or eggs are simply released into the surrounding medium and must find each other by chance. It is most efficient when the male inserts the sperm within the female body close to the point of origin of the egg.

Once contact between the gametes occurs, a chain of chemical and physical events begins. First the sperm become sticky and clump together (agglutinate) on the egg surface. Actually it is necessary for many sperm to stick to the egg surface in order to break down the layers of material that in many species form external envelopes to the egg cell. Eventually, one sperm comes close enough to the egg cell membrane that it makes direct contact. The tip of the sperm head shows quite clearly a special structure, called the **acrosome** (Fig. 11.3). During sperm development, the acrosome is derived from the Golgi apparatus. When they come into close proximity with an egg, sperm show an "acrosome reaction"; the acrosome filament extrudes from the head end of the sperm. This filament makes the initial contact between sperm and egg and holds the sperm in place at the point of contact. It is possible that the acrosome filament may contract, pulling the sperm to the egg and bringing about the actual contact between sperm and egg cell membrane. Enzymes associated with the sperm acrosome then digest the egg cell membrane so that the sperm may enter. It is interesting that this enzyme (hyalurinodase) is the same enzyme secreted by many bacteria when they infect healthy cells.

The egg membrane also responds physically to contact. It rises up to meet the sperm head, forming what is known as a **fertilization cone** (see Fig. 11.3). Formation of the fertilization cone is followed by a general lifting off of the outer egg membrane all around the cell, until a layer (the **hyaline layer**) is formed, separating the inner membrane from an outer membrane (the **fertilization membrane**). The inner membrane (the **vitelline membrane**) forms the new boundary layer of the egg cytoplasm.

The sperm cell maintains contact with the vitelline membrane so that once the sperm cell membrane is broken, the sperm nucleus is released within the egg. The sperm centriole divides to provide the spindle for the first cleavage division. Precisely what happens to the female egg cell's

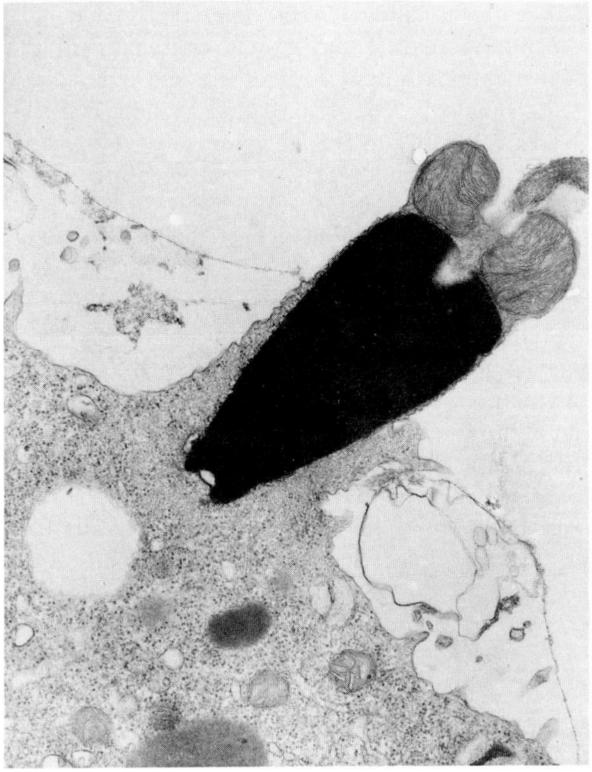

Fig. 11.3 *Electron micrograph (×17,600) of sperm in the process of penetrating the egg cell membrane. The acrosome, shown as the dark pointed object at the head of the sperm, enables the sperm to penetrate the egg membrane. Once penetration has occurred, the egg cell membrane lifts up to engulf the entering sperm. (Photo courtesy Dr. Everett Anderson.)*

centriole is unknown. There is evidence that it is present, because if the egg is caused to divide parthenogenetically, it forms a spindle of its own.

It should be emphasized that the egg contributes considerably to the initiation of development, just as the sperm does. There is, of course, a genetic as well as a structural or physiological contribution from the egg. The egg's genome governs all the early stages of development, for the maternal gene's *m*RNA is present in the egg cytoplasm at the time of activation. The influence of the maternal genome persists until approximately the time of gastrulation, when the male genome first begins to come into play. Evidence for this notion is derived from analysis of early development in hybrid embryos between two strains of the same species. In these cases, chemical products during blastula formation are almost exclusively of the sort characteristic of the maternal species. More light is thrown on the process by studies of early development in eggs that lack a nucleus and yet are activated by exposure to dilute acid. A series of unfertilized eggs were submitted to certain physical conditions that caused numbers of cells to be cleaved in half. One-half of each egg contained a nucleus, the other half lacked a nucleus. The resulting half cells were then separated into two batches, the enucleated and the nucleated. Both types of cells were activated, the nucleated ones by sperm, the enucleated ones by exposure to dilute acid. The response of the eggs was then determined by measuring the change in the rate of protein synthesis. This criterion was chosen because it is known that during normal fertilization protein synthesis increases rapidly shortly after the activation of eggs.

The results are shown in Fig. 11.4. Data for a control (unaltered) egg are included on the graph for comparison. The graph shows that there is very little difference in the initial stages of protein syn-

Fig. 11.4 *Changes in rates of protein synthesis after activation of whole eggs and half-eggs with or without a nucleus. Note that whether the activated egg contains a nucleus or not, the pattern of protein synthesis remains virtually the same.*

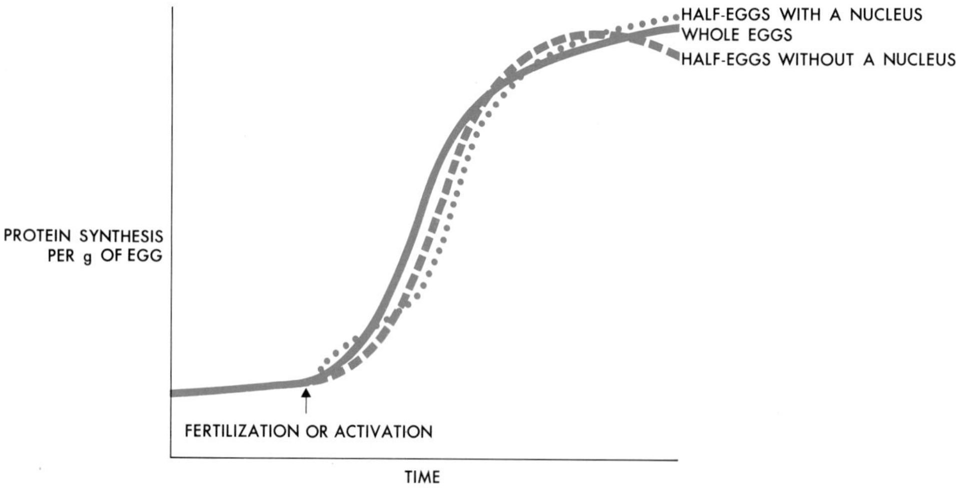

thesis between eggs that possess and those that lack a nucleus. Realization of this fact has led to the notion that *m*RNA for the initial protein synthesis following activation exists in the egg cytoplasm in a masked, or inactive, form. It appears to be transcribed from maternal cell DNA back in early stages of oögenesis. Of course, the half-eggs without a nucleus cannot exist for long—soon they become inactive and eventually die. But the fact that early protein synthesis occurs in the absence of any nucleus shows that this process has already been programmed by the maternal genes, a coding that must take place long before the egg is ready for actual fertilization.

11.4 Early Embryonic Development

The early development of an embryo is almost entirely epigenetic. This fact presents a very complicated situation to the embryologist. If the embryo were completely preformed, its development would be little more than simple growth in size. However, in epigenetic development, cellular differentiation and specialization must take place at precisely the right time and place if the individual is to be normal. Work on embryology has therefore been directed toward the problem of just *how* complex and specialized regions of an organism can arise from unspecialized cellular regions—for this is precisely what occurs.

In most higher organisms there are three such cellular regions, called the **primary germ layers.** The primary germ layers are the first distinguishable areas within the early animal embryo. They will give rise to the tissues and organs of the adult animal. The three primary germ layers are the **ectoderm** (outer skin), **mesoderm** (middle skin), and **endoderm** (inner skin). Table 11.1 shows some of the parts of the adult body that arise from each germ layer.

Cleavage

As we have just seen, embryological development generally begins with fertilization. The fertilized egg, or zygote, immediately becomes a beehive of biochemical activity. It soon undergoes its first mitotic division **(cleavage).** The resulting two daughter cells divide again, producing four. Each of these divides, giving eight cells. The eight soon become sixteen. As can be seen, early cleavages result in a

Table 11.1
The Three Primary Germ Layers and Some Body Parts That Arise from Them

Ectoderm	Mesoderm	Endoderm
Skin epidermis	All muscles	Lining of digestive tract, trachea, bronchi, and lungs
Hair and nails	Dermis of skin	Liver
Sweat glands	All connective tissue, bone and cartilage	Pancreas
Entire nervous system, including brain, spinal cord, ganglia, and nerves	Dentine of teeth	Lining of gall bladder
Nerve receptors of sense organs	Blood and blood vessels	Thyroid, parathyroid, and thymus glands
Lens and cornea of eye	Mesenteries	Urinary bladder
Lining of nose, mouth, and anus	Kidneys	Urethra lining
Teeth enamel	Reproductive organs	

geometric increase in cell number. Later, the increase in numbers is less precise. Eventually, however, a hollow ball of many hundreds of cells is formed.

Not all organisms show the same cleavage patterns. The amount of yolk present in an embryonic cell often determines the way in which it divides. Yolk is composed of very dense material. Thus yolk-laden cells divide more slowly than those containing less yolk. In the bird's egg, the yolk does not become divided into the individual cells at all. Three different egg cell types, based on amount and location of yolk, are shown in Fig. 11.5.

The hollow ball of cells formed by cleavage is called a **blastula**, the cavity inside the blastula is called the **blastocoel**, and the blastula's cells are called **blastomeres** (see Fig. 11.6). It is interesting to note that the blastula, composed of many blastomeres, is often no larger than the original zygote from which it developed. There exists a critical relationship between cell size and the amount of cell membrane available for respiratory exchange. The unfertilized egg has a relatively low rate of metabolism; respiratory exchange is low. Once it is fertilized, however, things change. The zygote immediately divides several times, and these cleavages bring surface–volume proportions to a point that will sustain the greatly increased metabolic activity accompanying development. Blastula formation also gives the embryo an adequate number of cells (essentially alike except for yolk content) for the first building blocks of the new organism.

Quite obviously the embryo cannot feed, so each cell must have its own source of energy and raw materials. Cleavage requires a great deal of energy for the mechanics of mitosis and the synthesis of nucleic acid (which must occur during the chromosome replication accompanying each division). The formation of a blastula is followed by a process called **gastrulation.** The result of gastrulation is to produce another hollow ball of cells, the **gastrula.** The gastrula often resembles the blastula in appearance. Unlike the blastula, however, the

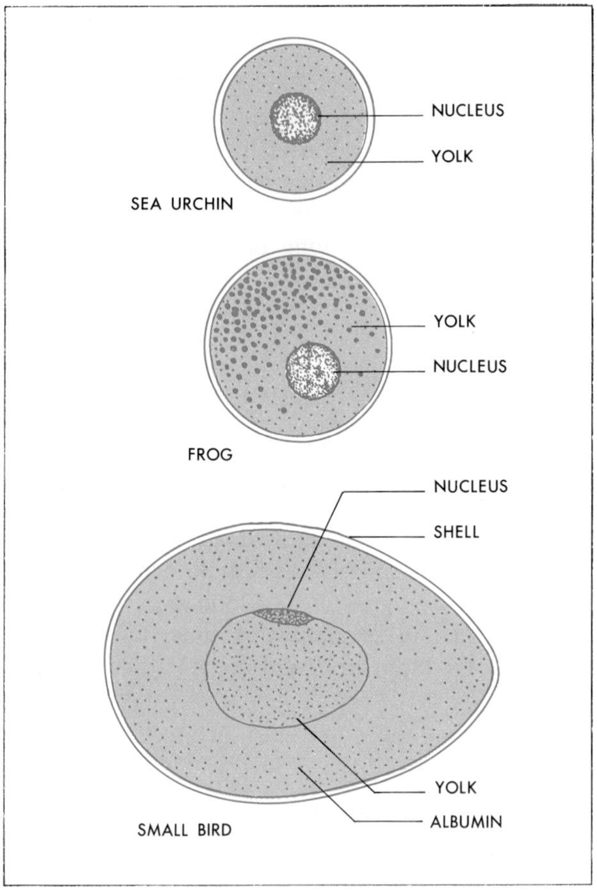

Fig. 11.5 *Three egg types showing differences in amount and distribution of egg yolk. Differences in yolk distribution affect the patterns of cleavage in the egg once it is fertilized. Yolk slows down cell division, so that those cells containing much yolk divide more slowly (and thus at any point in time appear larger) than cells containing less yolk. The yolk in the sea urchin and frog eggs can be apportioned into individual cells by cleavage; that in the bird's egg cannot, because of its massive volume in relation to the egg cell cytoplasm.*

Fig. 11.6 *Cleavage patterns for eggs with two different amounts and distributions of yolk: (a) sea urchin and (b) bird. In both cases, cleavage produces a rounded ball of cells, the blastula. In the bird, because of the vast amount of yolk, the blastula resembles a flattened, hollow disc (the blastodisc) lying on top of the yolk. Note that cleavage occurs without any significant overall increase in size between the zygote and the blastula. The cytoplasm of the zygote is partitioned out into successively smaller units, the embryonic cells, or blastomeres.*

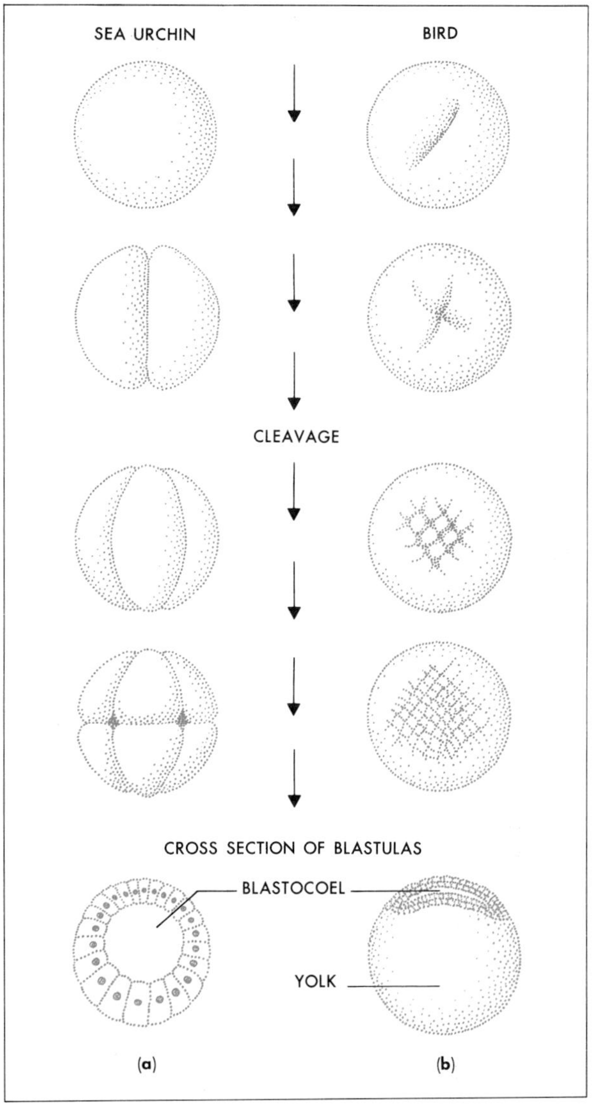

early gastrula has at least *two* layers of cells, instead of only one.

Gastrulation

Gastrulation differs considerably from one animal to another (see Fig. 11.7). Once again, yolk is an important determining factor. If there is very little yolk, as in the sea urchin, the side of the blastula simply pushes in, and the blastocoel is slowly obliterated. A new cavity, the **gastrocoel,** or **archenteron** (primitive gut), is formed. If there is a little more yolk, as in the frog blastula, invagination occurs more toward the top, or **animal** pole, of the blastula. The dense yolk-laden cells are concentrated at the bottom, or **vegetal** pole region. Finally, in yolk-laden eggs (birds, reptiles), invagination is greatly modified. It involves only those cells resting on top of the yolk mass.

An analogy may show how the yolk influences gastrulation. Imagine squeezing a soft rubber ball. If the ball is empty, one side can be pushed against the other. This might correspond to an *Amphioxus* blastula with little dense yolk. If the ball is half-filled with sand, only the top portion of the ball can be pushed in. This is similar to the intermediate yolk content of the frog's blastula. However, the ball almost completely filled with sand has only a small pinch of rubber at the top that can be moved. This ball is similar to the yolk-laden eggs of birds. Invagination must be considerably modified from that which occurs in blastulas containing less yolk. However, no matter how gastrulation takes place, the net result is roughly the same; *three primary germ layers are formed.*

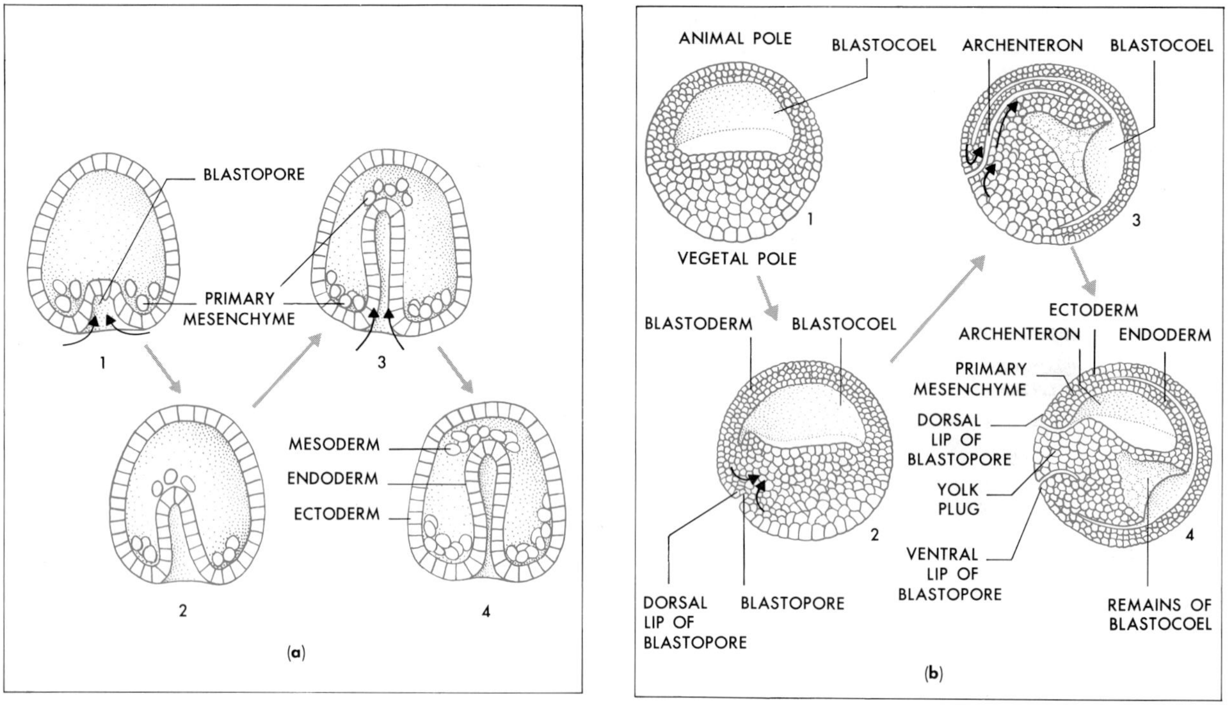

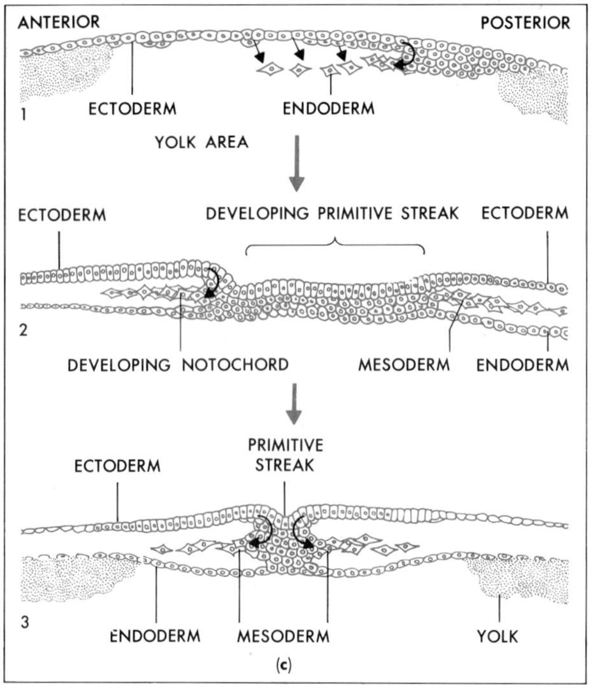

Fig. 11.7 *Comparison of gastrulation in three organisms: (a) sea urchin, (b) frog, and (c) chick. In all cases invagination takes place largely through the migration of cells (morphogenetic movements) inward at a point called the blastopore (in the sea urchin and frog) or the primitive streak (in birds). Cell migrations are indicated in each drawing by dark arrows. The result of gastrula formation is to produce first a two-layered embryo with an ectoderm and endoderm, followed by the migration of certain cells to form a middle layer, the mesoderm. The blastopore will become the anus in the adult, the archenteron the gastrointestinal tract. During gastrulation, the old blastocoel is largely obliterated.*

Blastula formation occurs without any significant increase in overall size. The cytoplasm of the original egg is merely portioned out into increasingly smaller compartments.

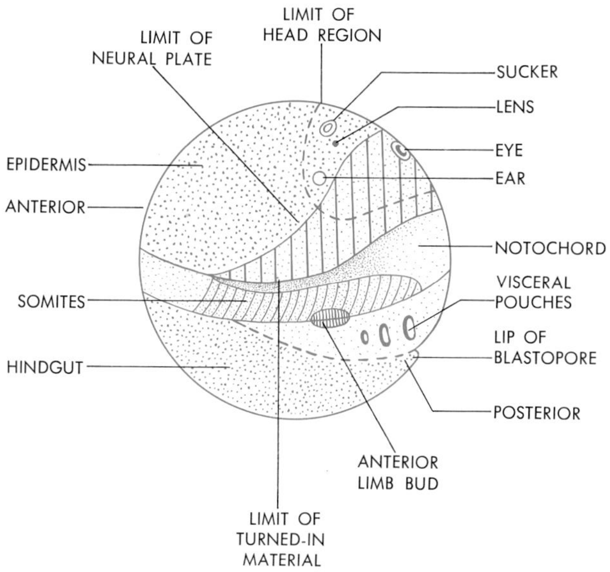

Fig. 11.8 *By the time the embryo is in the late blastula or early gastrula stage, it is often possible to predict what tissues and organs will be formed by many embryonic regions. (After Vogt.)*

Let us continue the analogy of the empty rubber ball. If you squeeze it so that one wall touches the other, you eliminate the inner cavity. This corresponds to the obliteration of the blastocoel. In doing so, however, your fingers form another cavity, corresponding to the archenteron. Note that this new cavity differs from the old one in two important respects. First, it opens to the exterior. This opening corresponds to the **blastopore.** Second, the new cavity is enclosed within a double- rather than a single-layered wall. Similarly, in the embryo the archenteron becomes surrounded by a double layer of cells, of which the outer layer is the ectoderm. With the formation of the mesoderm (by segregation from the endoderm in amphibians), the three primary germ layers are established. In *Amphioxus,* the mesoderm gradually spreads downward to lie between the endoderm and ectoderm (Fig. 11.7). The result is an embryo with three primary germ layers from which all the later organ systems can develop. In fact, by the time of late blastula or early gastrula, many regions of the young embryo are beginning to be determined for a specific fate in the adult. Figure 11.8 shows some of the predictive fates for cells in various regions of the late blastula.

The key to the process of gastrulation, and to many of the patterns that follow in later development, is the phenomenon of **morphogenetic movement.** During invagination of the blastula, for example, certain groups of cells near the region of the dorsal lip begin to move inwardly, literally migrating down into the indentation that forms the blastopore itself. Later, as the various germ layers organize themselves and begin differentiation into specialized tissues, additional cell migrations occur. These migrations are collectively called morphogenetic movements.

Gastrulation produces three primary germ layers, ectoderm, mesoderm, and endoderm, from which all the later tissues and organ systems of the adult organism are derived.

11.5 Causes of Differentiation: Embryonic Induction

In the normally developing embryo, the period following gastrulation includes formation of specific tissues, organs, and systems. At the gastrula stage the basic outline of the adult organism is already laid down. The blastopore will become the future anus, and the archenteron the future gastrointestinal tract. The organization of the gastrula into an embryo involves a variety of processes: cell growth, differentiation, and migration.

What makes a tissue? It might seem that this question was answered earlier. The primary germ layers (the ectoderm, mesoderm, and endoderm) give rise to the very specialized tissues and organs of the body. Yet we are still left with the question of *how* the primary germ layers give rise to these parts. What factors, for example, cause the descendants of certain endodermal cells to become parts of the intestinal lining, while others end up in the thyroid gland? At the end of gastrulation, a flattened plate of ectoderm cells—the neural plate—forms as a result of the movement of cells from more lateral regions. Depression of the center of the neural plate and the folding of its edges results in the formation of a neural groove. Eventually, the folds come together and fuse, forming the neural tube.

What initiates this process? How would experimental embryologists approach such a question? The neural plate does not begin its development until the gastrula has fully formed. At that time, the infolding of the blastopore is complete and the inner cell layers of endoderm and developing mesoderm come to lie close to the ectoderm layer (see Fig. 11.9, part 1). Where developing mesoderm touches ectoderm appears to be the site of neural plate differentiation. This observation leads to an hypothesis: neural plate development is initiated by contact between ectoderm and underlying mesoderm of the archenteron roof. Would the neural tube then form from ectoderm that has been isolated from an embryo considerably before it begins neural-tube formation? To find out, we remove ectodermal regions destined to form neural tubes from some frog embryos and culture them in a separate medium.

This experiment asks an important question. It

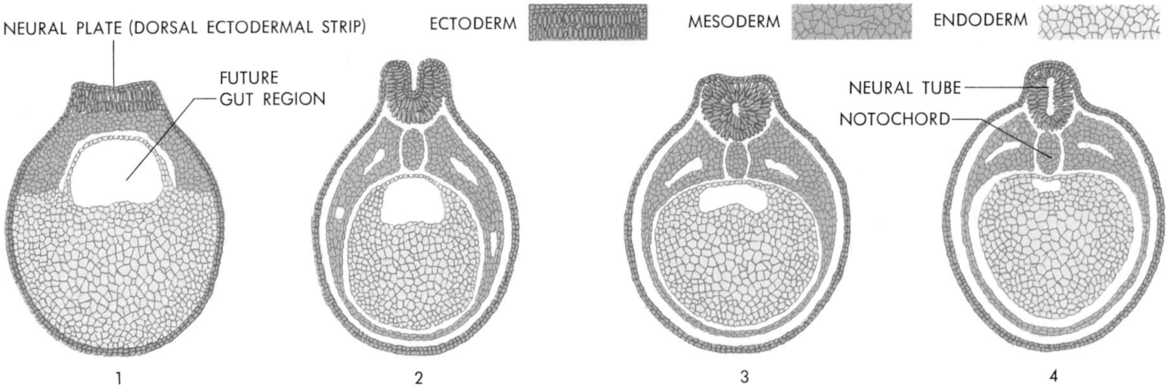

Fig. 11.9 *A cross-sectional view of neural tube formation in the frog embryo. Note that the neural plate sinks inward and rolls into a tubelike shape. Later, cartilage and bone will form around the notochord and neural tube to make up the spinal column.*

seeks to determine whether the development of a neural tube is caused by intrinsic factors within the ectodermal region involved, or by extrinsic factors contingent upon the location of the ectodermal region within the embryo. The results of the experiment seem to provide a definite answer. In no case are neural tubes formed by the isolated pieces of ectoderm. The embryos from which the ectodermal pieces were surgically removed do not develop a neural tube either, presumably because they were deprived of cells that would ordinarily have given rise to this structure.

It seems evident, then, that extrinsic factors from the surrounding environment of the embryo's body mass are responsible for the formation of a neural tube by the ectodermal region. The next question to be posed is: From what region of the embryo do these extrinsic factors originate? A logical choice is the mesoderm, since the neural-tube-forming ectoderm rests on top of it. An experiment can be performed to test an hypothesis proposing the mesoderm as the source of these extrinsic factors. We can hypothesize that if the mesoderm contributes extrinsic factors to the overlying ectoderm to cause it to form a neural tube, then mesodermal cells, wrapped in a sheet of ectoderm and isolated from an embryo at the same stage as in the previous experiment, will cause the formation of neural tissue by this ectoderm. Such an experiment was performed by the embryologist Johannes Holtfreter in the 1940s. The results were as predicted; the sheet of ectoderm enclosing the mesoderm differentiated into neural tissues.

The Primary Organizer

One of the great experimental embryologists was the German Hans Spemann (1869–1941). Working with Hilde Mangold and using amphibian embryos (salamander and frog), Spemann hypothesized that the mesoderm *induces* the ectoderm to differentiate into the neural tube; he called this action of the mesoderm on the ectoderm *embryonic induction.* Spemann observed that there were many inductive processes at work, and he visualized the embryo as being built by progressively less specific inducing tissues, or organizers. Somewhere in the chain of induction there had to be a tissue responsible for organizing the entire organism.

In the newly fertilized frog and salamander egg, a region called the *gray crescent* appears. Spemann turned his attention to cells located on the dorsal lip of the blastopore, which is formed just below the area of the gray crescent. Because of these cells' close association with the gray crescent, he considered that they might be the primary organizers he was looking for.

They were. Taking blastopore lip cells from one salamander embryo, he transplanted them into the presumptive belly tissue of another. Rather

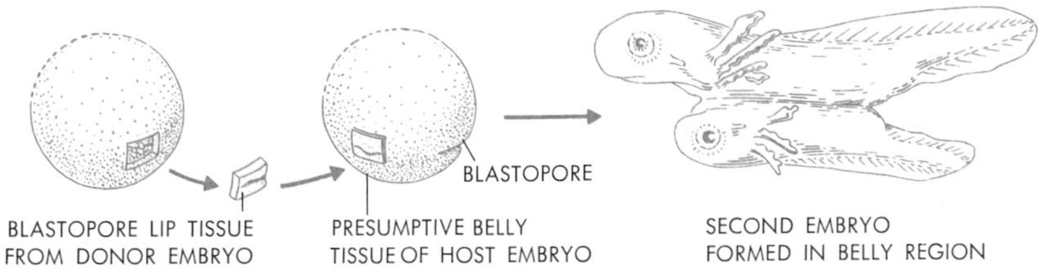

Fig. 11.10 *This experiment, for which Spemann received a Nobel prize, showed the dorsal lip of the blastopore to have primary organizer capabilities.*

than a brain or an eye lens, a *complete new embryo* was induced, joined to the other like a Siamese twin (Fig. 11.10).

From this point on, embryologists began to picture development of the embryo as a series of inductive processs, with each member of the series being essential for the development of the members that follow it. Thus, for example, the eye lens would not form if the midbrain were removed; the midbrain would not form if the head mesoderm were removed, and finally, nothing would form if the primary organizer, the dorsal lip of the blastopore, were removed.

Still, the nature of the primary organizer and the precise manner by which induction was carried out remained to be explained.

Chemical Induction

In the development of the neural tube, the mesoderm acts as an organizer. How?

If an underlying organizer tissue is separated by a thin piece of impermeable material from the tissue it is supposed to induce to differentiate, no differentiation takes place. Thus, one may hypothesize that a chemical is involved in induction, or that physical contact between cells is necessary for differentiation.

To test this hypothesis, two embryologists, V. C. Twitty and M. C. Niu, used the hanging-drop technique of tissue culture designed by the embryologist Ross G. Harrison (Fig. 11.11). Into half of their hanging-drop cultures, Twitty and Niu placed organizing mesoderm; the other cultures, serving as controls, received no mesoderm. After a few hours, the mesoderm was removed from the experimental cultures, and ectoderm was placed into both experimental and control cultures. In the experimental cultures that held mesoderm, the ectoderm differentiated into neural tissues. In the control cultures, no differentiation took place. It seemed obvious that the mesoderm released some substance into the culture medium that induced differentiation in the ectoderm, and that physical contact between cells was not necessary.

In examining the relationship between the inducing and the induced tissues, Oscar E. Schotté transplanted a cell region that would ordinarily form frog flank tissue to the mouth region of an older salamander embryo. The presumptive flank skin tissue of the frog formed a mouth. However, the mouth was distinctly one of a *frog* larva, and *not* one of a salamander! The organizer, then, seems to give only general instructions as to what kind of structure is to be formed. The genes of the tissue being induced determine what style of structure will be built.

More recent experiments have emphasized that embryonic induction involves two-way reciprocal interaction between inducing and induced tissue. The original inducer influences the induced tissue to differentiate. The differentiated tissue can now act

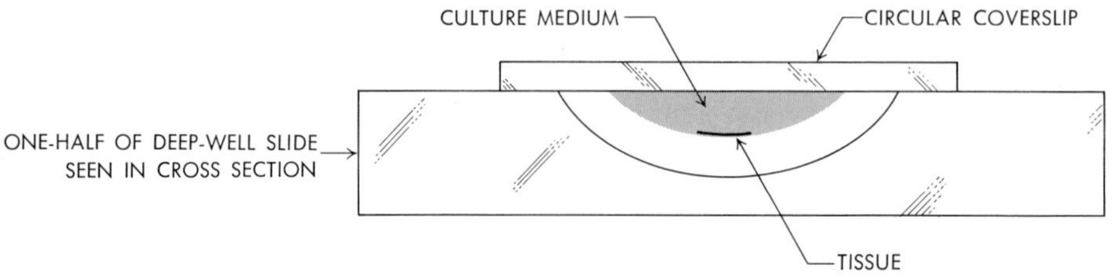

Fig. 11.11 *The hanging-drop technique, developed by embryologist Ross G. Harrison.*

as a secondary inducer on other, nondifferentiated tissues. Or, it can reciprocally induce further differentiation in its own original inducer.

Regeneration

Regeneration is the replacement of lost, injured, or worn-out parts by an organism, and the processes involved in it are similar to those that occur in embryonic development. Cellular division, migration, morphogenesis (the development of form), differentiation, and specialization are all examples of developmental phenomena common to both. Among the advantages of studying regeneration rather than embryonic development are that animals capable of regeneration are often quite hardy and the animal on which the limb is regenerating is comparable to a complex culture medium in which an embryo would be grown.

In general, the lower an animal on the scale of complexity, the greater its powers of regeneration. In addition, experimental evidence supports the generalization that the capacity of an animal to regenerate a lost limb declines as the age of the animal increases. Why should this be so? In *Xenopus laevis,* the South African clawed toad, a few hours after amputation of the hind limb, migrating epidermal cells form a protective cap over the wound. Damaged and dead cells are demolished and removed. Finally, and most important, a small group of unspecialized regeneration cells appear. These cells are the beginning of a cone-shaped mass of unspecialized cells, collectively called the **blastema.** The appearance of a blastema is typical and universal in limb regeneration. It is from the cells of the blastema that the new limb originates. Younger animals seem capable of more regeneration than older ones because as the animals get older, the number of regeneration cells that contribute to blastema formation is smaller (relative to the mass of the stump). The blastema is therefore smaller, and there are fewer cells available to produce the new limb. Also, the regeneration cells themselves are less "embryonic" in appearance. Research seems to indicate that an animal's capacity for natural regeneration decreases as the degree of tissue differentiation increases.

Does the fact that certain animals (such as frogs) do not regenerate lost limbs mean that their cells have completely lost their ability to do so? Dr. Marcus Singer hypothesized that the low ratio of nerve tissue to other tissues in the frog's limb (as compared to its relatively high ratio in the salamander's thin limb) might be responsible for the difference in the regenerative abilities of the two animals (see Fig. 11.12). He reasoned that if the differences between the regenerative ability of frogs and salamanders are attributable to differences in mass ratio between neural and other types of tissues in their limbs, then experimentally increasing the amount of neural tissue supplying a frog's limb might be sufficient to induce regeneration after amputation. Work done by Dr. Singer in 1954 confirmed this. Later, other investigators showed that regeneration could be induced in adult frog limbs under a variety of conditions, demonstrating that neural tissue was not the only factor involved.

Control of Differentiation

Genetic change? Do irreversible genetic changes occur with differentiation? If so, then the genetic capacity (sometimes called the **totipotency**) of a fertilized egg nucleus might be different from one in a later stage of embryonic development.

Experiments performed in the 1950s by R. Briggs and T. J. King tested this idea. Briggs and King devised the delicate experimental technique shown in Fig. 11.13. Nuclei were transplanted from the cells of older frog embryos into frog eggs from which the nuclei had been removed (enucleated eggs). Briggs and King first transplanted nuclei from blastula cells into enucleated eggs. Normal embryos resulted. Evidently, then, the transplanted nucleus was still capable of acting as a "general practitioner" of development. It could still direct (assuming it does so) the differentiation of a com-

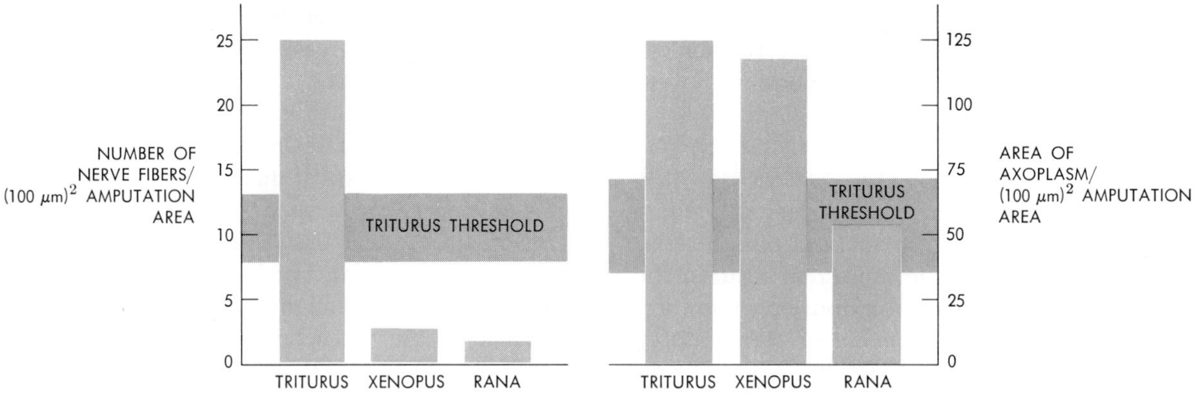

Fig. 11.12 *An hypothesis proposing that the number of nerve fibers per unit area accounts for the presence or absence of limb regenerative ability nicely accounts for the ability of the salamander* (Triturus) *to regenerate and the frog's* (Rana) *inability to do so. But this hypothesis is contradicted by the South African clawed toad* (Xenopus), *which* can *regenerate (left graph). Singer's hypothesis, proposing that the ratio between the cross-sectional area of the nerve and the cross-sectional area of the limb determines regenerative ability, has greater success. Note that the frog is on the threshold of regenerative ability, which could account for the success of Singer and others in inducing forelimb regeneration in this animal.*

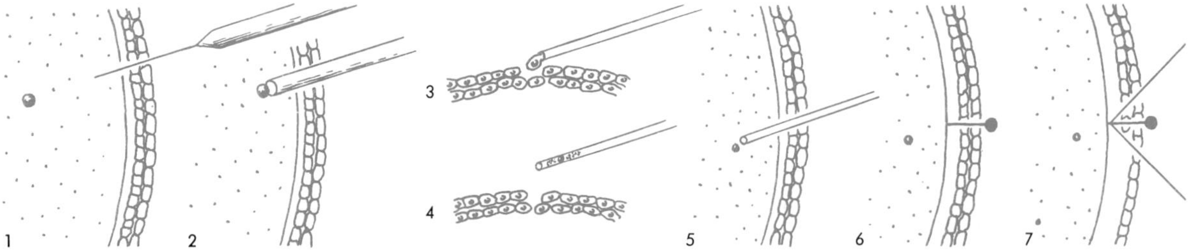

Fig. 11.13 *The Briggs and King operational technique. (1) The egg is pricked, causing the nucleus to move to the surface. (2) The nucleus is removed with a micropipette. (3, 4) An ectodermal cell from a blastula is drawn into the micropipette, whose diameter is so small that the cell membrane is ruptured. (5) The cell nucleus, with most of the cytoplasm removed from it, is injected into an enucleated egg. (6, 7) Snipping the protrusion prevents egg cytoplasm from escaping.*

plete embryo, rather than merely that part of an embryo it would have acted on if it had remained in the blastula. The experiment was repeated with cell nuclei transplanted from early gastrulas into enucleated eggs. Again, normal development occurred. However, when nuclei from cells of *late* gastrula stages were transplanted, abnormal development occurred.

Do these experiments support the hypothesis that a genetic change occurs in cells as they differentiate? It may be tempting to think so, but we must not jump too quickly. Consider what a "genetic change" involves. It means a chemical change within molecules of DNA—a mutation. Yet experimental evidence to date strongly indicates that mutations occur entirely at random. It is hard to base an hypothesis explaining such a highly organized process as cellular differentiation on a purely "chance" process. Besides, development is a regular and repeated *series* of events, which makes its occurrence by chance still more unlikely.

Have the nuclei from later stages in embryos lost their ability to control normal development? From these experiments, it would seem so. Yet in 1962, Dr. J. B. Gurdon and his associates transplanted to host eggs the nuclei from endodermal cells of *Xenopus laevis* donors ranging in age from late gastrula to the free-swimming tadpole stage. In some of the experimental host eggs, complete and normal development to the adult stage was obtained. In connection with the results obtained by Briggs and King it would seem, therefore, that nuclei might undergo functional but reversible differentiation (Fig. 11.14). The regeneration experiments discussed earlier showed that nuclei from the cells of adult amphibians are able to direct the differentiation necessary to grow a new appendage. If genetic change is assumed, then such cells must have undergone the proper random "unmutation." Surely this stretches the imagination too far.

Loss of DNA? It is possible that differentiation might occur because of a selective loss of DNA from cells during development. Perhaps, for example, cells of a plant destined by their position to become tracheids might lose all the DNA but that required for this particular developmental pathway. It was shown in 1950, however, that the quantity of DNA is constant in various fully differentiated cells of corn and *Tradescantia,* including those of the root, leaf, petal, and stamen hair, thus contradicting this hypothesis. Further, the amount of DNA present in the haploid cells of these plants (such as the pollen grains) was found to be precisely half that of the diploid cells. This result is to be expected, of course, since haploid cells possess an n rather than a $2n$ number of chromosomes.

Still more evidence against an hypothesis proposing that cells lose part of their DNA during their growth and development is provided by cell and tissue culture experiments. The results of this experiment demonstrate quite conclusively that cells do not lose any of their DNA during differentiation. The pith tissue of tobacco is fully mature and its cells are completely differentiated and specialized. Normally they would not again divide and grow. Yet Hildebrandt and Vasil's results (1965) show that the same cells can be stimulated by experimental manipulation to grow, divide, and produce a mass of cells that can be induced to grow into a whole plant. If the pith cells had lost some of their DNA during their differentiation, it seems highly unlikely that they would have been able to regenerate the entire tobacco plant.

Gene suppression? There is another possibility, however. Abandon the hypotheses proposing either genetic change or loss of DNA during development. Assume that the nuclei of all cells retain their genetic potential to control complete and normal differentiation, but suppose that certain genes are *suppressed,* or prevented from expressing themselves, and that only those genes used to perform the task at hand are allowed to remain functional.

This idea poses another question. Where and how does this gene suppression originate? Genes

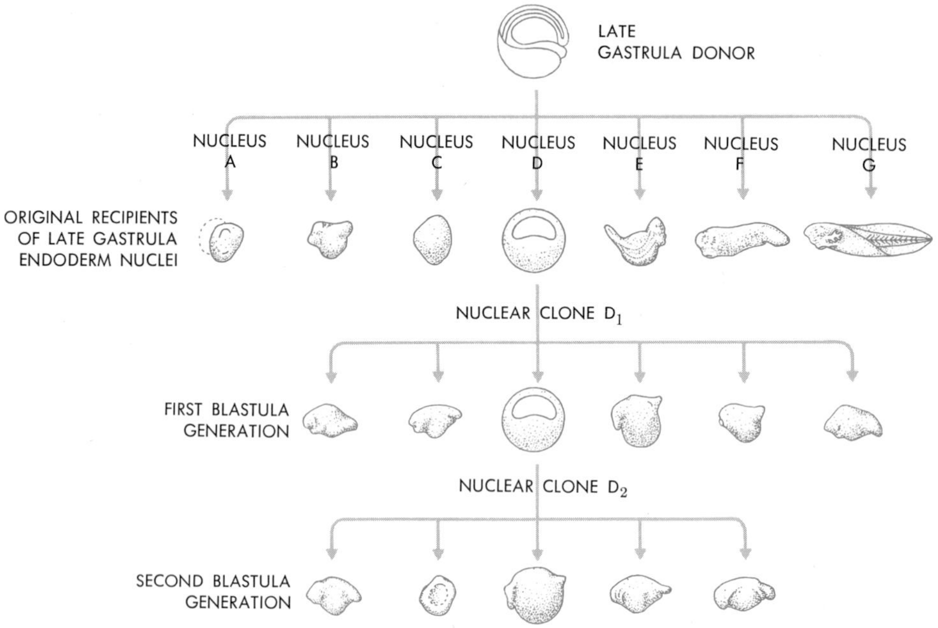

Fig. 11.14 *Nuclei removed from various late gastrula endoderm cells vary widely in their ability to control embryonic development when injected into enucleated eggs. Nucleus A, for example, seems to have lost its ability to control normal development, while nucleus G still retains the ability and causes the development of a complete embryo. However, when descendants (clones) of any one particular nucleus are injected into enucleated eggs, they yield embryos that are uniform in the stage of development they attain. This indicates that the genes controlling development are "turned off" in cell nuclei as development proceeds.*

don't have minds of their own; they don't "know" when to and when not to act. Is it not possible that elements in the cytoplasm are responsible? The fact that an enucleated egg does not develop does not mean that only the nucleus plays a role in its development, any more than the disabling of a car by removing its wheels indicates that the wheels are responsible for causing the car to move.

The logical conclusion is one we might well have been prepared for. Just as environmental factors must be considered in dealing with developing embryos and regenerating limbs, so must the nuclear environment (i.e., the cytoplasm) be considered, as well as the inherent factors residing within the nucleus.

11.6 Genes, Regulation, and Development

In the development of the chick, the heart appears about 24 hr after incubation. The cells that form the heart migrate from other places into the heart-forming region.

Heart-muscle myosin (the muscle protein) can

be distinguished from skeletal-muscle myosin by certain chemical tests. Presumptive heart cells can be recognized by chemical analysis of their proteins before they reach the heart region or have taken the shape of cardiac muscle cells. In other words, differentiation can be detected by looking for changes in the composition of cellular proteins.

The experiments of Beadle and Tatum, using the mold *Neurospora crassa,* demonstrated a one-to-one relationship between genes and enzymes (recall that all enzymes are proteins). A segment of DNA produces a certain polypeptide needed for the production of a certain enzyme. Each enzyme, in turn, is responsible for catalyzing a particular chemical reaction, perhaps one step in the production of a pigment. If the gene is absent, mutated, or suppressed, the proper enzyme is not produced. The individual produces no pigment and is an albino. Note the following series of reactions:

$$A \overset{\text{enzyme } a}{\rightleftharpoons} B \overset{\text{enzyme } b}{\rightleftharpoons} C \overset{\text{enzyme } c}{\rightleftharpoons} D \overset{\text{enzyme } d}{\rightleftharpoons} E$$

Suppose that compound E is needed by the organism. The rate at which the reaction of the series proceeds is proportional to the rate at which E is removed from the reaction site.

Suppose that the gene responsible for producing enzyme *c* is prevented from expressing itself, i.e., coding the construction of enzyme *c*. No compound D is formed and thus no compound E. Compound C, unusable by the organism, accumulates. If C is poisonous in large quantities, the organism may be damaged or die. Since C is accumulating, the reaction series reverses, and the concentrations of B and A also increase. Suppose, now, that compound E is an important structural protein needed for cell differentiation. Suppressing one or all of the genes responsible for coding enzymes *a, b, c,* or *d* would prevent differentiation.

A brief review of current thought about gene action and protein synthesis is in order here (for more details see Chapter 10). The gene, a part of the DNA molecule, passes on its inherited message, coded in the order of its nitrogenous bases, to messenger RNA (*m*RNA). Messenger RNA then passes from the nucleus to the ribosomes. There the code is translated during protein synthesis, using amino acids collected and brought to the ribosomes by transfer RNA (*t*RNA). Gene activity, then, involves chemical activity. The slowing down or stopping of this chemical activity will effectively suppress the gene. Indeed, chemical reactions are known in which the products have a slow-down or depressant effect on the reactions that produce them. If, for example, in the reaction series

$$A \overset{\text{enzyme } a}{\rightleftharpoons} B \overset{\text{enzyme } b}{\rightleftharpoons} C \overset{\text{enzyme } c}{\rightleftharpoons} D$$

product D should affect the rate at which enzyme *a* can catalyze the reaction converting A to B, then D is also indirectly controlling the rate of its own synthesis. Substance D at one end of the line is having a "feedback" effect upon the first step in the series. A parallel feedback situation applies to development. It was established earlier that embryonic cells are sensitive to their surrounding environment as well as to their own internal one. This can be summarized diagrammatically as follows:

$$\text{GENES} \rightleftharpoons \text{CYTOPLASM} \rightleftharpoons \text{ENVIRONMENT}$$

The genes influence the cytoplasm. The cytoplasm, in turn, has a "feedback" influence on the genes, turning them on and off when needed. In turn, cells influence other cells and thus the composition of their environment.

The Jacob–Monod Model

In what way might this gene regulation take place? How are the genes within a cell controlled so that some of them act and others do not? Some early clues were provided during the late 1950s by experiments conducted by Jacques Monod and François Jacob at the Pasteur Institute in Paris. Jacob and Monod found that the bacterium *Escherichia coli* does not synthesize beta-galactosidase (an enzyme that catalyzes the hydrolysis of lactose sugar into galactose and glucose) in significant quantities unless lactose is present in the cell. If lactose is supplied, however, the enzyme is synthesized from amino acids already present inside the bacterium. *E. coli* can also synthesize the amino acid arginine, using raw materials from its environment. If arginine is supplied, however, the formation of an enzyme needed by *E. coli* to produce arginine is inhibited. The bacteria stop making arginine and use that which is available from the environment. In both the lactose and arginine cases, *E. coli* is being "told" by its environment to synthesize its own enzyme for breaking down lactose or not to make an enzyme for producing arginine, etc. Might not the same principles of interaction account for the changes in potentiality shown by the differentiating cells of an embryo or the stump of a regenerating limb?

It seems clear that the environment of the bacterium containing the lactose or arginine molecules is influencing the activity of the genes—simply turning them on and off when needed. Later, Jacob and Monod found that the mechanism of gene control is far more complex than originally thought. They established that the ability of *E. coli* to break down lactose may depend on at least four genes. One of these (gene 1) may provide the information for making the enzyme beta-galactosidase, while gene 2 may code for the synthesis of another enzyme involved in the absorption of lactose by the bacterial cells. A third gene probably controls the synthesis of yet another enzyme necessary for lactose utilization. All three of these genes are known as "structural" genes, since they determine the structure of enzymes (proteins).

Various genetic studies (i.e., by mapping the bacterial chromosome) showed that these three structural genes are probably located close to one another on the DNA of the bacterium. In addition, Jacob and Monod uncovered evidence that the activity of all three of these structural genes may well be controlled by a fourth gene, called an operator gene, located near gene 1. The entire four-gene system is known as an **operon.** When the operator gene is "on," the three structural genes are active; when it is "off," they are inactive. Through the analysis of various *E. coli* mutations, Jacob and Monod discovered that the operator gene may itself be controlled by the combined action of two agencies: a regulator gene, located some distance away from the lactose operon, and various chemical-inducing substances, such as lactose or arginine. The regulator gene is thought to control the manufacture of a regulator substance capable of combining with the operator gene and switching off its control over the three structural genes (see Fig. 11.15). When this occurs, the operon is prevented from making the *m*RNA required for synthesis of the three enzymes involved in lactose utilization. When an inducer substance such as lactose is supplied to the bacteria, however, it appears to combine with the regulator protein and prevent it from switching off the operon. When this happens, then enzymes for breaking down lactose are synthesized, using the *m*RNA formed by the three structural genes of the operon. If arginine is supplied to the bacteria, however, these small molecules may react in some manner with the regulator substance, causing it to switch the operator gene off more effectively, and blocking the synthesis of the enzyme required for arginine production.

The Jacob–Monod concept of gene regulation (for which they shared the Nobel prize in medicine for 1965) is a purely hypothetical model to explain how genes may act in development. Since first proposed in 1961, the hypothesis has undergone slight changes as more experimental evidence

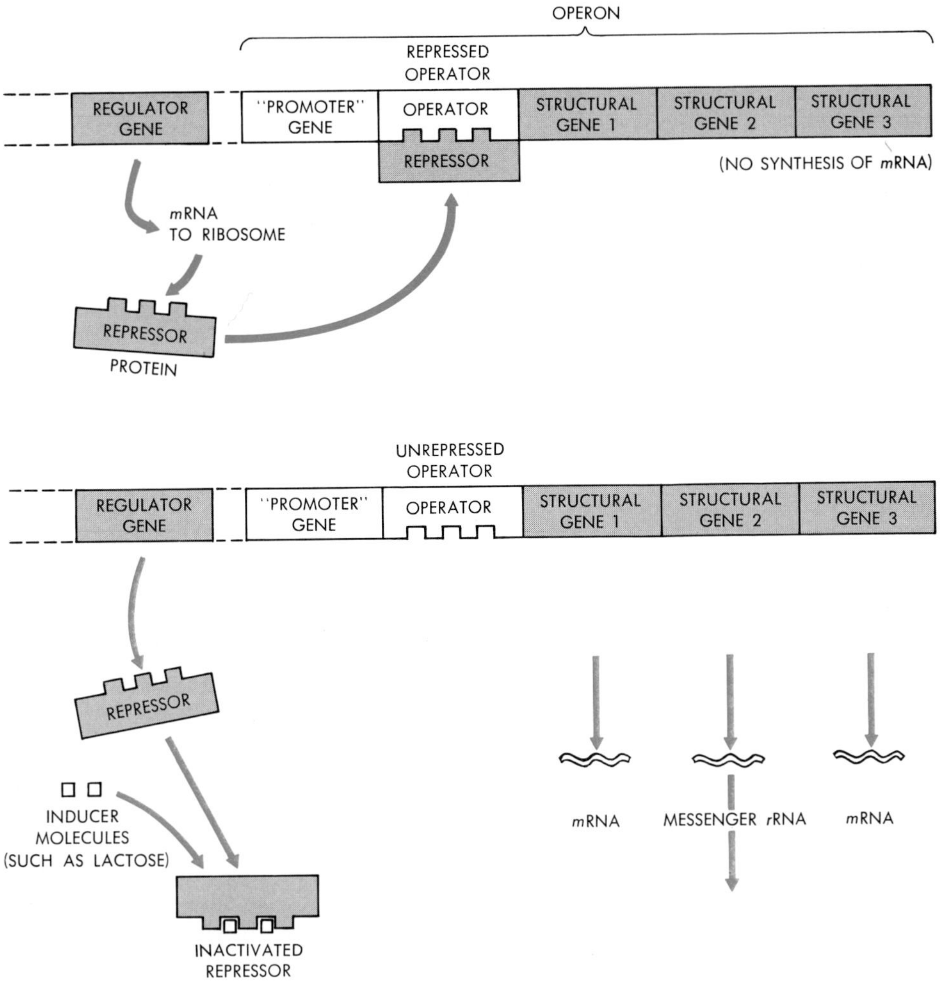

Fig. 11.15 *The Jacob–Monod hypothesis of gene regulation. The three structural genes (forming an operon) are all under the control of an operator gene. When the operator is "on," the operon is active; when it is "off," the operon is inactive. Each operator gene, in turn, is controlled by a regulator gene, which produces a repressor substance capable of switching off the operator gene. Inducer molecules modify the repressor and prevent it from switching off the operon; thus, the messenger RNA can still be synthesized. The "promoter" gene is the site where RNA polymerase must attach to begin transcribing the* mRNA *molecules for expression of the structural genes. If the operator gene is blocked by a repressor, the RNA polymerase is physically blocked from initiating transcription.*

has become available. The hypothesis has been very useful in explaining many observations of gene control of enzyme synthesis in bacteria.

A Positive Feedback System

However, the Jacob–Monod model cannot explain some phenomena observed with the lactose operon. For example, if the medium in which bacteria are growing already has an abundant supply of glucose, addition of lactose will not cause production of β-galactosidase and the other enzymes associated with lactose metabolism. Since glucose is the most widespread sugar in nature, cells would have no need to produce a second set of enzymes to metabolize lactose as long as glucose was available. Thus, the lactose-metabolizing system is repressed as long as glucose is present—a phenomenon known as **catabolite repression** since it is suppression of a system involved in catabolic—i.e., breakdown—pathways.

In addition to lactose, another substance is required to de-repress the lactose operon. This substance is cyclic 3′, 5′ AMP (cAMP). RNA polymerase cannot attach to the promoter genes of the lactose operon unless cAMP is present. As long as glucose levels are high, cAMP levels are low (most of the AMP has been converted to ATP when glucose is plentiful), thus ensuring that the lactose operon remains repressed even if lactose is present. When the glucose level falls low, electron transport slows down and AMP accumulates because ATP is being used faster than it is produced. The enzyme adenyl cyclase converts AMP into cyclic 3′, 5′ AMP. RNA polymerase is able to bind the promoter genes only after forming a complex with cAMP and the cAMP receptor protein (CRP).

The cAMP–CRP system functions quite differently from lactose as a signaling device to the lactose operon system. The lactose system as described so far functions as a *negative* feedback system. The regulator gene produces a repressor, which represses expression of an operon. Lactose derepresses the system by deactivating the repressor. The function of lactose is basically negative: it prevents an inhibitor from having its effect. But lactose does not do anything positive with regard to the operon itself. However, cAMP and CRP function as a *positive* feedback mechanism. They are necessary elements for the functioning of the RNA polymerase. This positive feedback system is diagrammed in Fig. 11.16.

Positive feedback mechanisms have turned out to be more common in bacterial gene regulation than negative feedback. For example, some operons that appear to function under negative feedback control lack repressor proteins. In these cases, the product of the regulator gene functions like cAMP–CRP: it is necessary for the functioning of RNA polymerase. The original Jacob–Monod model has thus been somewhat expanded and modified in the light of new findings about positive feedback.

What about gene regulation in organisms more complex than bacteria? During the 1950s, the geneticist Barbara McClintock of the Carnegie Institution of Washington found that grains of corn (*Zea mays*) have many spots of color of various sizes and frequencies in different varieties. Through

---

The Jacob-Monod model for the lactose operon is an example of negative feedback regulation of genes: lactose stimulates lac operon activity by deactivating an inhibitor. Cyclic 3′, 5′ AMP and its associated cAMP repressor protein are positive feedback regulators; their presence makes possible the attachment of RNA polymerase to the promoter gene of the lac operon.

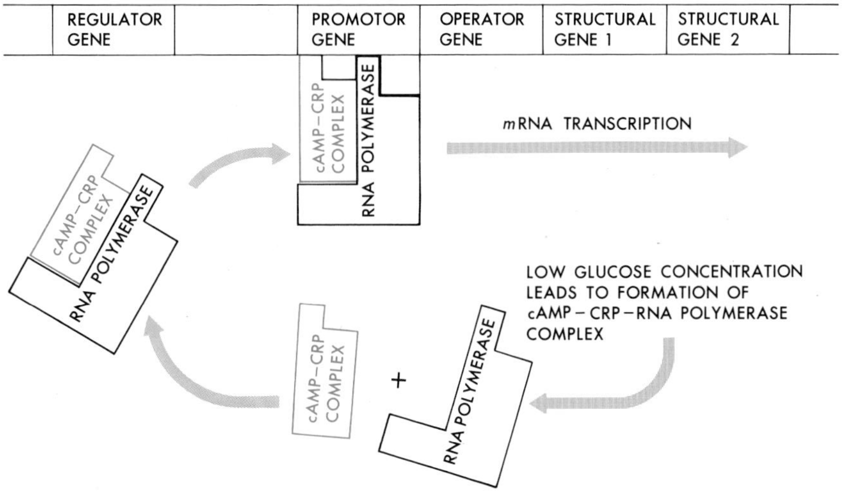

Fig. 11.16 *Schematic diagram illustrating the function of a positive feedback system in gene control. Low levels of glucose in the cell make possible the formation of a complex between cAMP (3′, 5′ cyclic AMP) and CRP (Cyclic AMP Receptor Protein), which in turn forms a complex with RNA polymerase. Formation of the complex with RNA polymerase makes it possible for the latter to bind to the promoter gene, thereby making transcription of* mRNA *possible. Since their presence is necessary for functioning of RNA polymerase, cAMP and CRP act as positive feedback molecules.*

ingenious genetic analyses, McClintock obtained evidence that these spots are due to the action of controlling elements that affect the genes for color. One of these elements directly controls the action of the structural gene (and thus is similar to the operator gene in bacteria). Another controlling element functions in a manner comparable to the regulator gene in bacteria. It is not at all certain, however, that the gene regulation system found in corn grains is sufficiently precise to control the orderly development of an entire multicellular plant. The gene control systems in corn produce randomly distributed spots of color instead of predictable patterns.

Conclusive evidence is still lacking for the existence in multicellular organisms of a gene control system as precise and orderly as the Jacob–Monod model. It seems wise to be cautious in extrapolating from a unicellular organism of procaryotic cell organization to multicellular eucaryotes. Yet it would seem reasonable to expect complex organisms to possess gene regulating devices at least as sophisticated as those of bacteria—though, of course, they may be of a different kind. Still the bacterial system is valuable as a working hypothesis for use in studies of the growth and development of multicellular organisms.

11.7 Evidence for Gene Regulation in Growth and Development

Is there experimental evidence that some genes are active while others are not?

At certain times during the development of insects, "puffs" or swellings along the chromosomes

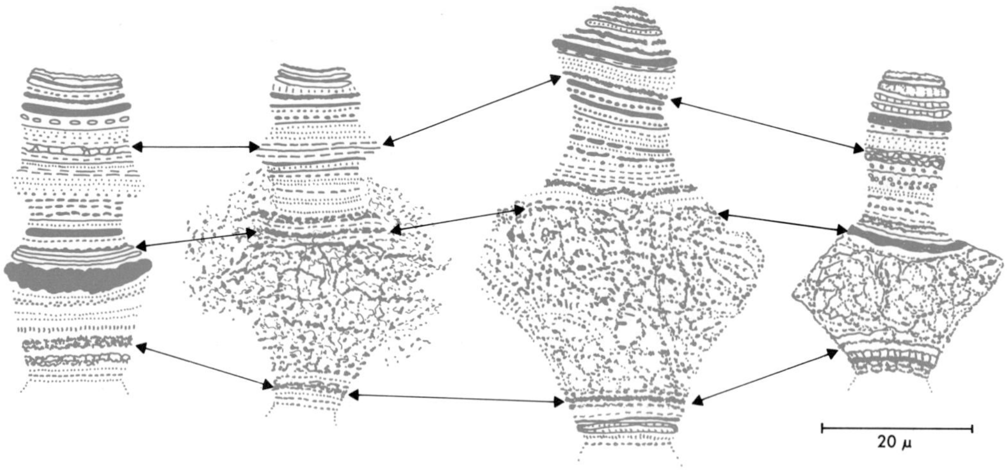

Fig. 11.17 *A portion of one giant chromosome from the midge larva* Rhyncosciara angelae *during different stages of larval development. The arrows and connecting lines indicate comparable bands. Changes in puffing (expansion) of particular regions indicate increase or decrease in activity of genes in those areas. Different genes appear to be active at different stages in the insect's development.*

can be associated with gene activity. The puff areas are thought to be regions of intense *m*RNA transcription. The nonpuff areas are thought to be inactive genes. The existence of puffs in particular regions of particular chromosomes appears to be both *tissue-specific* and *time-specific*. Tissue-specific means that normally, in adult tissue of a particular type, the same chromosomal regions appear puffed, suggesting that these regions are permanently active in that tissue. Generally, only a very small percentage of genes appear to be active in any one type of tissue. Specific puffs appear at particular times in given chromosomal regions during embryonic development. These puffs come and go as insect development (metamorphosis) proceeds. Fig. 11.17 shows the changing patterns of chromosome puffs during stages of metamorphosis in the midge, *Rhyncosciara angelae*. Puffing in insect chromosomes can be induced by hormones *in vitro*. The fact that insect metamorphosis is guided by changing patterns of hormone production suggests strongly that changes in gene activity may be either triggered or suppressed by the appearance and disappearance, respectively, of certain hormones.

Similar findings have been made in the developing oöcytes of amphibia. A few years ago workers discovered that the so-called "lampbrush chromosomes" found in cells of insects, a number of other organisms, and especially in amphibian oöcytes, were surrounded by brushy fibers (Fig. 11.18). Extensive chemical studies have revealed that these brushy fibers are DNA, each long central fiber being one double-stranded DNA molecule. Extending from each fiber are loops, now known to be DNA—presumably in the process of being transcribed. Since the oöcyte is known to be very active in synthesizing proteins necessary for the early stages of development, it is clear that the extended fibers and loops represent selected genes being transcribed. Studies of these chromosomal loops

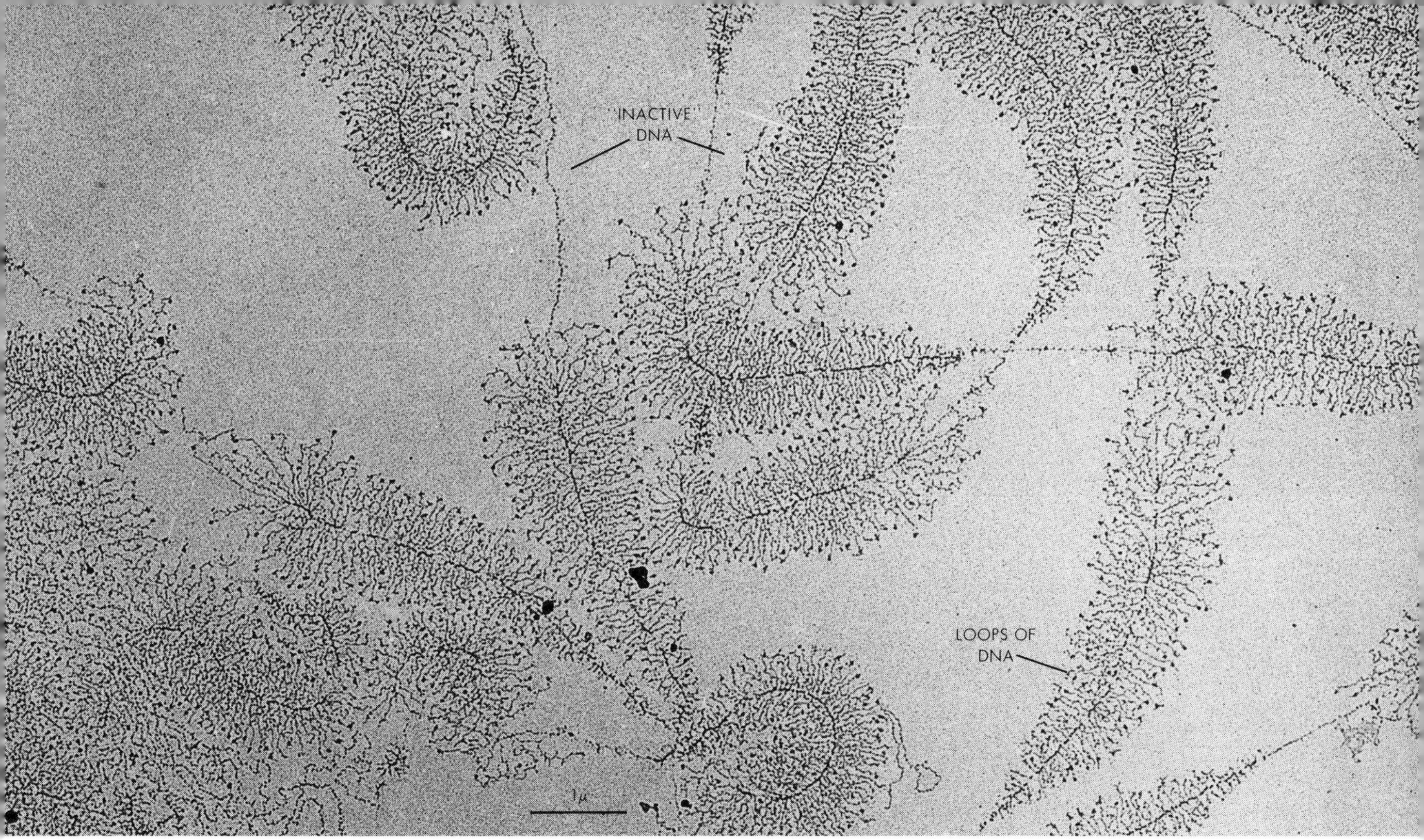

Fig. 11.18 *Electron micrograph of lampbrush chromosomes in the amphibian oöcyte. The particular genes shown here are in the nucleolus and code for ribosomal RNA (rRNA). The long horizontal axis is DNA. The brushy fibers extending from it are rRNA molecules being transcribed. Where no brushy fibers extend from the DNA, no transcription is assumed to be occurring. (Electron micrograph courtesy O. L. Miller, Jr. and Barbara R. Beatty, Biology Division, Oak Ridge National Laboratory; ×25,000) [From O. L. Miller, Jr. and Barbara R. Beatty, "Visualization of nucleolar genes," Science 164 (May 23, 1969), pp. 955–957. By permission of O. L. Miller, Jr. and Barbara R. Beatty, Biology Division, Oak Ridge National Laboratory.]*

and chromosome puffs suggest strongly that genes in higher organisms are selectively turned on and off during development. The mechanism for this switching process is still a matter of conjecture.

Much of the evidence for gene activation and deactivation during growth and development has come from experiments using plants in which the antibiotic actinomycin D was used to suppress the synthesis of *m*RNA. Without a continuing supply of *m*RNA, the production of enzymes and other protein molecules ceases. By this method, *m*RNA has been demonstrated to be involved in many aspects of growth and development. It has been shown, for example, that the cotyledons of peanut seeds begin synthesizing *m*RNA at the onset of germination, and within a week the content of *m*RNA per cotyledon doubles. Treatment with actinomycin D, however, severely impairs the synthesis of *m*RNA by the cotyledons. Actinomycin D unites with the guanine of DNA and prevents the RNA polymerase molecule from using the DNA chain in the synthesis of *m*RNA. Actinomycin D has also been shown to inhibit growth of soybean hypocotyls through suppression of *m*RNA synthesis. This antibiotic has been found to inhibit the synthesis of an enzyme (amylase) by germinating seeds of barley. Sensitivity to actinomycin D suggests that the genes for RNA synthesis are being inactivated in some manner.

SUMMARY

1. Developmental biology comprises the study of all regular changes within the life history of the organism from fertilization through embryonic development to growth, aging, and death. In addition it studies such processes as the immune response, aging, and cancer. Embryology is only one aspect of developmental biology.
2. The central problem of developmental biology is that of cell differentiation: how cells that are alike at a very early stage become progressively different, ending up as highly specialized cells such as muscle, nerve, or red blood cells. Much of the history of the field has been devoted to trying to understand the causes of differentiation.
3. Embryonic development can be said to begin with formation of gametes: sperm from primary spermatocytes and eggs from primary oöcytes in the germ tissues of the parents. Fertilization of the egg by the sperm, however, actually initiates the process of embryonic development. Many sperm are required to ensure that one actually penetrates the egg cell membrane. Once this has happened, the egg cell surface undergoes certain physical changes, usually ensuring that a second sperm does not get through. The egg membrane lifts up, forming an outer hyaline layer around the cell. The outermost boundary of this layer is the fertilization membrane; the innermost layer closest to the egg cytoplasm is called the vitelline membrane.
4. Activation of the egg follows fertilization. A certain number of the egg's genes have been transcribed into *m*RNA and translated into protein prior to fertilization; after fertilization this material is activated, which leads to intense metabolic activity. Thus maternal genes determine the characteristics of the initial stages of development.
5. Early embryonic development passes through the following stages: single-celled zygote, 2-cell, (each cell is called a blastomere), 4-cell, 8-cell, etc. to the formation of the blastula, a hollow ball of unspecialized cells surrounding a cavity, the blastocoel. Invagination of the blastula (at the blastopore) eventually produces the three-layered gastrula composed of ectoderm (outer layer), mesoderm (middle layer), and endoderm (inner layer). These are called the primary germ layers.
6. Each primary germ layer gives rise to certain groups of tissues in the adult organism: ectoderm to skin, nervous system, lens and cornea of eye, lining of nose, mouth and anus, hair, nails, and sweat glands; mesoderm to muscles, connective tissue, blood and blood vessels, kidneys, and the reproductive organs; endoderm to lining of the digestive tract, liver, pancreas, various endocrine glands, the bladder, and the lining of the urethra.

7. Early divisions of the egg cell are called cleavages. Patterns of cleavage are determined largely by the amount and distribution of yolk in the egg cell. The presence of a great deal of yolk slows down cleavage. In eggs such as the frog's, with only a moderate amount of yolk concentrated at one end, a blastula forms with larger blastomeres at the end where the yolk was concentrated. In the bird's egg, where there is a massive amount of yolk, cleavage is restricted to the cytoplasmic area that resides as a disc at the top of the egg.
8. Gastrulation occurs by morphogenetic movements of cells from the outside of the blastula to the inside.
9. Differentiation of cells of the various germ layers into tissues and organs occurs primarily by a process known as embryonic induction. The formation of the nervous system is an example of induction. Ectodermal cells lying just to the dorsal side of the blastopore come into contact with mesodermal cells. Contact induces ectodermal cells to begin elongation to form the neural plate. Soon cells of the neural plate begin to buckle or fold inward. Eventually buckling occurs to such an extent that the two ends of the plate meet, forming the neural tube. The primary induction responsible for this sequence is contact between archenteron roof (the inducer) with cells of the overlying ectoderm (the induced). Without this contact, no neural plate develops.
10. Induction usually starts a chain of events known as cell determination. The initial inducer is called the primary inducer; once induced, a differentiating tissue may itself become a secondary inducer, etc. German embryologist Hans Spemann found that the tissue of the dorsal lip of the blastopore was the primary inducer for differentiation of the whole embryo, and he called it the "organizer." The dorsal lip serves to initiate all subsequent differentiation.
11. Once tissue has started differentiation, determination becomes progressively more specialized. Most tissue is not irrevocably determined from the moment it comes into contact with an inducer. The initial direction of determination may be only generalized. Later, as secondary and tertiary induction occurs, the cells become more specialized and their differentiation less reversible.
12. The causes of induction are still largely unclear, especially at the molecular level. It is known that induction can only begin after contact between inducer and induced tissue. However, if ectodermal tissue is placed in a medium in which underlying mesoderm has been allowed to incubate but has been removed, the ectoderm begins to form neural plate. Inducer tissues appear to produce a substance or substances that diffuse into induced tissues and initiate differentiation. To date, no one has identified what such a substance or substances might be.
13. Induction appears to selectively stimulate the genes present in the induced tissue. The inducer determines the general direction of differentiation. The induced tissue responds within the limits or capabilities of its own genome. Frog mesoderm can induce differentiation in salamander ectoderm, but the induced tissue responds genetically like salamander, not frog.
14. Embryonic induction is a reciprocal process. Inducer causes changes in the induced tissue. The induced tissue can then "induce" changes in the inducer. Inducer and induced tissue thus undergo differentiation together by a constant interaction.
15. Regeneration is the process by which an organism replaces a lost, injured, or worn-out part. Human beings constantly regenerate new skin, but we cannot regenerate fingers, arms, or legs. Many lower animals, such as frogs and salamanders, can regenerate whole appendages.

Some observations have been made regarding ability to regenerate:

a) The more highly differentiated the tissue, the less likely it can regenerate.
b) The older the organism, the less ability its various tissues and organs have for regeneration.
c) In vertebrates, regenerative ability of a tissue appears to be related in some way to amount of innervation of that tissue.

16. As regeneration of a limb bud begins, blastema cells appear at the wound site. Blastema are unspecialized cells that congregate at an area where regeneration is occurring; they ultimately differentiate into the specialized tissues of the regenerating organ. Differentiation of blastema cells has been useful as a model system for studying many aspects of the general differentiation process.

17. Experiments with frog eggs, tobacco, and other organisms indicate that highly specialized cells from embryos or adults still retain the ability to form a whole new organism. These cells are genetically totipotent; they still carry all the genetic information for forming a complete adult. Differentiation appears to be a process of selectively turning off (or on, as the case may be) particular genes.

18. The Jacob–Monod model of enzyme induction in bacteria has been useful as a possible way of visualizing differentiation at the genetic level. This model accounts for the fact that the enzymes involved in lactose metabolism in *E. coli* are not synthesized unless lactose is present in the medium. Lactose "turns on" the genes for enzyme synthesis. Jacob and Monod hypothesized that there are several kinds of genes: regulator genes that produce a repressor substance (protein), operator genes to which the repressor can reversibly bind, structural genes that code for specific enzymes active in lactose metabolism, and promoter genes where RNA polymerase attaches to the DNA to initiate transcription. The promoter, operator, and structural genes associated with production of a particular protein are called an operon. The regulator gene controls the whole operon. When repressor binds to the operator gene, transcription of the structural genes is blocked. Appearance of a "signal," in this case lactose molecules in the medium, serves to inactivate the repressor protein. With no repressor able to bind to the operator, the operator becomes free and transcription of the structural genes can take place. This system is an example of *negative* regulation, or de-repression, since the function of lactose is only to inactivate a repressor.

19. Positive regulation also exists. Cyclic 3′,5′ AMP (cAMP) and its associated cAMP repressor protein (CRP) are necessary components for transcription to take place, even if lactose is plentiful. CRP and cAMP activate RNA polymerase, making it possible for the latter to bind to the promoter gene and start transcription. The amount of cAMP is a function of the amount of glucose present in the cell (glucose metabolism produces energy for converting AMP into ATP; hence the more glucose, the less cAMP). As long as glucose is present in abundance, the bacterial cell does not produce enzymes for breaking down lactose, even when the latter is available. CRP and cAMP are positive regulators, since their presence is necessary for one component, RNA polymerase, to function.

20. There is no direct evidence suggesting that the Jacob–Monod model is specifically applicable to differentiation in cells of higher organisms. It is only a tempting possibility. Recent evidence, involving the study of chromosome puffs and RNA synthesis in chromosomes of amphibian oöcytes, does suggest at least that during development different regions of specific chromosomes become selectively active in transcribing RNA. Some data point to the pos-

sible involvement of histones, a protein component of chromosomes in higher organisms, in gene regulation.

21. The study of development has left more unanswered than answered questions. The basic problem of cell differentiation, at the tissue as well as the cellular level, is still largely unsolved.

EXERCISES

1. How can the amount of yolk available in a fertilized egg determine the course of embryonic development?
2. Describe the changes that occur in an egg immediately after fertilization.
3. How does the amount and distribution of yolk in the blastula affect the process of gastrulation?
4. In what several ways does the major course of embryonic development after gastrulation differ from that which precedes it?
5. What were the results of Spemann's famous experiment? Explain the concept of embryonic induction. In terms of the direction this work gave to embryological studies, why would you say it was extremely important?
6. Explain briefly how inducing-agent substances might function. Do they seem to control directly the development in cells to which they diffuse, or do they act indirectly by affecting the genetic information of those cells? What evidence can you cite to back up your answer?
7. Observe Table 11.3.* It refers to an experiment in which nuclei were transplanted

Table 11.3

		Normal cleavage		Complete embryos	
Source of donor nucleus	**Total eggs injected**	**Actual number**	**%**	**Actual number**	**%**
Blastula	204	116	57	69	34
Early gastrula	135	52	38	20	15

from a donor embryo into an enucleated egg. What do these data indicate? What changes seem to occur in the nucleus as embryonic development occurs? How can you account for this?

8. Spinal cord and notochord cells induce cartilage formation in amphibian embryos. They will also induce cells to form cartilage when cultured outside the embryo. However, they will only induce cartilage formation in cells that would ordinarily form cartilage. Does this fact support or refute the conclusions drawn from Schotté's experiment?

* Taken from Barth, Lucena J., *Development, Selected Topics.* Reading, Mass.: Addison-Wesley, 1965, p. 58.

9. How would you account for the following generalizations? Do you think these generalizations are completely valid ones?
   a) The less complex the animal, the greater its power of regeneration.
   b) The older an organism, the less its power of regeneration. (Explain your answer in terms of the turning on and off of genetic information within cells.)
10. There has been much controversy in recent years about the use of X-rays for medical examination of pregnant women. Some authorities claim that this harms the developing fetus, while others claim that the effects are so slight that the medical advantages of such examinations far outweigh any disadvantages. Table 11.4 compares stillbirths (in which the fetus is born dead), congenital defects (in which the child lives but has some malformation), and healthy children in X-rayed and non-X-rayed mothers. Would you conclude for or against the theory that X-rays cause definite harm to the fetuses of pregnant women?

Table 11.4

	X-rayed mothers	Non-X-rayed mothers
Stillbirths	1.403%	1.222%
Congenital defects	6.01	4.82
Normal children	80.42	83.23

11. How might studies on embryonic development (in both plants and animals) possibly give important information about cancer?

SUGGESTED READINGS

General Descriptive and Experimental Embryology

Balinsky, B. I., *An Introduction to Embryology,* 3d ed. Philadelphia: W. B. Saunders, 1970. A thorough, well-illustrated, classic introductory text in embryology.

Ebert, James D., and I. M. Sussex, *Interacting Systems in Development,* 2d ed. New York: Holt, Rinehart and Winston, 1970. An excellent modern discussion of the main outlines of development. Focuses especially well on tissue and cell interactions at the molecular level.

Trinkaus, J. P., *Cells into Organs.* Englewood Cliffs, N.J.: Prentice-Hall, 1969. A well-written, summary review of theories of how cells form into organs in embryonic development.

Wessel, N. K., *Tissue Interactions and Development.* Reading, Mass.: Addison-Wesley, 1977. A compact discussion of the main processes of development and how they are brought about.

Regeneration

Bryant, Susan V., "Regeneration in amphibians and reptiles," *Endeavour* 29 (January 1970), p. 12. A short review article describing the current status of studies on limb-bud regeneration. Contains good illustrations (some in color) of developing tissues.

Genetic Aspects of Development

Davidson, Eric H., *Gene Activity in Early Development.* New York: Academic Press, 1969. A well-documented study of the genetic and chemical events associated with early stages of development. Particularly good in describing organization of the egg and localization of control substances.

TWELVE

HUMAN REPRODUCTION, DEVELOPMENT, AND SEXUALITY

12.1 Introduction

Despite the beauty and fascination that the processes of human reproduction, development, and sexuality have always held, few subjects have been so widely misunderstood or fraught with so much superstition, myth, and ignorance. Indeed, only recently have human sexuality and reproduction come to be considered proper spheres for biological research. It is no exaggeration to state that more has been learned in the past 30 years on these subjects than in the previous 3000!

In this chapter all aspects of human reproduction and sexuality will be considered, including the anatomy and physiology of the male and female reproductive systems, sexual intercourse, embryonic development, contraception, and the phenomena of sexual identity and sex roles.

12.2 The Male Reproductive System

The human male reproductive system consists of several organs, including the penis; the testes, contained in a sac, the scrotum; seminal vesicles; ejaculator ducts; and various lubricating glands (Fig. 12.1). An internal diagram of these organs is shown in the sectional view in Fig. 12.2.

The **testes** serve two very important functions. First, they produce sperm and thus are the organs necessary for male fertility. They are also endocrine glands, producing the male sex hormone, **testosterone.** The sperm-producing part of the testis consists of a series of highly coiled tubes, the **seminiferous tubules.** The seminiferous tubules converge into about a dozen ducts leading into the **epididymis,** a much-coiled collecting tube that lies on the upper part of each testis. Along the lower surface of the testis the epididymis connects to the **vas deferens,** or seminal duct, which leads upward, curves over and behind the front arch of the pelvis, eventually joining the urethra coming from

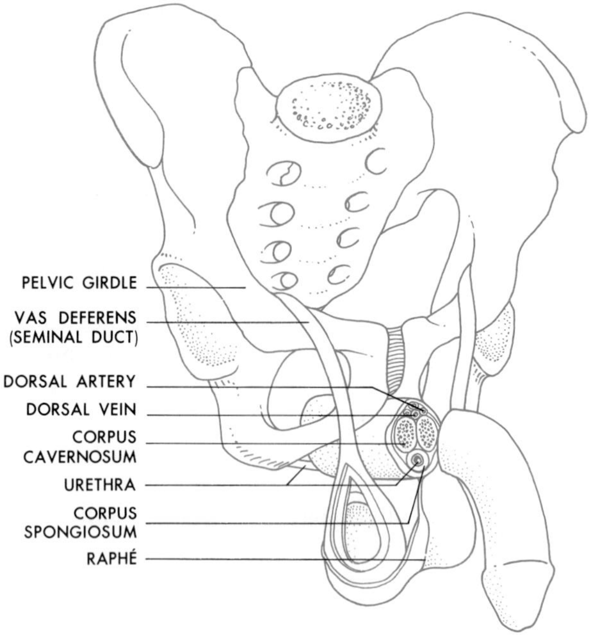

◀ **Fig. 12.1** *The human male reproductive organs shown in general location within the lower pelvis.*

Fig. 12.2 *Longitudinal section of the male reproductive organs.*
▼

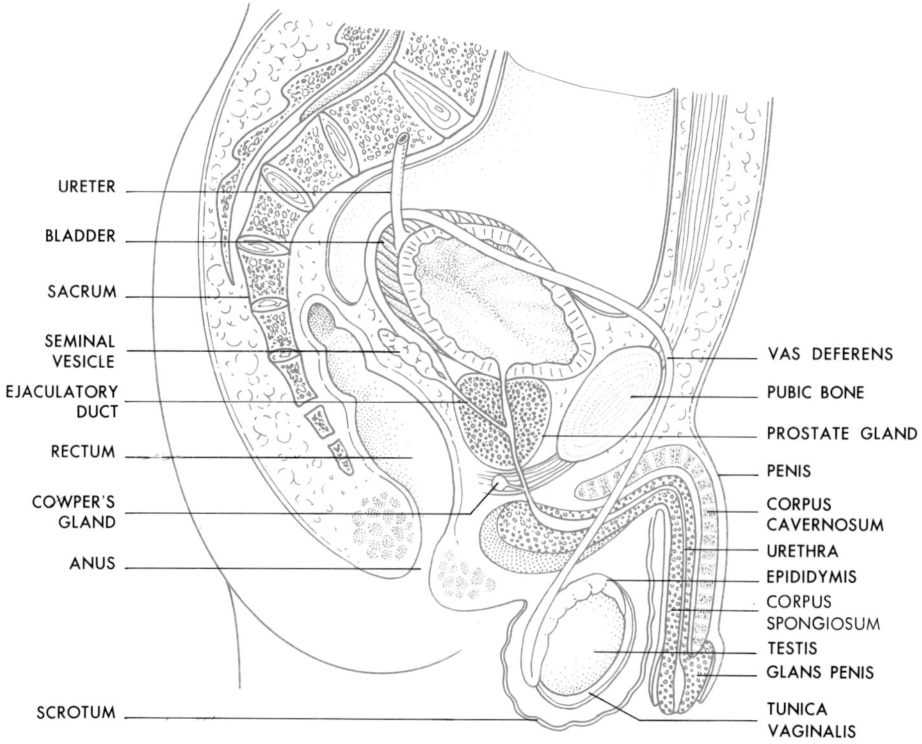

the bladder. From this point of juncture the two tubes follow a common course to the exterior through the penis.

At several points along the vas deferens and urethra, ducts leading from various glands pour secretions into the sperm suspension(semen). These glands include the paired **seminal vesicles,** the **prostate gland** (in human males the two prostate glands are fused to form a single gland; in other mammals there are two prostates), and the paired **Cowper's glands.** The seminal vesicles and Cowper's glands contribute a mucous alkaline secretion, while the prostate adds a thin milky fluid with a characteristic odor. This entire mixture is known as the seminal fluid. The seminal fluid contains glucose and fructose that provide energy sources for the sperm, acid–base buffers, and mucous materials that lubricate the passages through which the sperm travel.

The **penis** is a cylindrical organ composed of three bodies of spongy tissue. Normally there is only a moderate flow of blood through this tissue. During sexual excitation, however, the capillaries open up, allowing greatly increased volumes of blood to enter the spongy tissue. This extends the length and diameter of the penis until it becomes firm, a state known as **erection.** As Fig. 18.1 shows, the three layers of spongy tissue making up most of the penis are not identical. The two side layers, the **corpora cavernosa** (singular, *corpus cavernosum*) are larger; the lower layer, the **corpus spongiosum,** is smaller and surrounds the urethra. The corpus spongiosum expands at the upper end of the penis to form the cap-like **glans penis.** The glans is covered by a fold of skin called the **foreskin,** which is often removed surgically in the process of circumcision shortly after birth. The foreskin serves as a protection for the glans penis. (Originally done for religious purposes, removal of the foreskin is still practiced routinely by many physicians on the grounds that unless the skin underneath the fold is kept clean, it may become infected.)

The mammalian penis, including that of the nonhuman primates, usually contains a bone, the **os penis,** which serves to give the organ part of its rigidity. The human male lacks this bone; the aforementioned flow of blood into spongy tissue alone makes the penis rigid. The effect of rigidity during sexual excitation is to allow the penis to be inserted into the female.

Spermatogenesis

Figure 12.3 shows a sectional drawing of the testis and a more highly magnified picture of a cross section through a seminiferous tubule. Within these tubules gametogenesis takes place. Spermatogenesis is the process by which, through mitosis and meiosis, functional sperm are produced. The actual process of sperm formation is diagrammed in detail in Fig. 12.3(c); the process is compared with oögenesis in Fig. 7.11. The starting cell, the "grand-parental" cell of the future sperm, is called a **spermatogonium.** One spermatogonium undergoes mitosis to produce two daughter cells, one a **primary spermatocyte** and a second cell that remains a spermatogonium. The primary spermatocyte begins the first meiotic division and divides into two **secondary spermatocytes.** The secondary spermatocytes undergo the second meiotic division to form four equal-sized **spermatids,** cells with a considerable amount of cytoplasm and a haploid nucleus. The

---

Human development includes all aspects of the sequence of biological and behavioral changes that characterize the human life cycle. Development of the individual begins with the formation of sperm and egg in the parents' reproductive organs and ends with death.

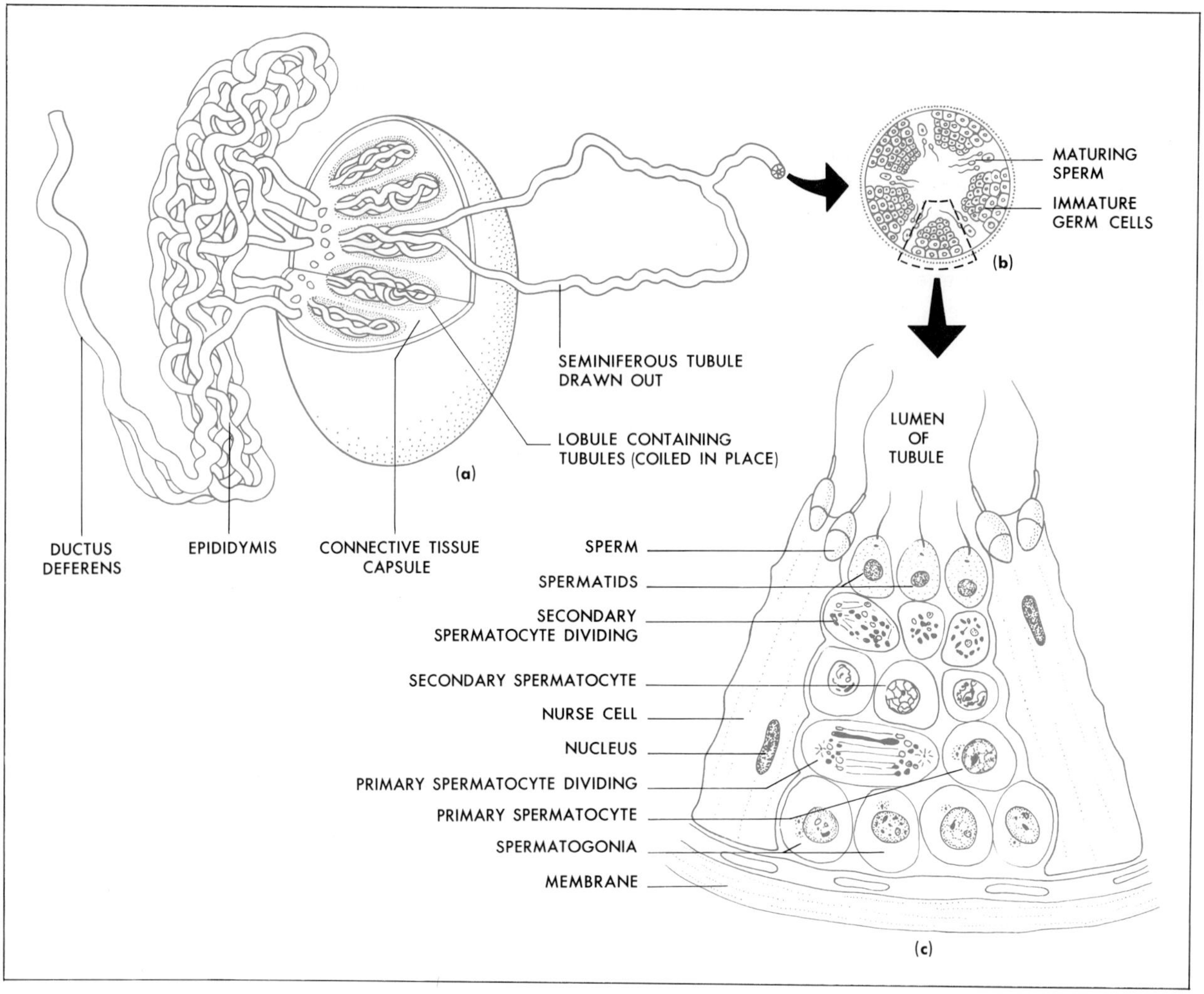

Fig. 12.3 *The human testis showing cross section of a seminiferous tubule and detailed view of sperm production. (a) One testis with epididymis viewed from the top, showing seminiferous tubules. (b) Cross section of one seminiferous tubule, showing cells that produce sperm. Newly produced sperm are pushed into the center of the tubule. (c) Detail of section of seminiferous tubule showing meiotic divisions as primary spermatocytes develop into spermatids and eventually functional sperm. Nurse or Sertoli cells help provide nourishment for the developing spermatogonia, spermatocytes, and spermatids. [(a) and (b) after* Biology Today, *New York: CRM/ Random House, 1972; (c) after Claude A. Villée,* Biology, *6th ed. Philadelphia: W. B. Saunders, 1972.]*

spermatid then undergoes a complicated process of growth and development (but no further cell divisions) to form the mature sperm.

During the transformation of the spermatid into a sperm, the nucleus shrinks in size and becomes the head of the sperm; at the same time the cell loses most of its cytoplasm. Several Golgi bodies congregate at the front of the sperm and form a point, the **acrosome,** which appears to help the sperm penetrate the egg cell membrane (Fig. 12.4). The two centrioles present in the spermatid migrate in opposite directions, so that one comes to reside at the front and the other at the rear of the developing sperm. The rear centriole gives rise to the filament of the sperm flagellum. Like all flagella, that of the sperm consists of two longitudinal fibers surrounded by nine pairs of fibers in the familiar 9 + 2 structure (see p. 96). The mitochondria of the spermatid move to a point where head and tail meet, becoming tightly packed into a thickened region known as the **neck** or **middle piece** (see Fig. 12.4). These mitochondria provide the energy for the beating of the flagellum once the sperm becomes functional and starts to move. As is true for so many aspects of developmental biology, the factors that initiate and guide the differentiation of spermatid into functional sperm are, to date, largely undetermined.

In the male of most mammalian species, spermatogenesis continues throughout the animal's lifetime, although there is some decrease in old age. This situation is in marked contrast to the female, who normally has at birth all the eggs she will ever possess as an adult (approximately 400,000). For most human females, only 400 of these will be ovulated and thus be available for fertilization. The male, on the other hand, releases an average of about 480,000,000 sperm in one ejaculation!

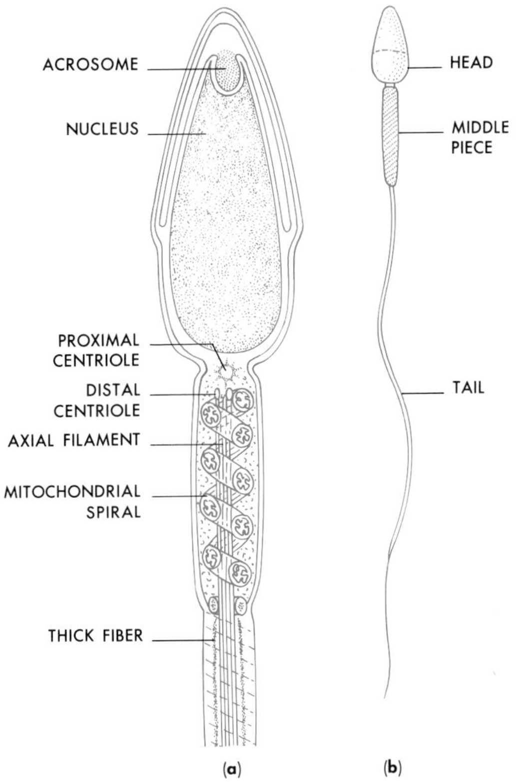

Fig. 12.4 *Typical mammalian sperm, showing general size (b) and internal detail (a).*

The testes are located in the scrotum. The sperm-producing cells and the spermatids cannot tolerate the higher temperature in the abdomen. Since it is more exposed to circulating air, the scrotum has a lower average temperature than the abdomen. However, testicular tissue is also sensitive to cold. Thus, to help in maintaining a constant temperature for the testes, the scrotum is furnished with a lengthwise set of involuntary smooth

While the human female ovulates one egg at a time, a male may release as many as 480,000,000 sperm in one ejaculation.

muscles. When the testes become cold, as for example after swimming, the muscles contract, drawing them up close to the abdominal cavity. When the testes become too warm, the muscles of the scrotum relax, lowering the testes away from the abdominal cavity.

Male **sterility** is the result of failure of the testes to produce functional sperm. Sterility may arise from a number of different causes. In some cases it may be due to the failure of the testes to develop normally during embryonic and later childhood or adolescent growth. In other cases it may be due to chromosomal disorders, such as the tetra-*XY* chromosome complement discussed in Chapter 9. Such a chromosome complement appears to impair normal meiotic processes and lead to lowered or nonexistent sperm production. Sterility should be distinguished from **impotence,** which is defined as the inability to copulate. In men, this is most commonly caused by failure to achieve erection; however, the term is applied to various conditions in women that also prevent copulation.

The Testes as Endocrine Glands

In addition to their sperm-producing function, the testes also serve as one of the body's important **endocrine glands.** Endocrine glands secrete substances known as **hormones,** "chemical messengers" which circulate through the blood and affect only the cells of certain "target" tissues. The target tissues are then signalled to carry out a specific function. As might be suspected, the testes produce the male sex hormone, consisting of several different hormones collectively called **androgens.** The most potent of the androgens is **testosterone,** which is a steroid with the following chemical structure:

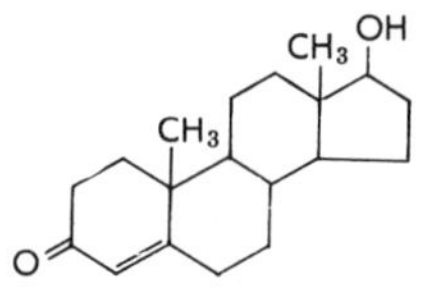

Other male sex hormones include androsterone and dehydroepiandrosterone. Both are less active than testosterone.

The male sex hormones control the development of **secondary sex characteristics.** Secondary sex characteristics are those phenotypic traits that differentiate males from females in any species but are not directly related to sex. In deer, for example, antler development is controlled by androgens, as is the large comb development in roosters. In humans, pattern of body growth, deepening of the voice, hair distribution, and general musculature are all controlled by androgens. The development of the actual reproductive tissue—for instance, the sperm-producing cells of the testes—is not controlled by the androgens, but by another group of hormones produced in the pituitary gland.

The androgens are synthesized from cholesterol, which is derived from acetate (acetic acid). Thus the production of sex hormones is directly

Male sex hormones are collectively called androgens; female sex hormones are collectively called estrogens.

tied to the citric acid cycle and connected to the general pathways of intermediary metabolism. In the testis, androgens are produced in the **interstitial** or **Leydig cells,** which are different from those producing the sperm. As an organ, then, the testes are differentiated into two quite different cell types: one producing functional sperm, the other producing several androgens. In addition to the interstitial cells, androgens are produced by another endocrine gland, the adrenal. The adrenal gland is located on top of the kidney in both sexes and in both secretes androgens. In most females, however, the masculinizing effect of the androgens is not enough to override the more powerful influence of the female sex hormones.

The process of **castration** involves removal of the testes. If castration occurs before **puberty**—the time when males undergo rapid differentiation of secondary sexual characteristics—the secondary sexual characteristics do not develop. The male grows to normal adult size, but the voice does not deepen and he never develops the male pattern of hair growth and distribution. Though generally showing some sex drive, he is, of course, sexually infertile. If castration occurs after puberty, the already-developed secondary sexual characteristics do not disappear (the voice does not suddenly become high-pitched, for instance). But the individual may tend to put on more weight and show a lessening of sex drive. Castration is routinely used in many domesticated farm animals (a "steer" is a young bull castrated before puberty) to produce leaner meat. It also makes the animals more docile and easier to handle.

12.3 The Female Reproductive System

The female reproductive system consists of the paired ovaries and Fallopian tubes, the uterus, vagina, and clitoris. The internal organs are located within the pelvic cavity (Fig. 12.5). Since it serves a greater variety of functions, the female reproductive system is more complex than that of the male. The following discussion of the anatomy of the female reproductive system is based on Figs. 12.5 and 12.6.

Internal Organs

The **ovaries** are paired glands about 3 cm long. Like the testes, they serve a double function, producing sex hormones as well as gametes. The ovaries are held in place within the body cavity by two pairs of ligaments. The mature egg released from the ovary passes into the abdominal cavity, where it is picked up by the **ostium,** the upper end of the fallopian tubes. The egg is guided into the ostium by the beating of cilia in the epithelial lining of the **Fallopian tubes.** The Fallopian tubes serve to conduct the egg down toward the uterus, and it is in the Fallopian tubes that the egg is usually fertilized.

It is an interesting anatomical fact that the ovaries do not make direct contact with the upper end of the Fallopian tubes. This means that an egg could escape into the region of the abdominal cavity around the ovaries. In rare instances, an egg has been fertilized just on leaving the ovary and has begun development within the abdominal cavity. Normal development cannot take place beyond a certain point, however, since the tissues of the abdominal cavity are not able to nourish the embryo in the manner of the uterus. However, at least a few instances have been reported where development of a full-scale embryo did occur within the abdominal cavity and a full-term baby was surgically delivered alive.

The Fallopian tubes end in small openings in the wall of the **uterus**, or womb, a pear-shaped organ just behind the bladder. The uterus has thick walls of smooth muscles and a mucous lining, the endometrium, filled with blood vessels. The function of the uterus is to provide nourishment for the developing embryo. Normally the cavity in the center of the uterus is almost closed. During pregnancy, however, it expands enormously to hold the growing fetus. The uterus terminates in a muscular neck, the **cervix,** which projects a short distance

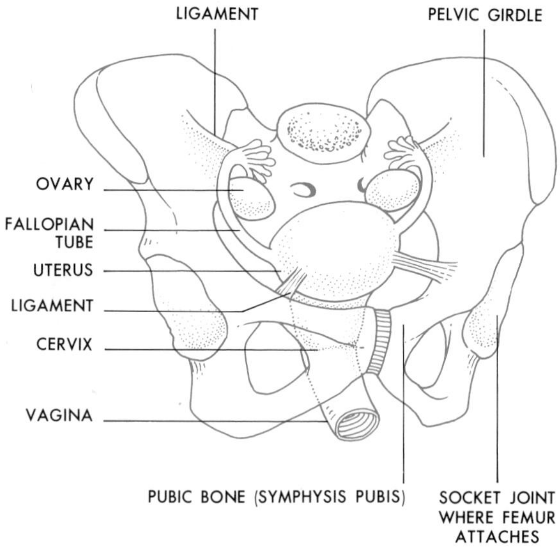

Fig. 12.5 *Location of the human female reproductive organs within the pelvis.*

Fig. 12.6 *Detailed cut-away view of the female reproductive system seen from the side.*
▼

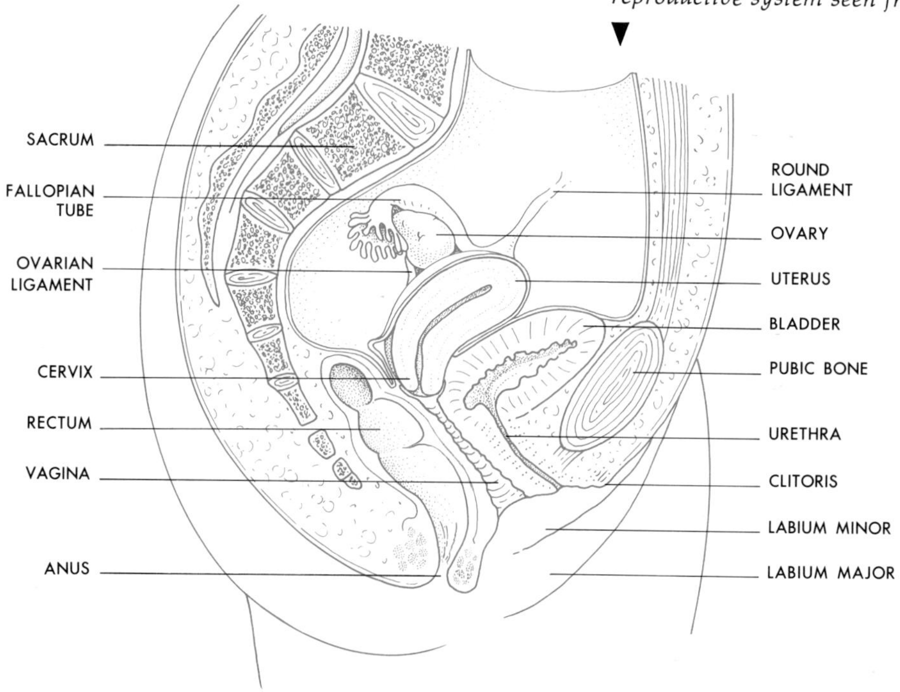

In the human being, as in all higher vertebrates, the testes and ovaries serve two functions: they produce germ cells (eggs and sperm) and they produce the male and female sex hormones. Two very different groups of cells are responsible for these functions.

into the **vagina,** a short muscular tube extending from the uterus to the exterior. The vagina serves a double function: it is a receptacle for the penis during copulation (and thus for sperm at ejaculation), and it serves as the birth canal when the baby is born.

External Organs

The external genitalia of the human female are known collectively as the **vulva** (see Fig. 12.7). The vulva consists of the **labia majora,** two folds of fatty tissue covered by skin; the **labia minora,** thin folds of tissue lying within the folds of the labia majora and concealed by them, and the **clitoris,** a sensitive structure composed of erectile tissue, which is homologous to the penis in the male. Like the penis, the clitoris becomes engorged with blood during sexual excitement. A large number of nerve endings are found in the spongy tissue of the clitoris, whose function appears to be to provide one of the areas responsive to sexual stimulation in the female.

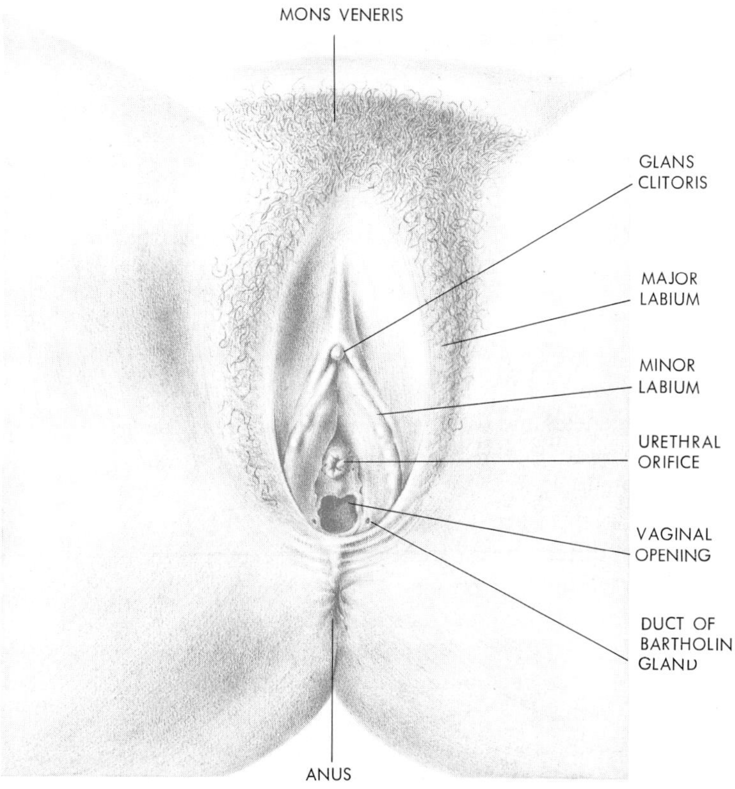

Fig. 12.7 *External female genitalia*

Just behind the clitoris is the opening of the urethra, which in women serves only for the passage of urine. Behind the urethra is the opening to the vagina, which is partly occluded by the **hymen,** a thin membrane composed of connective tissue. The hymen is usually opened by the first sexual intercourse, but various forms of exercise or such simple activities as riding a bicycle can also bring this about. Opening into the vagina on either side are certain glands that produce a mucoid fluid whose basic function is unknown.

Oögenesis

Eggs are produced in the ovary in structures known as **Graafian follicles** (see Fig. 12.8). In a cross section of an ovary a number of follicles can be observed in various stages of development. As the follicle is maturing, the egg undergoes meiotic divisions. The follicle consists of a series of cells that line the wall and surround the egg cell. When the egg is ready the follicle bursts, a process known as **ovulation.** Usually the egg completes its final meiotic stage, including extrusion of the **polar body,** after ovulation.

Oögenesis has been discussed in considerable detail in Chapter 7. In all the basic meiotic events it is identical to spermatogenesis. The chief difference between the two lies in the fact that starting from a primary oöcyte, only a single egg is produced, as compared to four sperm from a single spermatocyte. In Fig. 7.11 the basic patterns of spermatogenesis and oögenesis are compared. These patterns apply to the human being as well as to other animals.

It is in the formation of the polar bodies that oögenesis differs most from spermatogenesis. In the first meiotic division, one of the daughter cells, called the **first polar body,** is smaller than the other. The larger cell, called the **secondary oocyte,** contains more cytoplasm from the parent cell, but both cells contain the chromosomes divided up during the first meiotic division. Each cell, the secondary oöcyte and the first polar body, then undergo the second

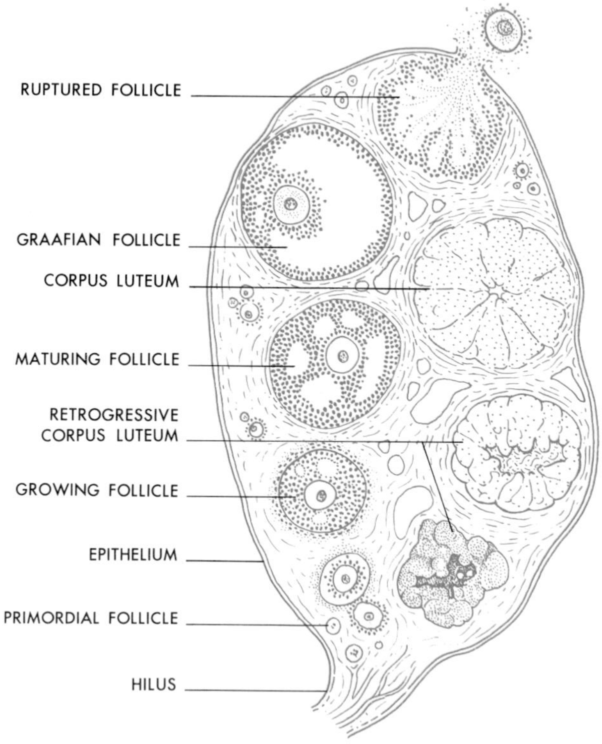

Fig. 12.8 *Longitudinal section of ovary, showing various stages in the progressive development of ova and Graafian follicles. The drawing is a composite in that the youngest eggs and follicles are at the lower left, with progressive development arranged chronologically in a clockwise pattern. In any normal ovary, follicles in different stages of development bear no spatial relationship to one another.*

meiotic division. Both daughter cells from the first polar body contain little cytoplasm and hence are also small. In addition, division of the secondary oöcyte produces one large cell called the **ootid,** containing most of the cytoplasm, and a small cell, one of the **second polar bodies.** Thus three haploid sets of chromosomes are ultimately discarded during oögenesis. A single haploid set is preserved within the oötid, which matures into an ovum, or egg cell.

The function of polar body formation during female gametogenesis is to produce an egg that has a large amount of cytoplasm. Even in animals such as the human being, where nourishment of the embryo is supplied early in life by the mother, a rich supply of cytoplasm in the egg is necessary for those first few days before the embryo becomes firmly attached to the uterine wall.

12.4 Sexual Intercourse

Sexual intercourse, or **coitus,*** involves a complex set of behavioral and physiological activities in the human female and male.

Genital response in the male begins with lengthening of the penis. As the penis becomes more engorged with blood, it becomes increasingly firm and erect. At the same time, nerve endings at the tip become increasingly sensitive to touch. The process of erection is achieved by blood filling the three areas of spongy tissue (vasocongestion). The scrotum also experiences some vasocongestion; through involuntary contraction of muscles surrounding the vas deferens, the testes are actually lifted.

Next the penis becomes slightly larger in circumference, especially around the tip, or **glans penis.** The testes are not only lifted up, but may also increase in size by as much as 50 or 100%. This phase is essentially the period during which full arousal is achieved and maintained as intercourse progresses.

The sex act culminates in orgasm. Prior to orgasm, lubricating fluid from Cowper's glands exudes from the tip of the penis. This fluid frequently contains active sperm, though not in high concentration. When orgasm is achieved, semen is forcefully pumped out of the epididymis and vas deferens and through the urethra, in a process known as **ejaculation.** During the first part of ejaculation, semen is pushed up into the base of the urethra by contraction of muscles all along the spermatic cord. The semen first collects at the base of the urethra in the **urethral bulb,** which can expand to two or three times its usual size. The sphincter muscles surrounding the bladder close, to prevent sperm being forced back up into the bladder. During the second part of ejaculation, seminal fluid from the prostate gland flows into the urethral bulb. Due to muscle contraction initiated by the continual rubbing of the penis, the sperm and various seminal fluids are forced out of the penis. Most of the sperm is ejaculated in the first portion of the semen. The remainder of the semen consists largely of various lubricating fluids.

In the female, genital changes include erection of the clitoris and vaginal lubrication by mucoid secretions. During excitement the vagina itself expands somewhat due to vasocongestion, and the uterus becomes partially elevated. As intercourse proceeds, rubbing of the clitoris maintains or increases the female's arousal. Areas of the labia minora and the outer region of the vagina become more engorged with blood and aside from the breasts and clitoris become the most sensitive areas for sexual stimulation. It is in these areas of the genitalia that repeated stimuli produce orgasm.

At orgasm the woman feels a series of short contractions in the labia minora and lower end of the vagina. The uterus then follows with irregular uterine contractions. Females can often experience several orgasms within a few minutes of one another. During orgasm, muscle tension increases,

* In nonhuman animals, sexual intercourse is generally referred to as copulation.

heart and breathing rate increase, blood pressure goes up, and many areas of the skin show a flush. The general overall feeling, as in the male, is that of explosive release.

12.5 The Female Reproductive Cycle

In addition to its role in the sexual act and gamete production, the female reproductive system must also serve another very important function for which there is no counterpart in the male. Should fertilization of the egg occur, the uterus must be able to receive and nourish a developing embryo.

In mammals, the embryo is nourished and protected during the early part of its development (approximately the first 9 months in humans) *within* the mother's body. Because development takes place within a specific organ, the uterus, it is referred to as **intrauterine development.** Intrauterine development is characteristic of mammals and occurs in no other group. In humans as in most mammals, the very young embryo becomes **implanted** in the uterine wall. About two weeks after conception, the embryo begins to develop a special organ, the **placenta,** that serves as the basic point of exchange for nutrients and wastes between the embryo and the mother's blood.

Estrus and Menstrual Cycles

Most mammals exhibit a cyclic recurrence of reproductive physiology and behavior. Periods when the female is ovulating and will copulate with males alternate with periods, often of long duration, when the female is not ovulating and will not copulate with the male. This cycle is called the **estrus cycle** and for most mammalian species is seasonal, with periods of ovulation and reproduction occurring once or twice a year. During this period the female is said to be in estrus or "heat." In human beings and a few higher primates, there is no seasonal estrus cycle, nor any particularly specific relationship between ovulation and sexual drive. The human female has an estrus cycle, known as the **menstrual cycle,** occurring on an approximately monthly basis. The name "menstrual" derives from the Latin word *mens* (month) and refers to the discharge of blood and tissue from the uterus that occurs approximately every 28 days in adult women. Because it is neither seasonal nor related to specific behavioral patterns associated with intercourse, the menstrual cycle is not viewed simply as the human version of estrus. Obviously, however, the menstrual cycle evolved from the estrus cycle shared by our mammalian ancestors.

Since the fertilized egg eventually becomes attached to the uterine wall, the lining cells must be prepared to receive and support the very young embryo. Support in this case requires that the endometrium be built up into a thickened layer richly supplied with blood vessels. Obviously to maintain the endometrium at all times at the thickness required to nourish a young embryo would require a considerable amount of energy. The mammalian reproductive system has evolved in such a way that the endometrial lining is progressively built up during the time that a follicle is ripening and the egg is being ovulated. If the egg is fertilized, the endometrium remains in its full state for implantation. If the egg is not fertilized, however, the endometrium breaks down and the lining is passed out of the uterus as the monthly menstrual flow. In human beings, this is accomplished over a three- to five-day period with the loss of both tissue and blood. Immediately after the old endometrium is

Males usually experience only one orgasm during each act of intercourse, whereas females may experience several orgasms in rapid succession.

sloughed off, a new lining begins to develop as a new egg is prepared for ovulation.

How does the female body control the cyclic events of estrus? How does it "know" when fertilization has taken place and thus keep the endometrium intact? The processes controlling the menstrual cycle are complex and involve many of the body's endocrine glands. They also involve a complex and interrelated group of feedback mechanisms that are among the most well-tuned of the body's control systems.

Events of the Menstrual Cycle

As pointed out previously, the ovaries are endocrine as well as gamete-producing organs. The hormones produced by the ovaries, in conjunction with the hormones produced by other endocrine glands throughout the body, are responsible for the events that compose the menstrual cycle. The chief difference between the menstrual cycle and other physiological feedback systems is the time elapsing before the feedback control shows its effect. This is partly because of the complex set of interacting hormonal systems involved and partly because of the nature of the hypothalamus and pituitary gland.

Two endocrine glands—one in the brain and another in the ovaries themselves—are involved in regulating the various stages of the female menstrual cycle. The gland in the brain, the **pituitary,** is a "master" endocrine gland for the whole body in both males and females. It produces a number of different hormones (see Fig. 12.9) which in turn affect numerous other glands. Of particular interest in the present case are its effects upon the ovaries. The pituitary, in turn, is regulated by the hypothalamus, a nerve center in the brain sensitive to hormonal, temperature, and other changes in the blood circulating through it. Thus the rate at which the pituitary secretes its various hormones is governed by signals from the hypothalamus. The hypothalamus, in turn, responds to subtle chemical changes in the blood. To understand how the menstrual cycle operates, we will follow a complete cycle, first without fertilization, then under conditions where fertilization has taken place. The hormones involved in regulating the menstrual cycle are summarized in Table 12.1. In considering this cycle, it is customary to refer to the start of menstrual flow as the "beginning" of one cycle. The onset of flow is thus labeled day 1. The hormonal, uterine, and ovarian events associated with the cycle are diagrammed in Fig. 12.10, upon which the following discussion will be based.

FSH and LH. The anterior lobe of the pituitary produces two hormones that act in different ways on the female reproductive organs. **Follicle-stimulating hormone (FSH)** is produced in large quantities during the first few days of the menstrual cycle (the period of menstrual flow) and stimulates one or more follicles in the ovary to begin enlarging and the ovum within it to begin oögenesis. As secretion of FSH by the pituitary begins to decline, secretion of a second hormone, **lutenizing hormone (LH),** begins to increase. LH contributes to maturation of the ovum, so that both LH and FSH are necessary for normal follicular development. In addition, LH contributes significantly to the triggering of ovulation and is the major stimulus for the changes the follicle undergoes as it forms the corpus luteum subsequent to release of the egg. When a correct mixture of FSH and LH is present in the blood the follicle ruptures, releasing the egg (ovulation). This usually occurs on about the fourteenth day of the cycle—that is, about 14 days after the start of the preceding menstrual flow. The mixture of FSH and LH that triggers ovulation varies from individual to individual and from time to time in the same individual, depending on the circumstances.

The changes occurring within the uterine lining during the first 14 days of the estrus cycle will now be examined. Under stimulation from FSH, the developing follicular cells become an endocrine organ themselves, secreting estrogens, the most im-

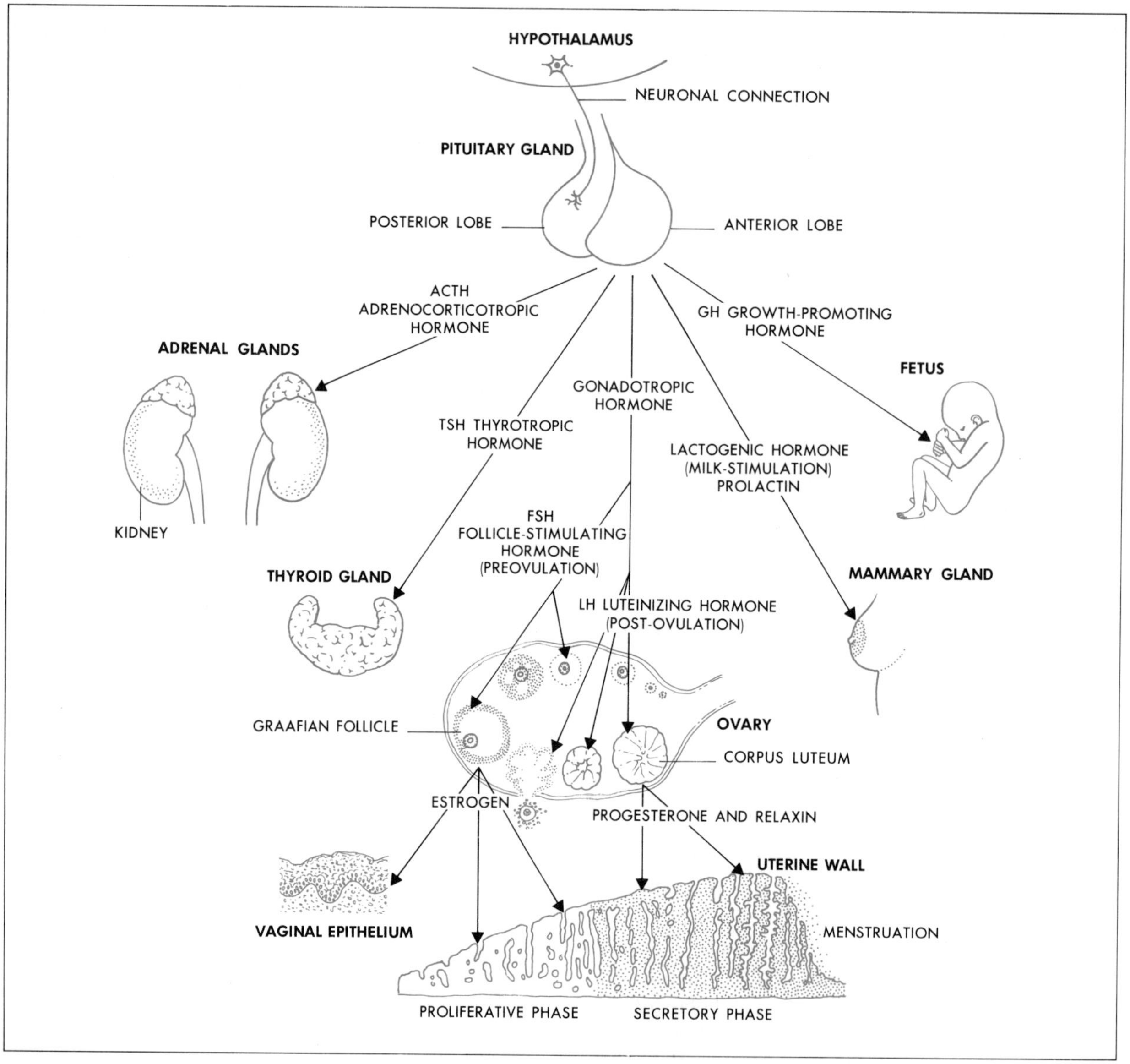

Fig. 12.9 *The pituitary gland is often called the "master" endocrine gland of the vertebrate body. This diagram illustrates why. Hormones produced in the pituitary control general body metabolism (by controlling production of adrenaline), growth (by controlling production of thyroid hormone), sexual development and maintenance (through production of FSH and LH), and various aspects of fetal growth and development (lactogenic hormone and growth hormone, GH). As the diagram illustrates, the pituitary is particularly important in regulating the female reproductive cycle and for controlling aspects of pregnancy. (After Roberts Rugh,* et al., From Conception to Birth. *New York: Harper and Row, 1971, p. 11.)*

Table 12.1
The Major Hormones Involved in the Menstrual Cycle of Human Females

Gland	Name of hormone	Abbreviation	Principal action (target organ)	Mechanism controlling secretion
Pituitary (anterior)	Follicle-stimulating hormone	FSH	Stimulates follicles of ovary to grow. (In males FSH stimulates development of sperm-producing cells.)	Blood level of estrogen inhibits FSH release; nerve impulses from hypothalamus stimulate release of FSH.
	Luteinizing hormone	LH	Stimulates formation of corpus luteum in females. (In males LH regulates production of testosterone.)	Testosterone or progesterone in blood inhibits LH release; hypothalamus stimulates LH release.
Ovary				
Follicle cells	Estrogens (estradiol, estrone, and estriol, listed in order of potency)		Stimulates growth of uterine lining. Guides development of female secondary sex characteristics at puberty.	Increase in FSH produces increases in estrogen secretion.
Corpus luteum	Progesterone		Promotes continued development of uterine lining; makes implantation possible; promotes development of mammary glands if pregnancy occurs; prevents maturation of new follicles and eggs.	Increase in amount of LH in blood; produces increase in progesterone secretion by corpus luteum.

portant of which is **estradiol.** Estradiol and other estrogens stimulate cell division in the uterine lining. As more estrogens are secreted into the blood by the follicle cells, the uterine lining grows thicker and thicker. By the time of ovulation, (around the fourteenth day) the uterine lining is about as thick as it will become by the end of the menstrual cycle. After the breakdown of the follicle following ovulation, growth of the uterine lining slows down temporarily, since of course the supply of estrogen secreted by follicle cells falls off. However, other events following ovulation ensure that the uterine lining will continue to grow so that it will be prepared to receive the zygote should fertilization occur.

Beyond ovulation, the level of FSH secretion

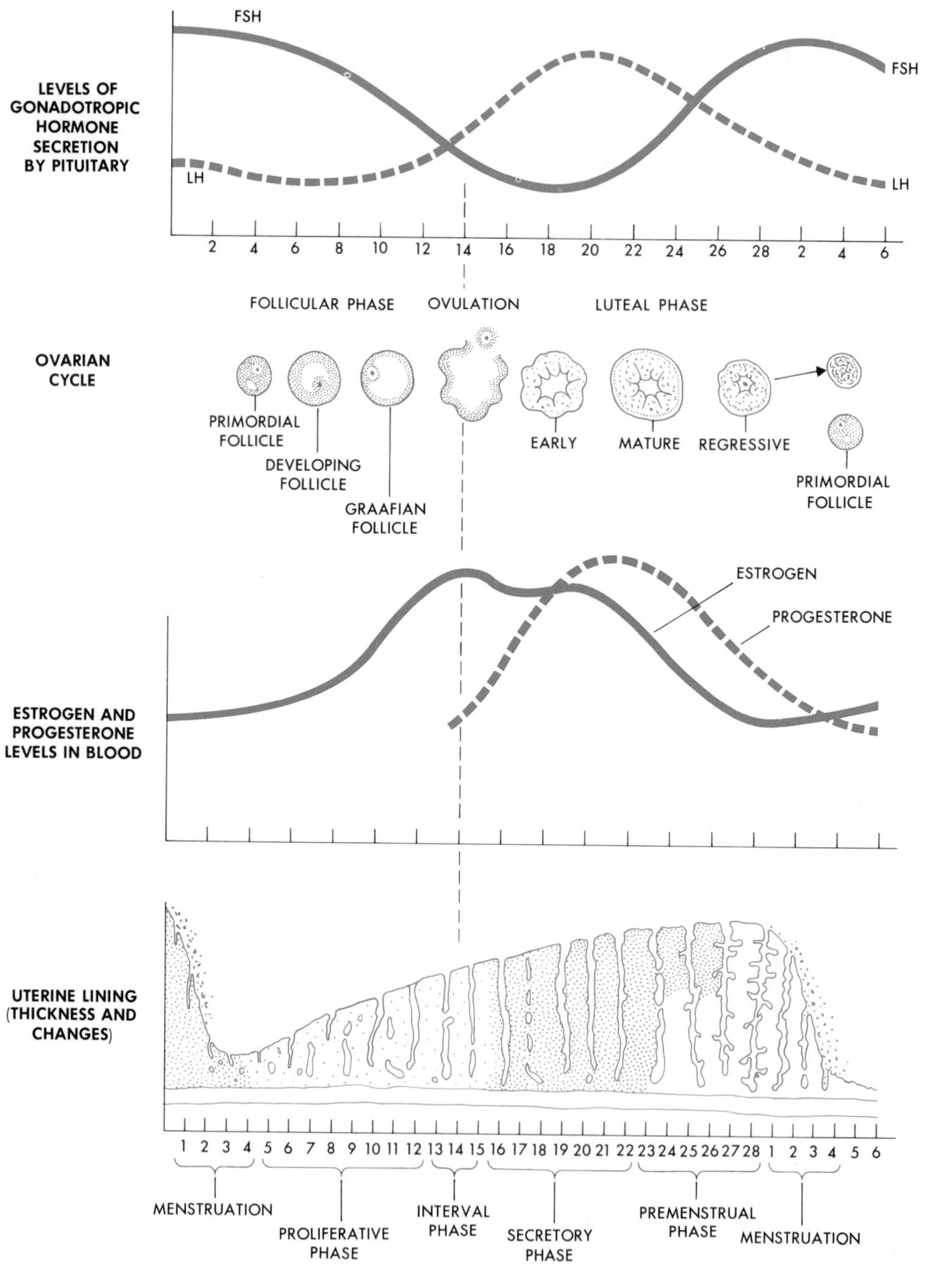
FSH
LEVELS OF GONADOTROPIC HORMONE SECRETION BY PITUITARY
FSH
LH
LH
2 4 6 8 10 12 14 16 18 20 22 24 26 28 2 4 6
FOLLICULAR PHASE
OVULATION
LUTEAL PHASE
OVARIAN CYCLE
PRIMORDIAL FOLLICLE
DEVELOPING FOLLICLE
GRAAFIAN FOLLICLE
EARLY
MATURE
REGRESSIVE
PRIMORDIAL FOLLICLE
ESTROGEN
PROGESTERONE
ESTROGEN AND PROGESTERONE LEVELS IN BLOOD
UTERINE LINING (THICKNESS AND CHANGES)
1 2 3 4 5 6 7 8 9 10 11 12 13 14 15 16 17 18 19 20 21 22 23 24 25 26 27 28 1 2 3 4 5 6
MENSTRUATION
PROLIFERATIVE PHASE
INTERVAL PHASE
SECRETORY PHASE
PREMENSTRUAL PHASE
MENSTRUATION

◀ **Fig. 12.10** *Diagram of the various events of the menstrual cycle, showing simultaneous changes in pituitary secretions, ovarian follicles, estrogen and progesterone levels in the blood, and thickness of the uterine wall, or endometrium.*

by the pituitary continues to fall, while that of LH continues to rise (see Fig. 12.10). Under the influence of LH, cells of the old follicle are transformed into an endocrine gland, the **corpus luteum.** Cells of the corpus luteum secrete progesterone, a hormone that stimulates the final stages of development of the uterine lining. Progesterone also stimulates the mammary glands to begin producing milk and prevents the growth of additional follicles and eggs. Thus the second half of the menstrual cycle, under direct guidance of progesterone produced in the corpus luteum, is characterized by development of the uterine lining to its fullest thickness.

If the egg is fertilized. . . . After the egg has been released from the ovary, the period of time during which it passes down the Fallopian tube and can be fertilized is limited—about 24 hr. Similarly, sperm deposited in the female vagina can usually live for a maximum period of about 48 hr. Thus the period of maximum fertility in human beings is only a brief span within the menstrual cycle. That period occurs approximately midway between successive menstrual periods. If the egg is fertilized, the corpus luteum does not break down, but continues to secrete progesterone almost until the time of birth. Progesterone is absolutely essential for retention of the young embryo by the uterine wall. If the corpus luteum is lost during the first two or three months of pregnancy, the uterine lining does not retain the embryo and a spontaneous abortion occurs. Progesterone produced by the corpus luteum also stimulates the mammary glands during the latter months of pregnancy. It prepares them for the action of prolactin, a hormone secreted by the anterior lobe of the pituitary, which stimulates certain mammary gland cells to secrete milk.

If not. . . . If the egg is not fertilized by about 27 days after the onset of the previous menstrual flow, the corpus luteum undergoes regression, thus limiting and finally ending the secretion of progesterone. Since the uterine lining requires progesterone for its continued maintenance, the fall of progesterone levels in the blood below a critical point initiates breakdown of the uterine lining. This breakdown starts the menstrual flow, at which point a new cycle can be said to have begun.

Menstrual flow would seem to be a traumatic demand on a woman's body. The amount of blood lost with each flow is about 100-200 ml (somewhat less than the fluid held in an average-sized drinking glass), a considerable drain on the body's resources. What prevents the amount of blood loss from being even greater than it is? As the progesterone level in the blood begins to fall, arterioles in the outer third of the uterine lining begin to constrict, cutting off blood supply to that region of tissue. This layer of the uterine wall begins to die and as it becomes

---

The menstrual cycle is a repetitive developmental phenomenon under the control of negative feedback systems. These systems operate between the ovaries and the pituitary via the hypothalamus.

---

detached, arterioles in the innermost layers of the endometrium also begin to close down. These latter arterioles have specialized muscle cells in their walls that cause the arteriole to coil up or contract longitudinally. This coiling tends to seal off the lumen of each arteriole. The resulting cessation of blood flow causes death of cells in the lower layers of tissue of the lining, allowing that layer also to be detached. The coiling of the arterioles also prevents excess hemorrhage. Blood flow in menstruation is not stopped by the clotting mechanism associated with wound healing. Although the loss of a certain amount of blood and tissue each month might seem wasteful to the body's resources, from an evolutionary point of view the primates have found the menstrual process adaptive. The increased chances for successful pregnancy obtained by preparation of the uterine wall to receive the embryo appear to more than offset the disadvantage to the female of losing blood or other body fluids.

Feedback Mechanisms

To understand the feedback system involved in the menstrual cycle, it is necessary to begin with the anterior lobe of the pituitary and its initial secretion product, FSH. At day 1 in the menstrual cycle FSH is being secreted in large quantities, yet it begins to fall off by day 3 or 4 (see Fig. 12.10). High levels of FSH stimulate growth of the follicle, which produces estrogen. Estrogen inhibits the production of FSH by the pituitary. Hence, as the follicle grows and produces increasing amounts of estrogen in the blood, FSH production by the pituitary decreases. As typical of so many feedback processes, the end product (estrogen) inhibits the first step in its own production (FSH secretion). The intriguing aspect of the menstrual system is that several such feedback inhibition loops interact to guide the entire cycle.

As FSH diminishes in quantity and estrogen increases in quantity, the pituitary responds by increasing its secretion of LH. The direct stimulus for LH secretion appears to be the increase in estrogens. As Fig. 12.10 shows, however, FSH reaches its lowest levels of secretion and LH its highest levels by about the eighteenth or twentieth day of the cycle. Thereafter both show reverse changes in production with FSH beginning to rise and LH beginning to fall. Similarly, production of estrogen begins to fall and that of progesterone to rise.

What controls these developments? The corpus luteum develops under stimulation of LH. As the corpus luteum grows, it begins secreting progesterone. Progesterone serves not only to guide the further build-up of the uterine lining; it also causes the hypothalamus to inhibit production of LH by the pituitary. This is yet another negative feedback loop built into the estrus cycle. The more progesterone the corpus luteum produces, the more it inhibits production of the very hormone (LH) necessary for continued maintenance of the corpus itself. After about the twenty-first or twenty-second day of the cycle, the hypothalamus causes the pituitary's output of LH to decline. At this point it might appear that the whole cycle would grind to a halt, since estrogen, progesterone, FSH, and LH secretion are all on the downswing (though some, such as FSH and estrogen, more than others). Recall that FSH production is inhibited by estrogen. Since estrogen secretion has begun to fall as early as the fourteenth day and continues to fall until the eighteenth day, its inhibiting effects on FSH secretion gradually decrease as time goes on. Secretion of FSH by the pituitary is thus "unchained" once again, and by about the nineteenth or twentieth day, the blood level of FSH begins to increase. In addition, the fall of progesterone production caused by its own inhibiting effect on LH production makes maintenance of the uterine lining increasingly difficult. By the twenty-eighth day, progesterone levels are low enough so that the uterine lining begins to break down. FSH production is now on the upswing again, however, reaching its peak by the first or second day of the new cycle, and the sequence of control events begins anew.

In pregnancy, these negative feedback systems are interrupted. Implantation of the fertilized egg in the uterine wall, which occurs about 9 days after fertilization, stimulates the corpus luteum to remain intact. The precise manner by which this occurs is unclear—whether through direct influence on the corpus luteum itself or through indirect stimulation of the pituitary gland to continue its secretion of LH. At any rate, retention of the corpus luteum ensures that progesterone production continues and the uterine lining is maintained. Shortly after implantation the developing embryo forms the placenta, which will play a major role in hormonal regulation in the pregnant mother. Within a month or so after fertilization (2 to 3 weeks after implantation), the placenta produces considerable quantities of estrogen and progesterone, along with chorionic gonadotropins, the general name for FHS- and LH-like hormones that stimulate the reproductive organs. These hormones are secreted by the placenta throughout pregnancy and essentially replace the pituitary and (to a lesser extent) the corpus luteum in the regulation of hormonal balances. The placenta secretes such an overdose of hormones that it overrides all the control mechanisms described earlier. Furthermore its tissue is not sensitive, as is that of the pituitary, to the inhibiting effects of estrogen and progesterone. Consequently, the feedback systems that operate during the menstrual cycle do not occur during pregnancy. The system thus maintains itself in a more or less steady state until just before birth. As we shall see later, in Section 12.6, various hormonal changes do eventually occur within this system to initiate the birth process.

12.6 From Conception to Birth

The human gestation period (period of pregnancy) averages 266 days (nine months), which is divided into three trimesters of three months each. During the first trimester the developing organism is usually referred to as an embryo; thereafter it is called a **fetus.**

The First Trimester

The newly fertilized egg begins cell division while it is still in the Fallopian tubes, being swept along the oviducts toward the uterus by the rhythmic beating of the cilia that line the Fallopian tubes. During its early cell divisions, the size of the embryo stays about the same; successive cleavages merely produce smaller cells. By the time the embryo reaches the uterus it is in the 32- or 64-cell stage. It is now a hollow sphere, the **trophoblast,** which is one cell thick on one side (this side will give rise to the membranes surrounding the embryo) and several cells thick on the other (this side will develop into the embryo proper).

The embryo lies free on the uterine surface for two or three days, then starts implanting itself in the uterine walls by secreting an enzyme that digests away part of the uterine tissue. Placenta formation begins with the penetration of the uterine wall by the **chorionic villi** of the trophoblast (Fig. 12.11). In its fully developed state, the placenta consists of a meshwork of capillaries derived from the embryonic circulatory system. The fetal capillaries penetrate the uterine lining tissue and lie close to the maternal circulatory system, but normally the blood of the two circulatory systems does not mix. The exchange of materials is accomplished by diffusion and active transport across capillary membranes. The embryo itself is surrounded by its **amniotic sac**—a fluid-filled membrane, the amnion, that is formed from the trophoblast.

Differentiation begins with the setting aside of germ cells for the eventual formation of reproductive organs. By the age of 30 days the embryo shows development of a **primitive streak** (longitudinal axis), on either side of which appear muscle segments called **somites.** From the somites will arise most of the voluntary muscles, much of the skeleton, and the connective tissue of the skin. Rudi-

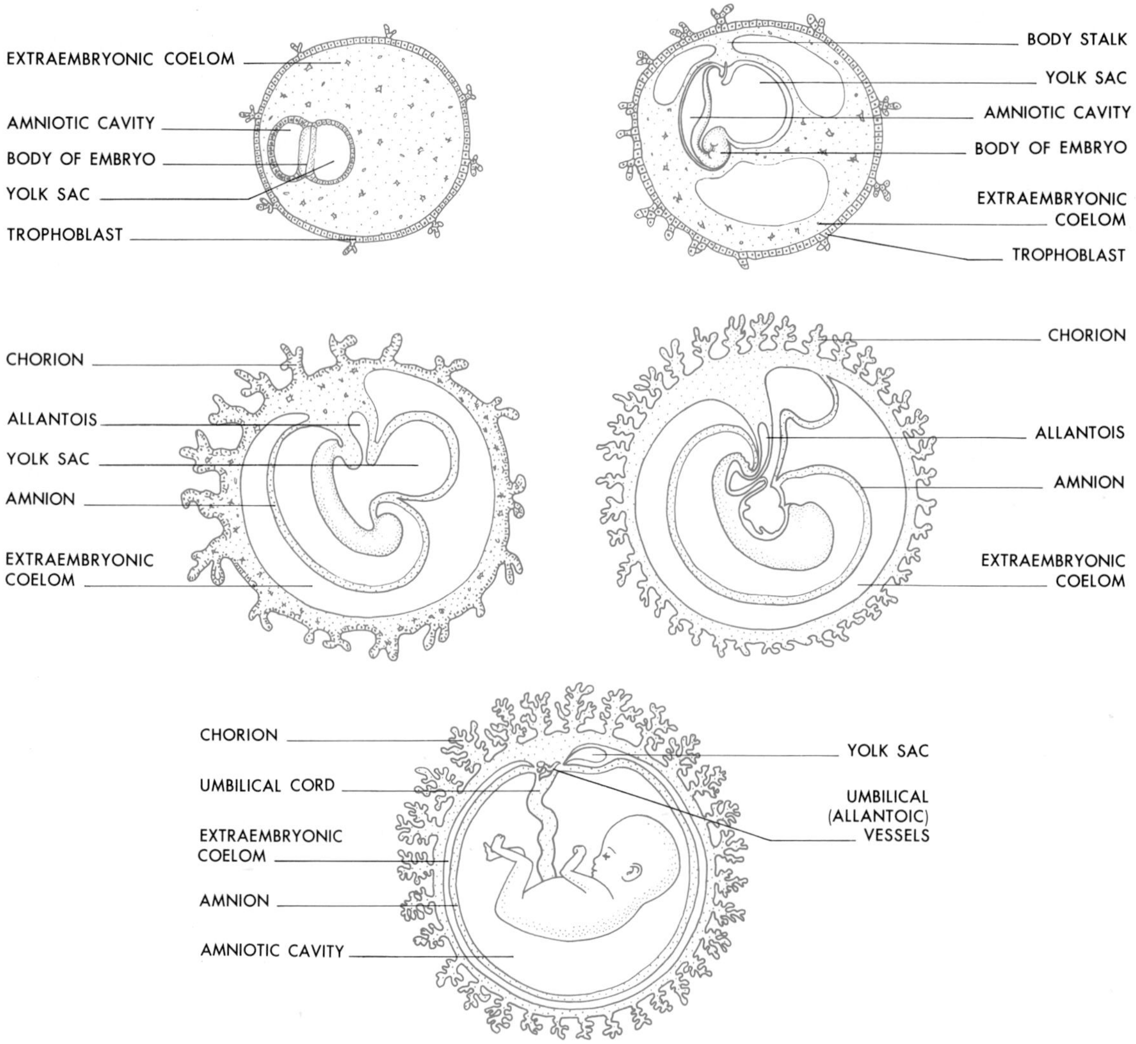

Fig. 12.11 *Development of the embryonic membranes. Surrounded by the amnion, the amniotic cavity gradually expands outward, filling up the cavity known as the extraembryonic coelom. The chorion forms outgrowths, the chorionic villi, that penetrate the uterine wall and will eventually form the placenta As the embryo grows and obtains more and more of its nourishment from the mother, the yolk sac decreases in size. The yolk sac does not serve as a food reservoir as in many other vertebrates, but rather appears to be largely a vestigial organ that develops only after the embryo starts growing. (Modified from B. M. Patten,* Human Embryology. *McGraw-Hill, 1968, p. 104.)*

ments of other organs and body parts are also visible, as summarized in Table 12.2. The embryo is now about one-third of an inch long.

During the second and third months the embryo gradually takes on a recognizably human form (see Table 12.2). By the end of the second month its length is about 3 cm (1¼ in.), and by the end of the third it is about 7.5 cm (3 in.).

The Second and Third Trimesters

During the second trimester, the fetal brain begins to form convolutions and thereby increases its surface area. The eyes are able to perceive light, but the ears apparently do not yet hear sounds. By the end of the fifth month the fetus is approximately 23–28 cm (9–11 in.) long and weighs about 225 g (one-half pound). It can kick, turn around, and float freely within the amniotic sac, since its only attachment is via the umbilical cord that connects it to the placenta. The placenta itself is still growing; the lungs are complete but the alveoli are nonfunctional. Although the sixth month is the earliest that any fetus can be born and have much hope for survival, at least one fetus has been known to survive after birth at five months—but it required to-

Table 12.2
Events in the First Trimester of Pregnancy

Month	Event	Approximate age of embryo
First	Fertilization occurs.	0
	Zygote divides into two cells (blastomeres).	30 hr
	Second division produces four-cell stage.	40 hr
	Fourth division produces 16-cell stage.	70 hr
	Embryo reaches uterus.	9 days
	Implantation is complete; placenta begins to form.	12 days
	Tube-shaped heart has become visible.	21 days
	Neural plate and neural tube have become visible.	24 days
	Four primary brain vesicles begin to form; somites appear (32 pairs); sense organs start to develop.	26 days
	Mouth appears; visceral arches ("gill slits") appear on each side of throat—destined to become Eustachian tube connecting throat to middle ear.	27–28 days
	Rudimentary lungs appear; primitive kidneys start to form; aminotic sac complete.	30 days
Second	Germ cells migrate to kidney region and become incorporated in primitive urogenital system.	31–45 days
	Reproductive tissues begin to form testes and ovaries.	46 days
	Additional somites are formed (total now is 49 pairs).	31–60 days
	Fingers and toes are visible.	60 days
	Bone tissue starts to become apparent in sections of skeleton.	60 days
Third	Embryo continues growth and differentiation; proportion of head length to total body length (formerly one-half) begins decreasing as trunk and limbs develop at faster rate.	61–90 days

tally artificial sustenance such as intravenous feeding, extra oxygen, temperature regulation, and so on.

The third trimester is characterized largely by further fetal growth and refinement of the nervous system and brain. About 85% of the calcium and iron in the pregnant mother's diet goes into the fetus for building the skeleton and synthesizing hemoglobin, respectively. Nitrogen is consumed in great quantities in building fetal proteins, especially those going into the nervous system.

During the eighth or ninth month of preg-

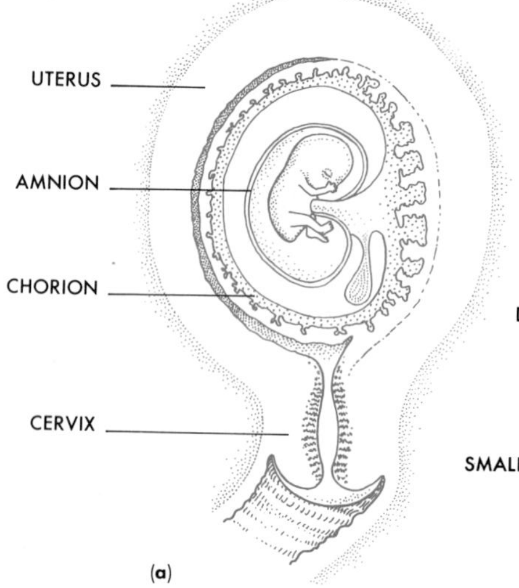

(a)

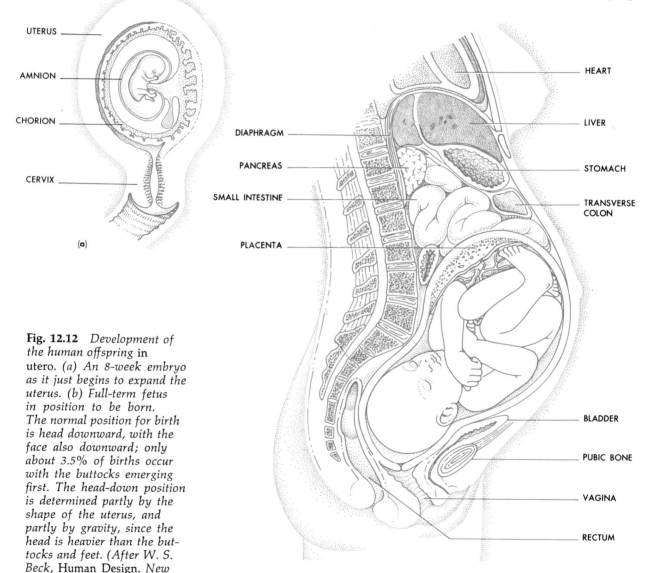

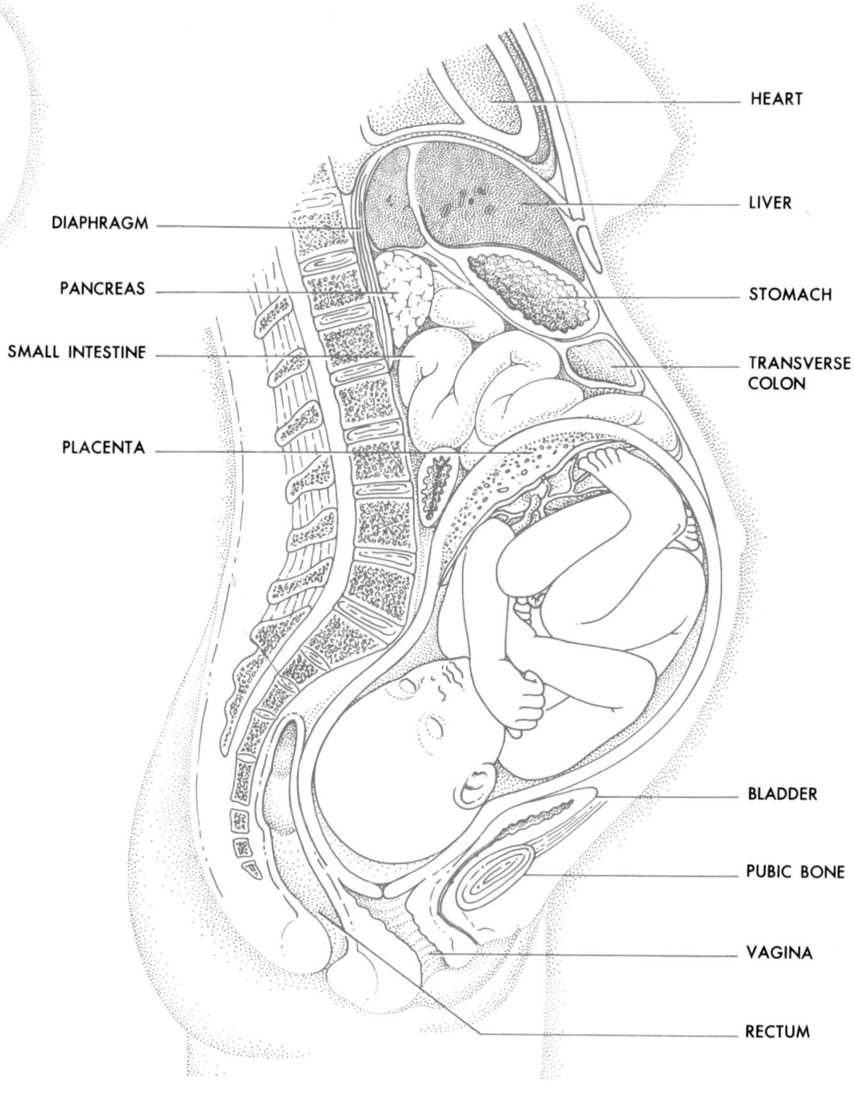

Fig. 12.12 *Development of the human offspring* in utero. *(a) An 8-week embryo as it just begins to expand the uterus. (b) Full-term fetus in position to be born. The normal position for birth is head downward, with the face also downward; only about 3.5% of births occur with the buttocks emerging first. The head-down position is determined partly by the shape of the uterus, and partly by gravity, since the head is heavier than the buttocks and feet. (After W. S. Beck,* Human Design. *New York: Harcourt Brace Jovanovich, 1971, p. 660.)*

nancy, the fetus "drops" into position for birth. It moves forward, with its head usually pointing downward in the uterus (Fig. 12.12). Then, late in the ninth month, the mother's "water" will "break" —this is, the amniotic sac will burst and release the fluid within it. Although this may occur a day or two early, it usually does not occur until after the onset of labor.

Birth (Parturition)

It is not clear what factors initiate the process of birth. During the seventh month, the growth and physiologic activity of the placenta slow down. Just prior to birth, the placenta exhibits regions of degeneration and breakdown of its capillary bed.

Complex hormonal changes also occur in the mother's body from the seventh month onward. Progesterone secretion begins to slow down while estrogen secretion increases. Immediately before birth, relatively large quantities of estrogens appear to be secreted, promoting contractility and irritability of the smooth muscles of the uterus. By the onset of labor, the hormone oxytocin is secreted by the posterior lobe of the pituitary in greatly increased quantities. Oxytocin causes contraction of the uterine muscles. In turn, stretching of the uterine muscles is known to trigger an increase in oxytocin secretion. Oxytocin can also be administered intravenously to induce labor artificially.

The process of birth begins with a long series of involuntary contractions of the uterine smooth muscle known as "labor pains." During labor the contractions become more frequent (every one to two minutes) and last longer (50–90 seconds), and the cervix dilates to make possible the passage of the fetus. During delivery the mother may push with her abdominal muscles to aid the uterine contractions in moving the baby out of the birth canal. After delivery of the baby, the final phase of labor involves contractions to expel uterine fluid and eventually the placenta with an attached portion of the umbilical cord. Far from being the end of development, the birth is only one important step toward continued growth and change that lasts a lifetime.

12.7 Controlling Human Fertility

For thousands of years, human beings have sought methods of controlling fertility to either increase or decrease the chances of successful pregnancy. Most attention in the recent past has focused on methods of preventing, rather than increasing, fertility. Within the past 25 years, various methods of preventing pregnancy have been developed, each with varying degrees of effectiveness and safety. And within the last decade, the introduction of the "pill" has promised a highly dependable, though not flawless, method of birth control.

Methods of birth control must balance a number of interacting and often opposing considerations. The first has always been effectiveness: To what degree can the selected method prevent pregnancy? Second, the most desirable form of birth control has always been one that is not permanent, where the effects are easily reversible. (Methods that involve permanent changes in the male or female gonads, although highly effective, are often not desirable on a large scale.) A third factor is safety of the birth control method for general use; those that have side effects dangerous to the health of the person using them involve risks most people are not willing to take. A fourth factor is a sociological one: For which partner, male or female, should fertility control devices be aimed? Not without reason, many women's groups have complained that the large bulk of birth control devices have been designed for women, not men. A fifth factor involves ethical problems which for centuries have influenced decisions about the use or development of birth control methods. Some people believe that it is not morally right to prevent or terminate pregnancy.

As already pointed out, birth control is obviously aimed at preventing the birth of a child. This can be effected in a number of places by a variety of means: at the behavioral level by preventing intercourse, at the general cellular level by preventing union of sperm and egg, at the physiological level by preventing implantation of the fertilized egg, or at the developmental level by aborting the young embryo. Various methods are currently in wide use at each of these levels. In general, the term contraception refers to all methods that prevent pregnancy while still allowing sexual intercourse to proceed. The term contraception does not extend to the practice of abortion, which is applied at a much later stage of the reproductive process.

Behavioral Methods of Preventing Fertilization

Obviously the most effective method of preventing union of sperm and egg is to avoid sexual intercourse altogether. In some religions abstinence from sex, or **celibacy,** is a custom practiced by the clergy and even some lay persons. Effective as celibacy is as a birth control method, it has simply not been a practical method for use on any large scale. Most people are not willing to avoid sex, even if they do not want children.

Some people use the **rhythm method,** involving abstinence from sexual intercourse during the time when it is believed that ovulation is taking place. However, since it is not always possible to be certain when ovulation is occuring, the rhythm method is generally regarded as the least effective method of birth control.

Another behavioral method for reducing the chances of fertilization is known as **coitus interruptus,** or withdrawal. This method means that the male withdraws his penis from the female shortly before he reaches orgasm. Withdrawal has several problems that make it a less than successful method of birth control. One is that sperm can be found in small quantities in some of the secretions from the penis prior to actual orgasm (in the lubricants released from Cowper's and other glands). These sperm may be concentrated enough to fertilize an egg. Another problem is that withdrawal is sometimes difficult to achieve before orgasm.

Mechanical Methods of Preventing Fertilization

Mechanical methods of birth control consist of various contraceptive devices that mechanically block the uniting of sperm and egg. There are two classes of such devices, those designed for use by the males and those designed for use by females. Both function on the same principle: they provide a barrier separating sperm from egg during the act of intercourse.

For the man. Most mechanical contraceptive devices for males fall under the general term **condom,** or penis sheath. This is a cap made of rubber or some other elastic substance placed over the end of the erect penis. The condom has a relatively high degree of effectiveness: it works between 75% and 95% of the time.

For the woman. There are a variety of mechanical devices for birth control used by the female, but only one, **the diaphragm,** works by actually blocking union of sperm and egg. The other mechanical devices for women work to prevent implantation of the fertilized egg and are thus not technically contraceptive agents.

The diaphragm is a rubber-covered ring designed to fit over the cervix and thereby block entry of sperm into the uterus. The diaphragm or cap forms a barrier so that sperm introduced into the vagina cannot get through the cervix and hence do not ever come in contact with the egg. The diaphragm by itself, however, is never effective in preventing union of egg and sperm. Sperm are highly motile and can usually get around the edge of the cap at some point. To increase the effectiveness of the diaphragm, it must be used in conjunction with a **spermicidal jelly,** which is rubbed around the inside of the diaphragm dome and along the rim.

Birth control methods include two basic types: those that prevent conception (contraceptive methods) and those that prevent development of the fertilized zygote. Among the former are condoms, diaphragms, and the pill. Among the latter are "morning after" pills, intrauterine devices (IUD's), and abortion.

Spermicidal jelly or foam can also be used by itself, applied within and around the walls of the vagina. Used this way it is sometimes referred to as vaginal jelly. Without the diaphragm, however, vaginal jelly has a poor record in preventing fertilization.

The Birth Control Pill

The birth control pill, or "oral contraceptive" as it is sometimes called, is a means of controlling ovulation by introducing synthetic forms of estrogen and progesterone into the female's body. Most commercial forms of birth control pills must be taken on a daily basis for three weeks; during the fourth week placebo pills containing no hormones are taken. During this time menstruation occurs. The cycle is then resumed with three more weeks of pills. The effect of the pill is to interfere with the normal feedback inhibition processes by which ovulation is triggered. By introducing higher-than-normal levels of estrogen and progesterone into the blood, the pill suppresses production of FSH and LH by the pituitary. Without release of the pituitary hormones, the normal cycle of follicular growth and ovulation do not occur.

The pill has many advantages over other forms of birth control. As long as the pills are taken as directed, it is virtually 100% effective. Moreover, it works continually and thus does not require interruption of the sexual act. It is extremely easy to administer by following a simple schedule, and pills are usually packaged to indicate a specific pill for each day of the month.

Increasingly in recent years, some doctors and medical physiologists have been questioning the overall influence of the pill on women's physiology. Up to 25% of the women taking the pill experience some dizziness, headache, nausea, vomiting, or some of the phenomena normally associated with pregnancy, including enlargement and sensitizing of the breasts and periods of increased irritability. Many women also experience an increase in weight after taking the pill for some time. A very grave problem is the increased rate of formation of blood clots in the arteries and veins (thromboembolisms). Although only a small percentage of pill users have actually experienced such clotting, the matter is a very serious one, since such clots frequently lead to death. A study in Great Britain indicates that the incidence of thromboembolism in the lungs and heart is 10 times greater among women who are taking the pill than among women who are not. Proponents of the pill point out, however, that the rate of death due to thromboembolism is about 3 per 100,000 users, whereas rate of death from normal complications in pregnancy and childbirth is about 25 per 100,000.

Control of Spermatogenesis

In recent years at least one method of chemical birth control has been developed for use with men rather than women. Compounds known as diamines can be administered orally to suppress spermatogenesis. However, they require eight weeks to become effective. If the drug is stopped, it takes another eight weeks for full sperm count to be restored. Obviously this long time lag is something of a disadvantage, though not a critical one.

Surgical Methods of Preventing Fertilization

An extremely effective method of female contraception is to tie off or cauterize (seal shut by heat) the Fallopian tubes. This process, called **tubal ligation,** can be carried out with only minor surgery. A small incision is made in the woman's abdomen, and with the help of an optical device known as a laparascope, the tubes are located and either cauterized or tied. Tubal ligation is a virtually certain method of birth control. It is, however, irreversible, and should be used only when the full implications of the procedure are clear to the patient, and the irreversibility of the process fully understood.

A surgical procedure to prevent sperm transport in the male which is becoming increasingly popular is the vasectomy. Vasectomy consists of making an incision in the scrotum, cutting each vas deferens (see Fig. 12.2) (the ducts through which sperm are transported from the epididymides to the urethra), and tying the cut ends. Since hormone secretions are not altered, sperm continue to be formed but they must be resorbed in the testis. The accessory sex glands—the prostate gland and seminal vesicles—continue to produce their secretions comprising the seminal fluid, so that ejaculations remain the same except for sperm content (which comprises only 15% of the semen by volume).

Among the advantages of vasectomy are (1) it is a relative simple operation (it can be performed in the physician's office with only a local anesthetic), and (2) libido remains unaffected or is improved because of the lack of worry over unwanted children. The major disadvantages of vasectomy include (1) the fact that it is a surgical procedure, and (2) that it is usually irreversible.

Post-Fertilization Birth Control: Inhibiting Implantation

"Morning after" pill. The so-called "morning after" pill has been developed for use immediately after intercourse. It contains massive concentrations of estrogens which produce overgrowth of the endometrium. If an egg has been fertilized in the Fallopian tube, by the time it reaches the uterus the endometrium is incapable of receiving it. Consequently, the zygote cannot become implanted and is passed out of the body. The "morning after" pill has never been very satisfactory, however. Despite its name, it must be taken for up to three days after intercourse to be effective. In addition, it has very unpleasant side effects, including vomiting, nausea, and excessive vaginal bleeding. Furthermore, if the "morning after" pill is taken after implantation has occurred, it generally fails to cause extrusion of the embryo. Worst of all, if the embryo is a female the massive doses of estrogen that the pill contains can cause vaginal cancer when the child reaches puberty.

Intrauterine devices. Certain physical objects called **intrauterine devices** (IUD's) can be inserted into the uterus by a doctor or other properly trained person to prevent normal implantation. A number of such devices exist in the shape of rings, coils, and loops. It is not completely clear how IUD's have their effect.

Once inserted, the IUD is a relatively permanent device that does not need to be inserted or removed prior to or after sexual intercourse. Like the pill, it allows sexual activity to proceed in a more spontaneous manner. The IUD can also be removed with relative ease by pulling the strings attached to it, which project into the vagina. Once inserted, an IUD involves no further attention on the part of the wearer, other than to check that expulsion has not occurred. IUD's are inexpensive to maintain and distribute and thus are easily accessible.

Despite the apparent advantages of the IUD, there are several disadvantages. One is that the device may be expelled unnoticed during menstruation. This possibility greatly reduces the effectiveness of the IUD. Another disadvantage is that insertion of the IUD by other than a skilled doctor or paramedical worker can lead to perforation of the uterine wall. This causes not only loss of effectiveness as a birth control device, but also the possibility that the IUD may work itself through the uterine wall into the body cavity. There it can block the normal peristaltic movements of the intestines and otherwise cause considerable damage to normal bodily functions. One of the chief objections to use of IUD's for birth control programs in economically depressed countries is that routine medical care is necessary for the IUD to be used as an effective and medically beneficial device. In many countries of the Third World, medical care adequate for use of IUD's does not exist. Under these conditions, use of IUD's is actually harmful rather than helpful. It is becoming clear that birth control cannot be isolated from the total social, economic, and medical environment in the countries involved. A given technique cannot be introduced into a culture as some sort of separate entity. Birth control involves questions of economics and philosophy that cannot be adequately judged by people from outside the culture in question. Those who live in industrialized countries need to reexamine their advocacy of mass distribution of birth control methods, including IUD's, sterilization, and the pill, in nonindustrialized countries.

Post-Fertilization Birth Control: Abortion

One method of preventing completion of a successful pregnancy is to artificially induce an abortion—the shedding of an embryo or fetus from the uterus prior to normal birth. Various methods of abortion are used at different stages in embryonic development. The issue of abortion has raised widespread and often heated philosophical, legal, and moral problems.

Birth control methods, at whatever point in the reproductive cycle they may be used, appear to have certain attendant risks. However, normal childbirth in itself involves certain risks for women. In comparing various methods of birth control in women, the disadvantages or dangers must always be weighed against the risks of becoming pregnant, as well as against the disadvantages and dangers of other methods of contraception. One meaningful comparison, obviously, is between the risks of using any particular birth control method and the risks involved in becoming pregnant. In many cases, the risks involved in birth control methods are far less than would be encountered on the average during pregnancy. This statement is not meant to suggest that pregnancy is a dangerous condition. It simply underscores the fact that pregnancy, like any situation that places stress on the human body, involves some risk. The vast majority of pregnancies, of course, lead to successful birth with no serious complications for the mother.

12.8 Male–Female Differences

Nowhere has the nature–nurture argument been more visible than in the question of whether male and female sex roles and patterns of behavior are innately or culturally determined. In the past, a large variety of behavioral and personality traits have been said to be innate in the male and female psyches. Men were supposed to be strong physically and emotionally, to seldom cry or show their inner feelings, and to be aggressive, competitive, and analytical. Women, on the other hand, were supposed to be physically and emotionally weaker than men, irrational, passive, and nonanalytical. In the late nineteenth century, anthropologists and biologists buttressed such ideas with measurements of skull size and shape. Some claimed that skull size and proportions made women behave more like children or the "savages" of Africa than like white European males.

It goes without saying that no one today accepts any relationship between skull size or shape and any kinds of intellectual or behavioral capacities. Yet the same type of argument resurfaces today in an only slightly modified form. Biologist Edward O. Wilson argues that women's position in society as passive homemaker and mother has probably been determined genetically and has been selected for during human evolution. As evidence for such a claim, Wilson cites a variety of animal species—not all of which, however, support his point. On the one hand, he points out that present-day anthropologists almost unanimously agree that in human societies the male is the dominant figure, even referring to this trait as "rampant machismo." Sheep, deer, antelopes, grouse, and elephant seals are a few of the species Wilson cites to support his contention that dominance and aggressiveness among males, coupled with passivity among females, is the general rule in nature.

On the other hand, however, Wilson also shows that in many other species, especially the "social insects" (bees, ants and wasps) and certain fishes, the female is the dominant sex; in many cases the males are virtual parasites serving only to inseminate the eggs. As Wilson himself points out, in some species there appear to be rigidly defined sex-role differences. In others, the sex-role differences are not so great, or are even reversed. The discontinuities and contradictions among the different species are simply too great to warrant any such generalizations as Wilson makes.

Many of the sex roles that biologists of the past have claimed might be innate are nothing more than socially conditioned stereotypes. Looking around in society today it is easy to see that most women, especially among those under 35, are not passive, irrational, or nonanalytical. Women have discovered that they are capable of doing other things well, if they choose, than occupying themselves at home, cooking, and caring for children. The personality traits traditionally assigned to women are largely stereotypes invented and advocated by men. Yet the women's movement has shown that women can defy the traditional stereotypes. If such behavior were genetically controlled, it should be difficult if not impossible to modify it easily. The fact is that over the past decade increasing numbers of women have begun to take their place in important positions in society, and the old stereotypes of "typical" female behavior are fading. In the face of such changes in actual behavior, it becomes increasingly difficult to maintain that specific sex-role patterns are innately ingrained in either men or women.

It is true, of course, that some biologically determined factors influence aspects of male and female sexual behavior. Factors that may influence sexual behavior include hormone differences, childbearing, and nursing. Hormonal fluctuation is considerably greater in the female body than in the male over any extended period of time. It has been amply demonstrated that hormonal changes can produce a variety of emotional responses in women. Some women become depressed just before or at the onset of menstruation. Others become hyperactive and restless around the time of ovulation. During pregnancy, mood changes have been correlated with physiological changes. There can be no doubt that some aspects of behavior are caused by such physiological elements. However, the kinds of behavior affected by hormonal levels are very generalized: depression, lethargy, irritability, etc. Specific behavior traits such as mechanical ability, competitiveness, or analytical skills do not appear to be governed by testosterone or estradiol.

Childbearing and nursing, as specifically mammalian traits, have had an important effect on differences in behavior patterns between males and females. Without question, pregnant women are less physically active, less agile, and tire more easily than men or nonpregnant women. And caring for a young child and nursing it requires intensive concentration of attention and energy. These facts have not been ignored in attempts to account for the origins of sex-role differentiation. Women have been said to be sedentary, passive and "homebound" because of the biological reality of childbearing and rearing. Unburdened by these biological realities, males have always been free to assume the aggressive role in society.

Such a scenario for the origin of sex-role differences sounds convincing. However, in opposition to it, anthropologists have pointed out that some human societies in the past have been "matriarchal"—dominated by women who performed many of the behavioral roles that in modern times have been taken over by males. In such societies, women do the farming or hunting and defend the community, while the men stay at home, cook, and look after all children except those still suckling. This suggests that there need be nothing innately determined about specific sex roles. Like other forms of behavior, it is largely learned within the very broad context of physical, physiological, and psychological makeup of the individual.

SUMMARY

1. The human male reproductive system consists of the penis, testes, seminal vesicles, ejaculatory ducts, and various fluid-producing glands. The testes serve two functions: production of sperm and secretion of the male sex hormone testosterone by special cells, the interstitial cells.
2. Testosterone is important in stimulating the development of the male sex organs during embryonic growth and the secondary sex characteristics at puberty. The general name for all male sex hormones is androgens.
3. The female reproductive system consists of the ovaries, Fallopian tubes, uterus, vagina, and clitoris. The clitoris is anatomically homologous to the male penis and functions as an organ of stimulation. The vagina is a receptacle for the penis during intercourse. The uterus contains the embryo during development and is the site of menstrual build-up every month. The Fallopian tubes connect the uterus with the ovaries which produce, between them, one egg per month. Like the testes, the ovaries are also endocrine glands, producing the female sex hormones collectively called estrogens.
4. Eggs are produced in the ovaries within structures called Graafian follicles. It takes a follicle about two weeks to complete the final stages of growth prior to release of the egg. After the egg is released, the old follicle becomes the site of production of the hormone progesterone. As an endocrine gland, the follicle is known as the corpus luteum.
5. The egg completes its meiotic divisions as it is passing down the Fallopian tube toward the uterus. As it makes this passage, the egg extrudes the third polar body. Fertilization, if it occurs, generally takes place in the Fallopian tubes.
6. Sexual intercourse generally involves sexual arousal by both partners followed by coitus, or insertion of the penis into the vagina. For both male and female, the basic responses involve: sexual arousal; increased sensitivity of the genitalia to touch; and orgasm, which in the male includes the ejaculation of sperm and in male and female an explosive feeling of release.
7. The estrus cycle in mammals consists of alternating periods of ovulation (during which the female will accept males sexually) and nonovulation (during which the female will not accept a male). In some higher primates and human beings, there is no strict estrus cycle, since the female will usually accept a man for intercourse at any time in her reproductive cycle. Human and anthropoid ape females have a 28- to 30-day menstrual cycle. During three or four days of this cycle, blood and tissue are given off as the menstrual flow.
8. During the first two weeks of the 28-day menstrual cycle, the egg is developing within the Graafian follicle, and the lining or endometrium of the uterus is building up a very rich capillary supply. If the egg is not fertilized, the endometrium breaks down, producing a loss of blood for approximately the last four days of the cycle. If the egg is fertilized, the

uterine lining does not break down and becomes the place where the young embryo implants.

9. The menstrual cycle is under hormonal and neural control. The master endocrine gland, the pituitary, is under neural control of the hypothalamus. The pituitary secretes two gonadotrophic hormones in both men and women: follicle-stimulating hormone (FSH) and luteinizing hormone (LH). In females, FSH stimulates the Graafian follicles to grow and the egg to mature. FSH also triggers the follicular cells to secrete estrogens (especially estradiol), which cause the uterine lining to start building up. High levels of estradiol cause the hypothalamus to inhibit secretion of FSH by the pituitary. As the level of FSH falls, ovulation is stimulated around the fourteenth day. Meanwhile, the pituitary is signaled by the hypothalamus to start secreting LH, which triggers the conversion of the follicle into the corpus luteum. The corpus luteum begins to secrete a hormone of its own, progesterone, which maintains the uterine lining. The progesterone level reaches its peak by about the twentieth day of the cycle, after which it declines. The decline is caused by a decline in the secretion of LH by the pituitary. By the twenty-eighth day of the cycle the progesterone level has fallen to such an extent that the uterine lining can no longer be maintained and begins to be shed, forming the menstrual flow. From the twentieth day or so onward, the pituitary begins secretion of FSH, which starts the cycle all over again.

10. The menstrual cycle shows all the characteristics of a negative feedback (feedback inhibition) process. FSH stimulates growth of the follicle, which produces estrogens; a high level of estrogen in the blood causes the hypothalamus to inhibit secretion of FSH by the pituitary. LH stimulates production of progesterone by the corpus luteum; a high level of progesterone in the blood causes the hypothalamus to inhibit secretion of LH by the pituitary.

11. The menstrual cycle shows a long-term (28-day) oscillation compared to the more usual short-term oscillation of a few minutes to a few hours for most other negative feedback systems. The long-term response of this system is a result of the fact that it does not begin to release neurotransmitter to the pituitary until well after the levels of estrogen or progesterone have begun to fall.

12. If fertilization occurs, implantation of the embryo in the uterine wall serves to keep the uterine lining from breaking down. Soon the embryo develops its organ of nourishment, the placenta, through which nutrients and waste are exchanged with the mother's blood. The placenta is also an endocrine organ, secreting estrogen and progesterone. These hormones negatively inhibit FSH and LH oscillation and serve to maintain the uterine lining throughout pregnancy (approximately 266 days, or 9 months).

13. The period of human gestation, or pregnancy, is divided into three periods, or trimesters, of approximately three months each.
    a) *First trimester:* The embryo becomes enclosed in its various membranes, becoming the trophoblast, by the end of the first month. The primitive streak appears, marking the anterior–posterior axis. The nervous system appears, as do the muscle segments and the rudimentary sense organs and the gill slits. The germ cells begin differentiation by the end of the first month, and by the middle of the second they migrate to form the future testes or ovaries. The bones become visible in outline during the second month, and the head shows rapid growth. During the first trimester, the future child is referred to as an embryo.

b) *Second trimester:* The brain begins to form by the fourth month, and by the end of the second trimester the future child looks like a miniature human being, with all the major body parts differentiated. From the fourth month on it is called a fetus.

c) *Third trimester:* The final stage is characterized largely by growth of the already-differentiated parts. As it reaches the end of term, the fetus "drops" into position within the uterus; the head is usually pointed downward and is normally the first part to emerge.

14. Birth is initiated by changes within the placenta, whose physiological activities begin slowing down from the seventh month onward. Hormonal and mechanical changes (such as crowding within the uterus) also appear to influence the onset of birth. Labor pains are contractions of smooth muscles in the uterine wall prior to birth. During labor, contractions of the uterine muscles become increasingly stronger and more frequent. The cervix also dilates to allow passage of the fetus.

15. Birth control methods are procedures that prevent the birth of a child. The methods can be grouped into two categories: those that prevent conception and those that prevent development of the fertilized zygote:

a) *Preventing Conception (Contraception):*

- Celibacy: avoidance of sexual intercourse.
- Rhythm method: periodic abstinence from intercourse around the midpoint between two menstrual flows (ovulation).
- Coitus interruptus (withdrawal): removal of the penis from the vagina prior to ejaculation.
- Mechanical device for the male (condom): a sheath over the end of the penis that prevents sperm from being released into the vagina.
- Mechanical device for the female (diaphragm): a rubber or plastic barrier placed at the cervix, mechanically blocking the passage of sperm into the uterus. It is usually used with spermidical jelly around the rim.
- Physiological device (the female birth control pill): an oral contraceptive that contains high levels of estrogen and progesterone, mimicking the hormonal levels characteristic of pregnancy. The pill prevents ovulation but allows menstruation to occur every fourth week.
- Physiological device (the male birth control pill): an oral contraceptive which, if taken
- regularly for eight weeks, suppresses spermatogenesis in men.
- Surgical methods (tubal ligation for the female, vasectomy for the male).

b) *Preventing Normal Development (Post-Fertilization Methods):*

- The "morning after" pill: a pill containing massive concentrations of estrogen producing an overgrowth of the endometrium. This overgrowth prevents implantation. The pill is taken immediately after intercourse.
- Intrauterine devices: rings, coils, loops, or other small physical objects that can be embedded lightly in the uterine wall, mimicking implantation of the embryo. This process prevents the implantation of further zygotes but allows normal ovulation and menstruation to occur.
- Abortion: artificially inducing expulsion of the embryo or fetus before it reaches full term.

16. While certain structures and functions of the male and female reproductive systems are biologically quite distinct, sexual behavior and attitudes appear to be determined by social rather than biological causes. Such aspects of sexual behavior as sex roles, sex identity, and the like also appear to be the result of social rather than biological development.

EXERCISES

1. How does the structure of the male and female reproductive system in humans work to ensure maximum chances of fertilization of the egg?
2. Compare and contrast oögenesis and spermatogenesis. Describe the divisional aspects of the processes and the final differentiation of each gamete type.
3. In terms of their structures and functions, compare and contrast the ovaries and testes.
4. On what evidence shown in Fig. 12.10 might the conclusion be based that high levels of estrogen *cause* FSH secretion by the pituitary to slow down?
5. Describe how the relationships between ovarian hormones and the pituitary–hypothalamus centers act as a negative feedback inhibitor.
6. Explain how the negative feedback system of the menstrual cycle is structured so as to produce a very long-lasting cycle.
7. What is the function of the embryonic membranes?
8. What advantages result from the pattern of intrauterine development? What disadvantages?
9. Women often suffer "menstrual cramps" just before or at the onset of menstruation. On the basis of the changes the uterine lining undergoes about this time, what might be the psychological and physiological events associated with cramps?
10. In many mammalian species that regularly have two or more offspring at a time, each fetus normally has its own placenta. Occasionally, however, two fetuses share the same placenta, which means that the circulatory systems of the fetuses are directly interconnected. If the twins are of opposite sex, the female always turns out to be a sterile form called a freemartin, while the male is sexually normal. Given the fact that in male embryos the production of testosterone by embryonic interstitial cells occurs earlier than production of estrogens in female embryos, how can you explain the freemartin case?
11. For what reasons might it be that the most concentrated research on birth control has always focused on devices and procedures for women rather than for men?
12. Many groups have lobbied aggressively against legalization of abortion on the grounds that it is murder. Other groups have worked for legalizing abortion, claiming that it is the right of a woman to decide whether she wants a child or not and that unwanted children create many problems for families. Opponents of abortion claim that the unborn fetus has rights too, and that no one should be allowed to decide whether it is to be born or not.
    a) Describe the ways in which the antiabortion argument displays elements of mechanistic philosophy (see Chapter 1).
    b) What is the central issue about which the two sides disagree? How might differences on that issue be resolved?
    c) Why is it difficult to derive definitive, all-encompassing answers to issues such as abortion?

SUGGESTED READINGS

As is the case with many relatively new areas of research, the literature on human reproduction, development and sexuality is plentiful and highly diverse in terms of both content and quality. We have listed here sources of a more standard nature. For the interested student, extensive bibliographies and information about other sources is available from the Sex Information and Education Council of the United States (SIECUS), 122 East 42d St., New York, N.Y. 10017.

Human Reproductive Physiology

Katchadourian, H. A., and D. T. Lunde, *Fundamentals of Human Sexuality.* New York: Holt, Rinehart and Winston, 1972. An excellent, clearly written book discussing many aspects of human reproduction from physiology to psychology.

Page, E. W., C. A. Villee, and D. B. Villee, *Human Reproduction.* Philadelphia: W. B. Saunders, 1972. A comprehensive volume that straightforwardly discusses all aspects of reproduction and sexual development.

Human Embryology

Patten, B. M., *Human Embryology,* 3d ed. New York: McGraw-Hill, 1968. A standard, quite thorough textbook on human embryology. Profusely illustrated, it serves as an extremely useful reference on all aspects of human embryonic development.

Rugh, Roberts, and Landrum B. Shettles, *From Conception to Birth.* New York: Harper and Row, 1971. This well-written, simple, and beautifully illustrated work is a most valuable source of information on human embryology and birth. There is virtually no other book available that assumes no previous knowledge of biology yet is so authoritative and interesting.

Human Sexuality

Churchill, Wainwright, *Homosexual Behavior Among Males.* Englewood Cliffs, N.J.: Prentice-Hall, 1971. A cross-cultural and cross-species account of the prevalence of homosexuality and various attitudes toward it. This book is a well-documented statement of the theory that homosexuality is not a sickness.

Goldberg, Steven, *The Inevitability of Patriarchy.* New York: William Morrow, 1973. A perfect example of an argument based on biological determinism, this well-written book maintains that the arguments of women's liberation run counter to biological reality. The author, a sociologist, is to the "innateness" theory of male domination what Robert Ardrey is to the "innateness" theory of human aggression.

Kaich, Dolores, *Woman Plus Woman.* New York: Simon and Schuster, 1975. Also printed in paperback by William Morrow, Inc., New York.

Masters, W. H., and V. E. Johnson, *Human Sexual Response.* Boston: Little, Brown, 1966. Aside from Kinsey's reports of 1948 and 1953, the work of Masters and Johnson is the most thorough-going and scientific account of human sexual behavior in the past twenty-five years. Their conclusions about every aspect of sexual behavior are well documented and revealing.

Pengelley, Eric T., *Sex and Human Life,* 2d ed. Reading, Mass.: Addison-Wesley, 1978. A simply written and concise general textbook of human reproduction, development, and sexuality.

THIRTEEN

THE ORIGIN OF LIFE AND THE EVOLUTION OF PROCARYOTIC AND EUCARYOTIC CELLS

13.1 Introduction

The origin of life is a problem of recurring interest in the history of biological thought. From the time of Aristotle to the present it has occupied the attention of the world's most prominent biologists. How did living material come into being in the first place? When and where did it originate? Did it originate only once, or have repeated origins taken place? Such questions as these have received a variety of answers in the past. One type of answer has been a religious one: special creation (see Section 14.2). Historically, this represents an early explanation to be advanced concerning the origin of life. A second type of answer is the idea of spontaneous generation (see Section 13.2). This idea, advocated by Aristotle, was accepted by most people, including some eminent naturalists, up until the middle of the nineteenth century. A third type of answer makes reference to the idea of **chemosynthesis.** Chemosynthesis attempts to explain the origin of life in terms of present-day chemical and physical laws. We shall examine the second and third of these ideas in some detail.

13.2 Spontaneous Generation

While the idea of spontaneous generation may seem farfetched to us today, it is not difficult to see why it gained many adherents. It is a commonsense sort of theory. If meat is left outside for a few days, maggots appear on and within it. A small pond is seen to have no life in it one day and to be filled with swimming tadpoles the next. For many years it was not known that maggots are larval flies, or that frogs lay eggs that develop into tadpoles. During those years, spontaneous generation seemed a plausible explanation for the sudden appearance of such organisms.

Yet the theory of spontaneous generation had weaknesses. It seemed to be applicable to only a

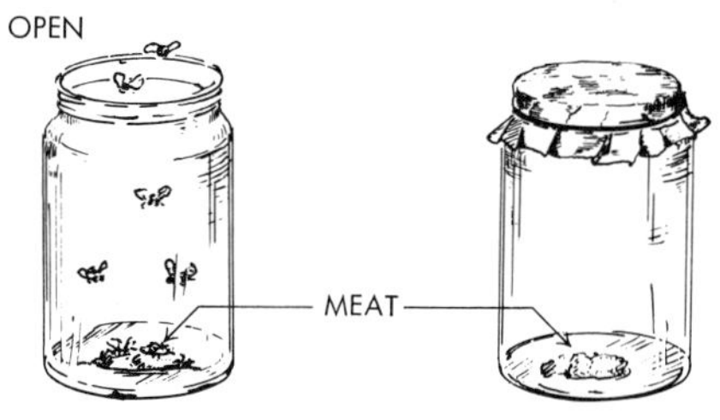

Fig. 13.1 *Redi's experiment. The left-hand jar represents the control group, the right-hand one the experimental group. When flies are prevented from coming back into contact with the meat, no maggots develop.*

few types of organisms. Although the late sixteenth-century alchemist Jan Baptista van Helmont (1577–1644) claimed that mice could be spontaneously generated from rags and bread in a dark corner, most of his contemporaries did not go that far. It was generally held that only lower forms of life—insects, tadpoles, certain kinds of eels—could be spontaneously generated. At the same time, it was recognized that chickens develop from eggs and calves from cows. Thus, for reasons within the nature of the theory itself, the idea of spontaneous generation became less acceptable as an effective explanation for the origin of life. Investigators turned to experimentation to test the hypothesis. And the experiments of Francesco Redi (1626–1697) and Louis Pasteur (1802–1895) provided conclusive evidence to refute the theory of spontaneous generation.

In 1668, Redi set up a very simple but conclusive experiment (Fig. 13.1). He placed samples of meat in two separate jars, one covered with foil, the other left open. The open flask served as the control, the covered flask as the experimental group.

Redi reasoned that if maggots arise by spontaneous generation, then they should appear on meat in both the covered and uncovered flasks. Redi observed flies hovering about both flasks. But even after several days, contrary to the prediction, maggots appeared only in the open flask. Hence the hypothesis of spontaneous generation, at least in this case, was invalid. (Recall from the Truth Table in Chapter 1 that if the prediction from an hypothesis is shown to be incorrect, the hypothesis must be incorrect.)

Despite Redi's work, many people continued to believe in spontaneous generation. In the latter half of the nineteenth century the French bacteriologist Louis Pasteur was able to demonstrate convincingly that another supposed example of spontaneous generation (microbes in decaying broth) was also invalid.

Pasteur began his experiments with the working hypothesis that all life comes from preexisting life (the concept of **biogenesis**). He placed beef broth in several flasks and boiled it. Boiling sterilized the broth. A day or two later, flasks that had been left open and allowed to cool could be seen to teem with microorganisms. Yet even months later, no microorganisms had appeared in the flasks that had been sealed immediately after boiling.

However, this demonstration was not enough to convince the proponents of spontaneous generation. They argued that boiling had altered the air in the flask so that spontaneous generation could not occur.

Pasteur met this argument by a new approach. He reasoned that if spontaneous generation requires the contact of the broths and fresh air, then boiled broth brought into contact with air cleared of all its floating particles should still generate microorganisms. To test this hypothesis, Pasteur did the following experiment. He boiled the broth as before. This time, however, instead of sealing the flask, he drew the neck out into a long, S-shaped

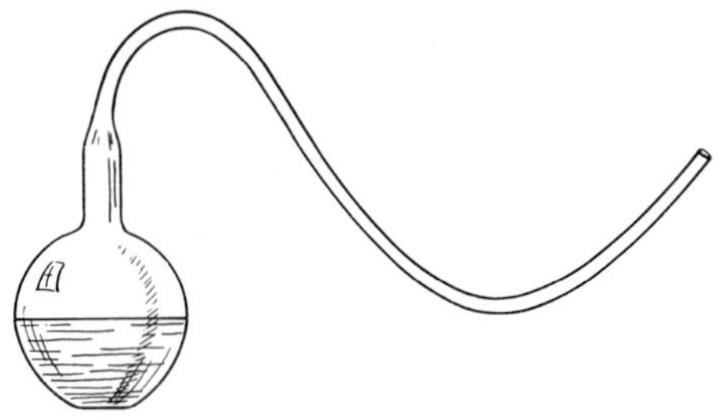

Fig. 13.2 *S-shaped flask prepared by Pasteur to show that spontaneous generation does not occur. Dust-borne microorganisms and bacterial spores are caught in the lower band and thus do not reach the broth. However, air is free to pass down the neck of the flask.*

tube (Fig. 13.2). Air would still be able to pass down the entire neck of the tube and come into contact with the broth, but the lower bend in the tube would serve as a trap for all dust particles and bacterial spores.

The experimental results showed that broth in the flasks with S-shaped necks did not decay. On examination, even months afterwards, there was no sign of microorganisms. Several of Pasteur's flasks are still, over 100 years later, on display in the Pasteur Institute in Paris—and their contents show no signs of decay!*

To make his point more effectively, Pasteur then took an S-shaped flask that had been sitting for several months and tipped it so that some broth ran into the lower bend of the neck. He then allowed the broth to flow back into the main part of the flask. Soon the broth in this flask showed signs of putrefaction. This seemed to be dramatic evidence that for spontaneous generation to occur, something more was needed than contact between broth and air. It appeared more likely that the dust particles (containing bacteria) and microorganism spores carried in the air caused the decay of broth.

Pasteur's experiments did not disprove the idea of spontaneous generation *in general;* he did not prove that such a process *never* occurs. Recall from Chapter 1 that such proof is impossible in science.

* They have been resealed to prevent evaporation. Pasteur did his original experiments in 1862.

What Pasteur did do was to show that one supposed example of spontaneous generation was invalid. However, this example was so widely accepted by proponents of the theory that Pasteur's work dealt a deathblow to the concept of spontaneous generation. Ironically, by showing that a common example of spontaneous generation was invalid, Pasteur banished the question of the origin of life from scientific investigation.

In more recent times a way out of this dilemma has been found. Chemosynthetic theories of the origin of life advocate a new variety of spontaneous generation. By claiming that organisms could arise from the nonliving environment through chemical means, they place the question of the origin of life on a rational, scientific basis. Let us examine how chemosynthesis provides an explanation for the origin and development of living organisms.

13.3 The Origin of Life by Chemosynthesis

Geological evidence has been interpreted to suggest that the early atmosphere of the earth may have contained water vapor, methane, ammonia, and perhaps some free hydrogen. These compounds provide the most essential elements found today in living organisms: carbon, hydrogen, oxygen, and nitrogen. In addition, phosphates of various sorts may have been present in the oceans. The presence of these elements set the stage for the origin of life.

The First Phase

The chemosynthetic hypothesis holds that proteins, carbohydrates, fats, and nucleic acids were formed from methane, ammonia, water, and hydrogen in the earth's atmosphere or on the surface of the oceans. The energy for these reactions is thought to have been provided by lightning and/or ultraviolet and cosmic radiation.

Probably the first molecules of any size to be synthesized were similar to present hydrocarbon chains (precursors of fatty acids); these can fold back on themselves to form ring structures. Also formed were simple amino acids, carbohydrates, and perhaps some small peptides. Probably only relatively small molecules accumulated in the oceans because the same energy used to build molecules could also be used to break them down.

Evidence: Stanley Miller, in the early 1950's, set up an apparatus like that shown in Fig. 13.3. By boiling and condensation, the ammonia, methane, hydrogen, and water were kept circulating in the apparatus. Two electrodes provided a periodic electric discharge into another flask. In the residue collected were several amino acids, one carbohydrate (succinic acid), and several other organic compounds. Other workers using Miller's procedure have produced simple components of nucleic acid and even ATP by varying the type and amounts of reactants used.

The Second Phase

The relatively small molecules were probably joined together to produce macromolecules: amino acids formed peptides; glucose and other carbohydrate units formed large sugar and starch molecules; hydrocarbon chains could have united with three-carbon sugars to yield primitive fat molecules. Possibly heat provided the energy needed to drive the reactions.

Evidence: Sidney Fox heated mixtures of amino acids for varying periods of time and found that dipeptides and even long-chain peptides were produced. These proteins were similar to those produced by living organisms. Cyril Ponnamperuma showed that sound energy (volcanic eruptions), the formation and collapse of bubbles in mud, and

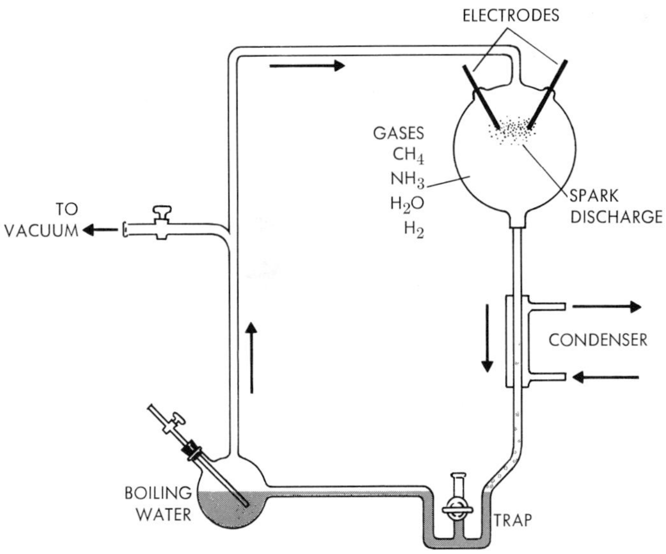

Fig. 13.3 *Apparatus designed by Stanley Miller for the circulation of methane, ammonia, water vapor, and hydrogen. Water is boiled in the flask at lower left. The products of chemical reaction are collected in the trap at lower right. Energy for chemical reaction comes from an electric discharge in the flask at upper right.*

high-energy radiation (X-rays, ultraviolet rays) could also generate complex molecules. Experiments showed that even under prebiotic conditions, amino acids do not join together purely randomly. Some order is involved. Furthermore, the polypeptides formed have a tendency to assume definite shapes (tertiary structure).

The Third Phase

A. I. Oparin proposed the coacervate theory: proteinlike substances in the early broth formed aggregates that tended to develop a simple membrane around them due to surface tension. It is possible that fatty acids may have collected at the surface (due to their polarity) and thus formed a primitive lipid membrane. These coacervates could have been similar to the precursors of the first living organisms.

Evidence: Oparin showed that mixtures of gelatin and gum arabic produced small spherules of these substances. Fox showed that thermally produced proteins form spherules similar in size to bacterial cells, and these were separated from the external medium by a surface layer. The spherules showed osmotic properties, and were capable of combining or splitting apart (Fig. 13.4).

The Fourth Phase

Nucleic acids became the major molecular "organizers" within aggregates, providing reproductive continuity and directing the immediate activities of the complex. Energy-capturing systems could be-

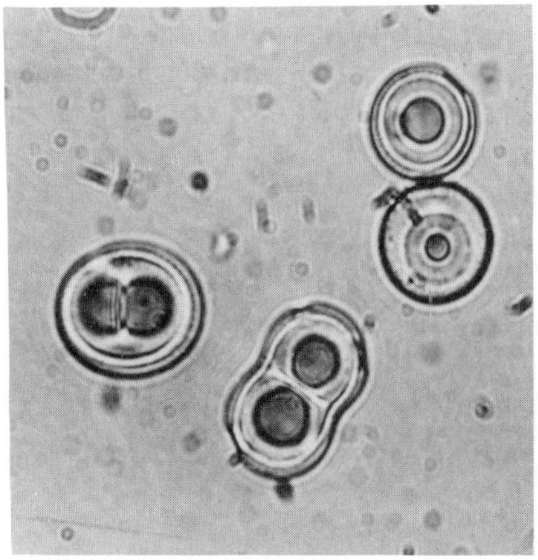

(a)

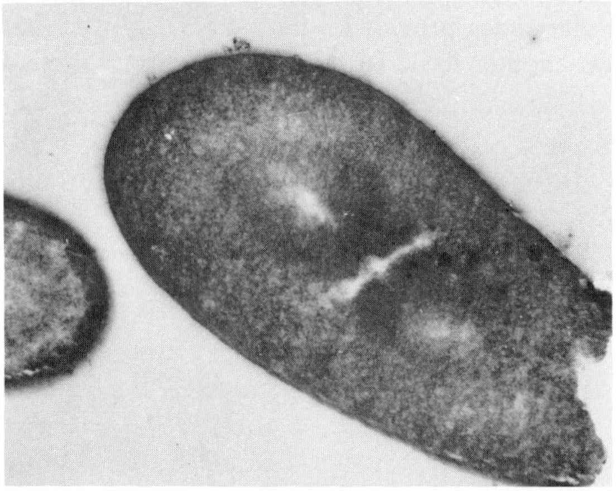

(b)

Fig. 13.4 *(a) Photograph showing both single and aggregate microspheres produced in Fox's laboratory at Florida State University. Note the regular appearance of the spheres. In general they resemble spherical bacteria. It is thought that such aggregates (mostly of protein) could represent the beginnings of organization that eventually led to life. (b) Proteinoid microspheres produced by Dr. Fox. Note the close resemblance to the cellular structure shown by some bacteria. (Photos courtesy Professor Sidney W. Fox.)*

come more efficient as they became an hereditary part of the molecular aggregates. By the end of this phase, the molecular aggregates had become what would be generally accepted as true living organisms.

The Fifth Phase

Evolutionary development began. Once continuity in the form of genetic control was introduced, natural selection (see Chapter 14) could come into operation. No doubt one of the first results of evolution was the increased efficiency in capturing energy from carbohydrate breakdown. It was probably at this time that the processes of electron transfer and storage of energy in phosphate bonds first appeared. The efficiency with which organisms could use energy sources increased, leading to an increase in their reproductive capacity. The number of organisms in the early oceans began to rise, and competition became more intense.

The Sixth Phase

Up to now, organisms were heterotrophs (dependent on an outside source of ready-made food); the biotic world had been running downhill because as time went on, the carbohydrate supply diminished. Competition ensued, and variations in the direction of carbohydrate manufacture using light energy were greatly favored. Thus the process of photosynthesis evolved and autotrophs appeared.

The first autotrophs may have resembled certain present-day bacteria that can use sunlight to produce carbohydrates. The appearance of photosynthetic organisms provided the biotic community with a balance that has remained ever since—while heterotrophs break down carbohydrates, autotrophs build them up.

Any hypotheses about the origin of life are at best tentative and subject to continual and frequent revision. Indeed, a fundamental aspect of the explanation outlined above has been called into question, for new geological evidence suggests to some workers that the chemical composition of the earth's early atmosphere contained very little ammonia or methane. Since ammonia is known to dissolve very rapidly in water, the presence of free ammonia in the atmosphere would be very unlikely. A new hypothesis proposes that a number of elements necessary for life (carbon, hydrogen, and nitrogen) were spewed forth into the atmosphere by volcanoes. Recent studies suggest that molecules such as amino acids can be produced by bombarding these elements with sunlight of a specific wavelength (2536 A). At the present time there is no crucial test to decide which of these two hypotheses better explains how life actually did arise on the earth. New evidence will have to come from studies on the chemical composition of the earth's oldest rocks. From a detailed knowledge of their composition, it may be possible to deduce what the early atmosphere of the earth was really like.

Perhaps the most difficult problem encountered by the chemosynthetic theory is to account for the transition from molecular aggregates lacking genetic control to organized forms having this control. How likely is it that unorganized molecular groups could form some type of spontaneous organization? How likely is it that nucleic acid could become associated with these molecules in a central role?

Oparin holds that the origin of life was neither a lucky accident nor a miracle, but the result of perfectly natural and ordinary scientific laws. In other words, given the conditions that existed on earth at the time, life was bound to arise and to evolve along the lines that it has. The basis of the development of life is the result of the development of hydrocarbons, and Oparin believes that the gradual evolution in the complexity of hydrocarbons is universal. There is some geological evidence to support this claim: the older the rocks, the simpler the hydrocarbons.

In Oparin's opinion, the problem of the "primal soup"—the sea containing dissolved organic substances, including simple polymers of amino acids and nucleic acids—has now been largely solved by the innumerable experiments of the sort performed by Stanley Miller. Such systems, however, do not exhibit any great degree of organization. To Oparin, a more difficult question is how *enzymes* first came into existence, since an enzyme's structure is adapted in a very specific way to its particular catalytic task. This appearance of adaptiveness in the simple aqueous solution of the primeval soup *before* the formation of such whole systems seems entirely improbable.

As seen earlier, to Oparin the best model to simulate the evolution of enzymes is the coacervate drop, an isolated part of the primeval soup separated from the solution by surface boundaries but able to interact with the environment. In one set of experiments, Oparin has made coacervate drops in his laboratory by polymerizing the nucleotide adenine in the presence or parallel formation of another polymer, usually histone. As soon as a certain degree of polymerization was reached, coacervate drops containing these polymers began to form. Once formed, the drops selectively absorbed and concentrated small molecules from the external medium and changed and incorporated them once they were inside the drop. Oparin refers to such coacervate drops as "protobionts." Under the influence of wave action, these protobionts can be broken up into "daughter" droplets, each with the attributes of its "parents." Once this has happened, of course, a sort of prebiological natural selection is free to operate, with the more "successful" drops (those with a greater catalytic ability to incorporate their surroundings) growing at the expense of the less successful.

Catalysts, which could only have come from the environment, thus become the telling requirement. The first of the catalysts must have been rather inefficient, like the industrial catalysts of today. However, through modification by the addition of organic molecules, their catalytic activity could have been improved a thousandfold. An enormous number of molecular alternatives must have arisen, of which only the most efficient remained. These may possibly be today's coenzymes, used now only as an aid, albeit a vitally important aid, to an enzyme's action.

Meanwhile, the random joining of amino acids could have led to primitive proteins with some form of active center. These would soon disappear, however, unless a further controlling mechanism evolved that could maintain their structure. Even at these early stages the polynucleotides, later to become DNA and RNA, might have had some organizing ability and, if they produced favorable catalysts, would have been selected. It is reasonable to assume that many millions of primitive proteins must have arisen; presumably the most successful remains today.

Oparin maintains that all these processes are subject to experimental testing despite the fact that we have no time machine to return and check them "in the field," and his laboratory is still actively engaged in research along these lines. Of course, such experiments can never *exactly* duplicate the way things actually must have happened when life first appeared on earth. But even approximate experiments are extremely helpful.

13.4 The Geological Time Scale

Establishing a date for such events as the origin of life is quite a difficult task. The earliest types of true organisms have left little or no fossil records. The earliest traces of life are the remains of primitive plants in strata that are at least three billion years old. Life must have originated at some time considerably before that, but to find out just when, we must find some evidence in the geological record. This is very difficult, so our estimates of when life may have originated are highly speculative.

Once fossil remains or even traces of life begin to appear in the strata, the history of life can be much more easily put on a definite time scale. Geological dating procedures make it possible to establish the age of various fossils and strata with a fair degree of certainty. One of the most prominent methods currently in use is based on the rate of **radioactive decay** of the isotopes of certain elements—for example, U^{238}, an isotope of uranium. Radioactive elements undergo transmutation to more stable forms by emitting subatomic particles, such as β-particles or neutrons. They may also emit radiant energy in the form of gamma rays. This "decay" procedure continues until all the atoms of a given sample have been transmuted to the stable form. For example, atoms of U^{238} undergo radioactive emission and decay ultimately to the stable form, Pb^{206} (lead).

The critical feature of radioactive decay, so far as dating procedures are concerned, is that it occurs at a steady, measurable rate. The rate differs for each isotope of each element. Yet the process is so regular that, for example, we can say it takes 4.5 billion years for half of the atoms in a sample of U^{238} to become Pb^{206}, or that it takes 15 days for half the atoms in a sample of P^{32} to decay to P^{30}. This period is termed the **half-life** of the isotope.

Many rock samples can be accurately dated by comparing the ratio of U^{238} to Pb^{206} contained in them. U^{238} is a relatively common isotope and is found in many rocks and fossils. The basic assumption of this method is that the older the rock, the less U^{238} it will contain and the more Pb^{206}. The oldest rocks known are estimated by this method to be more than 3.2 billion years old. The age of the earth is estimated to be between four and five billion years. A general scheme of geological time, with indications of the history of life in each period, is shown in Table 13.1.

For any accurate idea about the appearance of true cells or living systems on earth, it is necessary to have fossil remains. Unfortunately, cells do not make the best fossils, since they have no hard parts (like shells or bones) that can be preserved easily. The problem is even more acute for flimsy associations of macromolecules like coacervates. In the past two decades, however, a field known as micropaleontology has shown remarkable progress. By very special techniques, light and electron microscopy have been applied to the study of thin sections from sedimentary rock. These sections show many very primitive cells whose impression has been left in mud now turned to rock. Several remarkably clear micrographs of "fossil" cells are shown in Fig. 13.5. They come from the so-called

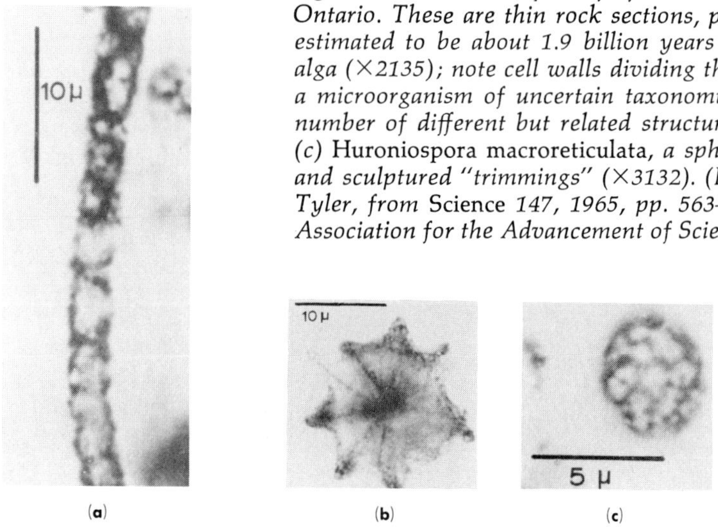

Fig. 13.5 *Three examples of "fossil" microorganisms from the Gunflint formation, Ontario. These are thin rock sections, photographed in transmitted light, of specimens estimated to be about 1.9 billion years old. (a)* Gunflintia grandis, *a filamentous alga (×2135); note cell walls dividing the filament into units. (b)* Kakabekia umbellata, *a microorganism of uncertain taxonomic relationships. This organism appears in a number of different but related structural forms in the Gunflint formation (×1763). (c)* Huroniospora macroreticulata, *a spheroidal, sporelike body showing a thick wall and sculptured "trimmings" (×3132). (Photos courtesy E. S. Barghoorn and S. A. Tyler, from* Science *147, 1965, pp. 563–575. Copyright © 1965 by the American Association for the Advancement of Science.)*

Table 13.1
Kinds of Life in Various Geological Eras*

Eras†	Periods†	Epochs	Aquatic life	Terrestrial life
Cenozoic 63 ± 2	*Quaternary* 0.5–3	*Recent* *Pleistocene*	*Periodic glaciation*	Humans in the new world First humans
	Tertiary 63 ± 2	*Pliocene* *Miocene* *Oligocene* *Eocene* *Paleocene*	All modern groups present	Hominids and pongids Monkeys and ancestors of apes Adaptive radiation of birds Modern mammals and herbaceous angiosperms
Mesozoic 230 ± 10	*Cretaceous* 135 ± 5		*Mountain building (e.g., Rockies, Andes) at end of period* Modern bony fishes extinction of ammonites, plesiosaurs, ichthyosaurs	Extinction of dinosaurs, pterosaurs Rise of woody angiosperms, snakes
	Jurassic 180 ± 5		*Inland seas* Plesiosaurs, ichthyosaurs abundant Ammonites again abundant Skates, rays, and bony fishes abundant	Dinosaurs dominant First lizards: *Archeopteryx* Insects abundant First mammals, first angiosperms
	Triassic 230 ± 10		*Warm climate, many deserts* First plesiosaurs, ichthyosaurs Ammonites abundant at first Rise of bony fishes	Adaptive radiation of reptiles (the codonts, therapsids, turtles, crocodiles, first dinosaurs, rhynchocephalians)
Paleozoic 600 ± 50	*Permian* 280 ± 10		*Appalachian Mountains formed, periodic glaciation and arid climate* Extinction of trilobites, placoderms	Reptiles abundant (cotylosaurs, pelycosaurs); cycads and conifers; gingkoes
	Pennsylvanian 310 ± 10	*Carboniferous*	*Warm humid climate* Ammonites, bony fishes	First insects, centipedes First reptiles; coal swamps
	Mississippian 345 ± 10	*Carboniferous*	*Warm humid climate* Adaptive radiation of sharks	Forests of lycopods, club mosses, and seed ferns Amphibians abundant, land snails

	Devonian 405 ± 10	*Periodic aridity* Placoderms, cartilaginous and bony fishes Ammonites, nautiloids	Forests of lycopods and club mosses, ferns, first gymnosperms Millipedes, spiders, first amphibians
	Silurian 425 ± 10	*Extensive inland seas* Adaptive radiation of ostracoderms Eurypterids	First land plants Arachnids (scorpions)
	Ordovician 500 ± 10	*Mild climate, inland seas* First vertebrates (ostracoderms) Nautiloids, Pilina, other mollusks Trilobites abundant	None
	Cambrian 600 ± 50	*Mild climate, inland seas* Trilobites dominant First eurypterids, crustaceans Mollusks, echinoderms Sponges, cnidarians, annelids Tunicates	None
Precambrian 4600		*Periodic glaciation* Fossils rare but many protistan and invertebrate phyla present First bacteria and blue-green algae	None

* From John W. Kimball, *Biology*, 2d ed. Reading, Mass.: Addison-Wesley, 1968.
† With approximate starting dates in millions of years ago.

"Gunflint" deposits in Ontario and are approximately 1.9 billion years old. In older rocks, very primitive bacteria-like structures have been dated at about 3.1 billion years.

Radioactive dating techniques indicate that the rocks in which the earliest fossil cells have been found are at least 2 billion years old. If cells existed that long ago, the events that formed the necessary foundation (roughly phases 1 through 4) must have been under way long before that. Even a conservative estimate could well set the formation of molecular aggregate at 3 to 4 billion years ago! Life has been developing on earth for a very long time.

A measurement scale based on periods of time like billions or millions of years is difficult to imagine. We can compare the whole history of life on earth to a 24-hr scale, the appearance of various forms being designated in terms of "time of day." If we set the origin of life at one minute after midnight, the first fossils do not begin to appear until 6 P.M. Even more striking, the age of mammals begins at 11 P.M., and humans appear at 11.59. The short history of our species on earth is some indication of the vast time over which the origin of life and evolution to its present stage are thought to have occurred.

13.5
Evolution of the Procaryotes

Many biologists consider the two procaryote groups, the bacteria and the blue-green algae, to be closely related. This knowledge does not, however, indicate their order of descent. Trying to work out precise evolutionary relationships is one of the problems that evolutionary biologists face. In this case, for example, there are two alternative hypotheses. The first is that if the bacteria and the blue-green algae evolved from a common ancestor, then they should possess many features in common. In recent years, this prediction has been shown to be true. The electron microscope reveals that the cell structure of these two groups of organisms is unique, differing markedly from that of all other organisms; this unique cell organization is called procaryotic (see Section 4.5). In addition, the cells of these organisms contain cytoplasm without vacuoles. The cytoplasm is resistant to heat, drying out, and osmotic shock to a degree not present in other organisms. Finally, the cell walls of organisms in both groups are constructed of a polysaccharide–amino acid long-chain macromolecule known as mucopeptide. The possession of so many features in common certainly adds support to a hypothesis that the bacteria and the blue-green

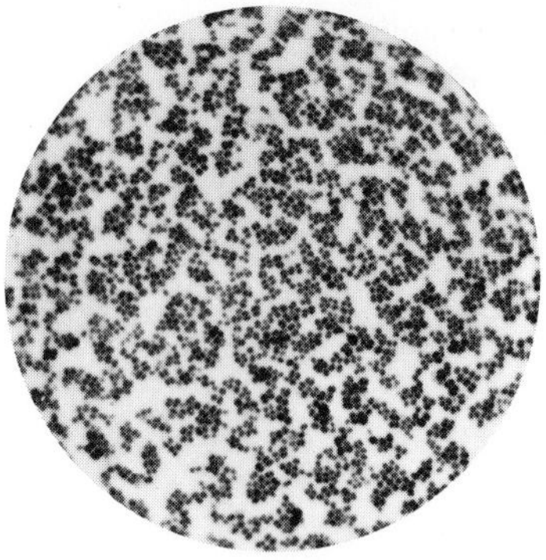

(a)

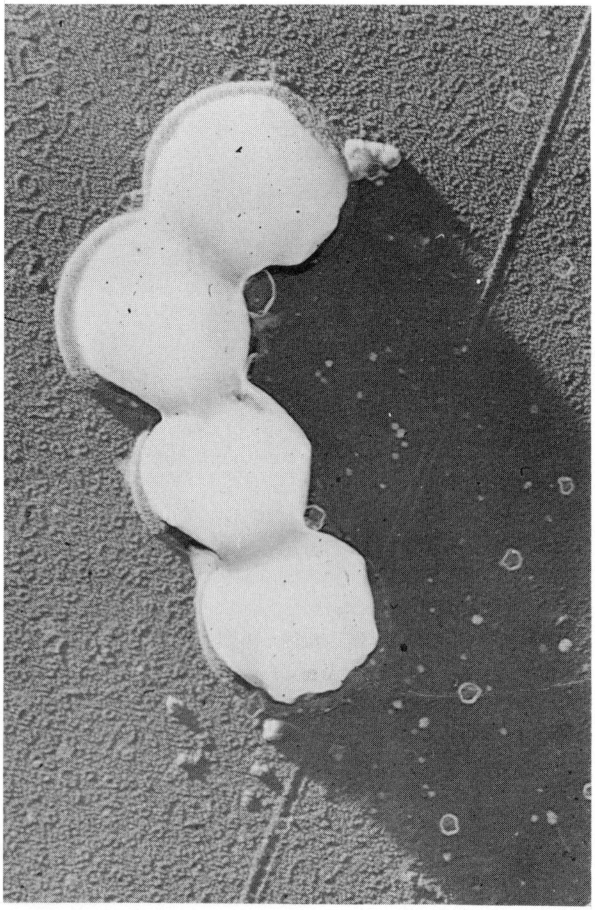

(a)

Fig. 13.6 *Light and electron micrographs of the three morphological types of bacteria. (a) Spherical or coccus form (top, light micrograph of* Micrococcus *×1000; bottom, electron micrograph of* Staphylococcus aureus *×55,000). (b) Rod-shaped or bacillus form (top, light micrograph of* Azotobacter *×1000; bottom, electron micrograph of* Bacillus subtilis *×80,000); (c) Spiral or spirillum form (top, light micrograph of* Spirillus volutans *×1232; bottom, electron micrograph of* Aquaspirillum bengal *×7700. [Light micrographs courtesy Turtox/Cambosco, Macmillan Science Co., Inc., Chicago, Illinois 60620. Electron micrographs: (a) courtesy Eli Lilly and Co.; (b) courtesy Thomas F. Anderson; (c) courtesy Noel R. Krieg and the International Association of Microbiological Societies.]*

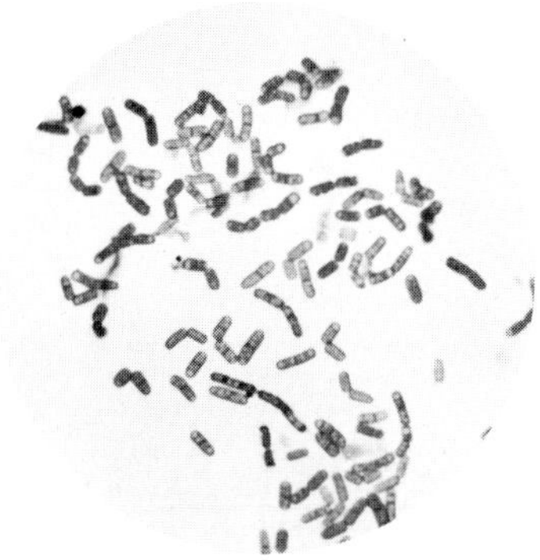

(**b**)

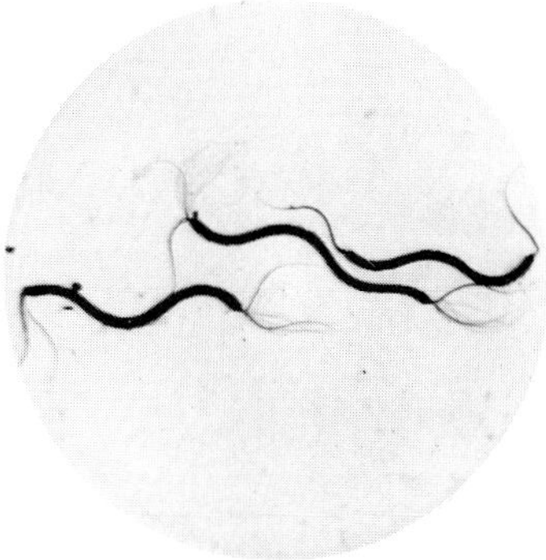

(**c**)

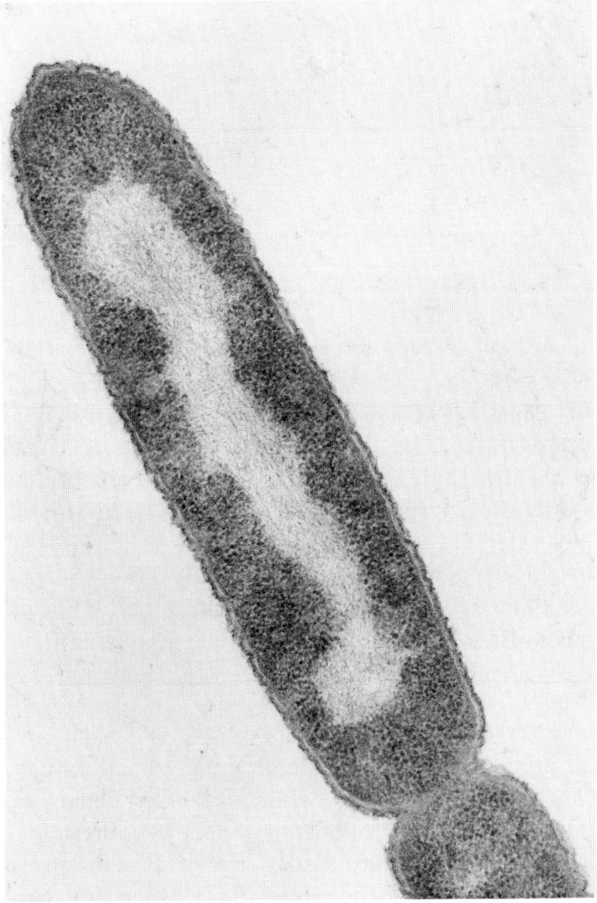

(**b**)

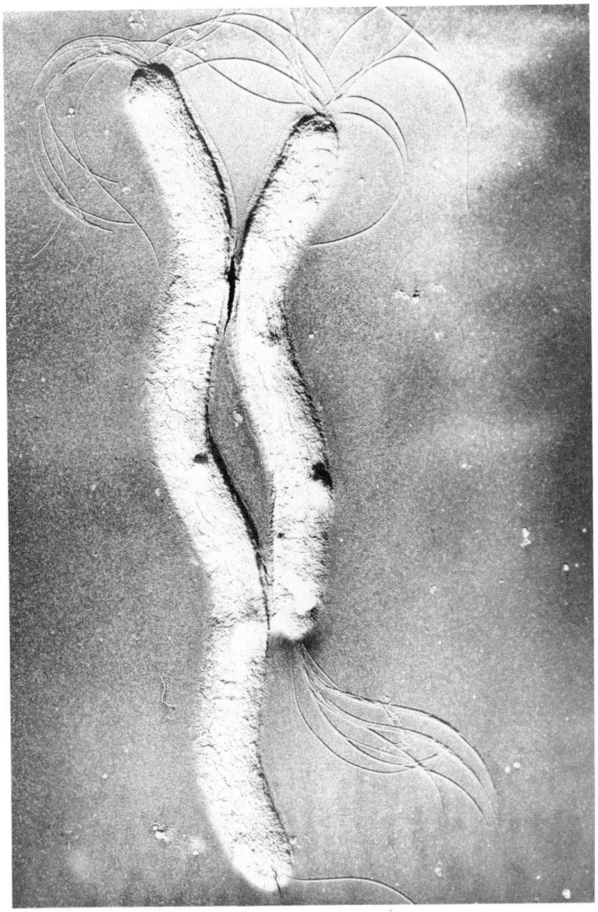

(**c**)

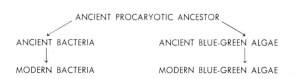

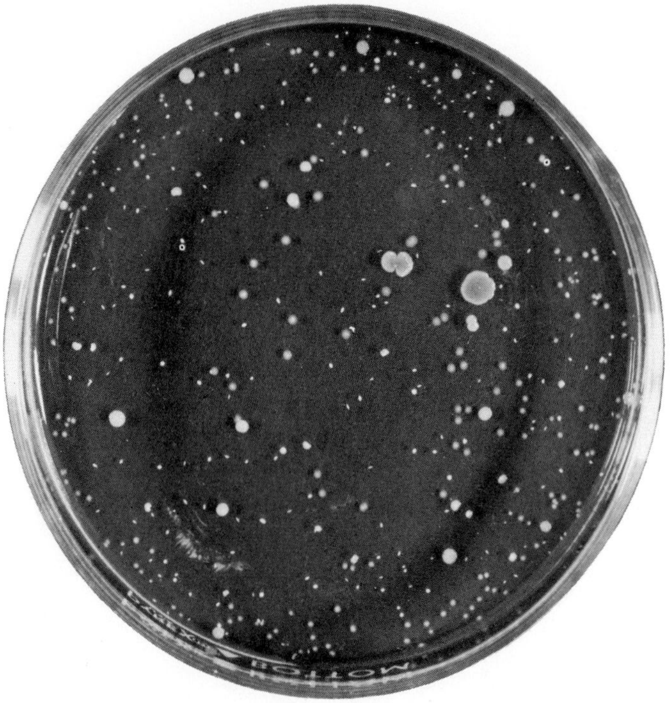

Fig. 13.7 *Bacterial culture plate (petri dish) containing several colonies. Each colony descended from a single bacterium, so all the cells in any one colony are genetically identical.*

algae arose from a common ancestor, since it would be quite unlikely that so many unusual features would arise independently on more than one occasion. One possible evolutionary scheme might be:

ANCIENT PROCARYOTIC ANCESTOR
↙ ↘
ANCIENT BACTERIA → ANCIENT BLUE-GREEN ALGAE
↓ ↓
MODERN BACTERIA → MODERN BLUE-GREEN ALGAE

An alternative hypothesis holds that the bacteria and the blue-green algae do *not* share a common ancestry. This reasoning is that if the bacteria and the blue-green algae each evolved independently, then we should expect to find an array of distinctive differences between the two groups. The blue-green algae are generally aerobic; that is, they must have oxygen to carry on their cell respiration. Many bacteria, however, are anaerobic and capable of using substances other than oxygen as the final electron acceptor. Nearly all the blue-green algae use chlorophyll *a* to capture light energy in photosynthesis. In this process, they utilize hydroxide ions (derived from water) as electron donors and produce free molecular oxygen. Only a few bacteria are photosynthetic; these lack chlorophyll *a* but do have other kinds of chlorophyll. Photosynthetic bacteria use molecules such as hydrogen and hydrogen sulfide as electron donors and do not release free oxygen (as do the blue-green algae and all other green plants). Finally, many bacteria possess flagella (Fig. 13.6), while blue-green algae lack these structures. Proponents of this hypothesis believe that these differences are great enough to preclude any close relationship between the bac-

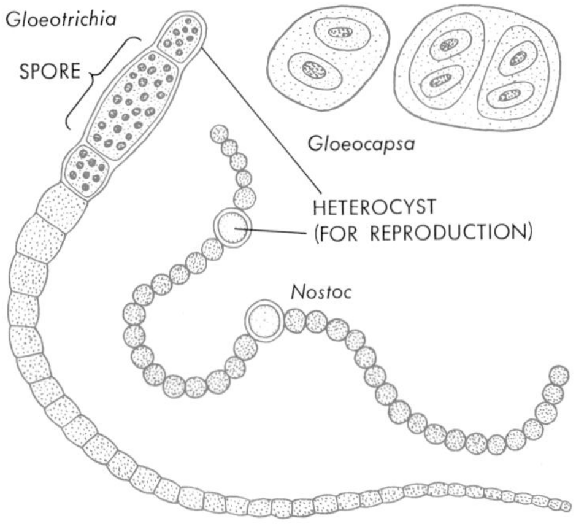

Fig. 13.8 *Four genera of the phylum Cyanophyta, the blue-green algae. All these organisms possess chlorophyll* a, *plus several other pigments. One of these pigments, phycocyanin, is responsible for the blue-green color characteristic of the cyanophytes. Reproduction is by fission largely; on occasion special spore-like cells, called heterocysts, can germinate to produce a new filament.*

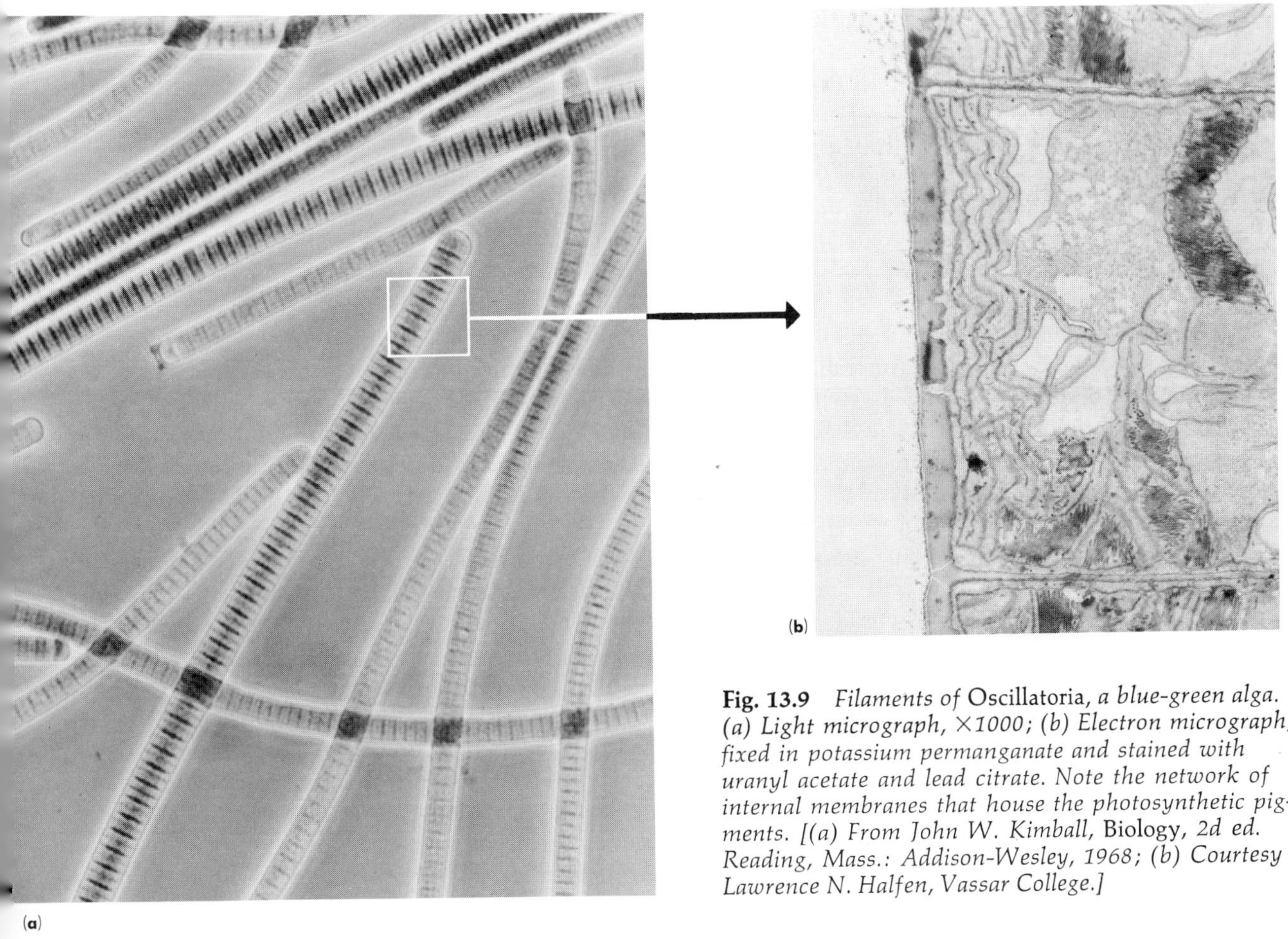

Fig. 13.9 *Filaments of* Oscillatoria, *a blue-green alga. (a) Light micrograph, ×1000; (b) Electron micrograph, fixed in potassium permanganate and stained with uranyl acetate and lead citrate. Note the network of internal membranes that house the photosynthetic pigments. [(a) From John W. Kimball,* Biology, *2d ed. Reading, Mass.: Addison-Wesley, 1968; (b) Courtesy Lawrence N. Halfen, Vassar College.]*

teria and the blue-green algae. They would diagram an evolutionary scheme as follows:

ANCIENT PROCARYOTIC ANCESTOR
↓
ANCIENT BACTERIA
↓
MODERN BACTERIA

ANCIENT PROCARYOTIC ANCESTOR
↓
ANCIENT BLUE-GREEN ALGAE
↓
MODERN BLUE-GREEN ALGAE

It is evident that each hypothesis is supported by an impressive body of information. As yet there is no evidence contradicting the predictions that follow acceptance of either of the two alternative hypotheses. To date, the fossil record is of little help in the matter; however, recent studies in paleobotany and paleobiochemistry show some promise of yielding information pertinent to the problem. In 1965, the paleobotanist Elso Barghoorn and his associates at Harvard University found organic remains interpreted as fossil bacteria in rocks from the northern shore of Lake Superior in Ontario, Canada. The remains have been dated as approximately 3 billion years in age. These investigators also discovered cellular microfossils similar in form to modern blue-green algae and green algae in one-billion-year-old rocks of central Australia. In 1966, they found a minute, rod-shaped, bacterium-like fossil, *Eobacterium isolatum*, in rocks from South Africa; these were dated at more than 3.1 billion years in age (Fig. 13.10). Later study of these same rocks revealed spheroidal microfossils interpreted as the remnants of unicellular blue-green algae. If these remains have been correctly identified as those of blue-green algae and bacteria, then these two evolutionary lineages represent the oldest known groups of organisms.

13.6 Origin of Eucaryotic Organisms

It is reasonable to hypothesize that the living bacteria and blue-green algae represent vestiges of an early, primitive stage in the evolution of cellular

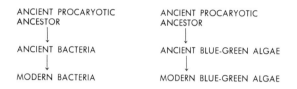

Fig. 13.10 *An electron micograph of a bacteriumlike fossil,* Eobacterium isolatum *(×66,000). (Photo courtesy Elso S. Barghoorn, Harvard University.)*

structure. Evidently the procaryotic cell was a fairly successful evolutionary experiment, for it has persisted in the descendants of the early procaryotes without significant fundamental change. At some time in the distant history of life, however, an evolutionary advance of the greatest importance occurred—the compartmented or eucaryotic cell came into existence. This event seems to have set the stage for the processes of evolution to generate the vast diversity of life with which the modern biologist is concerned.

Did the evolution of eucaryotic organisms take place only once? Or did it occur many times? If the former alternative is correct, then all living organisms with the eucaryotic types of cell structure, including all animals and the vast majority of plants, have evolved from a common ancestor. If the latter alternative is true, then the similarities in cell organization among the various groups of eucaryotes are the result of convergent evolution and do not indicate any close degree of relationship among the organisms.

Again the fossil record offers little help in resolving the problem. The oldest known fossils of

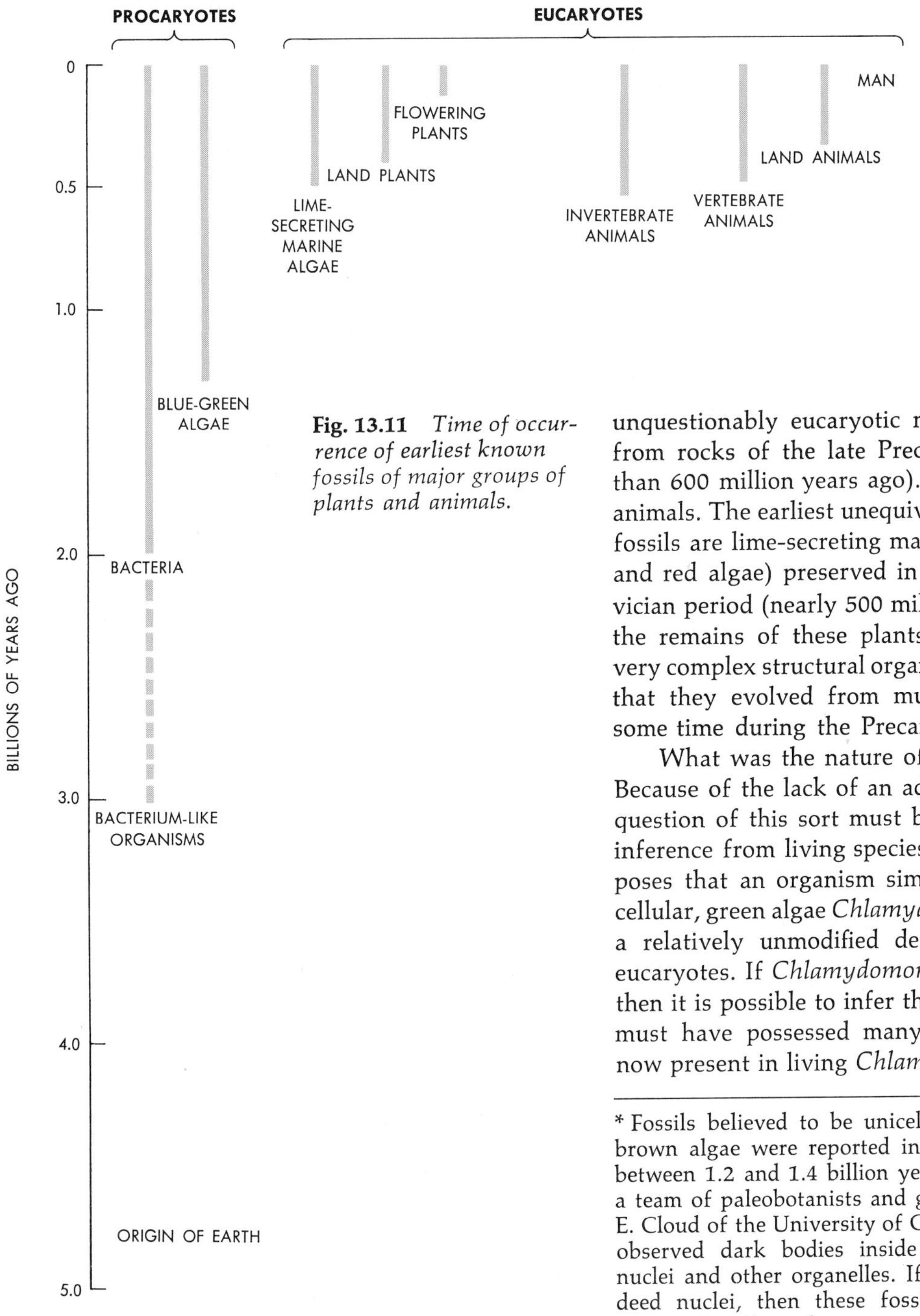

Fig. 13.11 *Time of occurrence of earliest known fossils of major groups of plants and animals.*

unquestionably eucaryotic nature (Fig. 13.11) are from rocks of the late Precambrian period (more than 600 million years ago). They are invertebrate animals. The earliest unequivocally eucaryotic plant fossils are lime-secreting marine algae (green algae and red algae) preserved in rocks from the Ordovician period (nearly 500 million years ago).* Since the remains of these plants and animals show a very complex structural organization, it seems likely that they evolved from much older ancestors at some time during the Precambrian period.

What was the nature of the early eucaryotes? Because of the lack of an adequate fossil record, a question of this sort must be answered largely by inference from living species. One hypothesis proposes that an organism similar to the living, unicellular, green algae *Chlamydomonas* may represent a relatively unmodified descendant of the early eucaryotes. If *Chlamydomonas* is indeed primitive, then it is possible to infer that the early eucaryotes must have possessed many of the characteristics now present in living *Chlamydomonas.*

---

* Fossils believed to be unicellular green and golden-brown algae were reported in 1969 to occur in rocks between 1.2 and 1.4 billion years old. The discoverers, a team of paleobotanists and geologists led by Preston E. Cloud of the University of California, Santa Barbara, observed dark bodies inside the cells that suggest nuclei and other organelles. If these structures are indeed nuclei, then these fossils represent the oldest eucaryotic cells yet known.

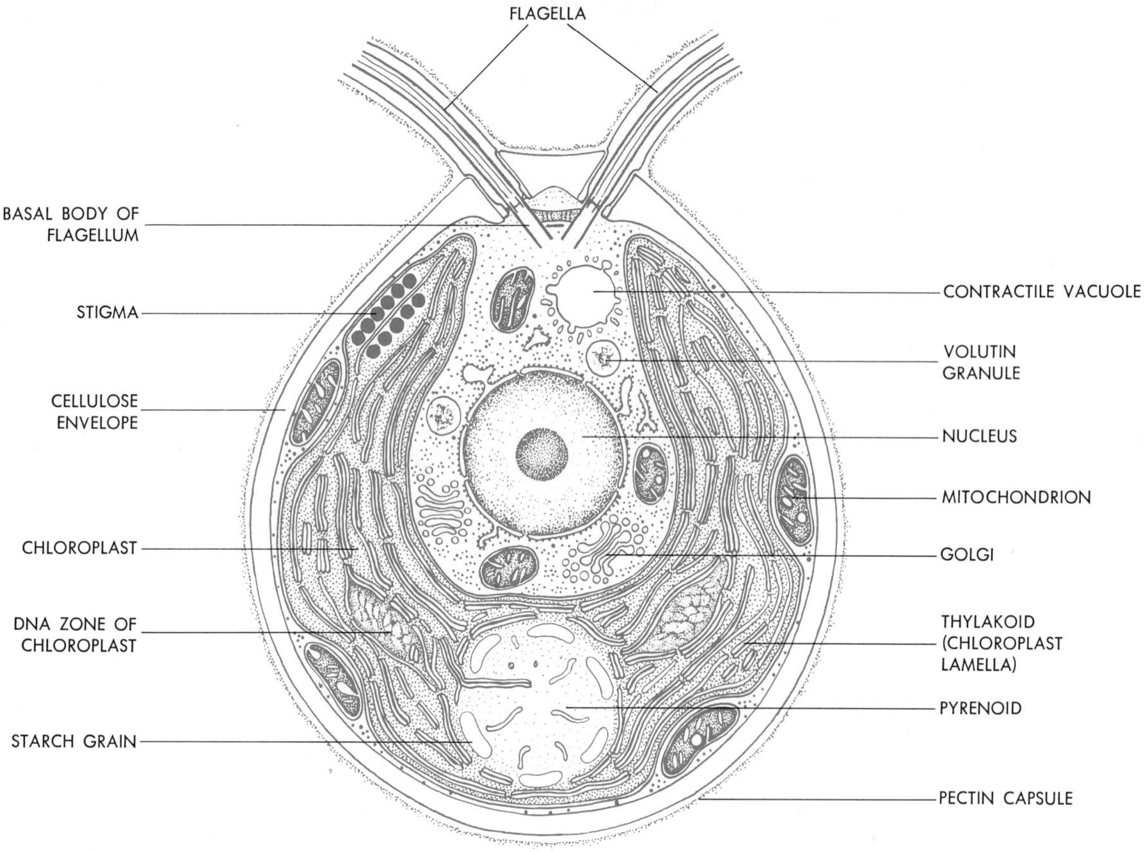

Fig. 13.12 *Diagram showing structure of the unicellular green alga* Chlamydomonas *as revealed by the electron microscope. Only one of the two contractile vacuoles is shown. Volutin granules are probably stored phosphates. (From Keith Vickerman and Francis E. G. Cox,* The Protozoa. *Boston: Houghton Mifflin, 1967, p. 20; first printed in Great Britain by John Murray, Ltd.)*

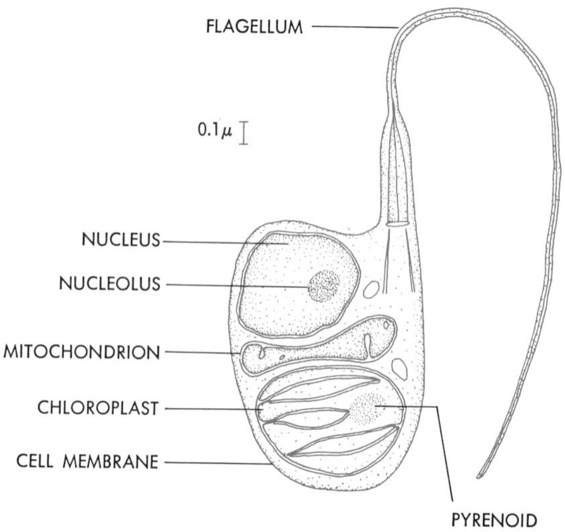

Fig. 13.13 *Diagram of* Micromonas, *a minute, unicellular alga of the sea. (After Peter R. Bell and Christopher L. F. Woodcock,* The Diversity of Green Plants. *Reading, Mass.: Addison-Wesley, 1968, p. 5.)*

The single cell of an individual *Chlamydomonas* (Fig. 13.12) is usually less than 25 microns long. Two flagella protrude through the cell wall at the anterior end. Lashing movements of these flagella propel the alga through the water. Inside each cell is a single large chloroplast, usually cup-shaped, with an eyespot near its rim or apex. The chloroplast usually obscures a nucleus that lies in the colorless cytoplasm. Near the anterior (flagellar) end are two or more contractile vacuoles. Endoplasmic reticulum, Golgi complex, mitochondria, and ribosomes are present in the cytoplasm.

Using *Chlamydomonas* as a model, we can postulate that the early eucaryotes were also unicellular, flagellated, photosynthetic organisms with at least minimum compartmentalization of their functions into organelles. The minimum structural requirements for a eucaryotic organism would appear to include at least one of the following: nucleus (with chromosomes), nuclear and cellular membranes, endoplasmic reticulum, Golgi complex, chloroplast, mitochondrion, and flagellum. It is likely that the early eucaryote may have been minute in size. Recently an organism closely fitting this description was found as a component of ocean plankton. Being only a micron or two long, this organism, the smallest known eucaryote, was named *Micromonas* (from two Greek words meaning "small single organism"). The existence of such a minute eucaryotic organism adds further support to the postulated primitive eucaryote. *Micromonas* (Fig. 13.13) seems even less evolutionarily modified than the larger and more complex *Chlamydomonas*. It is, therefore, conceivably far more primitive.*

The appearance of the various organelles characteristic of eucaryotic cells all have to do with the more efficient use of energy in the living cell. Thus compartmentalization may have arisen at the same time as the increased concentration of oxygen in the earth's atmosphere, which made aerobic respiration possible.

13.7 Origin and Adaptive Significance of Multicellularity

If the diversity of plants and animals is to be explained by the hypothesis of evolution by natural selection, there must be good reasons for the retention of any distinguishing features that a species manifests. Acceptance of the hypothesis necessitates belief that at any particular point in its history, a species demonstrates the sum total of adaptive changes preserved (or at least not eliminated) by natural selection. Thus, any structural or functional characteristic incorporated into a successful organism must have some selective value—or at least be of no selective harm. If it did not meet these criteria, the feature could not have evolved.

In general, the fossil record supports the contention that structurally complex organisms originated later than simpler forms. It is certainly reasonable to assume that complex biological systems were preceded in time by simpler ones. It is also consistent with the hypothesis of evolution by natural selection to assume that certain groups of organisms received some selective advantage in becoming more complex. Note that we are not saying that an increase in complexity is a general trend in evolution, acting uniformly in all phylogenetic lines.

* There is a possibility, of course, that both *Micromonas* and *Chlamydomonas* owe their relative simplicity and small size to an evolutionary reduction from much more complex ancestors. Even if this were the case, these organisms can still serve admirably as phylogenetic models for the postulated early eucaryotes.

The evolution of multicellularity was favored because it allowed for cell specialization. Specialized cells can work cooperatively to extract energy more efficiently from the environment.

Supplement 13.1

THE ORIGIN AND DIVERSIFICATION OF EUCARYOTES

If it is postulated that the original eucaryote may have been something like the unicellular, flagellated, green alga *Chlamydomonas,* then how did the evolutionary transitions to the many kinds of eucaryotes take place?

In the absence of a complete fossil record, this question can only be answered by inference from a comparative study of living species. About 10 years ago, two botanists at the New York Botanical Garden, Richard M. Klein and Arthur Cronquist, published an extensive and comprehensive review of the phylogeny of algae and fungi. In this essay, the authors hypothesized that the green algal order Volvocales may provide models for the initial stages in the diversification of the early eucaryotes. According to their hypothesis, the present-day Volvocales represent relatively unmodified descendants of a *Chlamydomonas*-like early eucaryote. The ancient Volvocales, therefore, may well be the common ancestral group that gave rise to nearly all of the major phyletic lines of eucaryotes.

Klein and Cronquist proposed that one interesting family of the Volvocales, the Polyblepharidaceae (Diagram A), may provide a good model for the ancestry of four phyletic lines of algae (the Euglenophyta, Pyrrophyta, Chrysophyta, and Phaeophyta), the fungi, and indirectly the unicellular animals (Protozoa). Most members of the Polyblepharidaceae lack a cellulose wall. They do possess a periplastic envelope, however, rigid in some species and flexible in others. Klein and Cronquist hypothesized that the organisms have lost the cellulose wall during their evolution but retain the genetic basis for cellulose synthesis (since cellulose appears in some of their presumed descendants). Contractile vacuoles and eyespots are present. Some species exhibit a tendency toward dorsiventrality (a flattening of the body) and asymmetry. Most species possess more than two flagella, while several have an inpocketing of the cell surface at the anterior (flagellar) end. Several

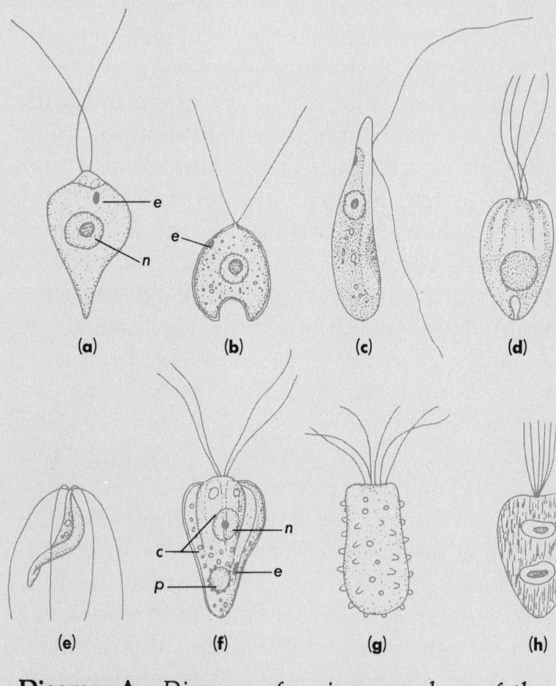

Diagram A *Diagram of various members of the family Polyblepharidaceae (Volvocales, Chlorophyta). (a)* Phyllocardium complanatum. *(b)* Furcilla lobosa. *(c)* Dangeardinella saltatrix. *(d)* Pyramimonas delicatulus. *(e)* Spermatozopsis exultans. *(f)* Pyramimonas tetrarhynchus. *(g)* Pocillomonas flos-aquae. *(h)* Polyblepharides singularis. c, *chloroplast;* n, *nucleus;* p, *pyrenoid;* e, *eyespot. (From F. E. Fritsch,* The Structure and Reproduction of the Algae, *Vol. 1. New York: Macmillan, 1935, p. 86.)*

of these features can be considered as foreshadowing attributes characteristic of one or more of these four groups of algae, fungi, and protozoa (see Diagram B).

A radical alternative to the Klein and Cronquist phylogenetic model for the evolutionary diversification of the early eucaryotes was elaborated in 1968 by biologist Lynn Margulis of Boston University (see Table 1). Margulis questioned the idea that the ancestor of all eucaryotic organisms was a simple, one-celled, green flagellate. She hypothesized that the ancestral eucaryote was a nonphotosynthetic amoebo-flagellate, not too dissimilar to some species of living flagellated amoebae (see Diagram C). According to Margulis' model, this organism possessed mitochondria and was heterotrophic in its nutrition. Much later in evolution, some of these heterotrophic amoebo-flagellates became autotrophic cells through the ingestion of photosynthetic procaryotes resembling certain blue-green algae. Eventually the ingested green organisms evolved into chloroplasts and the resulting photosynthetic eucaryotic organism became the ancestor of the eucaryotic algae and all other green eucaryotic plants (Diagram C).

Margulis further hypothesized that the fungi evolved not from various algal groups as suggested by Klein and Cronquist, but from the heterotrophic amoebo-flaggellate postulated to have been the ancestor of all eucaryotes. According to this hypothesis, therefore, the living fungi would be more closely related to the living protozoan animals than to the green plants (Diagram C).

Phylogenetic models should help the biologist make predictions. The symbiotic theory proposed by Margulis enables several predictions to be made. For example, all cells containing chloroplasts should also possess membrane-enclosed nuclei, and, at the same time, chloroplasts should contain DNA different from the DNA of the nucleus and the mitochondria. These predictions, as well as others that arise from Margulis's hypothesis, may challenge evolutionary biologists to carry out research designed specifically to test this phylogenetic model.

Diagram B *A phylogenetic chart summarizing the hypothesis of Klein and Cronquist on the evolutionary diversification of eucaryotes. (Adapted from R. M. Klein and Arthur Cronquist,* Quarterly Review of Biology *42, 1967, p. 109.)*

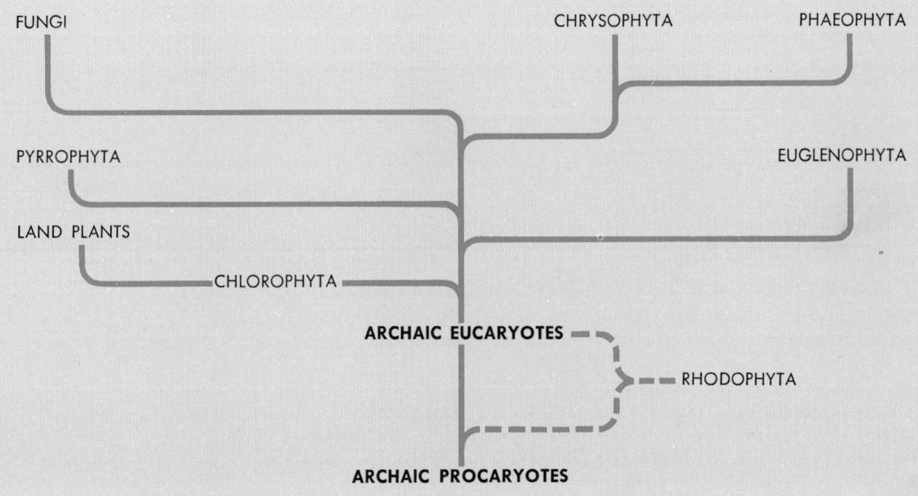

Table 1*

Klein and Cronquist model: assumptions	Margulis model: assumptions
1. The basic dichotomy between organisms of the present-day world is between animals and plants.	1. The basic dichotomy between organisms of the present-day world is between procaryotes and eucaryotes.
2. Photosynthetic eucaryotes (higher plants) evolved from photosynthetic procaryotes (blue-green algae, "uralgae").	2. Photosynthetic eucaryotes (higher algae, green plants) and nonphotosynthetic eucaryotes animals, fungi, protozoans) evolved from a common nonphotosynthetic (amoebo-flagellate) ancestor. There is not now, nor was there ever, an "uralga."
3. The evolution of plants and their photosynthetic pathways occurred monophyletically on the ancient earth.	3. The evolution of photosynthesis occurred on the ancient earth in bacteria and blue-green algae; higher plants evolved abruptly from procaryotes when the heterotrophic ancestor (2 above) acquired plastids by symbiosis.
4. Animals and fungi evolved from plants by loss of plastids.	4. Animals and most eucaryotic fungi evolved directly from protozoans.
5. Mitochondria differentiated in the primitive plant ancestor.	5. Mitochondria were present in the primitive eucaryote ancestor when plastids were first acquired by symbiosis.
6. The primitive plant differentiated the complex flagellum, the mitotic system, and all of the eucaryote organelles.	6. Mitosis evolved in heterotrophic eucaryotic protozoans by differentiation of the complex flagellar system.
7. All organisms evolved from a primitive ancestor monophyletically by single steps.	7. All procaryotes evolved from a primitive ancestor by single mutational steps; all eucaryotes evolved from a primitive eucaryote ancestor by single mutational steps. Eucaryotes evolved from procaryotes by a specific series of symbioses.
8. Morphological, biochemical, and physiological characters are useful in classification of Thallophytes.	8. Only total gene-based biochemical pathways resulting in the production of some selectively advantageous markers are reliable "characters" in classification; morphology is useless in most procaryotes.
Predictions of the two models	
Nothing predicted; no consistent phylogeny possible, many predicted organisms not found, for example "uralgae"; no correlation with fossil record possible; no presentation of phylogeny as a function of time elapsed is possible.	Major biochemical pathways predicted; consistent phylogeny constructed; biological discontinuity at Precambrian boundary predicted.

* This table by Dr. Lynn Margulis compares the Klein and Cronquist hypothesis concerning the evolutionary origins of organisms with the Margulis hypothesis. Note the reliance on the ability to predict as a major reason for selecting one model over the other. From Lynn Margulis, "Evolutionary Criteria in Thallophytes: A Radical Alternative," *Science*, September 6, 1968, pp. 1020–1022.

Diagram C *A phylogenetic chart illustrating an hypothesis proposed by Lynn Margulis on the origin and diversification of organisms. (Courtesy of Lynn Margulis, Boston University.)*

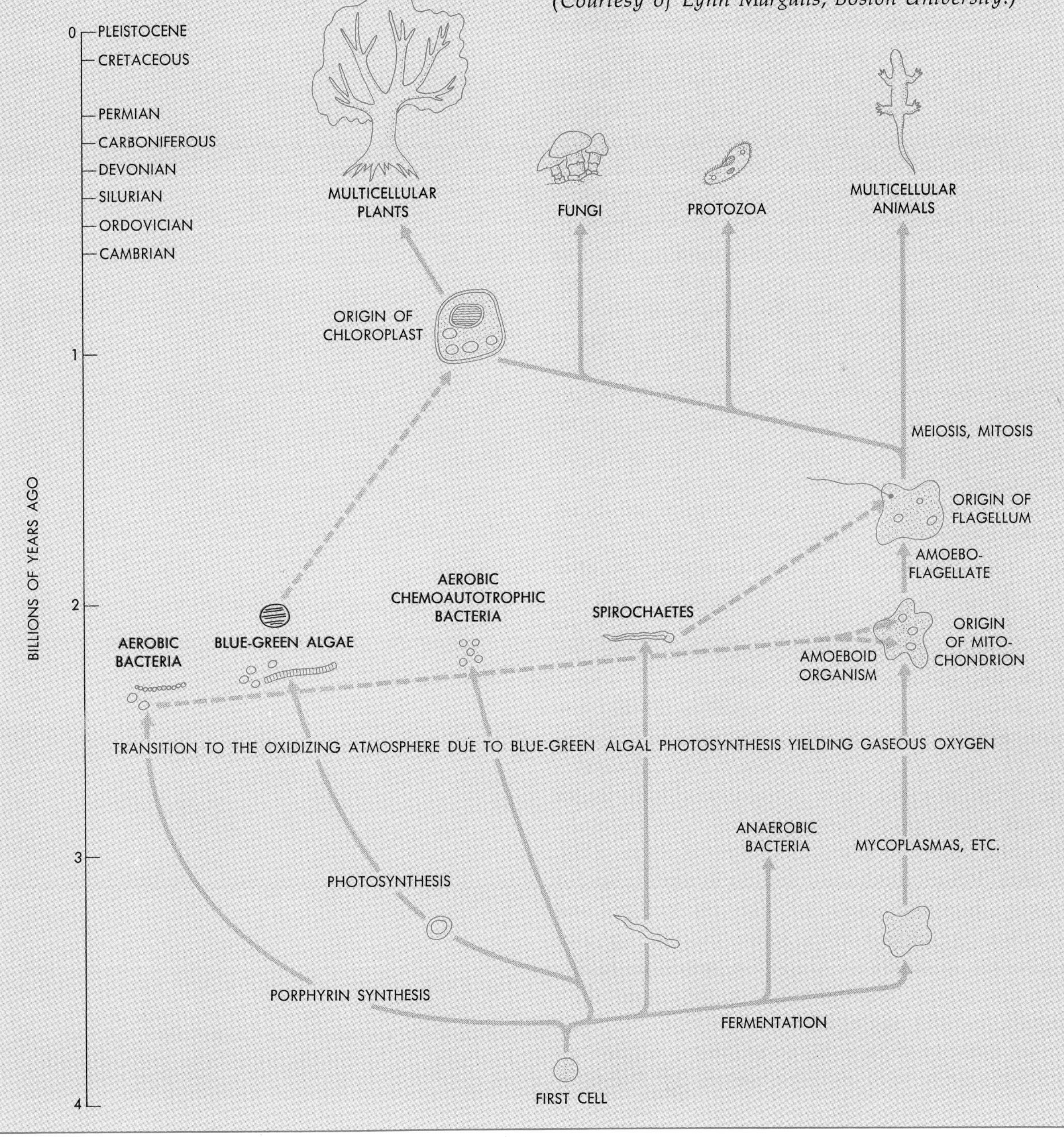

Indeed, in some organisms (e.g., leaves of *Equisetum*), a *decrease* in complexity is found. We are simply stating that *for most groups* increase in complexity has been favored by natural selection.

Consider the evolution of multicellularity. Multicellular organisms must surely have been preceded by unicellular organisms. What selection pressures favored the adoption by some groups of a multicellular state? Examination of such forms reveals one possible answer. The multicellular state allows for cell specialization (some cells capture sunlight and synthesize food, some play a role in reproduction, some protect the organisms, some anchor it, and so on). The result is a corresponding increase in the ability of the organism to exploit its environment and increase its own chances for survival.

Concerning when and how multicellularity evolved, biologists can only speculate. Complex multicellular animals were already firmly established by the beginning of the Cambrian period, some 600 million years ago. Algae with highly differentiated multicellular bodies had evolved into at least two main groups (the green and the red algae) by the Ordovician period, some 500 million years ago. Thus the fossil record is obviously of little help. We must be satisfied to examine living organisms, such as green algae, and try to draw reasonable and fruitful inferences about the nature of the first multicellular organisms.

It seems reasonable to hypothesize that the multicellular state originated through the aggregation of separate cells into a colony. Several surviving species of green algae demonstrate likely stages in this evolution. *Chlamydomonas*, for example, generally exists in a unicellular motile state (Fig. 13.14a). When conditions become unfavorable for activity, however, each cell loses its motility and becomes aggregated with other individuals into temporary associations. Upon the return of favorable conditions, the individual cells regain their flagella and the aggregation breaks up.

A somewhat later stage in the evolution of multicellularity may be represented by *Palmella*,

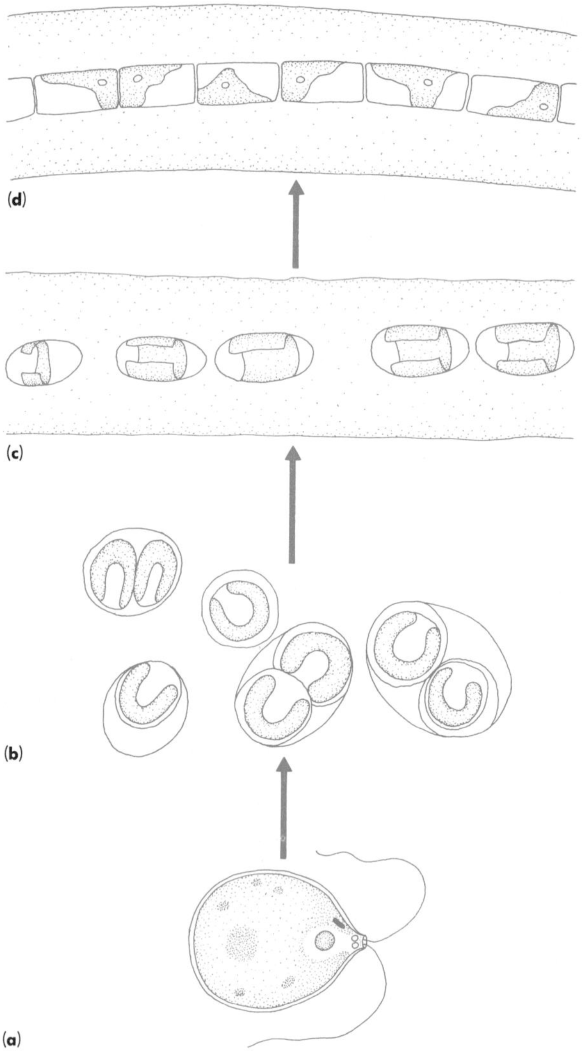

Fig. 13.14 *Diagrams (not drawn to same scale) illustrating one hypothesis concerning the origin of the multicellular condition. (a)* Chlamydomonas. *(b)* Palmella. *(c)* Geminella interrupta. *(d)* Geminella minor.

Supplement 13.2

THE ORIGIN OF BACTERIA, MITOCHONDRIA, AND CHLOROPLASTS

A number of similarities have been noted between bacterial, mitochondrial, and chloroplast structure and function. These similarities have suggested the intriguing hypothesis that both the cell organelles may have arisen from various types of procaryotic cells that became incorporated into the structure of primitive eucaryotic cells. What is the evidence on which this suggestion is based?

Mitochondria and chloroplasts are roughly the same size as bacteria and have a similar membrane structure. The inner membrane of chloroplasts resembles the cell membrane of blue-green algae in that it is the site of photophosphorylation (blue-green algae do not have chloroplasts, only infoldings of the cell membrane to which chlorophyll is attached). Similarly, the inner membrane of the mitochondrian resembles the cell membrane of bacteria in that both are sites of oxidative phosphorylation.

A number of antibiotics exist that inhibit the growth of procaryotic cells by interfering with their protein synthetic pathways. These same antibiotics interfere with protein synthesis within the mitochondria and chloroplasts. They do not inhibit protein synthesis in the cytoplasmic ribosomes (those bound to the endoplasmic reticulum) of eucaryotes. Conversely, drugs that inhibit protein synthesis by cytoplasmic ribosomes do not inhibit mitochondrial or chloroplast ribosomes.

Both chloroplasts and mitochondria contain their own DNA and ribosomes; the DNA is very similar in its general structure to that found in bacteria. The ribosomes are not bound to membranes and are a slightly smaller size and density than cytoplasmic ribosomes of eucaryotes.

The permeability characteristics of bacterial, mitochondrial, and chloroplast membranes are very similar. They also resemble the eucaryotic cell membrane very closely.

Putting all these observations together, it is not difficult to speculate on how procaryotic cells could have become incorporated into eucaryotic cells. If the original procaryotes were devoured by the primitive eucaryotes in a process similar to endocytosis (engulfing), then the similarities of the outer membranes of eucaryotes, procaryotes, mitochondria, and chloroplasts would be conveniently explained. If the procaryotes were not digested away, they could have taken up an independent existence within the cytoplasm of the larger host cell. Such occurrences are by no means rare in the biological world. Bacteria live in the digestive tract of all mammals and are essential to mammalian physiology. A number of small invertebrate animals are known to incorporate algal cells into their body (so that the animals actually appear green); and at least one invertebrate, a slug, has been discovered recently that incorporates chloroplasts directly into some of its cells! There is ample evidence to suggest that foreign bodies can be taken into cells and remain intact after incorporation.

Over time, as primitive procaryotes continued their existence within eucaryotic cells, they lost a number of their general functions and became more and more specialized. The DNA in mitochondria no longer contains genetic information for the electron transport system (such as the cytochromes or NAD); this information is coded in the eucaryote's nuclear DNA. Mitochondrial DNA appears to code only for the structural proteins of mitochondrial membranes and other components. Today mitochondria and chloroplasts are dependent on eucaryotic cells for their existence. They have lost their ability to live independently.

The procaryotic origin of chloroplasts and mitochondria is only a speculative hypothesis. It may never be possible to demonstrate it with any certainty. Nonetheless, it does bring together a number of interesting observations that otherwise appear disconnected and unrelated. Such is at least one function of any theory.

which forms loose colonies of an irregular shape (Fig. 13.14b). A still closer approach to the multicellular condition is represented by *Geminella*, an alga in which numerous cells are loosely lined up within a cylindrical mucilaginous sheath, forming a simple filament. In certain species the cells lie at some distance from each other (Fig. 13.14c), while in others they are sufficiently close for their ends to touch (Fig. 13.14d). The latter arrangement would appear to be only a slight step away from that present in such filamentous genera as *Spirogyra* and *Ulothrix*, whose cells are attached to each other with a cementing material.

Another hypothesis on the origin of multicellularity maintains that the multicellular condition arose through the adherence of daughter cells after cell division. Some evidence in support of this hypothesis is provided by observations of the early growth of many filamentous algae. In these plants, the unicellular motile reproductive cells are released into the water, where they swim around for a time before coming to rest on a rock, stick, or another alga. Upon alighting, these cells fasten themselves to the substratum by secreting a sticky material. Following cell division, the two daughter cells remain together. Further division produces a row of cells.

The multicellular condition may have originated by both methods hypothesized here, depending upon the particular group or evolutionary line. After all, the transition to multicellularity has taken place many times in numerous groups, among the procaryotes as well as the eucaryotes.

SUMMARY

1. Several theories exist to explain the origin of life (they are the same kinds of theories that also explained the diversity of life:
   a) Special creation (see Chapter 14).
   b) Spontaneous generation: The idea that living beings simply come into existence out of nowhere under certain conditions.
   c) Chemosynthesis: The idea that all components of earliest living systems were synthesized by ordinary chemical and physical processes in the very early history of the earth.
2. The most widely accepted theory of chemosynthesis maintains that the early atmosphere of the earth contained water vapor, methane, ammonia, and free hydrogen. These gases contain all the elements essential for life as we know it: carbon, hydrogen, oxygen, and nitrogen.
3. One version of the chemosynthetic theory sees life originating in six phases:
   a) First Phase: Random synthesis of organic molecules such as amino acids, carbohydrates, and small peptides. The work of Stanley Miller showed that these could be produced from water vapor, methane, ammonia, and hydrogen if energized by an electric current or discharge.
   b) Second Phase: Formation of macromolecules (dipeptides, long-chain peptides) from free amino acids, sugars, etc. Sidney Fox has shown that heat could cause such combinations. Fox and others found that some order is involved in the joining together of amino acids thus synthesized, and that the polypeptides formed tend to assume definite shapes.
   c) Third Phase: Association of proteinlike macromolecules into aggregates surrounded by a primitive lipid membrane. Factors such as polarity could have been involved in forming such associations. Oparin's coacervate theory is one way of accounting for the tendency to form higher-level associations among macromolecules. Fox has

shown that thermally produced proteins will form spherules about the same size as bacterial cells; the spherules can combine with one another or split.

d) Fourth Phase: Appearance of molecular "organizers" within aggregates. These organizers, nucleic acids, controlled the internal organization of the aggregates and the division of internal material when a droplet split in two.

e) Fifth Phase: The beginnings of evolutionary development. After the introduction of an organizer molecule into living systems made heredity possible, continuity could be guaranteed between generations, and natural selection could operate. Among the first processes selectively favored must have been those involved with more efficient methods of harnessing energy.

f) Sixth Phase: Introduction of photosynthetic mechanisms. Up until this phase all "organisms" were heterotrophic. With the evolution of photosynthesis, autotrophism became possible. The appearance of autotrophs meant that life could become self-sufficient, depending only on solar energy to keep the system going. A general summary of these phases is given in Fig. 13.15.

4. One of the important developments in the early history of life was the introduction of catalysis. The first catalysts were probably no more than inorganic ions (such as iron) incorporated into coacervate droplets or proteinoid spherules. Association of inorganic ions with small peptides enhanced catalytic function—hence the origin of enzymes.

5. The time scale for the origin of life on earth covers over 2 billion years (possibly 4 billion).

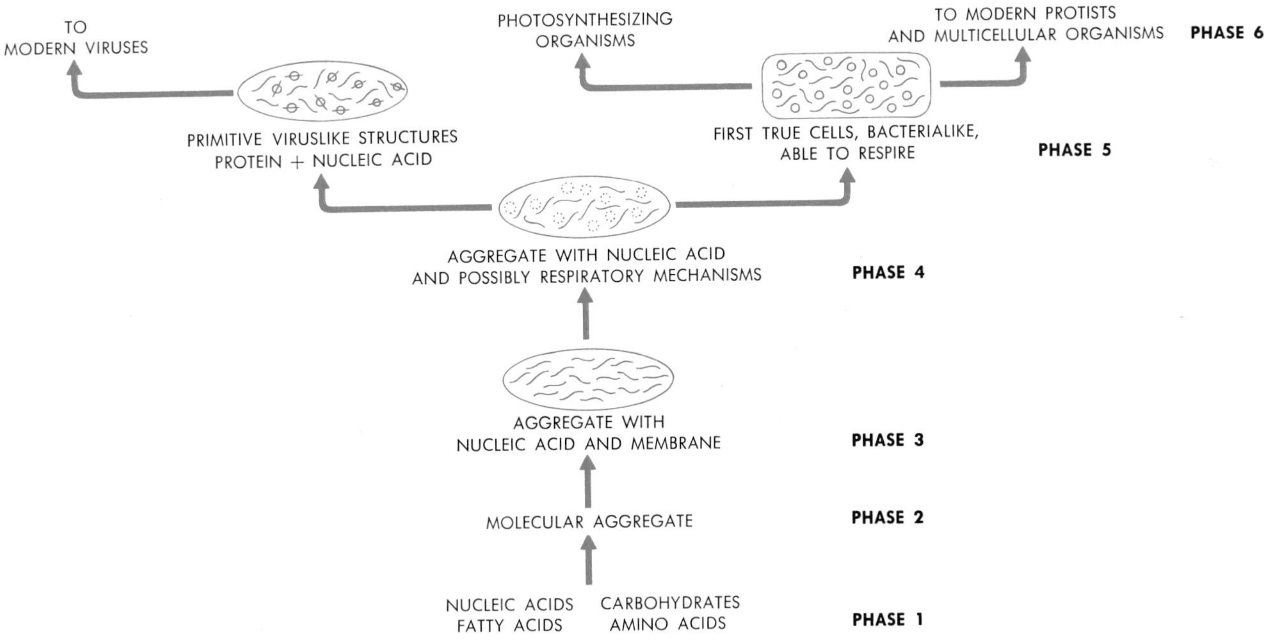

Fig. 13.15 *Summary of possible stages of development in the early history of life on earth.*

The first true fossil remains of primitive cells are at least 1.6 billion years old. Obviously the more primitive organisms and molecular aggregates must have preceded the appearance of cells by many millions or billions of years.

6. Modern blue-green algae and bacteria have many features in common. Both are single-celled organisms lacking a true, membrane-bound nucleus. Both lack most of the larger cell organelles such as mitochondria, Golgi bodies, and lysosomes (though both have ribosomes). These similarities have led to the hypothesis that both modern forms evolved from a common ancestor billions of years ago.
7. However, critics of the common ancestor theory point out that blue-green algae and bacteria are quite different in some basic features. Blue-green algae are all aerobic, while many bacteria are anaerobic, using substances other than oxygen as the final electron acceptor. Most blue-greens carry out photosynthesis using chlorophyll *a*; only a few bacteria are photosynthetic at all, and those that are use other chlorophylls than *a*. Many bacteria possess flagella, while none of the blue-green algae do. These differences have led to the alternative hypothesis that blue-green algae and bacteria had a completely separate evolutionary history. To date no evidence can rigorously distinguish between the two hypotheses.
8. Eucaryotic organisms, such as the one-celled green alga *Chlamydomonas*, possibly originated from procaryotic forms. *Chlamydomonas* is photosynthetic, having true chloroplasts and mitochondria, and yet moves about rapidly by means of flagella. A very primitive eucaryote, *Micromonas* is much smaller than *Chlamydomonas* and may represent an intermediate form between procaryotic and eucaryotic cells.
9. Multicellular plants were undoubtedly favored by natural selection over purely unicellular forms. As in animals, multicellularity allows for the development of specialized tissues and organs. Multicellularity may have originated by the sticking together of daughter cells after division. Such a process would by itself produce nothing more than a string or clump of similar cells, such as is found in the filamentous green or blue-green algae of today. Even in such simple systems, some specialization of cells can be observed. Certain cells in a filament of blue-green algae differentiate and become the reproductive cells. As long as cell groups adhere together, specialization is possible.

EXERCISES

1. Explain briefly how living, cellular forms could have originated from simple organic molecules such as amino acids, carbohydrates, lipids, and nucleic acids.
2. Figure 13.16 is a graph taken from the experiments of Stanley Miller described in Section 13.3. Considering his work and the graph, answer the following questions.
   a) Would it be reasonable to conclude that ammonia is incorporated into the formation of amino acids in Miller's experiment? Why or why not?
   b) The curve for concentration of amino acids is plotted on a different scale of magnitude from that for concentration of ammonia. What is the difference in magnitude? Why should this be the case? Draw the curve representing change in concentration of amino acids as plotted on the same scale as that for ammonia.
   c) Methane concentration is not shown in Fig. 13.16. If it were, what shape do you hypothesize the line would have? Why?

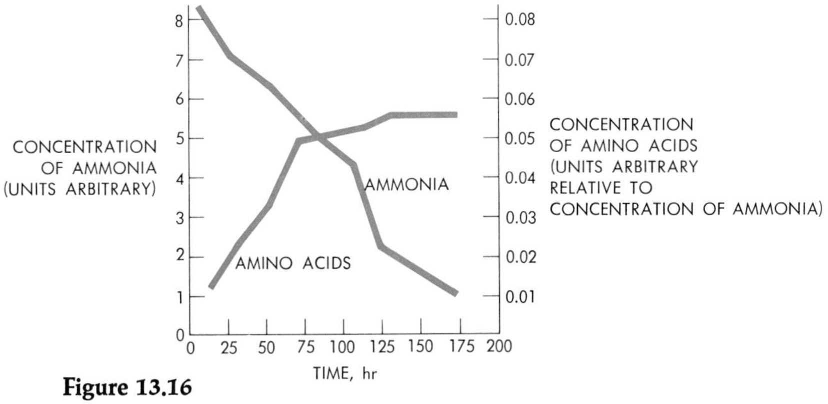

Figure 13.16

3. Dr. Sidney Fox performed a number of experiments that were a continuation of the work of Stanley Miller. He heated a dry mixture of a number of amino acids to about 90°C. Analysis of the results showed that a number of polypeptides had been formed. He also found that if he heated the amino acids in the presence of phosphoric acid, polypeptides were formed at a temperature as low as 71°C. The polypeptides produced in this manner could be broken down by specific enzymes from animals.

   In addition, bacteria could use these polypeptides for food.

   a) What does this experiment show about the idea that organic substances could have been formed on the primitive earth by random action? Are living organisms essential for the synthesis of organic molecules?
   b) From a thermodynamic point of view, what is the significance of the fact that polypeptides form at a lower temperature in the presence of phosphoric acid?
   c) What evidence from Fox's experiments suggests that the proteins formed by experimental methods are similar to native protein found in living organisms?
   d) Would it be logical to conclude that proteins formed on the primitive earth (before any life appeared) were similar to the proteins that later became incorporated into the living matter of organisms? Why or why not?
4. Identify the selection pressures that may have been operating at each of the points of development listed below:
   a) Formation of a membrane around nucleic acid–molecular aggregate complexes.
   b) Development of a definite respiratory process.
5. In what ways did Redi's and Pasteur's experiments provide evidence that spontaneous generation does not occur? Were their experiments designed to support or disprove an hypothesis?
6. Why are hypotheses that explain the origin of life on earth by saying that spores were brought to the earth on meteorites unsatisfactory?

SUGGESTED READINGS

Bernal, J. D., *The Origin of Life.* Cleveland: World Publishing Co., 1967. An extremely well-written account of current hypotheses about the origin of life. Assumes no special background and discusses philosophical, biological, and biochemical aspects.

Fraenkel-Conrat, H., "Rebuilding a virus." *Scientific American,* June 1956, p. 42. An interesting article dealing with the problem of how viruses are constructed, taken apart, and put back together again. The TMV and HR hybridization stories are presented in detail in this article.

Keosian, John, *The Origin of Life.* New York: Reinhold, 1964. A popular account with accurate and well-presented scientific details. Less stimulating than Bernal's treatment because it deals with few of the broader philosophical questions.

Margulis, L., *Origin of Eukaryotic Cells.* New Haven: Yale University Press, 1970. Discusses the serial symbiosis explanation for the evolution of eucaryotic from procaryotic cells.

Oparin, A. I., *Genesis and Evolutionary Development of Life.* New York: Academic Press, 1968. A useful summary of some of the more recent thinking of one of the pioneers in the field of chemosynthesis. This book replaces Oparin's older classic, *The Origin of Life on Earth,* first published in 1938.

Orgel, L. E., *The Origins of Life.* New York: Wiley, 1973. As the preface states, "This book is not written for professional biologists or chemists, but rather for college or advanced high school students and general readers who have a limited background in chemistry or biology." The author emphasizes the role of natural selection even during the earliest phase of the origin of life.

Wald, George, "The Origins of Life," in *The Scientific Endeavor.* New York: Rockefeller University Press, 1964. A short and very clear exposition of aspects of the chemosynthetic theory. The same paper is also available in *Proceedings of the National Academy of Sciences (U.S.)* 52 (1964), p. 595.

FOURTEEN

THE PROCESS OF EVOLUTION

14.1 Introduction

The concept of natural selection is undoubtedly one of the most important generalizations in modern biology. It was introduced to the world in comprehensive form in 1858 by the English naturalists Charles Darwin (1809–1882) and Alfred Russel Wallace (1823–1913). Darwin's book *On the Origin of Species by Means of Natural Selection* (generally referred to as *The Origin of Species*) of 1859 had significant effects on the intellectual world of the mid-nineteenth century. *The Origin of Species* contained a massive amount of data supporting the concept of evolution in general and the theory of natural selection in particular. No matter how much religious leaders and biologists might have disagreed with Darwin, they could not ignore his work.

Although Darwin's theory contained some gaps and loopholes when it was published in 1859, it has stood the test of time. Today more than ever, evidence shows that the fundamental principles of variation and natural selection that Darwin proposed are valid. The present chapter will discuss the mechanism of natural selection as it is viewed by modern biologists and the ways in which it accounts for the origin of species. First, however, we must stop and ask what a species is.

The Problem of Species Definition

The definition of a species was far simpler in the past than it is today. In the eighteenth century, it was believed that species were fixed and unchanging. Therefore, it was thought that detailed descriptions of plants and animals would apply equally well at any point in time, past, present, or future.

However, as ideas of evolution became widely accepted and paleontological work brought to light many forms with no counterpart among living organisms, it became obvious that the old idea of unchanging species was incorrect. Today, we think of living species as merely the top branches of a con-

Biologists have long attempted to devise a single criterion, such as ability or inability to interbreed, for distinguishing between species. But no single criterion has ever been successful in defining species across the animal, plant, and microbial groups.

stantly growing tree (see Fig. 19.5). If we were to cut across the branches at any level, we would see the species distribution as it existed at that time.

The procedure followed in early classification schemes was to describe a given species in terms of a detailed anatomical study of a few "representative" specimens of that species. During the eighteenth and nineteenth centuries, examples of new and undescribed species discovered in remote areas of the world were sent back to professional **taxonomists** whose job it was to describe and classify the organisms. After examining a new organism, the taxonomist would decide whether it was merely a subdivision of an existing species or was different enough to be rated as a new species. The number of specimens the taxonomists studied, however, was usually limited to one or two of each species, and the study did not take them outside the laboratory or workroom.

Later taxonomists took to the field to study specimens and thus were able to study species not only on a larger scale, but also on the basis of a different criterion—the presence or absence of crossbreeding. As the following example illustrates, breeding patterns reveal species distinctions that would be overlooked in a simple anatomical examination.

A number of different species of the small fruit fly *Drosophila* live in the vicinity of Austin, Texas. Many of these species are quite similar anatomically, so much so that it would take a real expert to note any distinguishing features. The differences are far less extreme, for example, than between a bulldog and a great dane, which are members of the *same* species. Yet, study of these flies in their natural habitat shows that none of these species (about 40) crossbreed with each other. Their physiological differences are great enough so that crossbreeding, or hybridization, does not produce viable offspring. If the older species concept had been ap-

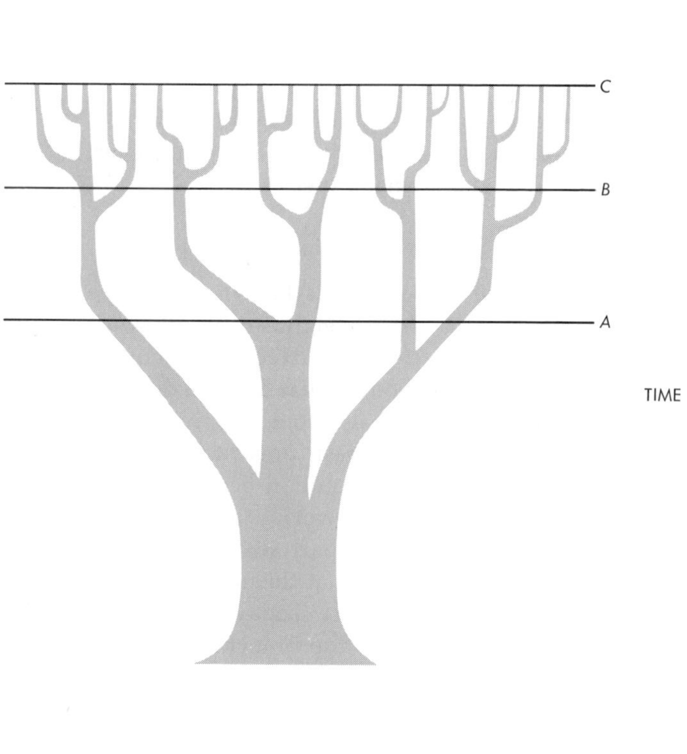

Fig. 14.1 *Schematic view of the evolution of a group of species from a single common ancestor. The tops of the "tree" at level C represent present-day species of one particular genus. Levels* B *and* A *represent different periods of time in the geologic past. The number of species was smaller at time* A *than at time* B. *Species and species groups are constantly changing in time, and modern taxonomy must allow for this.*

plied to this problem, and a few flies studied anatomically under the microscope, the number of and distinction between the species would probably not have been determined. By observing the organisms in relation to their environment, a more reliable grouping of the flies into species is obtained.

The use of interbreeding as the sole criterion for species differentiation nevertheless presents certain difficulties to the modern biologist, who wishes to develop a species concept applicable to all groups in the animal and plant kingdoms. First, although interspecific breeding seldom occurs in the animal kingdom, it occurs rather frequently among plants. Two parent plants capable of producing hybrid offspring may frequently have to be considered separate species because they differ sharply in anatomical and other details. Taxonomists feel that if the plants do not usually interbreed in nature, a few cases of hybridization do not require those plants to be considered the same species.

An example of the reverse side of the coin is found in the animal kingdom, in the case of the grass frog. There is a continuous distribution of this amphibian from Vermont to Florida. Despite the fact that adjacent populations can and do interbreed, the Vermont variety cannot mate with the Florida variety if the two are brought into contact in one or the other environment. However, if specimens from Florida are first conditioned to the water temperature in which Vermont frogs live (or vice versa), mating occurs. How, then, do we classify the Vermont and Florida varieties of the grass frog? Are they separate species? On the basis of the plant example discussed in the preceding paragraph, we might be inclined to say yes, since the two frogs generally do not mate in nature. However, taxonomists *do* consider the two varieties to be members of the same species (*Rana pipiens*). The reason is that because each geographic variety can mate with the variety adjacent to it, genetic characteristics can spread completely throughout the population, from Florida to Vermont.

Another difficulty with the criterion of interbreeding as a basis for a species concept is presented by microorganisms. Many one-celled species do not reproduce by sexual means. It is obvious that they cannot be divided into species on the basis of whether or not they interbreed. Such a criterion is meaningless. Yet species do exist among microorganisms as definitely (in anatomical and physiological terms) as among sexually reproducing forms.

These are just a few of the problems encountered in attempting to arrive at a satisfactory species definition. It is now possible to look at some of the ways in which modern taxonomists have come to view the species and some of their attempts to grapple with attendant problems.

Biologists think that closely related species have diverged from a common ancestor. Divergence can be seen in various stages throughout nature today. What may be only varieties of a single species at one point in time may become distinct species at another. For example, all housecats (Siamese, Persian, Angora) are today considered members of one species, *Felis domestica*. In the distant future, however, they may diverge enough to be considered members of separate species. The species concept developed by modern biologists tries to account for this dynamic quality of species.

The most workable species concept is that which considers the species as an active population of organisms with many anatomical, physiological, and behavioral characteristics in common. In many cases, physiological or behavioral differences can distinguish species where anatomical ones cannot. This way of looking at species may be called the **multidimensional species concept.** It is based on the idea that *no single criterion* is sufficient to define a species. In accordance with the multidimensional concept, the taxonomist tries to take all an organism's aspects—morphology, behavior, physiology, and reproductive patterns—into account in asking whether two groups of organisms represent the same or different species. Let us examine some of these criteria.

Behavioral criteria. Differences in behavior patterns have become one such important criterion in modern animal taxonomy. For example, differences in the mating behavior of the three- and nine-spined stickleback fish are a species-distinguishing characteristic. There is nothing else to prevent them from mating, for when their eggs are artificially inseminated, they produce healthy hybrid offspring. But the three-spined male has one pattern of courtship behavior (a form of swimming "dance") that lures only three-spined females. The nine-spined stickleback has a slightly different "dance," one that attracts only nine-spined females. It is interesting to note that the dances of both males are similar at the start. Therefore, a three-spined male can hold the attention of a nine-spined female until the moment his behavior deviates from the pattern of her species. As soon as it does, she loses interest. In this manner, hybridization is prevented from occurring naturally.

Another example of the use of behavior patterns in taxonomy can be seen in Fig. 14.2. Here, analysis of sound records of cricket calls provides the basis for separating the members of a single genus into different species. Cricket calls are emitted by males to attract females. A female will respond only to a call that closely resembles the pattern for her species. Mating does not occur between species whose patterns differ greatly from each other.

Biochemical criteria. Biochemical differences are an important means by which species may be distinguished. For example, some types of protein, such as pigments or the blood protein hemoglobin, vary in their composition from species to species. The composition of proteins can be determined by the process of electrophoresis and the electrophoretic patterns for the same proteins in different species can then be compared. Since this technique

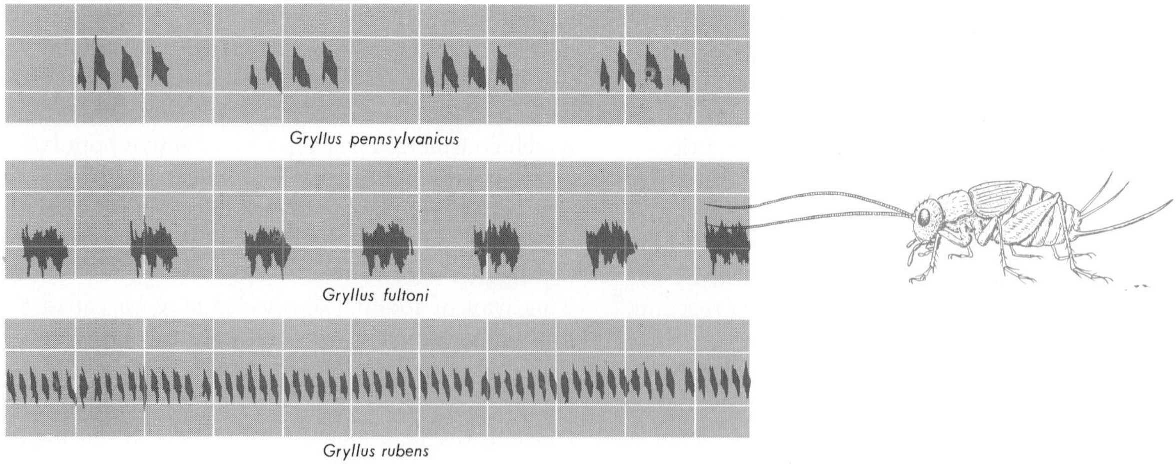

Fig. 14.2 *Sound patterns recorded for three species of the common field cricket. Such techniques reveal that six different species of field cricket exist in the eastern United States alone. The use of traditional taxonomic procedures had earlier led taxonomists to believe that all the crickets in the western hemisphere were of a single species.*

gives quantitative results, the exact differences between two samples become apparent (see Fig. 14.3). Such biochemical studies are especially useful in work with microorganisms, notably bacteria. Two bacteria may appear completely identical anatomically. However, if the sum total of biochemical differences is great enough, they may be considered separate species.

Immunological criteria. Immunological techniques have been rather widely used in taxonomy. As a defense against infection, an organism's body reacts negatively to the introduction of foreign protein and produces a reaction. The more unlike its own protein the foreign protein is, the greater the organism's reaction; thus the chances of success of a transplant operation are greater if a close relative, especially an identical twin, is used as the donor.

The immunological reaction thus has obvious taxonomical significance. For example, although the blood proteins (albumins) of human and rat show a relatively high degree of immunological reaction, far less reaction occurs between the blood proteins of human and monkey, and still less between those of human and ape. Indeed, so little reaction is shown between the blood of human beings and that of gorillas and chimpanzees that Dr. Morris Goodman of Wayne State University in Detroit has stated his belief that these animals should be reclassified and put into the family with human beings (Hominidae) rather than the family of apes (Pongidae).

The studies of blood albumin are nicely borne out by comparisons of the blood hemoglobins of the same organisms. It turns out that human and chimpanzee have identical hemoglobin; the hemoglobins of human and gorilla differ by only two amino acids and those of human and monkey by twelve. Drs. A. C. Wilson and V. M. Sarich have used this information to suggest that human beings and the apes diverged from the monkeys about 30 million years ago and that humans diverged from the ape around five million years ago. There is by no means total agreement on this matter, however, and since Wilson and Sarich's technique is based on the assumption that the mutation rate in any fixed interval of time is constant, there is not likely to be agreement until other techniques either confirm or deny their results. But such work does have obvious taxonomical significance as a means of determining degrees of relationship between living organisms.

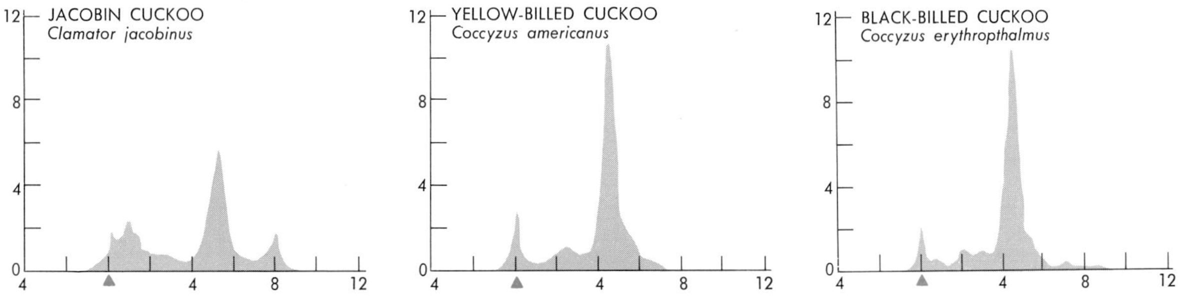

Fig. 14.3 *Electrophoretic patterns of egg white proteins. Note particiularly the similiarities between the species of a single genus. (Based on photos in R. E. Alston et al.,* Biochemical Systematics. *Englewood Cliffs, N.J.: Prentice-Hall, 1963. Courtesy Charles G. Sibley.)*

Today's species concept is multidimensional. It attempts to define species by a number of criteria, such as anatomy, ability to interbreed, ecological role, behavioral traits (animals), geographic distribution, and biochemical characteristics.

Similarity of DNA. A relatively new technique for analyzing interrelationships between species involves comparison of their DNA for similarities. The technique is based on the well-established belief that DNA determines the characteristics of living organisms. The more closely related two organisms are, the more the DNA of one should resemble the DNA of the other. The DNA molecules of one species are embedded in a porous agar. Fragmented DNA molecules of the other species are labeled with a radioactive isotope and percolated through the agar. The more similar the two types of DNA, the more combinations will occur between the labeled and unlabeled DNA fragments, and the fewer radioactive DNA fragments will pass through the agar to be collected again. If the two species are not closely related, very few labeled DNA fragments will be picked up by the unlabeled DNA in the agar. Instead, most of the labeled DNA fragments will pass through the agar and be collected. The technique is represented diagrammatically in Fig. 14.4.

DNA comparison has confirmed many relationships already indicated by investigations based on anatomical or physiological similarities. DNA from the bacterium *Escherichia coli* picks up most of the

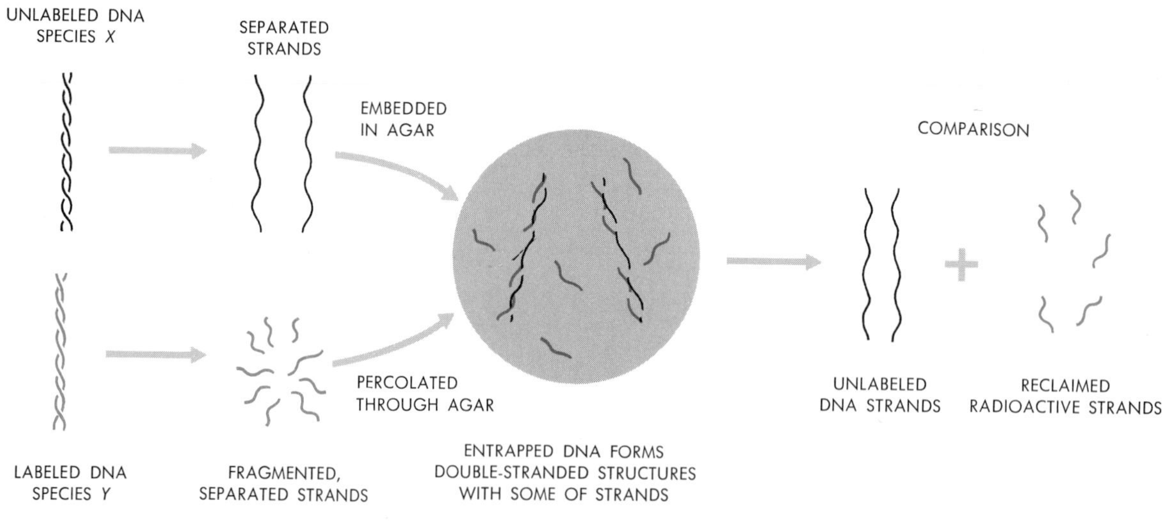

Fig. 14.4 *Diagram showing how comparison of DNA similarities can establish or confirm relationships between different groups of organisms. The DNA shown in color has been labeled with a radioactive isotope. By measuring the amount of radioactivity present in the reclaimed DNA shown on the right, the degree of relationship is established. The more DNA reclaimed, the more distant the relationship between the two groups of organisms.*

labeled DNA from other bacteria but almost none from vertebrates. About 75% of labeled DNA from mice is trapped by rat DNA, while only 60% combination occurs between the DNA of mice and hamsters. Human DNA shows 25% combination with mouse and calf DNA, but only 5% with DNA from a salmon.

With these ideas in mind about the nature of species, we can now proceed to understand how species have evolved over time.

14.2 The Evidence for Evolution

There are three general kinds of explanation for the wide diversity of form in the living world. The first is the idea of **special creation,** which holds that species arose in the past as the result of a supernatural act. At the time of creation, these species possessed exactly the same characteristics they do today. The second kind of explanation is that of **spontaneous generation.** According to this idea, many species arise from previously nonliving matter, as tadpoles from mud or maggots from decaying meat. A third type of explanation is the theory of **transmutation of species** (also called evolution), which holds that over the course of time, one species has given rise to another. All modern theories of evolution incorporate some idea of transmutation, though they may differ as to the mechanism by which one species may actually be modified to become another.

The problem in dealing with the concept of special creation is that it is inevitably linked to belief in a divine being.* Its adherents, therefore, though insistent that special creation be taught in biology courses, are not willing to submit it as a legitimate scientific hypothesis for objective testing. Such a mixing of science and religion is always unfortunate, for it makes an objective weighing of the evidence pro and con virtually impossible; the emotional attachment is simply too strong.

At times such emotional attachment can lead to tragedy. A pertinent example is that of Vice-Admiral Robert Fitzroy. Fitzroy had been captain of the *H.M.S. Beagle,* on which Darwin had sailed in 1831 on a five-year voyage which was to provide him with his evidence for evolution by natural selection. Fitzroy was an ardent believer in special creation, and Darwin had found him a valuable asset on the voyage as a source of opposing opinions against which to test the ideas that the naturalist was beginning to formulate in his mind.

Thirty years later, in Oxford, the famous debate between Bishop Wilberforce, representing special creation, and Thomas Huxley, representing Darwin's viewpoint, took place. A high point of the debates occurred when Wilberforce asked Huxley if it was through his grandmother or his grandfather that he claimed to have descended from the apes. Remarking privately to some friends that "the Lord hath delivered him into my hands," Huxley rose to reply. He stated that he would prefer to be descended from an ape than from a cultivated man who prostituted the gifts of culture and eloquence to the service of prejudice and falsehood. In brief, Darwinism triumphed. But seated in the audience was Vice-Admiral Fitzroy. He rose to his feet and, full of rage and waving his Bible in the air, shouted his regrets at having taken Darwin aboard the *Beagle* for the voyage. As long ago as 1836 Darwin had written to his sister Susan, "I often doubt what will be his [Fitzroy's] end. Under many circumstances I am sure it would be a brilliant one, under others I fear a very unhappy one." On Sunday morning, April 30, 1865, Robert Fitzroy committed suicide. (We do not *know* whether this was a result of the challenge his ideas of special creation faced.)

The Oxford debates had a rerun in the United States in the mid-1920s with the notable Dayton, Tennessee, trial of John Scopes, a high school biology teacher. Scopes was brought to trial for teaching evolution. The prosecution engaged the

* There is nothing wrong in this belief, of course; however, since the existence or nonexistence of a divine being is not subject to scientific testing, an hypothesis based on its existence cannot be a scientific one.

services of William Jennings Bryan, three-time unsuccessful candidate for the presidency and a noted special creationist. The defense countered with the brilliant defense lawyer, Clarence Darrow. The resulting "debate" was similar to the one at Oxford; in the words of the press, "Darrow made a monkey of the man." Scopes was given a minimum fine and a suspended sentence. Bryan died five days later.*

Unlike special creation, the idea of spontaneous generation *is* amenable to scientific investigation and can be experimentally shown to give rise to false predictions. For example, if meat broth spontaneously generates bacteria, and if boiling meat broth kills all previously existing bacteria, then boiled meat broth *should* still spontaneously generate bacteria. When French biologist Louis Pasteur tested this hypothesis he found that the evidence contradicted his expectations. Therefore, according to the truth table (see Fig. 1.4), he could reject the hypothesis with some certainty, at least with respect to spontaneous generation of bacteria.

Eventually the ideas of special creation and spontaneous generation lost ground. In the early and middle nineteenth century people became more favorably disposed to the idea of transmutation. To a large extent this was due to an increase in various types of evidence, such as the findings of paleontology (the study of fossils), taxonomy, and embryology. Evidence from various sources now strongly supports the belief that transmutation of species does occur.

Evidence from Taxonomy

One kind of evidence comes from the area of taxonomy, that division of biology dealing with the classification of organisms. As was pointed out in Section 14.1, it is sometimes extremely difficult to draw a distinct line between two closely related species. As naturalists accumulate more evidence about any species groups, it becomes increasingly obvious that (1) not all individuals in a single species are exactly alike and (2) between two quite distinct species there is often a graded series of intermediate forms. Such evidence supports the idea of evolution from a common ancestor. It is difficult to account for these data on the basis of immutability and special creation.

Evidence from the Fossil Record

A second kind of evidence comes from the fossil record. Studying the sequence of fossils from ancient to more recent rock layers discloses several important items of information to the paleontologist. First, the sequence shows, in general, an increase in diversity as well as complexity of fossil forms. Slight modifications in fossil forms from one stratum to another seem indicative, in many cases, of the slow modification of forms over time. While this evidence does not exclude the ideas of special creation or spontaneous generation, it certainly supports the idea of transmutation. Second, some fossil forms have no living relatives; these forms are said to be **extinct.** The phenomenon of extinction shows that some species have been unable to perpetuate themselves. By extension, if old species die out, cannot new species originate from previous ones? Given the alternative theory of special creation, it would be necessary to suppose that the Creator whimsically allowed some species to perpetuate themselves while causing others to die off. The fact that there are many more extinct species than species presently alive on the earth would indicate that the Creator's whims must have been rather frequent. And third, the fossil records sometimes show forms intermediate between two presently living types. For example, it is now well established that birds and reptiles diverged from a common ancestor. The most convincing item of information on this score was the discovery of *Archaeopteryx*, a fossil form intermediate between

* Somewhat later Bryan College, seemingly committed to the special creation "hypothesis," was founded in Dayton—an act comparable, perhaps, to the founding of a "Napoleon Tech" at Waterloo.

Supplement 14.1

EVOLUTION AS AN HYPOTHESIS

On April 30, 1973, Senate Bill 394 became law in the state of Tennessee. The bill requires that all textbooks used for teaching biology within the state of Tennessee must give equal time and emphasis to theories of the origin of species that are alternatives to the concept of evolution. The law refers not only to secular theories such as spontaneous generation, but also to religious accounts such as that found in Genesis. Urged on largely by religious groups, similar bills have been, or are currently (1976) being considered by the legislatures in a number of other states. Only Tennessee's, however, has become law. Tennessee's former anti-evolution law, under which school teacher John Scopes was tried in 1925, was repealed in 1967. By and large biologists have felt that a demand such as the one formalized by the Tennessee Legislature is both unnecessary and unreasonable.

What can this controversy indicate about the nature of science and the process of accepting or rejecting hypotheses in any area of rational thought? Let us keep in mind first that the bill refers to the teaching of the theory, or hypothesis, of *evolution,* rather than to Darwin's (or anyone else's) hypothesis for how evolution is brought about. Evolution is one hypothesis to account for the universally accepted observation that there are many different species of animals and plants inhabiting the earth. Other hypotheses attempt to account for these same observations. One is the idea of "special creation": species have been separately and specially created by a supernatural power, as described in Genesis. There are two ways to approach the issue of whether special creation and evolution (transmutation of species) are both legitimate scientific hypotheses and should thus be included together in textbooks. One way is to examine the logic of the two hypotheses; the other is to ask in what way or ways each corresponds to people's daily experience.

Logic

With regard to logic, anti-evolutionists argue that most modern biology textbooks present the hypothesis of evolution as an accepted *fact* and thus do not emphasize its hypothetical nature. This criticism has some justification, in that too often hypotheses are presented in science as if they had no controversial elements. In biologists' view, however, evolution is the best hypothesis developed to date to account for the origin of widespread diversity in the living world.

On what basis can such a claim rest? Biologists prefer the hypothesis of evolution to belief in special creation because it is testable. The hypothesis of evolution can be tested against the available fossil record. By definition, special creation puts the issue out of the range of testability. Creation occurred once in the time before human beings existed; it occurred by supernatural processes of which by definition we can never have full knowledge. Thus, to accept the notion of special creation requires an act of faith. While acceptance of the notion of evolution requires faith, too, it is faith of a sort that the human mind can rationally evaluate by examination of sensory data.

Another reason most biologists prefer the hypothesis of evolution to that of special creation is that evolution accounts for more of the observed facts about living organisms than special creation. For example, Huxley's remark that "God must have been inordinately fond of beetles" points up the difficulty of imagining a Creator creating separately the several hundred thousand species of them on earth today. Or consider the observation that many species of organisms show a large number of anatomical and physiological characteristics in common. By the theory of special creation, one must assume that the Creator was either lazy or simply liked certain anatomical and physiological schemes and thus used them repeatedly. Such assumptions require a further stretch of faith. Or consider the observation that fossil remains often show some linear progression over time, from more ancient species to their modern counterparts. An inseparable part of that series of observations is the fact that all the more ancient forms have become extinct—they have ceased to exist. By the theory of special creation, it would have to be assumed that the

Creator came to dislike certain species in the past and allowed them to become extinct. Such an assumption implies imperfection or caprice on the part of the Creator. One way out of such a dilemma was to assume, as did numerous eighteenth-century naturalists and religious zealots, that the Creator placed fossils in the strata as decoys, to lead arrogant human beings astray if they attempted to unravel the mysteries of creation!

The hypothesis of evolution can account for each of the above observations in a far less complex and capricious way. Similarity of body plans for a group of organisms can be explained as a result of descent with modification from a common ancestor. Extinction and continuity among certain fossil lines can be explained as a natural outcome, in each circumstance, of the modifiability of species and their environments. Special creation requires that new assumptions be invented specifically for each new set of observations.

Evolution can account for each set of observations by logical extensions of its most basic premises, that species are modifiable and pass some of their modifications on to their offspring.

Experience

More important, in each case where the hypothesis of evolution requires further subhypotheses or extensions, those extensions are in agreement with people's everyday experience or knowledge from other fields. It does not require a new and particular set of assumptions with no basis in experience. For example, we all know that no two organisms of the same species are alike, but that each has its own individual differences. We also know that weather and environmental conditions change in areas over time; few people would doubt the geological evidence of an ice age that capped much of the northern hemisphere with glaciers. We also know from agricultural experiments that selective breeding for certain characteristics (such as high milk production in cows) produces considerable change over a number of generations. All these observations from people's everyday experience lend support to the idea of species modification with descent.

Given the fact that most biologists today have a logical reason for preferring the hypothesis of evolution to that of special creation, is there any justification for opposing the inclusion in textbooks of the theory of special creation as an equally viable hypothesis? Anti-evolutionists argue that this attitude on the part of biologists is dogmatic and does not give students a chance to come to grips with all sides of the issue for themselves. Anti-evolutionists also claim that biologists indoctrinate students with their side of the story, systematically excluding the other side.

Biologists have responded to such criticism in several ways. Some argue simply that special creation is a religious belief and thus has no place in a scientific textbook. Even were special creation true, since it cannot be tested, it is beyond the realm of science. These arguments are the most practical ways to approach the controversy from the scientific side, but they avoid one of the most intriguing issues that the controversy raises.

The long-standing debate between biologists and religious advocates over evolution is a reflection of two very different world views. It boils down to a debate between rational and nonrational philosophy. It is not a debate that can be settled in a few words, nor can either viewpoint be conclusively demonstrated to the satisfaction of its opponents. The rational view of the world is based on philosophical materialism, as described in Chapter 1. It maintains that people's ideas are developed from their concrete daily experience rather than being "revealed" to them from some unknown and unknowable source. A fundamental premise of the rational, materialist view of the world is that all phenomena are eventually understandable. While not claiming that everything is already known or that there will ever be a time when everything is known, the rational view argues against the existence of unknowable forces, or beings, or spirits at work in the universe.

In contrast, the nonrational view of the world, as reflected in most traditional religions and in secular movements such as astrology, is essentially idealist. It maintains that people's ideas come from nonmaterial experience, such as divine revelation or inner "energies" of some unspecified sort. The nonrational view maintains that some aspects of human experience are by nature unknowable and that all events in the world do not conform to the same basic laws. The nonrationalist view is unwilling to see all aspects of the world explained in rational terms. The origin of species has historically been one realm in which nonrationalists have argued most vehemently against the introduction of rationalist methods. No doubt this is because findings in this area are related to the origins of human beings themselves.

The rationalist and nonrationalist views of the world are in fundamental conflict. The very methods of the one are considered inapplicable by advocates of the other. The two can be reconciled only by partitioning off human experience into two categories: an area in which rational thought prevails and an area in which nonrational thought prevails. The fact that in the past many scientists have held strong religious beliefs indicates that some people are able to partition off their experience in this way. Yet the history of the past 300 years has shown gradual reduction in application of nonrational methods and gradual increase in the application of rational methods to all areas of human experience. The reason is that the rational methods of observation, logic, and testing of ideas against experience have proven to be most effective for people, not just scientists, in dealing with all aspects of the world around them. As earlier parts of this book have tried to suggest, this method is not peculiarly "scientific." It is a method applicable in all areas of human experience, from psychology to history to art.

Because belief in special creation is based on nonrational methods, most biologists do not think it should be included on equal footing with the idea of evolution in textbooks or courses. They argue that to include it as an equally valid way of explaining the origin of species would be philosophically untenable and revert to a way of thinking that is outmoded in the present world.

these two groups (Fig. 14.5). *Archaeopteryx* had teeth and a long tail (characteristic of reptiles), and it also had feathers (characteristic of birds). Such genuine "missing links" are strong evidence for transmutation of species.

A noticeable feature of the fossil record is that it contains distinct gaps. The lines of demarcation between adjacent layers of rock are often quite sharp (Fig. 14.6). For many years (especially in the eighteenth and nineteenth centuries) these gaps posed problems for theories of evolution. Many times the fossils in one layer were totally different from those in the layer just above or beneath it. In the eighteenth and early nineteenth centuries, these gaps were thought to be the result of great cataclysms that swept the earth, destroying all or nearly all the life forms. It was supposed that after each successive cataclysm, the earth was repopulated with new forms by acts of special creation. This view of the geological record was called **catastrophism.**

More recent geological evidence has shown why such gaps occur. Changes in the level of the earth's surface cause regions to be raised or lowered from one period to another. This elevation of fossil-laden sediments exposes them to such eroding forces as wind and water. Thus whole layers, along with the fossils they contain, may be lost forever. Since such layers may represent vast periods of time, considerable evolutionary history will be missing. Furthermore, a layer of sediment deposited when an area is under water will be quite different in appearance and composition from a layer deposited when the area is dry. The fossil forms will obviously be different as well.

Evidence from Comparative Anatomy

A third kind of evidence is that of comparative anatomy. Comparative anatomy is the study of similarities and differences between the anatomical structures of two or more species. Reasoning from the hypothesis that evolution does occur, we can say that if organisms descend from ancestral forms

The fossil record shows that many species of the past are now extinct. It also shows that many of those species have modern counterparts with intermediate forms progressing up the geological strata.

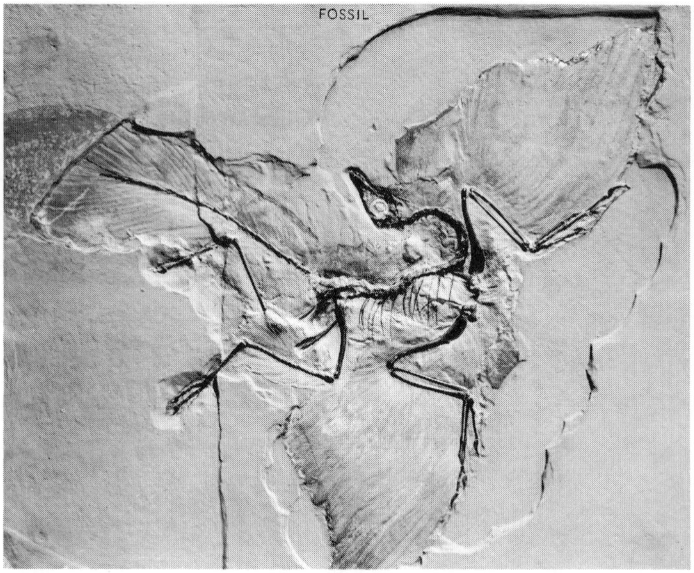

Fig. 14.5 *Fossil of the primitive bird* Archaeopteryx, *which clearly represents the evolutionary development of modern birds from primitive reptiles. Note the long tail and the claws on the wing, which are reptilian characteristics, contrasted with the feathers and the general structure of the appendages, which are avian features. (Courtesy American Museum of Natural History.)*

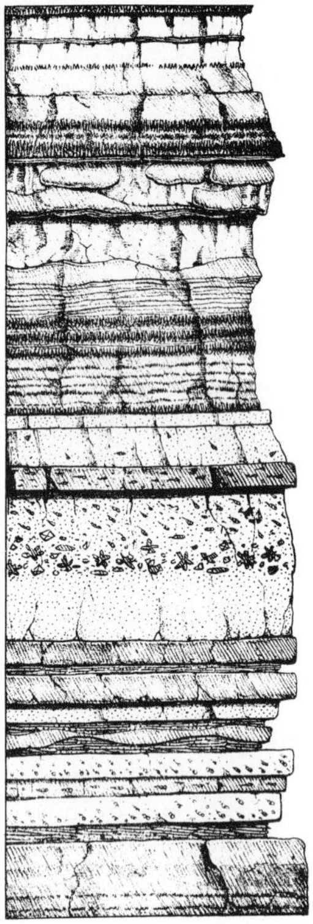

Fig. 14.6 *An illustration from a study in paleontology and geology published early in the nineteenth century, showing the distinct breaks in strata, with similar changes in their fossil content, that led many naturalists to the idea of catastrophism. Distinct divisions between layers were taken to indicate great upheavals by fire or flood.*

Comparative anatomy demonstrates the many similar patterns of form that exist in the animal and plant kingdoms. One way of explaining the existence of similarities of pattern is through descent with modification from a common ancestor.

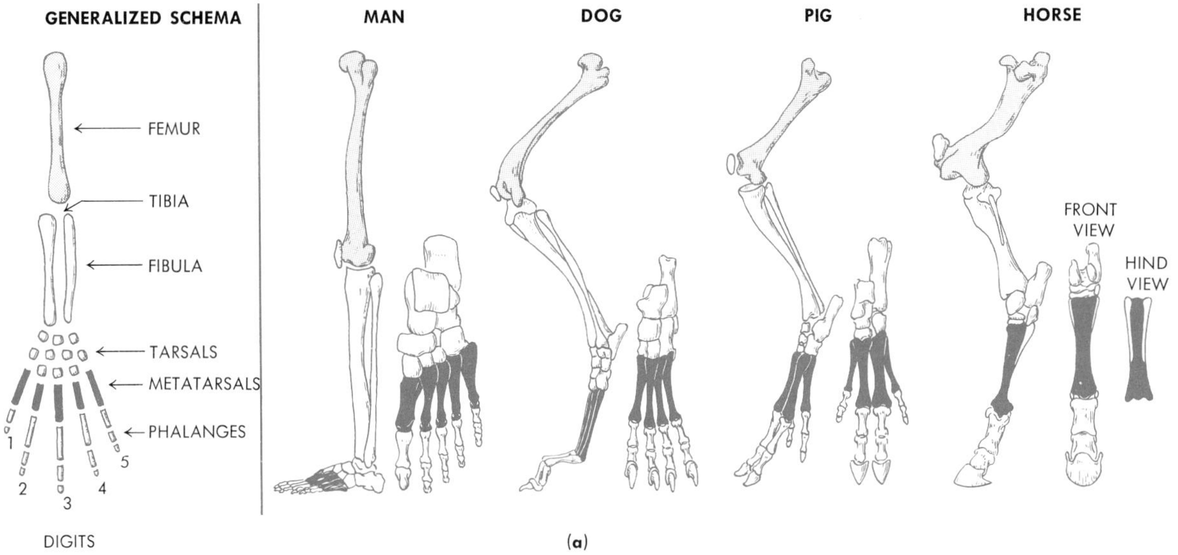

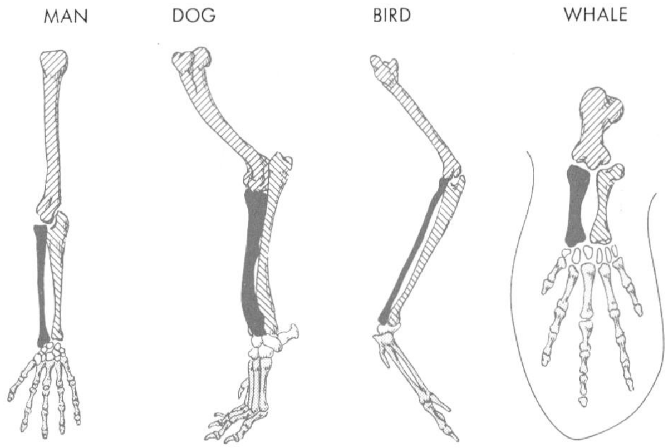

Fig. 14.7 *Homologous bones in the forelimb and hindlimb of several vertebrates. (a) Hindlimb homology. To the left is a generalized scheme for the vertebrate limb. (b) Forelimb homology. Note that though the forelimbs of these four vertebrates are all used for quite different functions, they nevertheless show a similar basic pattern. (Adapted from* Life: An Introduction to Biology, *by G. G. Simpson, C. S. Pittendrigh, and L. H. Tiffany, copyright © 1957 by Harcourt Brace Jovanovich, Inc. Used by permission of the publisher.)*

by modification, then we should expect to find large groups of animals (or plants) that are anatomical variations on the same theme. This prediction says, for example, that because the human, bat, pig, and dog are all mammals, the bone structure of their appendages should have many features in common, despite some obvious differences. The work of comparative anatomists has shown that this is indeed the case (Fig. 14.7a). Although the forelimb of each is modified for a different function (the human's for grasping, the bat's for flying, the pig's and dog's for walking), the similarities in bone structure are still evident (Fig. 14.7b).

Comparative anatomy emphasizes the difference between **homologous** and **analogous** structures. Homologous structures are those that have basic

Homologous structures are those that are derived from the same basic embryological structures in different species; the adult structures may or may not perform the same function (such as the arm of a person and the front leg of a horse). Analogous structures are those that do not have any embryological history in common but serve the same function in the adult (such as the wing of a butterfly and the wing of a bird).

anatomical features in common but perform different functions. For example, the wing of a bat, the leg of a horse or dog, and the hand of a human being show homologous bone structures. Analogous structures, on the other hand, are used for similar purposes but are of different origin and are not necessarily built on the same anatomical plan. The wing of a bird and the wing of a butterfly are analogous but not homologous, since they do not have any basic structural units (bones, etc.) in common.

The idea of evolution from a common ancestor is strongly supported by the study of homologous structures. In cases in which different animals have started with the same ancestor, subsequent generations show slight modifications of the basic anatomical plan. Homologous structures thus give an indication of the degree of divergence between two related forms.

The existence of homologous organs, or the existence of similarities in the structure of various body parts, does not refute the idea of special creation. There is just as much reason for believing that the Creator would choose a common plan for building His creatures as that He would choose different plans for each. However, one of the results of studies in comparative anatomy is the discovery of **vestigial structures.** Vestigial structures are those that appear in the organism in a seemingly functionless role. The human appendix, hip bones in snakes, and rudimentary legs in whales are all vestigial structures. It is easy to explain the existence of such structures by the idea of descent. They represent useless or inoperative parts against whose existence natural selection is working. By the concept of special creation, however, it is difficult to explain why such parts exist.

Evidence from Genetics

Evidence from the field of genetics has made important contributions to the idea that transmutation of species occurs. Although details of the genetic interpretation of natural selection will be discussed in a later part of this chapter, some mention of this important line of evidence will be made here. It has been known since the early years of this century that genes are quite constant and are inherited on a statistically predictable basis. Generally speaking, the constancy of genes tends to keep species from changing. However, mutations do occur, introducing new heritable variations into a population. Experiments under carefully controlled laboratory conditions show that genes can have different degrees of "fitness" to any given environment. If selection is practiced by the biologist, either directly (by allowing only organisms with one particular genotype to breed) or indirectly (by placing organisms in environments that will favor one genotype over another), it is possible to alter the genotype of a laboratory population in a very few generations. Experimental evidence for change of genotypes through selection gives strong support to the idea of transmutation in general and natural selection in particular.

Evidence for evolution comes from other areas

as well—from embryology, for example, and from the study of the geographic distribution of animals and plants. It is not necessary to go into details in these cases, since the preceding examples have provided ample evidence for the origin of species by evolution from preexisting forms. The question that arises now is: By what mechanism does evolution occur? There have been many attempts to answer this question in the history of biology. Charles Darwin's concept of natural selection has provided the most satisfactory explanation (see Fig. 14.8).

Fig. 14.8 *Charles Darwin (1809–1882), an English naturalist and author of* On the Origin of Species by Means of Natural Selection, *a work that laid the foundation for modern evolutionary thinking. (Photo courtesy the Burndy Library, Norwalk, Conn.)*

14.3 Darwinian Natural Selection

The basic process of evolution by natural selection can be summarized by the following observations and conclusions.

Observation 1: All organisms have a high reproductive capacity. A population of organisms with an unlimited food supply and not subject to predation could quickly fill the entire earth with its own kind. Under optimal conditions, all organisms have an enormous capacity for population growth.

Observation 2: The food supply for any population of organisms is *not* unlimited. The growth rate of the population tends to outrun the growth rate of the food supply.

Conclusion 1: The result must be continual competition among organisms of the same kind (organisms that have the same food requirements).

Observation 3: All organisms show heritable variations. No two individuals in a species are exactly alike.

Observation 4: Some variations are more favorable to existence in a given environment than others.

Conclusion 2: Those organisms possessing favorable variations will be better able to survive in a given environment than those that possess unfavorable variations; thus, organisms with favorable variations will be better able to leave more offspring in the next generation.

Selection in the Animal Kingdom

To demonstrate how natural selection operates in nature, let us consider the following case. For over a century, two varieties of the peppered moth have been known to exist in various parts of England.

▲ **Fig. 14.9** *On a lichen-encrusted tree trunk in non-industrial regions of England, melanic forms of the peppered moth stand out conspicuously. The light form of the moth, however, is barely discernible (center). Under such conditions, the melanic form is subject to heavy predation. (Photo courtesy Dr. H. B. D. Kettlewell.)*

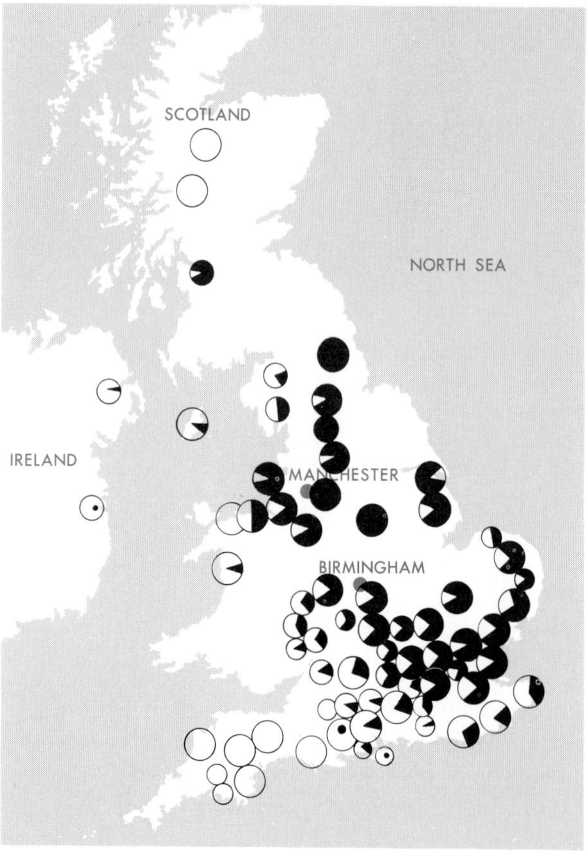

◄ **Fig. 14.10** *Distribution map of forms of the peppered moth in England. Each circle represents a population sample taken in a given area. The colored portion of each circle indicates the percentage of melanic forms found in the area, while the white portion indicates the percentage of light-colored forms. (Reprinted with permission. Copyright © 1959 by Scientific American, Inc. All rights reserved.)*

Fig. 14.11 *On a tree trunk darkened by soot from industrial areas, the light-colored form of the peppered moth is much more visible than the melanic form. In this situation, the melanic form has greater survival value than the lighter variety. (Photo courtesy Dr. H. B. D. Kettlewell.)* ►

One form, *Biston betularia,* is light-colored with small dark spots irregularly scattered over its wings and body. The other, *Biston carbonaria,* is much darker, due to the presence of the pigment melanin. This latter is often called the "melanic" form.

In the past, samples collected in the field showed that the light form was far more common than the dark. This was explained by showing that the light form was protectively colored. **Protective coloration** is the biological counterpart of military camouflage. Organisms that are protectively colored inherit certain patterns of pigmentation that allow them to blend with their backgrounds. On tree trunks covered with lichens,* light varieties of the peppered moth are perfectly camouflaged, but the darker form stands out prominently (Fig. 14.9). The dark form is thus more subject to predation by birds than the light form.

Nevertheless, in the past hundred years the number of melanic moths has been observed to increase quite drastically. In some areas the dark form has almost totally replaced the light. A map (Fig. 14.10) comparing distribution of melanics and light-colored moths shows that the darker form predominates in the industrial centers and the regions to the southeast. Investigation of these regions shows that the smoke and soot from factory cities has significantly darkened the natural background of these regions. In addition, prevailing southwesterly winds have deposited a good deal of soot in the eastern regions, which themselves are not industrial. This has been a very important factor in the evolutionary history of the peppered form.

On a darker background, the light-colored moths that had previously been well camouflaged became much more visible (see Fig. 14.11). As a result, these forms were put at a distinct disadvantage. Being more easily detectable, the light varieties became subject to greater predation from birds. The number of light-colored moths able to reach maturity and reproduce decreased proportionately. The genes for light color consequently reached fewer members of the next generation than before the change in background.

It might be hypothesized that the melanic form of the peppered moth arose originally by a mutation of the gene controlling coloration. This mutation was disadvantageous on the original light-colored background, and natural selection was unfavorable to it. However, through the change in the environment due to industrial growth, the mu-

* Lichens are greenish-gray growths found covering many rocks and trees. A lichen is actually a combination of an alga and a fungus. The moths often rest on the lichen-covered tree trunks during the day.

The evolutionary process produces change in species through two phenomena: the occurrence of inherited variations and the action of natural selection.

tant was favored. The result was that the melanic genes spread through the population. At the present time the melanic form has almost replaced the lighter form in certain areas, indicating the greater selective advantage of the former. The above hypothesis can become the basis of a prediction. If natural selection favors darker moths in industrial regions and lighter moths in nonindustrial regions, then releasing equal numbers of light and dark forms in both regions and recapturing the survivors after a period of time should show definite changes. For example, in the industrial region, far more dark forms should be recovered than light forms; in the nonindustrial areas the reverse should be true.

An experiment was designed in the early 1950s to test whether the dark form was indeed selectively favored in industrial regions as compared to nonindustrial regions. Dark and light forms of the moth were raised in the laboratory and released in roughly equal numbers into each type of environment. Before release, the moths were carefully marked with cellulose paint, a different color being used for each day of the week. In this way, it was possible to determine when a given moth had been released and how long it had survived. Other moth species were also released. (Why?)

After the experiment had proceeded for a certain period of time, large lanterns were set up in the woods at night and the surviving moths were recaptured. Table 14.1 shows the results, tabulated as percentage of the moths recaptured. These data suggest that more of the darker forms survive in the industrial region and more of the light forms in the nonindustrial region. It is apparent that the differences in the pigmentation of the moths are adaptive to their respective environments. The above test therefore supports the hypothesis that natural selection favors the dark form in industrial areas and the light form in nonindustrial areas.

Table 14.1

	Percent of moths recaptured	
Area	**Dark form**	**Light form**
Dorset woods (nonindustrial)	6.34	12.0
Birmingham (industrial)	53.2	25.0

This example of industrial melanism illustrates how a previously unfit organism (the dark form) has been rendered *more* fit by a change in the environment. This change likewise caused a well-adapted form (the light-colored moth) to become *less* fit. The mutant gene for melanism did not occur in response to the environmental change. As the experiments of Luria and Delbrück on bacteria indicate, mutations are not produced in this way. The mutant gene had always existed in a small percentage of the population. Spread of the gene was possible only when changed conditions favored those organisms that carried it. In Darwinian terms, it is the nature of the environment that acts as the "selecting agent."

Selection in the Plant Kingdom

Another example of the action of natural selection in nature was studied in a pasture in southern Maryland. The owner planted a mixture of grass and clover (*Trifolium*) seed. After the planting was completed, the owner installed a fence through the middle of the field, dividing it into two parts. One side was cut for hay a few times during the summers, while the other half was heavily grazed by livestock. Three years later, a botanist dug up plants of the various species of grass and clover and transplanted them to his experimental garden. After a period of growth under uniform conditions, a high proportion of the grasses and clover from the grazed half of the field produced a dwarf, rambling growth. However, the grass and clover plants taken from the ungrazed half were erect and vigorous (Fig. 14.12). When seed was raised from these plants, the differences were found to be inherited.

What is a reasonable explanation for these observations? In the field where livestock had grazed for three years, only those plants small and low enough not to be eaten were able to produce seed. However, in the field where the plants were mowed occasionally for hay, the tall grasses and clover were able to produce seed. In this field selection favored the tall plants.

14.4 Direction of Natural Selection

The evolutionist George Gaylord Simpson has defined selection as anything producing systematic, inheritable changes in populations between one generation and the next. This definition requires that there be direction in the way natural selection operates. The genetic makeup of the population may change in one direction, with respect to certain genes and genotypes. Such **directional selection** may favor one nonaverage or extreme phenotype,

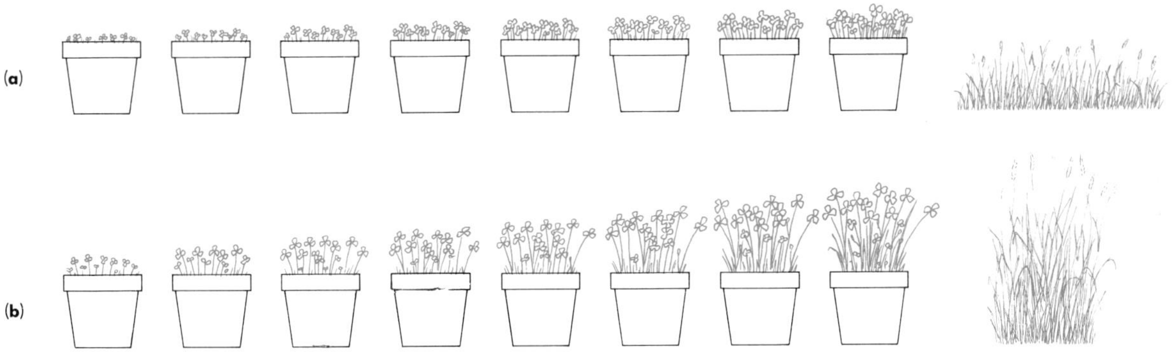

Fig. 14.12 *Diagrams illustrating the action of natural selection on the growth form of different plant species. (a) Plants transplanted from grazed field and grown under uniform conditions. Note dwarf growth habit. (b) Plants transplanted from ungrazed field and grown under uniform conditions. These plants are generally much taller.*

As Darwin emphasized in The Origin of Species, *the only criterion for success in an evolutionary sense is differential fertility: the ability to leave behind more offspring than other members of one's own species.*

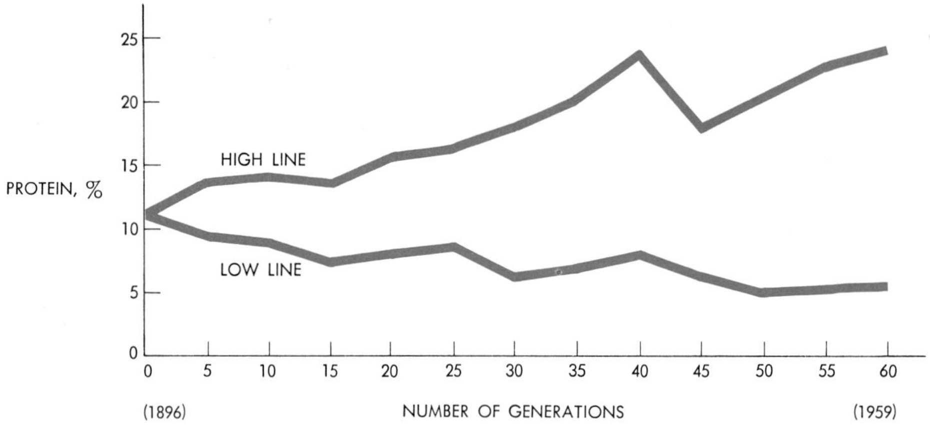

Fig. 14.13 *An example of directional selection for the protein content in grains of corn during 60 generations. (Data from C. M. Woodworth, E. R. Leng, and R. W. Jugenheimer, "Sixty Generations of Selection for Protein and Oil in Corn."* Agronomy Journal *44, 1952, pp. 60–65.)*

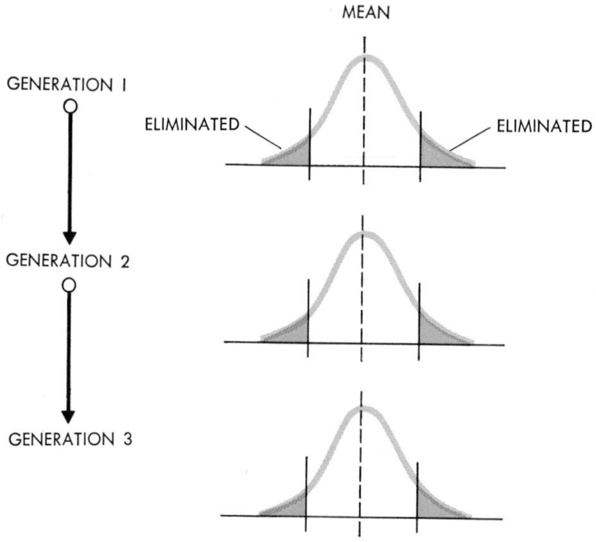

Fig. 14.14 *Normal curves of distribution illustrating the concept of stabilizing selection. Note that the extreme variant individuals on either end of the curve are eliminated from the population each generation.*

The net outcome of variation and selection is to produce organisms better adapted to their particular environments. Adaptation means increased effectiveness in using environmental resources and ultimately manifests itself in increased production of successful offspring.

---

pushing the population of phenotypes (and their genotypes) in the direction of this particular phenotype. This has happened, of course, in human domestication of crop plants and animals. For example, agronomists at the University of Illinois selected corn for high protein content in the grains of corn plants for 60 generations (Fig. 14.13). Each year the ears highest in protein were chosen to plant for the next generation. Throughout most of the 60-year period, there was a steady increase in the protein content of the grains from an average of 10.9% in the original population to an average of 19.4% in the sixtieth generation after the selection began. While this experiment was going on, the agronomists also selected for low protein content in another line originating from the same population of corn. The protein content of this corn steadily dropped from the starting average of 10.9% to 4.9% after 60 generations of selection. Thus artificial selection resulted in different directions of change in the populations: the grain of one line of corn increased in protein content while the other decreased. The agronomists were able to shift the population average for protein content of the corn from an original value of about 11% to over 19% for the upward and about 5% for the downward selection. The populations were responding genotypically to the selection of the phenotypes. This is directional selection.

But natural selection may also act to bring about no change in direction; in other words, natural selection may act to keep a population genetically constant. Such **stabilizing selection** occurs whenever the population has reached a high degree of adaptation to its particular habitat and the environment remains stable for long periods of time. By favoring the average or normal individuals and eliminating the extreme variants, stabilizing selection keeps the population genetically constant (Fig. 14.14). Botanists can infer that stabilizing selection occurs in plant species. This seems to be the most likely explanation for the existence of numerous species in relatively stable habitats for millions of years without appreciable change. For example, the fossil record shows that the giant sequoia (*Sequoiadendron giganteum*) of the Sierra Nevada Mountains in California has remained stable for millions of years.

14.5 The Population Concept of Evolution

One of the most important contributions to modern evolutionary theory has come from the field of genetics. Genetics has shown how variations arise and how recombination of characteristics takes place, and it has provided the means of treating evolution on the *population level.* The latter contribution is of great importance to a thorough understanding of the mechanisms of natural selection.

Population genetics is the field of biology that studies the genetic composition of whole animal or plant populations. The population itself, rather than individual cells or organisms, becomes the basic unit of biological study.

How is it possible to describe the genetic makeup of an entire population? This is not quite as difficult as it may sound. In the first place, the genetic characteristics of the population can be treated as the sum of the genotypes of all the individual members. Thus we may speak of the **gene pool** of a population. The number and kind of each allele found in the population determine the unique characteristics of the gene pool.

In the second place, population genetics uses statistical methods. Individual variations in organisms do not enter the picture unless a given variation becomes statistically significant. Only if enough organisms turn up with the same variation in a given generation will this variation become apparent as a definite and nonrandom change in the composition of the gene pool. Such methods of analysis are particularly important in studying the genetic changes involved in the evolutionary process.

By operating on individual organisms, natural

selection ultimately brings about a change in the composition of the gene pool to which these individual organisms contribute. The gene pool of a population is divided up each generation and parceled out to new individual members (the offspring). Inevitably new mutations or combinations of genotypes occur that express themselves as specific individual phenotypes. As a result of natural selection, certain genotypes will leave more offspring than others. Certain genes will be passed on to the next generation in greater numbers, others in less. The composition of the gene pool will be changed accordingly.

Genetic Equilibrium

Two opposing factors are at work on the gene pool of a population. One is natural selection, which tends to alter the composition of the gene pool from one generation to another. The other factor is expressed in the concept of **genetic equilibrium**, which holds that under very specific conditions the ratio between various alleles in a population tends to remain constant from one generation to the next. This concept applies regardless of the proportions of the genes in the initial population. The idea of genetic equilibrium was introduced into biology in 1908 by G. H. Hardy, an English mathematician, and Wilhelm Weinberg, a German physician; hence it is usually referred to as the Hardy–Weinberg law. Although it is sometimes thought by nonbiologists that recessive genes must eventually be wiped out of a population, according to the concept of genetic equilibrium this is not necessarily the case. For example, despite the fact that blue eyes are recessive to brown, the number of blue-eyed people in the human population remains relatively constant from one generation to another.

We mentioned earlier that genetic equilibrium would be maintained in a population *only* under very specific conditions. These conditions are as follows:

1. *Mating in the population must be random.* That is, each phenotype must have an equal chance of reproducing. If this were not the case—if one male phenotype were more acceptable to the females than another—then the favored males would leave more offspring. For example, if female birds of a given species always choose the males with red plumes, the genes producing this phenotype will spread throughout the population.

2. *All matings must yield, on the average, the same number of offspring.* Again, if this were not the case—if mating between two particular phenotypes yielded consistently fewer offspring—then the alleles of those phenotypes would decline in number with each generation.

3. *The population must be sufficiently large that chance variations in a small number of organisms do not affect the statistical average.* In small populations the effects of random mutations or immigration or emigration of a few organisms can produce statistically significant changes in gene ratios. By analogy, a failing grade in a class of three affects the class average far more than in a class of thirty.

4. *The mutation rate must have reached its own equilibrium.* This means that the frequency with which A mutates to a is equaled by the frequency with which a mutates back to A. As a result, the overall ratio of A to a will not change from one generation to the next.

These factors represent *idealized conditions* seldom met in natural populations. Yet, to the extent that they do occur, they tend to check or balance the opposing forces of selective mating, natural selection, small population size, and mutation.

The Hardy–Weinberg Law

The Hardy–Weinberg law is a cornerstone of the modern population approach to evolution. It can be expressed in quantitative terms that emphasize the exact nature of genetic equilibrium. In its simplest form, the law speaks to the case of a single gene locus with two alleles, *A* and *a*. There are three possible genotypes represented by a two-allele system: *AA* (homozygous dominant), *Aa* (heterozygous), and *aa* (homozygous recessive). The Hardy–Weinberg law states that the frequencies of these genotypes will remain constant from generation to generation, as long as the conditions outlined in the last paragraph prevail. In other words, the Hardy–Weinberg law states that:

> The frequency of *AA* + the frequency of *Aa* + the frequency of *aa* always equals 1.00 (or 100%).

By definition, the three genotypes possible will always represent 100% of the combinations in the population. Whatever the original frequency of each genotype, the proportions will remain the same from one generation to the next. A corollary of this expression is that the frequencies of the individual alleles (either *A* or *a*) will remain the same from one generation to another.

Customarily, the Hardy–Weinberg law is stated in terms of the arbitrary symbols p and q. Each symbol stands for one of the two alleles in a two-allele system (either for *A* or for *a*). The homozygotes are written mathematically not as pp or qq, but as the *squares* of each symbol. Thus, if $p = A$ and $q = a$, then $AA = p^2$ and $aa = q^2$. The heterozygotes would thus be written as pq. The mathematical formulation of the law is:

$$p^2 + 2pq + q^2 = 1.00$$

To understand the logic behind the Hardy–Weinberg law and its quantitative formulation, consider the case of random mating in a population of diploid organisms, as shown in Fig. 14.15. The basic hereditary pattern is Mendelian. If the original population, represented by the five organisms shown, consists of 49% *AA*, 9% *aa*, and 42% *Aa*, what will be the frequencies of these genotypes in the next generation? Further, what are the frequencies of the individual alleles *A* and *a*, as opposed to their frequencies in various homo-

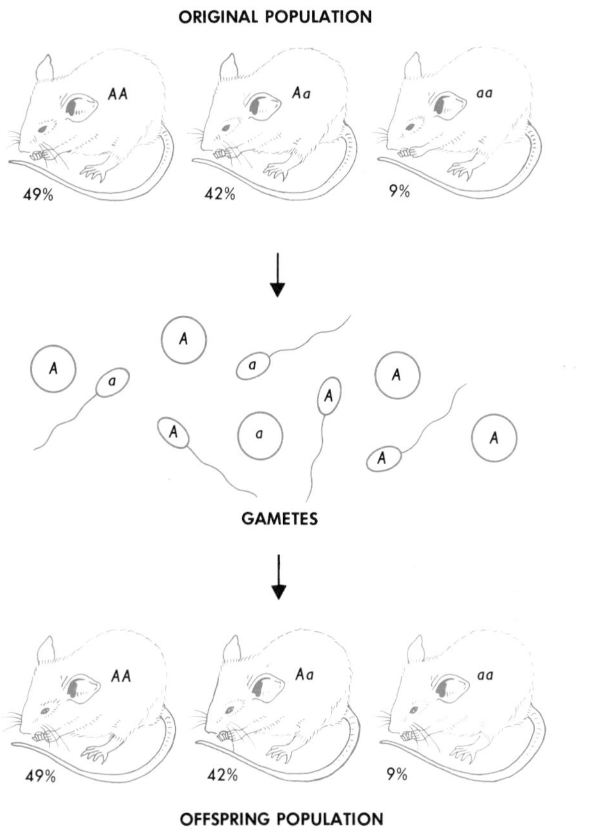

Fig. 14.15 *If 49% of the original parental population has a gentoype of* AA, *42% a genotype of* Aa, *and 9% a genotype of* aa, *the Hardy–Weinberg law states that the same precentages will be found in the next generation. With 49% of the original population* AA, *42%* Aa, *and 9%* aa, *the frequency of the* A *allele is 0.7, and that of the* a *allele 0.3. Thus, of the total gametes produced by the parental population (10 gametes are shown here), 7 are* A *while 3 are* a. *Random combinations of these alleles will yield pairings in the same proportions as existed in the original population.*

zygous and heterozygous combinations, in the parental and offspring populations?

First, to calculate the gene frequencies for each allele, we must translate percentage of each particular genotype in the population into gene frequency. This can be done in a simplified way. Suppose that the original population consists of 100 organisms. According to the given distribution of genotypes, 49 would be *AA*, 42 *Aa*, and 9 *aa*. For simplicity, let us assume that each organism produces 10 gametes. The distribution of each allele would be as follows:

	GAMETES *A*	GAMETES *a*	TOTAL
49 ARE *AA* AND PRODUCE	490	—	490
42 ARE *Aa* AND PRODUCE	210	210	420
9 ARE *aa* AND PRODUCE	—	90	90
	700	300	1000

This assumes that *A* and *a* are not sex-linked and that the usual 1:1 sex ratio holds. As the table shows, there are 700 *A* alleles, or a ratio of 7:3, distributed in the population's gene pool at the time of mating. The *frequency* of the *A* allele is thus said to be 0.7, while that of the *a* allele is 0.3.

The probability (the expected frequency) of getting the same genotype in the next generation can be calculated once the frequencies of each allele in the parental population are known. This can be done with the aid of the simple Punnett square (see Section 8.4). The mathematical probability of any two alleles coming together in a mating can be calculated as the product of the frequency of each allele in the population, as shown in the following Punnett square:

		MALE GAMETES *A* (0.7)	MALE GAMETES *a* (0.3)
FEMALE GAMETES	*A* (0.7)	*AA* (0.49)	*Aa* (0.21)
FEMALE GAMETES	*a* (0.3)	*Aa* (0.21)	*aa* (0.09)

Thus, the probability of getting the genotype *AA* is given as 0.7×0.7. The probability of getting the genotype *Aa* is given as $2 \times (0.7 \times 0.3)$; and the probability of getting genotype *aa* is given as 0.3×0.3. Thus the probability of getting genotype *AA* is 0.49; that of getting genotype *Aa* is 0.42; and that of getting genotype *aa* is 0.09.

It should be apparent that the values given for each genotype are identical to the original percentages of each genotype in the parental population. Expressing all this by the Hardy–Weinberg formula, we thus can write:

$$p^2 + 2pq + q^2 = 1.00$$
$$(0.7 \times 0.7) + 2(0.7 \times 0.3) + (0.3 \times 0.3) = 1.00$$

It should be apparent that this formula is simply an expansion of the binomial equation.

The Hardy–Weinberg law and its mathematical expression are an extremely useful approach to population genetics and its relationship to evolutionary problems. Using the formula, it is possible to determine the frequency of alleles in a population by sampling a limited number of organisms. If, for example, the gene *A* is completely dominant over *a* in the foregoing example, then the frequency of homozygous recessives (the only ones whose phenotype automatically indicates the genotype) can be used to calculate the frequency of each allele. If the frequency of *aa* is found to be 9 out of 100 (frequency = 0.09), then we can calculate that the frequency of *a* is 0.3 (since $0.3 \times 0.3 = 0.09$). This also establishes that the frequency of *A* is 0.7, since the frequencies of $p + q$ must equal 1.0. If *A* is incompletely dominant over *a*, the task is even simpler, since the phenotypes of homozygous dominant, homozygous recessive, and heterozygote will be different. Each can then be counted for sampling purposes.

If in sampling gene frequencies in one generation we note a shift in the frequency for the next generation, we can look for conditions in the population that might be working against the main-

The Hardy–Weinberg law states that gene frequencies in a population will remain constant from one generation to the next as long as certain conditions are met: the members of the population mate randomly, the offspring of all combinations are equally successful, the population is large enough to avoid sampling error, and the mutation rate has reached equilibrium.

tenance of genetic equilibrium. These conditions might be nonrandom mating or natural selection acting upon one or another genotype in the offspring. In some cases both conditions may be at work simultaneously. Whatever the case, the failure to maintain a Hardy–Weinberg equilibrium indicates that some evolutionary process is at work. Fundamentally, evolution is nothing more than a nonrandom shift in gene frequencies in successive generations. In the next section we will consider a specific example of how evolution produces a shift in gene frequencies.

14.6 Natural Selection and Changes in Gene Frequency

Although many factors serve to upset the genetic equilibrium of a population, perhaps none is so important as the combined effect of mutation and natural selection. How this may occur in natural populations is illustrated by the human condition known as sickle-cell anemia. This condition occurs primarily, but not exclusively, among blacks and affects the oxygen-carrying capacity of red blood cells. The biochemistry and genetics of sickle-cell anemia have been carefully studied in recent years. It is apparent that mutation of the gene for normal hemoglobin produces a slightly modified portion of the hemoglobin molecule. The mutation may be represented as:

$$Hb^A \longrightarrow Hb^s,$$

where Hb^A stands for normal hemoglobin and Hb^s for sickle-cell hemoglobin. When oxygen tension in the blood gets low, Hb^s molecules fold up, or collapse. As a result, the red blood cells, composed of almost 100% hemoglobin, also collapse. The change in shape of the cells is a diagnostic trait of the disease (Fig. 14.16). When this folding-up occurs, it is very difficult for the affected cells to pick up oxygen in the lungs and transport it to the tissues. In many cases, reduced vigor and death are the results of sickle-cell anemia.

Allowing *s* to represent the sickle-cell gene and *S* the normal allele, let us see how natural selection acts on this gene in a population. With these two alleles, three combinations are possible: *ss* (individuals with sickle-cell anemia), *Ss* (individuals with the sickle-cell trait, a mild form of the condition), and *SS* (individuals with normal hemoglobin). The sickle-cell alleles show incomplete dominance, so each of the genotypes also shows a characteristic phenotype.* Persons homozygous for the

* Individuals with the *ss* genotype have all of their hemoglobin in every RBC as the sickle type. Individuals with the *Ss* genotype have half sickle hemoglobin in every cell; the other half is normal. Individuals with the SS genotype have all normal hemoglobin.

Fundamentally, evolution is nothing more than a nonrandom shift in gene frequencies in successive generations.

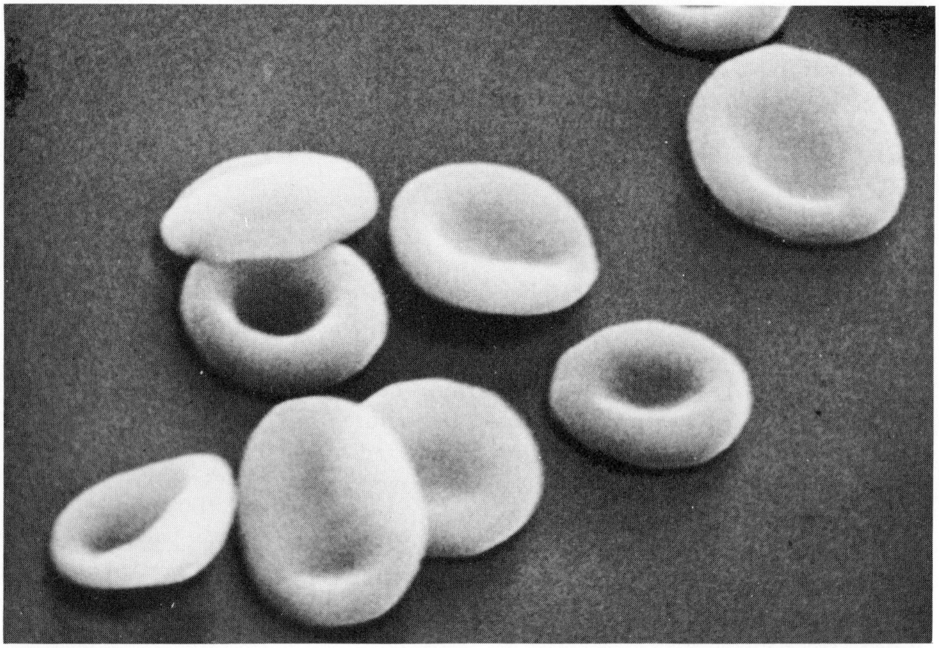

(**a**)

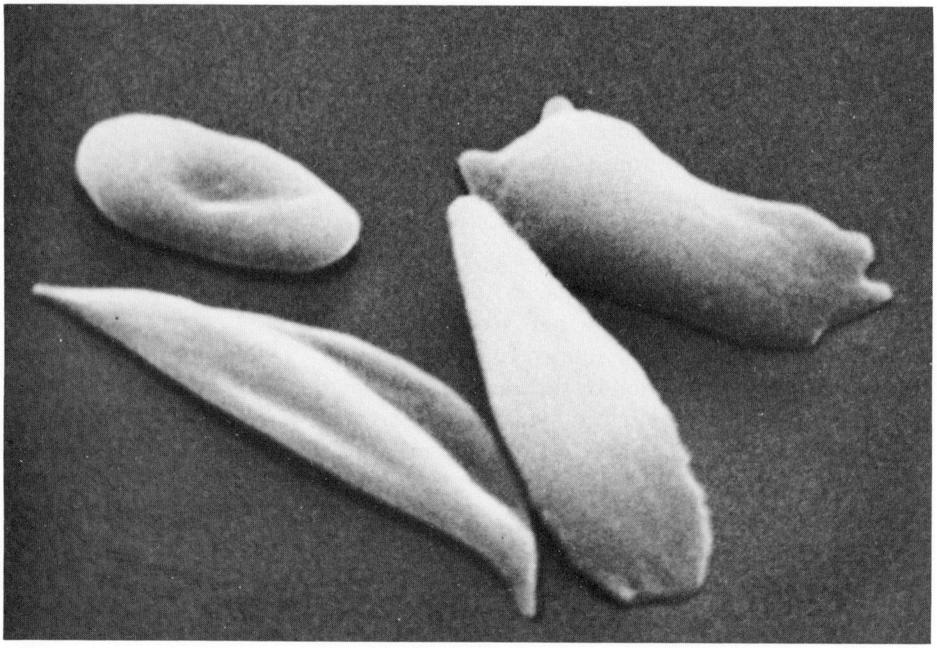

(**b**)

◄ **Fig. 14.16** *(a) Normal red blood cells showing basic concave circular shape (magnification ×6750). (b) Red blood cells that have sickled due to low oxygen tension. The deformed shape is a result of collapse of hemoglobin molecules; such cells have a greatly reduced ability to transport oxygen. (Scanning electron micrographs courtesy Dr. Marion I. Barnhart, Wayne State University School of Medicine. Originally published in R. M. Nalbandian, ed.,* Molecular Aspects of Sickle Cell Hemoglobin, Clinical Applications. *Springfield, Ill.: Charles C. Thomas, 1971).*

sickle-cell gene generally die very young and leave few offspring, if any. Persons heterozygous for the sickle-cell trait usually live to sexual maturity, since they are affected by this condition only in cases of extreme exercise or high altitudes.

The frequency of these three genotypes among American blacks has been studied by J. V. Neel and others. It is known that the frequency of the *s* allele is 0.05, while that of the *S* is 0.95.* According to the Hardy–Weinberg equilibrium, we should expect these frequencies to remain constant from generation to generation. However, natural selection is highly unfavorable to individuals of the *ss* genotype. It has been estimated that a child born with sickle-cell anemia has one-fifth as much chance as other children of surviving to sexual maturity. This indicates that roughly 80% of the *ss* genotypes fail to survive beyond infancy or childhood.

Natural selection thus limits the number of mutant *s* genes passed on to the next generation. If the frequency of *s* is 0.05 in one generation, then it will be reduced by about 16% in the next generation; the frequency will thus decline from 0.05 to 0.042 in one generation. (The figure of 16% is taken from calculations based on medical records of mortality rate among *ss* individuals.) The theoretical "cutting down" effect of natural selection in this instance is represented diagrammatically in Fig. 14.17. Note that the *rate* at which the mutant *s* allele is eliminated from the population becomes slightly lower in each generation. The fewer *s* alleles in the population, the less chance they have of coming together to form the lethal *ss* homozygote. Thus selection against a particular gene can reduce the frequency with which it occurs in the population. However, since the rate of decrease becomes less each generation, a very long time is generally required to effectively eliminate an unfavorable gene from the population.

The case of sickle-cell anemia illustrates very well how gene frequencies vary due to differences in environmental conditions. The frequency of the *s* gene is much greater in certain parts of Africa than in other parts of the world. Investigation has shown that the areas in which the sickle-cell gene is most frequent are also the areas of high malaria incidence. It was then discovered that the heterozygous carrier (*Ss*) individuals possessed selectively greater resistance to malaria than the homozygous nonsickling individuals.† Thus in areas in which malaria exists, the sickle-cell trait has a high adaptive value to the individuals who possess it. The dis-

* In other words, the *s* gene makes up 5% of the gene pool for this allele, and *S* about 95%.

† The way in which the sickle-cell trait appears to confer resistance to malaria seems to be something like the following. The infectious phase of the malarial protozoan enters the red blood cells as a parasite. The metabolic activity of the protozoan uses up oxygen, thus reducing the oxygen tension within the cell (i.e., the amount of oxygen). Now, molecules of sickle-cell hemoglobin collapse when the oxygen tension gets low, thereby causing the entire red blood cell to fold inward, assuming the sickle shape. Cells which have thus collapsed are, along with the parasitic protozoans within them, more readily destroyed by phagocytes. In this way the malarial parasite is prevented from spreading throughout the bloodstream.

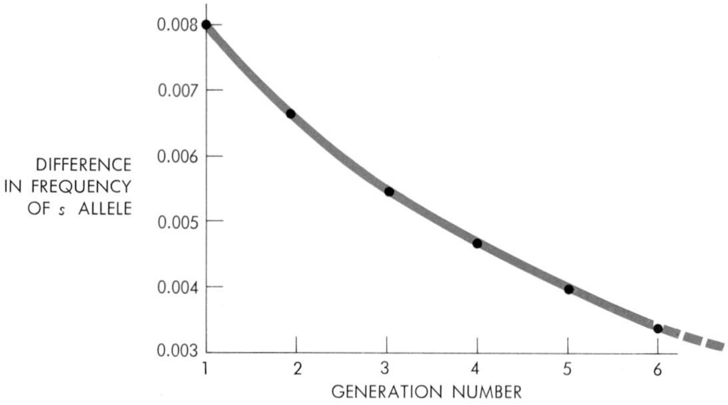

Fig. 14.17 *Graph showing the difference in frequency of s allele in successive generations.*

advantage of the sickle-cell trait is more than compensated for by the greater resistance to malaria.

Like melanism in moths, then, the sickle-cell trait is selectively disadvantageous under one set of conditions and advantageous under another. Therefore, natural selection tends to preserve both the *s* and *S* alleles in the population. Either homozygote is at a decided disadvantage in malaria-infested regions. Superiority of the heterozygote therefore keeps the frequency of either allele from declining to a very low level.

On the genetic level, the existence of sickle-cell hemoglobin is a result of a single mutation in the genes responsible for synthesizing the β-chain of hemoglobin. The α-chain of every hemoglobin molecule is normal. The mutant β-chain has the substitution of valine for glutamic acid in one position along the lengthy polypeptide chain. Since the DNA triplet codon for glutamic acid is cytosine–thymine–thymine (CTT) and that for valine is cytosine–adenine–thymine (CAT), it is reasonable to conclude that the mutation from normal to sickle-cell hemoglobin comes about by the substitution of a single base pair, adenine for thymine, in DNA. This small substitution produces an enormous impact on the individual's chances for survival, and on the course of evolution in a population.

It is not known whether this mutation occurred only once in the past, or a number of times independently. Given the smallness of the change, it is likely to have occurred repeatedly throughout human history. Only in populations that inhabited an environment in which the heterozygous condition was favorable was the mutant codon preserved.

14.7 Sexual Selection

In later editions of *The Origin of Species*, Darwin gave a great deal of emphasis to the idea of sexual selection. According to this concept, the female or male (depending on the species) selects a mate on the basis of certain distinct criteria, such as colora-

The case of sickle-cell anemia shows that adaptiveness of traits can only be understood in the context of all aspects of the population's environment. What is adaptive in one environment is nonadaptive or maladaptive in another.

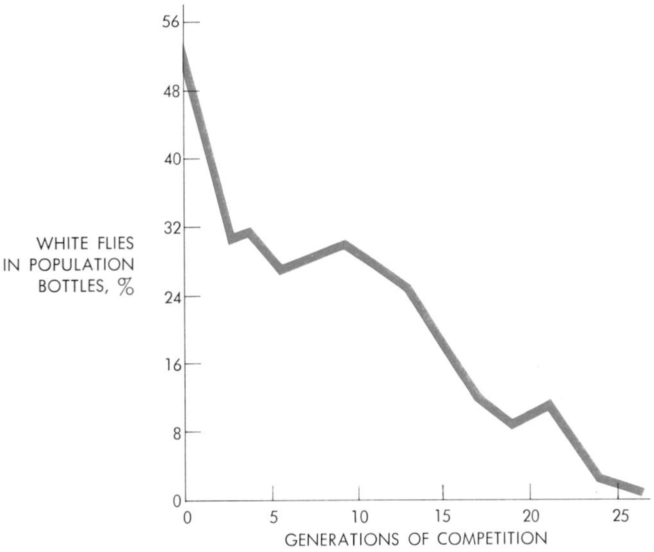

Fig. 14.18 *Graph showing the result of sexual selection in a culture bottle of* Drosophila. *Since females prefer red-eyed males to white-eyed males, the red-eyed males leave greater numbers of offspring, on the whole, in each generation. The culture bottle originally had about equal numbers of red-eyed and white-eyed males; the decline in percentage of the white-eyed genotype can be easily discerned.*

tion or behavior. Darwin held that the male or female of a species would consistently select a mate that possessed certain specific characteristics. Sexual selection is just a special case of natural selection. As a separate idea, however, it helps to explain how characteristics such as the gaudy plumes of the peacock or the extravagant tail feathers of the bird of paradise could be of adaptive value.

In such cases, two opposing selection pressures are at work. Sexual selection tends to favor any modification that increases the attractiveness of one sex for the other. Natural selection working for survival, however, favors any modification that renders the organism less visible to its enemies. If the disadvantage of being more easily spotted by predators is compensated for by the increased productivity resulting from bright coloration, the latter characteristic will be favored. Genes for coloration will therefore increase in the population. However, the moment the balance swings the other way (the moment the animal becomes too easy a prey), genes reducing coloration will be favored.

A vivid example of sexual selection in a controlled laboratory situation can be seen from the data plotted in Fig. 14.18. At the beginning of this experiment, a population of *Drosophila* was set up in which the initial percentages of genes for white eyes (w) and normal eyes (W) were known. This gene is sex-linked, so by and large only males will show the white-eye trait. *Drosophila* females will not mate as readily with white-eyed males as with red-eyed ones. In other words, they show a *preference* for red-eyed males. Sexual selection operated in this case to reduce the number of w genes passed on to each successive generation. The percent of individuals bearing the gene correspondingly declined over a number of generations.

14.8 Speciation

Speciation is the process by which one or more new species arise from previously existing species. Biologists identify two basic types of speciation: **phyletic** and **divergent.** The basic processes of variation and selection are identical for both types. The differences lie in the number of new species that arise from a single ancestral species. In phyletic speciation, a single ancestral species gives rise to a single descendent species. In divergent speciation, a single ancestral species gives rise to two or more descendent species.

Phyletic Speciation

Phyletic speciation can be diagrammed schematically as follows:

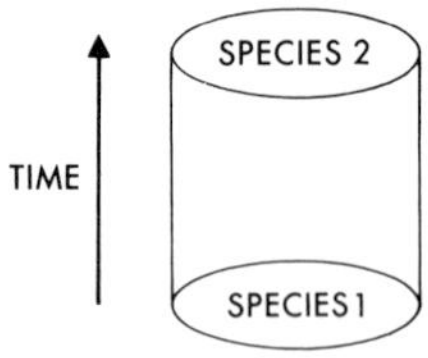

Its essential feature is that throughout the period of time that variation and selection are taking place, the population as a whole remains intact. That is, the population represented in species 1 undergoes gradual variation and selection—a shift in gene frequencies—over time. Because all parts of the population are in contact with one another, gene flow remains possible throughout the population at each stage along the way. This means that essentially no part of the population varies at a much greater rate than any other part. In time, the species as a whole may have changed so much that it is regarded as a new species. If species 2 could then come back in contact with the species 1 (which is, of course, impossible, since species 1 as such does not exist any more), the two would be unable to interbreed. One old species has given rise to one new species—the basic characteristic of phyletic speciation.

Divergent Speciation

Divergent speciation is diagrammed to the right. The basic feature of divergent speciation is that the ancestral species does not remain intact but rather fragments into distinct subpopulations. In the case diagrammed, there are three such subpopulations. Furthermore, for divergent speciation to occur, each of these populations must become isolated from the others geographically, so that gene flow between them is impossible. Each subpopulation then undergoes variation. Since it would be highly improbable that each subpopulation would have the same number and kind of variations occurring at the same time, and since each subpopulation inhabits slightly different geographic regions and is subject to slightly different selection pressures, the changes in gene frequencies over time would be expected to be different within each subpopulation. Thus over time the genetic constitution of each subpopulation will *diverge* from that of the others. The ultimate result will be that each population will become so different from the others that interbreeding will be impossible even if they are brought back together. New species will have been created.

The essential feature of divergent speciation is geographic isolation among subpopulations. Geographic isolation prevents the exchange of genes. If genetic exchange occurs even to a limited extent between subpopulations, divergence will never occur to the point of producing new species. Because the original subpopulations share a common genetic background, interbreeding remains possible for some time after geographic separation takes place. Only when gene flow between the subpopulations is prevented for a considerable time can enough

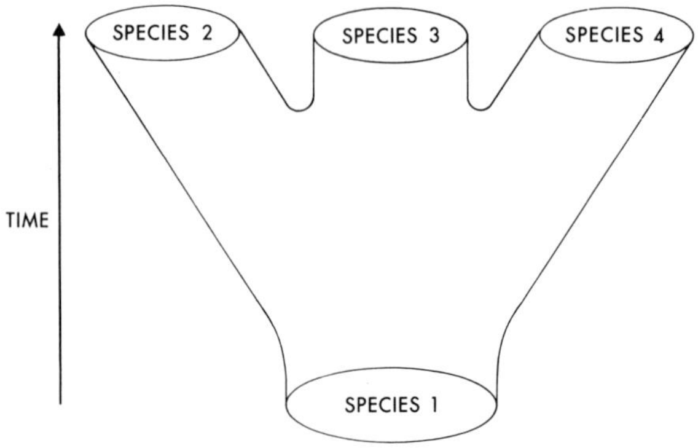

divergence accumulate to prevent the subpopulations from interbreeding should they ever come back into physical contact. Thus, geographic isolation must occur for divergence to begin and it must continue long enough for a new species to evolve. The new species is the result of the accumulation of small, quantitative changes over long periods of time.

Isolating Mechanisms

Let us consider an example of how geographic isolation comes about and how it produces divergent speciation. Two populations of tuft-eared squirrels inhabit the north and south rims of the Grand Canyon (Fig. 14.19). These squirrels are very similar, yet there are visible differences between them. The two squirrel populations are usually considered separate species, for, although some hybridization is possible in the laboratory, little or no gene flow exists between the two populations in nature.

It seems clear that both the Abert squirrel (southern rim) and Kaibab squirrel (northern rim) arose in the past from the same population. This was apparently a freely breeding population that inhabited the entire region now represented by the north and south rims. Widening of the bed of the Colorado River and the subsequent formation of a deep canyon created a geographic barrier that effectively prevented gene flow between the separate groups of organisms.

Not all geographic barriers are as spectacular as the Grand Canyon. A marshy area between two field habitats can also be an effective barrier. The question is: How does isolation produce a new species? Surely the separation of the original population of tuft-eared squirrels by the Colorado River did not automatically produce two species.

Once an original population becomes divided into two isolated portions, the kind of variations that occur in each will inevitably be different. When the effect of these variations begins to become visible, we say that evolutionary **divergence** has begun. Divergence does not represent the origin of a new species. It is only the first stage. If for some reason the two populations are brought back together after only a few generations, the accumulated anatomical and physiological differences are usually not great enough to prevent interbreeding.

When divergence has continued for many generations, however, the accumulated differences may make interbreeding impossible. Once this stage is

Fig. 14.19 *In nature, the Abert squirrel (left) and the Kaibab squirrel (right) are not found living in the same habitat as shown here. The Abert squirrel inhabits the southern rim of the Grand Canyon, the Kaibab squirrel the northern rim. Anatomical differences indicate that these two populations diverged from a common ancestor prior to being geographically isolated by the canyon. (Adapted from* Life: An Introduction to Biology, *by G. G. Simpson, C. S. Pittendrigh, and L. H. Tiffany, copyright © 1957 by Harcourt Brace Jovanovich, Inc. Used by permission of the publisher.)*

reached, we say that **speciation** has occurred. It is obvious that there can be no clear-cut dividing line between the end of divergence and the beginning of speciation. The terms are merely convenient to designate broad periods of evolutionary development.

Speciation in the tuft-eared squirrels probably started as a result of geographic isolation. In general, two or more species arise from a previous species only after some sort of physical separation of the original population into smaller groups. Once geographic isolation has allowed divergence and speciation to occur, a number of other isolating mechanisms come into play. These prevent the two species from interbreeding, should the populations ever come back together. Not all these mechanisms operate in any one case, but they do serve to strengthen the boundaries between the species. Isolating mechanisms are most evident as such when two similar species are living together in the same geographic area. Several isolating mechanisms are common in nature.

Seasonal isolation. Although two populations of organisms may inhabit the same geographic region, their reproductive periods may occur at different seasons of the year. As a result, members of one population cannot mate with members of the other, even though physical contact between the two is possible. This type of isolation is particularly common in plants, but it is also found in certain types of insects and snails.

Ecological isolation. If two populations are capable of interbreeding, yet do not do so because they live in different ecological niches, they are said to be ecologically isolated. For example, two populations of the deer mouse *Peromyscus* are capable of interbreeding in the laboratory. One population comes from the forest and the other from a nearby field habitat. In nature, each population remains almost completely within its own niche. The forest population rarely if ever enters the field, and vice versa. In this situation, divergence has reached the point where the two populations are close to being completely separate species.

Physiological isolation. Probably the first differences that begin to appear in geographically isolated populations are physiological. On the basis of the ideas that genes produce enzymes, slight changes in genetic code may show up first as changes in enzyme molecules. These changes need not be great enough to produce noticeable outward (phenotypic) effects, and they will not usually prevent interbreeding should the populations come back together. However, if the changes are in some way connected with the chemistry of the gametes themselves or of the seminal fluids, fertilization may be hindered. In sea urchins, for example, biochemical differences allow the sperm of one species to have greater success with eggs of its own species. Such physiological differences prevent or greatly inhibit successful hybridization.

Behavioral isolation. In many animals, specific behavioral traits are known to be inherited. Particularly important to evolution are those behavioral patterns connected with mating. In certain birds, fish, and insects, the male and female of a given species perform highly specific and elaborate courtship rituals. Male birds may strut about in a precise pattern, displaying their plumage in hopes of attracting a particular female. Once the female becomes interested, the male begins a new behavioral ritual that ultimately leads to copulation. If either animal fails to perform the expected behavior at the right time, the other may not respond further, and mating is not successfully completed.

That these patterns are genetic, i.e., built into each organism from the outset, has been shown by raising birds in complete isolation. In such cases, there is no opportunity for young birds to learn the ritual from older members of the species. The fact that such isolated birds still perform according to the exact behavioral pattern of other members of the species supports the hypothesis that proposes a genetic basis for such activity.

That behavioral patterns can serve as isolating mechanisms between species is shown by the courtship of several species of fish. The male three-spined stickleback (genus *Spinachia*) builds a nest attached to water plants. Then, by performing a series of zig-zag swimming motions, he induces the female of the species to lay eggs in the nest. These swimming motions are highly specific. If the male does not perform the ritual precisely, the female immediately loses interest and swims away.

In many of the rivers in which three-spined sticklebacks live, there also exists a second species, the nine-spined sticklebacks. When males of this species court females, the behavioral pattern is somewhat different. Instead of performing the zig-zag dance, male nine-spined sticklebacks bounce along vertically in the water, bobbing up and down in front of the female. The number of bounces and the position of the male are extremely important in bringing about the egg-laying response in the female. In both species the behavior of the male sets up a reflex action in the female. In some way not yet understood, this reflex pathway in both male and female is determined by the genetic makeup of the individual. If a male three-spined stickleback courts a female nine-spined stickleback, the female fails to show the egg-laying response. The female seems to lose interest as soon as the male begins the unfamiliar zig-zag dance. Such divergence in behavioral patterns, which probably first arose from geographical isolation, serves as an effective isolating mechanism even though geographical isolation is no longer in effect. Divergence from some ancestral stickleback allowed two courtship patterns to develop; speciation has taken place, and interbreeding has become difficult or impossible.

14.9 Migration and Genetic Drift

Migration of organisms from one population to another may introduce new alleles into a population's gene pool. If these new alleles have less survival value than the ones already present, they will not spread through the population. Thus the original gene frequencies of the population will not change. However, if the new alleles are in some way more adaptive, they will tend to spread in the population. As the frequency of these genes increases, the frequency of another allele must decline. In this way, migrations *into* an established population can cause a shift in gene frequencies.

A somewhat different situation may result when a group of organisms migrate *away* from a population into new, unoccupied territory. For example, mammals are thought to have originated about 165 million years ago from a migration of small reptiles out of the lowland marshes to the highlands. This **founder population,** as the emigrants are called, was a minority group from the parent population of reptiles. As is generally the case with founder populations, they were individuals who could not compete successfully in the original environment. The gene pool and the gene frequencies of a founder population are apt to be somewhat different from the parent population. Often the effects of a new environment are to increase this difference. If isolation of the founder population from the parent population is maintained, the founder population may eventually emerge as a different species.

Gene frequencies may also shift if the size of a population becomes very small. When there are few mating organisms, the effect of a mutation or chromosomal aberration and recombination may be greatly increased. If a mutation occurs in one organism in a population of 50, its frequency from the start is obviously much greater than if the mutation took place in a population of 1000 or 10,000. Often, the result is to establish certain alleles in a much greater frequency than would occur in larger populations. This tendency is often called **genetic drift.**

To a certain extent, genetic drift depends on chance. A particular allele may become fixed in a population through a chance mutation occurring at one particular time. Or it may become established because, by chance, a certain group of organisms

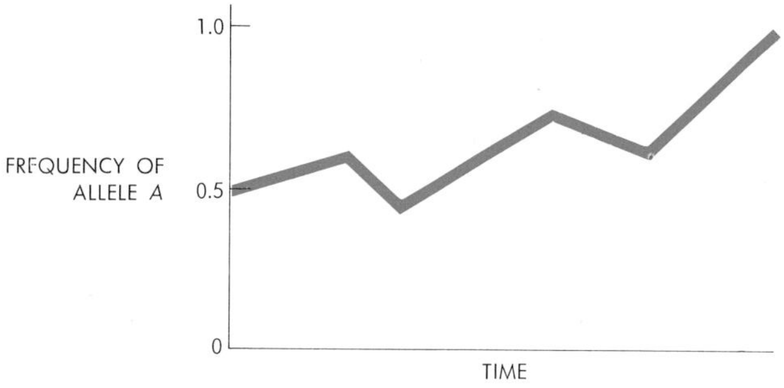

Fig. 14.20 *Graph showing the changes in frequency of a given allele over the course of time. The random fluctuations at various points in time represent the great changes that occur in small populations. There is no particular "direction" to genetic drift. In small populations, however, accidental death of even a few organisms, or the occurrence of a new mutation, may produce a definite shift in frequencies.*

in the population survives while another group does not.

Figure 14.20 shows a change in frequency of a particular gene in a small population. This graph shows that in the course of time gene *A* completely eliminates its allele *a* by reaching a frequency of 1.0 (i.e., it occurs in 100% of the population). This may be considered an example of genetic drift. By "fixing" certain genes in the population, genetic drift tends to reduce the variation that can occur.

There is debate among biologists as to whether genetic drift plays much of a role in the majority of evolutionary situations. There has been a tendency in the past to explain many changes in gene frequency by ascribing them to genetic drift, when in reality the factor responsible was natural selection. However, populations do exist in which the number of reproducing organisms is less than 50. For example, at last count (1962) there were 32 whooping cranes in existence. In a population of this size genetic drift could be an important factor in evolution. It is certainly fair to say, however, that in the evolution of average animal and plant populations (which contain many thousands of individuals), genetic drift is usually negligible.

14.10 Adaptive Radiation

All organisms are adapted to particular environments, but some are better adapted to a specific environment, or environmental niche, than others. When a population becomes too large, it is inevitable that some organisms must either adapt to new niches or perish. Those that can move into some area of the environment hitherto unoccupied will have a better chance of surviving. Natural selection favors those organisms that can exploit (take advantage of) new niches in which there is less competition. For example, if the organism has some chance variation that allows it to use a food source that other members of the population cannot use, it can exploit this new niche. The organism has taken advantage of an evolutionary opportunity. Indeed, nature places a premium on the ability of organisms to make the most of new opportunities.

When a species is introduced into a new area, it tends to spread out and occupy as many different habitats as possible. In time, such a single species will give rise to a variety of forms, each adapted through natural selection to a given niche. This evolution from a single ancestral species to a variety of types is called **adaptive radiation;** it results from the tendency of organisms to exploit unoccupied ecological niches. In a new niche, with reduced competition, an organism can produce more offspring. Variations enabling those offspring to exploit the niche further will be favored by natural selection. As a result, each population will become more and more adapted to a particular way of life. The ancestral population will give rise to a number

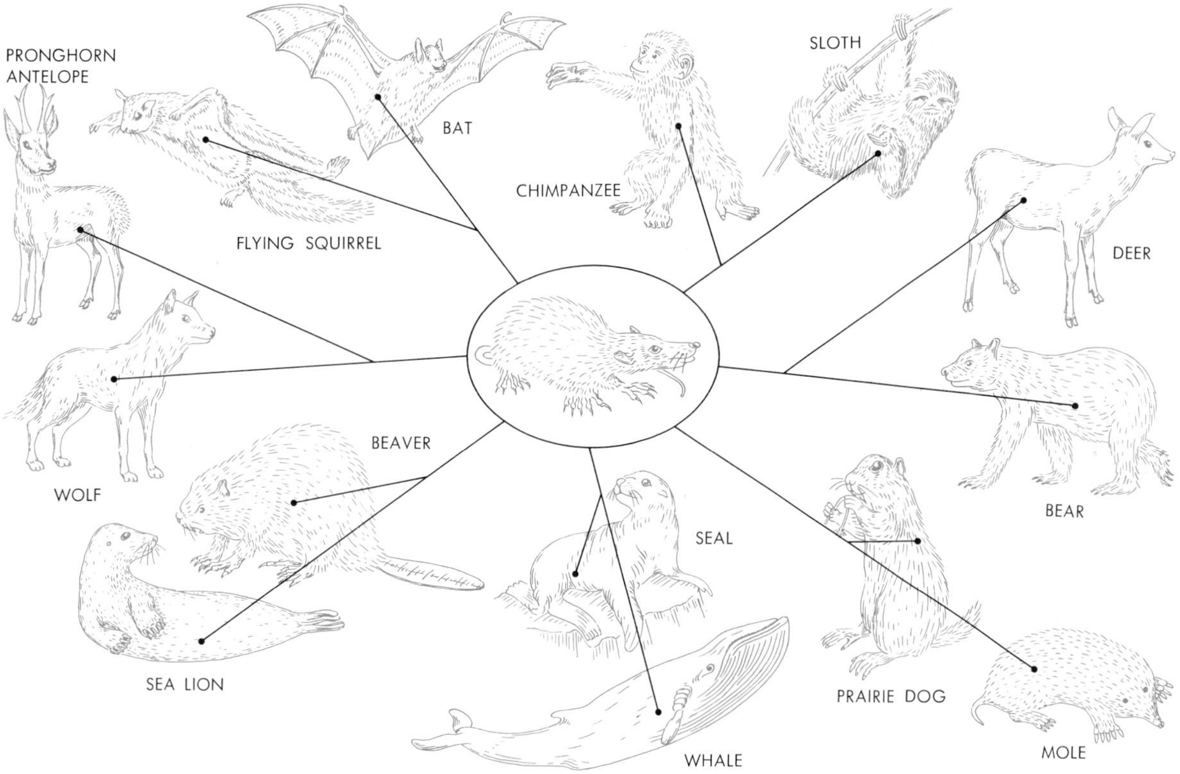

Fig. 14.21 *Adaptive radiation in mammals. All the many forms of mammals living today are hypothesized to have developed from an ancestor similar to the organism pictured at the center of this diagram. By invading and exploiting varied ecological niches, mammals have become very widespread.*

of different populations, each with its own special adaptations. A diagram showing adaptive radiation in the placental mammals appears in Fig. 14.21. This represents a very extensive radiation, one that has been occurring for long periods of time. As a result, the modern forms are considerably different from the primitive mammalian ancestor.

Just how does adaptive radiation take place? In 1832 Darwin visited the Galapagos, a group of islands just off the western coast of South America. There he collected some specimens of ground finch belonging to the subfamily Geospizinae. Although Darwin himself did not draw any special conclusions from study of these birds, later investigators have found in them excellent sources of data on adaptive radiation and other evolutionary principles.

Geological evidence shows that the Galapagos are of rather recent volcanic origin (having been formed about 1,000,000 years ago) and have never been connected to the mainland. As soon as vegetation developed on the islands, immigrant animals were free to move in and occupy the various niches.

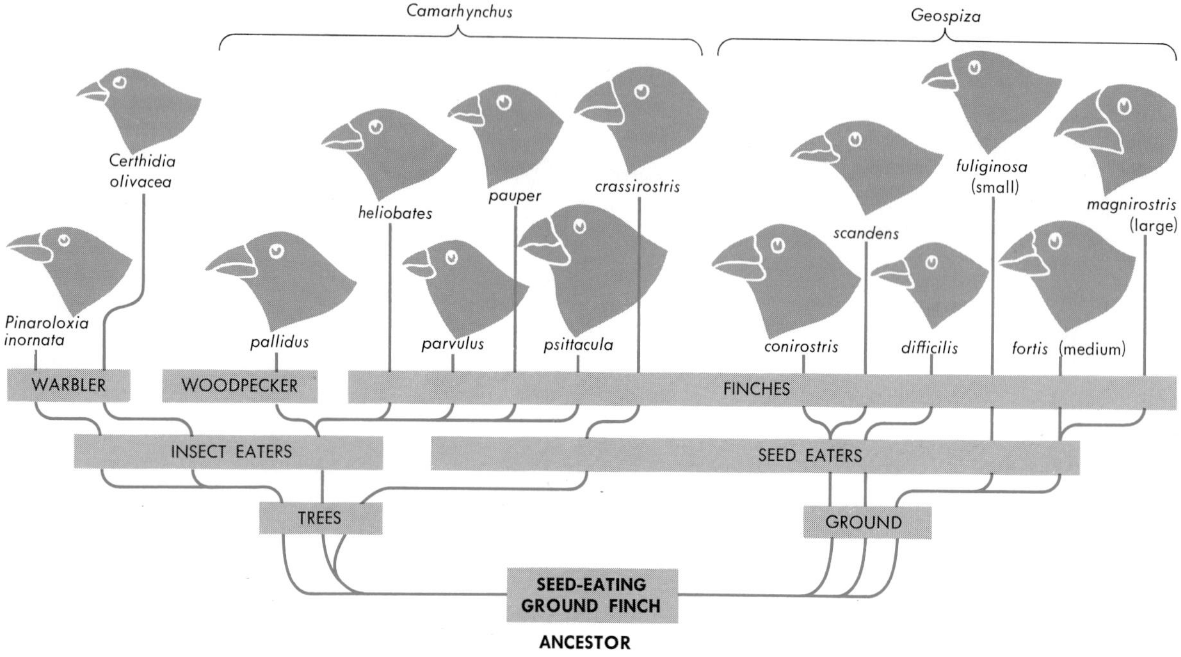

Fig. 14.22 *Evolutionary history of the finches on the Galapagos. This diagram illustrates the basic radiation that occurred at each point. The first radiation was between the two major habitats: ground and tree. Within each of these areas, subradiations have taken place.*

It is hypothesized that the present-day Galapagos finches are descended from a single type of finch that immigrated to the islands from the mainland sometime in the distant past. All attempts to find any such a mainland ancestor have, however, been unsuccessful. Perhaps this is because the ancestral finch has become extinct, or perhaps it is because the Galapagos finches have diverged so far from the ancestral traits that their relation to the mainland form is unrecognizable. In the absence of any conclusive evidence, the former possibility seems more likely.

An evolutionary diagram of the adaptive radiation of the finches throws light on the probable way in which divergence has occurred (Fig. 14.22). The fourteen species of finch living today can be roughly grouped into two categories: the ground finches (genus *Geospiza*) and the tree finches (genus *Camarhynchus*). In addition, there is a single "warbler" species and one isolated species peculiar to the most distant island of the group.

Figure 14.22 indicates that the ancestral finch population split up into ground-dwellers and tree-dwellers soon after migrating to the Galapagos. Variation and selection produced adaptations not only in foot structure, but also in the shape and size of beaks. Seeds form the major diet of ground finches, while the tree finches feed mostly on in-

Contrary to the popular conception of Darwinian selection, the outcome of evolution is not to increase but to decrease competition between species.

sects. Competition for food in these major niches produced further adaptive radiation within each area; all six species of ground-dwelling finches have slightly different diet preferences, even though all may be classified as seed-eaters. For example, three of the six species have differences in beak size and shape that make them specialized for specific sizes of seed (*G. magnirostris*, large seeds, *G. fortis*, medium seeds, and *G. fuliginosa*, small seeds). It is interesting that these three forms are found together on nearly all the Galapagos islands. Because these species have adapted to specific sizes of seed, competition for food between them is reduced. This subradiation occurs because organisms are able to exploit numerous unoccupied niches. The other three species of ground-dwellers, on the other hand, combine seed diets with flowers or cactus pulp.

Among the tree finches, one type feeds on seeds while the rest feed on insects. Like the ground finches, the tree finches have shown further radiation, especially in size and shape of beak. An especially interesting adaptation has occurred in one species of tree finch, *Camarhynchus pallidus*. This bird exploits essentially the same niche inhabited by woodpeckers in other geographic regions. Although it lacks the hard beak and long tongue of true woodpeckers, this finch has learned to use a cactus spine or small stick to probe insects from cracks in trees or from under the bark. In this case, an adaptation in behavior has allowed the organism to exploit an unoccupied niche.

On the South American mainland, no species of finch has been able to undergo the great adaptive radiation seen on the Galapagos. It seems clear that on the mainland most of the available niches are already occupied by other types of birds. The mainland forms lacked the "ecological opportunity" that makes adaptive radiation possible. The founder population that inhabited the Galapagos, however, had little competition from other birds. As a result, its members could reproduce rapidly and spread out to occupy innumerable niches. Adaptive radiation is, then, partly dependent on the availability of niches into which organisms can move.

14.11 Adaptation and Survival

The concept of natural selection implies that all surviving variations will in some way be more adaptive than those that do not survive. The mechanism for evolution by natural selection is the development of adaptations. When organisms are unable to adapt to changing environmental conditions, **extinction,** not evolution, is the result.

However, some organisms adapt so well to a specific niche that when conditions change they are hopelessly unfit for a new environment. It has been hypothesized that some of the large dinosaurs of about 230 million years ago were too specialized. They lived in swampy regions, depending on water to buoy up their massive bodies. When climatic conditions changed and the region dried out, these organisms were unable to move into new territories. Nor could they survive long enough for chance variations to allow them to exploit some aspect of the changed environment. As a result, they became extinct. This hypothesis is based on the fact that one important characteristic of a species is its ability to adapt to changes in the environment. If an organism cannot adapt, its evolutionary future is likely to be very dim. In evolution there is a distinct balance between specialization to one niche and flexibility to move into a new niche if necessary. Organisms can be viewed as evolving between two

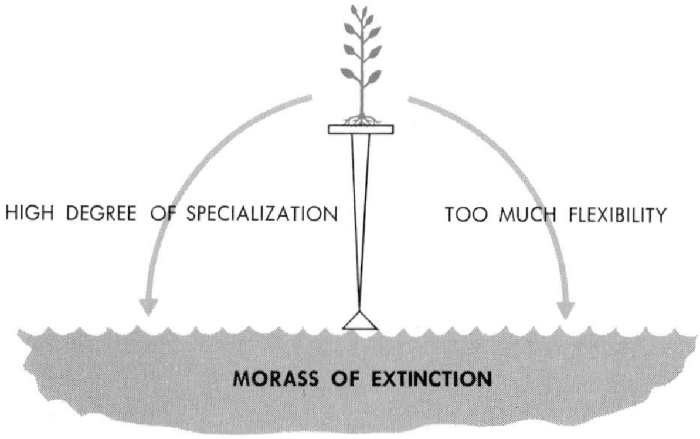

Fig. 14.23 *Organisms evolve between two opposing tendencies. On the one hand is the advantage a species gains by specializing to a very high degree. But this can lead to lack of flexibility and hence, if conditions change, to extinction. On the other hand is the advantage gained by being able to adapt to a changing environment. An organism that never specializes, however, may be unable to compete with its more specialized contemporaries under given conditions—and thus may likewise pass into extinction.*

opposing forces: specialization and adaptability (Fig. 14.23). Natural selection ultimately weeds out those who go too far to either side. The fact that in the history of life so many species have become extinct emphasizes how difficult it is for organisms to maintain a balance between these two pressures.

Writers who forget the importance of adaptability in the survival of new variations frequently speak of evolution as proceeding in a given direction, or for a particular "purpose." The ideas of purpose and direction have long been associated with the concept of evolution. Darwin's work, more than that of anyone else, showed that chance, not purpose or direction, is the basis of the origin of species. There is no evidence that evolution has any preconceived purpose, or that organisms evolve in a predetermined direction toward any absolute, ideal form. They simply undergo random variations that make them more or less adapted to given environments. Evolution along a given line, from ancestral to present form, takes place because the environment itself does not remain absolutely constant. The variations that survive are those that best adapt the organism to its environment.

SUMMARY

1. There are mainly three types of explanation for the wide diversity of living organisms in the world today.

   a) Special creation holds that each type, or species, was specially and separately created by a supernatural power at one time in the past.

   b) Spontaneous generation holds that living beings come into existence under the right organic or inorganic conditions, such as the generation of maggots from decaying meat. Spontaneous generation of each species can have occurred once in the past, or may be thought of as continually occurring whenever the conditions are right.

   c) Transmutation of species (evolution) holds that over many generations one kind of species gives rise to another through some process of variation; the evolutionary explanation postulates a genetic link, through history, between various species of animals.

   Neither special creation, a religious belief, nor spontaneous generation postulates any genetic continuity between the different species existing today. By and large biologists prefer the transmutation concept since it is most in agreement with a wide range of observations about past and present species.

2. Evidence favoring the transmutation concept comes from: (a) geology—the existence of fossils that often show continuity of development through geological strata from more ancient forms to their modern-day counterparts; (b) comparative anatomy—the existence of similar structural patterns among a number of different species; (c) taxonomy—the fact that different species can be arranged in groups showing a cluster of relationships, as if all had descended from a common ancestor; and (d) genetics—the fact that although organisms replicate themselves faithfully most of the time, they can undergo slight but distinct variations in heredity known as mutations; these mutations are the "raw material" of evolution.

3. Darwin's theory of natural selection is an attempt to explain how transmutation of species (evolution) could have come about. The essential elements of the theory are:

   a) All species tend to reproduce at a rate that exceeds the replenishment rate of the available food supply.

   b) Competition is keenest between members of the same species with very similar needs.

   c) All individuals in a population have slight inherited variations.

   d) Some of these variations enable the organism to compete better for limited resources, while some variations are a handicap.

   According to Darwin, those organisms with more favorable inherited variations would leave more offspring and thus more copies of themselves. The favorable variations would eventually eliminate or greatly reduce the less favorable variations. This process is known as natural selection, since environmental conditions continually select the more favorable from the less favorable variations.

4. Evolution is a continual process: no group of organisms is ever 100% adapted to its environment; hence, there is always the tendency and potential for greater adaptation. In addition, environments are continually changing: traits that are adaptive today may not be adaptive tomorrow. The relationship between adaptiveness of traits and environmental changes is illustrated by the cases of industrial melanism in moths and sickle-cell anemia in humans.

5. Evolution is nothing more than a shift in gene frequencies over a series of generations. The shift is not random, but occurs in some directional way. A population can be said to be evolving whether it is becoming better adapted (increasing in numbers) or becoming extinct (decreasing in numbers). To speak of direction in evolution does *not* imply purpose or ultimate goal.

6. The modern concept of evolution is based on understanding population dynamics. The genetic composition of a population is known as the gene pool, the figurative arena into which each organism casts its gametes during sexual reproduction. As the "pool" conception implies, it is theoretically possible for a gene coming from any one individual in the population to become combined with its counterpart from any other individual, as long as random mating is the rule.

7. A cornerstone of the population concept of evolution is the Hardy–Weinberg law of genetic equilibrium. This law states that the frequencies of any two alleles in a population will tend to stay the same from one generation to the next, whatever the original frequencies might be. Equilibrium is maintained only under a set of special conditions: (a) mating in the population is random; (b) all matings must yield the same basic number of offspring (i.e., no type of mating is preferred to another); (c) the population must be large enough to avoid sampling error; and (d) the mutation

rate must have reached its own equilibrium. If gene frequencies are found to be shifting from one generation to another, one or more of the above conditions are not being met, and evolution is occurring.

8. Darwin emphasized the concept of sexual selection as a special aspect of natural selection. Sexual selection refers to the process by which males and females of a species choose one another during mating. Certain anatomical or behaviorial features were thought by Darwin to be "more attractive" to one or the other sex and hence to influence mating choices. By this concept Darwin accounted for the origin of such bizarre structures as the male peacock's tail, the bright colors of birds of paradise, and the elaborate appendages seen in many species of tropical fish.

9. There are two ways in which species undergo evolution: phyletic speciation and divergent speciation. Phyletic speciation refers to the condition where one species, acting more or less as a single population, undergoes change over many generations; eventually it produces a modern form so different from the ancestral form that if the two could come back together, they would be unable to interbreed. The human population has demonstrated an essentially phyletic evolution. Divergent speciation involves the splitting up of an ancestral population into two or more subpopulations geographically isolated from one another in such a way as to be unable to mate and thereby exchange genes. After a long enough period of separation, the two populations diverge so much through variation and selection that even if they come back together they cannot interbreed. For divergent speciation to occur, geographic isolation between subpopulations is essential for a long period of time. Geographic isolation is always the first step in divergence of one population into two or more species.

10. Once divergent speciation has begun, there are many forms of isolation which act to keep species inhabiting the same area from interbreeding. These include: (a) seasonal isolation—two species reproduce at different times of the year; (b) ecological isolation—two species capable of interbreeding live in very different niches; (c) physiological isolation—copulation is possible but the sperm and egg of the two species are incompatible; and (d) behavioral isolation—applicable to animals among whom specific behavioral cues (such as courtship movements) are required to bring about copulation. All these isolating mechanisms develop as a result of an initial period of geographic separation between subpopulations; they then serve to reinforce the isolation even when the two descendent populations come back together.

11. In very small populations of organisms, a phenomenon known as genetic drift is observable. Genetic drift is nothing more than sampling error, which can allow certain genes to take over the population, coming to exist in 100% of the organisms, simply by chance. Genetic drift is not a major force in evolution in large populations, but it can be important in small ones or in subgroups of a large population that migrate to establish new territories.

12. Adaptive radiation is the process by which an initial population of one species undergoes considerable divergent speciation in an environment where there are available ecological niches. A common example of adaptive radiation is shown by the marsupials of Australia. An even more geographically widespread example is the mammals, who have adaptively radiated from a single common ancestor into a wide variety of niches distributed throughout the entire world. Darwin's finches represent an example of adaptive radiation in a limited environment.

EXERCISES

1. Distinguish between evolution and natural selection.
2. What is the value of comparative anatomy to the study of evolution?
3. In Darwinian terms, we say that the "fittest" organisms survive. How do we determine what is meant by "fit"?
4. Explain the process of natural selection in terms of the melanic moth.
5. What is the role of isolation in evolution? Can a population of organisms of the same species give rise to two separate species if no isolation occurs? Give reasons for your answer.
6. Occasionally individual fruit flies are born with shortened, stubby wings, the so-called "vestigial wing." The condition is inherited. These organisms cannot fly and do not survive in nature. Design an experimental environment that would selectively favor these flies over winged ones.
7. One criticism of Darwinian natural selection centered on the evolution of such adaptive features as protective coloration. An American biologist named MacAtee undertook a thorough study of the contents of the stomachs of birds. He noted whether the insect forms found there were of the variety supposed to be "protected." His data showed that, indeed, there were large numbers of "protected" insects eaten by birds. He thus became an outspoken critic of the idea of adaptive coloration. He claimed that there was no adaptive value in being camouflaged and that natural selection did not produce adaptively colored individuals.
   a) Is MacAtee's reasoning valid when applied to the mechanisms of natural selection? Why or why not?
   b) What further type of statistical or quantitative information is necessary before you could agree or disagree conclusively with MacAtee's findings?
8. Another argument against the evolution of protective coloration states that because many animals survive well without it, protective coloration is not essential to survival. Hence such coloration is not adaptive. What is the fallacy in this argument?
9. Figure 14.24 shows the different rates at which tadpoles grow under different conditions. Curve A shows the growth of an uncrowded population of tadpoles all roughly the same age. Curve B shows the growth of a similar population, but in a tank in which older tadpoles were living. The available food per tadpole was the same in both cases. Scientists attempted to determine how the presence of older tadpoles could check the growth rate of smaller ones. Water from a tank in which older tadpoles had lived was placed in a tank with young tadpoles. The growth of these young tadpoles was inhibited.
   a) What hypothesis can you suggest to explain how the presence of larger tadpoles can inhibit the growth of smaller ones?
   b) How might this inhibition be of advantage to the survival of a population of frogs in a pond? It is obvious that this is of advantage to the larger tadpoles themselves. But since the inhibitory mechanism has obviously evolved over long periods of time, how is it advantageous to the whole population? (There are several possible ways.)

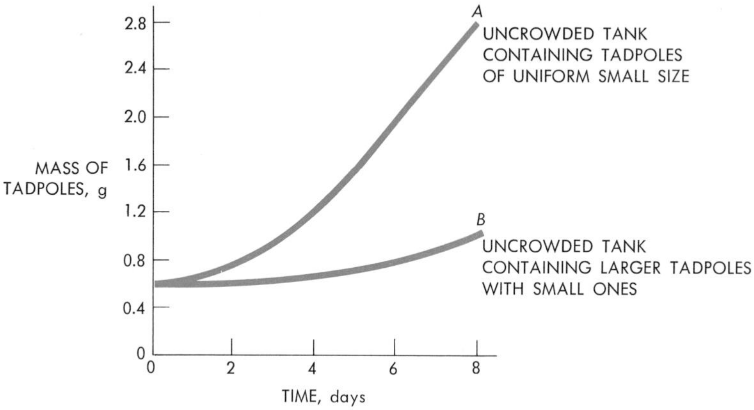

Figure 14.24

10. Two individuals heterozygous for a mutant gene become isolated. They leave two progeny. What are the chances that the mutant gene will be absent in the progeny and thus be lost (at least until a fresh mutation occurs) from this tiny subpopulation?
11. Explain why the application of population genetics to evolution by natural selection has been so useful in understanding the latter concept.
12. The situations described below are examples of specifically inherited genetic traits of populations, retained over many generations, that could hardly be adaptive to the survival of the individuals of the population. Yet the fact that these traits, involving the death or infertility of individuals, have survived in the population indicates that they must have some adaptive value. Offer an hypothesis explaining how this could be.
    a) When population crowding in groups of lemmings or locusts becomes intense, definite hormonal and biochemical changes are known to occur and cause mass migrations of hundreds of thousands of individuals from the population either to new territory or, in the case of lemmings, to the sea. In either case, the phenomenon is a result of more than merely forcing out the weaker individuals; it is a definite physiological characteristic that is somehow controlled genetically.
    b) Ant, bee, and wasp colonies have many sterile individuals who never reproduce during their entire lifetime.
13. Explain what is meant by the term "genetic drift." What conditions are necessary for genetic drift to have any influence on evolution?
14. Explain how the Galapagos finches seem to have developed from a single common ancestor. Discuss the stages by which this could have occurred. Include in your discussion the role played by the following: (a) adaptive radiation, (b) competition, (c) isolation (of various sorts), (d) available ecological niches.
15. To what extent has the human species solved the problem of specialization versus flexibility? Has our particular solution posed any additional problems?
16. Twenty-five percent of a population of cavefish show an albino phenotype. If albinism is known in this species to be due to a single recessive gene, estimate the

frequency of the gene in the population. What is the frequency of the heterozygous genotype?

17. The two alleles at one locus in a population of moths were observed to consist initially of 70% *A* and 30% *a*. *A* is the dominant allele, and *a* the recessive, for controlling color of body. After a severe frost and before another generation could be produced, the genotype frequencies were found to be as follows: 55% *AA*, 44% *Aa*, and 1% *aa*.
    a) Determine the genotype frequencies for the initial population before the frost.
    b) Account for the change in genotype frequencies.
    c) What relationship does such a change have to the process of evolution?
18. How would it be possible to "develop" a breed of moth that would be able to digest something that the species does not presently digest? What biological (evolutionary and genetic) principles would be involved?

SUGGESTED READINGS

The History of Evolutionary Theory

There have been volumes written on the history of our modern ideas of evolution, especially on the work and influence of Darwin. The best historical sources are:

De Beer, Gavin, *Charles Darwin*. New York: Thomas Nelson, 1964. This book combines the biology of Darwin's theory with an accurate and readable historical treatment. It is well illustrated and written clearly for the nonspecialist.

Moorhead, Alan, *Darwin and the Beagle*. New York, Harper & Row, 1969. A beautifully conceived and illustrated book, describing Darwin's voyage as naturalist aboard H.M.S. Beagle. Illustrations are all taken from contemporary publications and documents.

Aspects of the Modern Theory of Evolution

Clarke, Bryan, "The causes of biological diversity," *Scientific American*, August 1975, p. 50. A very well-illustrated article; its main point is that diversity within a species is not only advantageous to natural selection but is maintained by natural selection.

De Beer, Gavin, *Atlas of Evolution*. New York: Thomas Nelson, 1964. A general survey of many aspects of evolutionary theory, with brief descriptions of such topics as speciation, geographic isolation and distribution, sources of variation, etc. Very well illustrated and laid out. For quick and graphic review. Does not focus to a large extent on population aspects of evolution.

Kettlewell, H. B. D. "Darwin's missing evidence," *Scientific American*, March 1959, p. 48. This description of Kettlewell's investigations on the melanic moth explains his experiments in detail and presents much of the original data.

Smith, Homer W. *Kamongo*. New York: Viking, 1973 (first written in 1932). The author, a biologist noted for his work in kidney physiology, here presents a beautifully written book that nicely contrasts essential differences between the scientific and religious viewpoints of nature. The format is a dialogue between a biologist and an Anglican priest.

Stebbins, G. Ledyard, *Processes of Organic Evolution*, 2d ed. Englewood Cliffs, N.J.: Prentice-Hall, 1971. One of the very best short expositions of evolutionary biology. It contains a good discussion of population genetics, along with more traditional aspects of plant and animal evolution.

FIFTEEN

ECOLOGICAL RELATIONSHIPS: COMPETITION FOR ENERGY

15.1 Introduction

A population is a group of like organisms (belonging to the same species or subspecies) which live sufficiently close together that all members of the group can theoretically interbreed. A species or subspecies of any animal or plant consists of one or more local populations that have a certain geographic localization. For example, the people living in St. Louis, Missouri, or Hartford, Connecticut, form two local populations of the species *Homo sapiens*. There are also populations of other organisms living in those same geographic localities—dogs, cats, beetles, oak trees, etc. Populations do not have rigid boundaries but consist of collections of like organisms inhabiting a more or less definable area.

All organisms live in nature in a close and delicately controlled relationship with a great many other organisms. All the living things in any given area (pond, lake, river, ocean, or forest) form part of a biotic community. In a biotic community the existence of each population, as well as that of each individual, is governed to some extent by the presence of all the others. The biotic community forms a web the structure of which depends on every single strand. If just one strand of this web is changed, the structure may assume an entirely different character.

Introduction of the Japanese beetle into America and introduction of the jackrabbit into Australia have had considerable effect on the native biotic communities. Introduction of non-native organisms often allows for rapid growth of the new species while some native species decline in abundance. Without natural enemies, the new species reproduces rapidly and can do extensive damage. Great numbers of jackrabbits in Australia, for example, grazed intensively on the vegetation. This affected a large variety of other organisms that relied on the vegetation for food. The natural history of many parts of Australia has been altered temporarily, or perhaps even permanently, by the rapid spread of rabbits after their introduction in the last century.

All organisms and the processes that occur in their environments are interrelated. Ecology is the branch of biology that studies these interactions.

The study of the various relationships between organisms and their environment is known as **ecology.** The environment of any organism includes two major aspects, the physiochemical environment and the biotic environment. The physiochemical environment includes the amount of light, moisture, wind, water current, pressure, temperature, and acidity and the presence or absence of various minerals. The biotic environment on the other hand, includes all the various living organisms with which an animal or plant comes into contact.

Organisms in a biotic community depend on each other directly or indirectly for day-to-day existence. But they also compete for the limited resources available in the environment. There is competition for food, minerals, water, sunlight, and space. In short, *there is competition for energy.* To maintain itself and reproduce, an organism needs a constant supply of energy. This energy comes from the environment. When plants compete for sunlight, they are really competing for radiant energy. When animals compete for food, they are really competing for a source of energy and an adequate mix of nutrients. A significant feature of this competition is that the total amount of available energy is *limited.* For example, seed-eating birds can grow and reproduce only so long as the supply of seeds (a source of energy) lasts. The dependence of organisms on a limited food supply is one of the major principles involved in understanding how ecological relationships develop and how they are maintained.

15.2 Interaction between Organisms

In their search for energy, organisms interact in various ways. These interactions are grouped into several categories: **competition, predation, parasitism,** and **symbiosis** or **mutualism.** Some of the most obvious interactions are the eating of animals by other animals and the eating of plants by animals. When there are too many herbivorous animals living in a particular region (as happens when grasslands are "overgrazed" by livestock), the plants may be greatly reduced in number or even completely eliminated. As a result, many former grassland communities are now deserts.

Competition

Interaction between organisms in the ecosystem often takes the form of **competition,** an interaction wherein individuals use the same resources, such as matter, energy, and space. Two kinds of competition are recognized in natural populations: intraspecific and interspecific competition. **Intraspecific competition** occurs between members of the same species. For example, many acorns beneath a parent tree of white oak may germinate and begin to grow. Only a few, however, manage to survive to reproductive maturity. Those individuals with the genetic makeup enabling them to make maximum use of the available light, water, and minerals usually survive; they can be said to be the more fit individuals. The less fit seedlings within the species are eliminated. Through intraspecific competition, the general character of the species may change over long periods of time (see Chapter 14).

Interspecific competition results when individuals of two different species living in the same general area have similiar ecological requirements. The frequent result of such competition is the decline and eventual elimination of one species and the complete predominance of the other. A conspicuous example of competition between individuals of different species is found in the southeastern United States. Many years ago, plants of

Japanese honeysuckle (*Lonicera japonica*) were introduced and spread over much of the region. This plant is a vine, and it sprawls over low-growing shrubs and herbs and climbs high into trees. The shrubs and herbs are often nearly eliminated. Even tall trees may be so densely covered with the vine that they are eventually killed.

Parasitism

Since by definition a **parasite** is an organism that obtains its energy and matter at the expense of another, grazing animals are actually "parasites" upon the plants they eat. Plants may also be parasitized by other plants. Many fungi enter the tissues of living green plants and obtain their energy by breaking down this living tissue. For example, the fungus *Puccinia graminis* lives one portion of its life cycle within the tissues of wheat and another portion in barberry plants, producing a disease known as "rust" (from the rusty color of the reproductive structures of the fungus). Some species of flowering plants even live as parasites on other flowering plants; the stems of dodder (*Cuscuta*) become attached to the stems of other herbs (that are photosynthetic) and obtain their food. During the course of evolution, such parasitic plants as dodder have lost the genetic ability to produce chlorophyll and thus cannot manufacture their own food by photosynthesis. Animals also parasitize other animals. For example the tapeworm eats the food directly from the gastrointestinal tract and the flea sucks fluids from its host.

Other Interactions

Interactions involving parasitism or competition might be viewed as negative interactions in that one of the participants is usually affected negatively. There are also many cases of positive interactions within communities. These interactions involve the mutual coactivity of two or more different species and are known as symbiosis, or mutualism. A good example of such mutualism is the association of nitrogen-fixing bacteria with leguminous plants (see Fig. 15.1). The interaction of two species of plants may be so intimate as to justify the botanist's recognition of the association as a

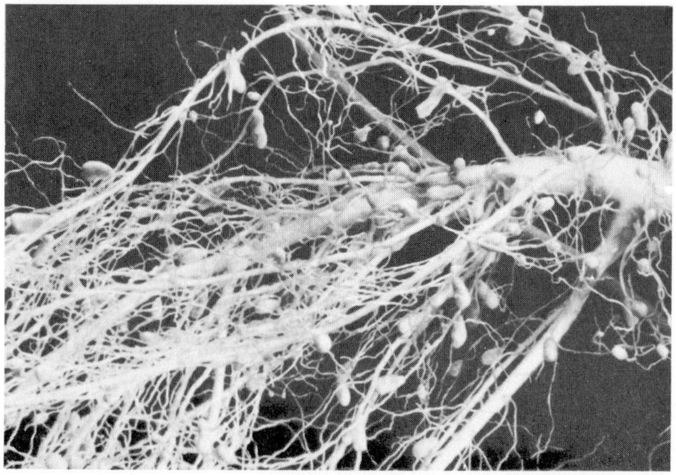

Fig. 15.1 *Nodules produced on the roots of crimson clover after inoculation of the soil with nitrogen-fixing bacteria. (Photo courtesy United States Department of Agriculture.)*

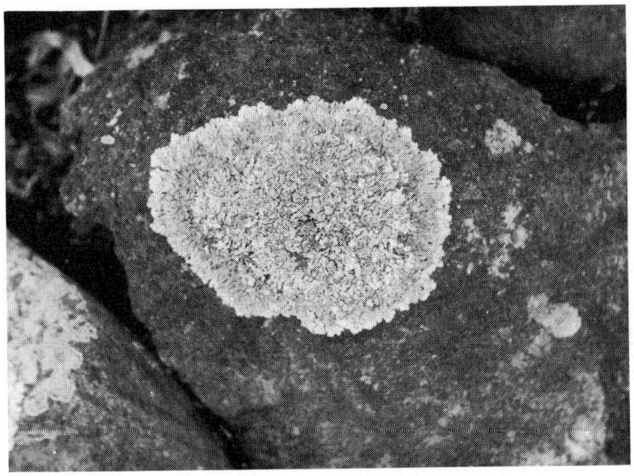

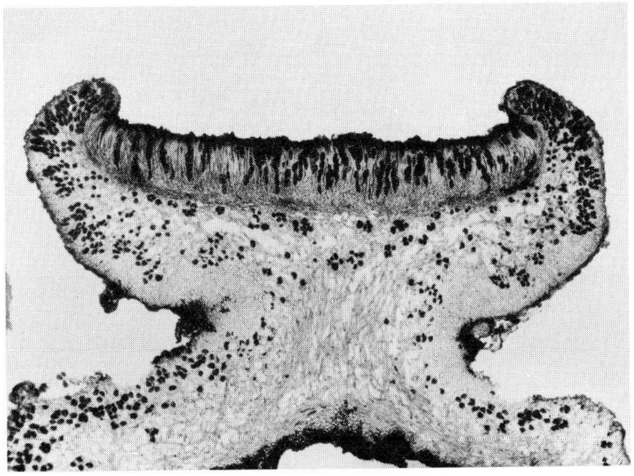

Fig. 15.2 *(a) Lichen growing firmly atttached to rock. (b) Structural organization of a lichen. The algae (dark round bodies) are interspersed among the fungus filaments. The vertically aligned dark bodies in the top center are ascospores, reproductive structures of the fungus. [Photo (a) from John W. Kimball,* Biology, *2d edition, Reading, Mass.: Addison-Wesley, 1968. Photo (b) courtesy Carolina Biological Supply Company.]*

single species; lichens, for example, consist of an alga and a fungus living in very close association (see Fig. 15.2). A lichen is a symbiotic interaction between a producer (the alga) and a consumer–decomposer (the fungus); both organisms are completely independent.

Other positive interactions occur between plants and animals that have evolved together in a mutually beneficial manner. For example, some leguminous trees (*Acacia*) have hollow thorns inhabited by ants that feed on sugary secretions produced by the trees. The ants patrol the tree branches and remove any other insects that might attempt to live on the tree and perhaps injure it.

During the hundreds of millions of years since plants and animals migrated from the seas to the land, other equally complex interactions have evolved. In one diverse group, the flowering plants, many of these interactions are involved with pollination and dispersal of seeds and fruits. Many species of plants are pollinated by insects; the insect obtains food in the form of pollen and nectar. This interaction has resulted in the evolutionary production of both elaborate flower structures and complex insect organs. In figs (*Ficus*), for example, the complex flowers of many species (there are several hundred different species) are pollinated by distinct species of wasps. Birds and other animals are attracted to the fleshy fruits of many plants. The seeds, resistant to the digestive processes of the animals that eat them, are carried from one habitat to another, where they are deposited in feces. The distance birds transport seeds may be much greater than windblown seeds would travel. The importance of wide dispersal relates directly to intraspecific and interspecific competition among different individual plants. Widely dispersed individuals are less likely to compete.

15.3 Interaction between Organisms and the Physiochemical Environment

In addition to their many interactions with one

another, organisms also interact with the physical and chemical environment where they live. Various physiochemical factors—sunlight, water, minerals, and pH, among others—have a strong effect on the kind of life that inhabits any particular area. This point can best be illustrated by considering several types of environments found throughout the world.

Various ecological systems on land have been distinguished in broad categories known as biomes. A **biome** is a very large community of species that inhabit a particular region, and it has a distinctive and easily observable physical appearance. Figure 15.3 shows four of the major biome types found within land areas of the earth. From the diagram, we can see that the differences between the clusters of organisms that comprise each biome type are obvious—particularly among the plants. For instance, the hot desert biome consists of a certain assemblage of plants that are all adapted to the same general physical conditions of high temperature and low moisture. The species of plants and animals inhabiting the Sahara desert are different from those inhabiting the Mojave desert; yet both regions would be spoken of as desert biomes. The physical conditions determine the types of organisms that inhabit each region.

Ecologists use the concept of the biome to describe a certain group of organisms and their interactions with the physical conditions of their environment. As a consequence, biomes are somewhat artificial categories. The assemblages of organisms they describe have distinct realities, but the borders of the biome may vary according to the purposes of the ecologist who does the describing. Like so many categories used in biology, the biome is not a discrete and rigid pigeonhole into which any particular geographic area can automatically be fit. Adjacent biomes of different types gradually fade into one another in such a way that it is often impossible to say exactly where one begins and the other ends. Moreover, the ecologist may expand or contract the limits of a particular biome being studied by making decisions about which organisms to include in the description. For instance, how far down under the soil does a grasslands biome extend? Many times, the depth is determined solely by whether the ecologist is interested in measuring subsurface organisms or equipped to do so.

The biome concept is useful in emphasizing one of the most important ideas in modern ecology: that the physiochemical environment determines the totality of life that can exist in any broad geographic area. The physiochemical environment operates primarily by "dictating" the kinds of plants that can exist; the kinds of plants operate to determine the kinds of animals that inhabit an area. If the major physical and chemical factors prevalent in a given region are known, a trained ecologist can predict fairly accurately the kind of biome that will be found there. Among the most important of the physiochemical factors are temperature (range as well as average) and amount of water. Each of the four biomes pictured in Fig. 15.3 is differentiated from the others primarily in terms of these two factors. As the graphs show, average temperature in a tropical rain forest stays about 80°F all year, but the amount of rainfall varies considerably from early March (1 inch/month) to early December (16 inches/month). On the other hand, the average temperature in a grassland biome in the United States varies considerably (from 25° to 70°F) from midwinter to midsummer, while the amount of precipitation is low all year (between 1 and 2 inches/month). Differences in these two factors produce markedly different types of biomes.

Of course, other physical and chemical factors are also important, such as presence of certain minerals, pH, type of soil (amount of clay, sand, etc.), and amount of wind and sunlight. Limitations in each of these other factors may modify the kind of biome that might be predicted for a given area on the basis of temperature and water conditions. For example, an area where temperature and rainfall patterns would otherwise produce grasslands might produce a semidesert biome if the soil were sandy and allowed rapid runoff of water. Biome

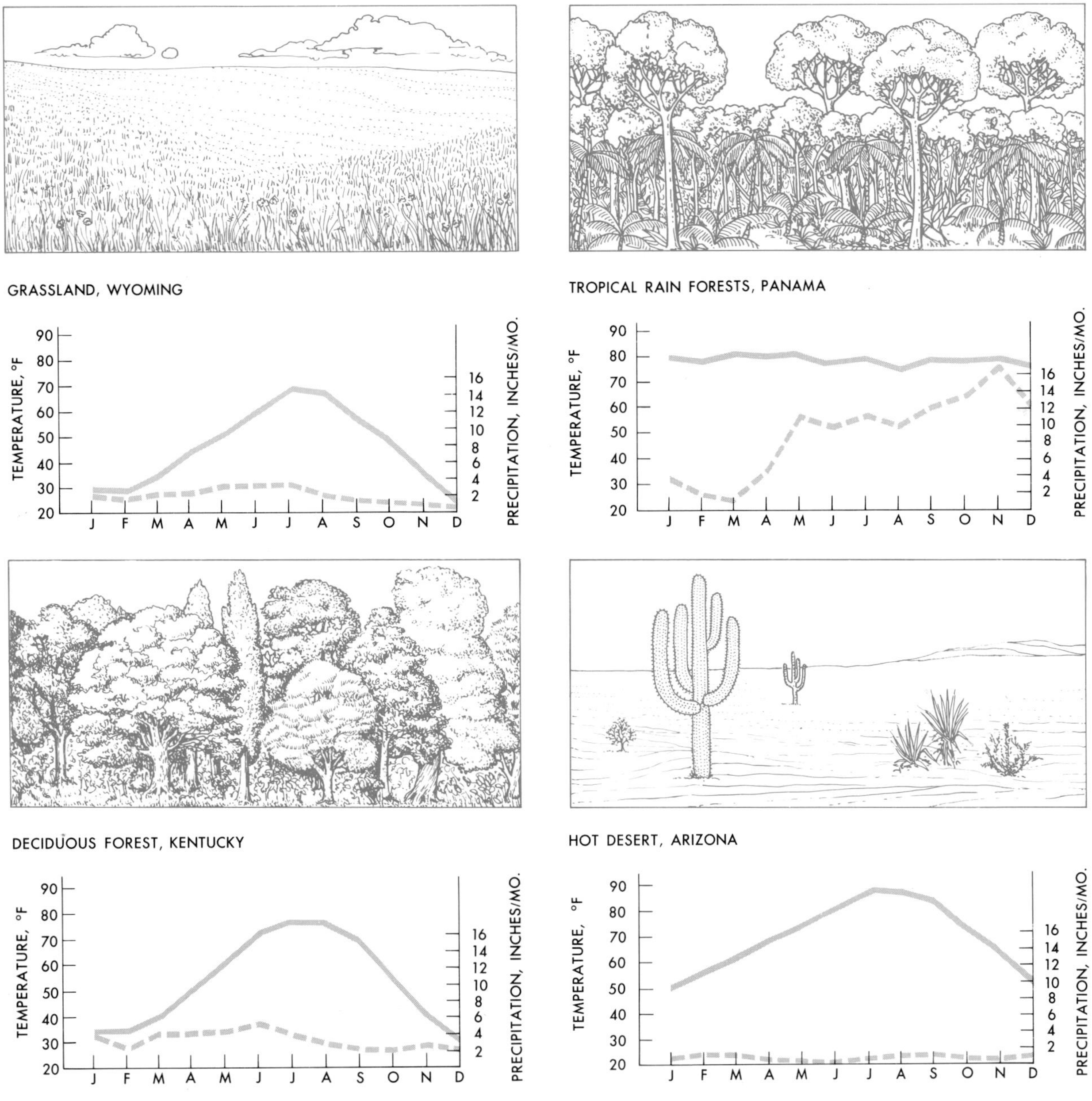

Fig. 15.3 *Four types of land biomes found throughout the major continents (except the arctic and antarctic regions) of the earth. Each biome type is largely determined by temperature and amount of rainfall per year. Graphs showing temperature and rainfall for each biome type accompany the drawings. On the graphs the solid line traces temperature, the broken line precipitation.*

development is subject to the principle of limiting factors. Although all factors except one might be right for determining a particular biome type, that one factor can influence what biome type actually exists. It cannot be emphasized too strongly that the physical factors in the environment are the ultimate determiners of the biotic community that can exist in any particular place.

A dramatic example of the effects of physiochemical environment on biome type can be seen by simply climbing up a high mountain on foot or in a vehicle (see Fig. 15.4). There is a strong parallel between the types inhabiting certain altitudes on the mountain, and those inhabiting certain latitudes on the surface of the earth as one proceeds from the equator northward. The ascent up the mountain presents us, in telescoped form, with a journey across a large segment of the earth's surface. At the bottom of the mountain (suppose for the sake of completeness that the mountain is in a fairly tropical climate to begin with) is the very dense tropical rain forest, with its high humidity, tall trees, and large number of species. A little way up the mountain the tropical rain forest gives way to a forest of deciduous trees (those which lose their leaves in winter), corresponding to the dominant forests of the north-temperate zone of the United States. Deciduous forests require less rainfall and humidity and can withstand lower temperatures than tropical forests. They also have a less extensive assemblage of species.

Proceeding up the mountain, or northward along the earth's surface, we encounter the coniferous forests, consisting of various spruces, pines, and firs. Conifers require still less rainfall and can withstand more severe cold than deciduous trees. They

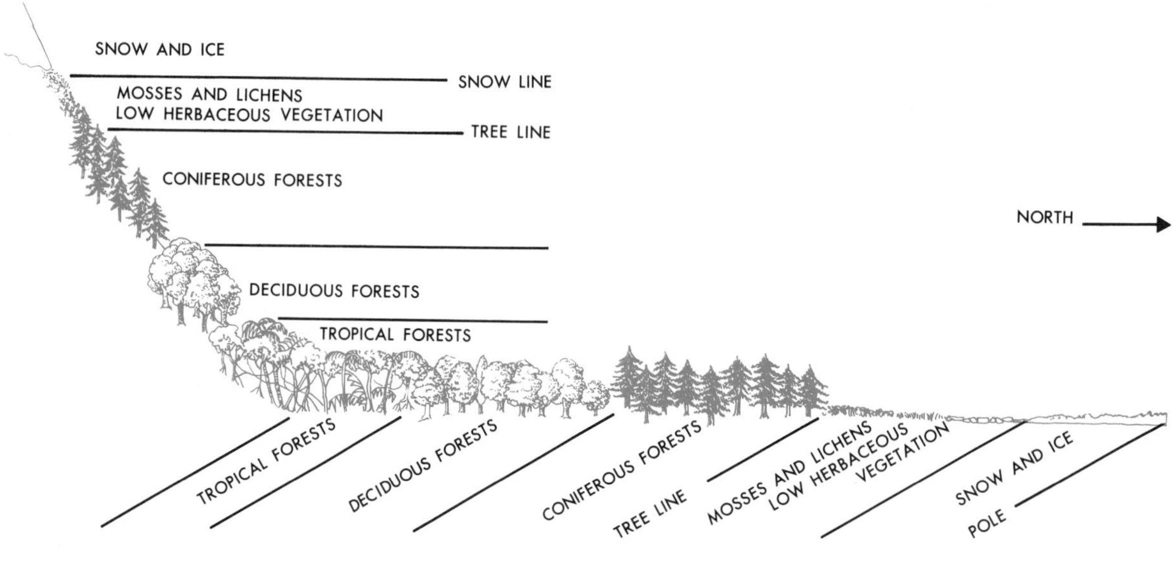

Fig. 15.4 *Biome types proceeding up a mountain correspond to biome types proceeding northward from the equator toward the pole. In North America, average temperature falls as one proceeds up a mountain or toward more northern latitudes. The different biome types found in each area of latitude or altitude are a consequence of temperature, moisture, type of soil, wind, and sunlight. These physical factors determine the total assemblage of life that can inhabit each particular area.*

can also live more readily in rocky and sandy soil, since they have taproots that penetrate deeply into the earth. Proceeding on upward, or northward, the coniferous forest begins to disappear. First the trees that remain are scrubby, showing the effects of high wind and cold temperatures. Finally trees disappear completely. On the mountainside we say we have reached the "tree line." The biome to occupy the next highest altitude, or northerly latitude, is often called the tundra. It is cold and windy, with vegetation limited to low-lying shrubs, mosses, and lichens. Tundra usually has a permanent stratum of frost, the **permafrost,** the exact depth of which varies according to latitude (or altitude). Because of the permafrost, when the tundra thaws in warmer weather, water cannot soak into the ground and run off. Consequently, tundras are marked during the summer with bogs and small lakes or pools. Lying still beyond the tundra, at the top of the mountain or around the arctic regions of the earth, is permanent ice and snow. Such areas are not necessarily lifeless, but contain algae, bacteria, and other organisms. In this respect the mountain top differs from its polar counterpart. While the snowy and icy peaks of many mountains do harbor some algae, in many cases the severity of wind conditions and low air pressure become limiting factors.

The distribution of biomes on the surface of the earth is a consequence of the various physiochemical (including climatic) factors characterizing each geographic area. As we will point out in a later section, the presence of living systems in a given area can, after some period of time, alter physiochemical conditions. There is thus a constant interplay between the physiochemical (abiotic) factors that determine the kind of life that can exist, and the life forms (biotic factors) that do exist.

15.4 The Growth and Regulation of Populations

Every type of organism has a **biotic potential,** or the ability to increase in numbers given ideal environmental conditions. Under such conditions, some types of organisms produce many more offspring than others. Jackrabbits, for example, have a very high biotic potential. They reproduce very rapidly under favorable conditions. On the other hand, populations of whooping cranes have a very low biotic potential. They produce only one or two offspring per parent every two years.

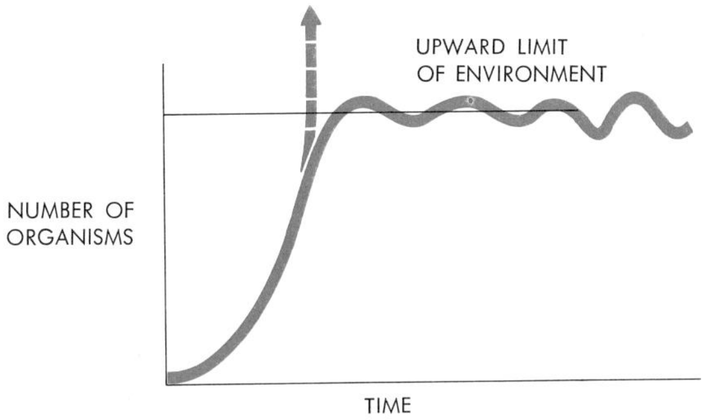

Fig.15.5 *Graph of generalized population growth.*

Opposing the effects of biotic potential is the **environmental resistance** a population encounters. Environmental resistance comprises all those factors in the environment that tend to prevent a population of organisms from multiplying at an unlimited rate: limitation of food supply, competition with other organisms, predation, or the effects of climate. All these limiting factors act together to restrict population growth.

Environmental resistance may be measured as the differences between the theoretical growth rate of a population under ideal conditions and the observed growth rate in nature. For example, given optimum conditions, a single pair of fruit flies could yield 3368×10^{52} offspring in a year! The fact that the fruit fly population does not increase at this rate is a result of such environmental resistance as limited food supply and physiological stress produced by low temperature.

The forms of environmental resistance discussed above are among the major checks to population size. If a population of organisms is introduced into a new territory where environmental conditions are favorable, the population grows rapidly. It is not that the rate of reproduction of any pair of individuals is increased, but simply that more offspring are able to survive. Figure 15.5 shows two possible population-growth curves. The solid curve shows the actual growth of the populations as observed in nature. The dotted line indicates the path the curve would take if there were no environmental resistance.

Competition as a Regulator of Population Size

As the number of organisms increases, the environmental resistance begins to increase proportionately. A decrease in the supply of food per organism reduces the number of offspring that survive. The growth curve begins to level off. The horizontal line running across the top of the graph represents the maximum number of organisms the environment can support. As the actual number of organisms in the population approaches this upward limit, environmental resistance becomes greater. When the curve levels off, the population maintains itself at about the maximum limit, called the **carrying capacity** of the environment, and the number of offspring that survive is very close to the number of organisms that die. The population has reached a **dynamic equilibrium.**

This dynamic equilibrium is not evenly maintained, however. The population size actually seesaws, first on one side of the upward limiting value, then on the other. When the population size goes over this limit, environmental resistance has exceeded the biotic potential. The death rate increases and the population decreases. When the line dips below the limiting value, biotic potential exceeds environmental resistance, and the size of the population is able to increase.

Predation as a Regulator of Population Size

Predation is a second factor that can limit population size. Just as the amount of any food-producing population determines the size of the population that feeds on it, the reverse is also true. In other words, the size of a food population itself is determined by the size of the population that preys on it. In a stable ecosystem, the relative ratios of prey and predator will not change much over the course of time. The graph in Fig. 15.6 illustrates dramatically how the sizes of prey–predator popu-

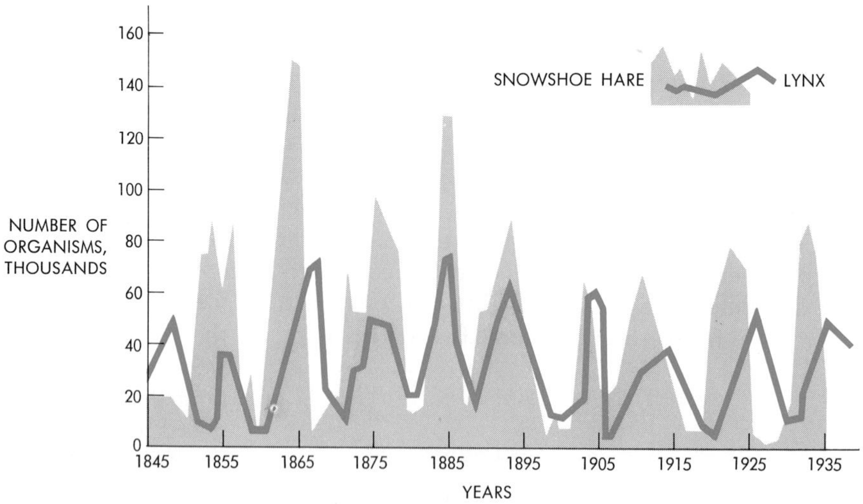

Fig. 15.6 *Cyclic changes in populations of snowshoe hares and lynx in Canada from 1845 to 1935. This graph is constructed from records of the number of pelts received per year by the Hudson's Bay Company. (After C. A. Villee,* Biology. *Philadelphia: W. B. Saunders, 1957, p. 577.)*

The regulation of population size, like the regulation of physiological systems, occurs through modulation of a dynamic equilibrium. Consequently the size of any population is constantly oscillating around a mean and never remains absolutely constant.

lations are completely interdependent. This graph shows estimated cyclic changes in population of lynx and snowshoe hares in Canada for a period of 90 years. Three things are important. First, note that on any one occasion the peak of the curve for the snowshoe hare (the prey) is always a good deal higher than that for the lynx (the predator). Second, note that the peak for the lynx population always occurs a little later than that for the hare population. As the number of hares increases, so does the number of lynx. This relationship results because more lynx kittens can survive when the supply of snowshoe hares is abundant. Third, note that when the number of hares decreases, so does the number of lynx.

Predation may damp out oscillations and allow the prey population to remain relatively stable. The result of interfering in a prey–predator relationship is often illustrated by the case history of Kaibab deer in Arizona earlier in this century. The Kaibab Plateau of Arizona was thought to sustain a healthy deer population that was kept from exceeding the supply of vegetation by predation from wolves, coyotes, and pumas. As a result of this predation, there were apparently never more deer than the vegetation could support. A well-meaning campaign was mounted that aimed to benefit the deer by killing off all the predators. The deer population was estimated to increase from about 4,000 in 1907 to 100,000 in 1924. In two winters over 6% of the herd was thought to die from starvation. The estimated size of the deer population declined until 1949 when the total number was back down to 10,000. This population was well within the reported carrying capacity of the environment as it had existed in 1907. Recently, a careful reexamination of the original data in the Kaibab deer case has cast doubt on its use as a "classic" illustration of the point in question. Nevertheless, as other studies do indicate, interference in a natural prey–predator relationship can be disastrous.

Density as a Regulator of Population Size

A third way in which the size of natural populations may be regulated is through what are called **density-dependent factors.** An example of density-dependent factors can be seen in the classic case of small rodents known as lemmings, which live chiefly in Scandinavia. When a lemming population becomes too large, great hordes of lemmings begin to migrate from the overpopulated region. This migration is impulsive. Lemmings on the march stop neither to rest nor to eat. Frequently they bypass regions in which they could easily settle. At some point in their migration, lemmings usually encounter a lake, river, or fjord. Ignoring the barrier, they usually plunge right into the water and drown. This famous spectacle has given rise to a great number of romantic and exaggerated stories of the lemmings' "march to the sea."

The mass migration of the lemmings can be understood as the result of several interacting variables including predation, competition, disease, climatic variations, and possibly physiological stress. All these parameters seem to be linked together in ways that produce cyclic fluctuations in the lemming populations so that peak densities occur about every 3 to 4 years, on a long-term average.

All the mechanisms operating on populations serve to keep population size at a relatively uni-

Habitat is an organism's home, the place where it lives. Niche is an organism's occupation, what it "does for a living" within the ecosystem.

form level of balancing environmental resistance and biotic potential. These built-in mechanisms, which maintain a steady-state carrying capacity for different populations, operate on a feedback principle similar to that active in physiological systems. Again, one of the most important and general features of living organisms is their internal control by feedback mechanisms.

15.5 Ecology and Evolution

At the beginning of this chapter we stated that all organisms in an ecosystem are in competition for the same basic requirement: a source of energy. Interspecific competition results when two different species living in the same general area have similar ecological requirements. We have said that a frequent result of such competition is the decline and eventual extinction of one species and the complete predominance of the other. This has led to a formal statement known as Gause's hypothesis, or the **competitive exclusion principle.** Gause's hypothesis states simply that two species with similar ecological requirements cannot successfully live together for more than a short time. The results of several of Gause's experiments are shown in Fig. 15.7. When one species of the protozoan *Paramecium* (in this case, *P. caudatum*) is grown alone in a nutrient culture, it shows a typical sigmoid or S-shaped growth curve. The same is true for another species, *P. aurelia* (Fig. 15.7a). When the two are grown together, however, *P. aurelia* eliminates *P. caudatum* in the course of about 11 days (Fig. 15.7b). They do not prey on one another. Both species compete for nearly all the same minerals and food supplies, and *P. aurelia* wins.

The evolution of interwoven ecological relationships depends on such competition. In heavily competing populations, any variations that tend to reduce this competition are selectively favored. If *P. caudatum* individuals are able to use some other

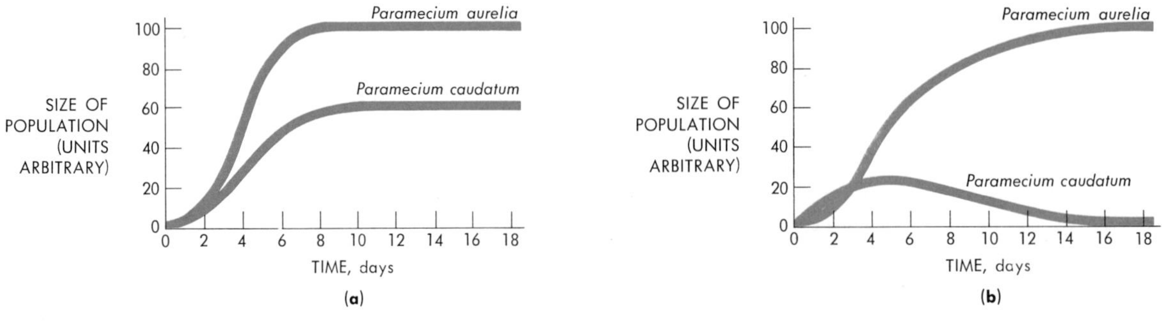

Fig. 15.7 *Typical growth curves for populations of* Paramecium aurelia *and* Paramecium caudatum *(a) when each species is grown separately and (b) when the two cultures are mixed.*

food supply than that monopolized by *P. aurelia,* competition will be reduced. This will increase the survival rate of those individuals of *P. caudatum* that bear the variation. In nature, organisms frequently invade unoccupied habitats simply to avoid intense competition. Once the organism is in a new habitat, any variations that allow it to exploit the available resources more efficiently will tend to be perpetuated. In this way, the genetic makeup of the population may slowly change, and it becomes well adapted to a new niche.

15.6 Ecological Niches, Habitats, and the Ecosystem

The concepts of habitat, ecological niche, and ecosystem are very useful in describing the various relationships between organisms.

The **habitat** is the place in which an organism lives in the biotic community. The term may refer to an area as large as the ocean or desert, or as small as the underside of a lily pad or the intestine of a termite. Ecologists sometimes speak of habitat as the "address" of an organism within the community. It is a place that contains suitable supplies of such required resources as food, water, nesting sites, etc.

The **ecological niche** occupied by an organism is harder to pinpoint than its habitat. The term "niche" refers to the role an organism plays within the biotic community. To what organisms does it serve as food; upon what organisms does it feed? What are the upper and lower limits to its toleration of changing temperatures? What minerals does it require from the environment? What minerals does it return to the environment? Answers to such questions help establish the exact niche an organism occupies. Just as an organism's habitat is spoken of as its "address" within the biotic community, its ecological niche is called its "profession." Unlike its habitat, an organism's niche is defined by all the physical, chemical, and biotic factors that influence an organism's maintenance and reproduction.

Because of the different ways they use the habitat, organisms may live together in the same general habitat yet have quite different ecological niches. Tidal pools near the ocean are a good example. They contain a wide variety of organisms: animals such as starfish and sea anemones and plants such as "seaweed" and smaller filamentous algae, all of which have roughly the same habitat. Yet within a single tidal pool the algae serve as producers, since they can manufacture carbohydrates by using energy from the sun, and the animals serve as consumers. They feed on smaller animals that ultimately feed on the plants. Thus the algae occupy a different ecological niche from the starfish or sea anemones.

Consider another example. In the shallow water along the shore of a lake it is possible to observe a large variety of water insects, all of which have the same habitat. Some of these, such as the "backswimmer" (genus *Notonecta*), serve as predators and feed on other small animals. Other backswimmers, of the genus *Corixa,* serve as decomposers and feed on dead or decaying material. Both organisms have the same habitat. Their ecological niches within that habitat, however, are quite different.

Every environment offers a large number of niches and habitats. In fact, although it tends to be similar for members of a given species, the environment of any two organisms is never precisely the same. While it is true that two species can occupy roughly the same habitat, in any given community they cannot occupy the same niche for very long. Occupying the same niche means competing on nearly every level of existence, and such competition generally results in the survival of one species and elimination of the other.

The term **ecosystem** has been used in several previous sections to refer to many interactions between biotic communities and their associated physiochemical environment. Any community or group of communities can be treated as an ecosystem. For example, the boundaries of a pond ecosystem can be defined in terms of the entire watershed, or only in terms of the shoreline vegetation, depending on the limits one wishes to draw. If a study deals with the nutrient dynamics of the entire pond, then the whole watershed of the surrounding terrestrial communities must be included. If the study is limited to the role of certain species of aquatic snails as grazers on particular plants, then the boundaries of the ecosystem will consist only of the shallow portions of the pond where those snails and plants live. Generally, ecosystems will be defined so that specific inputs and outputs can be identified, and the boundaries will be determined in a manner that allows for measurement of these flows. Thus all ecosystems are abstractions; they can be as large and complex as our abilities to measure them.

15.7 Community Structure and Energy Flow

Members of a biotic community can be classified according to their mode of obtaining energy. They may be producers, primary consumers, secondary consumers, tertiary consumers, or decomposers. These are arbitrary but convenient designations for major groups of organisms that differ greatly in their sources of energy. The **producers** are primarily green plants that use solar energy to convert inorganic substances into living tissue. While most bacteria use chemical energy derived from mineral substances, some bacteria are also photosynthetic. These green plants and specialized bacteria may differ greatly in their modes of reproduction, longevity, and size, but they share the same attribute of converting nonliving sources of energy into living substances which, in turn, provide energy for the consumers and decomposers.

Generally **primary consumers** are herbivores, organisms that eat only plants. These "vegetarians" may be microscopic in size and consumers of small algae, or they may be very large mammals who forage on grasses and shrubs. For example, the small "water flea" *Daphnia major*, is an herbivorous microcrustacean. These "zooplankton" swim and filter floating algae in suspension in the upper, well-lighted water of a lake. The **secondary consumers** are carnivores that consume the herbivores. Similarly, the **tertiary consumers** are other carnivores that consume the secondary consumers. These distinct groups interact to form a **food chain.** Several interconected groups produce a **food web** (Fig. 15.8). *The movement of matter through food webs is cyclic,* because the **decomposers** (fungi and some bacteria) convert dead organic tissue of producers and consumers back into mineralized matter that can once again be recycled through the food web.

Energy Flow

The movement of energy is not cyclic but unidirectional; it flows from the producers to the consumers and decomposers. Energy must continually be added to the food web in order for the recycling of matter to continue. In some communities, where there is no direct input of solar energy, the source of energy is the organic debris and breakdown products from other communities that *are* dependent upon solar energy. The dark inner caverns of cave communities and the very deep waters in the ocean receive no direct solar energy, but they do obtain energy indirectly by using the debris or by-products raining down on them from other communities exposed to the sun. In these specialized "dark-adapted" communities, the producers may be greatly reduced or completely absent.

The structure of a community can be depicted graphically by counting the number of individuals or weighing the mass of all members of each species. From an energetic point of view, these groups are called **trophic levels.** The producers are the **autotrophs,** or the first trophic level, while the various consumers are the **heterotrophs.** Each group constitutes a distinct trophic level. The diagram is called a **pyramid of numbers** (Fig. 15.9a) or a **pyra-**

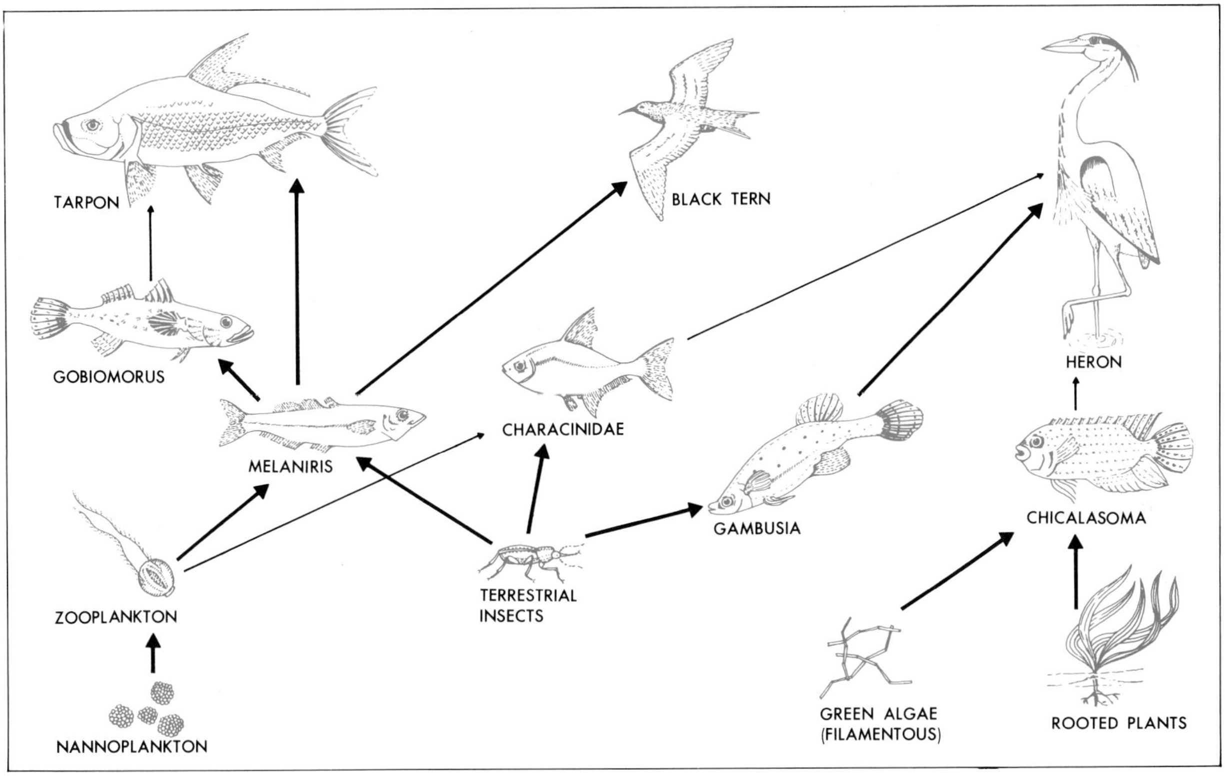

Fig. 15.8 *Generalized food web of Gatun Lake in the Panama Canal. Heavy arrows indicate connections between major consumers and producers, while thin arrows denote connections of minor energy flows. (After Zaret and Paine, "Species introduction in a tropical lake,"* Science *182, 1973, pp. 449–455.)*

Fig. 15.9 *Pyramids of numbers (a), mass (b), and productivity, or energy (c) for communities or organisms in an experimental pond. Productivity was estimated from the rate of phosphorus uptake. All three show the essential characteristic of less available mass or energy the higher up a food or energy web.*

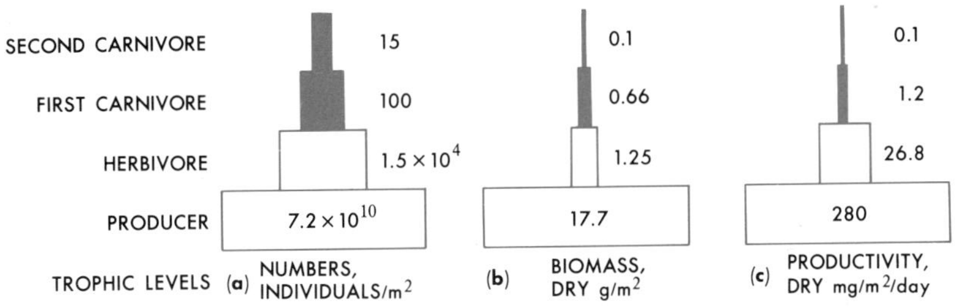

mid of mass (Fig. 15.9b). If the mass is converted to units of energy, then the diagram is termed an **energy** or **productivity pyramid** (Fig. 15.9c). In many communities, the structure is steeply truncated; the greatest mass and energy content occurs in the base among the producers. However, it is possible to have an inverted pyramid of mass if the flow of energy is very rapid and the size and life cycle of the producer organisms is reduced relative to the consumers. For example, the amount of microscopic algae in the open water of the ocean is often very small relative to the grazing herbivores. The only way this low abundance of algae can supply adequate energy to the herbivores is if there is a rapid "turnover" of algae: the algae reproduce quickly and are eaten quickly by the consumers. In such cases there is an inverted pyramid of mass, *but there is always a typical pyramid of energy* (Fig. 15.10).

In no community can a flow of energy greater than that available in the first trophic level persistently occur in the upper trophic levels. In fact, in all communities decreasing amounts of energy are available to the organisms in the upper trophic levels. This decrease occurs because all organisms are somewhat inefficient in their metabolic use of energy and all lose energy through the generation of heat. At each step up the pyramid, the organisms are able to use only 10 to 20% of the total energy available from the preceding level (although a few species may be more highly adapted and much more efficient than others, none have been observed to use more than 60%).

Why does each transfer of energy extract only a small fraction of the total potential energy from the previous level? The decrease in available energy along a food chain is in agreement with the first and second laws of thermodynamics. The first law holds that energy cannot be created or destroyed, but only changed in form, in ordinary chemical reactions. None of the energy in the universe is lost. The second law states that the total amount of *usable* energy in any system tends to decrease with time. This is because no transformation of energy is 100% efficient. In a given transformation, some energy is always converted into heat.

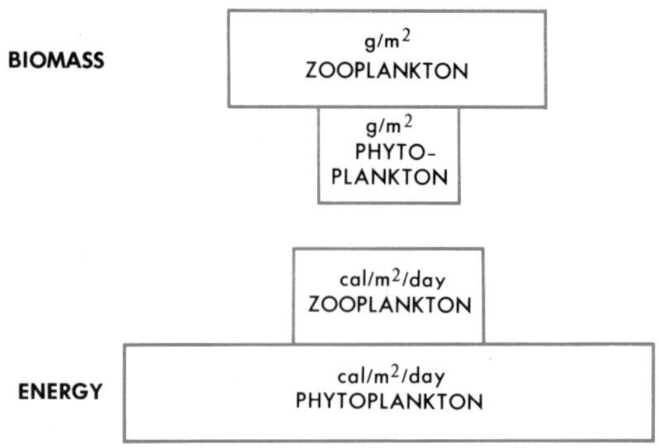

Fig. 15.10 *Inverted pyramids of mass in the sea. The values are in grams per square meter. A small crop of phytoplankton (plants) supports a larger crop of zooplankton (animals) and bottom animals that presumably all derive their nutrients from the small mass of plants. The low concentration of microscopic organisms occurs because they "turn over" quickly, so that the rate of energy flowing through the phytoplanktonic producer level is fast enough to be capable of sustaining the large biomass of animal trophic levels that feed on them. This illustrates the limited value of information derived solely from measurements of standing crop biomass or yield for studies of energetics. (After Odum 1959, from data of Riley 1956 and Harvey 1950.)*

A food chain involves a series of energy conversions. With each transformation of energy from prey to predator between trophic levels, there is a loss of usable energy. When a cow eats grass, for example, a great deal of the bulk consumed does not contribute to the nourishment of the cow. Much of it is indigestible and is returned to the environment in feces. In addition the cow must move, reproduce, and carry on a great many other energy-expending activities. This means that much of the energy value of the carbohydrate contained in the grass must go to support such functions. Only a small amount of energy is actually transformed into structural parts of the animal that another organism (such as a human) later eats. This low efficiency is some indication of why the competition for energy among many groups of organisms is often so severe, especially if they all occupy roughly the same consumer level on the food chain.

Three important principles emerge from this discussion of trophic dynamics. First, to be complete and self-containing, any food chain must always have photosynthesis at the beginning and decay at the end. In generalized form, then, a food chain may be represented as:

PHOTOSYNTHETIC ORGANISM ⟶ HERBIVORE ⟶ CARNIVORE ⟶ ORGANISM OF DECAY

Energy must be constantly supplied from the outside if the food chain is to keep operating.

Second, in general, the shorter a food chain the more efficient it is. Conversely, the more steps in such a chain, the greater the waste of energy that results. Third, the size of any population is ultimately determined by the number of steps in the food chain. With the decrease in useful energy at each step along the chain, very little energy is available for a population of quaternary consumers. The size of a population of quaternary consumers is typically less than that of tertiary consumers; a population of tertiary consumers is smaller than one of secondary consumers, and so on.

Movement of Materials

What happens to the materials consumer organisms take into their bodies? Some may move from one ecosystem to another, while others are only removed temporarily. Animals are continually returning various substances to the environment by the elimination of their waste products. And when an animal or plant dies, the materials within its body are acted on by organisms of decay—bacteria and various other decomposers. These organisms break down the complex proteins, fats, and nucleic acids of the dead organism and release many of the components back into the environment. At the same time, the decomposers are gaining their own nourishment. Decomposer organisms thus fill an extremely important niche within an ecosystem. Through the metabolic activity of decomposers, vital organic materials are prevented from remaining locked up in the bodies of dead organisms.

Consider a freshwater lake, a typical ecosystem (Fig. 15.8). Solar radiation is the driving force that keeps any such system going. This energy is harnessed by the producers: rooted water plants and free-floating algae. Small organisms feed on the algae and are eaten in turn by larger forms. If these larger forms have no natural enemy, they ultimately die and are then acted on by organisms of decay, which break down the large organic molecules and absorb the substances they need. The rest is returned to the water. As a result, all the materials that plants withdraw from the water in the course of their normal metabolism are returned. Besides producing carbohydrates, plants often synthesize complex compounds that animals are unable to produce on their own. Thus consumers obtain from producers not only a source of energy, but also many of the substances essential for their own growth and development.

Some plants also produce toxic substances which inhibit growth by their competitors or repel consuming species. These intricately evolved chemical interactions can be disrupted when human-made compounds are introduced into natural ecosystems (see Supplement 15.1).

Competition exists at all levels in an ecosystem. In a lake, algae on the surface of the water and plants around the shore compete for available sunlight and for the mineral nutrients in the water. Zooplanktonic copepods compete for available algae, while fish compete for available copepods, and so on. All forms of competition are simply attempts to obtain a source of energy.

Supplement 15.1

CONCENTRATION OF TOXINS IN FOOD WEBS

An understanding of energy flow is necessary to understand how certain synthetic compounds move through natural food webs. Initially many people assumed that the continued use of artificial pesticides would not pose problems for wildlife because the total amount introduced was relatively small and would be rapidly diluted. Yet as more data became available, a startling pattern emerged. Small doses accumulated and became concentrated in the bodies of consumer species at the uppermost trophic levels. (Remember that fewer consumers at high trophic levels consume many more organisms at the next lower level, and so on.) The poisons are not excreted but accumulate, with deadly effect, in the tissues of higher organisms. In some areas excessive amounts of pesticides were applied and entered aquatic ecosystems, either through indirect runoff and rainfall or through direct application.

One of the surprising outcomes of the introduction of these synthetic pesticides was the long distances they were transported from the application sites. Pesticides were found in birds, fish, and seals in Antarctica, for example, though the nearest place these chemicals were used was thousands of miles away. Oceanic and atmospheric currents had carried pesticides far beyond the localized sites of their application. Evaporating water vapor and dust from croplands can contain relatively high doses of pesticides which are then redeposited when rain falls in other locations. It is estimated that 40 tons or more of DDT reach England through rainfall each year. Because many pesticides, such as dichlorodiphenyltrichloroethane (DDT), are not broken down by oxidation or microbial metabolism, they can persist in the environment for many years. These compounds can enter the material cycles of the global ecosystem and reside in certain tissues of consumer species. Concentrations of these synthetic toxins can increase over a long period of continued ingestion if the compounds are not excreted. One study of DDT in Lake Michigan showed that even though concentrations were relatively low in the lake sediments (0.0085 parts per million), there was 48 times higher concentrations in the aquatic invertebrates (0.41 parts per million). The fishes feeding on these invertebrates showed a twenty-fold increase in DDT concentration (ranging from 3.0 to 8.0 parts per million) in their bodies. Predaceous birds that fed on these fish accumulated as much as 3,177 parts per million of DDT in their fatty tissue. These high concentrations in the uppermost consumers were found to have very toxic effects. A number of different predatory birds showed seriously impaired ability to reproduce because DDT and other synthetic chemicals apparently interfered with the formation of egg-shell. There were high mortality rates among the young birds because very thin egg shells frequently broke before the birds were ready to hatch.

One of the first ecologists to report these effects was Rachel Carson. Her book *The Silent Spring* set off a major effort to document further the indirect effects of pesticides and eventually led to some global regulations on the use of synthetic chemicals. A new impetus was given to exploring ways of controlling pests by biological means, such as maintenance of natural predators that consume noxious insects and rodents and use of natural hormones to change reproductive behavior. Ecologists have found that many species use complex chemical signals to attract their mates or to defend themselves from competitors and predators. Currently, a great deal of research focuses on how these natural chemicals are synthesized and broken down in food webs so that their effectiveness can be maintained in balancing predator–prey interactions without leading to the disastrous results of the introduction of synthetic compounds that may destroy a

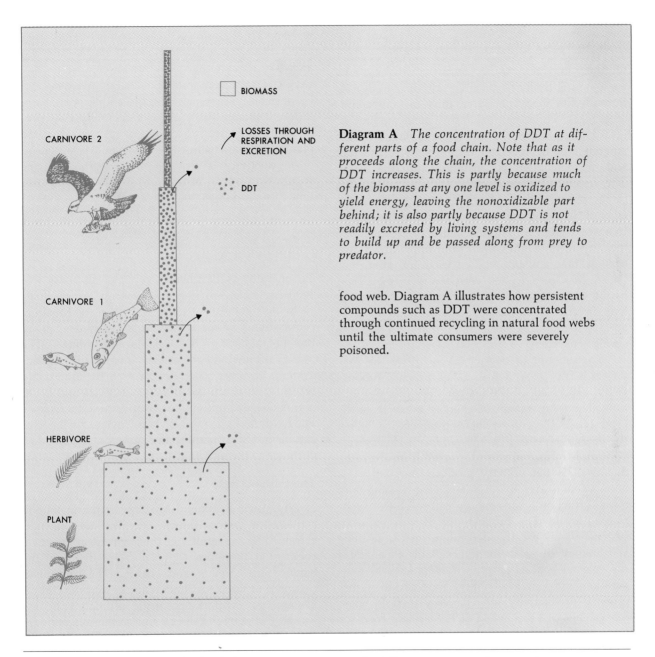

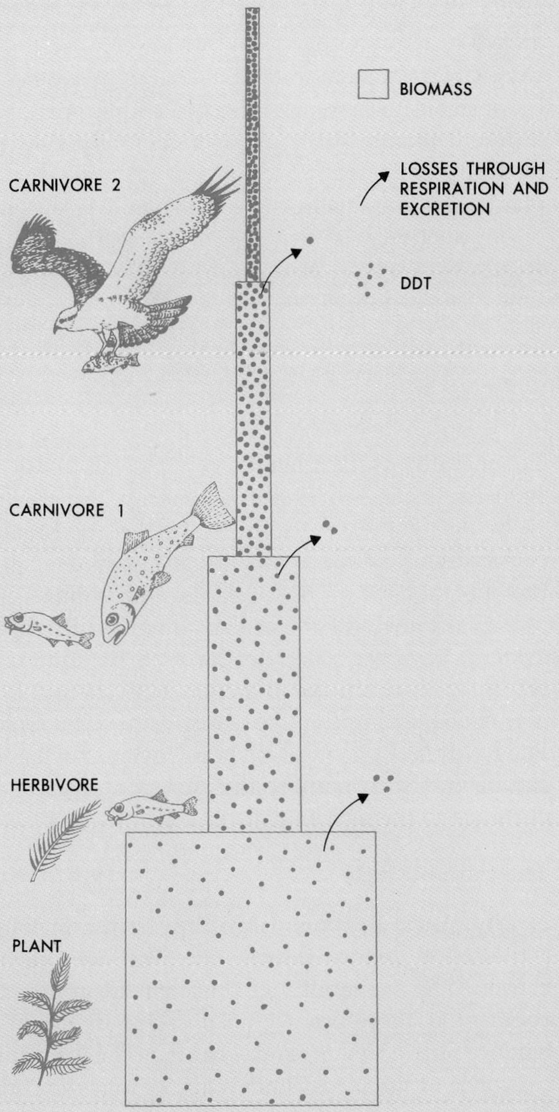

Diagram A *The concentration of DDT at different parts of a food chain. Note that as it proceeds along the chain, the concentration of DDT increases. This is partly because much of the biomass at any one level is oxidized to yield energy, leaving the nonoxidizable part behind; it is also partly because DDT is not readily excreted by living systems and tends to build up and be passed along from prey to predator.*

food web. Diagram A illustrates how persistent compounds such as DDT were concentrated through continued recycling in natural food webs until the ultimate consumers were severely poisoned.

The movement of materials through an ecosystem is cyclic. The movement of energy is unidirectional.

15.8 Productivity as a Measure of Energy Flow

One way in which very different ecosystems can be compared is by measuring the rate of energy flow through the biological community on a per-unit-area basis. For example, if the same amount of solar energy falls on a fresh-water spring, a certain amount will be reflected from the water surface and the remainder will penetrate into the water. The amount of energy actually taken up by plants in the spring will be dependent on the transparency of the water and the abundance of the plants. In a clear spring with high densities of green plants, much of the sunlight will be intercepted and a portion of that amount will be used for photosynthesis. In a turbid pond with few plants growing in the water, the flow of energy into this ecosystem directly from the sun will be relatively less on a per-unit-area basis than that entering the spring ecosystem.

The amount of energy taken up by the producers in any ecosystem is called **primary productivity,** expressed in calories per unit area per unit time. The total amount of energy taken up by the producers is the **gross primary productivity.** When the energy expended by the producers during respiration is measured and subtracted from the total, the result is **net primary productivity.** Only this latter value is available to the consumers, and only a small fraction of this energy will reach the upper trophic levels. The productivity of different ecosystems varies greatly, depending on the latitude and amount of solar energy available throughout the year, as well as on other factors such as temperature, nutrient availability, and rainfall. Gross primary productivity may be very high in some ecosystems, but their yield of harvestable food for humans may still be very low. The ecological measure of energy flow does *not* refer to the quality or type of material being produced. In other words, the ecologist measuring "productivity" of a cornfield might include all of the corn plant (not just the ears of edible corn) as well as the associated weeds. This complete compilation would then stress the relative productivity of the cornfield as an ecosystem so that it would be compared with, say, a forest community. Of course, it would also be possible to compare just the edible portions of different plants, but ecologists generally view productivity from the standpoint of all the associated plants and animals in the ecosystem. Measurement of productivity is also useful for comparing the stages of a given ecosystem as it develops over time (see Section 15.10).

15.9 The Cyclic Use of Materials

To operate for any length of time, an ecosystem requires a constant input of energy. As we have seen, that energy is captured in the food-making processes of green plants. It is released again in the metabolism of both plants and animals. The matter (atoms and molecules) involved in an ecosystem, however, does *not* have to be continually replenished from the outside. The chemical elements composing living organisms may be recirculated within a given ecosystem. Several such cycles can be traced in nature, and they are of great importance to the maintenance of life.

The Carbon Cycle

Nearly every compound involved in the metabolic activity of living things contains the element carbon. The availability of this element in the environment is therefore a crucial factor in the maintenance of animals and plants. The continued existence of any ecosystem requires that the carbon "locked up" within organisms ultimately be returned to the environment, unless there is periodic input from new materials. Thus atoms of carbon in the global ecosystem are passed around the carbon cycle.

A carbon cycle is pictured in Fig. 15.11. Let us begin with free carbon dioxide in the atmosphere. The level of atmospheric carbon dioxide is

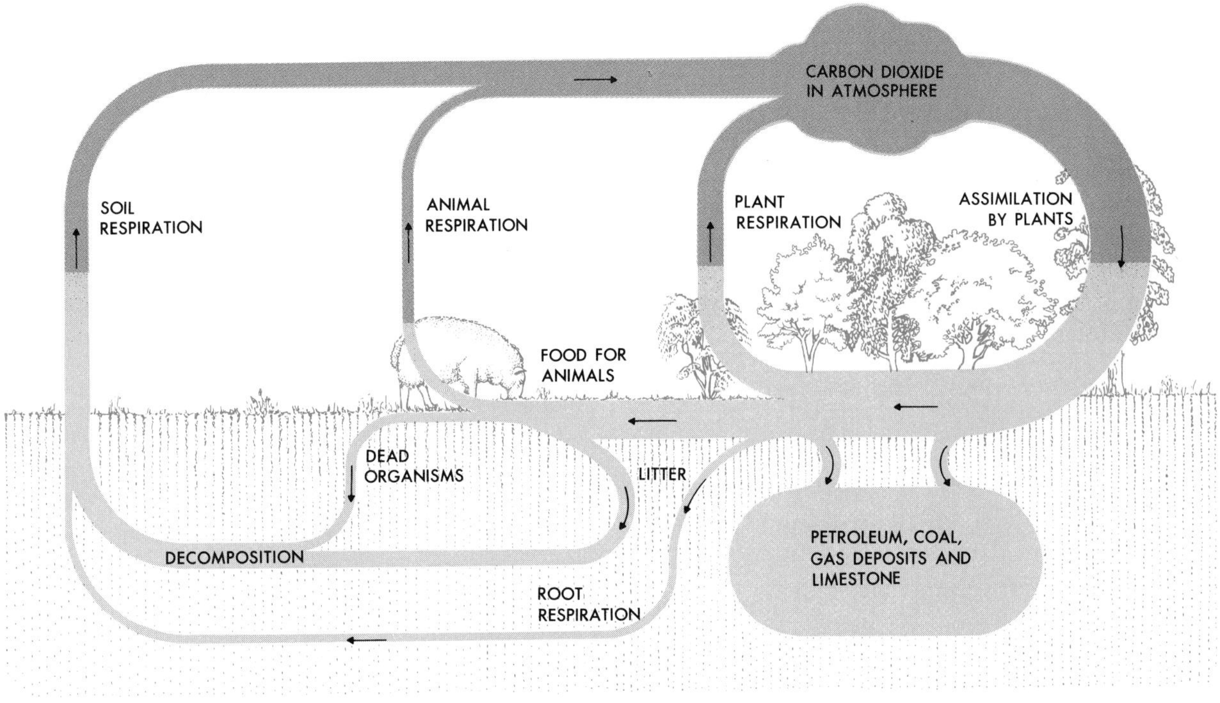

Fig. 15.11 *Diagrammatic representation of the carbon cycle. Carbon from the atmosphere, in the form of carbon dioxide (CO_2), is incorporated into carbohydrate in the process of photosynthesis. Carbohydrate is oxidized for energy by plants, animals, and microorganisms. This process returns CO_2 to the atmosphere. Some carbon is taken out of circulation for varying periods of time in what can be called a "carbon sink" (lower left) as coal, petroleum, gas, or limestone. In time, much of the carbon locked in the carbon sink will be returned to the atmosphere as CO_2 through one or another form of oxidation.*

maintained by animals and plants, both of which release this gas as an end product of respiration, the process of releasing energy. Besides releasing carbon dioxide, plants have the ability to use carbon dioxide in the manufacture of carbohydrates.

The carbohydrates in plant material can follow one of several courses. If the plant is eaten by an animal, the carbohydrate is burned in the animal tissues to yield energy and release carbon dioxide back into the atmosphere. Some carbon may pass out in the waste products of animal metabolism, such as urine or feces. The carbon in these waste products, as well as that which is part of the animal material at death, is acted on by organisms of decay. One end product of these decay processes is CO_2, which is released back into the atmosphere.

On the other hand, the plant may simply die before being devoured by an animal and in this case its organic substances (carbohydrates, fats, proteins, nucleic acids, vitamins, etc.) are acted on directly by the bacteria and fungi of decay. Sometimes, however, the plant materials may undergo a quite different process at death. By being deposited at the bottom of lakes in tightly packed layers, covered over with mud or other organic debris, and subjected to great pressure, plant parts may be turned into coal. This process requires very long periods of time and normally occurs less frequently than decomposition by bacteria. The carbon locked in coal is taken out of the global ecosystem for a long period of time. In due course, however, it is returned to the atmosphere as carbon dioxide through burning or weathering.

In addition to coal, there is a second very large "reservoir" for carbon that recycles very slowly. Calcium carbonate (limestone) is precipitated chemically in tropical oceans and in hardwater lakes and streams, and a chalky material is deposited as sediment. In addition, $CaCO_3$ is also produced biologically by microorganisms such as algae and foraminifera and by large reef-building invertebrates, as well as many vertebrates. These shells, bones, and microscopic bodies accumulate over long periods and are gradually transformed into sedimentary rock. Massive outcrops of limestone may contain billions of fossils. A close examination of a single piece of blackboard chalk reveals many different species preserved in these ancient carbonates that were deposited in former shallow seas. Carbonic acid is a major link in the carbon cycle as it forms from carbon dioxide in the atmosphere being dissolved in rain water. Through the slow weathering of limestone by carbonic acid, this soft rock weathers to release calcium and bicarbonate ions into rivers, lakes, and oceans.

In all these ways, carbon taken from the atmosphere by plants in photosynthesis is ultimately returned. In the course of time, a single atom of carbon in an ecosystem may have existed in a variety of compounds in many different organisms.

The Nitrogen Cycle

The element nitrogen is no less essential to life than carbon. In living organisms, nitrogen is found chiefly in amino acids and proteins. Since these molecules are constantly being built up and broken down in normal metabolic activity, it is essential that new sources of nitrogen be always available to an organism. Nitrogen is cycled from environment to organism back to environment by one of several paths (Fig. 15.12).

Four types of bacteria are involved as key parts of the nitrogen cycle. Before considering the details of the entire cycle, it will be helpful to examine the specific nature of each of these bacteria.

Nitrogen-fixing bacteria. These bacteria live in the soil and on the roots of leguminous plants (plants that bear their seeds in pods, such as beans or peas), in little swellings known as **nodules** (see Fig. 15.1). They also occur in several other types of plants, either in root nodules or in dense colonies in leaf tissues. Nitrogen-fixing bacteria have the ability to take free nitrogen gas from the atmosphere and convert it into soluble nitrates (compounds containing NO_3 such as potassium nitrate, KNO_3). Because they are soluble, nitrates can be taken into the roots of higher plants. Nitrogen-fixing bacteria and some types of algae can make use of atmospheric nitrogen.

Putrefying bacteria. Putrefying bacteria are found chiefly in the soil and in the mud at the bottom of lakes, rivers, or oceans, where they break down animal and plant proteins, converting them into ammonium compounds such as ammonium phosphate $(NH_4)_3PO_4$. These compounds are released into the soil or water where they can be acted on by other types of bacteria.

Nitrosofying bacteria. Nitrosofying bacteria (genus *Nitrosomonas*) act on ammonium compounds such as those produced by the process of putrefaction. By various chemical processes, these compounds are converted in **nitrites,** molecules containing NO_2. Like nitrates, nitrites are soluble.

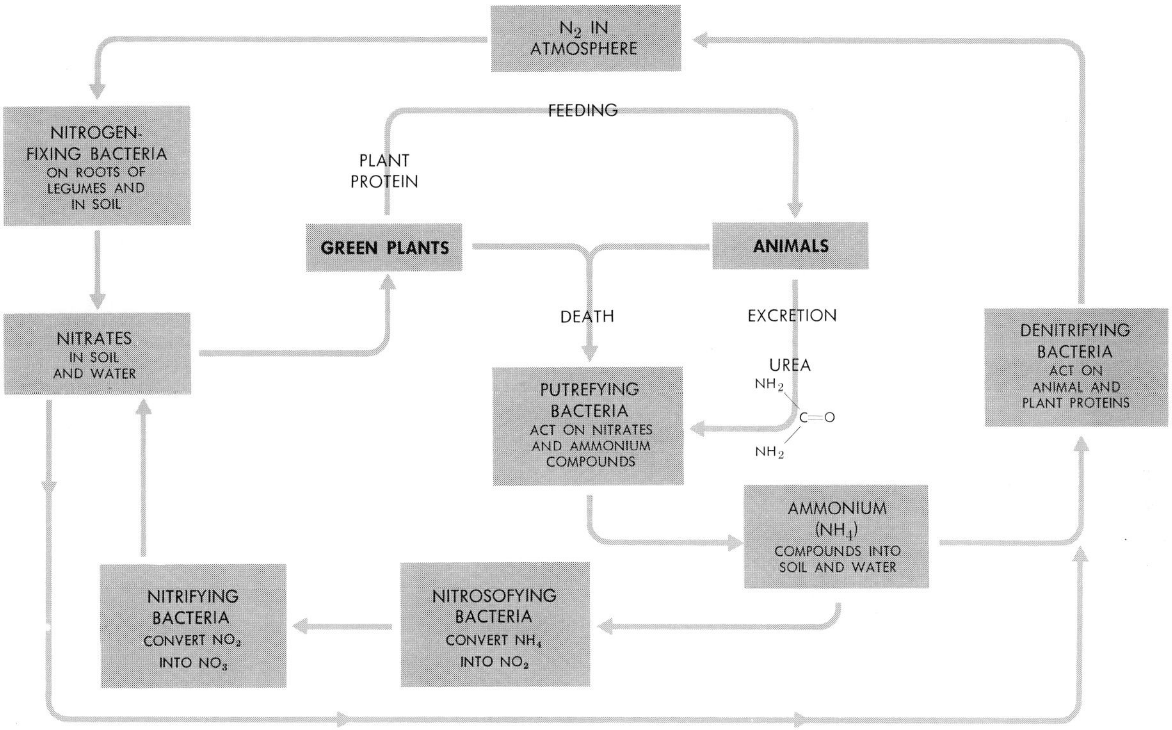

Fig. 15.12 *The nitrogen cycle. Nitrogen from the atmosphere is converted into a biologically usable form (that is, rendered soluble as nitrates) by nitrogen-fixing bacteria found on the roots of leguminous plants. From there, the nitrates can follow a number of paths and pass through a series of conversions. Denitrifying bacteria convert nitrates and ammonium back into atmospheric nitrogen, to complete the large cycle. Nitrates are used in the synthesis of amino acids by bacteria, green plants, and fungi. Animals get their nitrogen by eating plant proteins.*

Nitrifying bacteria. These bacteria (genus *Nitrobacter*) are able to convert nitrites, produced by nitrosofying bacteria, into nitrates.

Denitrifying bacteria. These bacteria convert either nitrates or ammonium compounds into molecular nitrogen (N_2). Thus the denitrifying bacteria serve as a means of returning molecular nitrogen to the atmosphere.

Consider now how all these components fit together to form the complete nitrogen cycle. It is convenient to begin again with the atmosphere. Molecular nitrogen composes about 78% (by volume) of the earth's atmosphere, but neither animals nor green plants use nitrogen in this form. It must first be converted into soluble nitrate compounds. This is accomplished largely by the nitrogen-fixing bacteria on the roots of leguminous plants. The nitrates produced there pass into the soil and are absorbed by the roots of plants. Once inside the plant, nitrates can be converted into amino acids that are the building blocks of proteins. In this way, atmospheric nitrogen becomes incorporated into protein. Or it may become converted into bacterial protein without passing into green plants.

The nitrogen in plant protein may now take two different routes. If the plant is eaten by an animal, the protein is broken down and reconstituted as animal protein. In higher animals, the breakdown of proteins yields the nitrogen-containing compound urea. Lower animal forms excrete excess nitrogen in other forms, sometimes as ammonia (NH_3) or as a compound known as uric acid. In any case, such nitrogenous wastes are acted on by either putrefying or nitrosofying bacteria. The ultimate result is that the nitrogenous wastes are returned to the cycle as either nitrates (if acted on by nitrifying bacteria) or molecular nitrogen (if acted on by denitrifying bacteria).

If the plant dies, on the other hand, the proteins may be acted on by putrefying bacteria (the bacteria of decay). This changes the protein into various ammonium compounds, as well as some nonnitrogenous waste products. The ammonium compounds thus produced can be acted on by either of two types of bacteria. They may be converted into atmospheric nitrogen by the action of denitrifying bacteria, or they may be converted into nitrite compounds by nitrosofying bacteria. Either of these processes occurs in soil or water, as bacteria act on ammonium compounds released by the bacteria of decay. The nitrogen taken from the green plants by herbivorous animals may be returned to the environment by the death of the animal. In this case, the putrefying bacteria are again the agents by which proteins are converted into ammonium.

The nitrogen cycle involves a number of organisms and a variety of pathways. There is no single nitrogen cycle; a group of nitrogen cycles all interact with each other. They ultimately ensure that no atoms of nitrogen are permanently withdrawn from circulation.

The nitrogen cycle illustrates the fact that animals are unnecessary for the operation of an ecosystem. Only green plants and bacteria are essential. Bacteria are needed because they can make use of the nitrogen in the atmosphere. Plants are needed because they can use sunlight to synthesize the organic compounds that provide the bacteria with a source of energy when they decay. Animals only enter the picture by feeding on plants. The nitrogen cycle (as well as the carbon cycle) would operate perfectly well without them.

These cycles are by no means the only ones that may be traced in nature. They serve to show the intricate relationships that exist among a wide variety of organisms. Each type of organism in an ecosystem requires many substances; carbon and nitrogen are only two of the most important. For a source of these substances, each type of organism has come to depend on the activity of one or all of the others. Plants, and thus animals, are completely dependent on nitrogen-fixing bacteria to convert atmospheric nitrogen into soluble nitrates. In turn, the nitrogen-fixing bacteria depend on the denitrifying bacteria to return nitrogen to the atmosphere as N_2. Such cycling is another example of the fundamental interdependence found within ecosystems.

15.10
Succession: Changes in Ecosystems through Time

Anyone who has planted a garden knows what happens if it is not cultivated: the desired plants are soon hidden by various kinds of weeds. When a farmer allows a cultivated field to lie unused, a crop of annual weeds generally grows on it during the first year and perennial herbs may appear the second. Gradually, however, the perennials are replaced as the **dominant species** by shrubs and trees.

Such a series of changes in plant species occupying a single region is called **secondary succession. Primary succession** occurs when aquatic habitats such as lakes and ponds are filled in with sediment and vegetation to become swamps and eventually forests. Another type of primary succession occurs when bare rocky surfaces are slowly covered by lichens and mosses that build up a soil layer that eventually supports shrubs and trees.

Primary Succession

Scientific study of ecological succession began during the seventeenth century with analyses of the development of peat bogs. In the eighteenth century, attempts were made to apply to burned-over and disturbed areas the principles gained from bog studies. During this research, the term succession was first used to refer to changes in the vegetation.

Around the turn of the twentieth century, much of the modern thinking concerning succession was developed through the pioneering efforts of several ecologists, including Henry C. Cowles of the University of Chicago. Cowles described in great detail the successional changes in the vegetation of the sand dunes at the southern end of Lake Michigan. Many centuries ago, the shore of the lake extended much farther south than at present. Over the years the lakeshore slowly retreated northward, leaving behind a series of progressively older sand dunes and beaches. Walking from the present-day lakeshore, Cowles observed a series of different plant communities (Fig. 15.13). Near the edge of the water there are no plants because of the destructive action of the waves in this large lake. Higher up on the beach, where the sand is dry in summer and frequently buffeted by the waves of winter storms, a few species of succulent annual plants manage to survive. Behind the beach, the sand dunes begin. The sand dunes are rigorous environments, very hot in the day and cold at night. Beach grasses survive on the dunes and actually help to secure them from the action of wind with their extensive underground stem (rhizome) systems. Various species of insects are the principal animals among the dunes.

Once the dunes are stabilized by the grasses, various species of shrubs, including cottonwoods, become established. The matted roots of these plants add to the stability of the dunes. On the slightly older dunes behind the cottonwood community, shrubs of other genera, including junipers and jack pine, flourish. Further back from the pine woods is a forest dominated by oak trees. Finally, several miles from the present lakeshore, Cowles observed forests of sugar maple and American beech growing in deep, rich soil.

Cowles interpreted his observation by hypothesizing that the series of communities represent different stages in ecological succession beginning with bare beach and culminating in a well-established forest of sugar maple and beech trees. The farther a given area was from the shore, the *older* a stage of development it represented. Space could be transformed, for the ecologist, into time.

Cowles's original outline of the subsequent steps these communities followed was essentially correct, but further study demonstrated that the

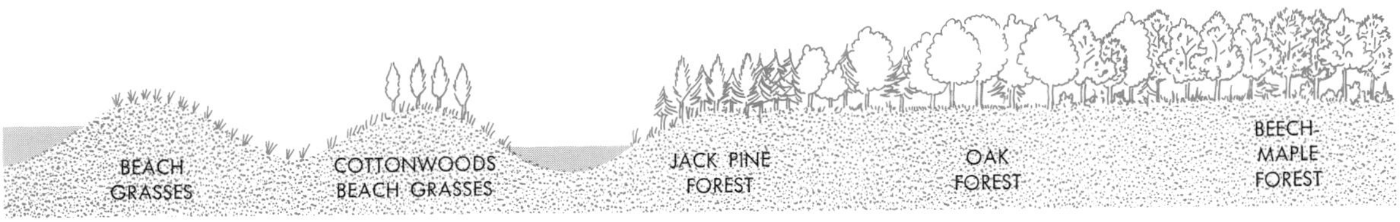

Fig. 15.13 *A highly diagrammatic portrayal of plant succession on the dunes at the southern end of Lake Michigan. (After Eugene P. Odum,* Fundamentals of Ecology. *Philadelphia: W. B. Saunders, 1959, p. 261.)*

dune successional sequence was more complex than Cowles had appreciated. Other investigators found there was no single linear sequence leading from pine to black oak to beech and maple. Instead a network of different assemblages could occur, depending upon specific characteristics of the soil, drainage, and biotic interactions.

Succession in a pond. The primary ecological succession of ponds has been extensively studied. The first plants to colonize a newly formed pond are the planktonic algae. Vascular plants such as pondweed and stoneworts (the macroscopic algae *Chara*) soon appear. Due to the accumulation of plant remains and soil washed in by rainwater from the surrounding land, the pond gets shallower with the passage of time. The filling process is hastened by the invasion of vascular plants whose shoots extend into the air above the water; cattails, bullrushes, and arrowhead are examples. With the appearance of the emergent plants, much of the former open water of the pond becomes a marsh. In shallower places, peat moss may produce a mat over the water surface that gradually adds to the filling of the pond with peat. As the land becomes better drained, herbs and shrubs such as willow, alder, and buttonbush colonize the site. With the passage of years, a forest composed of various tree species, including red maple, elm, and white pine, begins to develop. This entire sequence of plant species can sometimes be observed as a nearly continuous series of zones encircling the pond. After many years, the pond disappears and a forest community predominates (see Color Plate II).

Succession on land. In this example, primary succession starts with open water and takes place relatively rapidly. For example, in northern Wisconsin and Minnesota, there are numerous forested bogs that were once open-water lakes formed as the glaciers retreated less than 10,000 years ago. However, primary succession that begins on bare rock surfaces, such as those on granitic rock outcrops in the Piedmont region of the southeastern United States, is particularly slow. Rock surface is an extremely dry, harsh environment; the water escapes quickly by runoff and evaporation. The first colonizers here are usually crust-forming lichens, such as the grey-green *Parmelia conspersa*. As soil is washed by rain or blown by wind into deep cracks and depressions in the rock, such plants as black moss, hairy-cap moss, lichens, and even grasses and herbaceous vascular plants become established. A gradually thickening mat of plants and soil slowly spreads over the rock, eventually covering the bare surface. On the more exposed sites, however, the destructive forces of water and wind remove the soil and its covering of plants, exposing bare rock again. Where the soil does accumulate to some depth, as in deep crevices, shrubs such as sumac and even trees may become established. The progress of succession is evident as a series of girdles of vegetation. At the outer margin of the mat are the early colonizers, and each later stage of succession is nearer the center, where the thickest soil is found.

Secondary Succession

The vegetation of an area may be destroyed by fire, grazing, cultivation, or road building. If the soil is not eroded away by rainwater or wind, revegetation of the area will take place relatively rapidly. The process of secondary succession on such artificially modified habitats varies considerably, depending on the slope and climate. One well-studied example occurs in abandoned farmland in Georgia and the Carolinas (Fig. 15.14). As soon as cultivation stops, the fields are colonized during late summer and autumn by several species of herbaceous plants, including horseweed and crabgrass. The horseweed lives through the winter as a dwarf rosette. The following spring growth resumes, producing a tall, many-branched plant that flowers during the summer. While the horseweed is growing to maturity other herbs, including white aster, invade the field

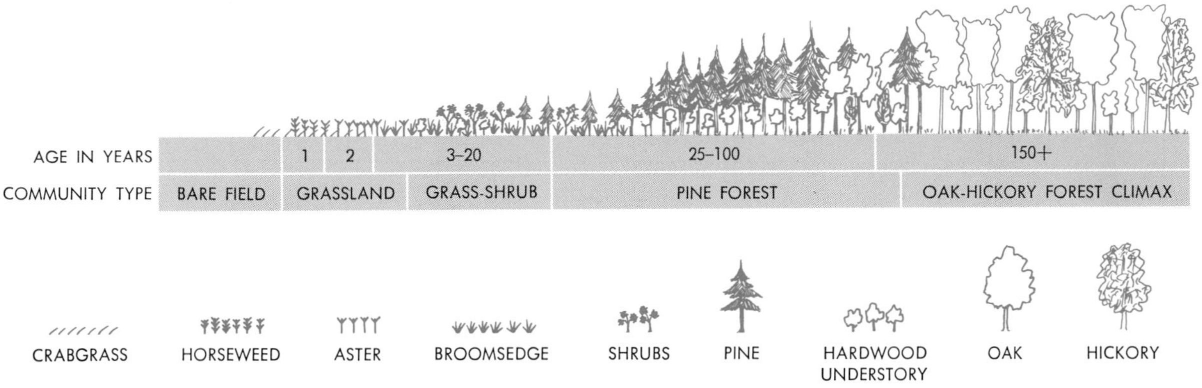

Fig. 15.14 *Plant succession following abandonment of crop land in the Piedmont region of southeastern United States. (After Eugene P. Odum,* Fundamentals of Ecology. *Philadelphia: W. B. Saunders, 1959, p. 263.)*

and develop flowers during the second summer. By the third summer a tall-growing bunch grass called broomsedge appears, replacing the aster and horseweed as the dominant plants.

By the time broomsedge appears, pine seedlings are noticeable, and within 5 to 10 years they form a forest where once the farmer grew corn and crops. Once the pines grow tall enough to cast a shade over the soil beneath their crowns, however, their own seedlings can no longer grow (young pine plants require almost full sunlight). Not only is there competition between the older pines and the seedlings for light, but there is also intense competition between these two growth stages for the available water in the soil. Thus there is little or no continued reproduction of the pines in this same area.

The seedlings of several other tree species *can* compete successfully in the shaded environment beneath the pines, however. Various species of oaks and hickory and broadleafed trees such as sweet gum are able to use the dim light for photosynthesis and to obtain water from deep in the soil (their roots extend several centimeters below the roots of the pines, which are confined mostly to the upper 20 cm or so). Thus sweetgum, oak, and hickory seedlings come up under the pines, and within 100 years after a field has been abandoned they develop into the dominant and overstory trees of the forest. During this interval of time, an understory of dogwood and red maple is also developing. Eventually the old pines begin to die and are replaced by oaks and hickories so that after about 150–200 years the forest consists mostly of broadleaf deciduous trees.

To observe the various states of ecological succession is one major problem: to *explain* why such successional changes occur is another. It seems clear that major changes in the vegetation of a region can only follow changes in the environment. Plants and other organisms living in a community modify their environment. For example, trees shade the ground beneath them, affecting soil temperature and humidity. Leaves fall to the soil surface and undergo decay. The resulting material affects the runoff of rainwater, soil temperature, and the formation of humus in the soil. These factors in turn affect soil development and change the quantity and type of available nutrients, soil pH, and aeration. These modifications of the environment by the organisms usually make it less favorable for themselves and more favorable for species that could only have survived there earlier with great difficulty. As organisms change the environment, they make it possible for other organisms to compete successfully with the established species. As has just been seen, the invaders sometimes even replace the original, pioneer organisms. Such behavior sounds like it would not be selected for evolutionarily. There is truth to this. The reason that such environmental changes occur is that they are unavoidable. Changing the environment becomes an inevitable consequence of the existence of life. Like it or not, organisms cannot help but change the place where they live. Permanence and lack of change do not exist in the natural world.

The environment may also be changed by forces other than living organisms. An overflowing stream, for example, may deposit fertile silt on bottomland. The bottom substrate level in a pond or lake may be raised because silt has been washed in, as often happens in bodies of water impounded by dams. The chemical content of the soil may change because of leaching. These and similar modifications of the environment are usually followed by changes in the vegetation.

Ecological succession eventually results in the formation of vegetation existing in a steady-state equilibrium with the soil, climate, and herbivorous animals. Although no natural community is static, such a relatively stable community has less tendency than the earlier successional stages to modify its environment in a way that is injurious to itself. The plants of such a "climax community" are able to perpetuate themselves because their seedlings can survive in competition with older plants. If environmental factors do not change appreciably, the climax vegetation will continue for centuries without being replaced by another stage. A climax community may change somewhat in the kind of plants that comprise it. Until about 50 years ago, for example, the climax forests of eastern North America contained abundant chestnut trees. Chestnut trees have now been virtually eliminated by a fungal disease.

In fact, the original climax vegetation that once existed over most of North America (and the entire earth) has been destroyed by humans. The normal stages in succession have been set back, modified, or stopped by such human activities as lumbering, grazing domestic animals, cultivation, urbanization, industrialization, and even radiation. In the settled regions of the earth, many of the plant communities that now exist do so not because of the natural process of ecological succession but because of deliberate human interference. People maintain such types of vegetation as crops, pastures, golf courses, and lawns, as well as wildlife preserves and managed forests, because of economic interests. These communities require some effort to maintain, however. Managed forests, for example, must be periodically thinned by cutting some of the trees.

15.11 Human Impact on Rates of Aquatic Succession

The earlier description of the filling in of a pond and formation of a forest habitat was based on the general pattern of succession that has been observed in many areas. The pattern of succession in larger aquatic ecosystems is much more complex, as is just now beginning to be understood. Until recently it was frequently stated that lakes undergo a natural process of nutrient enrichment as a result of the filling of their basins with sediments. This "aging" process was considered inevitable and was only accelerated when additional nutrients were supplied to lakes. When a lake basin was originally formed, for example by the glacial scouring action of a large ice sheet or by the natural damming of a river valley following a landslide, it was thought to be relatively poor in nutrients. Thus plant and animal production was relatively low and the water free of dense algae populations or abundant shoreline vegetation. Such deep nutrient-poor lakes were termed "oligotrophic" and were contrastasted with shallow, nutrient-rich "eutrophic" lakes. The process of succession in which deep lakes were slowly filled with sediments and increased concentrations of nutrients was called "eutrophication."

Recently two things have happened. First, the ecological term eutrophication has become widely used and given a broader set of definitions. Most of these meanings have stressed the end product of nutrient enrichment: the high production of "nuisance" plants such as blue-green algal scums. Eutrophication has taken on a negative connotation in newspaper articles and other reports. Although high concentrations of some nutrients can lead to excessive algal growths, production of toxins (by some blue-green algae and dinoflagellates), deoxygenation of the lake, and massive fish kills, these results need not occur if the ecosystem functions properly and continues to support an integrated food web. Indeed many European and Asian fish farmers add high amounts of nutrients to their lakes and ponds to increase fish production. Many of these managed systems are among the most productive and useful ecosystems in the world. The problem of excessive algal growth results from mismanagement of the entire ecosystem, not just from nutrient enrichment. If fish are not killed off by algal or human-made toxins, and if the reproductive stock are not removed at too rapid a pace, then nutrient enrichment can lead to higher production of nontoxic algae that are consumed by the fish. This increased nutrient and energy flow results in increased yields of fish.

The second recent change has happened as a result of further basic research on historical records of ancient and human-made lakes. Whereas ecologists initially thought that most lakes have gone through a transition from oligotrophy to eutrophy, it is now apparent that some lakes begin as eutrophic and become oligotrophic. They start out nutrient-rich and become progressively poorer in nutrients. This decline in productivity has been documented in many human-made lakes and may occur for a number of different reasons. Other lakes are known to have remained oligotrophic for extremely long periods and show no signs of gradual eutrophication until greatly disturbed by human activities. It is important to reject the conclusion that all lakes eventually fill in and undergo eutrophication and that any increase in nutrient inflow is therefore "natural." Some very deep lakes such as Lake Tahoe and Lake Superior showed no "natural" signs of eutrophication until sewage or industrial wastes were discharged into their basins. Evidence also demonstrates that the directions of eutrophication can be reversed: once limiting nutrients are removed from a well-fished lake, it can return to an oligotrophic condition. Professor G. E. Hutchinson of Yale University has concluded from his studies of lake histories that there is no support for any theories of *inevitable* development of a lake from an oligotrophic to a eutrophic condition. Each lake must be viewed a distinct ecosystem. Although some general tendencies are apparent, it is difficult to predict exactly how and at what rate succession will proceed in any lake basins.

15.12 Ecology, the Environment, and Economics

Most people are familiar with the "environmental movement" that began to flourish between the late 1960s and the early 1970s. By 1968 it had become increasingly apparent to many people in the United States that, if not checked, pollution of lakes, rivers, the oceans, and the atmosphere of certain regions of the earth (usually highly industrialized areas) could make the whole planet uninhabitable. Central to the growth of the "environmental movement," as it came to be called, was the study of ecology. As we have seen in this chapter, ecology is the general study of the interrelationships among organisms and the physicochemical environment in which they exist. A new breed of ecologists, the environmentalists, studied ecology in order to predict how certain human activities would affect large-scale ecosystems. The environmental movement emerged as an attempt to apply ecological principles and understandings to the specific kinds of problems human beings were creating in their daily activities.

However, the environmental movement was something more than the academic study of ecology. It included not only an attempt to understand the sources of environmental problems, but also the attempt to do something about them. Environmental activists made surveys and traced pollutants to their sources. Angry environmental groups picketed and demonstrated at industrial sites, power plants, politicians' offices—anywhere that they could spot a problem—and sometimes achieved a positive response. In the United States, recognition of increasing pollution problems eventually led to federal action, including establishment of the Environmental Protective Agency and passage of the Clean Air Standards Acts of 1970 and 1971. State legislatures also passed a variety of bills designed to clean up air and water and preserve land from the worst forms of degradation.

However, pollution itself turns out to be only part of the total spectrum of environmental problems modern industrial society faces. As many environmental groups pointed out, the demands of an ever-expanding production system were causing excessive wastes on the one hand and excessive depletion of future resources on the other. Moreover, it was noted that the problems of pollution and resource depletion were not evenly spread over the earth's surface. Pollution proved to be a problem for large industrial and relatively wealthy countries of the western hemisphere. Depletion of resources occurred mostly in the poorer countries of Asia, Africa, and Latin America, where financial interests of private corporations in the wealthy countries controlled the rate at which many natural resources were used. There was an irony, environmentalists pointed out, in the fact that resources such as copper or petroleum were being taken from the ground in relatively poor countries (such as Chile or Venezuela, respectively) and used at such an enormous rate in rich countries (such as the United States or Britain) that the latter were becoming polluted. Approaching the situation in a truly ecological fashion, environmentalists followed the flow of materials and energy over the whole planetary ecosystem—from where they were obtained to where they were used. The problems of pollution in one locality could not be separated from the problems of resource use in another.

One result of this concept is the recognition that environmental issues are intimately related to economics: human "rules" about how resources are to be controlled and distributed within a society. The pollution problems of the industrialized countries are often a direct consequence of the uneven distribution and use of the world's resources. For example, the United States has 6% of the world's population but consumes between 40% and 50% (higher by some estimates) of the world's natural resources (metals, petroleum, etc.). If the entire population of the earth used resources at a similar rate, the world's ecosystem could sustain neither the accumulated waste products nor the depletion of natural resources for very long. Recognizing this fact has caused increasing numbers of environmentalists to insist that a prerequisite to solving our ecological problems is reordering our economic priorities in a more rational way. This suggestion has meant not only readjusting rates of production and consumption in the wealthy countries to a level that the world ecosystem can sustain, but also redistributing wealth so as to put the consumption levels of all people on a more equal footing.

The relationship between ecology and economics suggests that environmental legislation in countries such as the United States treats only half the problem. The effect of such laws as the Clean Air Act is to require that industries clean up their waste-disposal processes—put filters on smoke stacks or pollution control devices on exhaust systems of automobiles. While this is important, it is relatively inefficient to make a mess and then spend time and energy to clean it up. In fact, as some critics have emphasized, the pollution control industry that has sprung up in the last 6 or 7 years itself creates an enormous amount of pollution! A much more efficient procedure would be to reduce consumption of raw materials. This would not only reduce the amount of pollution, but would also alleviate the burden on the world's natural resources.

SUMMARY

1. All organisms live in a closely interconnected web of relationships with each other and the physicochemical environment. Alteration of any single strand of this web may have profound effects on all other strands.
2. Ecology is the study of the various relationships between organisms and their environment.
3. All organisms compete to one degree or another for the limited supply of resources available. Ultimately all resources—food, water, living place—are reduced to one or another form of energy. All competition is thus competition for a form of energy. Competition is a result of the finite energy supply in any ecological system.
4. Interaction among organisms as they compete for energy takes several forms:

   a) Direct competition, where one organism gets something that another does not.

   b) Predation, where one organism consumes another, using the prey's bodily material (biomass) as a source of building materials and energy.

   c) Parasitism, where one organism consumes part of the biomass of another without directly killing it (at least immediately).

   d) Symbiosis (or mutualism), where two organisms exploit one another in such a way as to mutually benefit both without killing or impairing either.

5. Biotic potential is the ability of a species to increase in numbers, given ideal environmental conditions. Environmental resistance is the combination of those aspects of the environment that prevent organisms from achieving maximum biotic potential (examples are limited food supply and disease). The interaction of biotic potential with environmental resistance determines the size of any population in nature. Regulation of population size is the result of several kinds of interactions:

   a) Competition: as more organisms of a particular kind reproduce, competition for limited food increases, killing off increasing numbers of organisms. Thus the population size oscillates around its upward limit, or carrying capacity, for a particular environment.

   b) Predation: as the prey population increases in size, it becomes easier for the predator to find and kill prey. This counterbalances the increased reproduction rate of the prey and keeps its population down. Meanwhile, as a result of the availability of more prey, the predator population can increase in size. But as prey become scarcer, the predators must compete more sharply, which limits their population size.

   c) Density: as a population increases in size, the density of organisms increases. This can trigger certain responses in individuals that cause them either to die or migrate or to refrain from reproductive behavior. So-called density-dependent factors operate among the members of the population itself.

6. Ecological interrelationships such as competition, predation, symbiosis, and parasitism have all evolved by natural selection.
7. Interspecific competition is competition among populations of two or more species. Its net effect is usually to make the two species less and less alike over time in terms of their ecological requirements. The effect of such competition is to further reduce the amount of competition between the species in any particular locality, leading to divergent speciation.
8. Intraspecific competition is competition among members of the same species. Its net effect is to slowly alter the gene pool of the population and produce phyletic speciation.
9. Within ecosystems, organisms occupy a habitat and an ecological niche.
   a) Its habitat is the place where an organism lives, such as in fresh or salt water, on a desert, under rotting logs, or high in the trees.
   b) Ecological niche refers to the *role* the organism plays within the community. For what organisms does it serve as food? What organisms does it eat? What minerals does it require from, and contribute to, the environment? Ecological niche is sometimes compared to a human being's profession in the economic community.
10. Organisms within an ecosystem obtain energy as producers, as primary, secondary, or tertiary consumers, or as decomposers.
    a) Producers are those organisms that can convert the sun's energy directly into carbohydrate through the photosynthesis of green plants and certain photosynthetic bacteria.
    b) Consumers (primary, secondary, tertiary, etc.) are those organisms that must obtain their energy by eating other organisms. Primary consumers are herbivores that eat plants directly. Secondary and tertiary consumers are various forms of carnivores that eat herbivores or other carnivores, respectively.
11. The movement of *materials* through an ecosystem is largely cyclic. That is, most material used by organisms is eventually returned to the ecosystem through natural processes such as respiration, excretion, and decomposition. The carbon, nitrogen, water, oxygen, and other cycles are familiar illustrations of this point.
12. The flow of *energy* through an ecosystem is not cyclic, but unidirectional. Energy flows from producers to consumers and decomposers. Energy keeps moving through an ecosystem because there is a continual supply entering from the sun.
13. Organisms in an ecosystem exist at various trophic levels of the system. The primary producers, or autotrophs, represent the first level. The various consumers, or heterotrophs, represent the various other levels. In terms of trophic level, the number of organisms decreases in passing from the first to successively higher levels. The same is true most of the time for actual mass of organic material, or biomass, and for energy. Thus trophic levels can be depicted in terms of "pyramids" of mass and energy. This means that there are fewer organisms, or less energy available, the further one goes up the trophic levels of an ecosystem. This observation is in accordance with the first and second laws of thermodynamics.
14. The various trophic levels can also be described as a food chain for any particular ecosystem.
15. The carbon cycle can be thought of as starting

and ending with atmospheric carbon dioxide (CO_2). This CO_2 enters the living world through photosynthesis, during which it becomes fixed into carbohydrate. The oxidation of carbohydrate during cellular respiration returns most of the carbon to the atmosphere as CO_2. Some CO_2 is given off from decomposition of dead organic matter, a process carried out by microorganisms. A small amount of carbon becomes locked into a "carbon sink" where it is trapped as coal, petroleum, natural gas, or limestone. This carbon is withdrawn from the ecosystem for some period of time, but it eventually gets oxidized into CO_2 (when people burn coal or gasoline, for instance).

16. The nitrogen cycle can be thought of as starting with atmospheric nitrogen. Nitrogen-fixing bacteria living in nodules on the roots of legumes (such as pea or alfalfa) can convert this atmospheric nitrogen into soluble nitrates. Nitrates enter the plant and are incorporated into protein. Protein from dead organisms can be broken down by the action of putrefying bacteria, which convert the protein into compounds. Ammonium compounds are acted on by nitrosofying bacteria that convert them into nitrites (NO_2 compounds), which are soluble. Nitrites can be converted into nitrates by nitrifying bacteria. Finally, nitrates and ammonium compounds can both be broken down into molecular nitrogen, which is released back into the atmosphere. The whole function of the nitrogen cycle is to convert relatively inactive molecular nitrogen into a form (nitrates) that is soluble and can easily enter the plant or animal cell. Nitrates, of course, are necessary for the continued production of amino acids and proteins.

17. Ecosystems can exist perfectly well without animals. All the cycles and processes can be carried out by plants alone. The same is not true for animals, however.

18. Ecosystems are not static relationships; they are constantly changing through long periods of time. The regular, even highly predictable, stages through which ecosystems grow and develop are called succession. Primary succession occurs when aquatic ecosystems become filled in and eventually turn into land. Secondary succession occurs when a land ecosystem undergoes developmental changes into another kind of land ecosystem (a sand dune to a grassland to a pine forest). Successional changes occur because each stage changes the environment and paves the way for a new grouping of organisms. The general pattern of secondary succession begins with an open area (sand or meadow) inhabited with grasses. The grasses are replaced by pines or other evergreens, whose seeds need a lot of sunlight to grow. As a pine forest matures it shades the ground: neither the grasses nor future pine seedlings can grow well. Seeds of hardwood trees such as oak or hickory can grow in shady areas, so these plants move in and eventually come to dominate the pines. Since the oaks and hickories can replenish themselves, they represent a relatively stable or "climax" community in the successful process.

19. The environmental movement is an attempt to apply ecological principles to understand how human activities (such as factory or automobile pollution) are altering the natural ecosystem in which we live. The environmental movement seeks to change the improper use of natural resources and the unnecessary piling up of waste materials. Environmentalists recognize that in order to accomplish this task they must understand the relationships between ecological and economic processes.

EXERCISES

Figure 15.15 represents a sealed aquarium. In the figure, C represents enclosed air above the water, B the water, D the animal life in the water, A the plant life, and E the soil at the bottom. At the time it was sealed, the aquarium contained the proper balance of plant and animal life. Answer Exercises 1 through 5 on the basis of this information.

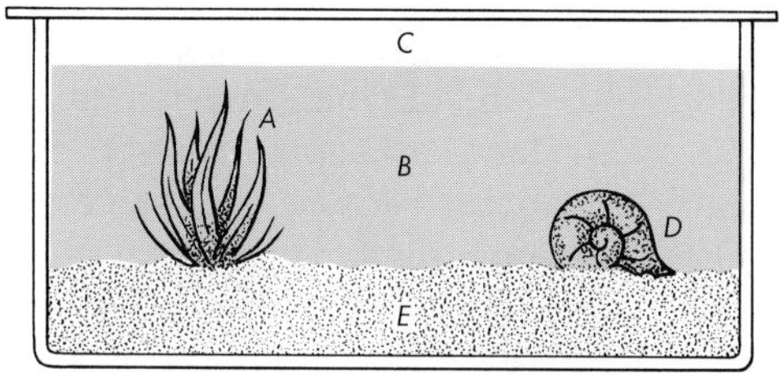

Figure 15.15

1. How long will it be possible for life to continue in the aquarium?
   a) Until the original oxygen supply in the air above the water is used, but not longer.
   b) Until the original supply of O_2 dissolved in the water has been used up, but not longer.
   c) At most, not more than two months.
   d) Until the original supply of nitrogen in the soil at the bottom is used up, but no longer.
   e) Indefinitely, as long as the sun shines regularly on the aquarium and the temperature stays above freezing.
2. At what point within the aquarium (A, B, C, D, or E) does energy first enter the cycle?
3. At what point within the aquarium is the oxygen supply replenished?
4. At what point or points is carbon dioxide given off as a waste product: A, A and B, B and D, A and D, D and E, or D?
5. At what point is carbon dioxide utilized as a raw material?
6. What is a food chain?
7. Why is there such a great decrease in the amount of available matter or energy along each step of a food chain? How does this relate to the second law of thermodynamics?
8. On the basis of the following assumptions, calculate the overall efficiency of the energy chain leading from the sun to you.
   a) Of the sun's energy, 99% goes unused.
   b) Of the 1% used by plants, only about 0.6% ends up in glucose.
   c) Half of this glucose is used up by the plant for its own life processes.
   d) Of the glucose reaching your cells, 60% is formed into ATP.

e) Your usage of this energy is 55% efficient.

Can you detect any "energy leakages" other than those listed here?

9. Can you suggest why the poorer nations of the world must necessarily be primarily vegetarian?
10. Distinguish between the terms "habitat" and "ecological niche."
11. What is meant by the terms "biotic potential" and "environmental resistance"? How do they relate to each other?
12. Distinguish between interspecific and intraspecific competition.
13. Nature's most unusual organisms are generally found in extreme environments (e.g., ocean bottoms, deserts, Arctic wastes). The grotesqueness and uniqueness of these organisms is generally anatomical, but it may also be physiological. What reasons can you give for this phenomenon?
14. Though all the forms living in any ecological area are seldom if ever closely related phylogenetically, they may show quite similar anatomical and even physiological characteristics. Explain why this should be the case.

An estimate has suggested that about 130 California jackrabbits eat as much forage on the open range as one cow. The cost of removing this many rabbits has been estimated at $47 (almost twice the value of a range cow). Thus a program to destroy the rabbits is not good economy. Furthermore, the rabbits help to aereate the soil of the range regions and make it more porous to water by their continual burrowing. Nevertheless, despite the apparent drawbacks, a campaign against the rabbits was launched. Since the beginning of this campaign it has been observed that there is a notable decrease in the number of hawks in the region. Hawks displaced from this region have moved into others, becoming a serious problem for chicken farmers. Stopping the rabbit campaign at this point, however, presents problems. With fewer hawks in the region, the rabbit population would begin to grow rapidly as soon as extermination ceased, since rabbits reproduce very rapidly and hawks reproduce very slowly. Thus, if the campaign were stopped, the rabbit population would begin to zoom upward, with no natural checks. Answer Exercises 15 through 17 on the basis of this information.

15. The food relationships described are best indicated by which of the following food chains?

a) Forage→rabbits→hawks→cattle→human beings

b) Forage ↗ chickens ↖ predator birds→cattle; Forage ↘ hawks ↗ predator birds

c) Forage ↗ rabbits→predator birds; Forage ↘ cattle→ human beings

d) Forage→rabbits ↗ chickens→predator birds; Forage→rabbits ↘ cattle→ human beings

16. If the rabbit population were allowed to increase by stopping the campaign against rabbits, what would be the first effect seen?
    a) An increase in the number of hawks.
    b) A decrease in the number of cattle.
    c) A decrease in the amount of forage available.
    d) An increase in the number of chickens in neighboring regions.
    e) A decrease in the number of cattle ranchers.
17. Observe the graph in Fig. 15.16. From studying these curves, explain which organism (A or B) is the prey and which the predator. How can you tell?

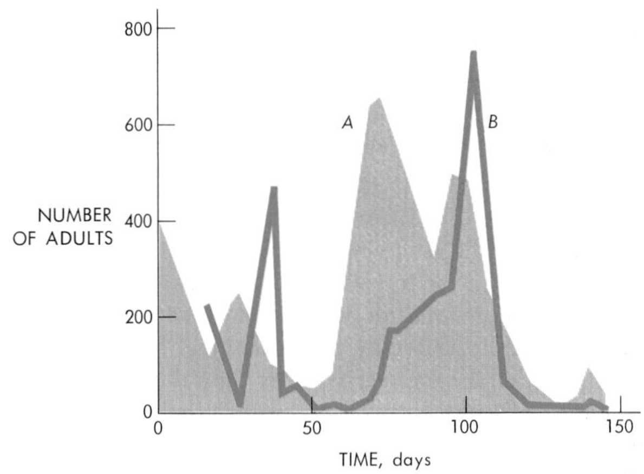

Figure 15.16

18. Observe the four graphs in Fig. 15.17. Both lots of *Tribolium castaneum* (one species of flour beetle) started with 20 adults but were kept at different temperatures. The two lots of *Tribolium confusum* were kept at the same temperature but started with different numbers of adults. Evaluate the following statements using letter symbols as follows: A, the statement is true according to the data given in the graphs; B, the statement is true according to an accepted biological principle, but it cannot be shown from these graphs; C, the statement is false according to the data given in the graphs; D, the statement is false according to an accepted biological principle, but this cannot be shown from these graphs; E, the statement cannot be judged on the basis of the data in the graphs.

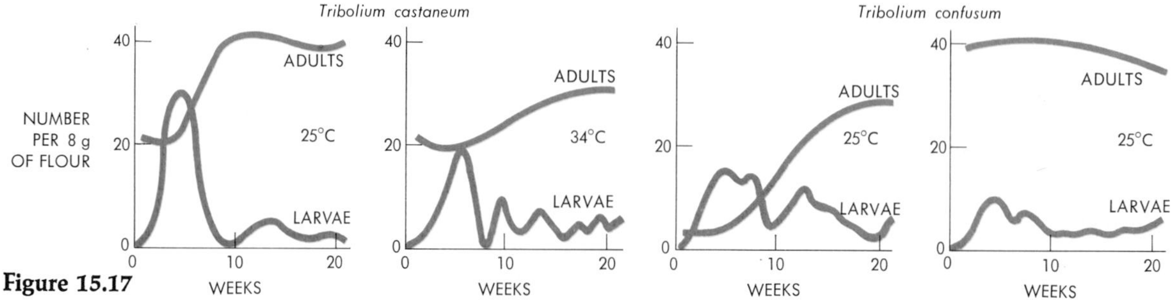

Figure 15.17

a) Adults of *T. castaneum* are more sensitive to an increase in temperature (above 25°C) than adults of *T. confusum*.

b) Adults of *T. castaneum* thrive better at 25°C than adults of *T. confusum*.

c) The larvae of *T. castaneum* are more greatly affected by a rise in temperature from 25°C to 34°C than are the adults.

d) The limiting value for the environment of these populations of adult *Tribolium* seems to be about 40 individuals.

e) There is no limit to the number of adults the environment can support, since growth rate always proceeds upward from the initial number.

f) Crowding organisms by starting a culture with an initially larger number of individuals has no effect on rate of growth of the population.

g) Starting a culture of *T. confusum* with 40 adults, as compared with 4, does not significantly affect the growth of larvae and pupae.

h) In general, the growth and development of all young organisms in a population is independent of the number of adults.

i) The number of *Tribolium* larvae in the population at any one point in time never exceeds the number of adults.

j) The growth in number of larvae as they reach their peak in any population occurs at a more rapid rate than the growth in number of adults as they reach their peak.

k) The curve for the number of adults in each culture eventually levels off because of a limited supply of food.

SUGGESTED READINGS

General Ecology Texts

Cheng, T. C., *Symbiosis.* New York: Pegasus, 1970. A review of the various kinds of symbiosis, written for the general reader.

Gosz, James R., Richard T. Holmes, Gene E. Likens, and F. Herbert Bormann, "The Flow of Energy in a Forest Ecosystem." *Scientific American,* March 1978, p. 92. A well-illustrated discussion of how a forest partitions solar energy.

Kormondy, E. J., *Concepts of Ecology.* Englewood Cliffs, N.J.: Prentice-Hall, 2d ed., 1977. An introductory text that is also available in paperback.

Turk, J. J., T. Wittes, R. Wittes, and A. Turk, *Ecosystems, Energy, Population.* Philadelphia: W. B. Saunders, 1975. A paperback for the general student. Introduces the concepts of ecosystems, energy transfer, and population growth. On the latter subject, the authors take a neo-Malthusian approach.

Environmental Issues

Commoner, Barry, *The Closing Circle.* New York: Bantam Books, 1972. A popular book by one of America's most vociferous environmentalists. Commoner lays blame for the majority of environmental problems in the United States on the economic system, particularly rampant industrialization. A challenging book that is easy reading.

Ehrlich, Paul R., and Anne H. Ehrlich, *Population, Resources, Environment,* 2d ed. San Francisco: W. H. Freeman, 1972. Focuses on a number of human environmental problems including population growth, food production, air, water, and pesticide pollution, and social changes involved with population stabilization. Most of the book deals with population growth in one way or another, viewed in a neo-Malthusian way.

Meadows, D. H., D. L. Meadows, J. Randers, and W. W. Behrens, *The Limits to Growth.* Washington, D.C.: Potomac Associates, 1972. A set of predictions about the rate of human use of natural resources relative to projected supplies. A controversial book claiming that the western industrial countries are rapidly exhausting the world's supplies of raw materials.

Ecology and Biogeography

MacArthur, R. H., *Geographical Ecology. Patterns in the Distributions of Species.* New York: Harper and Row, 1972. A specialized book that discusses a subject not covered in this chapter. The subject, an extremely important one in ecology, is the origin and distribution of different species on earth in terms of the ecosystems of which they are a part. It treats such subjects as the relationship between the stability of ecosystems and the number of species comprising them.

Human Population

Barclay, William, "Population control in the Third World," *NACLA (North American Congress on Latin America) Newsletter* 4 (No. 8, December 1970), p. 1–18. A brief analysis of population control from the socioecologists point of view. Contains an analysis of what people and groups in the United States strongly support population control programs in other (especially Latin American) lands.

Ehrlich, Paul R., *The Population Bomb.* New York: Ballantine, 1968. Though an older work, this book is a classic that launched the new wave of neo-Malthusian arguments about the future catastrophe of overpopulation.

Mass, Bonnie, *Political Economy of Population Control in Latin America.* Montreal, Editions Latin America, 1972. Although it is not always well organized, this short booklet presents some criticisms of the major neo-Malthusian arguments as they apply to Latin America. Written from a socioecological perspective.

Succession

Horn, Henry S., "Forest succession," *Scientific American,* May 1975, p. 90. A relatively short but worthwhile account of modern views of the nature of succession.

SIXTEEN

ANIMAL BEHAVIOR

16.1 Introduction

The movement of an appendage can be explained in terms of the contraction of certain key muscles; the contraction of these muscles, in turn, is brought about by the contraction of striated tissue of the muscle. These tissues contract because the muscle cells of which they are composed contract; the muscle cells contract because their myofibrils contract. The contraction of the myofibrils is explained by the hypothesis proposing that protein filaments called actin and myosin slide over each other to cause contraction. What causes this sliding? Investigations are still proceeding to attempt explanations on the molecular, atomic, and subatomic levels of investigation.

Yet even given 100% success in investigating muscle contraction at the most microscopic level, we still intuitively feel there is something more. We know, for example, that the muscle did not just contract on its own in moving the appendage; a nerve impulse triggered the action. What triggered the nerve impulse? The brain, most likely. But what triggered the brain? Probably it was an outside stimulus perceived by the organism—a noise perhaps, or the sighting of a predator. In other words, the entire system of levels of investigation has now been wrapped up into one package: **animal behavior.***

Early in this century the field of animal behavior became divided into two prominent schools of thought. One school of thought was given great impetus around 1924 with the publication of J. B. Watson's book, *Behaviorism*. The leading figures in this group, termed "behaviorists," were mostly American experimental psychologists. The second school of thought, mostly European, began around 1909 with the work of Jacob von Uexküll and O. Heinroth. Later, in the 1930s, important refine-

* Plants, lacking nervous systems, are not organisms that exhibit what one would generally call "behavior."

Ethology is the scientific study of animal behavior.

ments and additions were made by Austria's Konrad Lorenz and Holland's Niko Tinbergen, who shared the 1973 Nobel Prize in Physiology and Medicine for their work in this field. Many scientists, mostly zoologists, became associated with this school of thought and termed themselves "ethologists." **Ethology** may be defined simply as the scientific study of behavior.

It is important to recognize the existence of both the "behaviorist" and the "ethologist" schools of thought, for the conclusions they drew from their respective experimental work were often widely divergent and even contradictory. The behaviorists were trained in psychology and were most (if not exclusively) interested in learning theory. Furthermore, this work was almost entirely done in the laboratory. The maze to be learned, the positive reinforcement of food or negative one of electric shock, the "Skinner box" in which an animal learns to press a lever or peck a button for food or drink—these were the classic tools of the behaviorists. For behaviorism, ultimate success was seen in terms of accurately describing an organism's behavior patterns after training; in turn, these descriptions would enable the experimenter to increase learning efficiency and the degree of predictability. Naturally the behaviorists preferred organisms that lent themselves well to such experimental studies: the white rat became the behaviorist's *Drosophila.*

The ethologist, trained for the most part in biology rather than psychology, approached and interpreted things quite differently. Instead of working with just a few species, often highly inbred for laboratory study, the ethologist studied large numbers of species under both laboratory *and* natural environmental conditions. The differences between a wild rat's behavior and that of the behaviorist's white rats are considerable but hardly surprising. The former is genetically heterogeneous and exposed constantly to highly nonuniform sequences of environmental stimuli, while the latter may be genetically homogeneous and exposed to as uniformly controlled sequences of stimuli as the human experimenter can devise. With such very different approaches to such widely differing subjects, differing conclusions were inevitable.

Not long after World War II, clashes between behaviorists and ethologists began. The behaviorists accused ethologists of ignoring the role of learning in behavioral experiments, of underestimating the role of environmental factors in influencing behavior, and of treating the word "instinctive" as both a description and an adequate *explanation* of behavior—and thus using the label too widely and too often. The ethologists returned the fire. The behaviorists were accused of unjustified extrapolation to other mammalian species from the highly atypical laboratory rat. They were also accused of ignoring the role of instinctive or genetic factors in influencing an organism's behavior, and of thereby assuming that the environment alone provided all significant factors.

Which side was correct in its accusations? Both. The ethologists, who tended to study mostly nonmammalian species (e.g., birds, reptiles, fish, and insects) were concentrating mainly on groups whose actions *did* appear largely or entirely instinctive rather than learned. Conversely, the behaviorists *were* generally too occupied with their mazes to look at organisms in the wild, whose behavior patterns understandably bore no resemblance to those appearing in the laboratory.

In the following years, ethologists became more careful how they used the term "instinct." In turn, behaviorists turned open eyes and ears to other species studied under different conditions. The conflict was ultimately a productive one.

In the early twentieth century, another approach to animal behavior received considerable attention. The Russian physiologist Ivan P. Pavlov (1849–1936) tried ringing a bell every time he fed some dogs. The dogs soon learned to associate the sound of the bell with food. After a sufficient number of times, Pavlov showed that the dogs would secrete saliva in response to the bell alone; the sight, taste, or smell of food was no longer necessary. Such dogs were said to be "conditioned." Pavlov's work attracted much interest. Many variations of conditioning experiments were attempted and were for the most part very successful. Furthermore, the fact that training programs based on the principles of conditioning enabled animals to learn quite complicated behavioral routines suggested that the phenomenon might also account for the complex patterns of behavior found in nature. Indeed, Pavlovian conditioning became virtually a third school of thought in the study of animal behavior.

16.2 Simple Behavioral Systems: The Rise of Instinct Theory

Though interesting, conditioning nevertheless provided no satisfactory explanation for many types of observed behavior. The young orb-web spinning spider *Araneus diadematus* Cl., for example, spins a perfect web, specifically characteristic of the species, on her first try—despite the fact that she has never seen her mother perform this remarkably intricate procedure. *A. diadematus* Cl. raised to maturity in glass tubes so small that movement is entirely restricted soon spin perfect webs when released but still kept isolated (see Fig. 16.1). The web-spinning feat of *A. diadematus* Cl. is performed every morning by the female, who devours the old web before she spins a new one. The orb web is characteristic of several species of spiders. It is an economical one to produce; the spider covers the greatest area with as little material as possible. All other things being equal, the larger the area of the web, the greater the probability of capturing flying prey.

The web-spinning behavior of *A. diadematus* Cl. provides an excellent subject for behavioral study. For one thing, the path taken by the spider in spinning the web is faithfully recorded in its strands. The webs can later be removed and prepared for photographing by spraying with glossy white paint, spread over a dark box, and illuminated from the side. Projection of the resulting films allows precise measurements of the web's proportions. The size, number, and regularity of the web parts reveal a great deal about the motor behavior of the spider that spins the web (see Fig. 16.2).

One such revelation is the fact that web construction occurs in distinct phases. Each is executed in a different pattern of movements. The radii and frame threads are spun first, in a way that is distinctively different from the way the spirals are built. When the radii-building phase has been completed, the spider goes into the next phase, working on the interconnections between the radii that form the spirals. This, too, is performed in a precise sequence of individual steps.

While the glass-tube isolation experiments *seem* to rule out the role of any learning or experience in web construction, the intricacies of the process certainly make it tempting to believe that there is surely some sort of reasoning involved. Teleological explanations are also tempting; the spider spins the web *in order to* trap insects most efficiently. Yet this hypothesis predicts that, if some of the web strands are destroyed, the animal will stop to repair the damage before proceeding any further with construction. This prediction is not verified. The spider does not stop to repair the damage, even if it is great enough to make the web completely useless for the capture of prey.

Faced with observations of this sort, the early ethologists turned to the concept of **instinct** for an

Fig. 16.1 *At left is shown the web of a nine-month-old female cross spider* Araneus diadematus Cl. *(body mass 115.5 mg), raised in the laboratory and allowed to build a web every day. Its littermate was prevented from building any webs by being isolated in a glass tube. Upon release (body mass 77.8 mg), the experimental animal built the web at right. Note that only the size, not the pattern, is the distinguishing feature. By the time this animal had spun a fourth web, the product of its spinning was the same as that of the control's. Further investigation revealed that the silk glands of isolated spiders slow to a low level of productivity as a result of lack of use, and this factor seems to account for the smaller web size. When the glands regain their normal rate of silk synthesis, the webs spun become the normal size. (Photos courtesy Peter N. Witt, North Carolina Department of Mental Health, Raleigh, N.C.)*

explanation. Instinctive behavior was seen as behavior coordinated in the central nervous system and not generally determined in form by external stimuli (although it might be elicited by such stimuli). *Differences* in the various kinds of central nervous systems (and thus the instinctive behavior coordinated therein) are, of course, to a large extent genetically determined. To the early ethologist, therefore, instinctive behavior was "inherited" behavior, i.e., genetically programmed into the organism, and thus the opposite of learned behavior. The instinct-versus-learning or "nature-versus-nurture" controversy was off and running.

Testing the Instinct Hypothesis

Working first with the "instinctive behavior" hypothesis, it is possible to proceed further with an analysis of spider web-building behavior. As Figs. 16.3 and 16.4 show, the size and construction pat-

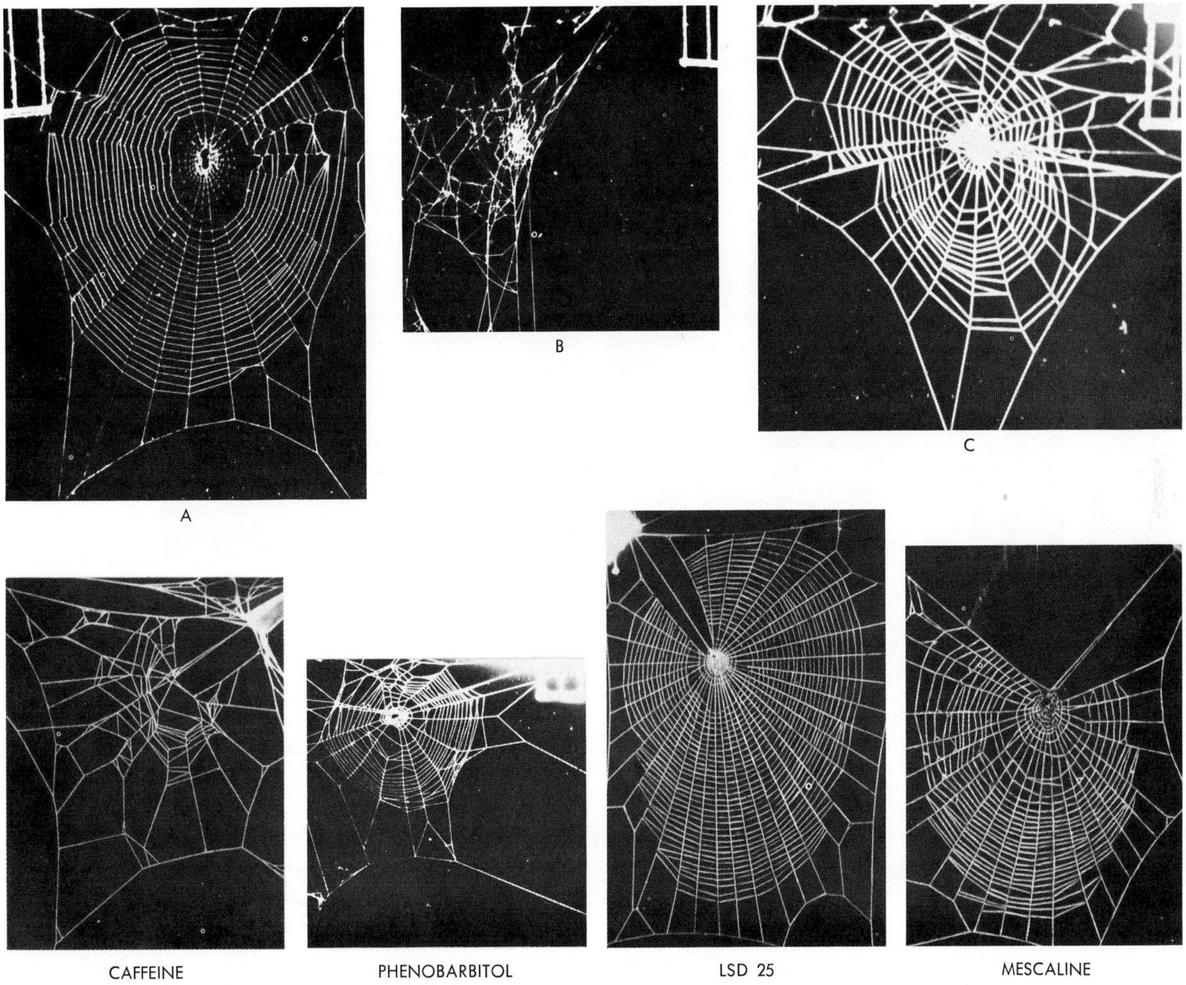

Fig. 16.2 *Spider web-spinning behavior is so characteristically precise that it provides a means to study the effects of drugs on behavior. At top are shown three webs of an adult female* Araneus diadematus *Cl. (body mass 157.9 mg) built on different days. Web A, the control, was built in approximately 20 min in the early morning. At 4 P.M., the spider was given 0.1 ml of sugar water containing 1 mg of dextro-amphetamine, also known as "speed." Web B was built approximately 12 hr later by the drugged animal; note that the web consists of only some remnants of a hub, a few irregular and frequently interrupted radii, and some erratic strands of sticky spiral. Web C was built 24 hr later and shows some signs of recovery. However, several more days were required to restore the web to normal. Since, as the other photos indicate, each drug produces a characteristic change in spider motor behavior (and thus in web design), the web pattern becomes a bioassay for drug identification. (Photos courtesy Peter N. Witt, North Carolina Department of Mental Health, Raleigh, N.C.)*

A B C

▲ **Fig. 16.3** *Webs built by the same spider at three different ages. Compare these with the information given in the graph in Fig. 16.4. The age of the spider that spun web A was a few days; web B, four weeks; web C, five months. (Photos courtesy Peter N. Witt, North Carolina Department of Mental Health, Raleigh, N.C.)*

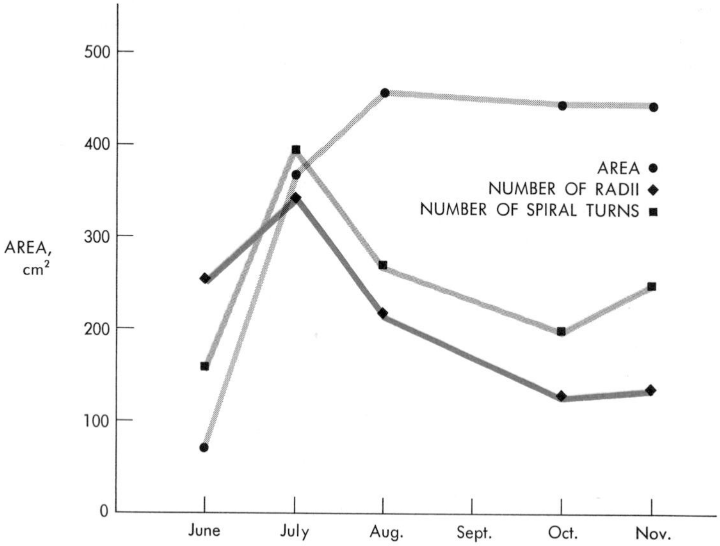

◀ **Fig. 16.4** *This graph shows the changes that occur in web design as the spider grows. Note that after an early increase the web area remains fairly constant. In contrast, the number of radii and spiral turns declines.*

Fig. 16.5 *The mass of this spider has been increased approximately 22% by the addition of a lead weight (arrow). The effects on web patterns are shown at right. Protein values per unit length for the experimental web B are nearly doubled. When the weight is removed, protein content returns to the level shown by control web A. (Photos courtesy Peter N. Witt, North Carolina Department of Mental Health, Raleigh, N.C.)* ▶

tern of the web change as the spider matures. Now a return to an hypothesis involving some learning or conditioning is tempting—e.g., the spider changes its web style as it "learns," with practice, to "perfect" the web. But this hypothesis is immediately contradicted: the first web spun by a glass-tube-raised spider is that of a mature, not a young, spider.

Some quantitative measurements can be made on the spider's web. For example, Dr. Peter N. Witt of the North Carolina Department of Mental Health in Raleigh has measured the amount of protein material going into each web. First the diameter, number of radii, and spiral turns are counted. From these data the thread length can be calculated. With the aid of a spectrophotometer, the precise amount of nitrogen each web contains can be calculated. The quantity of nitrogen, in turn, gives a direct measure of the amount of protein. If it is assumed that all the thread in one web is of uniform thickness, the quotient of nitrogen content and thread length provides a measure of thread thickness for any one web. Thread thickness, in turn, gives a measure of the amount of material per unit length.

The results of these studies show that after a certain time in the spider's life, the amount of web material used (as measured by nitrogen-content analysis) remains the same. So does web area. However, *later webs are built with shorter and thicker thread.* The end result is wider meshes.

Why does the spider build later webs with stronger thread? This intriguing question can be posed in two ways. One way is *functional* and requires demonstration of a reduced reproductive rate or increased mortality as a result of failure to build a stronger web. But the question can also be posed as a *dynamic* one; we ask, instead, what are the immediate causes and mechanisms that promote spinning thicker web thread. If this second alternative is taken, the simplest and most obvious explanation is that the spider is growing and thus increasing its body mass. Therefore a stronger web, with shorter and thicker strands, is required to support the spider's greater mass.

First, the "fact" of a relationship between body mass and web thickness must be established. The following hypothesis leads to a prediction: If the mass increase of a spider causes thread thickness and mesh width changes, then increasing an adult spider's mass with lead weights might result in even thicker thread and a web with even wider meshes. As Fig. 16.5 illustrates, the prediction is

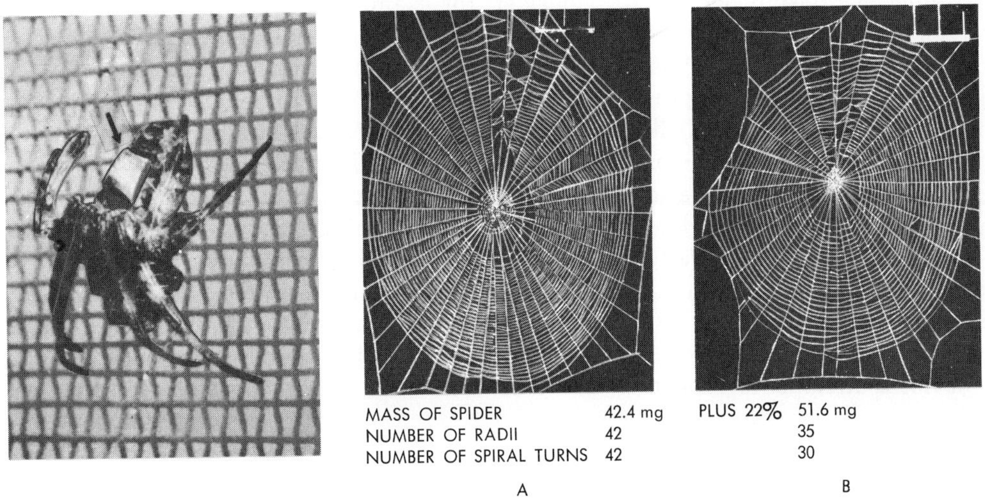

verified, the hypothesis supported. But this hypothesis merely establishes a fact; an explanatory hypothesis based on the facts is still needed to suggest how the web-building behavior pattern may be governed.

To hypothesize that the spider "knows" it must increase thread thickness or risk breaking the web and falling is clearly both anthropomorphic and teleological; the animal is credited both with reasoning ability and with anticipation of possible future events. But such an hypothesis is unnecessary. Instead, an hypothesis proposing a sort of computerlike circuitry in the spider's glandular and nervous system can be proposed. Some signal must communicate to the central nervous system the amount of silk that is ready in the gland for thread production. Body mass then operates to determine thread thickness. Thus the combination of the amount of silk available and body mass indirectly determines the length of the thread the spider can spin. In turn, the length of thread used to spin the web influences web size. From all these data, the mesh width of the future web can be established in advance and programmed into the system. By doing the radii first and determining the angle between them, the spider automatically predetermines the web pattern. In the end, it will have covered the largest possible area with a web strong enough to support its own body.

Note here the close interrelationship between biochemical processes (i.e., web material manufacture in the spider's silk glands) and motor behavior (i.e., the web-building process). The speed of protein synthesis determines the filling of the silk glands. This, in turn, regulates the motor behavior of the spider in building the web pattern. Note also the elimination of both anthropomorphic and teleological characteristics from the explanation; clearly neither are needed. Finally, note that the hypothesis is a complicated one, with many underlying assumptions that must be tested. But if the hypothesis is complex, the behavior it attempts to explain is still more so. Although a simple hypothesis is always preferable to a complex one, oversimplification is a danger when one is dealing with complex phenomena.

"Innate" as Opposed to "Instinctive" Behavior

It may now be quite evident to the reader that to label a behavior as "instinctive" explains no more about the phenomenon than to refer to it as "learned." For a time, there was a tendency among ethologists to label any behavior pattern in which the role of learning could not be readily demonstrated as "instinctive." Unfortunately, the term "instinctive" had in the past been applied to so many different conceptualizations of behavior (e.g., "I stepped on the brakes instinctively") that ethologists found great difficulty in freeing their own concept of the term from unwanted connotations. Partly in an attempt to give the ethological concept of instinctive behavior its own uniqueness, the term **innate behavior** was coined. Merely changing a name does not accomplish much by itself, of course. But it may often help to catalyze new thinking about a subject, or it may cause one to look at it in a new perspective.

"Releaser Mechanisms" as Innate Behavior

Consider, for example, the behavior of the female digger wasp, *Ammophila.* Within a few weeks of summer, she mates with a male, digs a nest hole, constructs cells within it to hold the future developing pupae, hunts and paralyzes with her sting such prey as caterpillars and spiders, puts the prey into the nest (where their state of suspended animation ensures that the young larvae will have fresh meat on which to feed before pupation), lays her eggs, and seals up the hole—a series of very complex behaviors carried out in a precise sequence. Since her parents died long before she was born and there is no time to "practice," it would *appear*

Few ethologists use the concept of "instinct" today. For behavior that is programmed into the organism the term "innate behavior" is more commonly used.

self-evident that her behavior could not be learned. Nor is it at all helpful to say it is "instinct."

In 1935, Konrad Lorenz introduced a hypothetical model to account for such highly stereotyped behavior. Lorenz pictured an organism as possessing a very specific receiving center in the central nervous system. He applied the term **innate releasing mechanism** (IRM) to this receiving center. Lorenz envisioned the innate releasing mechanism as being triggered by some specific behavioral patterns from the environment; these he termed "releasers." With this hypothetical model, the wasp's behavior can be seen as a sequence of highly coordinated responses (mating, nest building, prey capture, etc.) put forth in response to some releaser. For example, the releaser for the example cited might be the appearance and approach of the male. This alone may be all that is needed, with the mating act, nest building, prey capture, and so on following in order much as a row of dominoes standing on end may topple in sequence if the first one is pushed over. A programmed behavioral sequence—the standing row of dominoes—has been set off by the releaser, causing the innate releasing mechanism to allow the push on the first domino.

Note that this sort of thinking allows for a regular scientific sequence of hypothesizing and testing of prediction; for example, the female wasp can be deprived of exposure to the male to see whether the rest of the sequence will follow (it doesn't). Obviously many different interpretations can be placed on this experimental observation, but the point is that at least some progress has been made. Merely labeling *Ammophila's* behavior as "instinct" gets us nowhere at all.

It is interesting to note that the concept of releasers has led to some practical side effects in pest control. The male mosquito, for example, is attracted to the female by the sound frequency she emits in flight. Here, sound is the releaser mechanism that elicits the male's approach response. The male mosquito will also fly toward an electric trap and be killed if a sound source duplicating the sound frequency emitted by the female is used as "bait." No sight or odor of the female is necessary; once the releasing sound stimulus is produced, the males must topple like the dominoes.

16.3
Animal Behavior as a Combination of Learned and Innate Responses

Consider a well-known experiment performed by Tinbergen and Perdeck on the feeding behavior sequence occurring between herring gulls and their chicks. The parent arrives at the nest, lowers its head, and points its beak downward in front of the chick. The chick then pecks at the bill, occasionally grasping it and pulling it downward. After a few repetitions of this pecking and pulling, the parent regurgitates the partially digested food. The chick then pecks at the food, breaking it apart and eating it (see Fig. 16.6).

Now it is possible to explain this parent–chick feeding sequence on the basis of releasors and innate releasing mechanisms (such as was done with wasp behavior), in which the behavior of the parent acts as a releaser to bring about the proper response on the part of the chick. This is precisely what Tinbergen and Perdeck did. By building cardboard models, which varied both in color of beak and in the position of a characteristic red patch on the lower jaw, they showed that certain features

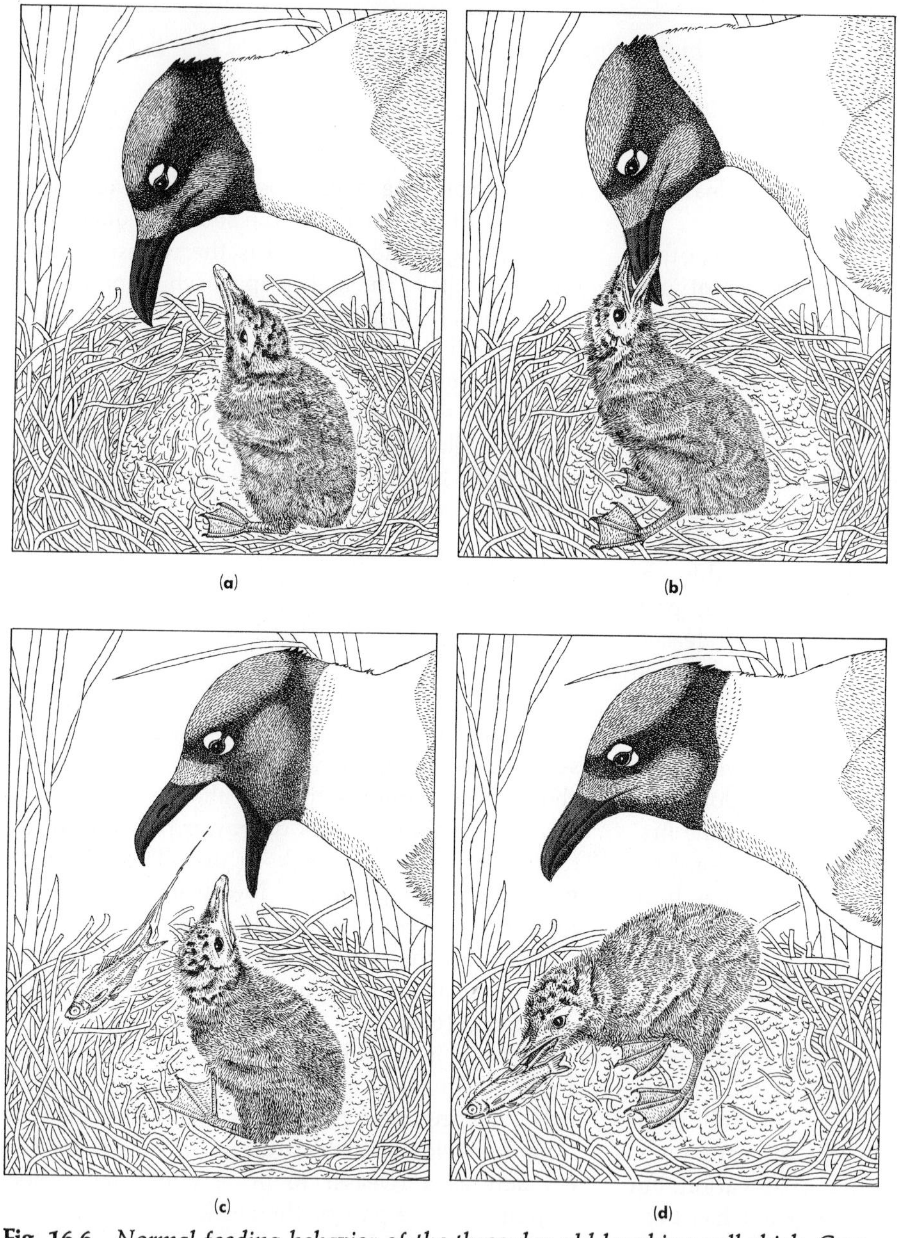

Fig. 16.6 *Normal feeding behavior of the three-day-old laughing gull chick. Components of the behavior include (a) "begging" peck at parent's beak when lowered, (b) grasping and stroking of beak, (c) regurgitation of partially digested food, and (d) the feeding peck. (From Jack P. Hailman, "How an Instinct is Learned."* Scientific American, *December 1969. Copyright © 1969 by Scientific American, Inc. All rights reserved.)*

(such as the shape of the head) were unimportant, whereas the shape of the bill, its motion, and the position and color of the red patch on the lower beak yielded significant differences in the experimenter's ability to elicit the chick's feeding response. All these observations are completely consistent with an hypothesis proposing that an interacting series of releasers and innate releasing mechanisms bring about the proper responses, and it again appears quite unnecessary to suggest the involvement of any learning component in the bird's behavior.

However, care must be exercised in accepting such an interpretation without reservation. In careful studies of the laughing gull, J. P. Hailman and his associates have shown that gull chicks raised in darkness (so that no visual stimuli and thus no pecking practice was possible) differed significantly in pecking accuracy from control chicks exposed to model heads from hatching. Furthermore, a significant increase in pecking accuracy was demonstrated by chicks reared in the wild. It was found that, while a certain improvement in pecking accuracy is achieved without practice, visual experience and practice are necessary if the animal is to attain full pecking accuracy. Finally, with continuing practice and maturation in the nest, the chicks gradually responded more to accurate models of the parent gull's beak and less to inaccurate ones. The results strongly suggest that in this case, and quite possibly others, what at first appears to be instinctive or innate behavior may actually be behavior in which learning is one of the causal factors producing the behavior. It seems likely that other sequences of behavior heretofore assigned entirely to the category of innate behavior can and should be reexamined.

The foregoing example suggests that we must use extreme caution in separating behavior into learned and innate components; perhaps we should even question whether such divisions between these two components are likely to be helpful. It will be necessary to return to this problem.

16.4 Motivation

Most if not all animal behavior is goal-oriented. This statement is not necessarily teleological; it is neither necessary nor fruitful to assume conscious behavior on the part of an animal in attaining a specific goal.* Yet it is obvious that an animal may orient itself toward a specific goal—say, food—quite strongly on one occasion and quite weakly on another. The degree of intensity of goal-oriented behavior, a degree highly influenced by an intervening variable, is termed **motivation.** Thus, if an animal is very hungry, it will be highly motivated toward the goal. If it has eaten recently, on the other hand, it will be far less motivated toward food. The time of the last feeding becomes the intervening variable in this case. The term **drive** is sometimes used to describe a specific motivation. Thus we speak of hunger drive, sex drive, thirst drive, etc. Some other examples of specific motivations are fear (leading, perhaps, to an escape drive), aggression, and so on.

A major problem in dealing scientifically with motivation is that it often resists quantitative measurement and must therefore be dealt with subjectively. This means that comparisons of drive intensities toward different goals are usually unprofitable; what does it mean, for example, to say that

---

* Though the statement is not teleological, what it asserts *is*, since teleological means "goal-oriented." There is a sticky semantic problem here. Evolution is not teleological, because those organisms that survive, survive, and that's all there is to it. Certain forms of behavior, on the other hand, are teleological because of their goal-oriented nature. The key is to diligently avoid the idea of *consciousness* in such behavior. Why? Because consciousness is an undefinable concept that cannot be scientifically tested. In an effort to avoid the connotation that an animal "knows" that reaching the goal will increase the probability of its own survival and that of its descendants, the term **teleonomical** has been suggested. By definition, the concept of teleonomy filters out both conscious and evolutionary factors and simply represents the goal-directedness that remains.

the hunger drive is stronger than the sex drive? How can the intensity of either drive be quantitatively measured? Assuming that some sort of quantitative scale *could* be established for hunger and sex drive, in what way could the scales ever be equated? Possibly the hunger drive can be placed ahead of the sex drive on the basis of the results of deprivation, one being fatal while the other is not. But most certainly this is an unsatisfactory way to compare motivation intensities.

As an example of goal-oriented behavior that has been subjected to some quantitative measurement, consider B. T. Gardner's work on the feeding behavior of jumping spiders. Jumping spiders do not spin webs but rather stalk their prey. The entire sequence of jumping spider feeding behavior can be broken down as follows:

A. Orientation: The spider sees its prey (e.g., a fly) and turns toward it.

B. Pursuit: If the spider is some distance away, it runs rapidly toward it, but slows down to a careful stalk as it approaches the fly. When it attains a certain crucial distance from the intended victim, it crouches and jumps upon it.

Gardner has shown that this sequence of steps in the overall hunting behavior of the jumping spider provides a scale by which the animal's motivation intensity can be gauged. A spider that has captured enough flies to satisfy its hunger will no longer orient itself to flies. If the animal has devoured a number of flies slightly less than that necessary to completely satisfy its hunger, it may orient itself toward flies, but do no more. At a certain level of hunger, however, the spider pursues after orientation and goes on to crouch and jump. Thus prey capture in the jumping spider provides an example of behavior in which an intensity scale can be correlated with units of time (i.e., the number of time units elapsed since the animal last fed).

Ethologists and psychologists also often quantify the motivational states in terms of the degree of deprivation. For example, Dr. Martin W. Schein of West Virginia University has shown that if the dust-bathing behavior of Japanese quail is prevented for about two or three days and then allowed to occur, the behavior is measurably higher than the standard for about an hour and then reverts to the standard. A deprivation of two weeks results in a predictable, still higher increment of behavior, and so forth. Unfortunately, not all bebehavioral sequences can be so neatly calibrated.

16.5 Imprinting

In 1890 ethologist Douglas Spalding noted that a baby chick that had not heard the call of its mother until eight or ten days of age would then refuse to recognize her at all, despite her coaxings. This observation was followed in 1910 by zookeeper O. Heinroth's report that young ducklings follow the first relatively large object they see moving. Heinroth labeled this rapid fixing of social preferences "prägung," the German term for "pressing" (as in the stamping of a coin). In 1935, Konrad Lorenz stressed the uniqueness of this form of learning, which he termed in translation **"imprinting."** Lorenz noted that, in geese, imprinting occurs rapidly, lasts a lifetime, and unlike other forms of learning does not seem to involve any "reward."

Imprinting is obviously of selective value, since the first moving and sound-emitting object a young bird is likely to see is its mother. Interestingly, however, young birds also become "imprinted" to objects such as toys, boxes with ticking clocks inside, and even the experimenter. Such objects, of course, bear little resemblance to the natural mother, and it is interesting to note that no matter what the object to which the animal becomes attached, the attachment is often a lasting one. Certain male birds "imprinted" to the human hand may prefer to attempt copulation with the hand than with a receptive female.

When first described, imprinting was thought to be permanent; for example, it was thought that a gosling imprinted to a human being would never be able to relate successfully to its own species. More recent work has shown that this is not always the case. It appears that it is mainly the existence of a critical learning stage which is the main characteristic of imprinting.

16.6 The Development of Behavior

The concept of the innateness of behavior, while not original with him, has perhaps been most succinctly stated by Konrad Lorenz. As his "psychohydraulic" model nicely illustrates, Lorenz visualizes the accumulation of energy for a specific innate act in a region of the brain specific for the control of that act. The act is not continuously performed, however, because it is blocked or inhibited. Only the appearance of specific stimulus patterns from the environment (e.g., the appearance of a fly in front of a hungry jumping spider, or the red belly of a rival stickleback male) causes the innate releasing mechanism to remove the inhibitor and trigger the outflow of motor impulses from the instinctive center to the muscles appropriate for that particular "innate" behavioral response.

Testing for Innateness

What major criteria must be met by a behavioral action if it is to meet the requirements of the "Lorenzian innateness" hypothesis? First, the behavior must be stereotyped and constant in form. Second, it must be characteristic of the species. Third, it must appear in animals raised in isolation from others. Fourth, it must appear fully formed in animals prevented from practicing it. The reader will recall several examples of behavior (e.g., web spinning in *Araneus diadematus* Cl.) that meet all these criteria.

But there are yet more examples. Domestic chicks, for instance, characteristically peck at objects, including food grains, soon after hatching. This behavior, like that of the jumping spider, can be broken down into highly stereotyped components—head lunging, bill opening and closing, and swallowing. This pecking behavior is stereotyped, characteristic of the species, appears in isolated chicks, is present at the time of hatching, and seems to appear without specific practice. Clearly, the Lorenzian criteria for the innateness of pecking behavior are adequately met.

But *is* pecking innate? In 1932, Z. Y. Kuo showed that the three-day-old chick embryo's head is passively moved up and down by the heartbeat. The head is also rhythymically touched by the yolk sac, which is moved mechanically by amnion contractions synchronized with the heart movements. At four days, the head bends *actively* in response to this touch. The bill begins to open and close. At about eight or nine days, swallowing of fluid forced into the throat by the bill and head movements occurs. By the twelfth day, a sequence of head movement, bill opening, and swallowing has been well established. Are these embryonic movements precursors of postembryonic pecking movements? Most certainly, as a result of these observations, the "innateness" of chick pecking must be viewed in a different light. Several alternative interpretations of the embryonic development of behavior (in this case, pecking) are now possible. A whole new field, that of the developmental biology of behavior, is opened up. Despite the fact that Kuo's experiments were performed several years ago, this area has yet to be adequately explored.*

* Kuo's papers were immediately challenged. It was claimed that the petroleum jelly he used to make the inner shell membrane transparent halted gas exchanges across the membrane and that the reactions he reported were the result of this interference. A sizable bibliography on the subject of chick embryonic behavior has been prepared by Dr. Viktor Hamburger of Washington University in St. Louis.

The pecking behavior of chicks is not the only example of behavior that at first seemed obviously innate and later less obviously so. Nest building and retrieving of young after they have been removed from the nest are also examples of behavior meeting the four requirements of the Lorenzian innateness hypothesis; rats raised in isolation will still perform these functions equally well. However, if isolated rats are raised in cages where the floor is of netting (so that the feces drop down out of reach) and the food is powdered (so that no pellets are available for manipulation), normal nest building and retrieval of young does *not* occur, presumably because the adults have had no opportunity to practice manipulation of objects. The nesting material is left scattered all over the floor of the cage and the young are simply moved at random from one place to another, rather than to any one particular nesting site.

The results of such experiments have many ramifications, but in particular they force us to take a long hard look at the classical isolation experiments. As D. S. Lehrman of Rutgers University points out, "an animal raised in isolation from fellow-members of his species *is not necessarily isolated from the effect of processes and events which contribute to the development of any particular behavior pattern*. The important question is not 'Is the animal isolated?' but '*From what* is the animal isolated?' " For example, song sparrows raised by a pair of canaries in a sound isolation room developed songs normal for their species. Thus it would *appear* that the song is innate; even the sound of the canary song did not affect its development. But a song sparrow deafened in its youth develops extremely abnormal songs. These results show that auditory monitoring of its own voice is essential for normal song development. The inherited potentiality to develop normal song without learning from other birds is not realized in the environmental absence of auditory feedback.

What has been concluded from isolation experiments in the past is rather similar to what Senebier concluded from his experiments on submerged leaves (see Section 6.3). Noting that no oxygen was evolved in the absence of carbon dioxide, he erroneously concluded that the carbon dioxide was the source of the evolved oxygen. In truth, the carbon dioxide was but one of several limiting factors. So it is with isolation experiments. They may well provide negative indications that certain environmental factors are not involved, directly at least, in the origin of a particular behavior pattern. But it lies beyond the isolation experiment's nature to provide positive indication that behavior is "innate" or, indeed, any information at all about what the process of behavioral development is composed of. Even behavior that seems to be most obviously innate may well be partly learned, since learning may emerge as a factor in behavior at the earliest of embryonic stages.

Indeed, it may well be that what is innate or inherited is not simply a matter of genetics. In other words, genes may not be the only factors involved in inheritance. Dr. Roger Williams, a biochemist from the University of Texas, has done studies on "individuality." He finds a hypothesis proposing genetic bases for individual differences sadly lacking, even in "lower" animals. For example, Dr. Williams found that rats inbred for 101 generations (a degree of inbreeding that should make their genetic makeup all but identical) still varied enormously in their behavioral patterns—and even in the biochemical composition of their urine, though they were given identical diets.

Casting about for a better source of "identicalness," Dr. Williams selected the nine-banded armadillo, *Dasypus novemcinctus*. This animal exhibits an unusual embryological development. The fertilized egg begins cleavage and undergoes implantation quite normally, but then produces four primordial buds. These buds are formed in two stages, and each will develop into a baby armadillo. Thus this animal always produces quadruplets, and since each is a descendant of the original fertilized egg, each contain identical genes. Yet, in 20 different features measured by Williams and Dr. Eleanor Shorrs (such as individual organ weight and biochemical characteristics), all varied widely within any given quadruplet set—and some varied as much as 140-fold.

Gene–Environment Interactions

If genetic differences are not the cause of this variation, what is? Williams has hypothesized that the *interaction* between genes and environment governs the intricate processes of cellular differentiation and may affect the extent to which each of the numerous types of differentiated cells proliferates. This hypothesis would account for the fact that one armadillo could have one organ, such as a liver, much larger than that of its genetically identical littermates.

Though it is still too early to judge, the potential implications of the armadillo results are profound. In essence, they seem to indicate that genes are not necessarily all-important in inheritance—a fairly revolutionary notion. Many deep-seated and important human characters, such as strength, fertility, body form, and intelligence, vary widely and continuously within the population. Obviously single genes fail to account for these variations, but there has been a tendency to assume that multiple-gene inheritance can do so. Even here, Williams' results with armadillos suggests another explanation. With regard to behavior, genes can determine very broad, general developmental directions. Any specific behavior develops as a result of the interaction between the organism's genetic potential and the environmental conditions to which the organism is exposed.

Two interesting implications of Williams' hypothesis can be mentioned. For one, it would seem that no matter how knowledgeable geneticists become about microbial genetics—and single-celled organisms such as *E. coli* are the main source of genetic information—we could still remain almost totally ignorant of how some of the most fundamental characteristics of mammals, for example, are inherited. Williams' hypothesis, if valid, also overthrows an assumption that for years was a mainstay of behavioral "nature versus nurture" experiments: that identical twins, because they have identical genes, must therefore have identical inheritance. Thus the countless numbers of identical-twin studies performed over the past decades may very well have been built upon an invalid foundation.

In summary, then, there is now a strong tendency in ethology to regard the old controversy over innate versus learned behavior as no longer a fruitful one to pursue. Only with complete recognition that the answer is probably age-specific is it considered a worthwhile goal to establish "how much" of a particular behavioral sequence is innate and "how much" is learned. In other words ethologists now recognize that the effects of learning and genetically determined differences in structural factors differ not only from component to component of the behavioral pattern, *but also from developmental stage to developmental stage.* Attention is therefore more apt to be directed toward analysis of the characteristics of each developmental stage and of the transition from one stage to the next. As Lehrman says: "The interaction out of which the organism develops is *not* one, as is so often said, between heredity and environment. It is between organism and environment! And the *organism* is different at each different stage of its development."

In systematically challenging the concept of innateness as viewed historically in their own field, ethologists such as M. S. Klopfer are challenging the concept of the gene as the repository of data or "blueprint" from which the organism is constructed. Their work suggests that the gene is rather an information-generating device that exploits the predictable and ordered nature of its environment. This view fits well with the model of gene action advanced by Jacob and Monod (see Section 11.6). In their study of insect development,

H. A. Schneiderman and L. I. Gilbert (1964) have shown that hormones whose synthesis can be traced ultimately to the action of particular segments of the DNA helix activate genetic transcription at other portions of the helix. As was seen in Chapter 11, the transcription products, in turn, may further exert a feedback and regulatory effect on development.

There are parallel situations at the cultural level. In discussing human development, E. Erikson (1968) writes: "The human infant is born preadapted to an average expectable environment. Man's ecology demands constant and natural historical and technological readjustment which makes it at once obvious that only a perpetual, if only ever so imperceptible, restructuring of tradition can safeguard for each new generation of infants anything approaching an average expectability of environment." Thus human behavior, as is the case with development of the embryo, is seen as the outcome of epigenetic rather than preformational processes. Epigenetic development, in turn, assures greater developmental stability, since an epigenetic system is buffered and self-correcting at several points. Like aggression, behavior is, in Klopfer's terms, a *process* not a noun, and the hope of finding an instinct on a chromosome is illusory.

We can conclude this section by asking, then, what *is* inherited? Some investigators have gone so far as to state that only the zygote is inherited, or that heredity is only a stage of development. But it is neither necessary nor fruitful to provide a rigorous definition of heredity here. Pragmatism must be ethology's guide. As Lehrman cogently points out,

> *To say a behavior pattern is inherited throws no light on its* development *except for the purely negative implications that* certain types *of learning are not directly involved. Dwarfism in the mouse, nest-building in the rat, pecking in the chick, and the "zig-zag dance" of the stickleback's courtship are all "inherited" in the sense and by the criteria used by Lorenz. But they are not by any means phenomena of a common type, nor do they arise through the same kinds of developmental processes. To lump them together under the rubric of "inherited" or "innate" characteristics serves to block the investigation of their origin just at the point where it should leap forward in meaningfulness.*

16.7 Genetic and Evolutionary Aspects of Behavior

In the honeybee, *Apis mellifera* L., certain strains are referred to as "hygienic." The name comes from the fact that if a larva dies within its enclosing cell, the workers uncap the cell and remove the corpse. Other strains are "unhygienic"; when a larva dies, its corpse is left to decay in what has now become its tomb. The descendants of hygienic bees all exhibit the same hygienic behavior pattern; those of unhygienic strains remain, like their progenitors, unhygienic.

Certainly these are sharply contrasting phenotypes, as distinct as Mendel's tall and short peas. And, as with Mendel and his peas, a cross can be made between the contrasting phenotypes. The result is an F_1 that is all unhygienic; the dead larvae are left untouched, their cells capped. Clearly, the unhygienic behavior pattern (or better, perhaps, the absence of the hygienic behavior pattern) is the dominant character here. On the basis of the results thus far, tentative genotypes can be assigned as shown below.

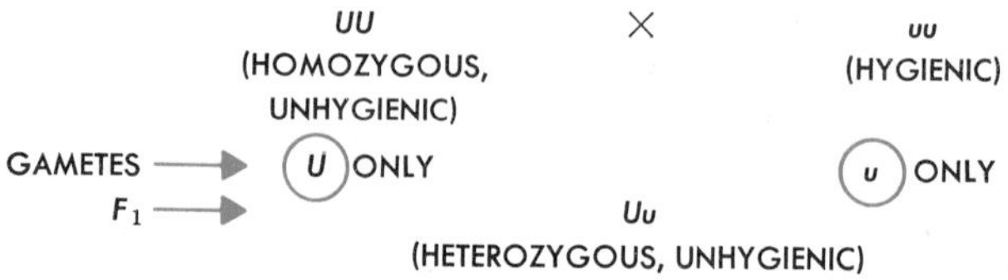

If a backcross is now made between the F_1 unhygienic hybrid, genotype *Uu*, and the recessive hygienic strain, genotype *uu*, Mendelian genetics

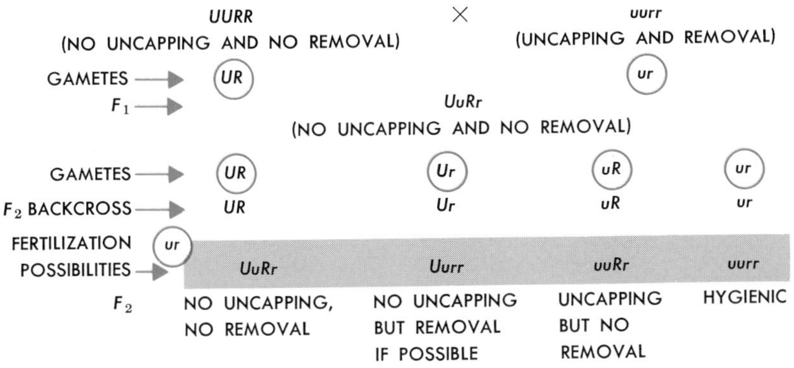

predicts a 50–50 distribution of unhygienic and hygienic bee colonies.

However, in 1964 the animal behaviorist W. C. Rothenbuhler showed that in 29 colonies of bees resulting from such a cross, the following distribution was obtained:

8 colonies: workers left cells capped and did not remove dead larvae (i.e., were unhygienic)

6 colonies: workers uncapped cells and removed dead larvae (i.e., were hygienic)

9 colonies: workers uncapped cells, but left corpses of larvae untouched

6 colonies: did not uncap cells, but would remove larvae if the caps were removed by the experimenter

It is reasonable to propose that two pairs of genes are operating here, not one, and that the data represent a 1:1:1:1 ratio. Each pair of genes controls one of the two steps in the hygienic behavioral sequence: (1) uncapping the cells and (2) removing the larvae. Thus the original cross of hygienic with unhygienic must be written as shown above.

In terms of Mendelian genetics, this situation works out most happily.* Not surprisingly, however, such neat results are rare in behavioral genetics. The example does show that *differences* in behavior can have a genetic basis and, further, that behavioral patterns can be broken down into their component parts, e.g., the hygienic behavior pattern into (1) uncapping and (2) corpse removal. Rothenbuhler's work and that of others with different organisms, seems to show clearly that the complexity of the *behavior* is not related to the mode of inheritance of *differences* in the behavior. The hygienic behavior, as we have seen, shows four phenotypes, but it appears to be controlled by gene segregation at only four loci. Yet a seemingly simple behavior such as the reaction of the fruit fly *Drosophila* to gravity (geotaxis) shows complex differences and appears to be influenced by many genes scattered across all the chromosomes. Geotaxis is a **polygenic character.**

Lovebirds are members of the parrot family. Within the genus *Agapornis* distinct evolutionary stages of nest building can be shown. In particular, two types of nest-building behavior are known. In one, strips are torn from leaves for nest-building material. The strips are then transported to the building site by being tucked underneath the rump feathers. In the other nest-building behavior pattern, however, the leaf strips are carried back to the nest one at a time in the beak or bill, not tucked into the feathers.

Dr. William C. Dilger is an ethologist who has carried out some intriguing experiments in orni-

* It is an interesting coincidence—but only that—that besides raising garden peas Mendel was an ardent beekeeper.

thology. Dilger wondered what would happen if he crossed "feather-carrier" lovebirds with "bill-carriers." Assuming complete dominance is involved, one would predict an F_1 of either all feather-carriers or all bill-carriers. With incomplete dominance, some sort of behavioral hybrid would be expected. The latter proves to be the case, and the result is a group of hopelessly confused young birds. All are completely incapable of building a nest, because they attempt a compromise between bill-carrying and feather-carrying—which means no carrying at all! A bird might begin to tuck a strip between its rump feathers, but then the bill-carrying "urge" would take over, and the strip would not be released until it was pulled back out of the feathers and dropped on the floor. Then the whole process would begin again.

Yet once again, a hypothesis proposing that nest-building behavior is entirely innate is contradicted; occasional successes were recorded when a hybrid managed to keep a strip in its beak after failing to tuck it into the feathers. After months of practice, success was achieved in more than a third of the trials. Two years later, nearly complete success was attained by the hybrid birds. What seemed to be strictly innate behavior revealed, upon further examination, considerable susceptibility to learning. The "genetic factors" were not entirely suppressed, however; even hybrid birds attaining complete success in nest building still made a preliminary movement toward tucking the strip underneath their feathers before flying off with it held in their bills.

If behavior patterns have a genetic basis, they must have considerable evolutionary significance as well. At least one example has already been discussed; the different mating calls of crickets serve to isolate species that may live in otherwise overlapping habitats (see Fig. 14.2). A somewhat similar example is found in lizards of the genus *Sceloporus*. During the mating season, subtle differences in head-bobbing movements (see Fig. 16.7) serve as courtship signals as well as species-

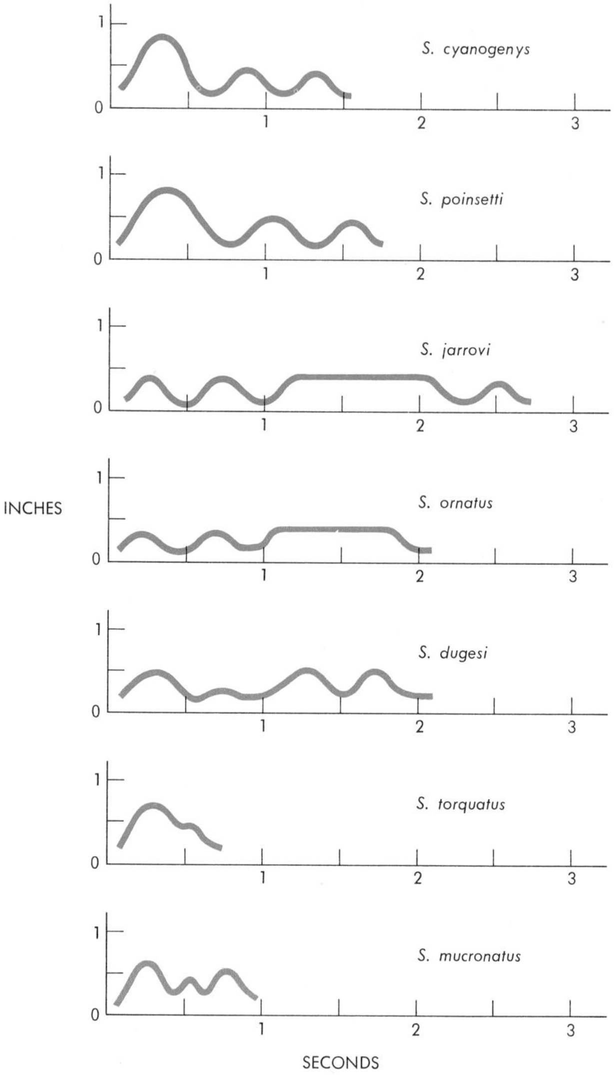

Fig. 16.7 *Heading-bobbing movements of some* Sceloporus *lizards. Movements of the head are represented as a line with height on the vertical axis and time in seconds on the horizontal axis. Note the variations in amplitude, speed, and length of the movements. (After Aubrey Manning,* An Introduction to Animal Behavior. *Reading, Mass.: Addison-Wesley, 1967. From Hunsaker,* Evolution *16, 1962, p. 62.)*

identifying signals. There are only two main groups of muscles involved in these characteristic movements, but even in this case a hypothetical model based on the concept of genes affecting nervous system thresholds can satisfactorily account for the remarkable diversity of head-bobbing behavior shown by these animals. Given more complex situations, involving many groups of muscles (more the rule than the exception), it is not difficult to picture how small accumulations of threshold changes might lead to changed behavior patterns which, in turn, might lead either to extinction or to further speciation.

Another example illustrating the evolutionary significance of behavior is pertinent here. Herring gulls remove the eggshells from the nest after their chicks have hatched. But what is the selective value of such behavior? The speckled coloring of the gull eggs before hatching gives them excellent protective coloration against the pebble-strewn background on which they are laid. The chicks, too, are protectively colored. The *inside* of the eggshell, however, is white. When the shell is broken at hatching, some fragments may lie with their inner surface exposed. This observation led N. Tinbergen to hypothesize that the white fragments might enable predators to locate the nest by sight (or even by the odor of the decaying portions of the remaining egg parts adhering to the shell fragments). This hypothesis leads to certain predictions: If removal of eggshells from the nest after hatching aids in protecting the young from predators, then nests with eggshells should be preyed upon more frequently than those without eggshells; and nests with eggshells left closer to them should suffer a higher degree of successful predation than nests with eggshells farther away. Tinbergen tested his hypothesis with experiments conducted in the field (outside the laboratory). Both predictions were verified. Thus the seemingly energy-wasting behavior of eggshell removal on the part of parent gulls was shown to have considerable selective value.

Indeed, it seems as though *behavioral changes may be of major importance in the process of evolution by natural selection.* While stressing the primary importance of geographical isolation in the formation of new species, Ernst Mayr states in his book, *Animal Species and Evolution:* "A shift into a new niche or adaptive zone is, almost without exception, initiated by a change in behavior. The other adaptations to the new niche, particularly the structural ones, are acquired secondarily."

16.8 Agonistic and Aggressive Behavior

The spectrum of behaviors ranging from overt attack to overt flight comprise what is called agonistic behavior. In general agonistic behavior involves ritualized interactions that avoid, rather than lead to, bloodshed.

This Land Is My Land

As has been stressed more than once in this book, the acceptance of scientific hypotheses is quite frequently related to particular historical periods.* One hypothesis wholly harmonious with the culture of its day was put forth to explain the fighting often observed among males of the same species during the breeding season. The reasons for this fighting seemed perfectly clear; the males were competing

* It can be correctly pointed out, of course, that the concept of ideas having "their time" is based on whether or not they are accepted; that is, if they are accepted, the time was ripe for this acceptance, and if not, the time was not ripe. The reasoning is circular.

for a female, with the strongest claiming the prize. The observation of such fighting was one fact among many that led Charles Darwin to his concept of "sexual selection," whereby nature allowed only the strongest to contribute to the species' future gene pool. This concept fit in well with the stereotyped views of manhood and womanhood fashionable in the Victorian era, as it also did with the capitalist ethic of competition as the most effective mechanism for ensuring progress.

In the late nineteenth century Henry Eliot Howard, an English businessman, began a detailed study of British warblers and moorhens. An avid bird watcher, Howard spent almost thirty years carefully studying warbler behavior and making copious notes. The result, in 1920, was a book entitled *Territory in Bird Life*. Its message was simple: it is *territory*, not females, for which the male birds compete.

Slowly at first, and then with increasing rapidity, Howard's idea gained acceptance. The aggressive behavior of other birds was observed in the fresh new light of **territoriality,** rather than sexuality, as the stimulus that provoked intraspecific fighting. According to the new territorial concept, the males fought for space—a tree, a meadow, a certain portion of a forest—with the winner taking a distinct and definable portion. The established territory, rather than any direct attribute of the male, became the attractant that won the female.

The basic concept of territoriality proved to have extensions beyond the class Aves (birds). On the Rio Piedras campus of the University of Puerto Rico lizards, not squirrels, scamper through the trees and along the sidewalks. The animals are strongly territorial; the experimental introduction of one male lizard into another's territory leads to characteristic warning gestures on the part of the owner which, if unheeded, lead to combat. When more than one male lizard is kept in captivity, the cage must be large enough to allow each male to have his own territory; if the cage is too small, the loser of the fight cannot retreat out of the disputed territory to safety, and his death is the result. Invertebrates too show territoriality, though less commonly. Male crickets, for example, fight vigorously to defend their areas against trespasser; indeed, cricket-fight matches are a popular sport in parts of the Orient. Most interesting, perhaps, is the contention of some researchers that territoriality can be demonstrated among some of our closest primate relatives, e.g., the howler monkeys of Central America. The trespassing of one troop of howler monkeys into an area claimed by another leads to a nonviolent but deafening vocal contest that continues until the trespassers retreat. The claim that howler monkeys are territorial animals is somewhat weakened, however, by the observation that howling contests may occur even when the troops are some distance from each other. It has also been shown that the "territories" of howler monkeys are not exclusive and that considerable overlapping occurs.

In many cases it is relatively easy to map the boundaries of the territory a particular organism considers its own simply by watching its behavior. It is far less easy to understand the ultimate functions of territoriality. An animal expends considerable energy establishing and defending a territory, which suggests that territoriality must have considerable selective value to those species exhibiting this behavior. One prevalent hypothesis proposes that territoriality serves to space out a population, and thereby helps to ensure that the available supply of food is not overexploited to the extent of jeopardizing the entire ecosystem.* But this cannot be the whole story. The spacing hypothesis nicely accounts for a hawk or eagle defending a square mile or so from being hunted over by other birds

* Here is another excellent example of how attempts at brevity and conciseness in writing can lead to teleological and anthropomorphic statements. It is hoped that the reader by now understands that animals don't space out to avoid jeopardizing the ecosystem. Rather, they spread out because their ancestors did, and their ancestors were ancestors because they carved out enough of their environment to reproduce successfully.

of prey, for the amount of food in the area available to these animals is certainly finite. But a pair of herring gulls will defend a nesting territory of only 18 square feet or so, an area from which they certainly get no food. The relationship among territoriality, spacing, and food supply may be an indirect one: the spacing of nests limits the size of the gull colony and thus controls the number of birds feeding in the area. This is a reasonable hypothesis from an ecologist's viewpoint, but it runs into difficulties when seen through the ethologist's looking glass. In general, seabirds do not seem to be food-limited; indeed, gulls appear to live in a superabundance of food. Most of their day is spent resting on beaches or mud flats with full bellies.

An hypothesis about territoriality with which ethologists feel more comfortable comes from Tinbergen's group, which has shown that nesting territories in the black-headed gull are closely related to predation: the first occupied and most vigorously defended territories in areas that suffer the least predation from foxes, crows, etc. Still another advantage stemming from territoriality, but unrelated to food supply, may be the ensuring of freedom from disturbance during mating pair formation.

Research on territoriality has established one basic fact: territory is different things to different species. A hawk or eagle defends a rather wide area in which it does all its feeding. Gulls, on the other hand, show no territoriality in feeding, and it is possible that what has been interpreted as territoriality in some gulls may be little more than nest defense, especially in those species in which the nesting "territory" is only two feet or so in diameter.

Agonistic Behavior and Aggression

Ethologist Aubrey Manning of the University of Edinburgh in Scotland points out that "the first essential for a territorial animal is that it be aggressive towards others of its kind." The defense of a territory can be accomplished only if the defender shows **aggression**—i.e., initiates or gives signs of initiating an attack—against any individual entering territory claimed by the defender. Note here an almost complete reversal in the use of the term aggression in ethology compared to its use in international conflict. In the latter case, the nation violating the territory of another is referred to as the aggressor. On the other hand, when the invaded nation attacks the trespassing nation's forces, the action is called defensive, not aggressive. As will be seen shortly, this differing use of the same term is but one of several difficulties encountered when we move from considering aggression in nonhuman animals to aggression in humans.

Clearly the study of aggressive behavior must be approached with caution. Merely to focus on the behavior patterns of an aggressive organism is not sufficient. Due consideration must be given the effects of this behavior on both the aggressor and the animal that is the target of the aggression.

Actually, there is a spectrum of behavior ranging in intensity from overt attack at one end of the scale to overt fleeing at the other. This spectrum of behavior is termed **agonistic behavior.** All displays of agonistic behavior (e.g., the raised and compressed threat stance of a territorial lizard) are combinations of motivations to attack and to flee, in various absolute and relative strengths (see Fig. 16.8).

Several years of ethological research on agonistic behavior have established certain "facts" concerning it. For example, the claim has been made that the production and release into the bloodstream of certain hormones is associated with aggressive behavior.* Time is also often a factor; only at certain periods, such as the mating season, do some animal species show agonistic behavior. Finally, it has been established that aggression may

* While there is little question that the release of certain hormones is associated with agonistic behavior, it *is* still possible to challenge the many cause-and-effect hypotheses stemming from the consequences of increased hormone output, and some ethologists have done so.

Fig. 16.8 *Agonistic behavior in mice. As a result of many encounters, the mouse at the left has assumed a dominant status over the one at the right, which adopts a characteristic defense posture of holding out the forefeet and moving only when attacked.*

be triggered by certain external stimuli. The male three-spined stickleback, for example, reacts violently to the visual stimulus of red on the underside of a rival male. That it is the color to which the animal reacts and not something else can be shown by experimentation. If presented with a crude model of a rival male with a red underside, the male attacks viciously. One case has even been reported in which males attempted to attack British mail trucks (which are red) driving by the window in which the aquaria containing the fish were kept. A more lifelike model, without the red underside, is not attacked. In parakeets and flickers, certain colorations or markings distinguish males from females. If the markings are changed experimentally, the males will attack their own mates (see Fig. 16.9).

The Function of Agonistic Behavior

Since the appearance of agonistic behavior in animals is fairly widespread, there must be some selective value to such behavior. In the case of species exhibiting territoriality, it is obviously necessary for defense. In territorial species, agonistic behavior makes territoriality possible and in turn territoriality may aid in spacing animals so as to prevent overexploitation of the environment.

In some species, the relationship of agonistic behavior to territory is a remarkably close and consistent one. In the stickleback, size and strength of the fighting males is far less important than which fish invaded the other's territory. Indeed, it seems possible to show a distinct relationship between proximity to a rival's nest and the number of battles "won" or "lost" (see Fig. 16.10).

However, while agonistic behavior may at times be closely linked to territoriality, it is by no means inseparably so. Many animal species are not territorial but exhibit agonistic behavior, and vice versa. In certain birds such as chickens, so-called **peck orders** are found (see Fig. 16.11). The members of a flock compete to establish a sort of class system, in which those higher in the social register get first choice of food, space, mates, etc. Once established, however, peck orders actually serve to reduce conflict, and attempts on the part of an animal to improve its standing in the order are relatively rare. Peck orders are known in primates, also, where they are usually referred to as **dominance hierarchies.** The dominant male of a baboon or rhesus monkey troop is easy to spot, not only by his own behavior but by the differing degrees of "respect" shown him by animals lower down in the social order. If he so desires, the dominant male is the one who drinks first at the water hole, who samples new food first, and who indicates when it is time for the troop to move to the safety of its sleeping quarters for the night.

Aggression in Humans

Aggressive behavior seems to have a genetic basis in many lower forms of animal life. But does the same situations hold true for humans? *Do* humans

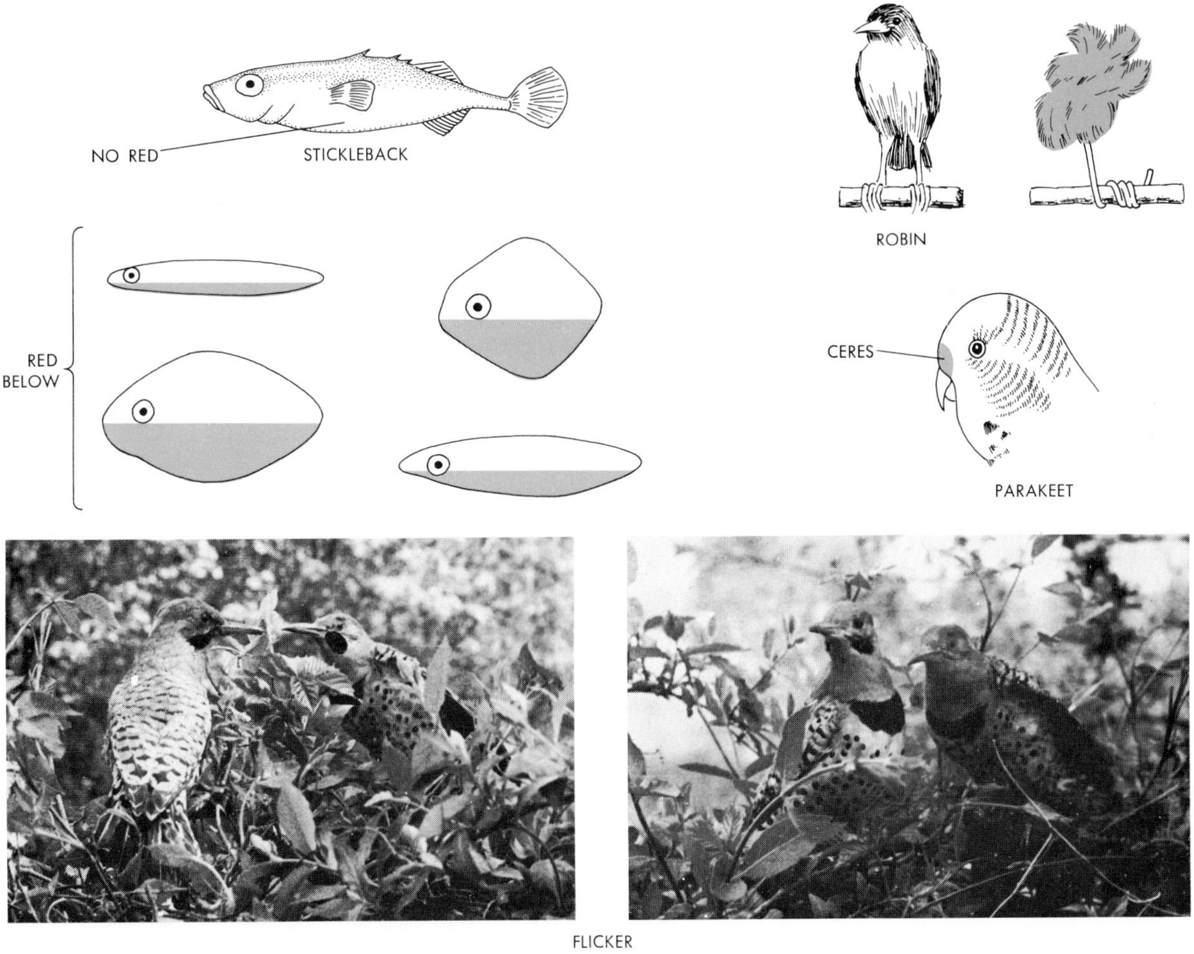

Fig. 16.9 *Four examples of "releaser" stimuli. Top left: The crude models of stickleback fish painted red on the underside are threatened by a male stickleback, while the true male at top with no red is ignored. Top right: The robin will ignore an immature male without the characteristic red breast, but may attack a tuft of red feathers. Middle right: If the characteristic color of the female parakeet's ceres is changed, her mate attacks her as if she were a rival male. Bottom: If the black spot or "moustache" of the male flicker* (Coleoptus aurates) *is painted on his mate (as shown at left), he attacks her as he would a rival male. When it is removed, he accepts her as his mate again. (Stickleback models after N. Tinbergen,* The Study of Instinct. *London and New York: Oxford University Press, 1951. Posed model photos courtesy Charles Tarleton, Wesleyan University.)*

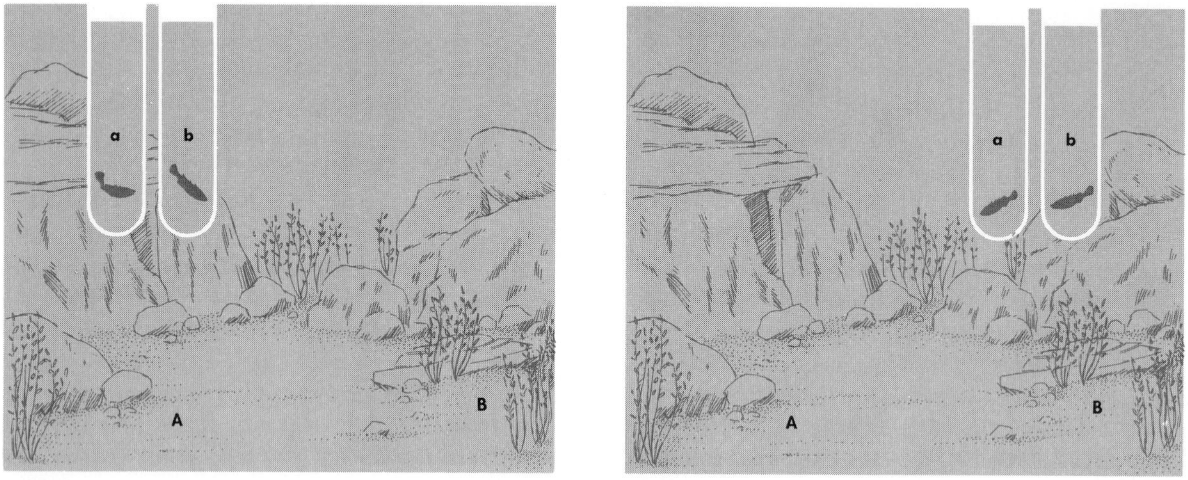

Fig. 16.10 *An experiment showing a relationship between types of behavior and territory. Male stickleback* b, *"owner" of territory* B, *is threatened by male* a *when brought into territory* A. *When the situation is reversed, male* a *is threatened by male* b.

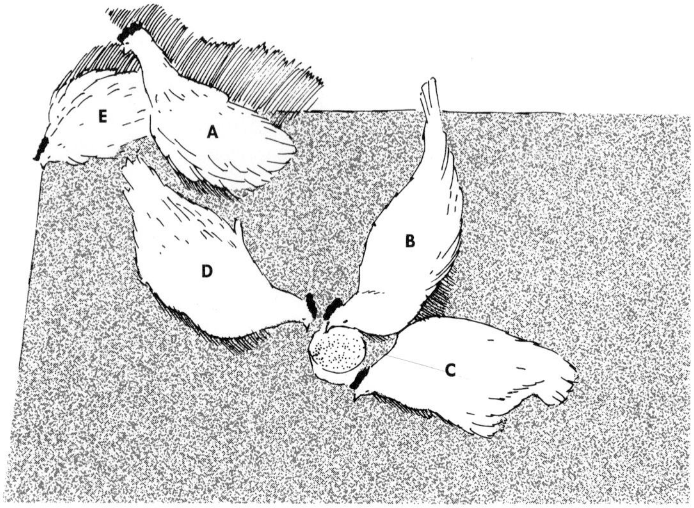

Fig. 16.11 *"Peck order" or dominance hierarchy in chickens. Once the hierarchy is established, physical conflict is minimized. Dominant hen A has just driven bottom-ranking hen E away from the food. While A is gone, B moves in and keeps C and D away by pecks and threats. When A returns, B will give way.*

Supplement 16.1

SEX AND THE DOMINANT MALE

In all discussions about social hierarchy and dominance among male animals, it is assumed that dominance confers a reproductive advantage on its bearers. Following Darwin's idea of sexual selection, it is assumed that the more sexually aggressive a male, the more frequently he will mate, and hence the more offspring he will leave behind. Making the further assumption that aggression is somehow controlled by genetic factors, it can be concluded that more genes for aggressive behavior (at least with regard to sexual behavior) will be passed on to the next generation than genes for nonaggressive behavior.

Recently this assumption has been criticized on several grounds. Irwin Bernstein of the Yerkes Primate Research Center at Emory University in Atlanta, Georgia, claims that the whole concept of dominance hierarchies, at least among primate males, is vague and ill-defined. Bernstein notes that various measures of dominance, as determined by different observers, do not agree with one another. If captive monkeys are ranked three times according to three different criteria—mounting, grooming, and agonistic behavior—different males come out on top each time. Thelma Rowell of the University of California, Berkeley, agrees with Bernstein. She claims that dominance hierarchies may exist only in the minds of the observers, who make primates compete in the laboratory or field for rewards such as food or water. Such procedures are unnatural and highly anthropomorphic. Primates seldom have to scramble for a limited supply of food in the wild. To create a human-like situation for animals, and then describe their reactions to it in such human terms as "dominant," "aggressive," "meek," "passive," etc., is not the same as observing natural behavior patterns of the animals involved. It has not escaped the attention of some critics that the concept of dominance hierarchies and the laboratory tests designed to support their existence are highly reminiscent of the social class hierarchies that characterize the very societies in which the biological concept of hierarchy was invented.

Not all ethologists agree with Bernstein and Rowell on this matter, however. Sandy Richards, of Cambridge University in England, reports that when captive rhesus monkeys are ranked according to Bernstein's criteria or other criteria, the same males come out as most dominant every time. Critics of Richards's work are quick to point out that her monkeys also lived in the very kind of unnatural circumstances that all observers agree lead to predictable competition for food and water. Further, Richards's groups consisted of a male and several females (rather than the more common multimale group) which could greatly influence the results.

An even more fundamental criticism of the older concept of dominance hierarchies has been made by Bernstein, Thomas Gordon, and Susan Duvall. They tested the assumption that the so-called aggressive or dominant males in a population have a selective advantage over less dominant males in terms of reproduction. Bernstein and others noted that in the past, primate ethologists were satisfied with counting the number of times a male monkey mounted a female as an indication of reproductive success. The greater the frequency of mounts, the more reproductively successful the male was judged to be. Several observations throw this standard of "success" into question. Lee Drickamer of Williams College has pointed out that observers often judge certain male rhesus monkeys to be high-ranking simply because they display the more conspicuous behavior. This was found to be true in judging reproductive success. The number of times a male mounts a female is not an accurate measure of reproductive success.

Bernstein and his coworkers set out to determine whether dominance hierarchies, however they are determined, really bear any relationship to reproductive success. They performed biochemical tests of paternity in a group of rhesus monkeys at the Yerkes Primate Center. They found that no direct relationship exists between dominance of a male, judged by a variety of different criteria, and the number of offspring the male produced. So-called nonaggressive males often left behind as many or more offspring than so-called aggressive males.

Bernstein and others further expressed the opinion that sexist biases may have crept into most past studies on this subject. Investigators had been assuming that the female was a totally passive participant in the reproductive process, willingly accepting whatever male made the most aggressive response toward her. Bernstein and his associates noted that females played an important part in the process itself, often rejecting the most aggressive, "dominant" males.

Similar results have been obtained observing baboons in the wild. Glenn Hausfater of the University of Virginia studied agonistic behavior among baboons in Kenya in order to test an hypothesis developed in the 1940s stating that dominant males have the first access to females who enter estrus. According to this hypothesis, if only one female in a group is in estrus at a given time, the top-ranking male will copulate with her. If two females are in estrus, the top two-ranking males would be expected to copulate with them, etc. During the first 400 days that Hausfater observed baboons, only one or no females were in estrus at any one time. If the older hypothesis were correct, he should have observed the vast majority of matings to be with the top-ranking male. The prediction was not fulfilled, however. The top-ranking male did not mate with three successive females who were in estrus, even though no other estrus females were present. At the same time, males who were not top-ranking did mate with the estrus females. It may be, Hausfater concluded, that the females chose the males, rather than vice-versa.

The observations of these various workers do not necessarily mean that dominance hierarchies never exist in primate social groups. They *do* indicate that the descriptive definitions of such hierarchies (1) may not be based on the same criteria that ethologists choose as important, (2) may not have much if any relationship to reproductive success, and (3) may be very fluid, changing from day to day under natural environmental conditions. It is obvious from these studies that our current picture of primate and other nonhuman animal social systems may be a primitive one at best, fraught with problems introduced by our own subjective biases and the difficulty of making solid, statistically significant observations under natural conditions. Such findings make it even more imperative that we avoid facile comparisons between animal and human societies.

fight and kill their own kind because of an innate tendency to do so? Or is our aggressiveness purely the result of past or present environmental influences? Or is some other explanation possible? The thought once prevalent in ethology and such related fields as psychology and anthropology can perhaps best be traced in the following two statements:

There can not be any doubt, in the opinion of any biologically minded scientist, that . . . aggression is in man just as much a spontaneous instinctive drive as most other higher drives.

Konrad Lorenz

. . . in the course of human evolution the power of instinctual drive is gradually withered away, till man has virtually lost all his instincts. If there remains a residue of instinct in man, they are, possibly, the automatic reaction to a loud noise, and in the remaining instances to a sudden withdrawal of support; for the rest, man has no instincts. . . . Evil is not inherent in human nature, it is learned . . . aggressiveness is taught, as are the forms of violence which human beings exhibit.

M. F. Ashley Montagu

Certainly these two quotations illustrate a very wide gap in thinking concerning the origin of aggression in *Homo sapiens.* It is a matter of much interest and concern which, if either, of these two schools of thought is correct. War seems always to have been part and parcel of the human heritage, and we stand now at a point in history where our next great war will very likely be our last.

If human aggressiveness is a highly unacceptable form of behavior in today's society, then clearly, like a disease, we should want to prevent it. But disease control is most effective when the causes for the disease are understood. One who believes human aggressiveness toward fellow humans to be innate may well recommend a quite different treatment for the condition from one who believes aggression to be entirely the result of environmental influences. So, indeed, do the two men just quoted differ in their approach to the problem. Lorenz, believing aggression to be an innate drive that must be expressed, recommends athletic events as a specific and harmless outlet for this aggressive drive. Montagu, on the other hand, would concentrate more on finding the proper physical and cultural environment for humankind, an environment in which aggressive behavior would not be learned.

SUMMARY

1. Biology has come to be organized into various levels of investigation, from the molecular to the population level. As a *concept,* however, the idea of levels of organization in living organisms can be misleading. In animal behavior, for example, the organism must be seen as a member of a population and as an individual organism, and of course the component parts of which it is composed affect its behavior.
2. The field of biology dealing with animal behavior is called ethology. The early development of the field was characterized by "heredity versus environment" or "nature versus nurture" debates.
3. Anthropomorphism—the ascribing of human characteristics to nonhuman organisms—and teleological explanations, which ascribe conscious purpose to an organism's actions, were early impediments to ethology. Their elimination put studies in ethology on a far more solid scientific footing.
4. The term "instinctive" behavior was a widely used and misused term in ethology and was later replaced with the term "innate." While both terms refer to behavior that is genetically based, the former term antedated the field of ethology and had acquired nonscientific connotations.

5. Lorenz's innate releasing mechanism (IRM) concept pictures a genetically based mechanism controlling a specific behavior activated only by exposure to the proper "release" stimulus. Thus, for example, the sight of a female turkey "releases" strutting behavior in the male; in turn, his strutting behavior "releases" her crouching behavior, and so on. While the IRM concept was extremely important as a scientific model that stimulated much important research, more recent studies have raised doubts about the "innateness" of some of the behavior elicited by the releasing stimuli.

6. Behavioral changes are now seen as being as important as anatomical changes in the process of evolution by natural selection.

7. Learning in nonhuman organisms spans a very wide range, from the duplication of "human" behaviors (e.g., abstract reasoning, symbolic communication) to the imprinting learning of animals such as birds. In the latter, each exposure to a stimulus object often (though not always) causes a lifetime attachment to the object, even when it may bear no resemblance to the natural object to which the young organism would ordinarily be exposed.

8. The studies of amateur ornithologist Henry Eliot Howard led to formulation of the concept of territoriality in organisms. Territoriality refers to the tendency of some organisms to defend a certain space (a tree, bush, plot of ground, etc.) from intrusion by others of their own species. Not all organisms are territorial, and even in those that are, the territorial boundaries are not always well defined. The precise reasons for the evolution of territoriality are unclear, though various hypotheses, ranging from spacing in accordance with food supply to protection from predation or disturbance during mating behavior, have been formulated.

9. A territorial animal must be aggressive towards another animal of its species that enters its territory if it is to successfully hold the territory. The aggression of one animal towards another often triggers a characteristic behavioral reaction in the organism being attacked (defensive posture, retreat, etc.) Aggression and the range of behaviors it elicits in the organisms fighting are collectively called agonistic behavior.

10. Historically, the hypotheses put forward for the causes of aggression have ranged from aggression's being "innate" to its being entirely the result of environmental factors. More recent work has emphasized the difficulty of satisfactorily delineating "aggressive" behavior in different species and casts doubt on the value of hypotheses based upon the concept of "innateness" for such a complex phenomenon as agonistic behavior.

11. "Peck orders" or "dominance hierarchies" are found in certain organisms (chickens, cows, baboons). In such systems, the population is seen as having a highly structured, "class" system, where those organisms higher on the ranking scale appear to get first access to food, space, mates, etc. More recent research, however, calls into doubt some of the interpretations of earlier investigators. For example, blood tests of newborn baboons indicate that as many or more are fathered by adult males or even juveniles, supposedly far down on the dominance hierarchy scale, than are fathered by the supposedly dominant male.

12. The concept of aggression as innate in humans has resulted from often unjustified or highly speculative extrapolations from aggressive behavior in lower animals. Ethologists such as Konrad Lorenz have often gone beyond their truly scientific data to express speculations with little or no scientific bases. Such specu-

lations have then been picked up and extended by writers who are not scientists, such as playwright Robert Ardrey.

13. Modern ethology looks upon behavior much as a developmental biologist looks upon the developing limb bud of a chick embryo—that is, as a system whose general features are genetically controlled but whose actual expression may be greatly affected by its internal or external environment. Just as exposure to a particular chemical at an initial stage of development may produce a deformed wing in the adult bird, so may aversive environmental stimuli or deprivation at critical times cause behavioral aberrations of a pathological nature. It is generally to the environment rather than the genome that we must look for means to ensure the potential for healthy behavior for all. The old "nature versus nurture" controversy is not a meaningful debate in terms of promoting progress in the field of ethology.

EXERCISES

1. Young Mediterranean cuttlefish (*Sepia*) raised in isolation will only attack and feed upon crustacea of the genus *Mysis* (probably their normal prey in this area of the Mediterranean). Propose hypotheses in the following three categories to answer the question: Why do young *Sepia* attack mysids?
   a) An anthropomorphic hypothesis.
   b) A teleological hypothesis.
   c) An ethological hypothesis (propose a test for this hypothesis).
2. Lorenz and Tinbergen (1937) caused the silhouette at the right to "fly" on a wire extending across a pen containing geese or various game birds. When the model in the illustration moved across the pen to the right, no particular attention was paid to it. But when it moved to the left, the birds crouched, piled upon one another in a corner of the pen, or otherwise showed alarm. Propose an hypothesis to explain these results and outline a test for your hypothesis.

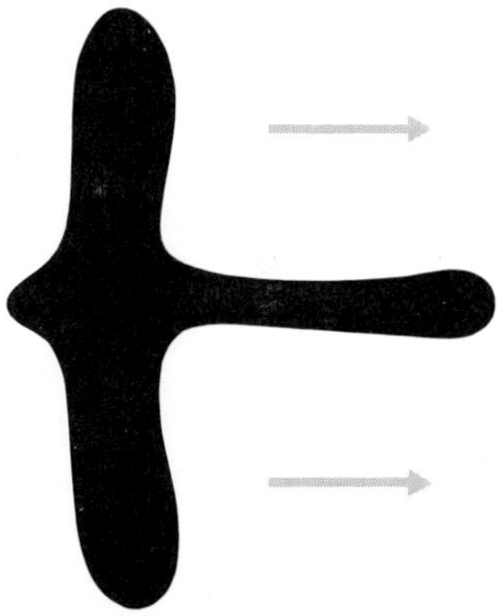

3. A mother turkey normally accepts and raises her own young with no problems. A hen turkey deafened several weeks before hatching of the eggs invariably kills all her young when they hatch. A hen turkey deafened 24 hours after hatching will raise her chicks normally. Propose a hypothesis to explain these observations and a test for this hypothesis.
4. A male and a female dove are put into adjoining cages separated by a glass pane. At a certain period of the year, the male "displays." Following completion of the display, the female builds a nest, lays infertile eggs, and sits on them. A control female, not exposed to a male, does not exhibit this behavior.

   Interpret or explain these observations. In other words, what do you think this experiment demonstrates?
5. Around 1900, there was a horse in Germany named Hans who, it was maintained, could carry out addition, multiplication, and division. He would respond to the question, "How much is 3 + 2" by pawing the ground five times. Hans could also

spell out words and sentences, pawing the ground an appropriate number of times for each letter of the alphabet. To recall that U is the twenty-first letter of the alphabet in the midst of spelling out a complicated sentence is a difficult task, yet Hans learned to do so in two years. Hans was studied by a committee of ethologists who verified his ability to do the things.

Propose a nonanthropomorphic hypothesis to account for Hans's feats, and propose a test for your hypothesis.

SUGGESTED READINGS

Alland, Alexander, *The Human Imperative.* New York: Columbia University Press, 1972. A book countering the ideas advanced by Robert Ardrey in the two selections listed below.

Ardrey, Robert, *African Genesis.* New York: Dell, 1961. The author, a journalist with considerable writing skill, makes the case for the innateness of human aggression toward fellow humans.

Aronson, L. R., E. Tobach, D. C. Lehrman, and J. S. Rosenblatt, *Development and Evolution of Behavior: Essays in Memory of T. C. Schneirla.* San Francisco: W. H. Freeman, 1970. An excellent collection of essays dealing with the evolution and development of behavior, behavioral processes, social behavior, and human behavior.

Dilger, William C., "The behavior of lovebirds," *Scientific American,* January 1962, p. 88. An interesting case study of how behavior affects divergence in a species and thus is related to its evolutionary position.

Hinde, R. A., *Animal Behavior,* 2d ed. New York, McGraw-Hill, 1970. A text for the serious student of behavior that provides a synthesis of ethology and comparative psychology.

Klopfer, Peter, and Jack P. Hailman, *An Introduction to Animal Behavior: Ethology's First Century.* Englewood Cliffs, N.J.: Prentice-Hall, 1967. An excellent study of animal behavior from the biological viewpoint. The book is especially good in its historical treatment of the subject.

Lehrman, Daniel S., "A Critique of Konrad Lorenz's Theory of Instinctive Behavior," *The Quarterly Review of Biology* 28, No. 4, December 1953, pp. 337–363. An excellent presentation of the case against the concept of instinct as used by Lorenz.

Lorenz, Konrad, *On Aggression.* New York: Harcourt, 1966. Lorenz's most controversial book, now available in paperback. Fascinating reading, but the case the author makes for the innateness of aggression seems not to be accepted by most ethologists.

———, *King Solomon's Ring.* New York: Crowell, 1952. A paperback book by a classic investigator in the realm of animal behavior. This delightfully written and excellent book covers many topics in animal behavior.

Manning, Aubrey, *An Introduction to Animal Behavior.* 3d ed. Reading, Mass.: Addison-Wesley, 1979. A good paperback introduction to ethology.

Tinbergen, Niko, "The evolution of behavior in gulls," *Scientific American,* December 1960, p. 118. An excellent case study in the evolution of behavior in several species.

Von Frisch, K., "Dialects in the language of the bees," *Scientific American,* August 1962, p. 78. An excellent article describing a communication pattern in honeybees. Comparison of the system in closely related varieties and species shows the divergence of communication patterns due to geographic isolation.

APPENDIX 1

CHIEF UNITS OF THE METRIC SYSTEM

Linear measure		Symbol	English equivalent
1 kilometer	= 1000 meters	km	0.62137 mile
1 meter	= 10 decimeters	m	39.37 inches
1 decimeter	= 10 centimeters	dm	3.937 inches
1 centimeter	= 10 millimeters	cm	0.3937 inch
1 millimeter	= 1000 microns	mm	
1 micron	= 1/1000 millimeter or 1000 millimicrons	μ	no English equivalents
1 millimicron	= 10 angstrom units	$m\mu$	
1 angstrom unit	= 1/100,000,000 centimeter	Å	

Measures of capacity			
1 kiloliter	= 1000 liters	kl	35.15 cubic feet or 264.16 gallons
1 liter	= 10 deciliters	l	1.0567 U. S. liquid quarts
1 deciliter	= 100 milliliters	dl	.03 fluid ounces
1 milliliter	volume of 1 g of water at standard temperature and pressure (stp).	ml	

Measures of mass			
1 kilogram	= 1000 grams	kg	2.2046 pounds
1 gram	= 100 centigrams	g	15.432 grains
1 centigram	= 10 milligrams	cg	0.1543 grains
1 milligram	= 1/1000 gram	mg	about .01 grain

Measures of volume		
1 cubic meter	= 1000 cubic decimeters	m^3
1 cubic decimeter	= 1000 cubic centimeters	dm^3
1 cubic centimeter	= 1000 cubic millimeters	cm^3
1000 cubic millimeters	= 1 milliliter (ml)	mm^3

APPENDIX 2

TAXONOMIC CHARTS

Many different classification schemes have been proposed for living organisms. An early one was the animal–vegetable–mineral system used by Linnaeus. More recent schemes have dealt only with animals and vegetables (plants). An examination of several textbooks reveals that at least four variations exist among modern schemes for classification. There are schemes involving two kingdoms (Plantae and Animalia), three kingdoms (Protista, Plantae, and Animalia), four kingdoms (Monera, Protista, Plantae, and Animalia), and five kingdoms (Monera, Protista, Plantae, Fungi, and Animalia). The scheme given in detail below is organized according to the five-kingdom plan, since that is the one used most frequently by biologists today.

The Five-Kingdom System

The five-kingdom system given here is adapted from R. H. Whittaker in "New Concepts of Kingdoms of Organisms," published in *Science* (the official journal of the American Association for the Advancement of Science), Vol. 163, 1969, pp. 150–160. Besides the Protista, this system recognizes the Monera, or procaryotic organisms, as a separate kingdom.

Each taxonomic system, whether it uses two, three, four, or five kingdoms, has its own logical basis and set of criteria for the grouping or separation of organisms. The five-kingdom system's internal logic is based on the criteria of levels of body organization and the different modes of nutrition (photosynthesis, absorption, or ingestion) utilized by the various groups.

The five-kingdom system is summarized in Fig. A.1. Note that the colored dividing line leaves the "plants" not only the larger group but also in possession of the hypothesized ancestral forms that gave rise to both the "plants" and the "animals." This arrangement reflects the fact that the early ancestral forms of both the "plant" and "animal" groups fall more logically into the "plant" world.

Kingdom Monera

Organisms with procaryotic cells, lacking nuclear membranes, plastids, mitochondria, or flagella with 9 + 2 strands; body organization unicellular, colonial, or mycelial. Diverse nutritional types, including absorption, photosynthesis, and chemosynthesis. Reproduction primarily asexual by cell division; some sexual recombination known in a few species. Motile by simple flagella or gliding, or nonmotile.

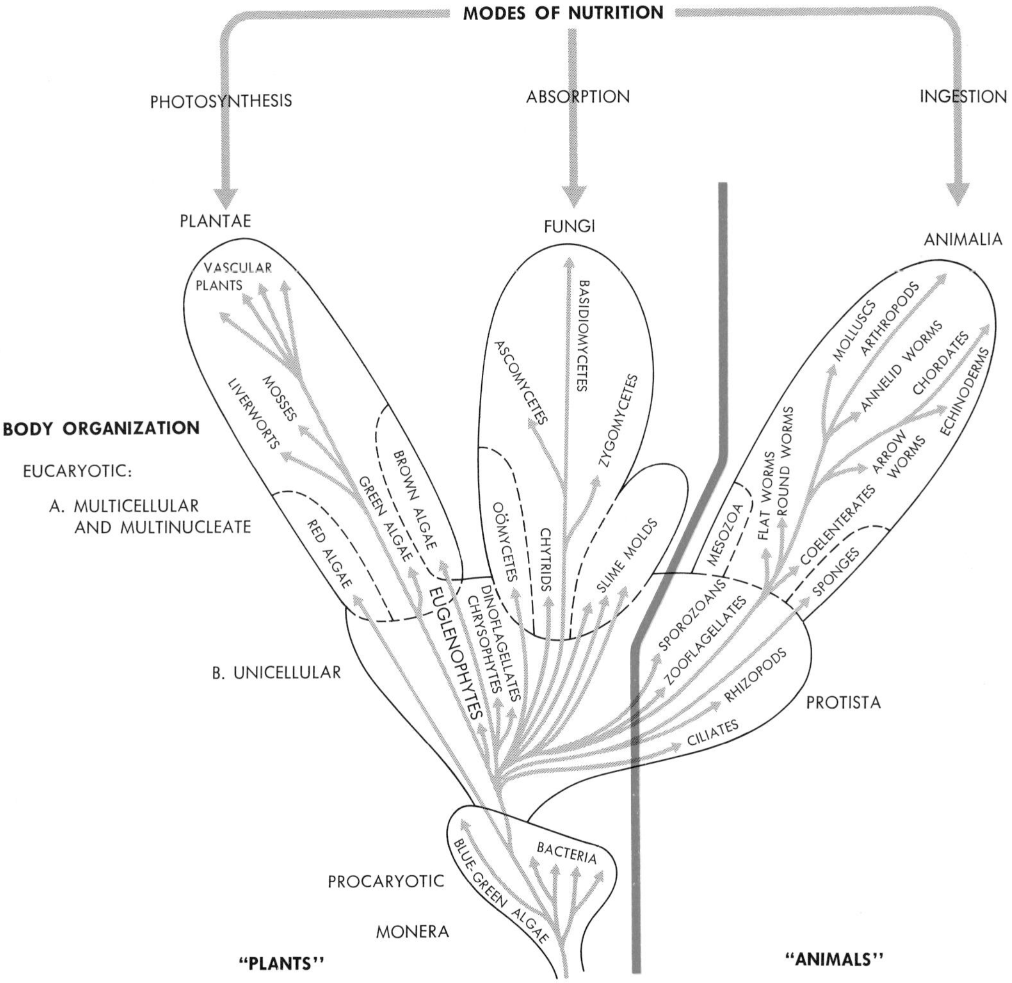

Fig. A.1 *This diagram illustrates the broad interrelationships of plants and other organisms. Five kingdoms, based on levels of body organization and modes of nutrition, are recognized in this classification system. The organisms traditionally called "plants" in the two-kingdom system are listed on the left of the diagram, the "animals" on the right. (Adapted from R. H. Whittaker, "New Concepts of Organisms."* Science *163, pp. 150–160, copyright 1969 by the American Association for the Advancement of Science.)*

PHYLUM 1. SCHIZOPHYTA. The bacteria. Many species motile by simple flagella; most absorb their food, a few are photosynthetic or chemosynthetic. About 1600 species, found almost anywhere conditions are favorable for growth.

PHYLUM 2. CYANOPHYTA. Blue-green algae. Blue color due to pigment phycocyanin, red color to pigment phycoerythrin: nearly all are photosynthetic, producing free molecule oxygen; flagella lacking, motility (if present) by gliding. About 1500 species, found in aquatic and marine habitats, in damp soil, on tree bark, and on rocks.
EXAMPLES: The genera *Nostoc, Gloeocapsa,* and *Oscillatoria*

Kingdom Protista

Organisms with eucaryotic cells (possessing nuclear membranes, mitochondria, plastids in the plant members, 9 + 2 strand flagella, and other organelles); primarily unicellular or colonial-unicellular organisms; diverse nutritional methods, including photosynthesis, absorption, and ingestion (in a few species); true sexual processes with nuclear fusion and meiosis present in most; motile by 9 + 2 strand flagella, or by other means.

PHYLUM 1. EUGLENOPHYTA. Euglenophytes. Flagellated unicells lacking cellulose cell walls, enclosed by a flexible or rigid pellicle; reserve food usually paramylum (a starchlike carbohydrate). About 300 species, found in stagnant fresh water or damp soil.
EXAMPLES: The genera *Euglena, Peranema, Astasia,* and *Phacus*

PHYLUM 2. PYRROPHYTA. Dinoflagellates. Unicells with a heavy cellulose wall (in many species) sculptured into plates, others naked (unarmored); cell has two furrows, one transverse and one longitudinal, each furrow containing one long flagellum. About 1000 species, mostly marine, a few in fresh water.
EXAMPLES: The genera *Geratium, Peridenium,* and *Gonyaulax*

PHYLUM 3. CHRYSOPHYTA. Golden algae and diatoms. Golden algae have unicellular, colonial, or filamentous bodies; some have elaborate internal siliceous skeletal structure (the silicoflagellates); other have heavily calcified rings, discs, or plates embedded in the cell wall (the coccolithophores); 300 species. Diatoms have cell walls heavily impregnated with silica and highly ornamented; over 5500 species. Both groups abundant in phytoplankton.
EXAMPLES: The genera *Ochromonas, Dinobryon, Synura, Pinnularia, Navicula,* and *Cyclotella*

PHYLUM 4. XANTHOPHYTA. Yellow-green algae. Plants yellow-green due to chlorophyll in plastids being masked by yellow carotenoid pigments; bodies mostly one-celled, a few being filamentous or tubular (coenocytic); widespread but relatively inconspicuous; grow in fresh waters and marine waters, on damp mud or moist soil; or as epiphytes on larger algae and other aquatic plants; over 400 species.

EXAMPLES: The genera *Characium, Tribonema,* and *Vaucheria*

Kingdom Protista also includes five phyla of protozoans and two phyla of funguslike organisms.

PHYLUM 5. GYMNOMYCOTA. The slime molds. Heterotrophic organisms most of which lack a cell wall but form sporangia at some stage in their life cycle. Most nutrition is by ingestion. Have amoeboid structure.

PHYLUM 6. PROTOZOA. Microscopic, unicellular animals, which sometimes aggregate in colonies. Some are free-living, others parasitic.

Kingdom Fungi

Feeding body usually composed of cobweblike threads (hyphae) that excrete digestive enzymes into the surrounding environment and absorb dissolved nutrients; decomposer organisms. Primarily multinucleate, with eucaryotic nuclei dispersed in a walled and often septate mycelium; plastids and mitochondria lacking. Little or no vegetative body differentiation; reproductive body of more advanced species composed of tissues. Primarily nonmotile (but with protoplasmic flow in the mycelium), living embedded in a medium or food supply. Reproductive cycles include both sexual and asexual processes.

Kingdom Plantae

Multicellular organisms with eucaryotic cells; cells walled and frequently vacuolate; with photosynthetic pigments in plastids. Nutrition mostly photosynthetic, a very few species being absorptive. Primarily nonmotile, living anchored to the substratum. Structural differentiation leading toward organs of photosynthesis, anchorage, and support and, in most species, specialized photosynthetic, vascular, and covering tissues. Reproduction in the plant kingdom is primarily sexual with cycles of alternative haploid and diploid generations, the haploid being greatly reduced in the more advanced members of the kingdom.

The plants have evolved on at least three separate occasions: the red and brown algae originated separately from the remainder of the plant groups.

PHYLUM 1. RHODOPHYTA. Red algae. Plants chiefly reddish due to the red pigment phycoerythrin; bodies vary from microscopic single cells through simple filaments to large plants over 10 ft long with some tissue differentiation due to the aggregation of filaments; compare favorably in complexity of symmetry and branching habit with that found in flowering plants; very complex sexual reproductive systems (but no flagellated cells of any sort); grow attached to rocks and larger algae along ocean shores from high in the intertidal region to depths of 360 ft, especially in the tropical seas; over 3500 species.

EXAMPLES: The genera *Porphyridium, Bangia, Porphyra, Nemalion,* and *Chondrus*

PHYLUM 2. PHAEOPHYTA. Brown algae. Plants brownish due to the xanthophyll pigment fucoxanthin; bodies vary from simple-unbranched filaments only 1 mm long to massive plants 200 ft long and with well-developed tissues (including food-conducting, sievelike elements similar to those of vascular plants); grow largely attached to rocks along ocean shores, often forming dense subtidal forests; over 1500 species.

EXAMPLES: The genera *Ectocarpus, Laminaria, Nereocystis, Fucus,* and *Sargassum*

PHYLUM 3. CHLOROPHYTA. Green algae. Plants bright green due to chlorophyll in plastids; body diversity is considerable, ranging from one-celled motile or nonmotile species to motile or nonmotile colonies, and from simple filaments to massive plants over 20 ft long (with tissues); wide diversity of life cycles and modes of sexual reproduction; abundant in both fresh waters and marine waters, on tree trunks, and on moist rocks, leaf surfaces, and soil; nearly 7000 species.

EXAMPLES: The genera *Chlamydomonas, Volvox, Ulothrix, Cladophora, Spirogyra, Ulva, Codium*

PHYLUM 4. BRYOPHYTA. Bryophytes. Multicellular terrestrial plants without vascular tissues or true roots; life cycle includes a conspicuous gametophyte with unbranched sporophyte permanently attached; sex organs (archegonia and antheridia) multicellular; reproduction by gametes and spores.

PHYLUM 5. TRACHEOPHYTA. Vascular, terrestrial plants with complex differentiation of organs into leaves, roots, and stem. The only motile cells are the male gametes of some species, which are ciliated. The vascular plants have well-developed conducting tissue for the transport of water and organic materials and minerals. The main trend in the evolution of vascular plants has been a progressive reduction in the gametophyte and in the evolution of the seed.

Subphylum 1. Lycophytina. Lycophytes. Vascular plants with microphylls (*i.e.*, "little leaf-like structures"). Extremely diverse in appearance. All lycophytes have motile sperm. There are five genera and about 1,000 species.

Subphylum 2. Spenophytina. Horsetails. Vascular plants with jointed stems marked by conspicuous nodes and elevated siliceous ribs; sporangia are borne in a strobilus at the tip of the stem. Leaves are scalelike. Sperm are motile. There is one genus, *Equisetum*, with about two dozen living species.

Subphylum 3. Pterophytina. Ferns, gymnosperms, and flowering plants. Although diverse, these groups possess in common the megaphyll, or large leaf, which in certain genera has become much reduced. About 260,000 species.

Kingdom Animalia

Eucaryotic multicellular organisms. Principal mode of nutrition is by ingestion. Animals lack the rigid cell walls characteristic of plant cells, and they are motile. Their reproduction is primarily sexual, with male and female diploid organisms producing haploid gametes that fuse to form the zygote. More than a million species have been described and some estimate the actual number at close to 10 million.

PHYLUM PORIFERA. The sponges, both freshwater and marine. The lowest of the many-celled animals, resembling in many respects a colony of protozoans. Body perforated with many pores to admit water, from which food is extracted. There are three classes, divided primarily according to the formation of the sponge spicules. Over 4000 species.

PHYLUM COELENTERATA (CNIDARIA). Radially symmetrical animals with a central gastrovascular cavity. Body wall consists of only two cell layers. In the outer layer are stinging cells (nematocysts. About 11,000 species.
EXAMPLES: Hydra, jellyfish, Portuguese man-of-war, coral, sea anemmones.

PHYLUM CTENOPHORA. The comb-jellies or sea walnuts. Move by means of eight comblike bands of cilia. About 80 species.

PHYLUM PLATYHELMINTHES. The flatworms. Bodies bilaterally symmetrical, flat, and either oval or elongated, with three cell layers. All have flame cells as excretory organs. True central nervous system. No skeletal or respiratory systems. About 15,000 species.
EXAMPLES: Planaria, liver flukes, tapeworms.

PHYLUM RHYNCOCOELA (NEMERTEA). The proboscis worms. Nonparasitic, usually marine animals, with complex digestive system and proboscis armed with hook for capturing prey. Lowest group on evolutionary scale with blood-vascular system; two-opening digestive system. About 600 species.

PHYLUM NEMATODA (ASCHELMINTHES, NEMATHELMINTHES). The roundworms. An extremely large phylum, characterized by elongated, cylindrical, bilaterally symmetrical bodies; live as parasites in plants and animals, or are free-living. About 80,000 species.
EXAMPLES: Vinegar eel, *Ascaris,* hookworm, pinworm, *Trichinella.*

PHYLUM ACANTHOCEPHALA. Spiny-headed worms. They are parasitic worms with no digestive tract and a head armed with many recurved spines. About 300 species.

PHYLUM CHAETOGNATHA. Arrow worms. Free-swimming planktonic marine worms, they have a coelom, a complete digestive tract, and a mouth with strong sickle-shaped hooks on each side. About 50 species.

PHYLUM NEMATOMORPHA. Horsehair worms. They are extremely slender, brown or black worms up to 3 feet long. Adults are free-living, but the larvae are parasitic in insects. About 250 species.

PHYLUM ROTIFERA. "Wheel animals" with circular rows of cilia around the mouth that beat with a motion suggesting rotation of a wheel. Well-developed digestive system. Smallest of metazoans. About 1500 species.
EXAMPLE: Rotifers

PHYLUM BRYOZOA. "Moss animals." Microscopic organisms, usually marine, form branching colonies. Have U-shaped row of ciliated tentacles by means of which they capture food. About 4000 species.

PHYLUM BRACHIOPODA. Marine animals with two hard shells, superficially like a clam. About 250 living species and 3000 extinct species.
EXAMPLE: The lamp shells

PHYLUM ANNELIDA. The segmented worms, with body cavity separated from the digestive tube; brain dorsal and nerve cord ventral; body wall contains circular and longitudinal muscles. About 8800 species.
EXAMPLES: Earthworm, leeches.

PHYLUM ARTHROPODA. Segmented animals with jointed appendages and a hard exoskeleton of chitin. Body divided into head, thorax, abdomen. About 765,000 species.
EXAMPLES: Crayfish, shrimp, barnacles, centipedes and millipedes, spiders, scorpions, insects.

PHYLUM MOLLUSCA. Unsegmented, soft-bodied animals, usually covered by a shell, and with a ventral, muscular foot. Respiration by means of gills, protected by fold of body wall (the mantle). About 110,000 species.
EXAMPLES: Clams, oysters, mussels, snails and slugs, squid, octopus.

PHYLUM ECHINODERMATA. Marine forms, radially symmetrical as adults, bilaterally symmetrical as larvae. Skin contains calcareous, spine-bearing plates. Possess unique water-vascular system. Respiration by skin-gills or outpocketing of digestive tract. About 6,000 species.
EXAMPLES: Starfish, sea urchins, brittle stars.

PHYLUM CHORDATA. Bilaterally symmetrical animals with a notochord, gill clefts in the pharynx, and a dorsal, hollow neural tube. Over 44,000 species.
EXAMPLES: Acorn worm, tunicates, *Amphioxus*, sharks, skates, fish, amphibians, birds, mammals, primates.

GLOSSARY

Abortion. Spontaneous or induced termination of a pregnancy before full term.

Acetabularia. A single-celled marine alga composed of a stalk and an umbrellalike cap.

Acid group. The carboxyl (COOH) group located on many organic molecules, especially amino acids and fatty acids. The acid group ionizes at physiological pH (pH = 7.4) to yield a proton in solution as a hydronium ion: $—COOH + H_2O \rightarrow —COO^- + H_3O^+$ (hydronium).

Acrosome. A caplike structure covering the head of the sperm cell; it appears to help the sperm penetrate the egg membrane.

Activation energy. The amount of energy necessary to initiate an exergonic reaction.

Active site. The portion of an enzyme molecule into which a given substrate fits. When the active site is blocked, the enzyme cannot catalyze a reaction with its substrate.

Active transport. The movement of molecules against a concentration gradient, requiring the expenditure of energy.

Adaptation. For an organism, any change (usually somatic) in its structure or function that allows the organism to better cope with conditions in the environment. For a species, any change (usually genetic, selected out by natural selection) that allows the species as a whole to better cope with its environment.

Adaptive radiation. The evolution by natural selection of a variety of types from one ancestral species.

Adenosine diphosphate (ADP). Adenine contains two phosphate groups, one of which is bonded to the other by a high-energy phosphate bond.

Adenosine triphosphate (ATP). A molecule consisting of a purine (adenine), a sugar (ribose), and three phosphate groups. A great deal of energy for biological function is stored in the high-energy bonds that link the phosphate groups, and it is liberated when one or two of the phosphates are split off from the ATP molecule. The resulting compounds are called adenosine diphosphate (ADP) and adenosine monophosphate, (AMP), respectively.

Adhesion. The tendency of molecules of different substances to stick together.

Aerobe. An organism that requires oxygen to carry on the process of respiration.

Aerobic respiration. A series of reactions for the breakdown of glucose in which the element oxygen serves as the ultimate electron acceptor.

Agglutination. The clumping of red blood cells when they are exposed to agglutinogens in blood of an incompatible type.

Agglutinogen. Blood substance that causes agglutination when introduced into blood of an incompatible type.

Aggressive behavior. Animal behavior in which attack is either initiated or threatened.

Agonistic behavior. The entire sequence of behavioral events in animals associated with aggressive behavior on the part of both the aggressor and the organism against which the aggression is directed.

Alkaptonuria. A relatively benign hereditary disease caused by an autosomal recessive gene. People afflicted with the disease cannot make the liver enzyme homogentistic acid oxidase, so that the intermediate compound homogentistic acid accumulates and is excreted in the urine. On exposure to air, homogentistic acid is rapidly oxidized to a dark brown color, an identifying feature of the disease.

Allantois. In bird and reptile embryos, an extraembryonic membrane for the storage of solid, nondiffusable nitrogenous wastes.

Alleles. Genes that occupy similar loci on homologous chromosomes but carry contrasting inheritance factors. For example, the gene for blue eyes in human beings is said to be an allele to the gene for brown eyes. Also, alleles are two or more genes capable of mutating into one another.

Allometry. Study of the relationship between the growth rate of a part of an individual and the growth rate of the whole individual.

Alternation of generations. A characteristic of the life cycle of certain plants in which a sexual generation alternates with an asexual generation.

Alveolus. An air sac in the lungs, thin-walled and surrounded by blood vessels. The hundreds of thousands of alveoli in each lung serve as the major vehicles for gas exchange in the mammalian body.

Amino acid. The basic structural unit of proteins, having the general formula:

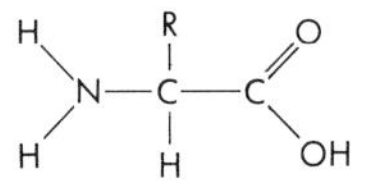

The name "amino acid" is derived from the fact that a basic amino (—NH_2) group and an acidic carboxylic group (—COOH) are attached to the same carbon skeleton. The R group varies from one amino acid to another, giving each amino acid its particular characteristics.

Amino group. The NH_2 group located on all amino acids that contributes certain basic (proton-accepting) properties to the molecule. The amino group is able to accept a proton from a hydronium ion in the following reaction: —$NH_2 + H_3O^+$ (hydronium) $\rightarrow$ —$NH_3^+ + H_2O$. At physiological pH (pH = 7.4), most animo groups are ionized in the NH_3^+ form.

Ammonia. A highly toxic and soluble waste product resulting from the deamination of amino acids. In aquatic animals, ammonia passes from the body almost continuously, so that a harmlessly low concentration is maintained. In terrestrial animals, ammonia is converted to other less toxic materials, such as urea, which can be safely stored in the body until excretion.

Amnion (amniotic sac). Transparent, thin, but tough membrane making up the sac that encloses and protects the embryos of mammals, birds, and reptiles. In humans the amniotic sac is the "bag of waters" containing amniotic fluid in which the fetus develops. It acts as a shock absorber and in other ways protects the fetus. In humans, spontaneous rupture of the amnion is usually an indication that labor is about to begin.

Amoeba. A one-celled organism of irregular shape that moves by extending part of its mass into temporary armlike extensions called pseudopods.

Anabolism. The build-up of more complex substances from simpler ones within a living organism.

Anaerobe. An organism that can carry on respiration in the absence of oxygen. Two types of anaerobes can be distinguished: facultative and obligate. Facultative anaerobes (such as yeasts) respire aerobically or anaerobically, depending upon environmental conditions. Obligate anaerobes can carry on only anaerobic respiration, regardless of whether or not there is oxygen in the environment.

Anaerobic respiration. A series of reactions involving the breakdown of fuel molecules (glucose) and the generation of ATP in the absence of oxygen. The end products of anaerobic respiration can be lactic acid or alcohol, depending on the type of cell.

Analogous. Term applied to body parts that are similar in function but not in structure, such as the wing of a bird and the wing of a bee.

Anaphase. The phase of mitosis characterized by the separation and movement of homologous chromosomes toward opposite poles of the dividing cell.

Androgens. General name for the various male sex hormones, or any substance that has "masculinizing" effects on an organism. Testosterone is the most common naturally occurring androgen. Androgens are produced in the testes of males and in the adrenal glands of both males and females.

Anemia. A state of deficiency of either the number of circulating red blood cells or the amount of hemoglobin in the red blood cells.

Animal pole. The surface of an egg close to the nucleus where the yolk density gradient within the egg is smallest.

Anthropomorphism. The assigning of human characteristics to nonhuman forms.

Anticodon ("nodoc"). Triplet of bases on transfer RNA complementary to the codon of messenger RNA.

Apoenzyme. A protein that forms an active enzyme system by combining with a coenzyme. The apoenzyme usually has the determining effect on the specificity of the enzyme complex.

Archenteron (gastrocoel). In embryology, the hollow interior of the gastrula stage forming a primitive gut.

Arteries. Tubular branching vessels that carry blood away from the heart to various other organs.

Arterioles. The smallest arteries in the circulatory system.

Asexual reproduction. Development of new organisms without the fusion of gametes. This may occur in plants by either spore formation or vegetative reproduction. Some animals may reproduce asexually by fission or budding.

ATP. See Adenosine triphosphate.

Autonomic nervous system. That portion of the central nervous system responsible for carrying out involuntary vital processes. The autonomic system is composed of two parts. The sympathetic system is responsible for integrating the body's many functions during an emergency. The parasympathetic system counteracts the effects of the sympathetic system. Both systems operate to some extent at all times, controlling such functions as the size of the iris diaphragm, salivary secretion, heart rate, peristalsis, and secretion in the stomach and duodenum.

Autoradiography (radioautography). A process whereby the location of radioactive materials is determined by use of photographic film. When a radioactive emission (such as a β-particle) hits a photographic film, it produces an exposure. The more particles, the "brighter" the exposure. In biology this process is especially useful in tracing substances throughout an organism. The organism is fed or injected with a substance containing radioactive atoms. Parts of the organism (a tissue section from the liver, or a leaf) are then exposed to film. Bright spots on the film reveal the distribution of the radioactive substance in the organ or tissue under observation. The exposed film is called an autoradiograph (radioautograph).

Autosome. Any chromosome that is not a sex chromosome.

Autotroph. An organism that can generate its own food supply from simple organic and inorganic elements and some external energy source such as sunlight. Green plants are autotrophs.

Average. The number that describes the sum total of a group of values divided by the number of values in the group (also called the mean).

Avogadro's number. The number of atoms or molecules in a gram atomic weight or a gram molecular weight: 6.023×10^{23} particles.

Bacteriophage (phage). A virus that attacks bacteria. The infecting phage causes the bacterium to produce a new generation of phages, destroying the bacterium in the process.

Binomial nomenclature. The system of naming in taxonomy introduced by Linneaus. Names consist of both the genus name and the species name for the organism. The human being is classified *Homo sapiens* (the genus is capitalized, the species is not, and both are italicized).

Biogenesis. The theory that all living things must be derived from other living things.

Biogenetic law. A nineteenth-century theory devised by Fritz Müller and Ernst Haeckel. This theory held that the stages of embryological development of a given organism repeat the evolutionary stages through which the species passed. This "law" is often stated "Ontogeny recapitulates phylogeny." At present most biologists question the validity of the biogenetic law.

Biomes. Large, easily distinguished community units arising as a result of complex interactions of physical and biotic factors. Grasslands or deciduous forests constitute two distinct biomes.

Biotic community. A varied aggregate of organisms

existing in a common environment, less extensive biologically or geographically than a biome. Division of labor or competition for food may be internal characteristics of a biotic community.

Biotic environment. The sum total of living organisms with which a given plant or animal comes in contact.

Biotic potential. The inherent power of a population to increase in numbers under ideal environmental conditions.

Blastocoel. The cavity inside the hollow blastula stage of the animal embryo.

Blastomere. A cell of the blastula in animal embryos.

Blastopore. The opening of archenteron to the exterior in the gastrula stage in animal embryos.

Blastula. The hollow, single-layered, ball-like structure forming the first identifiable phase of embryonic development in animals. The blastula is the same size as the original zygote, but it is the result of multiple cell divisions.

Blood. A fluid connective tissue composed of living cells and a nonliving matrix, the plasma. The blood carries oxygen, food, and waste products through the body.

Botany. The study of plants.

Bulk transport. The transportation of a large collection of molecules at one time across the plasma membrane. The processes of phagocytosis and pinocytosis are examples of bulk transport.

Calorie. The amount of heat required to raise the temperature of one gram of water from 14.5 degrees to 15.5 degrees Celsius.

Calvin-Benson cycle. A cycle comprising the dark reactions of photosynthesis. It is a major part of the biochemical pathway by which green plants reduce carbon dioxide to sugars.

Camera lucida. An instrument, usually used in conjunction with a microscope, which by means of mirrors or a prism projects the image of an object onto a plane surface. In biology it is used to make outline drawings of objects viewed under the microscope.

Capillaries. The smallest blood vessels in the vertebrate body, having walls one cell thick.

Carbohydrates. Compounds such as sugars, starches, cellulose, glycogen, etc., containing carbon, hydrogen, and oxygen, generally in a ratio that can be expressed as $(CH_2O)n$. Carbohydrates are primary energy foods.

Carnivore. An organism whose diet consists of meat. Such organisms usually display structural adaptations for meat-eating, such as sharp claws and/or teeth.

Carotenoids. Various colored pigments found closely associated with the chlorophyll in green plants and believed to be accessory to the photosynthetic process.

Catabolism. The degradation of complex organic compounds to simpler ones within living organisms or cells.

Catalyst. Any substance that lowers the activation energy of a system, allowing a given chemical reaction to proceed more rapidly. In living systems, enzymes are the main catalysts.

Catastrophism. A late eighteenth- and early nineteenth-century idea accounting for the changes in flora and fauna indicated by fossil records. According to this idea, from time to time great catastrophes destroyed all life on earth, and after each cataclysm a new special creation populated the earth with new forms of life.

Celibacy. Abstinence from sexual intercourse.

Cell. A discrete mass of living material surrounded by a membrane. The basic structural and functional unit of life in nearly all types of organisms.

Cell division. The splitting of a parent cell into two daughter cells. This process consists of two separate phenomena: division of the cytoplasm (cytokinesis) and division of the nucleus (mitosis). The events of the nuclear mitosis follow a regular pattern of four phases: pro-phase, metaphase, anaphase, and telophase. Between succeeding nuclear divisions the nucleus is in interphase.

Cell membrane. The phospholipid–protein bilayer forming the outer surface of every cell. The membrane is flexible, almost fluidlike, and it regulates the nutrients entering the cell and the waste products or secretions leaving the cell.

Cell plate. A cytoplasmic figure formed during plant cell mitosis at the site where a new cellulose partition will be synthesized to separate the two daughter cells.

Cell wall. A rigid structure composed of cellulose, surrounding plant cells.

Cellular metabolism. The total processes in which food and structural materials are broken down (catabolism) and built up (anabolism) within the cell.

Cellulase. An enzyme capable of splitting cellulose into its monosaccharide components.

Cellulose. A large, insoluble polysaccharide of repeating β-linked glucose molecules. Cellulose is the major component of plant cell walls.

Centriole. A small, deeply staining cytoplasmic structure with a 9 + 0 complex of microtubules. It is thought that the centriole performs a function in cell division; however, many higher plants, which seem to have no centrioles, still manage cell division.

Cervix. A "neck" of the uterus, which protrudes and opens into the vagina.

Chemical bonds. The forces of attraction that hold two or more atoms together in a molecule. Formation of chemical bonds is thought to be due to rearrange- of electron clouds.

Chemosynthesis. Synthesis of organic compounds using energy derived from other chemical reactions. For example, the chemosynthesis of carbohydrates or proteins occurs in living cells using energy from the oxidation of foodstuffs.

Chiasma (plural, chiasmata). Figure formed by the intertwining of chromosomes during prophase I of meiosis. Crossing over can occur during chiasma formation.

Chlorophyll. A molecule based on the same ring structure (porphyrin) as hemoglobin, but with magnesium replacing the central iron atom. Chlorophyll is found in all green plants and gives them their color. The molecule functions in photosynthesis by absorbing specific wavelengths of sunlight. It is now known that light raises electrons of the chlorophyll molecule to higher energy levels. As the electrons return to their original level through a series of acceptor molecules, ATP is generated to serve as the direct energy source for reducing carbon dioxide to carbohydrate.

Chloroplast. A small plastid present in the cells of green plants. The chloroplast contains chlorophyll, which is essential for the photosynthetic activities of the plant.

Chorion. In intrauterine development, the outermost membrane surrounding the embryo.

Chorionic villi. In mammals, projections of the chorion that extend into the uterine wall. These are outgrowths of the embryonic sac (chorion), and they provide the basis for surface exchange of materials between the mother and the very young embryo.

Chromatid. A term applied to each of the two parts of a chromosome after replication as long as these parts remain connected at the kinetochore.

Commensalism. Relationship between two different species of organisms in which one derives benefit and the other suffers no harm.

Competition. Struggle between organisms for the necessities of life. There are two types of competition; intraspecific (between members of the same species) and interspecific (between two or more different species).

Competitive exclusion principle. Also known as Gause's hypothesis, it states that two species with similar ecological requirements cannot successfully coexist.

Competitive inhibitor. A substrate that reversibly combines with the active site of an enzyme and lowers the capacity of the enzyme to interact with its regular substrate.

Complementation. In genetics, a test to determine whether two mutations occur within the same gene or within the same cistron. If two strains have the same mutant phenotype, the mutation may occur within the same gene or in a different gene altogether. In the complementation test, two mutant chromosomes are introduced into the same cell. If mutations occur within the same gene, the mutant phenotype will still be expressed. If the mutations are in different, nonallelic genes (or different cistrons), the wild-type phenotype can be expressed, since each chromosome contains (or "complements" for) the defective gene in the other.

Conditioned reflex. A behavior pattern learned through repetition of a sequence of events.

Conjugation. A physical association and exchange of materials, leading to reproduction in certain organisms such as the green alga *Spirogyra* and the protozoan *Paramecium.*

Control group. In a biological experiment, the organisms maintained under "normal" conditions to serve as a basis for comparison with the experimental group of organisms, in which some variant condition has been introduced.

Corpora cavernosa. Two distinct tracts of spongy erectile tissue, situated longitudinally along the length of the human penis surrounding the single, central area of erectile tissue, the corpus spongiosum.

Corpus luteum. The yellow, glandular structure in mammals that develops from an ovarian follicle after the egg has been discharged (ovulated). The corpus luteum is the site of progesterone production and secretion. If the egg is fertilized, the corpus luteum persists throughout pregnancy; if the egg is not fertilized, the corpus luteum deteriorates beginning about the third week of the menstrual cycle.

Corpus spongiosum. A central tubular mass of erectile tissue in the human penis surrounding the urethra and situated within the outer two masses of erectile tissue, the corpora cavernosa.

Cortex. In plants, the storage tissue of the root or stem. In animals, the outer area of an organ, such as the kidney or brain.

Cowper's glands. A pair of glands that lie at the base of the erectile tissue of the penis and contribute the final liquid component, a mucous alkaline secretion, to the seminal fluid.

Cristae. Projections into the central matrix of a mitochondrion, produced by the repeated invagination of the inner mitochondrial membrane and serving to increase the membrane surface area within the mitochondrion.

Crossing over. Exchange of chromosome segments between maternal and paternal chromatids during tetrad formation.

Crucial experiment (critical experiment). An experiment that distinguishes between two hypotheses by contradicting the prediction from one hypothesis while supporting the prediction from the other. It is possible to have a crucial experiment only if the two hypotheses are mutually exclusive.

Cutin. The waxy secretion from leaf epidermal layers that forms the leaf cuticle.

Cytochrome. A molecule in the respiratory assembly with the characteristic porphyrin ring structure found also in hemoglobin, myglobin, and chlorophyll. By contrast with hemoglobin, in cytochrome the central iron atom is easily oxidized and reduced. This allows the cytochrome to pass electrons in the electron transport chain.

Cytokinesis. Cytological changes, usually occurring along with mitosis, through which the cytoplasm of one cell is divided to form two cells.

Cytology. The study of cells.

Cytoplasm. All the liquid colloidal material in the cell that is enclosed within the plasma membrane, excluding that of the nucleus. Cell organelles reside in the cytoplasm of the cell.

Deamination. The removal of the amino group ($—NH_2$) from an amino acid by chemical oxidation.

De-differentiation. The reversion of a cell from a condition of specialization to a nonspecialized, embryonic type of state, as is often the case in cancer cells.

Deductive logic. A process whereby a conclusion is reached by proceeding from a generalization to specific instances.

Degradation. The process of breaking down complex molecules to simpler ones, generally accompanied by a liberation of energy.

Dehydration synthesis. The joining together of small units (such as amino acids or glucose) into a single, large molecule by the elimination of water. One of the units contributes the H^+, the other the OH^-.

Deletion. The loss of a segment of a chromosome during crossing over, resulting in a certain phenotypic deficiency in the developing organism. Such aberrations can provide significant clues in the mapping of gene loci.

Density-dependent factors. Factors operating on individuals to limit or reduce a population when it reaches some critical size. Density-dependent factors include physiological or behavioral changes in individuals which cause migration, reduced mating practices, etc.

Deoxyribose. The five-carbon sugar in DNA (deoxyribonucleic acid).

Desiccate. To dry out by losing water (desiccation).

Dialysis. The process whereby compounds or substances in a heterogeneous solution are separated by the difference in their rates of diffusion through a semipermeable membrane. For example, if a solution of sodium chloride and albumin is placed inside a dialysis bag immersed in water, the sodium chloride ions will diffuse outward about twenty times faster than the albumin. Thus, after a short time, the solution inside the dialysis bag will be mostly protein.

Diaphragm. A contraceptive device used by the female for birth control. The diaphragm is a rubber-covered ring designed to fit over the cervix, preventing sperm from entering the uterus.

Diatom. A form of marine alga living within a tiny, silicon-containing shell. Diatoms produce huge amounts of organic materials by carrying on photosynthesis with the brown pigment fucoxanthin.

2, 4-D (2, 4-dichlorophenoxyacetic acid). One of a

Dictyosome. A stack of membranous vesicles similar to the Golgi complex and found in certain animal and higher plant cells. The function of dictyosomes is uncertain, but at least in plants it appears to be related to synthesis of polysaccharides.

Differential fertility. A measure of the success of an inherited variation in terms of its effects on the reproductive capacities of an organism. A variation that increases reproductive capacity is considered a successful one.

Differentiation. The structural or functional changes that occur in cells during the embryonic development of an organism.

Digestion. The enzymatic breakdown of food from large molecules into small ones capable of entering the bloodstream, and eventually the cells, by a process of absorption. Important organs of the digestive system include the stomach, pancreas, gall bladder, and small intestine.

Diploid. Term applied to a cell that contains a pair of each type of chromosome. The number of chromosomes usually given for an organism is the diploid number; hence the human being has 46 chromosomes (23 pairs).

Divergent speciation. Evolutionary process in which a single ancestral species gives rise to two or more descendant species.

Dominance hierarchies. "Peck orders" in animals, in which social hierarchies serve to maintain order. A baboon troop manifests a peck order.

Dominant. In genetics, a term used to refer to a gene that always expresses itself over its recessive allele in the heterozygous condition.

Dynamic equilibrium. A state in which the concentration of reactants and products in a chemical reaction remains constant, though not necessarily in equal quantities over time.

Ecology. The study of the relationship between plants and animals, and their environment.

Ecosystem. All the interacting factors, both physical and biological, forming a biotic community.

Ectoderm. In an embryo, the outer germ layer giving rise to the epidermis, the neural tube, and the epithelial lining of vertebrates.

Ejaculation. During the male orgasm, the forcible ejection of semen from the genital tract.

Electrophoresis. In biochemistry, a process used to separate different kinds of organic molecules from each other in a mixture. Electrophoresis takes advantage of the differences in overall net electric charge on different kinds of molecules. When a mixture of molecules is placed on a moist surface (such as a gel) through which an electric current passes, molecules with an overall net positive charge move toward the negative pole (cathode) and those with a net negative charge move toward the positive pole (anode). Rates of migration also vary: molecules with more negative charges migrate toward the positive pole more rapidly than those with less negative charges, etc. In this way different types of molecules tend to separate out together at different regions along the gel or other surface.

Embryology. The study of the structural and functional development of an organism during its early life.

Embryonic induction. The ability of one type of embryonic germ layer to trigger or specifically influence the differentiation of another germ layer (which usually lies in direct contact with it).

Embryonic region. The growth region in roots, located just behind the root cap, where new cells are produced by rapid cell division.

Endergonic. Term applied to those chemical reactions that result in an overall increase in energy among the formed products and hence the storage of energy. Photosynthesis is an endergonic reaction.

Endocrine system. All the hormone-secreting glands in the body.

Endoderm. In an animal embryo, the innermost germ layer, which gives rise to the lining of the gut.

Endoplasmic reticulum. In cells, a maze of membranes in the cytoplasm, at places continuous with the nuclear envelope. The endoplasmic reticulum may serve to increase the surface area of the cell and thus aid in exchange of material.

Energy barrier. The amount of energy that any non-excited atom or molecule must gain in order to become "excited" and enter into a chemical reaction.

Environmental resistance. Factors in the environment that oppose or limit the increase in numbers of a given population.

Enzyme. A protein the synthesis of which is controlled and directed by a specific gene. Enzymes act as catalysts, directing all major chemical reactions in the living organism.

Epididymis. A single, complexly coiled tube lying on top of the mammalian testes in which sperm is stored.

Epigenesis. In embryology, the idea that an entire organism develops from an originally undifferentiated mass of living material.

Equilibrium phase. The stage where growth rate in a population of cells or organisms has leveled off, so that the appearance of new cells or organisms just equals the disappearance of old ones.

Erythrocyte. Red blood cell. Erythrocytes contain hemoglobin (hence their red color) and serve as oxygen carriers.

Ester bond. The anhydrous bonds in lipids formed by removing a hydroxide from the carboxyl group of a fatty acid and a hydrogen from an alcohol group of glycerol.

Estrous cycle. The recurrent, restricted periods of sexual receptivity in the nonhuman mammalian female, marked not only by egg production but also by increased sexual drive.

Ethology. The study of animal behavior.

Eucaryotes. Cells characterized by true nuclei bounded by a nuclear membrane. The cells of all protozoa and higher animals, most algae (except blue-greens) and higher plants are eucaryotes (they are eucaryotic).

Eugenics. The attempt to improve the human genetic stock by encouraging breeding of those presumed to have "desirable" genes (positive eugenics) and discouraging breeding of those presumed to have "undesirable" genes (negative eugenics).

Eutrophication. The successional process leading from an aquatic system with low productivity through increasingly greater productivity to the ultimate development of a terrestrial system on the same site.

Excretion. The removal of the waste products of metabolic activity. In higher organisms, the blood bathes each cell and carries away waste. The waste material is removed from the blood by the kidneys, the sweat glands, and the lungs.

Exergonic. Term applied to those chemical reactions in which the end products have less energy than the reactants. Exergonic reactions give off energy. Respiration is an exergonic reaction.

Experimental group. In an experiment, the objects or organisms whose environment is altered so that their resulting responses may be studied.

Exponent. An integer written slightly above and to the right of a number to indicate how many times the number is to be multiplied by itself in the given expression; for example, $x^2 = x \cdot x$, $4^3 = 4 \cdot 4 \cdot 4$.

Extinct. No longer present in the world population of organisms.

Extrapolation. The calculation of a value or prediction of an event beyond a given series of values or events, based on observing the trend up to a certain point (for example, extrapolating tomorrow's weather based on the trend of the weather over the past 5 days).

Facilitated transport. The movement of substances across a cell membrane from an area of higher concentration to one of lower concentration more rapidly than would occur by simple diffusion, but without the expenditure of energy.

Fallopian tubes. In humans and other mammals, the name for the oviduct, a connecting pasageway by which the ova (eggs) from the ovary are carried to the uterus. In human beings, fertilization usually takes place in the Fallopian tubes.

Fats. Lipid compounds composed of glycerol and fatty acids. Fats are energy-rich compounds, often stored in adipose (fat) tissue.

Fatty acid. Organic molecule composed of a long hydrocarbon chain and terminal acid (carboxyl) group.

Feedback mechanism. A self-regulating mechanism within all homeostatic systems. Part of the output of the system is cycled back into the system itself in order to regulate further function and output, as, for example, in a thermostat–furnace system.

Ferredoxin. An iron-rich protein found in photosynthetic organisms. During the light reaction, this compound is capable of accepting free electrons and passing them through a reducing system to generate reduced NADP (that is, NADPH) and ATP.

Ferredoxin-reducing substance (FRS). Hypothesized primary electron acceptor after light absorption by chlorophyll. FRS passes electrons from chlorophyll to ferredoxin.

Fertilization. In sexual reproduction, the union of the male (sperm) and female (ovum) gametes to form a diploid cell (the zygote) capable of developing into a new organism.

Fertilization membrane. A protective membrane that surrounds the egg once it has been fertilized by a single sperm and prevents multiple fertilizations.

Fertilizin. A carbohydrate and protein substance produced by the egg, capable of causing the agglutination of sperm or the binding of sperm to egg.

Fibrin. A long protein polymer composing the fibrous part of blood clots. Fibrin is produced from the soluble blood protein fibrinogen by the action of the enzyme thrombin.

First law of thermodynamics. A physical principle which states that matter is neither lost nor gained during ordinary chemical reactions. This law may more accurately be stated as the Law of conservation of matter and energy: the sum total of matter and energy in the universe remains constant.

Fission. A rapid and efficient method of reproduction, found in many microorganisms, which involves the splitting by mitosis and cell division of one cell into two, each of which is genetically identical to the other.

Flagellum (plural, flagella). A long, whiplike extension of cytoplasm from unicellular organisms such as *Chlamydomonas*, or most animal sperm. An outer membrane encloses a highly structured matrix which surrounds microtubules arranged in the familiar 9 + 2 pattern found also in cilia (a 9 + 0 arrangement is found in centrioles). The function of flagella is movement of the cell.

Follicle-stimulating hormone (FSH). A gonadotropin, one of the hormones secreted by the anterior pituitary gland which in mammalian females stimulates growth of ovarian follicles into estrogen-secreting glands (*i.e.*, corpus luteum). In males FSH stimulates spermatogenesis.

Food chain. The transfer of energy from one organism to another, starting with the primary producers (photosynthetic organisms), or through primary consumers (herbivores), secondary consumers (carnivores), and decomposers.

Food web. A complex set of interactions within an ecosystem in which any given species usually serves as both prey and predator. All living forms in the ecosystem have a variety of functions. Thus, there is no single one-way street of food consumption hierarchies.

Foramen ovale. An opening in the fetal heart allowing blood to pass directly from the right to the left atrium, by-passing the pulmonary transit.

Foreskin. Also called the prepuce, the fold of skin covering the head of the penis in human males.

FRS. See Ferredoxin-reducing substance.

Furrowing. The infolding of the cell membrane by an animal cell during telophase.

Gamete. Male or female reproductive cell containing half the total number of chromosomes (that is, the haploid number) for any given species. These germ cells are formed by the process of meiosis (reduction division) from diploid cells.

Gametophyte. The small, photosynthetic, haploid stage in the life cycle of lower plants such as mosses and ferns. The gametophyte contains male and female gamete-producing organs.

Gastrula. The embryonic stage produced by gastrulation. A hollow structure generated by infolding of the blastula and consisting of two germ layers.

Gastrulation. The process of embryonic development, produced by infolding of the blastula to form the next embryonic stage, the gastrula.

Gause's hypothesis. The principle of competitive exclusion, which states that two species with similar ecological requirements cannot successfully live together for any length of time because of their competition for all the basic requirements for life.

Gene. A part of the hereditary material located on a chromosome. The term "gene" was first used by Johannsen to mean something in a gamete which determined some characteristic of an adult. The gene concept has been refined to the idea of a single gene as the source of information for the synthesis of a single polypeptide.

Gene pool. The total genetic makeup of a particle population, consisting of all the alleles existing in that population at any given time, regardless of their proportions. This concept provides a way of looking at the possible genetic changes that may occur in a freely-breeding population from one generation to another. If 100 organisms (composing a hypothetical population) have 10 genes each (with 2 alleles for each gene), then the gene pool of that population consists of 2000 genes.

Genetic code. The linear sequence of bases along a DNA molecule which in turn determines the sequence of amino acids in a polypeptide chain. The code itself consists of triplet groups, each specific three bases coding one amino acid. It has been found that more than one triplet can code for the same amino acid; thus the term "degenerate" is used in reference to the genetic code.

Genetic drift. A condition in which one allele becomes fixed in a population. Genetic drift is a major factor only in very small populations of organisms.

Genetic equilibrium. The maintenance of a more or less constant ratio between the different alleles in a gene pool from generation to generation (often stated as the Hardy–Weinberg Law).

Genotype. The genetic makeup of an organism; what alleles it actually contains and can pass on to its offspring.

Geographic isolation. The physical division of an original population into geographically separate groups. Such isolation is usually followed by divergence and eventually speciation.

Germ plasm. The term developed by August Weismann to describe the reproductive cells (of testes and ovaries) which directly produce the gametes for the next generation. Weismann visualized the germ plasm as being immortal, having continuity with the germ plasms of the preceding and succeeding generations through the process of sexual reproduction. Changes in somatoplasm (body cells) do not affect the germ plasm.

Germinal ridges. Precursors of the primordial gonads in human and most mammalian embryos. Germinal ridges develop as thickenings of the mesodermal epithelium lining the primitive coelom.

Gestation period. The period required for a mammalian embryo to develop from fertilization until birth.

Glans penis. The central mass of erectile tissue at the tip of the penis in human males.

Glycerol (glycerine). A compound with the formula:

$$\begin{array}{c} CH_2OH \\ | \\ CHOH \\ | \\ CH_2OH \end{array}$$

which combines with fatty acids to form fats.

Glycolysis. The initial stage of respiration, involving the breakdown of glucose to pyruvic acid. In aerobic respiration, glycolysis yields pyruvic acid for the citric acid cycle. In anaerobic respiration the pyruvic acid is converted to lactic acid (as in bacteria and the muscle cells of higher organisms), or into ethyl alcohol (as in yeasts).

Glycosidic bond. The anhydrous bonds of carbohydrates formed by removing a hydrogen from an alcohol group of one sugar and a hydroxide from an aldahyde group of the other. Glycosidic bonds hold together the subunits (such as glucose) making up long-chain polysaccharides.

Golgi complex (Golgi body). A cluster of flattened, parallel, smooth-surfaced membranous sacs found within the cytoplasm. The Golgi complex appears to function in isolating, packaging, and transporting molecules out of the cell.

Gonadotropins. General name for the group of hormones produced and secreted either by the pituitary gland of both sexes or by the placenta during pregnancy in the mamalian female. Gonadotropins act as a stimulator of the reproductive organs; their increased production by the pituitary at puberty is responsible, in part, for the growth of the ovaries or testes.

Graffian follicles. Vesicles in the ovary in which the (eggs) are produced. One egg is produced per follicle. After ovulation, a follicle becomes transformed into the corpus luteum, producing progesterone which builds up the uterine lining.

Grana (singular, granum). Dense stacks of membranes that are part of the lamellar system within chloroplasts. Grana were the first structures discovered within the chloroplast by means of an electron microscope. Chlorophyll molecules are attached to the lamellae and the grana.

Gray crescent. A surface region in the newly fertilized egg of frogs and salamanders from which the orientation of the developing embryo can be ascertained.

Growth. Enlargement in size of an organism, due to an increase in the number of cells, enlargement of cells already present, or both.

Habitat. The surroundings in which an organism resides—the organism's "address" in the biological community.

Half-life. The amount of time required for half the atoms of a given radioactive sample to decay to a more stable form.

Haploid. A condition in which an organism (or a single cell) bears only one copy of each gene. Most higher organisms have two copies (alleles) of the gene for any given character and thus are called diploid. Many microorganisms, however, such as bacteria, *Paramecium,* and most algae, are haploid during most of their life span.

Hardy–Weinberg law. The generalization that the frequencies of both genes and genotypes will remain constant from generation to generation in a large, freely-breeding population in which there is no selection, migration, or mutation. The Hardy–Weinberg law is stated mathematically as the expansion of the binomial, where one member of a pair of alleles is designated p, and the other q: $p^2 + 2pq + q^2 = 1.0$. This mathematical expression means that the frequencies of the alleles symbolized as p and q remain constant (at whatever their initial values) from generation to generation as long as the conditions mentioned above prevail. The Hardy–Weinberg law is sometimes referred to as the law of genetic equilibrium.

Heme. A complex molecular structure (a porphyrin ring) in which a central ion is capable of undergoing repeated oxidation and reduction. The heme structure is the basis of such important biological molecules as hemoglobin, myoglobin, and the cytochromes.

Hemoglobin. A red, iron-containing protein pigment in erythrocytes that transports oxygen and carbon dioxide.

Herbivore. An organism whose diet consists exclusively of vegetation.

Hermaphroditism. A state characterized by the presence of both male and female sex organs in the same organism. An individual organism with both male and female sex organs is called a hermaphrodite.

Heterogametes. Two gametes, structurally dissimilar, capable of fusion to form a zygote. The sperm and egg are examples of heterogametes.

Heterogamy. The condition in which gametes are differentiated into two distinct forms (generally male and female).

Heterosexuality. Sexual attraction and/or intercourse between members of the opposite sexes.

Heterotroph. An organism that depends on its environment for a supply of nutritive material to build up its own organic constituents and also for general energy requirements.

Heterozygous. Term for the condition in which the two members of a pair of genes located on homologous chromosomes and influencing a given characistic are different.

Histogram. A graphical representation of statistical data showing frequency distribution by means of a series of rectangles. A bar graph is an example of a histogram.

Histology. The study of tissues.

Homeostasis. The dynamic equilibrium processes which maintain a relatively constant internal environment in the face of variations in the external environment.

Homologous. In anatomy, the term applied to body parts that are similar in structure but not necessarily in function, such as the arm of a human being and the front leg of a horse.

Homologous chromosomes. The pair of structurally similar chromosomes within a diploid cell which carry inheritance factors influencing the same traits.

Homosexuality. Sexual attraction and/or intercourse between members of the same sex. Female homosexuality is sometimes referred to as lesbianism.

Homozygous. Term for the condition in which the two members of a pair of genes located on homologous chromosomes and influencing a given characteristic are identical.

Hormone. A chemical substance produced in small quantities in one part of an organism and profoundly affecting another part of that organism. Chemically, hormones may be proteins (insulin), steroids (estrogens), or small metabolites (thyroxin).

Hunter's syndrome. An inherited disorder of mucopolysaccharide metabolism (less serious than Hurler's syndrome) which allows the accumulation of certain mucopolysaccharides in connective tissue. No corneal clouding is seen, and very few examples of dwarfism. The gene for Hunter's syndrome is an *X*-linked recessive.

Hurler's syndrome. An inherited disorder of mucopolysaccharide metabolism in connective tissue, resulting in a serious malformation of the skeleton, including dwarfism. Deafness and clouding of the cornea also occur. Results from lack of production of the enzyme glucose-6-phosphate dehydrogenase, thus allowing a build-up of mucopolysaccharides. The gene for Hurler's syndrome is an autosomal recessive.

Hydrolysis. The chemical breakdown of a larger molecule into smaller units by the addition of water.

Hydrostatic pressure. In the circulatory system of animals, pressure created by the pumping action of the heart. Hydrostatic pressure aids in forcing water out of the capillaries and into interstitial spaces.

Hymen. A thin membrane that often partially encloses the vaginal opening. The hymen has no known function.

Hypothesis. A tentative explanation suggested to account for observed phenomena.

Immunology. The field of biology dealing with the process whereby organisms develop a chemical resistance within their bodies to various types of foreign substances (bacteria, viruses, pollen, etc.).

Imprinting. The rapid fixing of social preferences; for example, the tendency of a duckling to follow the first moving object it sees after hatching.

Inclusion. Any of a number of nonliving structures occurring within the cytoplasm of a cell.

Inductive logic. A process whereby a conclusion is reached by proceeding from specific cases to a generalization.

Induction, embryonic. See embryonic induction.

Innate behavior. Ethological term that replaces the concept of instinct. In general, innate behavior is that which is presumably inherited to a large degree.

Innate releasing mechanism (IRM). Hypothesized receiving center in the central nervous system which is triggered by some external stimulus and elicits a specific sequence of behavioral events.

Instinct. Term applied to behavior that is primarily "genetic" in nature and seems less amenable to change through learning. Most behaviorists prefer to use the term "innate behavior" instead of "instinct."

Internal environment. The conditions generated inside an organism by the functions and interactions of cells, tissues, organs, and systems.

Interphase. The longest individual phase of mitosis, often considered the resting phase between two active cell divisions, during which the genetic material of the cell is being duplicated.

Interpolation. Filling in a value on a graph between two given values.

Interstitial cells. In the mammalian testis, cells that produce and secrete the male sex hormones.

Intrauterine device (IUD). Birth control device made of soft, flexible plastic or metal such as copper. It is not known precisely how the IUD works, but it seems to interfere with the implantation of the young embryo in the uterine wall.

Inversion. The process of chromosome breakage and rejoining in such a way that a whole segment breaks from a chromosome and is replaced in reverse order. Thus, if the sequence of genes on a normal chromosome were represented as abcdefghi, and two breaks occurred, one between c and d and the other between g and h, the middle segment defg might turn and rejoin in reverse direction, so that the sequence of genes on the chromosome would read: abcgfedhi.

Invertebrates. Organisms characterized by the lack of a notochord. Invertebrates compose approximately nineteen animal phyla.

IRM. See Innate releasing mechanism.

Isogametes. Two gametes which are morphologically alike and capable of fusion to form a zygote. Isogametes are not visibly differentiated into male or female forms.

Isolating mechanisms. Differences introduced by speciation and divergence during geographic isolation of two populations which prevent interbreeding should the populations come back together. Such mechanisms include seasonal isolation, ecological isolation, physiological isolation, and behavioral isolation.

Karyotype. Characterization of the chromosome complement of an individual organism with regard to the number, size, and shape of chromosomes present. Karyotypes are usually displayed as a series of chromosomes lined up in a specific order, showing each of the chromosome pairs present (including extra or missing chromosomes).

Kilocalorie (Calorie). The amount of heat required to raise the temperature of one kilogram of water from 14.5 to 15.5 degrees Celsius. The kilocalorie is used to measure the metabolism (energy turnover) of animals.

Kinetic energy. Energy of motion; energy in the process of doing work.

Kinetochore (centromere). The region of the chromosome at which the chromatids remain connected following chromosome replication and to which the spindle microtubules appear to be connected during cell division.

Krebs cycle (citric acid cycle). A series of reactions in the oxidation of pyruvic acid in which large amounts of energy are released. The hydrogens from pyruvic acid supply electrons for the electron transport chain.

Labia. In human female genitalia, the thin, pink folds of epithelial tissue that lie lateral to the vaginal opening. Homologous to the male scrotum. There are two labial structures: labia minora and labia majora.

Lamarckism. An evolutionary theory proposed by Jean Baptiste Lamarck in the eighteenth century, consisting of several basic ideas: Because of new physical needs, new structures arise and old structures are modified; use or disuse of parts causes variation in a structure; and acquired characters can be transmitted to succeeding generations.

Lamellae. Membranous sacs stacked parallel to each other within the chloroplasts of green plant cells. There are two types of lamellae: stroma lamellae and grana lamellae. Together they are the site of the light reactions of photosynthesis.

Larva. An intermediate free-living form of some organisms. In this stage the organism eats extensively and stores food. It then undergoes metamorphosis into the adult form, using the stored food as an energy source.

Leukocyte (also **leucocyte**). A white blood cell.

Leukoplast (also **leucoplast**). A colorless plastid, thought to serve as a cytoplasmic center for the storage of certain materials, such as starch.

Linkage. The location of two or more genes on the same chromosome, so that the two characteristics are passed on together from parent to offspring.

Lipids. A group of organic compounds including the fats and fatlike compounds and the steroids.

Logarithm. The power to which a given number (the base) must be raised in order to equal a certain number. Ten is the most commonly used base number. The logarithm of 100 to the base 10 is 2. This can be written: $\log_{10} 100 = 2$, or $10^2 = 100$. The base number is given as a subscript following the term "log."

Logarithmic phase. The phase of most rapid increase in size for a population; that is, the area of exponential increase on a sigmoid curve.

Luteinizing hormone (LH). A hormone secreted in vertebrates by the anterior lobe of the pituitary gland of both sexes. In females LH stimulates formation of the corpus luteum; in males it regulates production of testosterone.

Lysins. A class of substances produced by sperm to dissolve the protective egg membranes and allow the sperm to enter and fertilize the egg.

Lysis. The chemical breakdown of a cell, usually under the influence of enzymes released by the rupture of a lysosome, or by reproduction of viruses within the cell.

Lysosome. A saclike structure containing enzymes that catalyze the breakdown of fats, proteins, and nucleic acids. The membranes of lysosomes protect the cell from being digested by its own enzymes (autolysis). Lysosomes also serve as defense mechanisms, ingesting and digesting foreign toxic agents within the cell.

Macromolecule. A large molecule bulit up from small repeating units. Cellulose is a macromolecule built of repeating β-glucose units.

Mapping. In genetics, the description of the physical order of gene loci on a given chromosome, as determined by frequency of chromosome cross-overs. The further apart two genes are on a chromosome, the greater the chance that crossing over will occur between them; thus the percentages of cross-overs can be used to calculate relative distances between genes. Chromosome maps established in this way can be correlated with actual chromosome structure by a variety of cytological techniques.

Mass spectrometry. A process that sorts streams of electrically charged particles according to their different masses by using deflecting magnetic fields. The device that accomplishes this, a mass spectrometer, generally consists of a long tube that generates a magnetic field. The particles of varying masses are passed through the tube and the degree of deflection by the magnetic field is recorded on a photographic film at the other end. Given the strength of the field and the amount of deflection, investigators can calculate the relative masses of the particles. Isotopes of various elements can be detected by mass spectrometry.

Matter. That which has weight and occupies space.

Mechanism. The philosophical view that life is explicable in terms of physical and chemical laws, and that the whole is equal to nothing more than the sum of its parts. Also called mechanistic materialism.

Meiosis. A process of cellular division which results in each of the daughter cells containing half as many chromosomes as the parent cell (that is, they are haploid). Meiosis occurs primarily in the formation of gametes (in sexually reproducing organisms) or in spore formation in organisms such as ferns or mosses.

Melanin. Any of a group of dark brown or black pigments occurring in the skin and other parts of the body.

Menopause. The period (usually between 40 and 50 years of age) in the reproductive life of human females when the recurring menstrual cycle including ovulation ceases. Changes in hormonal balance and control within the female body also take place.

Menstrual cycle. A cyclic event approximately 28 days in duration, in the human female. The function of the cycle is to prepare the endometrium (the lining of the uterus) for implantation of a young embryo should fertilizaton occur. If a fertilized egg is not received, much of the endometrium is sloughed off and the cycle starts again.

Mesenchyme. An embryonic tissue derived from the mesodermal layer of an animal embryo. Mesenchyme appears in the embryo as a mass of scattered angular or pointed cells with long, protruding processes. In vertebrates the mesenchyme forms the connective tissues (bone, cartilage, etc.).

Mesoderm. In vertebrate embryos, the third germ layer, lying between ectoderm and endoderm and giving rise to connective tissue, muscle, the urogenital system, the vascular system, and the lining of the coelom.

Messenger RNA (*m*RNA). A strand of RNA synthesized in the nucleus of a cell with one DNA strand as a template. Thus *m*RNA has a base sequence directly complementary to the base sequence on the DNA molecule. After its formation, *m*RNA migrates from the nucleus to the cytoplasm, where it becomes associated with the ribosomes.

Metabolism. The sum total of chemical and physical processes within the body related to the release of energy by the breakdown of chemical fuel and the use of that energy by the cells for their own work.

Metaphase. The phase of mitosis characterized by the lining up of the chromosome pairs along the equatorial plate of the cell.

Metaphysics. A method of thinking which tries to go beyond the present physical reality and postulate the unknown. In metaphysics, phenomena are regarded as final, immutable, and independent of one another. Metaphysical thinking is generally contrasted to scientific thinking.

Microfilament. A minute protein fiber in the cytoplasm of certain cells; its function appears to be providing support and shape to the cell.

Microtubules. Tiny intracellular tubes composing such diverse structures as the spindle apparatus and the centrioles.

Minimal medium. A medium containing only those elements absolutely essential for the growth of a particular microorganism, and which the organism cannot synthesize itself. Generally, minimal media contain a carbohydrate source, various inorganic salts, and sometimes a growth factor such as biotin.

Miscarriage. An accidental or spontaneous abortion. Miscarriages can occur anywhere from the trophoblast stage through the eighth or ninth month of pregnancy.

Mitochondria. Cytoplasmic organelles of a characteristic structure containing the enzymes for glycolysis and for the citric acid cycle.

Mitosis. The series of changes within a cell nucleus by which two genetically identical daughter nuclei are produced.

Mode. In a set of data, the item or group of items that appears most often. For example, when a set of grades for a whole class is compiled, the mode is the single grade, or group of grades, that the largest number of students obtained.

Mole. A gram molecular mass (often called gram molecular weight) of a substance; the molecular mass of a substance in grams. One mole contains Avogadro's number of particles (6.023×10^{23}).

Monera. The kingdom that includes all those organisms that lack nuclei in their cells; the monera are all procaryotes.

Monocytes. One of three types of leukocytes; large, highly mobile cells whose main function is phagocytosis.

Morphogenetic movement. Cell movements (migrations) that change the shape of differentiating tissues in an embryo.

Mosaic theory. An embryological idea most often ascribed to Wilhelm Roux, which holds that certain regions of the egg are designated to become specific parts of the organism.

Muellerian ducts. Tubules that develop in the mammalian embryo just lateral to the Wolffian ducts and are the precursors of the Fallopian tubes, uterus, and part of the vagina.

Multiple alleles. Sets of alleles that contain more than two contrasting members for a given locus.

Mutualism. See Symbiosis.

Mutation. An inherited structural or functional variation of an offspring in relation to its parents. Mutations are due to a change in the chemical structure of DNA, the molecule bearing hereditary information. Once a mutation has occurred, it is transmitted to future generations.

Negative acceleration phase. The period of decreasing growth rate in a population, following the exponential increase that occurs during the logarithmic phase.

Negative feedback. A mechanism of self-regulation whereby a change in a system in one direction is converted into a command for a change in the opposite direction. A means of helping maintain a biological system in dynamic equilibrium.

Neural crest. During mammalian embryogenesis, cells that become detached from the forming neural tube, move laterally, and then move dorsally between the closed tube and the overlying ectoderm. The spinal and cranial nerve ganglia are developed from neural crest cells.

Neural tube. In vertebrate embryos the tube formed along the dorsal surface, produced by the infolding of a large mass of ectodermal tissue. The neural tube develops into the entire central nervous system and parts of the peripheral system as well.

Niche. The ecological position of an organism—the organism's "occupation" within the biological community.

Nitrogen-fixing bacteria. Bacterial organisms capable of drawing nitrogen from the atmosphere and converting it into soluble nitrates. These nitrates can then be used by plants.

"NODOC". See Anticodon.

Nodules. Swellings on the roots of certain leguminous plants where nitrogen-fixing bacteria reside in a symbiotic relationship with the plant.

Nondisjunction. The failure of a pair of chromatids to separate at metaphase, creating an abnormality of chromosome number in both daughter cells. Nondisjunction usually leads to deformed offspring.

Normal distribution curve. A bell-shaped curve, more or less symmetrical, with the greatest value in the center (mean) and with values decreasing equally on both sides. Also called a normal curve.

Notochord. A rod-shaped body located dorsally and serving as an internal skeleton in the embryos of all chordates and in the adults of some; replaced by a vertebral column in vertebrates.

Nuclear membrane. The unit membrane that separates the nuclear material from the surrounding cytoplasm in a cell. The nuclear membrane is not continuous, but rather is broken at different intervals by nuclear pores that provide a physical passage between the nucleur and surrounding cytoplasm.

Nucleic acid. A polymer composed of ribose sugar rings and phosphate groups, with organic bases of thymine, guanine, cytosine, adenine, and/or uracil. Both DNA and RNA are nucleic acids.

Nucleoplasm. The living material within the nucleus.

Nucleotide. The molecule formed from the combination of a purine or pyrimidine, an appropriate sugar (ribose or deoxyribose), and a phosphate residue. Nucleotides are the basic units of nucleic acid structure.

Nucleus (atomic). The dense area in the central part of an atom where all protons and neutrons are located.

Nucleus (cellular). A body found in nearly all cells which contains the hereditary information of the cell and acts as the control center of cell function.

Oligotropic (lake). Lacking in biological productivity; not yielding much biomass.

Oögenesis. The process by which haploid female gametes (eggs) are produced.

Organ. A unit composed of various types of tissues grouped together to perform a necessary function. The liver and a plant leaf are examples of organs.

Organelle. A small body appearing within the cell cytoplasm, with a characteristic structure and a definite, though perhaps not always clearly defined, function. Cytoplasmic organelles include mitochondria, ribosomes, the Golgi complex, and the endoplasmic reticulum.

Orthogenesis. An erroneous conception, originating in the nineteenth century, that evolution progresses in a given, straight-line direction. The evolution of the antlers of the Irish elk is often cited as an example of orthogenesis.

Osmosis. The passage of a solvent from a region of greater concentration to a region of lesser concentration through a semipermeable membrane.

Ostium. The upper end of the Fallopian tubes where eggs released from the ovaries enter for transport to the uterus.

Ovary. In plants, the basal portion of the pistil which encloses the ovules. The ovules, in turn, contain the female sex cell. After fertilization, the ovary becomes the fleshy part of the fruit and the ovules become the seeds. In animals, the organ (usually paired) of the female which produces the ovum, or egg.

Oviparous. Term applied to organisms that lay eggs in which the embryo continues to develop for some period of time, deriving nourishment from the yolk.

Ovoviviparous. Term applied to organisms whose young develop within the body of the mother but derive most or all their nourishment from the egg yolk.

Ovulation. The release of an unfertilized egg from the ovary. In human beings ovulation involves the discharge of a mature ovum from a Graafian follicle of the ovary.

Ovum. The female reproductive cell containing the haploid number of chromosomes, derived by meiosis from a diploid germ cell. The ovum is fertilized by the sperm, producing the zygote.

Oxidation. A type of chemical reaction involving the loss of electrons. Frequently, but by no means necessarily, the element oxygen is involved.

Oxidative phosphorlyation. The process whereby electrons removed from substrate molecules are passed through the electron transport chain (cytochromes), in such a way that their potential energy is coupled to the formation of ATP (from ADP and inorganic phosphate).

Oxygen debt. The amount of oxygen required to oxidize the excess lactic acid accumulated in muscle cells during strenuous exercise.

Oxytocin. A hormone produced in the hypothalamus and secreted by the pituitary gland, regulating uterine contractions.

Parasite. Organism that derives its food from another species of organism by living in or on the host organism, usually to the detriment of the host.

Parasitism. A type of heterotrophic nutrition found among both plants and animals. An organism engaging parasitism lives in or on the body of a living plant or animal (host) and attains nourishment from it.

Parthenogenesis. The development of an egg, without fertilization, into a new individual. Parthenogenesis occurs naturally in some organisms (such as aphids, rotifers, bees, and ants) but can be induced artificially in higher forms (such as frogs) by chemical or physical stimulus to the egg.

Partial pressure. The pressure exerted by a given component gas in a mixture of gases.

$$\text{Partial pressure of gas A} = \frac{\text{Volume of gas A}}{\text{Total volume}} \times \text{Total pressure of entire sample}$$

"Peck order". The establishment of social hierarchies, first noted in domestic chickens.

Penis. The male sex organ through which sperm are ejaculated.

Peptide bond. The bond, formed by dehydration synthesis (elimination of water), which links together two amino acids.

Peroxisome. A cytoplasmic cell organelle containing enzymes for the production and decomposition of hydrogen peroxide.

pH. Symbol for the logarithmic scale running from 0 to 14, representing the concentration of hydrogen ions or protons (actually hydronium ions) per liter of solution. On the pH scale, 7 represents neutrality, the lower numbers acidity (acids), and the higher numbers alkalinity (bases).

Phagocyte. A cell in the body capable of engulfing particles from the surrounding medium into its own cytoplasm for enzymatic breakdown (phagocytosis). Phagocytes are found in large numbers lining the walls of lymph node sinuses; they destroy bacteria that have entered the body and been picked up by the lymphatic system. Leukocytes and macrophages are examples of phagocytes in man.

Phagocytosis. The engulfing of microorganisms, other cells, or foreign particles by a cell. For example, phagocytosis occurs when an amoeba engulfs its prey or when a white blood cell engulfs a bacterium.

Phenotype. In genetics, the outward appearance of an organism, as contrasted with its genetic makeup (genotype).

Phlogiston theory. A widely accepted seventeenth-century theory of combustion, which held that the substance phlogiston was contained in all combustible bodies and was released from these bodies upon their burning (producing "phlogisticated" air). This theory remained popular until the later eighteenth century, when experimental work by Black, Priestley, Lavoisier, and others led to the oxygen theory of combustion.

Phosphoglyceric acid (PGA). An intermediate product in carbohydrate metabolism, composed of a 3-carbon backbone with a phosphate group attached to one (usually position 3) of the carbons:

$$
\begin{array}{c}
\text{O} \\
\parallel \\
\text{C} - \text{OH} \\
| \\
\text{H} - \text{C} - \text{OH} \\
| \\
\text{H} - \text{C} - \text{O} - \text{PO}_3 \\
| \\
\text{H}
\end{array}
$$

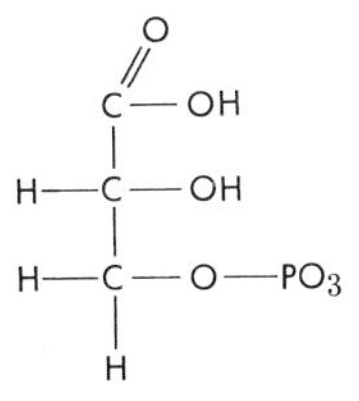

Phospholipids. An important structural part of cellular membranes, the phospholipids contain phosphorous, fatty acids, glycerin, and a nitrogenous base.

Phosphorylation. The addition of a phosphate group (such as H_2PO_3) to a compound, as in oxidative phosphorylation of ADP (producing ATP) during respiration.

Photophosphorylation. The process whereby phosphorylation is coupled to the transport of electrons that have been moved to higher energy levels by the absorption of light energy (that is, ATP production during the light reaction of photosynthesis).

Phyletic speciation. The process of evolutionary change in which one population gives rise, over time, to another single population genetically different from the original.

Phylogenetic chart. A diagram showing the evolutionary relationships among a group of species, or within a single species. The so-called "family tree" is an example of a phylogenetic chart.

Phylogeny. The study of the evolutionary history of species.

Physical environment. All the elements surrounding an organism, excluding other living organisms.

Pinocytosis. The process by which materials can be taken into the interior of a cell without passing through the plasma membrane.

Placenta. A structure created by the fusion of the chorion from the young embryo with the wall of the uterus. Respiratory, excretory, and nutritional functions of the fetus are carried on by exchanges across this structure. The placenta also secretes hormones regulating certain aspects of fetal development. Presence of a placenta is characteristic of all mammals except the marsupials.

Plasma. A protein-containing fluid, the liquid portion of blood.

Plasma membrane. A lipoprotein of a definite structure which surrounds and contains the living matter within a cell. The membrane has three layers: two outer protein surfaces surrounding an inner core of lipid. The polar part of each lipid molecule is associated with the protein on the surface, while the nonpolar portion points into the middle of the "sandwich." Average plasma membrane diameter is 75 Å.

Plasmagene. A term often applied to a gene of cytoplasmic rather than nuclear origin.

Plastids. Small bodies occurring in the cytoplasmic portion of plant cells. Plastids are classified according to color.

Polar body. The small daughter cells produced during meiotic divisions of the oöcyte. From the first meiotic division one polar body is produced; from the second meiotic division three polar bodies result. A single primary oöcyte thus gives rise to one large, mature ovum and three small polar bodies. The polar bodies contain the extra sets of chromosomes produced by meiosis. They contain very little cytoplasm, however, since most of this is reserved for inclusion in the single ovum.

Polarization. Term applied to an unequal distribution (separation) of charged ions, producing an electric potential. Cell membranes are said to be in a state of polarization when they have a greater concentration of positive ions on the outside than inside, or *vice versa*.

Polygenic character. A quantitatively variable phenotypic trait determined by interaction of numerous genes.

Polymorphism. The existence within a single species of members showing many different, but distinct and recurring forms; for example, the drone, queen, and workers occurring in the honeybee.

Polyribosome (polysome). A cluster of connected ribosomes, usually arranged along a strand of *m*RNA.

Polysaccharides. Complex carbohydrate molecules built up from simpler sugar units (such as glucose) into long-chain polymers. Polysaccharides are the major constituents of the cell walls and capsules of various microorganisms.

Polysome. See Polyribosome.

Polyspermy. The fertilization of a single egg by more than one sperm at a time.

Population genetics. The application of genetic principles to a large number of breeding organisms.

Positive acceleration phase. The first section of a sigmoid growth curve for a population, where the system described is just beginning to increase.

Positive feedback. Biologically speaking, an abnormal state in which a change in a system in one direction serves as a command for continued change in that same direction. This can create a severe physiological imbalance leading to the death of an organism.

Potential energy. Energy capable of doing work.

Precipitate. An insoluble product of a chemical reaction in a solution.

Predation. The process by which one species uses another species for food.

Preformation. The idea that an already formed, miniature individual exists within the egg and merely increases in size during embryological development.

Primary germ layers. The first distinguishable areas within the developing embryo, namely the ectoderm, mesoderm, and endoderm. These areas give rise to the tissues and organs of the mature organism.

Primary spermatocyte. The cells in the testes that derive from spermatagonia and ultimately undergo meiosis to produce haploid sperm.

Primitive streak. A longitudinal groove that develops on the surface of the embryo of fishes, reptiles, birds, and mammals. Formation of the streak is a consequence of the movement of cells and the formation of mesoderm. The primitive streak marks the future longitudinal axis of the embryo.

Procaryotes. Cells that have no true nucleus and no nuclear membrane. Procaryotes lack membrane-bound subcellular organelles such as mitochondria and chloroplasts. Bacteria and blue/green algae are example of procaryotes.

Progesterone. A hormone produced and secreted by the corpus luteum of the ovary. The corpus luteum promotes continuous development of the uterine lining and development of mammary glands if pregnancy occurs. If pregnancy occurs it also prevents maturation of new follicles and eggs.

Prophase. The first visible phase of mitosis, marked by the condensation of the chromosomal material into chromosomes and the disappearance of the nuclear membrane.

Prosecretin. The inactive form of the hormone secretin, which can be converted into the active form by the action of dilute HCl.

Prostate gland. A glandular and muscular organ surrounding the urethra just below the bladder in the male mammal. The prostate secretes a major portion of the seminal fluid.

Protective coloration. Patterns of surface pigmentation that blend with the environment, allowing an organism to remain unobserved and therefore camouflaged from predators.

Protein. A complex organic molecule composed of amino acids joined in specific sequence by peptide bonds. Proteins serve both structural and enzymatic functions.

Protista. One of the five kingdoms in the system outlined by Whittaker. Protista includes unicellular and some colonial organisms.

Pseudopod. An extension of the streaming cytoplasm of amoebae or other one-celled organisms which gives them their irregular shape. In the amoeba, pseudopods function as a means of locomotion and as the tool for the intake of food from the environment.

Purine. One of a class of nitrogen-containing compounds including the bases, adenine and guanine, important components of the nucleic acids DNA and RNA. The two-ringed purine structure is:

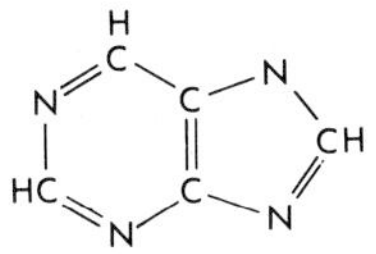

Pyrimidine. One of a class of nitrogen-contaning compounds including the bases cytosine, thymine, and uracil are found in RNA. The single-ring pyrimidine structure is:

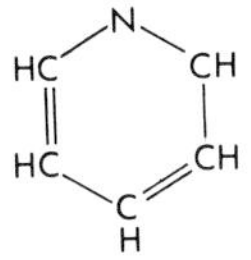

Pyruvic acid. The final product of glycolysis, with the formula:

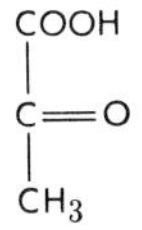

Quantasome. One of the individual membranous structures arranged in columns within a granum. Molecules of chlorophyll are aligned on the quantasomes.

Quantitative. That which can be measured or described in some definite and precise (generally numerical) form.

Quantum. A tiny energy packet in which light travels.

Quantum theory. The model which holds that light is composed of tiny energy packets (quanta, or photons) which are given off by any light emitter and travel intact through space.

Radioactive decay. The decrease in mass of certain unstable elements by emission of elementary particles, continuing until a stable isotopic form has been reached.

Range. The distribution of a group of numerical values describing the spread between the lowest and highest of the included values.

Recessive. In genetics, a term referring to the relative lack of phenotypic effect of a gene in the presence of its dominant allele. Thus the gene for blue eyes is said to be recessive to that for brown in the human population, because when one allele for blue eyes and one for brown eyes are present in the individual, the blue condition is masked and the individual has brown eyes.

Recombination. In genetics, the formation of new genotypes (a combination of genes not present in either of the parents) in offspring due to independent assortment of genes and chromosomes during gamete formation.

Reduction. A chemical reaction involving a gain of electrons.

Reduction division. A cell division during which the number of chromosomes in each daughter cell is reduced to one-half that found in the parent cell. This is accomplished by nuclear division without previous chromosome duplication.

Regeneration. The developmental process by which a lost part of an organism (such as an organ, part of an organ or tissue, the skin, etc.) is replaced after damage. Repair of wounds and regrowth of an amputated salamander's arm are examples of regeneration.

"Releasers". Environmental stimuli that trigger an innate releasing mechanism.

Respiratory assembly. A series of complex molecules (including cytochromes), found on the inner membranes of mitochondria, and capable of oxidation and reduction. Such assemblies accept electrons from reduced acceptors during the citric acid cycle and pass them along to the final acceptor, oxygen. The enzymes for oxidative phosphorylation are also components of the respiratory assembly.

Response. A change in behavior on the part of an organism (or tissue) as a result of some chemical or physical change in the environment.

Reversible reaction. A reaction system where reactants and products are interconvertible. If left to themselves, reversible chemical reactions reach an equilibrium point where just as much reactant is being converted into product as product into reactant at any time. Nearly all chemical reactions are to some degree reversible.

Rh factor. An antigen found on the red blood cells of certain human beings (designated Rh+). The Rh factor acts as an antigen, so that individuals with this factor cannot donate blood to individuals who normally lack it (Rh− individuals). The factor derives its name from having been first discovered in the rhesus monkey.

Rhythm method. A type of birth control relying upon abstinence from sex during the time of suspected ovulation and fertile period of the female.

Ribonucletic acid (RNA). A complex, single-stranded molecule consisting of repeating nucleotide bases:

adenine, guanine, cytosine, and uracil. At least three types of RNA are known, all of which are involved in transcribing the genetic code into protein. Messenger RNA (*m*RNA) carries the genetic code for amino acid sequence from the DNA to the ribosomes; soluble or transfer RNA (*t*RNA), of which there is a specific type for each amino acid, carries each amino acid to the ribosome where it is incorporated into protein in a specific place; ribosomal RNA (*r*RNA) is found only in the ribosomes, and its function is not yet known.

Ribosomal RNA (*r*RNA). A type of ribonucleic acid found as part of ribosomes. The function of *r*RNA is not well understood, though it appears to have some role in protein synthesis.

Ribosomes. Small particles found either free in the cytoplasm or attached to the outer surface of the endoplasmic reticulum in the cells of all eucaryotes and many procaryotes. Ribosomes contain high concentrations of ribonucleic acid (RNA) and are centers of protein synthesis.

Roentgen. A standardized unit measuring the amount of energy delivered in X-ray (or gamma ray) beams. One roentgen is the amount of radiation which under ideal conditions (0°C and 760 Hg), liberates 2.083×10^9 ion pairs per cubic centimeter of air.

Rudimentary phallus. Genital folds in both male and female embryos. In males, the rudimentary phallus develops into the penis. In females, the phallus grows only slightly and develops into the clitoris.

Scrotum. A pouch at the base of the penis containing the testes and parts of their spermatic cords.

Secondary spermatocyte. The daughter cell(s) produced by the first meiotic division of the primary spermatocytes during spermatogenesis.

Secretin. A hormone secreted by the cells of the duodenum under the stimulus of hydrochloric acid from the stomach. Secretin in turn causes the pancreas to secrete certain digestive enzymes into the duodenum.

Segregation (of alleles). The separation during gametogenesis of paired factors influencing a single condition.

Semen. A secretion of the male reproductive organs, composed of the spermatozoa and liquid secretions of various other glands.

Seminal vesicles. The portion of the male reproductive duct in which sperm are stored prior to copulation.

Seminiferous tubules. Tubules within the testes where the male sperm are produced. Each testis contains about 1,000 highly coiled seminiferous tubules.

Semipermeable (differentially permeable). Term applied to a membrane that allows some substances to pass through while prohibiting the passage of others.

Sex chromosome. The chromosomes, commonly referred to as *x* and *y*, whose presence in certain combinations determines the sex of an organism.

Sickle-cell anemia. A hereditary disease caused by a mutant form of hemoglobin. Under low oxygen tension in the blood, red blood cells containing sickle-cell hemoglobin collapse, assuming a half-moon, or sickle shape. The disease is mild in the heterozygous form but greatly shortens lifespan in the homozygous dominant form.

Simple sugar. A molecule composed of a single five- or six-carbon sugar.

Smallpox. A severe, infectious, viral disease now controlled by a cowpox vaccine.

Soluble RNA. See Transfer RNA.

Somatoplasm. The term used by August Weismann for all cells of an organism except reproductive cells ("germ plasm"). In each generation, the somatoplasm is derived from the germ plasm of the preceeding generation, but is distinct from it in that changes in somatoplasm (body cells) will not be passed on to the next generation.

Somites. Paired, block-like masses of mesoderm arranged in a longitudinal series along the side of the neural tube of the embryo. Each somite will form one vertebra and its associated muscles.

Special creation. An account of the origin of life and its diverse forms by some act of divine creation.

Specialization (of cells). The change in cell capability from the performance of a wide range of functions to concentration on one activity or set of activities.

Speciation. The process by which the accumulated effects of variation within a population make cross-breeding between two given organisms difficult or impossible.

Species. The smallest unit of taxonomic classification, referring for the most part to a group of individuals capable of breeding among themselves. Species are defined by morphological, ecological, physiological, and biochemical criteria.

Sperm. The male reproductive cell containing the haploid number of chromosomes, produced from a diploid germ cell by meiosis (reduction division). The sperm fertilizes the egg, producing the diploid zygote.

Spermatids. After the second meiotic division in spermatogenesis the secondary spermatocytes undergo development into spermatids, which in turn develop directly into functional sperm.

Spermatogonium. The starting cell, or the "grandparental" cell, of future sperm. A spermatagonium undergoes mitosis a number of times, eventually producing a primary spermatocyte. The latter undergoes meiosis to eventually produce four sperm.

Spermatogenesis. The process by which haploid male gametes (sperm) are produced from diploid primary spermatocytes.

Spontaneous generation. A concept according to which living organisms develop from nonliving matter.

Standard deviation. A statistical calculation defined as the square root of the variance. A means of showing the limits within which all items of a distribution should occur relative to the mean. The greater the standard deviation, the wider the "normal," or bell, curve of distribution for the set of data.

Static equilibrium. A state of balance in which there is no activity.

Statistical analysis. The use of mathematics to determine whether deviations from a pattern as predicted by an hypothesis are significant.

Stimulus. Any physical or chemical change in the environment which brings about a change in activity on the part of an organism (or portion of an organism, such as isolated tissue).

Substrate. The molecule upon which an enzyme acts during an enzyme-catalyzed reaction.

Succession. Ecological development that begins in a habitat or area not previously occupied by the given community. Primary succession refers to the invasion by one community (for example, grasses) of a previously unoccupied area (such as a sand dune along the shore of a lake). Secondary succession refers to the replacement of one community (such as grasses) by another community (such as conifers and scrub vegetation).

Symbiosis (mutualism). A relationship between two species of organisms in which both derive benefits from the other.

Syllogism. A logical scheme or analysis of a formal argument, consisting of three propositions (called respectively the *major premise* (major hypothesis), the *minor premise* (minor hypotheses), and the *conclusion*. For example: every virtue is commendable; generosity is a virtue; thus, generosity is commendable.

Synthesis. The process by which larger molecules can be built up from smaller molecules or atoms.

System. An association of independent organs throughout the body for the performance of a necessary body function. Some systems in higher animals are the circulatory, digestive, muscular, skeletal, and excretory systems.

Systematics. A term originally used to refer to the study of different systems of classifying animals and plants. Today the term "systematics" is used more or less synonymously with the term "taxonomy."

Taxonomy. The science of classification; in biology, this refers to the classification of organisms into kingdom, phylum, class, order, family, genus, and species.

Teleology. Assigning purpose to an action, such as saying the cell takes in calcium ions "in order to . . ."

Telophase. The final phase of mitosis, in which the cytoplasm of the dividing cell is cleaved and two daughter cells are formed.

Territoriality. The tendency of some organisms to defend a section of space surrounding them and/or their family.

Testes. The male gonad; the glandular organ in which male gametes (sperm cells) and sex hormones are produced.

Testosterone. An androgen; a hormone produced in the interstitial cells of the testes of males and responsible for the characteristic changes associated with puberty.

Tetrad. In meiosis, the four-part structure resulting from the duplication of each pair of homologous chromosomes.

Theory. A term sometimes synonymous with a hypothesis that has undergone verification and is applicable to a large number of related phenomena.

Thermodynamics. That branch of physical science which deals with heat as a form of energy. It is concerned with such problems as the exchange of energy (measured always in respect to the gain or loss of heat from a system) during chemical or physical processes.

Thylakoid discs. Flattened pancake-like "sacks" with double-layered walls, formed from the inner membrane system of the chloroplast. The lamellae and grana inside chloroplasts are composed of thylakoid discs.

Tissue. An aggregate of similar cells bound together in an ordered structure and working together to perform a common function.

Transamination. The stepwise series of reactions in which the amino group from one amino acid type is transferred to an intermediate substance, thus producing another type of amino acid.

Transfer RNA (soluble RNA). A type of RNA in the

cytoplasm of which there are at least 20 varieties, one specific for each amino acid. Transfer RNA (*t*RNA) unites with its specific amino acid and draws it to the ribosome during protein synthesis.

Translation. An aspect of protein synthesis in which the genetic information coded on messenger RNA is used to specify the order of amino acids in a polypeptide. Translation occurs in the ribosomes, where mesenger RNA, ribosomal RNA, and activated amino acids meet and peptide bonds are formed.

Transmutation (of species). An older term for the idea that over a long period of time new species arise through modification of old species.

Trophic level. Levels of nourishment in a food chain. A food chain is the transfer of energy from its ultimate source in plants through a series of organisms each of which eats the preceding organism.

Trophoblast. In mammals, the thin-walled side of a blastocyst which gives rise to the placenta and the membranes that surround the embryo.

Unit membrane. The model that sees cell membranes (plasma, nuclear, etc.) as consisting of a mosaic of small "sandwiches"; that is, phospholipid molecules between an inner and an outer layer of protein.

Uterus. The womb; the hollow muscular organ of the female reproductive tract in which the fetus undergoes development.

Vacuolar membrane. A unit membrane structure that separates the contents of a vacuole from the surrounding cytoplasm.

Vacuole. A bubble-like structure surrounded by a membrane, occurring in the cytoplasm, and serving as a reservoir to hold food and waste products.

Vagina. In mammals, a muscular tube that extends from the uterus to the exterior and serves both as a receptacle for the sperm during coitus and as a birth canal when the fetus completes its development.

Variance (s^2). A numerical calculation describing the extent of dispersion of data around a mean.

Vas deferens. A long tubelike structure at the top and in back of the testes, in which sperm are stored and through which they pass out through the penis.

Vegetal pole. The surface of an egg opposite the animal pole, on an axis running through the nucleus. The yolk density within the egg is greatest at the vegetal pole.

Veins. Vessels which carry blood from the various organs to the heart.

Vestigial organ. A structure in a degenerate state that remains in an organism but has little or no present function. The appendix in human beings is vestigial.

Virus. A noncellular, submicroscopic particle composed of a protein coat surrounding a nucleic acid core. Viruses can reproduce only inside living cells (eucaryotes or procaryotes).

Vitalism. The view that life is an expression of something above and beyond the chemical and physical interactions of a group of molecules.

Vitamins. Chemical substances, required in only trace amounts, that are thought to aid enzymes in catalyzing specific chemical reactions.

Viviparous. Term applied to organisms whose embryos develop within the body of the mother and derive their nourishment from the mother.

Vulva. The external female sex organs (genitalia).

Wavelength. The distance between a given position on one wave and the same position on the following wave. Wavelength is often symbolized by the Greek letter lambda (λ).

Wave theory of light. The model that depicts light as demonstrating all the properties of wave motion, analogous to waves on the surface of water.

Wolffian ducts. A pair of tubules which, in conjunction with the mesonephric ducts, form the primitive urogenital system of the male mammalian embryo. Both male and female embryos develop Wolffian ducts, though in females these ducts degenerate by the beginning of the third month.

Work. An indirect measure of the energy required to move matter a given distance.

Yolk. Fatty compounds and proteins stored within the egg that serve as the first food source for the developing embryo.

Yolk sac. An extraembryonic membrane in many kinds of eggs. It functions to gradually supply food material from the yolk to the developing embryo.

Zoology. The study of animals.

Zygote. A diploid cell, the product of fertilization formed from the union of male (sperm) and female (egg) reproductive cells (gametes).

INDEX

 Page numbers in italics refer to illustrations.

Page numbers in italics refer to illustrations.

Page numbers in italics refer to illustrations.

Page numbers in italics refer to illustrations.

Page numbers in italics refer to illustrations.

Page numbers in italics refer to illustrations.

Page numbers in italics refer to illustrations.

Page numbers in italics refer to illustrations.